56th American Association of Swine Veterinarians Annual Meeting 2025

Be The Pig's Champion

San Francisco, California, USA
1 - 4 March 2025

ISBN: 979-8-3313-1636-5

Printed from e-media with permission by:

Curran Associates, Inc.
57 Morehouse Lane
Red Hook, NY 12571

Some format issues inherent in the e-media version may also appear in this print version.

American Association of Swine Veterinarians
830 26th Street
Perry, IA 50220-2328
USA

Phone: 515-465-5255
Fax: 515-465-3832

aasv@aasv.org

Additional copies of this publication are available from:

Curran Associates, Inc.
57 Morehouse Lane
Red Hook, NY 12571 USA
Phone: 845-758-0400
Fax: 845-758-2633
Email: curran@proceedings.com
Web: www.proceedings.com

Proceedings of the 56th AASV Annual Meeting

Table of Contents

Howard Dunne Memorial Lecture

Alex Hogg Memorial Lecture

Angela Baysinger Memorial Lecture

Sunday: Research Topics

Sunday Session #1: Student Seminar

Sunday Session #2: Industrial Partners

Sunday Session #3: Industrial Partners

Sunday Session #4: Industrial Partners

Poster Session: Veterinary Students

Poster Session: Research Topics

Poster Session: Industrial Partners

Monday: Be the Pig's Champion

Monday #1: PEDV Elimination

Monday #2: Pig 101 on Pig One-on-One

Monday #3: Global Hot Topics

Tuesday: Influenza Insights

Howard Dunne Memorial Lecture

The challenges and opportunities of becoming the pig's champion

Clayton Johnson, DVM

In remembrance and honor of
Howard W. Dunne

Dr Howard W. Dunne received his veterinary degree from Iowa State University in 1941. He was in engaged in private practice for one year and industry for four years before entering graduate school at Michigan State University. After receiving his PhD in 1951, he spent two years as deputy chief of veterinary microbiology at the Chemical Corps Biological Laboratories in Camp Detrick, Maryland. In 1953, he joined the faculty of Pennsylvania State University, where he headed the veterinary science research program and the large animal diagnostic laboratory. Dr Dunne was internationally known for his work on swine diseases, particularly hog cholera and the swine reproductive viruses then referred to as the SMEDI syndrome. Dr Dunne authored more than 90 scientific publications, and is perhaps best known for initiating and editing the textbook *Diseases of Swine,* which is still widely considered the "bible" of swine disease.

In 1971, the American Veterinary Medical Association presented Dr Dunne with the XII International Veterinary Congress Prize to recognize his contributions to the international understanding of veterinary medicine. At the 1974 Congress of the International Pig Veterinary Society (IPVS), Dr Dunne was elected president of the 1976 IPVS Congress, to be held in Ames, Iowa. His sudden death at the age of 61 in September of 1974 saddened the swine veterinary community, and the 1976 IPVS Congress was dedicated to his memory.

At the suggestion of Dr Allen Leman, a financial gift made by the 1976 IPVS Congress organizers to the fledgling American Association of Swine Practitioners was combined with additional association funds to endow this prestigious annual lecture in remembrance and honor of Dr Howard W. Dunne.

HOWARD DUNNE MEMORIAL LECTURERS

1978	Michael Muirhead	1990	Stanley Curtis	2002	Tim Blackwell	2014	Daryl Olsen
1979	W. J. (Bill) Smith	1991	Roy Schultz	2003	Lonnie King	2015	Gregory Stevenson
1980	N.C. Nielsen	1992	Joseph Connor	2004	Robert Desrosiers	2016	John Harding
1981	Allen Leman	1993	Camille Moore	2005	Scott Dee	2017	Jeff Zimmerman
1982	Michael Wilson	1994	Robert Morrison	2006	Gregg BeVier	2018	William R. DuBois
1983	Edward Bohl	1995	Rodney Johnson	2007	Tom Burkgren	2019	John T. Waddell
1984	Richard Ross	1996	Paul Yeske	2008	Timothy J. Loula	2020	Bret D. Marsh
1985	Kent Kislingbury	1997	Barbara Straw	2009	John U. Thomson	2021	Jerome O. Geiger
1986	Ralph Vinson	1998	Mike Daniel	2010	James F. Lowe	2022	Angela Baysinger
1987	Gary Dial	1999	Larry Rueff	2011	Locke A. Karriker	2023	Egan Brockhoff
1988	Thomas Alexander	2000	David Reeves	2012	R. Dean Boyd	2024	Joel Nerem
1989	Steven Henry	2001	Max Rodibaugh	2013	Craig Rowles		

https://doi.org/10.54846/am2025/1

The challenges and opportunities of becoming the pig's champion

Clayton Johnson, DVM
Carthage Veterinary Service, LTD, Carthage, IL

Presenting the 2025 Howard Dunne Memorial Lecture is an incredible honor. However, the apprehension and consternation that follows accepting this responsibility is overwhelming! I not only have an obligation to uphold the impressive standards set by Dr Dunne, but also the considerable task of following an exceptional list of legends, mentors, and colleagues who have come before me. I have spent my entire life involved in pig production, visiting the farms of my father's clients, working on the farm of Ron and Carolyn Myers, learning early in my career from The Maschhoffs and continuing to learn later in my career now at Carthage. I am grateful to everyone who has contributed to my journey. I am also most fortunate to have been a member of this wonderful association, and I want to thank all the AASV staff and fellow members – you have all helped me more than you will ever know. If my thoughts in any way provide you with a new insight or inspire something that makes you a better swine veterinarian, a better champion of the pig, please know I am humbled and rewarded.

Introduction

I want to sincerely thank Dr Locke Karriker for the opportunity to author these proceedings and present on the topic of becoming the pig's champion. It is an honor and a privilege to give the Howard Dunne lecture at the 2025 AASV Annual Meeting. I always leave our annual meeting inspired to improve my performance as a swine veterinarian. I hope in some way the thoughts and messages I share both in written and verbal communication provide you with inspiration to improve your skills, broaden your perspective, and help you persevere through the myriad of obstacles that present themselves when supporting effective swine health management. To quote former AASV Director and Howard Dunne lecture author and presenter, Dr Tom Burkgren, "AASV is a good but not great organization. There is so much more we can do." These words inspire me daily – acknowledging our collective positive attributes and efforts while also embracing the reality that we can and must improve our impact on swine health and well-being. Being good but not great is a paradox and paradox is complicated, difficult to communicate and hard for our mind to embrace. But we must embrace this reality – we are good but not great, there is so much more we can do.

In preparing for this lecture, I embarked on a journey of reading every Howard Dunne proceedings written and available electronically to our membership (**https://www.aasv.org/foundation/endowed-lectures/howard-dunne/**). It proved to be an endeavor I found endlessly rewarding, and one I challenge all our membership to complete. All Howard Dunne lecture proceedings since 1999 are available for reference via the AASV website and all provide tremendous insights into the thoughts, opportunities and concerns previous AASV thought leaders have shared. It was striking to me how similar the opportunities and concerns of our current day resemble those of previous years. Opportunities such as leveraging technology and information systems to provide veterinary services across larger populations of pigs have been discussed in this lecture for decades. Concerns about the ability to create shared expectations regarding animal welfare between producers, consumers, and veterinarians fill these proceedings. My greatest learning from these historical proceedings is that our present reality is not unique, nor should it be viewed as a scenario with clear objectives that can be achieved to declare mission accomplished. Rather, we should see today's challenges and opportunities as an institutionalized mission, one that we must strive to improve, every minute, every day, every week and every year. Our challenges and opportunities as a swine veterinary profession should be seen as a journey, not a destination, and we must recognize that we are either getting better or worse in every moment, but never remaining static. We must assess our current situation and try to improve it, but we should never feel we can move on from our challenges and opportunities, nor should we feel the lack of final victory over these timeless challenges is in any way a failure of our efforts.

It is imperative that swine veterinarians be viewed as the pig's champion by swine producers, swine processors, pork retailers, pork consumers and regulatory authorities. Being seen as a credible champion of the pig's health and well-being builds trust with clients, influences critical decisions and maintains our professional reputation. Our credibility is fundamental for establishing and maintaining trust with the farmers and producers that make the most important decisions impacting animal well-being. Without credibility, the swine veterinarians' recommendations are not likely to be adopted, which could harm the pig. Simply put, credible veterinarians are more likely to influence key decisions and all stakeholders in the pork supply chain will be more inclined to adopt recommendations from veterinarians they trust. Being seen as a credible champion of the pig is critical for our ongoing professional reputation. This reputation opens opportunities for collaboration and public engagement. Maintaining our reputation is essential to ensure future generations of veterinarians enjoy the same decision-making influence and trust essential to serving as the pig's champion.

While objective measurement of pig health and well-being is tremendously complicated and difficult, assessing who is seen as the pig's champion is not. There are three main stakeholders in the assessment of who is seen as the pig's champion and those stakeholders are swine producers, pork consumers, and regulatory agencies overseeing the production of pigs and pork. Our brutal reality is several of these stakeholders do not see swine veterinarians as the pig's champion at the time of this publication.

What is a champion?

While it may seem trivial, I feel it important to create a shared definition of the word champion. For us to become the pig's champion, we must certainly know what that word means and have a shared definition that focuses our collective resources toward that effort. Webster defines champion as:

1. A winner of first prize or first place in competition; one who shows marked superiority
2. A militant advocate or defender
3. Warrior, fighter
4. One that does battle for another's rights or honor

While I respect that my interpretation is far from perfect, I would propose that our definition of champion for the purposes of being the pig's champion must revolve primarily around definition 4, and that we must represent the interests of the pig in all decisions. While our veterinary services are not at all militant in nature, I do think we should embrace the spirit of definitions 2 and 3, and that we must be an advocate for the pig's well-being and fight for health management strategies that optimize its health. In doing so, we optimize the pig's welfare and clearly demonstrate to pork consumers that we represent what's in the pig's interest. Lastly, I feel definition 1 has a place in our discussion. Like all veterinarians, swine veterinarians rarely have the sole decision-making power regarding animal health management strategies. Our clients face the same challenges as all animal owners, having to choose between health management strategies that have different advantages and disadvantages. Each option represents unique costs, outcomes and timelines and rarely do we find one option being agreed upon by all stakeholders as best. Health management opportunities, like all opportunities, are infinite, yet time, money, and resources are very finite. Veterinarians may not have sole decision-making power, but we should be highly influential in health management decisions. This is where we must strive to show marked superiority. In fact, I think a fair assessment of whether we are effectively being the pig's champion is to ask ourselves: Are we seen as the most credible source of information inputs and data points assessed as part of the health management decision making process? If we get feedback from swine owners or pork consumers that they view others as more credible in this space, we must embrace the reality that we aren't seen as the pig's champion in that moment. Being the pig's champion is a journey and not a destination, if we're not seen as the pig's champion in that moment, it is our duty to make changes to better represent the pig's well-being. Simply put, we must look inward and adjust our approach when we aren't highly influential in health management decisions.

I would humbly propose that defining what it means to be the pig's champion is to effectively represent the interests of the pig's welfare, researching, creating, proposing and supporting the implementation of practical solutions for all obstacles that negatively impact pig health and well-being. We MUST influence the decisions being made regarding pig care. In any situation where we find ourselves sidelined and not able to credibly influence change, we should assess the potential root cause (poor communication, lack of follow up, lack of practicality, etc) and adjust our behaviors and actions accordingly.

Assessing our ability to be the pig's champion

It is both poignant and ironic that we hold this meeting in the state of California, which recently passed very specific regulations regarding swine housing requirements aimed at improving animal welfare for animals produced all over the United States. Sow housing has been a topic of conversation within our industry for decades and has been a point of concern for many previous Howard Dunne lecture speakers. AASV has had a position statement on sow housing for most of my career. Despite all our discussion and research of the topic, swine veterinarians had at best, minimal influence on the creation of California's swine housing regulations. While our position statement includes "There are advantages and disadvantages to any sow housing that should be considered by weighing scientific evidence and veterinary professional judgement," the reality is our collective swine veterinary professional judgement was not seen as credible enough to strongly influence this process. The largest state in our union overwhelmingly excluded us from the process of defining pig housing standards. This is our reality, and we must learn from it and adapt our tactics to be more influential going forward.

I know these last few statements are terribly discouraging; trust me, they're as discouraging for me to write as they are for you to read. That being said, all feedback is a gift, and we can learn from the fact that our opinions weren't considered important or credible. It is easy for us to hide from this feedback, we can surround ourselves with those who will tell us this is just some vegan activism program at work. We can hide from the truth that we weren't seen as credible in defining sow housing standards, but we can't hide from the impact of these standards going forward. We know both animal well-being and pork availability have been adversely impacted by Prop 12 regulations. I often use the flippant line of "I spend a lot of time trying to figure out how to keep sows alive in 'welfare friendly' housing situations," and that's my way of escaping the reality that I am not seen as a credible advisor to those welfare friendly housing definitions. To be the pig's champion we must first embrace our current reality and ask ourselves, why aren't we seen as credible, and what can we do to retake the position of having markedly superior influence on improving swine health and well-being going forward. Remember, being the pig's champion is a journey, not a destination. We must now strive to better represent the interests of pigs we see negatively impacted by this and other well-meaning but ill-informed regulatory and retail sales programs.

Unfortunately, there is additional concerning evidence about our status as the pig's champion. Veterinarians are challenged with a unique burden of protecting both animal and human health. These goals can be occasionally adversarial with each other, with decisions regarding antimicrobial use causing a common conundrum of prioritizing the pig vs prioritizing the judicious use of antimicrobials. Indiscriminate use of antimicrobials increases the risk of antimicrobial resistance that could have both a negative impact on future pig health as well as the potential for spill over events into human health. Failure to use antimicrobials in the spirit of antimicrobial stewardship will likely be seen as having an immediate and unnecessary negative impact on pig well-being by our animal caretakers. This is a terribly difficult paradox to navigate for the practicing swine veterinarian, but one we must routinely embrace and attempt to optimize, despite its challenges and risks. For most of my career, carbadox, an antibiotic used to control severe animal health challenges, has been threatened by the United States Food and Drug Administration (FDA). FDA has used several different avenues to remove carbadox as an option for disease management and has always been met by significant concerns from swine veterinarians that animal health and well-being will be negatively impacted and the use of antibiotics that are important to fighting human infections would increase as an unintended consequence to the removal of carbadox. I have participated in multiple discussions with the FDA and have found each of them to be wholly unrewarding. In panel discussions with FDA staff charged with making decisions regarding carbadox, I have never once been asked a question, nor has there been any actual discussion between FDA and practicing veterinarians. We are given an opportunity to make a statement, it's entered into the record, and then the conversation is over. It is clear they do not value swine veterinarians' opinions and perspective in this process, rather they ask for our public opinion in the same way they would any US citizen. Simply put, swine veterinarian input into antimicrobial regulatory decisions is a check-the-box activity. Our input isn't valued by the FDA because it's not seen as credible. Again, this is as painful for me to write as it is for you to read, but we can only improve if we embrace our current state – the FDA does not see swine veterinarians as champions of the pig. We must ask ourselves why this is the case, and then work to address the perceived and/or real concerns that limit our influence.

Why are swine veterinarians not embraced as the pig's champion?

I believe the primary reason swine veterinarians are not viewed as being credible is because we're not viewed as being independent. Let's start with the FDA's feedback. While my notes above highlight our inability to influence a single antimicrobial, swine veterinarians have been involved in discussions with the FDA regarding antibiotic use for decades. When I was a younger veterinarian, the use of antibiotics for growth promotion was a huge topic. Howard Dunne lecture proceedings from years past shared concerns of "what would happen to pig health if we lost these tools." I certainly respect the observations that antibiotics used for growth promotion likely had unintended benefits to swine health and well-being, but it is impossible to look at the current state of swine health and well-being and make the case that ceasing to use antibiotics for growth promotion has created challenges we weren't able to manage. In hindsight, I feel that the process of removing antibiotics used for growth promotion in swine had an unintended impact, not directly on swine health, but rather it negatively impacted the view of consumers and regulatory officials that swine veterinarians represent an independent and credible voice. When I look at this from the FDA's perspective, I can appreciate that they see the veterinarian as simply representing the economic concerns of our swine producer clients, not honestly representing the health concerns of our patients, the pigs. We are viewed more and more as the pig farmer's champion rather than the pig's champion. This lack of independence from the pig farmer hinders our influence with both regulatory agencies as well as pork consumers.

Let's transition this discussion to California and Proposition 12. At the core of sow housing discussions has always been the widely held consumer beliefs that food animals should be able to participate in normal behaviors throughout their life. Tim Blackwell gave the Howard Dunne lecture in 2002, providing us with a scholarly review of the study of ethics. Tim summarized that "ethics is the discussion of considering right versus wrong when considering outcomes from multiple and differing perspectives." Sow housing is an excellent example – should a lactating sow be confined to prevent crushing mortality to her piglets? This housing decision has no perfect answer, and the best management practice depends on whether you view the outcome through the lens of the piglet or the lens of the sow. Tim reviewed the four categories of ethics: Traditional ethics, popular ethics, religious ethics and legal ethics. To briefly review, traditional ethics has dominated livestock production since the beginning of animal husbandry. We take care of the animals, and the animals take care of us is an easy phrase for us to embrace and for all industry stakeholders to support. This one line sums up most decisions for animal care in livestock operations. The popular ethic is difficult for the swine industry to embrace as it often leans toward an impractical view of livestock production that involves intense manual labor with little progressive thought or technological intervention. The phrase "fish should swim and birds should fly" is often used to summarize this ethic. Consumers embrace this perspective and envision a diversified farm with a red barn on a hill and a variety of livestock species sharing a small ecosystem. One can quickly see how traditional ethics and popular ethics do not always reach the same conclusion about what's best for the pig. The controversy surrounding gestation stalls to house pregnant sows was the exact topic Tim wrote about in 2002 when he highlighted that while the swine industry may desperately want to adopt the principles of the traditional ethic, it is the popular ethic that is the most powerful and influential of all the ethics. Why is the popular ethic so powerful? Quite simply, if the popular ethic perseveres long enough, it becomes a legal ethic. What is popular with the public for enough time, eventually gets enacted as the law of the land and pig producers must follow these laws if they want to continue to participate in the modern pork supply chain.

As I reflect on our inability to influence the process that resulted in the creation and adoption of Proposition 12, it is impossible to ignore the similarities to the antibiotic discussion with the FDA: Veterinarians were not able to influence the process or outcome because they were not seen as independent and credible. Sow housing has been a point of contention for decades. Sow housing has been an opportunity for veterinarians for decades. Swine veterinarians should be the undisputed welfare experts of humane practices for livestock production. Sow housing represented, and still represents, an opportunity for veterinarians to lead. Leadership isn't popular, and we must embrace the fact that to regain our position of influence, we'll have to start by regaining some level of independence. Swine veterinarians should see welfare discussions as opportunities to lead, opportunities to define acceptable and practical standards, but we'll only be seen as credible if each stakeholder group including swine producers, pork consumers, and food regulators feel their most important concerns are addressed. We've spent decades ignoring the elephant in the room with sow housing, that pork consumers overwhelmingly want sows to have the ability to turn around. Until our housing recommendations include that expectation, we can only hope that the popular ethics perspective embraced by most pork consumers in the US changes, and unfortunately hope has proven to be a poor tactic. If swine veterinarians do not include the simple recommendation that a sow should be able to turn around in our position statements, we will not be viewed as a credible source of influence by the average pork consumer and we certainly won't be viewed as the pig's champion.

What are the barriers to being the pig's champion?

While the barriers to being the pig's champion are highly variable and situationally dependent, there are several common themes to focus on. The first limiting resource that comes to mind is always the most important – our time. The last several decades have brought new demands for time including novel disease emergence and/or introduction as well as the increased expectation of veterinarians to provide oversight for industry audit and training programs. Unfortunately, these increased demands for time have not come with any additional human capital infusion to execute them and as such, practicing veterinarians find themselves regularly in triage situations, focused on putting out the most urgent disease fires as efficiently as possible. Collectively, we have too much ambulance time. As a profession, we have far too many hours consumed with long drives to see hopeless swine respiratory disease (SRD) situations and ultimately administer broad spectrum antimicrobials. To use a human hospital analogy, we have too many resources dedicated to the emergency room section of the hospital and not enough resources dedicated to preventative medicine.

A second common barrier to being the pig's champion involves our relationship with the pig farmer. There are times where we as veterinarians must make a difficult choice – be the champion of the farms' financial outcomes or be the champion of the pig's health and well-being outcome. In these situations, the challenge of pushing back on producers is overwhelming. For staff veterinarians working for a production company, it's natural to have concerns you'll get disciplined and/or fired if you don't toe the company line and repeat commonly accepted dogma within your organization. Staff veterinarians often report to non-veterinarians and as such, may have the complex task of influencing technical decisions made by individuals who don't have scientific training and the ability to separate good science from pseudoscience pedaled regularly from visiting salespeople and other producers they speak with at industry events. For private practice veterinarians, the concerns are similar: you're nervous you'll get fired if you challenge the client. While pseudoscience is a challenge throughout our industry, it is an even bigger challenge in the private practice situation, where the decision maker is almost never technically trained to evaluate scientific recommendations and make decisions based on the best available evidence. The reality is, almost every pig producer can always find another veterinarian who'll tell them what they want to hear. I've lived in both worlds, serving as both a staff veterinarian and a private practice veterinarian and I know each of these pains quite personally. The saying "the pigs are easy, but the people are hard" is apropos in these situations.

Another common obstacle to being the pig's champion is the lack of (or lack of access to) objective and empirical scientific data to support health management recommendations. The gap in data availability is even further amplified when we move from health management recommendations to animal well-being recommendations. Within health management, our challenges are primarily the lack of active research into swine disease management and/or the fact that most research that does exist is private and owned by individual producers or industry partners. How can you call out pseudoscience peddlers when you yourself have limited quality scientific information to support your perspective? Historically, the swine industry relied on academic institutions and extension services to conduct and share learnings from animal research, but the resources available to continually fund this model have dwindled. As such, we much look to alternative funding sources for future health management research. Within animal well-being, our primary challenge is much greater – we simply don't have metrics that are universally agreed upon to accurately characterize animal well-being. Before we can begin to consider who will conduct animal well-being research, we must define and align upon meaningful metrics to describe animal well-being. Once we have appropriate means of measuring animal well-being, then we can proceed with improved methods to generate and share objective data.

A final common challenge encountered in championing the pig is the fact that we lose experienced practicing veterinarians to non-practicing veterinarian roles all too frequently. The loss of well trained and highly influential swine practitioners to other roles within our profession after a significant training investment in farm level development results in on farm veterinary work mostly being done by the least experienced within our ranks. The root causes of these career transitions are often similar, swine veterinarians, like many working adults want a change in work life balance and less travel. This is particularly true of veterinarians actively raising children – we may have had plenty of flexibility to live the life of a practicing swine veterinarian when we graduated and in the immediate years thereafter, but as our family situation changes, we also find ourselves needing to adjust our values and time management accordingly. There are no shortage of industry and academia jobs that facilitate this transition, and inevitably there comes a time in most practicing veterinarians' career where they stumble across a non-practice related job that they can't refuse. I don't mean to suggest that industry and academia jobs don't provide value, quite the opposite; they play a critical role in championing the pig. I'm simply highlighting the reality that the veterinarians actively filling the vet roles that show up on a pig farm and hold the veterinarian-client-patient relationship (VCPR) are often young and inexperienced swine veterinarians. Whether we like to admit it or not, the lack of experience is often tied to a lack of credibility in the eyes of the producer, who is the ultimate decision maker in managing their animals. Simply put, we have a career ladder model in which a well-trained swine veterinarian who's finally in a position to champion the pig's well-being is often lost to one of the many other industry roles. In the eyes of most pig farmers and caretakers their definition of a swine veterinarian is a relatively recent graduate who's worked in the industry for a shorter amount of time than they have. As such, more and more producers don't see us as a collective group that's capable of solving complex problems because their interactions with swine veterinarians are often with the least experienced swine veterinarians amongst our membership.

What are the opportunities to be the pig's champion?

I can appreciate that after reading the preceding sections you may be uncomfortable and overwhelmed. Action is the antidote for anxiety, so let's figure out where we can get started! Don't overthink it – focus on helping the pig! Swine veterinarians are uniquely positioned to be the pig's champion. No other technical support functions visit the farm and its caretakers as frequently as swine veterinarians do. Those farm visits give us tremendous insights into what's possible, practical and likely to be effective. We have perspective, breadth of knowledge and experience from our farm visits that put us in an enviable position to champion the pig. How do we do that? First and foremost, force yourself to see every decision through the lens of the pig. Start small, and use this perspective when making basic, tactical recommendations on farms. If these were your animals and you wanted to optimize their health and well-being, what would you do in this situation? You must value other considerations, but before you begin the complex process of looking for the best outcome from multiple different viewpoints, focus on the best outcome from the viewpoint of the pig. Once you've created an action plan based on what's best for the pig, then begin to consider the financial impact of your recommendations. Most farms we work with are businesses, and businesses must return a sustainable profit to be viable – as such we cannot ignore the economic impact of our action plan. Fortunately, the old saying "take care of the pig and the pig will take care of you" is generally true, and often the best outcome for the pig is also the best outcome for the farms' financials. You must be thoroughly versed in understanding the costs and revenues ramifications of your recommendations, but you shouldn't feel constricted to only recommend the action that optimizes the economic outcome. There are times where we need to champion the well-being of the pig because it's the right thing to do as the pig's champion, even if it's not the most optimal financial outcome. We also must respect the pig owners' decisions in these situations – like all animal owners they have a limited number of resources, and they must deploy these resources strategically.

For the practicing swine veterinarian, we must be more efficient and productive with our time. We must be ruthless with our focus. Our reality is we have a shrinking pool of on-farm veterinarians accompanied by an ever-increasing demand for our services and neither of those trajectories is likely to change soon. Focus starts with setting goals, a process that's hard to do when you're constantly chasing sick pigs on ambulance visits. Create long term goals as well as short term goals, think of long-term goals as yearlong or multi-year efforts and short-term goals as the weekly and monthly milestones that progress us toward the big goals. End each week with a review of your short-term goal progress and take time to create a plan for the next week. If you don't have a plan, you'll easily fall into the trap of jumping into the minutia to start crossing off tasks, tasks that may not be the most important things to work on. Focus on reverse time management, think through what time sinks regularly prevent you from focusing on the pig. Which meetings can you get out of? What relationships constantly require your time but rarely provide value? What email list serves can you unsubscribe to? What nonproductive screen time can you avoid? Don't get caught in the efficiency trap – we all love to check off to-do lists but we inevitably end up doing minutia level work just to make ourselves feel productive – focus on the big problems, even if they aren't immediately going to be checked off your list. If you need lists, try to prioritize your tasks using the rule of the 4 Ds – for each task either Do it (just do it right now), Defer it (schedule it to be done later), Delegate it (send it to someone on your team and tell them to work on it) or Don't do it (communicate with other stakeholders that you're not going to complete this task). As a swine veterinary community, we need to be more results oriented and less action oriented. The culture of our profession is to work hard and that's a noble culture to have, but it's not necessarily productive. The culture of our profession must embrace the fact that working smarter is better than working harder.

I feel strongly that pig veterinarians must advocate for a new model of veterinary services, one that allows some level of independence for veterinary service and one that values practicing veterinarians having time available to think through proactive solutions to chronic problems vs spending the bulk of our time triaging and treating the symptoms of those chronic problems. We must advocate for veterinarians to be judged not just based on how many finishers they got through during a day, but also based on how proactively they manage health problems vs triaging them. Technology is our friend in this endeavor, hardware and software improvements have made telemedicine very accessible, and we must advocate for allowance to use these tools to minimize the time commitment involved in ambulance visits as well as regulatory visits simply to fulfill certificate of veterinarian inspection (CVI) needs. Technology is also our friend in disease prevention – the ability to utilize technology improvements and innovation to both manage disease with novel vaccine platforms as well as prevent infection with gene edited pigs represent breakthrough opportunities to eliminate current time commitments for practicing swine veterinarians. Technology is similarly our friend in disease control – using technology to training and empowering caretakers to make simple decisions regarding disease outbreaks such as simple diagnostic and decision tree tools can reduce the need for ambulance visits. The practicing swine veterinarian working in a world where virtual CVI visits (where a valid VCPR is already in place) and decreased ambulance visits because chronic disease situations instigated by porcine reproductive and respiratory syndrome (PRRS) infections are prevented would have a tremendous increase in time available to research and think through through optimal disease management plans. Further, the decreased travel needs may facilitate better retention of practicing swine veterinarians because this new working situation will allow for a more family-friendly working environment. That improved retention of on-farm vets will inevitably increase our credibility in the eyes of pig producers and caretakers. The bottom line is swine veterinarians need to put effort into working ourselves out of the ambulance and regulatory visits, rededicating this time to more strategic endeavors.

Both AASV membership and AASV leadership must consider formal advocacy efforts, beyond the scope of our professional organization's role today. Currently, the mission of AASV is education, but all the education in the world won't help practicing swine veterinarians improve the well-being of the pig if we continue to spend the bulk of our time triaging chronic disease and performing time consuming, low value regulatory functions. The needs of our profession have changed and the tools available to our profession have improved yet we aren't able to fully utilize them optimally. We need to advocate for changes that facilitate improved swine veterinary practice, and if swine veterinarians themselves don't champion that effort, we can't expect anyone else to do it for us. A very tangible example of where we need to advocate is sow housing, the debate on best management practices for sow housing is far from over. Swine veterinarians have much to contribute to this topic, yet our contributions will be minimal as long as they are individual in scope. A well-resourced organization focused on advocating for the priorities of swine veterinarians would have great credibility in defining, proposing and building support across industry stakeholders for practical and science-based sow housing recommendations. Such an organization would be able to similarly organize and lead efforts not just specifically into sow housing, but into whatever new priorities lie in store for the future of swine veterinary practice. Scope expansion into advocacy infers a resource need, and this is where most of us need to step up – we must support AASV with additional funds to accomplish this additional work. We must show up at the meetings, even when they're in California! We must show up not just to learn, but also to donate to the worthy cause of advocating for better working environments for our practicing swine veterinarians.

Our last and most important opportunity in being the pig's champion is developing our communication and influence skill set. From the farm to the highest levels of regulatory authority, we must embrace a more extroverted communication style than most of us are comfortable with. The old joke "the outgoing veterinarian is the one who looks at your shoes instead of their own while they're talking" comes to mind. Swine veterinarians of all generations must embrace soft skills as much as we embrace work ethic. There are numerous tools and techniques that teach us improved communication processes. Most involve understanding the natural tendencies of others and tailoring your communication style in that moment to best match what is well received by your audience. Communication is like any other behavior – we all have natural tendencies we want to prioritize but our actions don't have to be based on our tendencies. Human beings have the will power to choose actions that are different from our tendencies. Our communications can be evaluated and where our natural tendencies aren't the optimal communication style in that moment, we can choose a different communication approach. Changing your communication behaviors isn't necessarily easy, but it's necessary right now. We must embrace that most of our profession is introverted by nature and yet overcome those natural tendencies toward sitting on the sidelines in larger group settings.

Our communication opportunities exist not only domestically but internationally. US swine veterinarians are respected throughout all global swine industries, and there is a huge opportunity for US based veterinarians to get more involved in global swine health challenges and opportunities. Look at this from the perspective of our welfare credibility, it's hard to champion for the well-being of our pig herds and sound genuine when you're only talking about pigs that live in your geographic area – if we really want to champion for pig well-being, we must take the noble approach of championing for all pigs, in all geographies. We shouldn't advocate for well-being and care standards in the US; rather we should embrace advocacy at the global level and champion improved well-being conditions for all pigs. Benefits from this effort exist certainly beyond just the well-being and care standards, as pathogens know no boundaries. The interaction of swine, humans and other livestock species around the world has created tremendous risk for swine health. Poor swine health plans have resulted in pathogen outbreaks and worldwide spread of African swine fever, porcine epidemic diarrhea, porcine deltacoronavirus, senecavirus A, haemophilus parasuis, PRRS and influenza, just to name a few – we must do a better job of communicating effective health management strategies globally to protect our domestic swine health status.

Stretch goals for AASV members

One of the many things I love about our AASV Annual Meeting is leaving with a sense of both urgency and motivation. To that end, I'd encourage **ALL** of our membership to consider how they can personally make contributions toward being the pig's champion. Specifically, I ask all of you to:

- **Review your time management and priorities from the lens of being the pig's champion.** Shed tasks and time commitments that get in the way of championing improved health and well-being for your patients.
- **Always highlight the intervention that optimizes the pig's health and consider long term financial impacts in addition to short term financial impacts.** Consider the financial impacts of each strategy, but don't highlight only the best economic outcome, also highlight the best health outcome when it may differ from optimizing economics. Make sure to frame economic considerations in both short term (making money now) and long term (making money in the future) so that reflection is given to the sustainability of the intervention strategy and how its impacts will be viewed from other stakeholders including pork consumers.
- **Increase your advocacy budget.** That increased budget could come from either you or your business's time and human capital, it could be financial contributions, or ideally it would be some combination of both. To be seen as the pig's champion we must spend more time talking to regulators, retailers and consumers about the most important challenges to pig health and well-being as well as what we're doing about them. Get even more engaged with not only AASV's advocacy efforts but also the advocacy efforts of the National Pork Producers Council and other public-facing industry organizations.
- **Practice your communication by sharing how proud you are to be a swine veterinarian, and most importantly, how proud you are to be the pig's champion!** Be willing to strike up a conversation with a stranger sitting next to you on the airplane and tell them how hard you work to protect pig health. Jump into that awkward conversation with your vegan relative during the family holiday get togethers and have a real discussion about your role as a swine vet in representing the best interests of the pig in all decisions. Simply put, educate those you interact with outside of the farm, and do it much more often than you do today.

Closing

My sincerest thanks again to Dr Locke Karriker for providing me with the opportunity to give this lecture and to Dr Howard Dunne for his timeless impact on our profession. I want to thank and recognize my family for their countless contributions, supporting me as I live out my dream of being a swine veterinarian. To my parents, Bill and Jane Johnson, thank you so much for believing in me and always telling me I could accomplish anything I put my mind and effort toward. You have both been tremendous role models to me throughout my life. To my sister and brother, Torie and Henry, thank you so much for the fellowship we've been able to enjoy throughout our upbringing and thank you both for taking on the honorable responsibilities of raising the next generation of our family. To my children, Nina and Jake, thank you for being who you are – tremendous young people who are exceptionally talented and a joy to be around. Lastly, to my wife, Ali, thank you for your never-ending support in my journey as a husband, parent and veterinarian. Swine veterinarians have unique burdens which inevitably impact our families. Being the pig's champion means the occasional late nights, early mornings, and unexpected travel needs – your willingness to let me embrace my role as the pig's champion has certainly come with sacrifice and I can never say THANK YOU enough for all you've done to support me.

To my fellow AASV members, it is with humble gratitude that I thank you for your mentorship and fellowship throughout my career. I hope my message has stimulated your thoughts and given you inspiration to embrace your role as a swine veterinarian, the champion of the pig's well-being. Be good to each other and help your fellow veterinarian out whenever and wherever you can. Be kind to our caretakers, they serve an essential role in protecting the well-being of the pig and they deserve our recognition, support, and respect. Be principled in your approach to veterinary medicine, set long term goals toward the biggest problems your pigs encounter, and collaborate relentlessly to improve our current solutions and recommendations. Be an idealist, imagine a world with less wasted travel time, less time spent agonizing over chronic disease challenges. Be brave when overcoming the inevitable obstacles you encounter in your journey toward solving the big problems we face. Set short-term goals that serve as milestones toward improving our biggest problems and communicate a vision of improved pig well-being to everyone you encounter. Commit to this process and know you'll be championing the pig every step of the way. I appreciate you all and wish you all the best on this journey!

Alex Hogg Memorial Lecture

Who gets to be the pig's champion?

Cara Haden, DVM

In remembrance and honor of Alexander Hogg

Alexander Hogg
1920-2006

A native of Scotland, Alex (pronounced "Alec") Hogg immigrated to the United States with his family when he was six years old. After reaching adulthood, he served in the US Navy during World War II before entering veterinary school at Kansas State University. He received his DVM degree in 1950, and was a mixed animal practitioner in Iowa for twenty years. Dr Hogg returned to college - this time, Iowa State University - to complete a Master's Degree in 1972. He spent the following years serving and educating swine veterinarians and producers in his roles as Nebraska Swine Extension veterinarian and professor at the University of Nebraska-Lincoln. Upon "retirement" from the university in 1990, Alex worked as a technical services consultant for MVP Laboratories. Ever an enthusiastic learner, he graduated from the University of Illinois Executive Veterinary Program in 1995.

An active and longstanding member of the AASV, Dr Hogg joined what was then the American Association of Swine Practitioners shortly after it was formed. He served as secretary/treasurer of the association from 1974 to 1976, and was elected president in 1979.

The many awards and honors bestowed upon him over the years included several presented by the AASV: The Howard Dunne Memorial Award in 1985, the Meritorious Service Award in 1996, and in 2005, the first-ever AASV Foundation Heritage Award for a lifetime of achievement in veterinary medicine.

ALEX HOGG MEMORIAL LECTURERS

2011 Joseph F. Connor
2012 Lisa Tokach
2013 John T. Waddell
2014 Mark J. Engle
2015 C. Scanlon Daniels
2016 Peggy Anne Hawkins
2017 Matthew Turner
2018 Rodger Main
2019 Deborah Murray
2020 William L. Hollis
2021 Jeremy Pittman
2022 James A. Kober
2023 Attila Farkas
2024 Chris J. Rademacher

https://doi.org/10.54846/am2025/2

Who gets to be the pig's champion?

Cara Haden, DVM
Pipestone, Independence, Iowa

Introduction

It is an honor to have the opportunity to present the 19th Alex Hogg Memorial Lecture this year. Thank you, Dr Karriker, and the 2025 American Association of Swine Veterinarians (AASV) annual meeting planning committee for this opportunity. Thank you to the entire AASV. I am so grateful to be a part of this industry. It is a blessing to have a career that I love.

I am on this stage today because many people in this room took the time to mentor and challenge me. I will always be grateful to the staff and professors at Iowa State University, including Darrin Madson, Kent Schwartz, Alex Ramirez, Derald Holtcamp, and Chris Mower. Thank you to Cameron Schmitt and the Pipestone team for hiring me, investing in me, and trusting me in leadership roles. I am especially grateful to Scott Dee for his mentorship, encouragement, and his belief in me. To Bryan Myers for the 10 years, he has spent so far working alongside me and teaching me what it looks like to be an excellent veterinarian. I am grateful to the welfare community for welcoming me, specifically Anna Johnson, Madonna Benjamin, Monique Paris Garcia and Carissa Odland. I am thankful to my husband, Michael, and our 4 children Flynn, Kennedy, Ford, and Wells. You are the most important people to me. Finally, I want to thank all the veterinarians who came before me to shape this wonderful industry and organization. That list includes Dr Alex Hogg.

Dr Hogg passed away prior to my entering the swine veterinary profession, and so it was an honor to spend time learning more about his life and his work as I prepared for this lecture. Dr Hogg was born in Scotland in 1920 and moved to the United States during his childhood. He loved this new country and even volunteered to serve in the US Navy during World War II. After returning from service, he received his DVM and began a lengthy career in service to the swine industry.

During his career he spent time as a practicing veterinarian, veterinary pathologist, swine extension specialist and even completed the Executive Veterinary Program course at the age of 75. The most famous title he carries is that of a "lifelong learner."

This title is an excellent summary of his life, but it is also an incomplete one. As I read about his life and talked to colleagues who knew him, what really stood out to me was the way he valued and cared for people. He had a love for teaching people. He proudly attended his daughter's sporting events and recognized and had an appreciation for the difficulties she faced as a woman early in her field. He had empathy for the people of eastern Europe after the devastation of World War II. His story is full of people that were blessed because of their encounters with him. That is fitting because today our focus will be on people. There will be opportunities at this conference to talk about pigs. For the next few minutes, we will honor Dr Hogg with a focus on people.

Dr Clayton Johnson did an excellent job this morning of defining what it means to champion the pig. It is a calling to "effectively represent the interests of the pig's welfare, researching, creating, proposing and supporting the implementation of practical solutions for all obstacles that negatively impact pig health and well-being".[1] It is a role veterinarians are uniquely trained and qualified to fill.

The question I have been tasked with exploring today is regarding the question of who. Who are the future veterinarians that will take on this role, and is our industry ready to welcome them? The second question I want to explore today is who gets to influence what it means for veterinarians to champion pigs?

We will start today with this idea of who are the next generation of swine veterinarians.

Who are the next generation swine veterinarians?

To determine the demographics of our future swine veterinarians, we can look at information regarding the populations already in, and those applying to veterinarian schools. The American Association of Veterinary Medical Colleges (AAVMC) publishes an annual data report covering student enrollment and diversity as well as several other university related statistics. According to their report, prior to 1985 there were more male veterinary students than female students.[2] In 1987 female students outnumbered male students for the first time.[2] That trend has continued. In 2023, the percent of female students was 83.2%.[2]

This trend is likely to continue as only 13% of veterinary school applicants were male in 2023.[2]

This shift is noticeable in the teaching staff as well. Over 50% of research faculty, and over 60% of clinical faculty were women as of 2023.[2]

Not only has there been a shift in gender, but also in racial and ethnic representation. In 1980, less than 5% of veterinary students were from underrepresented groups. That number grew to 25% in 2023.[2]

These changes are making their way from schools and into the veterinary profession. According to the American Veterinary Medical Association (AVMA) market research statistics 76.5% of veterinarians in clinical practice are women, and 62.7% in public or corporate employment are women.[3]

The American Association of Swine Veterinarians (AASV) salary survey would suggest the swine industry has been slower than the small animal industry to see this shift. As of the 2023 results, 38.6% of the respondents were female and 61.4% male.[4] Amongst the respondents we see few women above 50 years of age responding. When we look at the 35-39 years of age category 50% of the respondents were female.[4] In the youngest category, swine vets less than 30 years, 72% of respondents were female.[4]

In 1987, the veterinary schools began the shift towards more female students, and it has taken almost 40 years for that change to be recognized in our industry.

One shift we have not discussed is the generational shift. Although the AVMA market research does not include age or generation information, it is safe to assume there are not significant numbers of boomers in the class of 2025. In fact, Generation Z, the generation born from 1996 to 2010, overtook boomers in the general workforce in 2024.[5] The veterinary industry will follow this trend.

In the past, the majority of veterinarians acting as the champion of the pigs have been white, male, and of the baby boomer and Gen X generations.

The future champions will be more diverse, more female, and will be represented by the millennial and Gen Z generations.

This represents a major change for our industry.

Part 1: Are we ready to welcome a new generation?

This change is coming. Are we ready as an industry to adapt to welcome this drastically different veterinarian?

The mostly white, male, baby boomer, and Generation X veterinarians grew and shaped our industry over their careers. They walked with their pig farmer clients as they made huge leaps to raise pigs inside. They were the ones who first recognized and diagnosed porcine reproductive and respiratory syndrome (PRRS). They helped their clients adapt to multi-site production. They developed the practice of swine population medicine, including disease eradication from large populations. Some of their triumphs are the eradication of pseudorabies, as well as controlling diseases such as rhinitis, mange, lice and *Actinobacillus pleuropneumoniae* (APP). They built the foundations of swine biosecurity including shower in and shower out, as well as air filtration.

They are the builders of modern swine medicine and should be incredibly proud of what they built. I am on this stage today as a part of this industry because people like Gordon Spronk, Tim Loula, Paul Armbrecht, Joe Connor, Max Rodibaugh, Tom Wetzell, Paul Yeske and many others built something compelling and meaningful.

As they grew and developed the modern practice of swine medicine, these practitioners shaped their industry to meet their needs and their goals. As they should have. Our modern industry today was built by them, and it was built for them, but the truth is they are not the future of this industry anymore.

The future swine veterinarians are vastly different and will need the opportunity to adapt and change the industry to meet their needs. This change needs to come sooner rather than later for the health of our industry.

One year ago, in the 18th annual Alex Hogg lecture, Dr Rademacher shared a disturbing trend of veterinarians leaving our industry. In phase one of his projects, his team looked at veterinarians leaving swine medicine entirely. He reported that those who are leaving are young.[6] They report travel, work life balance, and an inability to "disconnect" from work as reasons for leaving.[6] Other factors Dr Rademacher reported included the solitary nature of the job, lack of advancement opportunities, and gender-related challenges.[6] Interestingly, only four out of eleven reported incomes as a significant reason for leaving.[6]

In phase two, Dr Rademacher looked at those leaving practice but remaining in the swine industry. They tended to be older veterinarians with more than ten years of experience. Overwhelmingly they reported work life balance and an inability to disconnect as factors for their leaving.[6] Again, interestingly, only 50% reported that an increased salary was part of the decision.[6]

The industry today was not built or shaped by the next generation of veterinarians, and so unsurprisingly, it does not meet their needs.

Dr Rademacher identified some key concerns. The next generation places high value in the ability to disconnect.[6] They place significant value on their life outside of their jobs and expect to be able to maintain both within balance.[6]

In preparation for this talk, I reached out to a considerable number of colleagues out of school fifteen years and less. Nearly every person talked about this idea of the challenge of finding balance. Again, the theme of too much road time, too much to do, and no way to disconnect came up repeatedly. Unfortunately, gender related challenges were also brought up by nearly every female practitioner I talked to.

The veterinarians of the future are leaning even further into these topics. Roberta Katz, a former researcher at Stanford's Center for Advanced Study in Behavioral Sciences, in an article for the Stanford report, sums up the Gen Z priorities. Members of Generation Z are used to change.[5] They are flexible and resilient and have an expectation that things will change. They have a need to learn lessons for themselves.[5] They care about causes. They do not value hierarchy as much as previous generations and prefer collaboration and teamwork.[5] They value their mental and physical health and to them this means work life balance.[5]

Changes to our profession will be needed if we want to maintain and recruit this next generation of veterinarians.

Long term/high effort

The small animal industry is much further into this effort than we are. They have made investments in emergency practices and hotlines which manage calls and cases outside of business hours. They have also developed significant relief veterinarian infrastructure. These changes have allowed small animal veterinarians to invest in their lives outside of veterinary medicine, while still caring for the needs of their clients.

Changes along these lines would represent a major change for our industry, and the younger generation in the room should evaluate this carefully. First, small animal clinicians have been able to do this while still meeting the needs of the client. We should not forget that our jobs rely on our abilities to meet the needs of our clients. The previous generation of veterinarians understood this well. This is how they earned a spot at the table shaping and leading in the swine industry.

If the next generation wants the same opportunities to lead in the swine industry, then the client's needs cannot be missed in the pursuit of work life balance. We must find a way forward that maintains both.

Second, the ability for small animal veterinarians to disconnect has not come without a price. The small animal industry has made a drastic shift away from veterinary owned practices. Small animal vets are finding better work life balance, but it is at the cost of ownership and control. Our younger members should consider this carefully.

One change that may be an easier step towards balance would be telemedicine. Telemedicine may be a tool to drastically reduce travel time, which has been cited as a concern for those leaving the industry. There are certainly challenges that need to be overcome, but I would urge our organization to pursue the opportunities in telemedicine further.

Another step that would take careful consideration but could have large payouts is the utilization of veterinary technicians. This is not a suggestion towards the idea of a mid-level practitioner, but of a highly trained, limited-in-scope technician. Our small animal colleagues utilize well trained technicians to allow them to focus their time more appropriately. Could swine veterinary technicians do the same for swine veterinarians?

These ideas need effort to develop and deploy. There are other shorter term and lower effort investments we could adopt more easily.

Shorter term/lower effort

Artificial intelligence is moving and adapting quickly. As an industry we should be finding and testing technologies that support a reduction in hours, in travel, and on-call time.

According to the bureau of labor statistics, in 2023, 67% of families with children 18 and under reported both parents working.[7] This is a change from the past and has made finding and securing high quality childcare a nationwide challenge. Our industry makes this even more difficult because our typical working hours do not match most childcare hours. A simple shift in farm and clinic schedules could reduce the challenge of childcare for the veterinarians of the future.

Although pay was not a major concern presented by Dr Rademacher, according to the AASV salary survey, a pay gap does exist in this industry. For the age category ranging from ages 40-49, female practitioners made $0.73 to each dollar earned by a male colleague.[4] For ages 35-39, female practitioners made $0.69 to each dollar earned by a male colleague.[4] For age 31-34, it was $0.63 to each dollar.[4] The excellent news for our industry is that this gap completely disappears for practitioners 30 years of age and younger.[4] If our goal is to recruit and maintain the next generation of veterinarians, we need to continue to prove as an industry that we equally value the work of female practitioners.

Another concern that is unique to female veterinarians revolves around barns consistently failing to provide items needed specifically for a female practitioner. There is an unfortunate common experience held by all female veterinarians of not having the items needed for them on farm. Again, if we understand the importance in recruiting the next generation, then our industry should clearly communicate to our clients the value of our female veterinarians by requiring they have what they need when they go to a farm.

An item more specific to the generational shift would be to look at our current industry activities. Generation Z and even millennials have quite different hobbies and activities in comparison with Generation X and the baby boomers. Our industry still focuses vastly on activities such as hunting, golfing and skiing, which are not activities being readily adopted by younger generations.

A final comment to close this section of my talk would be regarding leadership. One of the best ways to shape our industry to fit the next generation, is to give leadership of our industry to the next generation.

My challenge to the older generation is to evaluate how you are transitioning leadership. What young veterinarians on your team are ready for leadership? If they are not ready, what role can you play in developing and coaching them?

My challenge to the younger generation is to not leave. Stay in this profession and be a part of the change you want to see happen. Engage in these conversations. Bring forward solutions. Be ready to step up and lead.

Conclusion

The future of our industry is in the hands of veterinarians that look very different from those that built this industry. If we do nothing to adapt and change, we will face significant difficulties in recruiting and maintaining the next generation of veterinarians. Instead, I challenge our industry to embrace change and to adapt the swine veterinary profession into the place where the veterinarians of the future want to be.

Part 2: Who gets to influence what it means to be the champion of the pig?

At this point, it is clear that veterinarians are uniquely trained and qualified to be the champions of pigs. The next question becomes, who gets to influence veterinarians regarding what it means to be the champions of pigs? Again, an obvious answer is the welfare professionals in this room. We will have an opportunity later this morning to hear from Dr Anna Johnson. She will do an excellent job highlighting the work of our welfare-focused colleagues.

The question that needs further discussion is regarding the pork consumer. Do they get to play a role? Do they get to influence what it looks like for us to be the pig's champion?

For years we have shaped our industry around production. Our efforts have vastly focused on improved performance with the goal of reducing the cost of producing pork. When markets experience oversupply, we rely on exports and reduced pricing to move additional pork. However, the most recent downturn in the market was different.

Here we will go back to the exact same discussion we had above regarding the transition of generations within the veterinary medical industry, but now looking at the transition of generations regarding pork consumers. With Generation X and baby boomer consumers, our model of focusing on production and pushing extra pork to the consumer via reduction in cost worked well. Unfortunately, millennials and Generation Z are hugely different generations. The old playbook is not working to increase pork consumption in these individuals.

The National Pork Board recently completed a large market study to understand and segment pork consumers. Their work has shown a significant drop in pork consumption year over year.[8] The drop in consumption falls along generation lines with millennials consuming significantly less pork compared to baby boomers.[8] Even more concerning, the consumption drops even further when comparing Generation Z to millennials.[8] The production focused approach of our industry is no longer a relevant approach for the modern consumer.

Instead, our industry must make a shift to a consumer centered industry. The day after our AASV meeting concludes, National Pork Board will be introducing a new consumer centered campaign to pig farmers. Later in the year it will launch in significant pork markets, including California.

Today we are sitting in a room in California, and I am telling you that our industry needs to shift to focus on the needs and wants of the Californian consumer. Our industry will not thrive if we cannot win over the millennial and Generation Z consumers of California.

At the exact same time, I am aware that the Californian consumer has no idea how food is produced in the United States. They are in no way qualified to determine how pigs are raised. So how do we move forward with both things being true?

What the swine industry needs are intelligent, highly educated professionals, with experience and expertise that can stand in this gap. People that can take what the consumer says they want, and work to find solutions that actually improve the welfare of the pig, while at the same time being executable by pig farmers.

We need people with critical thinking skills and the ability to plan and execute research trials. We need people with training in pig behavior and welfare. We need people that have a seat at the table to advise and assist pig farmers. We need the people in this room, swine veterinarians, to stand in the gap.

I have been in a unique position within our industry now for several years because of social media. I post videos highlighting the way we raise pigs in the United States. Although I do not get feedback from consumers necessarily, I do get feedback from the public on a regular basis.

These comments have helped me to understand how little the public truly knows about how food is produced in our country. These comments have also taught me that despite their lack of knowledge or legitimate understanding, the public can still play a role in helping me as a veterinarian to improve animal welfare.

I grew up in the swine industry, so raising sows on concrete seemed very normal and unproblematic to me. Then I posted to social media and experienced many comments regarding the viewers' concerns for pig comfort when being raised on concrete. Because of the constant comments, I became more aware of stiff sows and sows looking painful when standing up as I was filming for videos. I also struggled with how to respond to some of the direct comments regarding sow lameness. I could not reply that lameness is not an issue in our industry with it being in the top three causes of death loss on most sow farms.

I found myself in a gap. The public was pointing out a problem, sow discomfort and lameness on concrete, for which I did not have a solution. This led to me working with rubber flooring and eventually installing it in a barn for a pilot project. The project was a success, which led to Pipestone, a rubber flooring company and three different universities working together to apply for a pork board grant. That grant was funded, and the project started collecting data on rubber flooring in November of 2024.

This rubber flooring project is a fitting example of finding a potential solution that may improve the experience of the pig, may improve health and production on farm, and may improve the public perception of how pigs are raised. If this project works out, it is a win for pigs, producers, and consumers.

Another comment I receive all the time is that pig farmers and veterinarians only care about production numbers and finances, and we do not actually care about pigs. This is false. However, this did start me thinking about the sorts of reports we look at on a regular basis. We look at multi-page production and financial reports every single week. We identify concerns and set goals to improve on the metrics we measure.

Our industry historically has not looked at pig care or welfare metrics in the same way. We will not drive continuous improvement if we do not measure. So, we started to measure and report welfare metrics. Our report looks at care for lame sows measured by the number of treatments and treatment success rates. We look at care for farrowing sows, again, measured by number of treatments and treatment success rates. We also look at a variety of other treatment-related metrics which help us to at least start to measure pig care and welfare in a quantitative way. Now that the report is in the packet, it is a part of the discussion. We drive continuous improvements around production, finances, and welfare. A solution that is good for pig outcomes, production, and public perception of our industry.

It was the public that brought to my attention how much we need a solution to the pain at castration and tail docking problem. I knew this was an issue with no clear and easy solution, and so it was easy to ignore until I started posting on social media. A comment I get on a regular basis is this "She says she cares about pigs, but off camera she is going to cut off their tails and tear their testicles out without anesthesia." The people posting have no idea of the complications of AMDUCA. They have no concept of the amount of labor involved with picking pigs up a second time. They are not able to understand the nuance and challenge of this issue.

Those things are true. At the same time, it is also true that they are not wrong. It does not make sense to say that I do care about pigs and at the same time support castration and tail cutting without anesthesia. Being on social media helped me to see the issue of pain control at castration in an entirely different light. I have again found myself in the gap.

If we do not stand in the gap, then others will. Proposition 12 is an example of what this looks like. We, as always, were focused on excellent production. As veterinarians we were in the trenches focusing on health and disease elimination and all the legitimate health issues that take up our precious time. The voters in California were focused on freedom of movement for the sows. No one stood in the gap to bring solutions forward that would work for both the producers, the pigs, and the people of California. So, the people of California wrote the solution and now we follow their guidelines because we need to sell pork in their state.

We do not have to wait for the next Proposition 12.

The minds in this room are the people who can and should develop, evaluate, and roll out solutions that work for producers, pigs, and the people of California. As Dr Johnson pointed out earlier this morning, this isn't something general practitioners have time for in their schedules right now. As an industry we need to make this a priority. We need to find a way for this organization to prioritize development of pig, producer, and public focused solutions.

It will be a shame if our expertise sits by the wayside while the unqualified public writes the guidelines on pig production. We need to stand in the gap.

Farrowing and lactating sow housing matters to the public. We should be the ones that put forth a solution. Pain control at castration and tail docking matters to the public. We should be the ones that put forth a solution. Barren environments and enrichment matter to the public. We should be the ones that put forth a solution. Antibiotic use for production animals matters to the public. We should be the ones that put forth a solution.

The swine industry needs our expertise, our experience, and our ability to solve complex problems. We can and we should stand in the gap.

Conclusion

The future of both the swine veterinary profession and pig farming relies on the ability of the people in this room to meet the needs of the next generation. As a veterinary profession, we must adapt to be able to recruit and maintain new veterinarians. As professionals in the pig industry, we must help our clients shift to a consumer-centered approach that meets the needs of the pig, the producer, and the public.

Thank you again for the opportunity to be a part of this morning's agenda. I am so grateful to have had this opportunity, and I look forward to continuing these discussions.

References

1. Johnson C. 2025 Howard Dunne Memorial Lecture. The challenges and opportunities of becoming the pig's champion. AASV Annual Conference Proceedings. 2025.

2. Annual Data Report 2022-2023. Washington, DC: American Association of Veterinary Medical Colleges; 2023 March p.1-67. Accessed August 24, 2024. **www.aavmc.org/about-aavmc/public-data/**

3. US Veterinarians Numbers. AVMA. December 31, 2024. Accessed October 13, 2024. **https://www.avma.org/resources-tools/reports-statistics/market-research-statistics-us-veterinarians**

4. Brown D, Canon A, Schulteis S, Snelson H, Webb S. 2023 Salary Survey Results. American Association of Swine Veterinarians. 2023. Accessed August 24, 2024. **https://www.aasv.org/wp-content/uploads/2023/12/SalarySurvey2023.pdf**

5. De Witte M. 8 Ways Gen Z Will Change the Workforce. Stanford Report. 2024. Accessed October 23, 2024. **news.stanford.edu/stories/2024/02/8-things-expect-gen-z-coworker**

6. Rademacher C. 18th Alex Hogg Memorial Lecture. Past, present, and future challenges for the swine veterinary profession. AASV Annual Conference Proceedings. 2024. **https://www.aasv.org/swine-information/?item=27190**

7. Employment Characteristics of Families-2023. Bureau of Labor Statistics. April 24, 2024. Accessed November 1, 2024. **https://www.bls.gov/news.release/pdf/famee.pdf**

8. National Pork Board. Consumer Value at Stake Study, 2023. Accessed November 1, 2024. **https://live.porkcheckoff.org/news/reaching-the-consumer-new-research-helps-identify-the-future-pork-consumer/**

Angela Baysinger Memorial Lecture

Advancing pig welfare together: Standing on the shoulders of Angela

Anna Johnson, BSc, MSc, PhD

In remembrance and honor of Angela K. Baysinger

Angela K. Baysinger
1967-2024

Angela was born in 1967 in Rantoul, Illinois. After living on several Air Force Bases across the country where her father was stationed, she grew up on a rural farm near Martinsburg, Missouri. It was there at a young age that she gained a true love of farm animals, especially pigs. She attended the University of Missouri where she earned a Doctor of Veterinary Medicine degree in 1992. After graduation, she moved to Sutton, Nebraska, where she practiced at Sutton Veterinary Clinic.

Dr Baysinger left practice in 1995 to pursue a master's degree in epidemiology at the University of Nebraska/USDA Meat Animal Research Center – Clay Center, Nebraska. While there she served as the Interim Extension Swine Specialist. After almost 20 years of various positions with animal health companies as a technical swine veterinarian and with Farmland Foods as Vice President of On-farm Food Safety, she accepted a position with Merck Animal Health. There, she first was a Technical Service Veterinarian and then later was the North America Animal Well-being Lead until the time of her death in 2024. Animal well-being was her passion, and she championed many educational conferences to help bridge livestock producers with consumers. This position led her to obtain a master's degree in animal welfare, ethics, and law from the University of Edinburgh in 2021.

Dr Baysinger was a member of the AVMA Animal Welfare Committee from 2013 to 2024. She served on the Board of Directors of the Professional Animal Auditor Certification Organization (PAACO) from 2004 to 2007 and again from 2011 to 2024. In early 2024, she was installed as president of the American Association of Swine Veterinarians, an association that she was involved with for many years and truly loved. She also contributed to science with peer-reviewed scientific journal articles and many abstracts over the years.

A few of Dr Baysinger's professional accomplishments and awards include: The Feather in Her Cap Award (2021), the AASV Meritorious Service Award (2021), the PAACO Service Award (2022), and the AASV Howard Dunne Memorial Lecture (2022).

More information about Dr Baysinger's personal history, her professional contributions to swine health and welfare, and her lasting impact on colleagues, mentees, and friends is shared in the AASV Member Recollection Video at **aasv.org/video/heritage/angela-baysinger/**.

https://doi.org/10.54846/am2025/3

Advancing pig welfare together: Standing on the shoulders of Angela

A. Johnson[1], BSc, MSc, PhD; K. Barger[2], BA, BS, DVM; C. Brezina[3], BS, DVM; T. Burkgren[4], BS, DVM; E. Byers Taylor[5], BS, DVM; G. Golab[6], BS, DVM; B. Johnson[7], BS; C. Johnson[6], DVM, MS, MSc; C. Kaster[8], BS, MS; S. Millman[3], BSc (Agr), PhD; M. Pairis-Garcia[9], BS, DVM, PhD, DACAW; D. Sjeklocha[10], BS, DVM; L. Tokach[11], BS, DVM; J. Waddell[12], BSc, DVM, MBA; S. R. Webb[13], BSc, MS

[1]Department of Animal Science, Iowa State University, Ames, Iowa; [2]KB Welfare Consulting LLC, Seneca, South Carolina; [3]Veterinary Diagnostic and Production Animal, Ames, Iowa; [4]Retired Executive Director, American Association of Swine Veterinarians, Perry, Iowa; [5]Merck Animal Health Technical Services, Swine, Lenexa, Kansas; [6]American Veterinary Medical Association, Schaumburg, Illinois; [7]Roots and Legacy, Manhattan, Kansas; [8]American Meat Science Association/Professional Animal Auditor Certification Organization, Kearney, Missouri; [9]Department of Population Health and Pathobiology, Raleigh, North Carolina State University; [10]Merck Animal Health Technical Services; Beef, Topeka, Kansas; [11]Abilene Animal Hospital, Abilene, Kansas; [12]Boehringer Ingelheim Animal Health USA Inc. Duluth, Georgia; [13]American Association of Swine Veterinarians, Perry, Iowa

Introduction and the Angela Baysinger Memorial Lecture

Dr Angela Baysinger (nee Wright) was born February 14, 1967, in Rantoul, Illinois and died March 8, 2024, in Bruning, Nebraska. These are the bookends of a life well lived and a foundational welfare legacy. In her honor the AASV has created the Angela Baysinger Memorial Lecture. The endowed lecture will be given annually by a veterinarian or animal scientist with expertise in animal welfare in swine, bovine, poultry, fish medicine, or a combination of these species. The objective for this lecture is the education of members on current work in animal welfare in the lecturer's current species with comparisons and contrasts to swine medicine if applicable. Secondly, the lecture should educate members about opportunities and challenges for the improvement of animal welfare, providing practical examples.

The purpose of this inaugural proceedings paper is to provide the depth, scope, influence and powerhouse that was Angela. Furthermore, we will reflect on the human attributes that made her honest, vulnerable, open and approachable. So, let's take a walk down memory lane and you will clearly see how Angela has left an indelible mark on us, the swine industry and swine welfare.

The tender years

Angela shared with Dr John Waddell her first boss and mentor that the OB doctor who delivered her was named Dr Mayhem! After living on several Air Force Bases where her father was stationed, Angela grew up on a rural farm near Martinsburg, Missouri. It was there at a young age that she gained a true love of farm animals, especially pigs. Angela became enthralled when she helped their veterinarian pull piglets from gilts. Angela noted "*This is where you saw the circle of life, birth, growth and death.*" In high school she told the school counselors she wanted to be a large animal veterinarian. The counselors told her that this was a man's job and Angela thought "*I'll be darned!*" She attended the University of Missouri Columbia where she earned her BS in Animal Science (1988), her DVM (1992) and an honor badge of "*swine goddess!*" After graduation, she practiced at Sutton Veterinary Clinic, a beef cattle and swine focused animal practice. Dr Waddell noted that "*Angela hired us, and to this day she was the most prepared new grad.*" When she first moved her possessions into the veterinary clinic garage "*She topped off the pile with a sawed-off 12-gauge double barreled shotgun. It was a not-so-subtle message of don't go messing with me or my stuff!*".[1] Angela left practice in 1995 to pursue a master's degree in epidemiology at the University of Nebraska/USDA Meat Animal Research Center Clay Center and served as the interim Extension Swine Specialist. After almost 20 years in various professional companies and positions, Angela joined Merck Animal Health as a Technical Service Veterinarian and later as their inaugural North America Animal Welfare Lead. Angela's dedication to continuing education and new skills was evident throughout her life, culminating in a master's degree in Animal Welfare, Ethics, and Law from the University of Edinburgh in her final years (2021).[1]

When you think of Angela and welfare what comes to mind?

Angela lived life with great enthusiasm, adventure, and like "*Someone left the gate open*".[1] At the beginning of writing this legacy tribute, the co-authors were asked to identify words that reflected Angela as a person and a professional (Figure 1A) and we also reviewed her memorial board created in her honor that contained posts from over 130 family, friends and colleagues (Figure 1B). From these themes we explore Angela's personal and professional legacy.

Impact – devotion to improving livestock and aquaculture welfare

Theme 1: Connections

To influence a story, you need to be entrenched in the story. Angela was a consummate story teller, and the stories always had purpose and meaning which in-turn evolved into extraordinary impact. While delivering Angela's eulogy, Dr Lisa Tokach noted that "*The thing that I always found most impressive about Angela was that she always had good connections. Having Angela on your side was like holding a wild card because she connects to everything.*" Let's consider the Angela connections.

Figure 1: Words to frame Angela the person and professional **A)** from the co-authors and **B)** the wider community.

If the connection was "professional", the words might be veterinarian, agriculturalist, politician, and Angela. Angela was a leader in our field. She served in numerous capacities on committees, as the Annual Meeting Program Planning Committee chair, and recently as the AASV president. She was awarded the Meritorious Service Award, delivered the Howard Dunne Memorial Lecture (2022), and too many other awards to name. Angela was selected to be part of the inaugural Swine Veterinarian Public Policy training program.[2] She mastered media training videos to then march on to Capitol Hill like a boss armed with an agenda, dressed to the nines, and ready to do battle for her livestock producers. Angela was a master at bringing the livestock and poultry industry silos together and getting them talking. The food industry is often so segmented and often have antagonist relationships. Whereas Angela was an ally.

If the connection was "protector", the words might be animal welfarist, diversity equity and inclusion (DEI) advocate, mother, and Angela. Angela worked in the animal welfare arena from all angles. She made swine, beef, poultry and aquaculture producers better at animal husbandry, retailers better at understanding what production agriculture was all about, and consumers more appreciative of the care and effort put into getting food onto their plates. She did this all while staying mindful of what the animal needed. Angela was a charter member and champion of the AASV DEI committee. She was the person who believed that no one eats alone at the lunch table. She would always introduce herself, find some common connection and introduce you to others who shared similar passions.

If the connection was "life-long learner", the words might be education, scientific author, mental health advocate, and Angela. Angela's learning never stopped with a DVM, two master' degrees, animal welfare certification boards and publications. More to come later!

If the connection was "adventurer", the words might be Harley-Davidson motorcycle rider, pilot, runner, and Angela. It was Angela's dad that got her started with motorcycles. As her husband, Jerry mentioned in her obituary, Angela loved getting some much needed "wind therapy" on her own Harley. There was no backseat riding!

If the connection was "collector", the words might be fashion, quirky friends, and Angela. Dr Tokach noted that when she first started interacting professionally with AASV, she stayed with black or gray blazers to blend in. Not Angela, she showed up in a pink tweed blazer looking like a fashion model.

Dr Tokach confessed that I thought she looked fabulous, and I felt like my wardrobe was always boring. Angela never judged, she just said, *"Swine sister, I have to introduce you to Stitch Fix!"* (Figure 2A and B).

Figure 2: Angela wearing fabulous glasses at AASV 2023 with **A)** Emily Byers Taylor and **B)** Lisa Tokach.

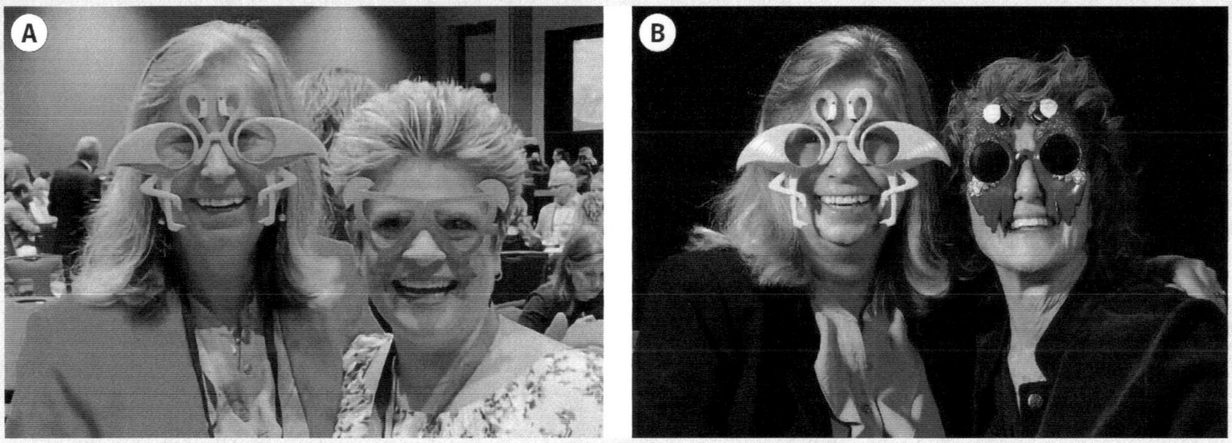

Theme 2: Developing audits and the creation of the Professional Animal Auditor Certification Organization (PAACO)

Back in 2002, Dr Anna Johnson joined the National Pork Board as the first Director for Swine Welfare and oversaw the Animal Welfare Committee, of which Angela was a member. This was a time of upheaval and uncertainty as to where the US pork industry was going with swine welfare. Producers were extremely wary of the word "welfare." Angela was one of the first soldiers to go out onto farms and converse with producers explaining to them what welfare was, how it could contribute positively to production, and that they were already doing a lot of good welfare practices. Examples that she used were vaccination programs to support good health, nutritionally sound diets appropriate to stage of production and using the right handling tools to move pigs.

Next came the checks and balances. Educational programs were acceptable to producers and with much coaxing, on-farm assessments began with the Swine Welfare Assessment Program. Angela was neck-deep in these communications to producers and all stakeholders. So many gaps were identified and filled, with Angela being involved in developing templates for standard operating procedures, identifying what type of on-farm records were needed, drafting practical methods to score animal-based measures in field conditions and so on. Angela was among those that recognized the value of integrating animal welfare criteria into the exiting Pork Quality Assurance program, of which producers were rightly proud. This resulted in the Pork Quality Assurance Plus Program that consisted of producer certification and on-farm site assessments. This updated and expanded program was a strategic move that gave producers a platform to articulate animal welfare in their own terms through the actions performed on-farm every day. The program continues to grow and as of December 2023, there are 20,427 sites that had a current Pork Quality Assurance Plus site assessment, representing an inventory of 65,360,736 pigs.[3]

The next progression after assessments were third party audits. Although in the late 1990s the Humane Farm Animal Care program was established and created the first ISO-certified swine welfare label and audit,[4] certain aspects of that audit such as space and housing requirements excluded most commercial producers. Hence, the National Pork Board began discussions in 2003 with SES: Environmental Audit Compliance and Sustainability group that desired to build and implement third party audits for the mainstream US swine industry.[5] Unfortunately, these discussions did not result in a formalized agreement and program because this company knew very little about pigs or animal welfare. These challenges led to conversation between Angela and Drs Paul Sundberg, Anna Johnson, John McGlone and Tom Burkgren about developing a specific auditor training program for animal welfare on a really hot summer day in Arizona at the American Society of Animal Sciences annual meeting! From these preliminary discussions, Angela and Tom spent a fair bit of time creatively doodling on the back of bar napkins and the Professional Animal Auditor Certification Organization (PAACO) idea was conceived.

Since then, PAACO has become the authority on animal welfare auditing in North America, providing high-quality training and certification credentials for auditors and audit tools. Those that serve as PAACO trainers are leaders who work in animal welfare and have extensive experience in their area of expertise as well as auditing. Angela was one of those icons. She stepped right in at the beginning and served as a board member for almost the entirety of PAACO's existence, taking only a short break for one term. She served on the Executive Committee twice and worked closely with both Mr Mike Simpson and Ms Collette Kaster during their tenures as Executive Director. She was highly active in putting together the meat plant welfare auditor training by working closely with the American Meat Institute (now Meat Institute) to create a robust audit for meat plants that met the expectations of customers such as McDonald's and Costco. She was known for being pragmatic and resolute to get the best possible standards into the audits. Angela helped train shadow auditors in the plant and had some stories! One time a trainee auditor noted they would have failed the plant. When Angela enquired as to why, they said *"There is metal sticking out in that pen."* The metal in question was the water drinker!

Angela saw merit and value in auditing. She was a stickler for correct auditor qualifications, training, practice and accountability. Dr Johnson noted that between Dr Jim McKean, ag lawyer and Angela an auditor connoisseur, every written word was scrutinized to be clear and transparent. *Is that a "should", "must," or "maybe"! Is it objective? Can it be measured to exactly the same standard every time by many people?*

In 2014, Mr Simpson approached Dr Suzanne Millman to chair the creation of PAACO's Common Swine Industry Audit (CSIA) curriculum committee. Together with colleagues they created educational materials and delivered the foundational CSIA auditor training. Angela served on the CSIA curriculum committee as the PAACO Board of Directors liaison, providing sage insights and relaying questions between the board, the task force that created the audit tool and the curriculum committee. Angela's enthusiasm, attention to detail and humor were noteworthy, and critical to the program's success. As the committee members beta-tested the CSIA on commercial farms, Angela graciously and effectively communicated our frequent but important questions for clarification *"Did the Task Force truly mean to say..."*. These important refinements strengthened the audit tool and the educational materials. Above all Angela radiated infectious energy that helped to keep the team going. During the first training event in Waseca Minnesota, Dr Millman recalls Angela bounding into the hotel breakfast room each day with a huge smile, glowing from her early morning run through the prairie and relaying tales of wild turkeys and cattle that paused to observe her trotting past! Such amazing energy and love for life! To date, there are currently 133 active certified swine welfare auditors.

Theme 3: Animal Welfare Assessment Contest and the American Veterinary Medical Association

In 2001, faculty members from Michigan State University (MSU; Heleski and Zanella) and Purdue University (Pajor) presented the idea of promoting animal welfare science to undergraduate university students by coupling it with the more traditional concept of livestock judging to the International Society for Applied Ethology. Two competitive grants were awarded to MSU for establishing the pilot Animal Welfare Assessment Contest (AWJAC) for undergraduate students in welfare assessment/judging. This event was held March 1, 2002 as one of the initial events of Michigan State's Agriculture and Natural Resources week. On competition day, 18 students from Purdue University, University of Guelph, University of Wisconsin and MSU tested their skills in assessment and oral reasons.[6] In 2008 a veterinary division was created with support from the American Veterinary Medical Association (AVMA) and species were expanded to include companion, laboratory, and exotic animals in addition to the traditional livestock and poultry focus. This further expanded to include a graduate student division. The AVMA became the primary organizer of AWJAC in 2014 and schools from across North America brought a total of 28 teams and 116 participants to compete.

Angela became involved in AWJAC during 2017 through her role at Merck Animal Health. In this role Angela welcomed and engaged participants and coaches at the networking receptions, supported the overall AWJAC platform and encouraged student participation when she was speaking at other events. Her biggest impact was in starting the inclusion of AVMA member veterinarians at the in-person AWJAC event, through Merck Animal Health sponsorship of targeted travel grants. Furthermore, her example and encouragement facilitated new sponsors, such as PAACO. The event continues to be well supported (Figure 3) and 95% of the 429 students surveyed valued the process, agreed that the event increased their knowledge about animal welfare science, and would recommend the AWJAC.

Consistent with Angela's sense of fun and mentorship, the evening social event at AWJAC evolved to require animal welfare experts and coaches to don light tiaras and other paraphernalia that lightened the atmosphere and helped students identify potential mentors and engage in low stakes networking to learn about career opportunities in the field.

While Angela's role with the AWJAC was limited to sponsor capacity, she did hold other responsibilities at AVMA. Angela served on the AVMA's Animal Welfare Committee (AWC) for 5 years, holding the roles of chair and chair-elect within that timeframe. During her time on the AWC she was instrumental in providing the AWC portion of the educational content for the 2023 AVMA Humane Endings Symposium and the 2023 Animal Welfare track of the AVMA Annual Convention. Angela was also the chair of the Swine Working Group for the 2024 edition of the AVMA Guidelines for the Humane Slaughter of Animals and did a large amount of work to assist in fully reformatting the document into species specific chapters while updating the literature, references and figures used in the guidelines.

Theme 4: Advancing Animal Welfare Together Symposium

No one was as good at breaking down silos and bringing people together to talk animal welfare as Angela! She pioneered the Advancing Animal Welfare Together Symposium (AAWT) in 2017. The purpose of AAWT was to bring together leaders in the fields of animal welfare and sustainability to broaden perspectives, leverage networks, and inspire innovative solutions for issues impacting the future of the industry. The event invites thought leaders in the cattle, poultry, swine, dairy and aquaculture industries and representatives from every link of the supply chain. The respect and reverence people had for Angela was evident in their willingness and desire to speak at AAWT, serve on her advisory council and/or attend the conference. *"As someone who plans events for a variety of clients, I can attest that this is not the norm! The appreciation people at the highest levels of the industry had for her and her work was evident in their willingness to give of their time and knowledge to her projects"* (Ms Becky Johnson personal communication; Table 1).

Angela was the decision maker in every aspect, from the location, to the topics and the speakers. Angela's industry connections made it possible to bring in the *"heavy hitters"* from all species and all aspects of the supply chain to challenge and inspire the industry toward progress. For example, at the inaugural meeting Dr Temple Grandin, a pioneer in animal welfare and handling, opened the symposium by emphasizing the importance of exploring future animal welfare advancements through the lens of lessons the industry has learned from the past. Other topics and renowned animal welfare experts included, animal welfare compliance: Drs Lily Edwards-Callaway, Colorado State University and poultry welfare; Dr Janice Swanson, MSU.

The symposiums were held in Dallas, Orlando, Des Moines and Calgary. At each event ~120 attendees were registered. In addition, Angela invited 5 to 10 graduate students each year and fostered connections for them with attendees, contributing to the future success of the industry (Figure 4A,B).

Figure 3: The AWJAC participant numbers by categories between 2017 and 2023.

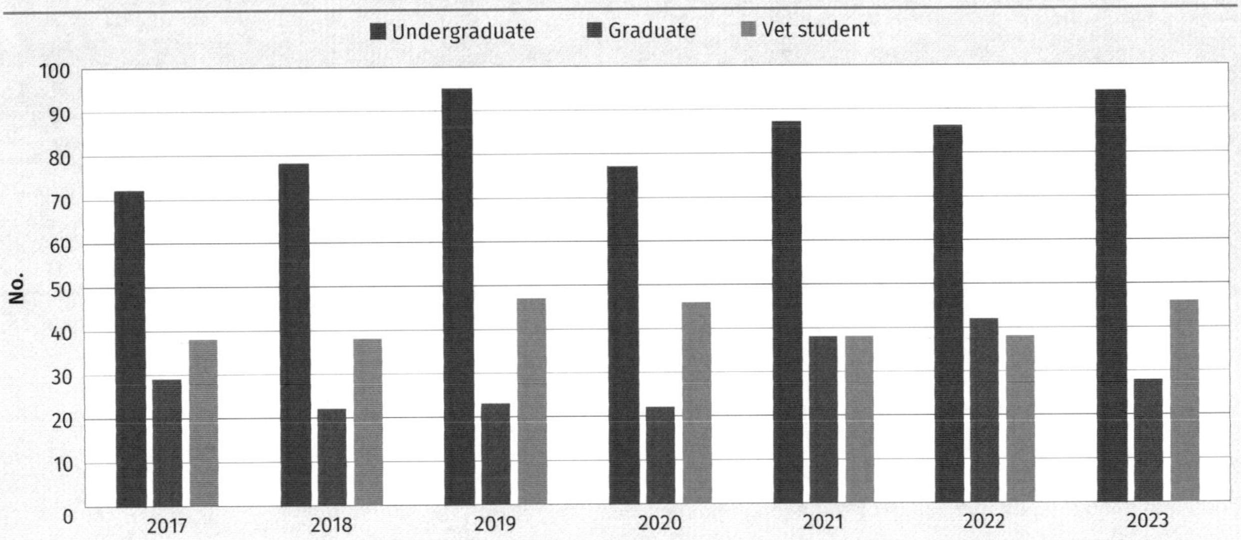

Table 1: AAWT feedback survey impact statements from 2023

Attendee	Quote
1	It's an honor to be invited and to attend the AAWT symposium. It's truly one of the best meetings that I've attended. Angela and her team do a remarkable job with the planning and execution of the event. I enjoy the wide scope of participants from "across the barnyard" for networking, learning and discussions. So many sectors are dealing with the same issues and it's great to be able to learn from each other. Thank you for hosting this!
2	As a producer, veterinarian, and lead of an animal health and welfare and sustainability research program, I thought the content would be very applicable. What I didn't realize was how constructive, collaborative, and invigorating this conference would be. Truly the best symposium I have attended in years.
3	Looking around the tables on the last night, seeing the number of lives Dr Baysinger has touched through her work was incredible. From students just beginning their careers to long time colleagues across multiple sectors and species, she has positively shaped agriculture for not only the animals, but for the people as well by curating and assembling individuals with unique perspectives and skill sets to stretch our beliefs and limits.
4	Merck and Angela are awesome partners with connecting industry members and academia. Appreciate the balance of audience and the tailored messages, but also the ability to network and converse freely.

Figure 4: A) Angela speaking at the AAWT symposium, Calgary, 2023. **B)** The Roots and Legacy team with Angela.

After Angela's passing, the members of the AAWT advisory council sought out the Merck team to ensure this symposium continues to encourage a safe forum for open dialogue and collaboration between producers, academicians, veterinarians, packers, grocery and restaurant corporations. The next AAWT will be held October 2025.

Theme 5: Doing the hard stuff

Livestock welfare can be challenging and it takes commitment, grit and determination to shoulder welfare projects in times of peace and hardship. Angela did not shy away from the hard stuff and embraced the idea that what is impossible today can become standard practice in the future. For example, Angela and Ms Kaster were instrumental in the implementation of CO_2 stunning in modern US pork plants and she published in the disciplines of euthanasia,[7] neonatal pain,[8] and depopulation.[9]

Specific to depopulation, COVID-19 tilted the globe into unknown waters. The swine industry tried numerous management interventions to slow growth, reduced sow breeding, conducted select euthanasia, targeted heavier slaughter weights, and opened more barns. These strategies provided a temporary stopgap but did not entirely resolve the backlog of market-ready pigs.[10] Next came depopulation using approved methods or techniques permitted under exceptional circumstances. Carbon dioxide supply was highly dependent on geographical location and the Midwest experienced severe disruptions.[11] The type and quantity of ammunition needed for gunshot were limited.[12] Use of penetrating captive bolt was difficult due to multiple devices needed to reduce overheating, reduce cleaning demand, and prevent shooter fatigue. Here came Angela with her team. She considered, researched, conducted, published and was a defender of ventilation shutdown+ (VSD+).

Angela appreciated that the barn yard came together and supported her VSD+ efforts. However, she also took the punches on the chin from stakeholders outside of the swine industry as well as from fellow swine veterinarians who did not agree with the method. Dr Kate Barger (2024, personal communication) attended an AVMA meeting with Angela. As they were crossing the street protestors were calling her terrible names for killing animals using VSD+. Dr Barger asked *"How do you keep your cool?"* and Angela's reply was *"I am not going to let them suck the oxygen out of the room and tell the story because they do not know the story. I know we did good, we are good people and I work with great people."* Dr Barger asked if she was concerned that protestors would seek her out at home and she replied, with a smile *"Well, I live on a pretty good-sized piece of land and can see people coming. Merck gives me protection and I am sure if they approached and saw me on my porch with my shotgun that they may think twice."* Dr. Barger and Angela formed an unbreakable bond during these difficult times, and maintained communication and support for each other when poultry were undergoing depopulation due to disease.

Theme 6: Learning new species and processes

Angela was keen to learn new topics, processes and species. Although she was a swine gal at heart, she expanded her repertoire to include poultry (International Poultry Welfare Alliance [IPWA], 2024),[13] beef and aquaculture. Angela was a board member of IPWA, representing allied industry, and was enthusiastic about the opportunity to connect the poultry supply chain to focus on welfare (Figure 5). With typical Angela flair and energy Dr Barger noted that Angela always encouraged IPWA members to collaborate on welfare initiatives, share their success and their challenges during committee meetings, and to continually discuss on how to improve welfare outcomes. With her swine welfare knowledge, she was able to contribute *"out of the box"* thinking to the poultry welfare discussion and at the same time eagerly wanted to learn more about chicken and turkey welfare.

Figure 5: Left to right: Collette Kaster, Angela, and Karen Christensen (Senior Director of Animal Welfare, Tyson Foods).

Angela was one of the first to work on the Processed Verified Program.[14] During this time, she had the opportunity to work directly with McDonalds, Dominos and others. This allowed her to understand the full value chain. Angela used her experiences and networks to create opportunities for discussions that had never been had. Towards the end of her career, Angela became immersed in aquaculture. She worked directly with the AquaCare365 program in reviewing and ensuring that the program was relevant to the Aquaculture industry, and she made it a priority to include fish welfare in the discussions at AAWT (**https://www.merck-animal-health-usa.com/aqua-care365**). It is remarkable the level of trust that Angela instilled among stakeholders, as well as a willingness to engage in tough conversations when they knew Angela was at the helm and ensuring safe spaces.

Theme 7: Paying it forward

A mentor is an individual with expertise who can help develop the mentee's career. A mentor has two primary functions; (1) the career-related functions that enhance the mentee's professional performance and development and (2) the psychosocial functions whereby the mentor establishes themselves as a role model and support system for the mentee. Both functions provide explicit and implicit lessons related to professional development and general work-life integration.[15] Angela had several mentors including Drs John Waddell, Cate Dewey, Lisa Tokach and Barb Straw, and in-turn she became a fabulous mentor. Angela learned and shared along the way. She built her community and village. She said that at the beginning of her career she was not one to ask for help, thinking that as a veterinarian you are supposed to know it all. However, as she matured this mindset changed, and she said *"I pick up the phone to my network and ask, what do you think?"* As Angela became a firm believer in the power of networking, she generously invited those she mentored into her circle and extended her network to help her mentees grow.

As one of the pioneer female swine veterinarians, Angela was a ground-breaker and shattered the glass ceiling for so many. Angela especially enjoyed mentoring younger veterinarians who added diversity into AASV and the swine industry, advocating for equity and inclusion. She was ecstatic that the swine industry and profession was drawing in more women (Figure 6). At the time of writing this paper, 10 of the 15 AASV committees are being chaired or co-chaired by women (boar stud, collegiate activities, communications, human health, safety and well-being, pig welfare, pork safety, PRRS, student engagement and, transboundary and emerging diseases).

Dr Emily Byers Taylor (2024; personal communication) noted *"Angela was my mentor and she paved the way for me and so many others to be successful."* Angela was strong, bold, and unafraid; and she was excellent at coaching and leading by example to encourage these qualities in her mentees as they discovered their own passions in the industry. Angela said *"Every time I was told you cannot do that I LOVED the challenge. Go ahead, draw the line, and watch me cross it."*

Dr Byers Taylor reflected on a time when she was struggling with confidence in a new role and spoke with Angela. Angela replied *'Don't you EVER let a man tell you to be LESS. You go out there and kick some ass!'* and that was all Dr Byers Taylor needed to pick her head up and give it her best. Dr Tara Donovan (2023; Dr A Baysinger, Heritage Video) remembered seeing Angela, fresh out of veterinary school being a successful and respected swine veterinarian. Dr Donovan was so impressed that she was inspired to become a swine veterinarian as well. Dr Wesley Lyons (2023; Dr A Baysinger, Heritage Video) noted she was always on my side, she supported me both professionally and personally and welcomed my family.

Dr Caleb Brezina (2024; personal communication) noted, *"[She] showed me the importance of striving to improve animal welfare, challenging where possible, compromising, when necessary, but always seeing compromise as a first step, not a final conclusion"* (Figure 7).

Indeed, it was Angela's work in animal welfare her involvement in the animal welfare judging competition and creation of student internships specific to animal welfare that inspired many,

Figure 6: Percentage of the 2024 AASV membership by gender (Dr Sue Schulteis).

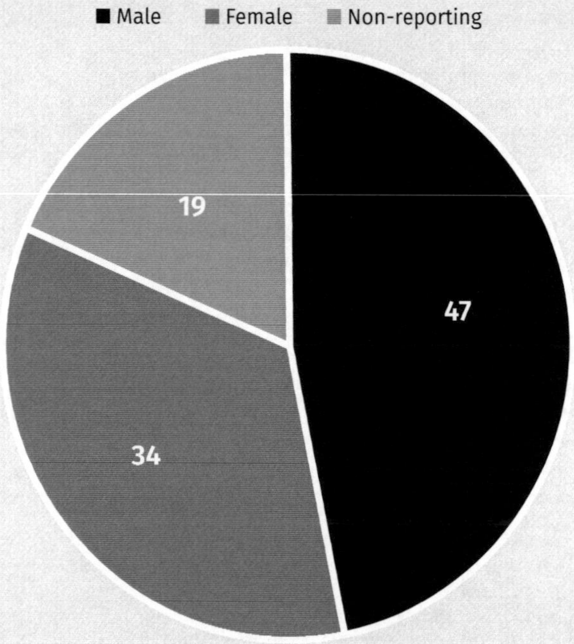

Figure 7: Caleb Brezina (center) won the top individual score for the veterinary division at the AWJAC, Colorado State 2019; with Angela (L) and Cia Johnson (R).

including Dr Byers Taylor to pursue and persist in careers focused on or including animal welfare. Numerous individuals owe their careers to the passion, determination, and vision Angela demonstrated in her everyday work. The world is undoubtedly a better place thanks to the mentorship of Angela. Her drive and guidance were a hallmark of her dedication to improving animal welfare and swine medicine.

Theme 8: Putting a human face on mental health

One of the most impactful stories Angela told AASV members in 2022 as she delivered the Howard Dunne Memorial Lecture was about her own struggles with mental health and suicide. Angela opened up to an audience of her peers and shared that she collected and lined up pills that she would take to induce unconsciousness, restrict respiration and result in cardiac arrest. As she prepared to take the pills the phone rang and it was her son. He had forgotten his homework and needed her to run it to school noting *"Look on the bright side Mom, see I do need you."* She picked up the pills and threw them away, she grabbed the assignment and took it to school. During the same day, three members of her friend group called her just to check in and chat, not knowing what had occurred earlier. After this she received mental health support, she received counseling and she rocked the world by learning how to fly. This message took enormous courage to deliver. It was raw, painful, hard, honest, and emotional. But in those few minutes of sharing, Angela stripped away the taboo of mental health and the incorrect assumptions that are attached to it.

Veterinarians are exposed to stressful occupational hazards including excessive workload, compassion fatigue, burnout and financial constraints.[16] This mental health crisis expands across all disciplines in veterinary medicine and is considered a crisis within the veterinary field.[16] Putting stigmas aside, Angela charged forward with the objective to acknowledge and put light onto depression, anxiety and burnout in the swine veterinary community. Another area of interest was to assess the mental health of swine veterinarians involved with depopulation events related to COVID-19 and the results were humbling. Of the 134 participants who completed an online survey, only 50% of participants felt comfortable discussing mental health with other veterinarians, 2% had thought about suicide in the past 12-months and 8% at some point in their career. One swine veterinarian had made plans in the past 12-months with five veterinarians saying they had made plans at some point in the past. A total of 12% of swine veterinarians were or had been on prescription medication for mental health, 20% were or had been an outpatient for mental health but 16% said they needed counselling in the past 12-months but had not received it. When the data was further broken down the results demonstrated that younger females were more likely to suffer from serious psychological distress and lower well-being than older male veterinarians. The results from this work indicated that the scale and scope of swine veterinary behavioral health issues are significant, and that action is needed to improve mental health outcomes associated with occupational hazards associated with this industry.[17] Angela championed this discussion by providing the data to allow the US swine industry to have an objective and honest conversation and to continue to keep the conversation going with her own personal perspective and experience. As Angela stated *"The most comforting words I have heard in my life are me too. In that moment you find out that your struggle is also someone else's struggle. You are not alone. Someone needs you."*

Angela's challenge to swine veterinarians and the American Association of Swine Veterinarians

After training and competing in a Tough Mudder race, Angela set a new goal to earn her pilot's license. She brought that same voracious spirit to her professional life. Once one challenge was met, there was always another that grabbed her attention and became her next area of focus. Her forward thinking and motivation for "W*hat's next?*" was inspiring to many. It is now up to us to carry on with that commitment and honor her legacy by identifying and taking on those next challenges.

1. Education. This begins with veterinary students. Many veterinary students are entering veterinary school with little to no livestock experience. This does not have to be a detriment to them or the industry, but we challenge that how veterinary schools introduce and expose students to agriculture curriculum may not be the most consistent and balanced. For example, only 4% of the North American Veterinary Licensing Exam questions were about pigs, compared to approximately 23% about dogs and cats and that 13 out of 29 veterinary schools only had one swine course. This leaves an enormous opportunity for veterinary schools to work with the swine industry to improve communication on what a swine career can be and look like, along with the opportunities and rewards.[18] Furthermore, a survey with third-year DVM students before and after completing an animal welfare course to assess their knowledge of key animal welfare topics, opinions regarding the inclusion of welfare education in the veterinary curriculum, and views on veterinarian responsibilities as advocates was conducted. Results indicated that this animal welfare course led students to view the inclusion of more animal welfare topics into the veterinary curriculum favorably and it improved their confidence in conducting research on animal welfare topics.[19]

- How can AASV support swine welfare educational exposure both within and after veterinary school?

2. Work-life integration. It has been reported that 62% of swine veterinarians who left practice either completely or took a technical/academic role did so because the new job meant less traveling and 69% reported that their work-life integration improved. Open comments included *"I can't disconnect from the job"*, *"I am on-call 24/7"*, and "*No flexibility to deal with family emergencies.*"[18]

- How can AASV become involved with new- and seasoned swine veterinarians to support retention?
- How can AASV help with continuing education in career and lifestyle planning?

3. Financial burden. In 2023 the mean starting salary in private practice for swine veterinarians was $92,641 compared to $120,317 for companion animal practioners. Approximately 51% of new veterinarian graduates carried a debt load of $150,000 or more.[20] The AASV Foundation offers student-debt relief grants to veterinarians at https://www.aasv.org/foundation/swine-veterinarians/debt-relief-grants/. AASV offers scholarships for participation in the MentorVet program (**https://www.aasv.org/2024/07/mentorvet-leap-scholarships-available-apply-by-aug-2-3/**). The first cohort of the participant-led, Early-Career Swine Veterinarian Development Program is nearing completion (**https://www.aasv.org/earlycareerdevelopmentprogram/**) and the Early Career committee has published an early career resource directory (**https://www.aasv.org/wp-content/uploads/2024/06/resource-directory.pdf**)

- How can the AASV organization and members make younger swine veterinarians aware of these opportunities?
- What other foundational support could AASV drive?

4. Paying it forward. Being an effective mentor takes time, practice and personalized communication. Knowing how someone communicates and learns can be assisted by using personality assessment tools such as the Myers-Briggs, DiSC Personality Test or the Predictive Index Behavioral Assessment.[21,22]

- How could AASV utilize these (and other) personality tools to match mentors and mentees better?
- How can AASV support continuing education to become an effective mentor?

5. Become a defender and challenger. When needed stand up for your profession and the swine industry when claims are blatantly untrue. However, do not be too quick to fall on the sword and always defend the status-quo. Let's be honest, some housing and production practices have dubious effects on pig welfare and we need to own that. It is critical as practitioners to keep listening and learning. We should always be thinking "*How can I do and be better today.*" The members of the profession are nimble and knowledgeable. However, we tend to lean into the "*We can't because...*" instead of "*Let's try because...*" This is going to become important when we talk to customers and consumers that ask curious and perhaps uncomfortable questions. They do not care if the pig has enhanced average daily gain or that we have reduced the wean-to-estrus interval. What they do care about is a happy pig!

- How can AASV forecast future welfare issues and be part of the solution?

Here are some issues: Growing pig osteoarthrosis, sudden sow death, hospital pen management and gilt development unit management, acknowledging pig affective state, mortality and cull animal transport.

- How can AASV defend sustainability when livability through the entire chain is questionable?

Working across the barn yard and perhaps the human medical field needs to be nurtured and utilized more. In certain aspects of production, we work more on the population - than individual level. But when needed can we hand on heart say we always find the individual pig that needs assistance quickly?

- How can AASV members begin to understand, implement and use tools in real time to identify individual pigs that need help?
- How can AASV support telemedicine and telehealth?
- How can AASV test and support products that use existing technology to help us complete on-farm welfare assessments?
- How can AASV identify technology used in one species and translate it to swine?
- How can AASV drive a gap analysis on missing technology that would help practitioners and assist in creating research trials to test the technology validity?

6. Advocate. Be proud to be the pigs' advocate and wear your DVM badge with pride. Try not to let the politics or marketing of the day sway you, and remember to listen and be part of the conversation.

- How can you as an AASV member become more involved with customers, consumers and legislators?

7. Flip the script. For decades the livestock industry has been on the defense and our messaging has been self-defeating. So, rethink, repackage and retool! Why do we keep using frameworks that show mostly the negative such as the Five Freedoms? Why do we not frame and bench mark data differently using a different framework such as the Five-Domains? [23,24] For example, instead of *"1.5% of our pigs on-farm have observed signs of severe lameness"* why not package this as *"98% of our pigs can walk easily. This results in a pig that is comfortable, not in pain, can explore their environment and interact in a positive way with other pigs and caregivers"*

- How can AASV become involved in improving the message throughout the livestock production chain?

8. Welfare in its own right. Ms Kaster and Drs Millman and Barger noted that Angela was an advocate for not placing animal welfare under the larger sustainability umbrella. The IPWA has five pillars with health and welfare standing separate from the efficiency and sustainability pillar.

- How can AASV voice their opinions on how swine welfare is categorized and prioritized by organizations in the future?

Angela's legacy

Angela's dedication to improving animal welfare has made a significant impact in the food-animal industry for both the animals and their human stewards, and her contributions will have long-lasting implications. Let us close with Angela's own words during her AASV legacy video. When asked *"What do you think your legacy should be?"* she simply but elegantly stated with her beautiful smile *"That I made a difference."* Dr Baysinger, Angela, colleague, consummate professional, animal advocate and our darling friend, please rest assured you excelled at this at every level, and we thank you.

References

1. AASV. American Association of Swine Veterinarians. 2024. Dr. Angela Baysinger Passes Away. **https://www.aasv.org/2024/03/dr-angela-baysinger-passes-away/**

2. NPPC. National Pork Producers Council. 2009. Veterinary Government Relations Externship. **https://nppc.org/get-involved/education-programs/veterinary-government-relations-externship/**

3. NPB. National Pork Board. 2023 Pork Quality Assurance numbers from Ms. D. Peebles. Director for certification programs. **https://www.porkcdn.com/sites/porkcheckoff/assets/files/NPBStaffDirectory2023_1689798761481.pdf**

4. HFAC. Humane Farm Animal Care. 2024. Humane Farm Animal Care standards (January 2018) Pigs. **https://certifiedhumane.org/wp-content/uploads/Standard_Pigs.pdf**

5. SES. SES Inc. environmental consulting services. **https://ses-corp.com/**

6. Heleski CR, Zanella AJ, Pajor EA. 2003. Animal welfare judging teams - a way to interface welfare science with traditional animal science curricula. *Appl. Anim. Behav. Sci.* 81(3):279-289.

7. Anderson KN, Allen KJ, Baysinger A, Benjamin M, Berger J, Claus JR, Greco BJ, Massie EA, O'Brien B, Ramirez A, Rendahl AK, Reues AA, Zhitnitskiy PE, Vogel KD. 2021. Relationship of tissue dimensions and three captive bolt placements on cadaver heads from mature swine (*Sus scrofa domesticus*) > 200 kg body weight. *J Anim Sci.* **doi. https://doi.org/10.1093/jas/skab327**

8. Baysinger A, Webb SR, Brown J, Coetzee JF, Crawford S, DeDecker A, Karriker LA, Pairis-Garcia M, Sutherland MA, Viscardi AV. 2021. Proposed multidimensional pain outcome methodology to demonstrate analgesic drug efficacy and facilitate future drug approval for piglet castration. *Anim Healt Res Reviews.* 22, 163–176. **doi. https://doi.org/10.1017/S1466252321000141**

9. Baysinger A, Senn M, Genhardt J, Rademacher C, Pairis-Garcia M. 2021. A case study of ventilation shutdown with the addition of high temperature and humidity for depopulation of pigs. *JAVMA.* 259:415–424. **doi: 10.2460/javma.259.4.415**

10. Johnson AK, Rademacher CJ, Eggers J, Gabler NK, Greiner LL, Kaisand J, Karriker LA, Millman ST, Patience JF, Ramirez BC, Schulz LL, Webb SR, Ross JW. 2021. Innovative strategies for managing swine during the COVID-19 pandemic in Iowa. *Transl Anim Sci.* **doi. https://doi.org/10.1093/tas/txab225**

11. US Energy Information Administration. 2021. Weekly Midwest (PADD 2) oxygenate plant production of fuel ethanol (thousand barrels per day). **doi. https://www. eia.gov/dnav/pet/hist/LeafHandler.ashx?n=PET&s=W_EPOOXE_YOP_R20_MBBLD&f=W**

12. Zent J. 2020. The great ammo shortage of 2020: when will it end? NRA American Riflemen. **doi. https://www. americanrifleman.org/content/the-great-ammo-shortage-of-2020-when-will-it-end/**

13. IPWA. International Poultry Welfare Alliance. **https://www.poultrywelfare.org/About-IPWA 2024**

14. USDA. United States Department of Agriculture. 2024. USDA Processed Verified Program. **https://www.ams.usda.gov/services/auditing/process-verified-programs/faq#:~:text=The%20USDA%20Process%20Verified%20Program%20%28PVP%29%20provides%20a,include%20standards%20requested%20by%20a%20customer%20or%20consumer**

15. APA. American Psychological Association. 2024. Introduction to mentoring: A guide for mentors and mentees. **https://www.apa.org/education-career/grad/mentoring**

16. da Silva C, Domingues-Gomes AA, Dos Santos-Doni TR, Antonelii AC, da Costa-Vieira RF, da Silva ARS. 2023. Suicide in veterinary medicine: A literature review. *Vet World.* 2023.1266-1276.Epub 2023 Jun 8. **doi: 10.14202/vetworld**

17. Baysinger A, Kogan LR. 2022. Mental health impact of mass depopulation of swine on veterinarians during COVID-19 infrastructure breakdown. *Front. Vet. Sci.* **doi. 10.3389/fvets.2022.842585**

18. Rademacher C. 2024. Past, present and future challenges for the swine vey profession Hogg Lecture AASV Annual Meeting Nashville Tennessee 2024. **https://www.aasv.org/10.54846/am2024/2/**

19. Johnstone ECS, Frye, MA, Lord LK, Baysinger AK, Edwards-Callaway LN. 2019. Knowledge and opinions of third year veterinary students relevant to animal welfare before and after implementation of a core welfare course. *Front. Vet. Sci.* **https://doi.org/10.3389/fvets.2019.00103**

20. AVMA. American Veterinary Medical Association 2023. AVMA report on the economic state of the veterinary profession. **https://ebusiness.avma.org/files/ProductDownloads/eco-state-of-profession-report-lr-secured-2022.pdf**

21. Peters L. 2023. Improving caretaker mental welfare through targeted on-farm euthanasia training programs. Accessed on October 20, 2024. **https://dr.lib.iastate.edu/entities/publication/03f1033e-a7da-4a9e-ba89-214461488447**

22. Yarian J. 2021. Investigating strategies in training swine caretakers for mental well-being and Mexican swine caretakers' attitudes, behaviors, and perceptions toward euthanasia." **https://dr.lib.iastate.edu/entities/publication/03f1033e-a7da-4a9e-ba89-214461488447**

23. Mellor DJ, Beausoleil NJ, Littlewood KE, McLean AN, McGreevy PD, Jones B, Wilkins C. 2020. The 2020 Five Domains model: Including human- animal interactions in assessments of animal welfare. *Anim.* **doi. http://dx.doi.org/10.3390/ani10101870**

24. Johnson AK, Rault, JL, Marchant JN, Baxter EM, O'Driscoll K. 2022. Improving young pig welfare on-farm: The five domains model. *J Anim Sci.* 100(1-15) **doi. https://doi.org/10.1093/jas/skac164**

https://doi.org/10.54846/am2025/4

Establishing best practices for oral fluid collection

Grzegorz Tarasiuk[1], DVM; Marta D. Remmenga[2], PhD; Kathleen C. O'Hara[2], DVM, MPVM, PhD; Marian K. Talbert[2], PhD; Sarah Mielke[2], MPH, PhD; Marisa L. Rotolo[3], DVM, PhD; Pam Zaabel[3], DVM; Danyang Zhang[4]; Jeffrey J. Zimmerman[1], DVM, PhD

[1]College of Veterinary Medicine, Iowa State University, Ames, Iowa; [2]USDA:VS:Strategy and Policy, CEAH, Fort Collins, Colorado; [3]National Pork Board, Des Moines, Iowa; [4]College of Liberal Arts and Sciences, Iowa State University, Ames, Iowa

Introduction

Disease detection and surveillance are crucial to the swine industry's ongoing improvement of pork production. An efficient surveillance approach that has gained widespread acceptance is the use of oral fluids for pathogen detection. The versatility of oral fluids is shown in their ability to detect a wide range of pathogens, including African swine fever virus (ASFV) and classical swine fever virus (CSFV) among many others.[1] Despite the widespread use of oral fluids, there is little critical data concerning sampling procedures. The premise of our research is that a better understanding of pig behavior is fundamental to optimizing oral fluid collection and, thereby, improving farm surveillance. In this study, we observed pig behavior in a variety of pen sizes over time (90 minutes) to determine the optimal oral fluid sampling time in field situations.

Materials and methods

We determined the cumulative percent of pigs that had contacted (chewed on) the sampling rope at any given time over a 90-minute observation period. The study was conducted on a commercial farm with 10- to 14-week-old cross-bred pigs using one rope per pen in pens of 30, 90, and 120 pigs (32 pens per size category). In each pen size category, 20 pigs were clearly marked with distinct colors and symbols so that they could be individually identified from a distance. Oral fluid sampling was recorded using cameras (GoPro® 7 cameras, GoPro Inc., San Mateo, CA) and the marked pigs' contacts with the rope were counted.

Results and conclusions

The aim of the analyses was to describe pig participation in oral fluid sampling as a function pen size and sampling time. The analysis was done in R (version 4.4.0. R core team 2024) using a generalized linear model. Pig participation results are presented in Table 1. Both sampling time and pen size significantly affected participation ($P < .001$).

These results have important implications at the barn level. Using a bootstrap simulation based on data from the 30-minute exposure study, and assuming oral fluid sampling from 6 pens in one barn, 6 oral fluid samples from pens holding 30, 90, or 120 pigs each would contain "contributions" from approximately 97, 213, and 237 pigs, respectively. Basing the bootstrap on the 60-minute exposure data, participation is estimated to increase to approximately 111, 257, and 292 pigs; based on the 90-minute exposure data, participation is estimated to increase to approximately 124, 298, and 346 pigs, respectively. This study has documented the value of oral fluids as a tool to efficiently collect diagnostic specimens from a large proportion of pigs in a given population and across pen sizes, and the significant impact of increasing sampling times to maximize pen level participation.

Reference

1. Munguía-Ramírez, B.; Armenta-Leyva, B.; Giménez-Lirola, L.; Wang, C.; Zimmerman, J.J. Surveillance on swine farms using antemortem specimens. In Optimizing Pig Herd Health and Production; Maes, D., Segalés, J., Eds.; Burleigh Dodds Science Publishing: Cambridge, UK, 2023; pp. 97–138.

Table 1: Pig participation in oral fluid sampling by pen size over a 90-minute observation period

	Cumulative participation (predicted percent, 95% prediction interval)			
Pen size	**Number ropes**	**At 30 minutes**	**At 60 minutes**	**At 90 minutes**
30 pigs	1	55% (15, 89)	64% (20, 93)	73% (27, 95)
90 pigs	1	37% (8, 81)	47% (11, 86)	57% (16, 90)
120 pigs	1	30% (6, 75)	38% (8, 81)	48% (11, 86)

https://doi.org/10.54846/am2025/5

Assessing the effect of pooling commonly used samples in breeding herds on the probability of influenza A virus sequencing

D. C. A. Moraes[1], DVM, MBA, MS; O. H. Osemeke[1], DVM, MS, PhD; T. Petznick[1,2], BS, DVM; E. Pratt[2], BS, DVM; M. A. Jones[3], BS; C. A. Otto[1]; M. N. Musskopf[1], DVM, MBA; C. Y. Wang[1], DVM; P. C. Gauger[1], BS, DVM, MS, PhD; G. Trevisan[1], DVM, MBA, PhD; G. S. Silva[1], DVM, MS, PhD; D. C. L. Linhares[1], DVM, MBA, PhD
[1]Department of Veterinary Diagnostic and Production Animal Medicine, Iowa State University; [2]ArkCare, Omaha, Nebraska; [3]NC State College of Veterinary Medicine, Raleigh, North Carolina

Introduction

Population-based samples such as udder wipes (UW) and family oral fluids (FOF) reduce diagnostic costs and improve the probability of pathogen detection by increasing the number of animals sampled. Nasal wipes (NW) are an individual sample type that can be used strategically for monitoring influenza virus A (IAV), obtaining sequences, and in situations where practitioners are interested in testing individual animals. Pooling samples allows cost reduction[1] and potential gains on the IAV probability of detection.[2] However, the effect of pooling IAV surveillance samples on the probability of successfully obtaining IAV hemagglutinin (HA) gene sequence is unknown. This study aimed to assess the effect of pooling (undiluted, 1:3, and 1:5) different sample types (FOF, UW, NW) on the probability of obtaining complete IAV sequences.

Materials and methods

Thirty-six PCR-positive IAV samples (12 FOF x 12 udder wipes x 12 nasal wipes) from an endemically infected breeding herd under surveillance were selected and pooled with PCR-negative samples in the following proportions: undiluted, 1:3, 1:5. Six samples of each dilution level for each sample type were tested for a total of 18 samples in each category. Category A included Ct values between 26-30, and Category B included Ct values between 30-34. A logistic regression model was used to characterize a complete or partial IAV sequence given the pooling level Ct value and sample type, using the lme4 package in the R Studio statistical program.[3] This study classified a sequence as partial if the HA1 portion of the HA gene could be generated. If the HA1 could not be generated, sequencing results were reported as negative.

Results

For FOF, 14 of the 18 samples had complete sequences in Category A, and 2 of the 18 samples had complete sequences in the undiluted level in Category B. For udder wipes, 7 of the 18 samples had complete sequences in Category A; in Category B, 2 of 18 samples had complete sequences in the undiluted level. For nasal wipes, 11 of 18 samples had complete sequences in Category A; in Category B, there were 2 complete sequences at the undiluted level. There was no statistical difference (P-value > .05) when assessing the IAV sequencing success by sample type at the dilution levels (Table 1).

Discussion and conclusion

Complete sequences were recovered from FOF and NW on both Ct categories, while for UW, complete sequences were only recovered when Ct was lower than 30. Overall, the success rate in obtaining complete sequences in Category A was consistent until the 1:3 dilution. The success rate to obtain sequences was much lower when the Ct value was higher than 30 across all sample types, indicating that pooling should be used cautiously in that scenario. Comparison of these pooled sample types is critical for guiding practitioners and veterinarians in selecting the most effective sampling approach by the Ct value for successful IAV sequencing in breeding herds.

References

1. Osemeke, O. H., de Freitas Costa, E., Almeida, M. N., Trevisan, G., Ghosh, A. P., Silva, G. S., & Linhares, D. C. (2022). Effect of pooling family oral fluids on the probability of PRRSV RNA detection by RT-rtPCR. Preventive veterinary medicine, 206, 105701.

2. Moraes, D.C.A et al. 2024. Probability of Influenza A virus RNA detection at different pooling levels for commonly used sample types in breeding herds. 55th AASV Annual Meeting 2024, United States, **https://doi.org/10.54846/am2024/12**

3. Team, R.C., 2021. R: A language and environment for statistical computing. Vienna, Austria: R Foundation for Statistical Computing; 2020.

Table 1: Influenza A virus complete and partial sequencing by family oral fluids, udder wipes, and nasal wipes

Dilution level	Family oral fluids				Udder wipes				Nasal wipes			
	A Ct 26-30 %, n[1]		B Ct 30-34 %, n[1]		A Ct 26-30 %, n[1]		B Ct 30-34 %, n[1]		A Ct 26-30 %, n[1]		B Ct 30-34 %, n[1]	
	Complete	Partial	Complete	Partial	Complete	Partial	Complete	Partial	Complete	Partial	Complete	Partial
Undiluted	0.79 (5/6)	0 (0/6)	0.3 (2/6)	0.50 (3/6)	0.35 (2/6)	0.65 (4/6)	0 (0/6)	0.50 (3/6)	0.7 (5/6)	0 (0/6)	0.20 (1/6)	0 (0/6)
1:3	0.79 (5/6)	0 (0/6)	0 (0/6)	0.35 (2/6)	0.35 (2/6)	0.65 (4/6)	0 (0/6)	0 (0/6)	0.65 (4/6)	0.20 (1/6)	0.20 (1/6)	0 (0/6)
1:5	0.65 (4/6)	0 (0/6)	0 (0/6)	0.20 (1/6)	0.50 (3/6)	0.50 (3/6)	0 (0/6)	0 (0/6)	0.35 (2/6)	0.50 (3/6)	0 (0/6)	0 (0/6)
Total sequences %, n1	0.78 (14/18)	0 (0/18)	0.11 (2/18)	0.33 (6/18)	0.39 (7/18)	0.61 (11/18)	0 (0/18)	0.17 (3/18)	0.61 (11/18)	0.22 (4/18)	0.11 (2/18)	0 (0/18)

[1] Number of sequences/total samples tested by each dilution level and sample type.

https://doi.org/10.54846/am2025/6

How many samples do you need for influenza A virus monitoring in farrowing rooms?

D. C. A. Moraes[1], DVM, MBA, MS; O. H. Osemeke[1], DVM, MS, PhD; P. C. Gauger[1], DVM, MS, PhD; G. Trevisan[1], DVM, MBA, PhD; C. A. Moura[2], DVM, PhD; G. S. Silva[1], DVM, MS, PhD; D. C. L. Linhares[1], DVM, MBA, PhD
[1]Department of Veterinary Diagnostic and Production Animal Medicine, Iowa State University, Ames, Iowa;
[2]Iowa Select Farms, Iowa Falls

Introduction

Influenza A virus (IAV) impacts the health and productivity of swine populations. Control of IAV starts by understanding the pressure of infection in weaning-age pigs. Determining the appropriate sample size for effective IAV surveillance in farrowing rooms ensures the detection of IAV and allows strategic implementation of control or elimination programs. Thus, the objective of this study was to determine appropriate sample sizes for monitoring IAV in farrowing rooms in breeding herds at different prevalence scenarios using different samples such as family oral fluids (FOF), udder wipes (UW), or individual pig nasal wipes (NW).

Materials and methods

The sample sizes were calculated using data from two studies. The first study compared IAV RNA detection by reverse transcription real-time polymerase chain reaction (RT-rtPCR) in matched FOF, UW, and NW samples from weaning-age pigs, where all samples were collected in one breeding herd on the same day.[1] A second study[2] compared the probability of IAV RNA detection in IAV-positive samples pooled with IAV-negative samples at different pooling levels (undiluted, 1:3, 1:5, 1:10). Sample sizes were calculated using Monte Carlo simulations performed in R statistical software on data from the studies described above.[3] The simulations assumed 30 stalls/farrowing rooms, IAV at 5%, 10%, 20%, and 50% litter prevalence, and a 95% confidence level to detect at least one positive sample.

Results

Sample size calculations to detect IAV at 5% prevalence using FOF were 13, 5, 4, and 3 pools for undiluted, 1:3, 1:5, and 1:10 dilutions levels, respectively. The sample size calculation to detect IAV at 5% prevalence using udder wipes was 16, 6, and 4 for undiluted, 1:3, and 1:5 dilution levels, respectively. In comparison, having a positive RT-rtPCR test with 95% confidence at 1:10 dilution was not possible. The sample size calculation for nasal wipes to detect IAV at the 5% prevalence was 54, 20, 14, and 10 for undiluted, 1:3, 1:5, and 1:10 dilution levels, respectively. For the prevalence at 10%, 20%, and 50%, a similar pattern was observed in the 5% prevalence, where population-based samples required fewer test samples than individual sample types. (Table 1).

Discussion and conclusion

For the same prevalence scenario, population-based sample types (UW and FOF) require fewer test samples than individual sample types (NW). The strategic use of population-based sample types, such as UW and FOF, reduces costs and enhances the probability of IAV detection in breeding herds. In addition, pooling should be considered in fixed testing budget scenarios to improve coverage. The sample size presented in this study was calculated based on room size from 30 stalls farrowing rooms. Future studies are needed to validate such numbers for rooms outside this range. This study offers new approaches to the practical utilization of diverse sample types and, thus, is a valuable and applicable resource for IAV, depending on the diagnostic testing goals of practitioners, veterinarians, and producers.

References

1. Moraes, D.C.A. et al. 2023. Use of Family Oral Fluid to detect Influenza A Virus in the breeding herd. 54th AASV Annual Meeting 2023, United States, **https://doi.org/10.54846/am2023/125**

2. Moraes, D.C.A et al. 2024. Probability of Influenza A virus RNA detection at different pooling levels for commonly used sample types in breeding herds. 55th AASV Annual Meeting 2024, United States, **https://doi.org/10.54846/am2024/12**

3. Team, R.C., 2021. R: A language and environment for statistical computing. Vienna, Austria: R Foundation for Statistical Computing; 2020.

Table 1: Sample size to detect IAV at a 95% confidence level in a 30-stall room at weaning

	Nasal wipes				Udder wipes				Family oral fluids			
Prevalence	Undiluted	1:3	1:5	1:10	Undiluted	1:3	1:5	1:10	Undiluted	1:3	1:5	1:10
5%	54	20	14	10	16	6	4	NP*	13	5	4	3
10%	28	12	10	8	13	5	4	3	10	4	3	3
20%	14	8	5	4	9	3	3	2	7	3	2	2
50%	5	3	3	2	5	2	2	1	5	2	1	1

* Not possible to detect at a 95% confidence level.

https://doi.org/10.54846/am2025/7

Effect of oral meloxicam administration to sows on piglet colostrum intake based on immunocrit, birth weights, and infrared thermography

K. Blaschko[1]; E. Kettelkamp[2], DVM; B. Payne[3], DVM; A. M. Betlach[2], DVM, PhD
[1]Gustavus Adolphus College, St. Peter, Minnesota; [2]Swine Vet Center, St. Peter, Minnesota;
[3]Veterinary Pharmaceutical Solutions, St. Peter, Minnesota

Introduction

Colostrum intake is essential for the early growth and survivability of piglets.[1] An enhanced understanding of management methods and interventions that may increase colostrum production and intake is important to optimize piglet performance. Research has shown that improving colostrum intake and quality is directly linked to piglet survivability, as adequate passive transfer of immunity from the sow is essential for their health.[2] A previous study showed that nonsteroidal anti-inflammatory drugs increase IgA and IgG colostrum levels, enhance IgA, IL-2, and IL-4 serum levels, and improve piglet and litter performance.[3] Meloxicam, a nonsteroidal anti-inflammatory drug (NSAID), is commonly used to alleviate pain and inflammation in postpartum sows.[4] However, minimal information exists on the effect of meloxicam on piglet colostrum intake. Moreover, challenges exist to assess optimal colostrum intake in real-time and under commercial settings. Therefore, this study investigated the effect of oral meloxicam administration to sows on piglet colostrum intake based on immunocrit, birth weights, and infrared thermography. In doing so, alternative methods for assessing colostrum intake were evaluated under field conditions.

Materials and methods

An experimental study was conducted on a 2,400 head commercial sow farm. Sixty dams and their corresponding litters (n = 30 per treatment) were randomly allotted to two treatment groups: 1) Meloxicam and 2) Control. Parity was balanced across treatments. Sample size was determined using a power calculation analysis, assuming an effect size of 0.2, standard deviation of 0.30, a power of 0.80, and an alpha of 0.05. Treated dams received compounded meloxicam 1.5% suspension (6mL; 90mg; orally) at the time of induction and 24-hours after. Five to six piglets were selected within each litter based on a pre-determined birth order. Body weights and thermal infrared surface body temperatures (FLIR®, FLIR Systems, USA) of enrolled piglets were taken before and after (161.5 ± 7.9 minutes) colostrum consumption. The thermal infrared surface body temperatures were taken from the abdominal area of the piglet. Blood was also collected from the fourth piglet in each litter after colostrum consumption for immunocrit testing via collection from the jugular vein using a vacuum tube and spinning it down to collect the serum. Immunocrit values were obtained through analysis of the separated piglet serum with a 40% ammonium sulfate solution (30 g $(NH_4)_2SO_4$), 15mL sterile water). The solution (50μL) was combined with 50μL of the serum, vortexed for 5 seconds, and collected in a capillary tube. Capillary tubes were then centrifuged for 10 minutes, and the precipitate was measured using a micro-hematocrit reading chart.

Colostrum intake (g) was calculated as described by Devillers et al, 2007.[5] Data was analyzed in R (version 4.2.3) using linear mixed regression models to evaluate the effects of meloxicam on piglet body weight, surface body temperature, and colostrum intake. Fixed effects of treatment and random effects of room were considered. Pearson's correlation coefficient was calculated to assess the linear correlation between body weight, colostrum intake, temperature, and immunocrit variables. Results were deemed significant with $P \leq .05$.

Results

A total of sixty dams and their litters were included in the study. Piglets born from sows that received meloxicam had increased colostrum intake (186 g; $P = .037$) (Figure 1) and final body weights ($P = .039$) (Figure 2) compared to pigs born from control sows. There was no difference in starting piglet body weights ($P = .540$) between treatments. There was no effect of treatment on piglet surface body temperatures ($P > .10$). Colostrum intake and immunocrit levels were positively correlated with post-colostrum body weight (r = 0.29-0.34, $P < .05$). No correlation was shown between body temperature and body weight, body temperature and colostrum intake, or body temperature and immunocrit levels ($P > .10$). Thermal images (FLIR One Camera) showed no visible differences in piglet surface body temperature.

Figure 1: Piglet colostrum intake (g) between control sow litters and meloxicam sow litters ($P = .037$).

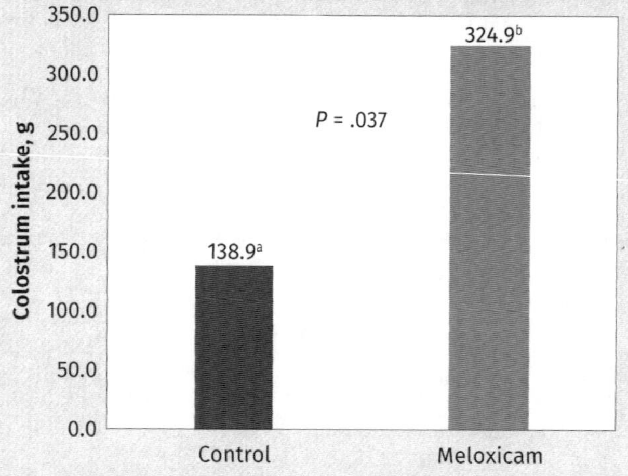

Figure 2: Piglet body weights (lbs) before and after colostrum consumption between control sow and meloxicam sow litters (pre-colostrum: $P = .540$, post-colostrum: $P = .039$).

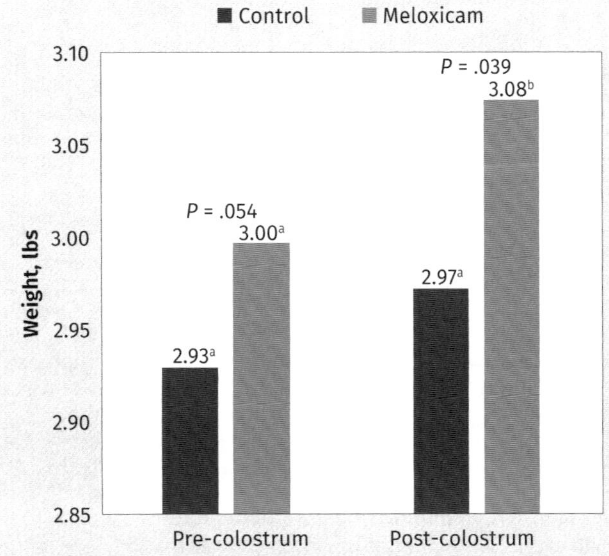

Discussion

The aim of this study was to investigate the effect of oral compounded meloxicam, a nonsteroidal anti-inflammatory drug, administration to sows on piglet colostrum intake. Overall, a benefit of providing meloxicam to sows on piglet colostrum intake was observed in this study, thus supporting findings from previous trials. Piglets born from sows that received meloxicam consumed more colostrum and had greater final body weights. These findings align with previous research that suggests NSAID treatment is beneficial to piglets and litter performance. Furthermore, administering meloxicam to sows may improve piglet health and growth, potentially leading to increased production efficiency and higher economic returns in a commercial setting.

Further research should be conducted to evaluate different timing and dosage strategies for meloxicam and its effect on sow and piglet performance. Under the conditions of this study, there was no correlation between piglet body temperature using thermal infrared technology on performance, colostrum intake, or immunocrit levels. Previous studies have shown that environmental factors may impact infrared temperature readings, resulting in variation, which was experienced during this study. Additional research on the application of thermal infrared technology as an assessment for colostrum management in field conditions is warranted to support these findings. Altogether, this study highlights the impact of administering meloxicam to sows with consequently positive benefits to piglet health, growth, and colostrum intake in a commercial swine setting.

References

1. Friendship RM. Importance of colostrum for newborn piglets. Ontario Ministry of Agriculture, Food and Rural Affairs.

2. Waddell J, Wallgren T. Piglet survival: from birth to weaning. In Practice, 41(2), 56-68.

3. Vallet JL, Miles JR, Rempel LA, Nonneman DJ, Lents CA. Relationships between day one piglet serum immunoglobulin immunocrit and subsequent growth, puberty attainment, litter size, and lactation performance. Journal of Animal Science, 93(6), 2722-2729.

4. Quesnel H, Farmer C, Devillers N, Hémonic A. Colostrum intake: influence on piglet performance and factors of variation. Livestock Science, 146(2-3), 105-114.

5. Devillers N, Farmer C, Dividich JL, Prunier A. Variability of colostrum yield and colostrum intake in pigs. Animal, 1:7, 1033-1041.

https://doi.org/10.54846/am2025/8

Evaluation of meloxicam residues in commercial sows, growing pigs and cull boars utilizing a sensitive diagnostic assay

D. Borts[1], MS, PhD; D. Schrunk[1], MS; L. Burns[1], MS; S. Radke[1], DVM, MS, DABVT; J. Brown[1], DVM, PhD; C. LeFevre[2], DVM; S. Williams[2], MS; J. Ellingson[3], DVM; N. Winkelman[4], DVM; T. Martin[5], DVM; P. Smith[6], PharmD, MBA; K. Bretey[6], DVM; B. Payne[6], DVM
[1]Iowa State University; [2]Carthage Veterinary Service; [3]AMVC; [4]Swine Services Unlimited, Inc; [5]Abilene Animal Hospital; [6]Veterinary Pharmaceutical Solutions

Introduction

Practitioners have a vested interest in international withdrawal periods of FDA-approved drugs and the American Association of Swine Veterinarians (AASV) commits resources to re-evaluate these withdrawal periods.[1] For prescribed extra-label and compounded drugs, it is the practitioner's responsibility to use the best data available to determine withdrawal periods.[2] This paper shares the results of residue studies in sows, growing pigs and cull boars. These studies provide necessary data for practitioners by measuring meloxicam residues in sows' and growing pigs' urine, liver and kidney tissues using a newly developed assay. The sow and growing pig studies were cross-sectional residue studies, and the cull boar evaluation was a case study.

Compounded meloxicam for livestock and poultry is available in multiple formulations from Veterinary Pharmaceutical Solutions (VPS, St. Peter, Minnesota). These include a 1.5% compounded meloxicam suspension for individual treatment of sows and boars, a 0.24% compounded meloxicam suspension for processing-aged pig individual treatments and a 0.6% compounded meloxicam solution for use through the drinking water of all ages. These formulations are compounded within a licensed pharmacy that is FDA and Boards of Pharmacy inspected, following the FDA Code of Federal Regulations Title 21 (21CFR), US Pharmacopeia (USP) Chapter 795 for Non-Sterile Preparations and Boards of Pharmacy regulations.[2] The active ingredients used are from FDA-approved final form products with assigned Abbreviated New Drug Application (ANDA) numbers which is a requirement for compounding for livestock and poultry.[2] In addition, veterinarian oversite is ensured as all compounded meloxicam is produced and sent under the direction of veterinarian prescriptions.

In October of 2024, the FDA issued a Dear Veterinarian letter indicating the concern of aspirin use in dairy animals. The conclusion was that aspirin being used was produced as an OTC monograph and not an FDA approved product nor was it compounded legally under 21CFR.[3] Repercussions from this letter throughout livestock and poultry veterinary groups have made veterinarians reassess the use of both acetylsalicylic acid and sodium salicylate as both drugs are produced as OTC monographs with only a National Drug Code (NDC) assigned and without FDA approval numbers such as ANDAs.[4-6] Many swine veterinarians and companies have looked to compounded meloxicam as a legal option for administering oral non-steroidal anti-inflammatories to swine. This makes these studies and data valuable to veterinarians writing prescriptions for compounded meloxicam.

Materials and methods

HPLC-MS/MS assays were developed and validated with respect to linearity, limit of quantitation (LOQ), selectivity, recovery, carryover, and inter- and intra-day accuracy and precision. From both previous work and a pilot study within this project, there is a known quick depletion of meloxicam in the tissues and thus achieving levels well below the FSIS Minimum Level of Applicability for meloxicam (MLA), 10ng/g, is possible in a couple days after administration.[7,8] The goal was to create a more sensitive assay to show a depletion curve, rather than only show negatives that would occur with a less sensitive assay which could be false negatives. The LOQs, 1ng/mL and 1ng/g, were created to have a 10x more sensitive assay than levels reported to be used by USDA FSIS slaughterhouse checks as the MLA.[7] All samples collected were frozen and stored at -80°C. Aliquots were tested for meloxicam residues via the developed assay for each study.

Study 1: Sow urine pilot study

This pilot study was used to determine sample collection timing for subsequent studies. Compounded 1.5% meloxicam suspension (90mg, 6mL, orally) was administered, twice, 24-hours apart, to sows (n = 7) peri-farrowing. Urine was collected (10-40mL) via free catch and tampon methods throughout an 11-day period post-second dose. A previous pilot experiment during this project (unpublished) was conducted showing there was no interference with the selected tampon used for collection (L. 100% organic cotton core tampon, Super, Proctor and Gamble).

Study 2: Sow study

Compounded 1.5% meloxicam suspension (90mg, 6mL, orally) was administered, twice, 24-hours apart, to sows (n = 30) peri-farrowing. Parity ranged from 1-6 and body condition scores (BCS) were 2-3. Though no weights were measured, lower parity sows represented a higher mg/kg dose while older parity sows represented sows that may have more chronic issues that would delay absorption and excretion. BCS 4+ were not included due to likely lowering of the mg/kg dose. Sows (n = 5/day) were cross-sectionally euthanized on days 8, 11, 14, 17, 20 and 31 via captive bolt. Urine (20-30mL), right liver lobe and right kidney were harvested.

Study 3: Growing pig study

Compounded 0.6% meloxicam soluble was delivered via drinking water for 7 days (targeting 0.2mg/kg/day, 2x normal dosing) to 8-week-old pigs (n = 24). Calculation of the disappearance of stock solution equated to 0.5mg/kg/day average daily disappearance, 5x normal dosing. Pigs (n = 4/day) were cross-sectionally euthanized on days 0, 2, 5, 8, 12 and 14 post treatment-cessation. Urine (5-30mL), right liver lobe and right kidney were harvested.

Study 4: Cull boar case study

Four cull boars were administered compounded 1.5% meloxicam suspension by practitioner protocol. Boar 1 was administered a single, 135mg (9mL) dose 6 days prior to euthanasia and found to have minor pleural adhesions and severe fibrinous pericarditis on necropsy. Boar 2 was administered a single, 135mg (9mL) dose 11 days prior to euthanasia and found to have severe peritonitis on necropsy. Boar 3 had splayed hind legs and was administered a single, 135mg (9mL) dose 18 days prior to euthanasia and there were no necropsy lesions observed. Boar 4 was administered three, 135mg (9mL) doses on days 24, 22 and 20 prior to euthanasia and had an open dewclaw wound but no other significant necropsy lesions.

Results

Study 1: Sow urine pilot study

Two days after last administration, most sows (5/7) had meloxicam residues in urine < 10 ng/mL level currently used to test residues in pork (R: 1.6-22.8 ng/mL). By day four, all sows had levels below the 10ng/mL FSIS MLA (R: < 1-2.1).

Study 2: Sow study

At all sampling points and tissues (Table 1 and Figure 1), all assay results were below the FSIS MLA. All urine samples were either below the LOQ or not detected. The highest liver concentration was 2.4ng/g (76% below MLA), 8 days after cessation of administration. The highest kidney sample was 3.3ng/g (47% below MLA), 11 days after cessation of administration.

Study 3: Growing pig study

At time 0, the last day of the 7 days of administration, all pigs were positive and had assay values that would have been violation results through the FSIS MLA of 10ng/g (Table 2 and Figure 2). Urine, liver and kidney results ranged from 9.4-12.4ng/mL, 38.4-76.7ng/g and 87.4-181.0ng/g, respectively. By 2 days after cessation of treatment, all results were well below the MLA (Figure 3). At this point and beyond, urine results were either below LOQ or not detected. Liver samples' maximum concentration was 1.3ng/g (87% below the MLA) at 2 days post treatment cessation and thereafter, all samples were either below LOQ or not detected. Kidney samples' maximum concentration was 3.2 ng/g (68% below the MLA) at 8 days post treatment cessation and thereafter, all samples were either below LOQ or not detected.

Study 4: Cull boar case study

All liver samples from the boars in this case study were either below LOQ or not detected (Table 3). Only 1 boar kidney sample was quantifiable at 1.3 ng/g, 87% below the FSIS MLA. The remainder were either below LOQ or not detected.

Discussion and conclusions

In the US, compounded products, such as meloxicam, do not have labeled withdrawal periods but internationally, where meloxicam is available for swine as a labeled product, there are 5-day withdrawal periods.[9,10] Internationally, the meloxicam minimum residue levels (MRLs) for swine are 65ng/g (liver and kidney) in Europe, 200ng/g for liver and 60ng/g for kidney in Canada, and 10ng/g (liver and kidney) in Japan.[11-13] While there are higher concentrations allowed in most countries, Japan's MRL for meloxicam is the lowest and the FSIS MLA is aligned with this low concentration. Having this low concentration as a stringent standard is important to protect US exports. The development of the sensitive assay to go beyond the MLA was important to prove the results would not only be non-violations but that they were well below the MLA.

Results for the two residue studies (sows and growing pigs) are in alignment with meloxicam withdrawal periods in swine for internationally labeled products, 5 days. Day 0 (last day of treatment) samples were the only samples that would be violation residues using the FSIS MLA. For the sows administered the 1.5% compounded meloxicam suspension, day 8 samples (earliest samples) and beyond were all below the MLA. For the growing pigs administered 0.6% compounded meloxicam soluble through the drinking water, day 2 samples and beyond were all below the MLA.

There are potential differences in pharmacokinetics and excretion of drugs in different sexes and ages within the same species.[14] Thus is it important to conduct studies in such categories of livestock species. The cull boar case study was not a controlled, rigorous residue study like the sow and growing pig studies. The results indicated there were no residues at or above the MLA for 6 days post administration and beyond in these male cull animals. There are opportunities to expand the residue case studies for other categories of swine such as body condition score 1 (BCS 1) cull sows and finishing pigs. Future work includes utilizing the FDA's statistical package for calculating withdrawal periods in the statistical program, R, to create a recommended withdrawal period.[15] Should a more sensitive assay be used for regulatory purposes in the future, lower detected concentrations would be violative.

The residue data compiled from these studies, with these specific formulations, will help practitioners better target withdrawal periods on their prescriptions. Other formulations, even with the same concentration of meloxicam, may have different outcomes than the results presented here. Use of compounded meloxicam is only legal to use in swine, other livestock and poultry by order of a veterinarian with a proper prescription including information such as specific animals to be treated, indication(s) of use, dosing instructions and withdrawal period. The compounding pharmacy must follow FDA, US Pharmacopeia and Boards of Pharmacy guidelines which includes inspections. Proper compliance with Boards of Pharmacy regulations is state specific for the state of the pharmacy and the location for administration to the animals.

Table 1: Study 2 sow residue data sorted by days post administration of the last dose

Sow number	Parity	Body condition score	Day post last dose	Urine, ng/mL	Liver, ng/g	Kidney, ng/g
1	3	3	8	0.5*	0.5	0.5
2	1	3	8	0.5	2.1	2.5
3	5	3	8	0.5	2.4	1.9
4	1	3	8	ND**	1.8	2.8
5	1	3	8	0.5	0.5	1.4
6	3	3	11	0.5	0.5	1.3
7	6	3	11	ND	1.4	3.3
8	1	3	11	0.5	0.5	1.9
9	1	3	11	0.5	0.5	1.4
10	1	2	11	0.5	0.5	1.1
11	1	3	14	ND	0.5	0.5
12	1	3	14	ND	0.5	0.5
13	1	3	14	0.5	0.5	0.5
14	2	3	14	ND	0.5	0.5
15	5	3	14	0.5	0.5	0.5
16	1	3	17	ND	ND	ND
17	1	3	17	0.5	0.5	0.5
18	2	3	17	0.5	0.5	0.5
19	4	3	17	0.5	0.5	3.0
20	1	3	17	0.5	0.5	1.3
21	2	3	20	ND	0.5	0.5
22	5	3	20	0.5	0.5	1.6
23	1	3	20	0.5	0.5	1.2
24	1	3	20	ND	0.5	0.5
25	1	3	20	0.5	1.2	1.1
26	1	3	31	ND	ND	0.5
27	3	3	31	ND	ND	ND
28	1	3	31	ND	ND	ND
29	2	3	31	ND	ND	ND
30	1	3	31	ND	ND	ND

* 0.5 used for the purpose of illustration of value below LOQ = 1.0 but above LOD

** ND = not detected

Figure 1: Study 2 sow residue concentrations, by tissues and days post administration.

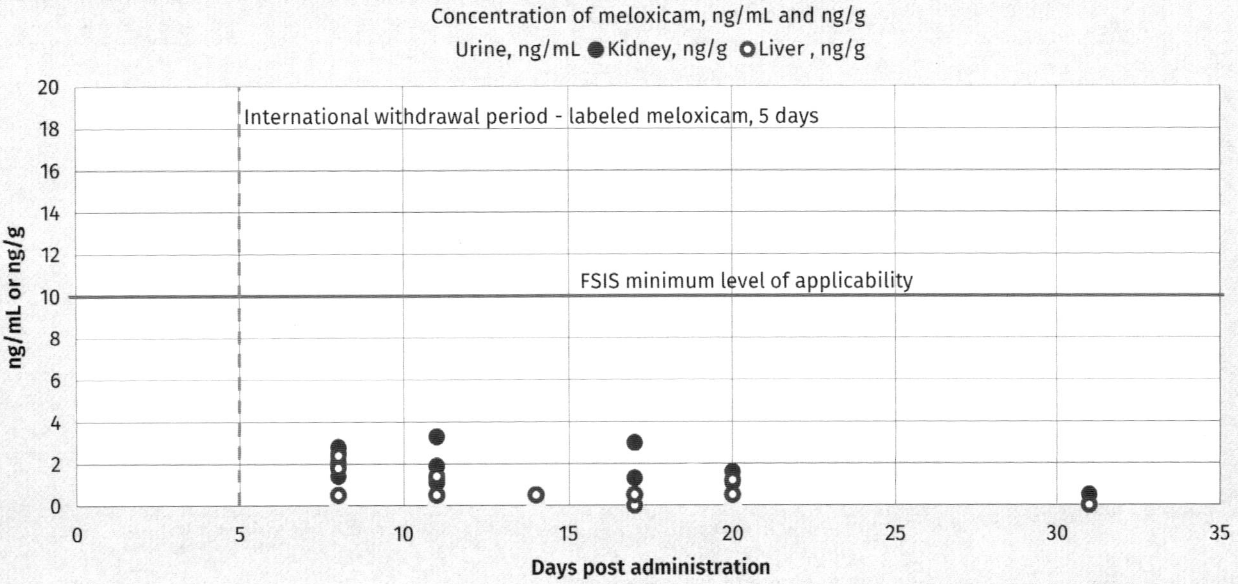

Table 2: Study 2 growing pig residue data sorted by days post administration of the last dose

Number of pig	Days post last dose	Urine, ng/mL	Liver, ng/g	Kidney, ng/g
1	0	12.4	61.7	119
2	0	12.3	38.4	87.4
3	0	9.4	76.7	181
4	0	no sample	56.3	111
5	2	0.5*	1.1	2.4
6	2	0.5	1.3	2.8
7	2	0.5	0.5	2.9
8	2	0.5	1.1	3.2
9	5	0.5	0.5	0.5
10	5	ND**	0.5	0.5
11	5	ND	0.5	0.5
12	5	ND	0.5	1.0
13	8	ND	ND	0.5
14	8	ND	0.5	1.1
15	8	0.5	0.5	1.1
16	8	ND	0.5	0.5
17	12	ND	0.5	0.5
18	12	ND	0.5	0.5
19	12	ND	ND	0.5
20	12	ND	0.5	0.5
21	14	ND	0.5	0.5
22	14	ND	0.5	0.5
23	14	ND	0.5	0.5
24	14	ND	ND	0.5

* 0.5 used for the purpose of illustration of value below LOQ = 1.0 but above LOD

** ND = not detected

Figure 2: Study 3 growing pig residue concentrations, by tissues and days post administration.

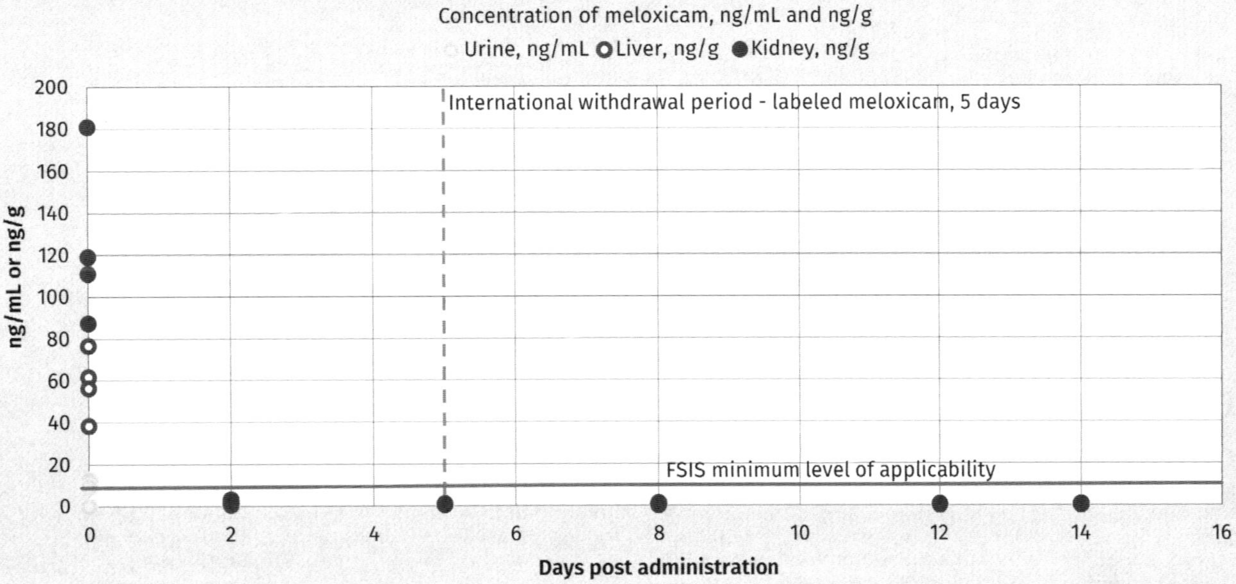

Figure 3: Study 3 zoomed in growing pig residue concentrations, by tissues and days post administration.

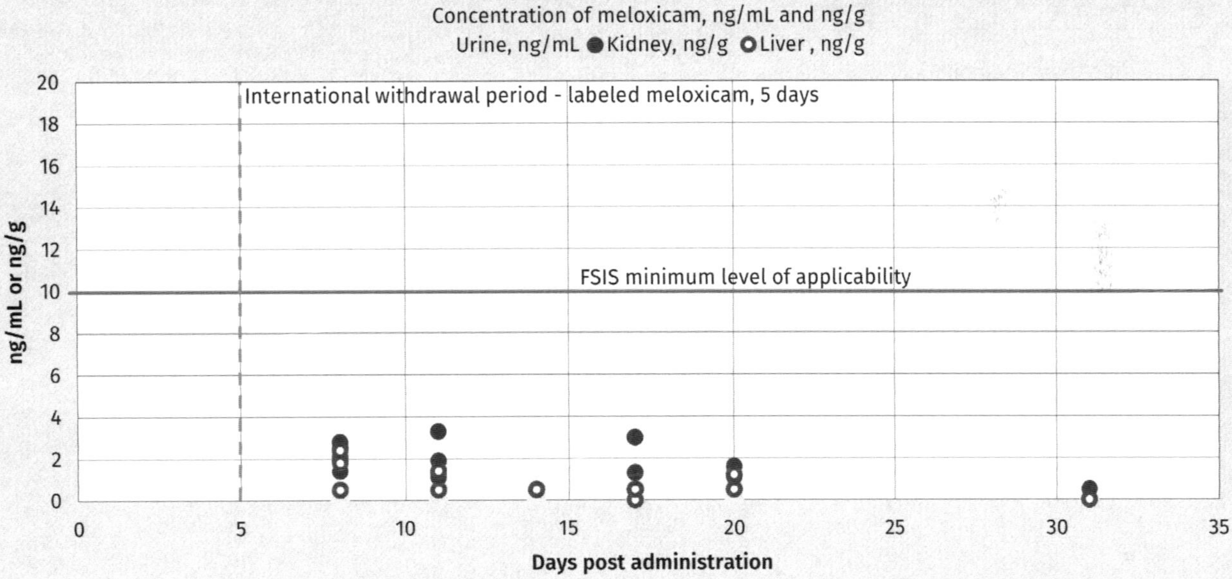

Table 3: Study 4 boar case study residue data sorted by days post administration of the last dose

Boar Number	Days post last dose	Liver, ng/g	Kidney, ng/g
1	6	0.5*	0.5
2	11	0.5	0.5
3	13	0.5	1.3
4	20	ND**	0.5

* 0.5 used for the purpose of illustration of value below LOQ = 1.0 but above LOD

** ND = not detected

References

1. AASV. Committee Plan of Work. March 2024. Accessed November 11, 2024. AASV Pharmaceutical Issues Committee Plan of Work, **https://www.aasv.org/wp-content/uploads/2024/09/2024_pow_pharmissues.pdf**

2. US National Archives. Code of Federal Regulations, title 21, chapter I, subchapter E, part 530, November 7, 2024. Accessed November 11, 2024. **https://www.ecfr.gov/current/title-21/chapter-I/subchapter-E/part-530**

3. FDA. Dear Veterinarian Letter regarding use of aspirin products in lactating dairy cattle. October 11, 2024. Accessed November 11, 2024. **https://www.fda.gov/animal-veterinary/product-safety-information/dear-veterinarian-letter-regarding-use-aspirin-products-lactating-dairy-cattle**

4. FDA.report. National Drug Code Directory. Aspirin Bolus-240 Tablet. Accessed November 11, 2024. **https://fda.report/NDC/46066-010#46066-010-56**

5. FDA.report. National Drug Code Directory. Sodium Salicylate 48. Accessed November 11, 2024. **https://fda.report/NDC/51072-038#51072-038-01**

6. FDA.report. National Drug Code Directory. List of aspirins. Accessed November 11, 2024. **https://ndclist.com/drug-index/substance/aspirin**

7. FSIS. USDA. Screening and Confirmation of Animal Drug Residues by UHPLC-MS-MS CLG-MRM1.08 – USDA-O.Pub.Health. Accessed September 19, 2024. Page 47, 49, 51. **https://www.fsis.usda.gov/sites/default/files/media_file/2020-09/CLG-MRM1.pdf**

8. Nixon E, Mays T, Routh P, Yeatts J, Fajt V, Hairgrove T, Baynes R. Plasma, urine and tissue concentrations of flunixin and meloxicam in pigs. BMC Veterinary Research (2020) 16:340. **https://doi.org/10.1186/s12917-020-02556-4**

9. European Medicines Agency (EMA). Metacam (meloxicam): An overview of Metacam and why it is authorized in the EU. March 2018. Accessed November 11, 2024. Page 4.

https://www.ema.europa.eu/en/documents/overview/metacam-epar-summary-public_en.pdf

10. Drugs.com. Metacam for swine (Canada). Accessed November 11, 2024. **https://www.drugs.com/vet/metacam-for-swine-can.html**

11. European Medicines Agency (EMA). Committee for Veterinary Medicinal Products. Meloxicam Summary Report (1). Accessed September 19, 2024. **https://www.ema.europa.eu/en/documents/mrl-report/meloxicam-summary-report-1-committee-veterinary-medicinal-products_en.pdf**

12. Government of Canada. List of maximum residue limits (MRLs) for veterinary drugs in foods. January 19, 2024. Accessed September 19, 2024. **https://www.canada.ca/en/health-canada/services/drugs-health-products/veterinary-drugs/maximum-residue-limits-mrls/list-maximum-residue-limits-mrls-veterinary-drugs-foods.html**

13. The Japan Food and Chemical Research Foundation. Maximum residue limits (MRLs) list of agricultural chemicals in foods. Pig kidney and pig liver. Accessed September 19, 2024. **https://db.ffcr.or.jp/front/food_group_detail?id=14900**

14. Canton L, Lanusse C, Moreno L. Rational pharmacotherapy in infectious diseases: Issues related to drug residues in edible animal tissues. Animals (Basel). 2021 Oct 1;11(10):2878. **doi: 10.3390/ani11102878**

15. U.S. Department of Health and Human Services, FDA, CVM, General principles for evaluating the human food safety of new animal drugs used in food-producing animals. May 2022. Accessed November 13, 2024. Page 28. **https://www.fda.gov/media/70028/download**

https://doi.org/10.54846/am2025/9

Prevention of *Streptococcus suis* disease using avirulent strain inoculation

Samantha J. Hau[1]; Nahuel Fittipaldi[2,3]; Servane Payen[2,3]; Daniel Grenier[3,4]; Daniel W. Nielsen[1]; Susan L. Brockmeier[1]; Marcelo Gottschalk[2,3]

[1]National Animal Disease Center, USDA, Ames, Iowa; [2]Groupe de recherche sur les maladies infectieuses en production animale, Faculté de médecine vétérinaire, Université de Montréal, St-Hyacinthe, QC, Canada; [3]Centre de Recherche en Infectiologie Porcine et Avicole (CRIPA), Fonds de Recherche du Québec - Nature et Technologies (FQRNT), Saint-Hyacinthe, QC, Canada; [4]Groupe de Recherche en Écologie Buccale (GREB), Faculté de Médecine Dentaire, Université Laval, Quebec City, QC, Canada

Introduction

Streptococcus suis causes severe systemic disease in pigs including septicemia, polyarthritis, polyserositis, and meningitis. It is an important cause of post-weaning mortality and imposes a significant economic burden on the swine industry. Currently, there are no commercially available vaccines to prevent *S suis* disease. There has been increasing interest into novel vaccine platforms, including inoculation with live avirulent or attenuated *S suis* strains. Live vaccines may provide protection through multiple mechanisms: competitive exclusion of pathogenic strains and immune stimulation. Additionally, some *S suis* strains also produce antibacterial compounds (bacteriocins) that inhibit the growth of other *S suis*. Bacteriocin production may provide a competitive advantage for these strains when colonizing the nasal cavity/tonsil which could increase their efficacy as a live vaccine. *S suis* strain 90-1330 is an avirulent bacteriocin producing, serotype 2 strain that has inhibitory activity against a panel of pathogenic *S suis* isolates.[1] This study examined the use of *S suis* strain 90-1330 as a live, intranasal vaccine and its efficacy against challenge with the virulent *S suis* strain P1/7.

Methods

Animal studies used cesarean derived, colostrum deprived (CDCD) pigs and an intranasal challenge model. The first study evaluated 90-1330 virulence, dosing, and protection 21 days after inoculation. Pigs were divided into three groups: low dose (~10^7 CFU/mL 90-1330), high dose (~10^9 CFU/mL 90-1330), and P1/7 control. Serum was collected on days 0 and 21 to evaluate antibody titers. On day 21 post-inoculation with strain 90-1330, pigs in the low dose and high dose group were challenged with strain P1/7 and survival was assessed. The second study investigated the role of the bacteriocin produced by strain 90-1330 and the role of competitive inhibition on protection. Pigs were divided into two groups: inoculated with strain 90-1330 or mock inoculated with PBS. Serum was collected on days 0 and 3 to evaluate antibody titers. Pigs were challenged 3 days post-inoculation with strain P1/7 and survival was assessed.

Results

The first study revealed strain 90-1330 was avirulent in CDCD pigs and could prevent *S suis* disease when challenged with a virulent strain 21 days post-inoculation. No pigs showed clinical signs of *S suis* following inoculation with strain 90-1330, while six animals in the P1/7 control group (6/8) developed clinical signs of *S suis* disease. Antibody titers to *S suis* strain 90-1330 rose in both the low dose and high dose groups prior to challenge (21 days post inoculation). Following challenge with strain P1/7, no pigs in the high dose group (0/9) and two pigs in the low dose group (2/9) developed *S suis* disease. The second study demonstrated that immune stimulation and antibody, not competitive exclusion or bacteriocin production, were important to the protective response. Pigs were still colonized at the time of challenge (3 days post-inoculation); however, antibody levels had not increased from the time of inoculation. Mortality and median survival time were similar between pigs inoculated with strain 90-1330 and control animals, indicating that the effects of strain 90-1330 on survival required additional time to develop and were likely associated with the development of a humoral immune response.

Conclusions

Inoculation with the avirulent *S suis* strain 90-1330 stimulated an adaptive response in exposed pigs that provided protection against disease with the pathogenic *S suis* strain P1/7. The protection generated by exposure to strain 90-1330 requires time to develop and is suspected to be antibody mediated. Although protection was not due to competitive inhibition or bacteriocin production in the high dose challenge model used in these studies, these mechanisms may contribute to protection when exposure doses are lower, as would be seen in the field. *S suis* strain 90-1330 should be further investigated as an intranasal vaccine as it provided significant protection against *S suis* disease with a single inoculation.

Reference

1. LeBel G, Vaillancourt K, Frenette M, Gottschalk M, Grenier D. Suicin 90-1330 from a Nonvirulent Strain of *Streptococcus suis*: a Nisin-Related Lantibiotic Active on Gram-Positive Swine Pathogens. *Appl Environ Microbiol*. 2014;80. **https://doi.org/10.1128/AEM.01055-14**

https://doi.org/10.54846/am2025/10

Mitigating economic losses from *Streptococcus suis* through data-driven decision-making

László Gombos[1,2], DVM; László Búza[3], DVM, PhD; Veronika Szűr[4], DVM; László Vargo[5], MSc, PhD, DSc
[1]Wittmann Antal Multidisciplinary Doctoral School of Plant, Animal, and Food Sciences, Széchenyi István University, Hungary; [2]Agrofeed Research and Development Division, Hungary; [3]Institute of Food Chain Science, University of Veterinary Medicine, Hungary; [4]Vadócdoki Research and Development Division, Hungary

Introduction

Streptococcus suis is a facultative pathogenic and zoonotic bacterium with global prevalence, affecting pigs of all age groups in industrial farming. The severity of the disease is influenced by various environmental factors, such as poor ventilation, high humidity, elevated dust and ammonia levels, and stress, as well as by improper management practices, including overcrowding, mixing pigs of different ages, and the absence of an all-in-all-out system. Co-infections with other bacterial or viral pathogens further exacerbate the condition.[1] The disease leads to reduced performance and increased mortality, posing a significant economic burden on swine production worldwide. With new regulations limiting the preventive use of antimicrobials in livestock and the absence of effective vaccines, controlling *S suis* infections has become increasingly challenging.[2] While antibiotics have traditionally been the primary tool for prevention and treatment, rising antimicrobial resistance has underscored the urgent need to reduce their use. This study aims to develop a data-driven approach to analyze farm-specific information on *S suis* and implement stabilization strategies to mitigate its impact effectively.

Materials and methods

A comprehensive three-year study (June 2020 to June 2023) was conducted on a farrow-to-finish farm that has been in operation for over 40 years, with 1,000 sows and a total of 126,000 newborn piglets tracked and monitored throughout the study. The farm has been free of PRRS. In 2020, the primary animal health challenges identified were *Actinobacillus pleuropneumoniae* (APP) (serotypes 5 and 13), *Lawsonia intracellularis* (ileitis), postweaning *Escherichia coli* diarrhea, and *S suis*. For more than 20 years, the sows have been routinely vaccinated against *Escherichia coli*, swine parvovirus, *Erysipelothrix rhusiopathiae,* and atrophic rhinitis. Piglets have been vaccinated against *Mycoplasma hyopneumoniae* and porcine circovirus type 2. In 2017, the farm initiated a strategy to significantly improve overall animal health and reduce antibiotic use. The introduction of vaccinations against APP (2017) and ileitis (2019), along with the optimization of biosecurity protocols, led to a notable improvement in the health of fatteners. Mortality dropped to below 1%, animal homogeneity improved, and the average daily weight gain increased by 60 g/day. Despite these improvements, *S suis*-related issues, including purulent and subclinical lesions, meningitis, and a 2% mortality rate per batch, remained a critical concern among weaners. Necropsies revealed that 80% of the examined piglets exhibited purulent, fibrinous lesions in body cavities and organs. It is important to note that cannibalism (tail and ear biting) was not observed on the farm. In 2020, the farm began developing a comprehensive protection strategy against *S suis*. As the first step, farm management needed to recognize that addressing this pathogen required a multifaceted and complex approach.

We collected and analyzed several key parameters, including:

1. Clinical symptoms: meningitis, arthritis, and septicemia (Table 1).
2. Pathological symptoms: Purulent lesions and septicemia indicators (Table 2).
3. Bacteriological results: The study primarily focused on meningitis and sudden death in suckling piglets and weaners, as these symptoms made the appearance and treatment of the disease unpredictable.
4. Laboratory testing for pathogen identification and antibiotic resistance: Routine aerobic bacterial examination of diagnostic samples was conducted on Columbia 5% sheep blood agar (Biolab, Budapest, Hungary) at 37 ± 1°C for 24–48 h under normal atmospheric conditions. Colonies of presumptive streptococci were sub-cultured and identified at the genus level using Gram-staining and basic biochemical tests. Species-level identification was performed using the API 20 Strep biochemical test kit (bioMérieux, Schaerbeek, Belgium). To detect *cps*-types, a multiplex PCR system was implemented,[3] with specific assays to discriminate *cps*-types 1/2 and 2, as well as 1 and 14, using a mismatch amplification mutation assay.[4] *Streptococcus suis* isolates without detectable *cps*-type were confirmed using a species-specific PCR.[5] Any isolates with uncertain or negative PCR results were sub-cultured and re-tested.

Key data points analyzed included:

1. Patterns of antibiotic use.
2. Bacteriological testing outcomes and resistance profiles.
3. Internal biosecurity measures.
4. Use of alternative products to reduce antibiotic reliance and pathogen-related losses, including monolaurin, symbiotics, and flavonoids.
5. Prevention protocols and technological improvements, such as:
   - By March 2021, strain identification was completed, and sow vaccinations began in May using a farm-specific autogenous vaccine (SUIBAC *S suis* serotype 1/2 and serotype 2). SUIBAC is a standardized autogenous vaccine designed for sows to enhance passive immunity in piglets. It contains a water/oil adjuvant, developed to induce high maternal antibody levels while maintaining vaccine safety. Manufactured under GMP conditions, each batch of the vaccine ensures consistent quality. Sows were vaccinated 8 weeks pre-partum, followed by a booster 4 weeks later. In subsequent cycles, sows received a single booster 4 weeks pre-partum.
   - Hygiene: Products and concentrations for cleaning and disinfection were revised.
   - Stricter disinfection protocols: Pens, equipment, and tools were subjected to more rigorous disinfection measures.
   - Personal hygiene: Increased attention to hygiene when handling animals, including the mandatory use of disposable rubber gloves.

Table 1: Clinical symptoms by age and antibiotic use

Age group	Clinical symptoms	Antibiotic use	Notes
Lactating sow	Metritis Shoulder lesions	Amoxicillin	Prophylaxis: all sows treated quarterly
Suckling piglet	Arthritis Sudden death	Amoxicillin → Ceftiofur →	Litter treatment Prophylaxis: injection
Weaner piglet	Sudden death Meningitis Septicemia	Amoxicillin → Ceftiofur → Enrofloxacin → Florfenicol →	Prophylaxis: 2 weeks after weaning in feed Individual treatment Individual treatment Metaphylaxis: in water
Fattener	Respiratory symptoms (*Actinobacillus pleuropneumoniae* and *Pasteurella multocida* infections in damaged lung areas)	Amoxicillin Enrofloxacin Tulathromycin	Individual treatment

Table 2: Pathological symptoms by age group

Age group	Pathological symptoms
Lactating sow	Catarrhal-purulent metritis, cystitis
Suckling piglet	Purulent arthritis, bacteremia
Weaner	Purulent arthritis, meningitis, septicemia
Fattener	Encapsulated, healed abscess (locus minoris resistenciae) Abattoir: Abscesses – with positive *Streptococcus suis* laboratory results

- Minimization of infectious transmitters and efforts to limit animal movement.
- Needle exchange policy: Needles were changed per litter or piglet.
- Colostrum management and fostering techniques: Colostrum quality in 1,738 breeding animals was evaluated twice using a Brix refractometer—once after the first liveborn piglet and again 24 h later. A Brix value of < 20 was considered low, 21–25 adequate, and > 26 high. Postpartum dysgalactia syndrome (PDS), suckling pig mortality, weaned piglets per sow, and fostering potential were individually recorded and analyzed.

Data analysis was conducted using a specialized animal health monitoring software (AB Kontroll; Agrofeed, Győr, Hungary).

Results

From 2020 to 2021, the study identified *S suis* serotypes 2 and 1/2 as the primary causes of severe clinical symptoms. Throughout the research, the pathogen consistently displayed sensitivity to the antibiotics used. However, alternative products showed limited efficacy, offering only minor benefits. The study also examined total mortality and antimicrobial treatments following weaning. Based on data collected in 2021 and 2022, the most effective interventions were:

1. Vaccination of sows with a farm-specific inactivated *S suis* vaccine.
Six months after the vaccination program began, the results were significant. Mortality associated with *S suis* ceased, and subclinical lesions decreased by 50%. By July 2022, one year after the start of vaccination, the use of critically important antibiotics (AMEG Category B: ceftiofur, enrofloxacin) was completely discontinued. By the end of 2022, prophylactic use of amoxicillin was also permanently stopped, and florfenicol usage for metaphylaxis across all batches decreased by 25%.

2. Revision of internal biosecurity measures.
The most effective biosecurity interventions included the consistent use of disposable gloves, stricter boot-washing protocols, and improved hygiene for pig-driving boards and other tools. While needles were replaced for each pig in three experimental groups, no significant improvement was observed in these groups. Similarly, minimizing animal movement in three groups did not yield significant results.

3. Enhanced colostrum management.
At farrowing, 98.8% of sows had a Brix value greater than 21, with 85.3% categorized as high (> 26). Twenty-four hours after farrowing, 80.1% of sows still had a value greater than 21, and 35.4% had a very high level, largely due to colostrum fostering. Among sows with an initial value of 21-25, only 41.7% reached a value above 21 after 24 h, and just 0.4% exceeded 26. For sows that initially measured above 26, 87.6% maintained a value above 21, with 47.9% remaining above 26. The number of weaned piglets per sow was higher in the 21–25 (13.4) and > 26 (13.6) Brix groups compared to those with values below 20 (12.4). Homogeneity among weaned piglets improved, and their average body weight increased by 10%, from 6.8 kg to 7.5 kg at 26 days of age. Sows with Brix values above 26 were able to raise 25% more piglets compared to those in the 21-25 or < 20 groups (Table 3). Total mortality after weaning decreased from 2% pre-vaccination to below 1% post-vaccination, with *S suis*-related mortality reduced by 90%. Antibiotic use in suckling and weaned piglets was reduced by 98.2%, and the use of critically important antibiotics associated with the pathogen was eliminated one year after the vaccination program began (Table 4).

Table 3: Results of colostrum measurements (n = 1738)

Measurement 1			Measurement 2		
> 21	1717	98.8%	> 21	1393	80.1%
> 26	1465	84.3%	> 26	616	35.4%
> 26 from > 21		85.3%			
21–25	252	→	> 21	105	41.7%
			> 26	1	0.4%
> 26		→	> 21	1284	87.6%
			> 26	615	47.9%
> 21		→	> 21	1389	80.9%
			> 26	616	44.3%

Table 4: Reduction in antibiotic use (2021–2023) and antibiogram results (2020–2023)

	Antibiotic use (mg)			MIC (µg/mL) – CLSI							
Antibiotic	2021	2023	Difference	2020		2021		2022		2023	
Amoxicillin	600,000,000	480,000	-99.9%	≤ 2/1	S	≤ 2/1	S	≤ 2/1	S	≤ 2/1	S
Ceftiofur	1,200,000	0	-100%	≤ .125	S	≤ .125	S	≤ .125	S	≤ .125	S
Enrofloxacin	780,000	0	-100%	= .25	S	= .25	S	= .25	S	= .25	S
Florfenicol	13,800,000	10,600,000	-23.2%	≤ 1	S	≤ 1	S	≤ 1	S	≤ 1	S
Total	615,780,000	11,080,000	-98.2%								

Summary of outcomes:

- 98% reduction in antibiotic use.
- Complete elimination of critically important antibiotics.
- 90% reduction in mortality and clinical symptoms.
- Improved biosecurity protocols, yielding additional benefits.
- 91% reduction in medication costs (from $83 to $8 per sow), showcasing significant economic and health improvement.

Conclusions

This case study underscores the critical role of detailed farm-specific analysis, combined with precise monitoring of *S suis* serotypes and antibiotic resistance patterns, in developing an effective protection strategy. The significant improvements in animal health highlight the importance of ongoing monitoring and data collection, enabling timely identification and resolution of emerging challenges to ensure the sustained success of implemented measures.

References

1. Vajda L, Kis IE, Kiss K, et al. *Streptococcus suis*: microbiology and laboratory diagnostics: Literature review – Part 1. *Magy Allatorvosok Lapja.* 2024;146:409-418.

2. Segura M, Aragon V, Brockmeier SL, et al. Update on *Streptococcus suis* research and prevention in the era of antimicrobial restriction: 4th International Workshop on *S. suis*. *Pathogens.* 2020;9:374.

3. Kerdsin A, Akeda Y, Hatrongjit R, et al. *Streptococcus suis* serotyping by a new multiplex PCR. *J Med Microbiol.* 2014;63:824-830.

4. Lacouture S, Okura M, Takamatsu D, et al. Development of a mismatch amplification mutation assay to correctly serotype isolates of *Streptococcus suis* serotypes 1, 2, 1/2, and 14. *J Vet Diagn Invest.* 2020;32:490-494.

5. Ishida S, Tien LHT, Osawa R, et al. Development of an appropriate PCR system for the reclassification of *Streptococcus suis*. *J Microbiol Methods.* 2014;107:66-70.

https://doi.org/10.54846/am2025/11

Effect of stocking density on F1 gilt grow-finish and reproductive performance

Christine Mainquist-Whigham[1,2], DVM, MS; Brady McNeil[1], MS; Amanda Cross[1], PhD, MBA; Ethan Stephenson[2], MS; Ashley Hartman[1,2], PhD; Tom Rathje[1], PhD, MBA
[1]DNA Swine Genetics, Columbus Nebraska; [2]Pillen Family Farms, Columbus, Nebraska

Introduction

Gilt development is one of the most critical, but most often overlooked aspects of a commercial sow system. As the United States swine industry has increased replacement rates, gilt developers have largely remained unchanged, thus reducing space per gilt. The impact of this reduced stocking density is not well understood in modern developing gilts and for producers to invest in additional space an increase in performance is needed. Therefore, the objective of this study was to evaluate the effect of different stocking densities on replacement gilts grow-finish and reproductive performance.

Methods

At 10 weeks of age, DNA L241 females (n = 824) were randomly allotted to finishing pens with 3 different stocking densities. Each stocking density treatment consisted of 12 pens of 0.71 m^2 (n = 360), 0.93 m^2 (n = 264), and 1.25 m^2 (n = 204). Weights were individually collected at 10, 24 and 31 weeks of age as well as each cycling event. Pen weights were collected at 16, 20, and 28 weeks of age. Backfat and loin depth were measured using an Exago Ultrasound between the 10th and 11th rib at 24 and 31 weeks of age and at each cycling event. Feed consumption was measured at the pen level using a Feed Logic system. Mortality date, reason and weight were recorded throughout the grow-finish phase. Structure scores were recorded at allotment and at 24 weeks of age. Lesion scores were recorded at 31 weeks of age on a 1-3 scale, with 1 being no lesions and 3 being deep cuts. Heat detection started at 20 weeks of age with fence line exposure for 2 weeks at 15 minutes per pen. Within pen boar exposure occurred from 22 to 31 weeks of age for a minimum of 10 minutes per pen. A blood progesterone assay was performed at 24 weeks of age on all females to determine true occurrence of females cycling. Additional assays were performed at 27, 29 and 31 weeks of age for females without a heat cycle by progesterone reading. Females were serviced on their 1st estrus if they were over 145 kg, otherwise on their 2nd or subsequent estrus once they were at least 136 kg.

Results

No significant differences were observed in mortality or structure scores by treatment. Weights were significantly different ($P < .05$) at 20 weeks for 0.71 m^2 at 90.0 kg compared to 0.93 m^2 and 1.25 m^2 at 94.1 kg and 94.0 kg respectively (Table 1). At 24 weeks of age, weight was different ($P < .01$) for all treatments and continued through 31 weeks of age, with 0.71 m^2 at 142.9 kg, 0.93 m^2 at 152.0 kg and 1.25 m^2 at 156.5 kg. Weight at first cycling were significantly different ($P < .05$) between 0.71 m^2 at 114.8 kg and 1.25 m^2 at 124.3 kg. Feed to gain was similar from 10 to 16 and 16 to 20 weeks of age, however, from 20 to 28 weeks of age 1.25 m^2 was significantly different ($P < .05$) at 3.39 compared to 3.83 for 0.71 m^2. Lesion scores were lower ($P < .05$) for 1.25 m^2 compared to both 0.93 m^2 and 0.71 m^2 with scores of 1.34 and 1.38 respectively. Cycling results determined by progesterone assay showed a 9% difference in cycling rate at 28 weeks of age for 0.71 m^2 at 64.0% cycling and 1.25 m^2 at 73.3% cycling. Comparing visual heat detection to progesterone assays resulted in an almost 7% difference in accuracy of heat detection with 0.71 m^2 at 75.72% and 1.25 m^2 at 82.3% accurate. Most of the inaccuracies, were females that were cycling by progesterone but missed by visual heat detection.

Table 1: Effect of stocking density on growth rate and reproductive weights on F1 gilts

	0.71 m^2	**0.93 m^2**	**1.25 m^2**
Week 10 weight (kg)	28.71	28.21	28.26
Week 20 weight (kg)	90.01^a	94.06^b	93.96^b
Week 24 weight (kg)	119.75^a	122.92^b	125.65^c
Week 28 weight (kg)	127.01^a	136.96^b	141.93^c
Week 31 weight (kg)	142.88^a	151.95^b	156.49^c
First cycle weight (kg)*	114.76^a	116.57^{ab}	124.28^b
First mating weight (kg)*	140.61^a	146.06^b	149.23^b

a,b,c Superscripts denote significant differences between treatments at ($P < .05$)
* First cycle weight and first mating weight do not include females that did not cycle by 31 weeks of age

Discussion

Growth rates in literature differ in the effect of stocking density on maternal gilt growth, however this trial only looked at growth up to 24 weeks of age with stocking densities between 0.77 and 1.13 m^2.[1] The current trial showed a significant growth effect measured at 20 weeks of age between 0.75 m^2 and the other treatments. This could partially be explained by genetic progress for growth, as there are 15 years between these trials. Results showed similar trends to the literature, as tighter stocking densities for gilts through the growing phase resulted in lower cycling rates but showed no difference for structure scores or mortality rates.[1] Lesion scores showed a similar trend to the literature, however stocking density resulted in a difference in lesions scores between 0.8 and 1.0 m^2 compared to the current trial only showing 1.25 m^2 as being a significant improvement.[2] To the authors knowledge this was the first trial of its size to evaluate visual heat detection compared to progesterone assays in relation to stocking density. Based on feedback from the heat detection team, the tighter stocking densities became challenging for the boars to appropriately check each gilt as there was not enough room for them to move around. This could explain part of the differences seen in accuracy of heat detection, as the tighter stocking densities had a lower accuracy of heat detection. While not a goal of this study, results from the literature show that faster growing gilts have higher total born and females that cycle earlier have better retention to P5.[3] By providing females with appropriate space, producers can increase the growth rates of their gilts and cause a higher percentage to cycle earlier. Overall, the 1.25 m^2 stocking density resulted in faster growth, better feed to gain, lower lesion scores, better cycling rates and a higher accuracy of cycling rates.

References

1. Young M, Tokach M, Aherne F, Dritz S, Goodband R, Nelssen J, Loughin T. Effect of space allowance during rearing and selection criteria on performance of gilts over three parities in the commercial swine production system. *J Anim Sci.* 2008;86(11):3181-3193. **https://doi.org/10.2527/jas.2007-0600**

2. Vermeer H, Dirx-Kuijken N, Bracke M. Exploration feeding and higher space allocation improve welfare of growing-finishing pigs. *Animals.* 2017;7(5):36. **https://doi.org/10.3390/ani7050036**

3. Roongsitthichai A, Cheuchuchart P, Chatwijitkul S, Chantarothai O, Tummaruk P. Influence of age at first estrus, body weight, and average daily gain of replacement gilts on their subsequent reproductive performance as sows. Livest. Sci. 2013;151(2-3):238-245. **https://doi.org/10.1016/j.livsci.2012.11.004**

https://doi.org/10.54846/am2025/12

PRRSV-2 variant classification: A dynamic nomenclature for enhanced monitoring and surveillance

K. VanderWaal[1]; N. Pamornchainavakul[1]; M. Kikuti[1]; J. Zhang[2]; M. Zeller[2]; G. Trevisan[2]; S. Rossow[1]; D. Linhares[2]; D. J. Holtkamp[2]; J. Herrera da Silva[1]; C. A. Corzo[1]; J. P. Baker[1]; T. K. Anderson[3]; D. N. Makau[1,4]; I. A. D. Paploski[1]
[1]University of Minnesota; [2]Iowa State University; [3]Virus and Prion Research Unit, National Animal Disease Center, USDA-ARS; [4]University of Tennessee

Introduction

Porcine reproductive and respiratory syndrome virus-type 2 (PRRSV-2) is one of the most sequenced viruses in the world,[1] largely because sequencing is used by animal health professionals as a tool for routine monitoring of virus circulation within and between farms. While phylogenetic analysis is the gold standard for interpretation of sequence data, practitioners and field epidemiologists often find it faster and more convenient to have a name in which they can refer to a group of genetically related sequences (ie, a genetic "variant") as part of everyday communication and outbreak investigations. Currently, the naming method used by the industry to discriminate between sequences is restriction fragment length polymorphism (RFLP)- typing,[2] sometimes in combination with an additional label corresponding to phylogenetic lineage.[3,4] However, lineages and sub-lineages are large and diverse, and are too coarse for on-farm disease monitoring, and using RFLP-types often leads to misleading conclusions (eg, viruses assigned to the same RFLP-type often are not genetically similar, and vice versa).[5-7] For example, RFLP 1-4-4, which is one of the most abundantly reported in the US today, occurs in seven different lineages.[8]

Here, we introduce the lineage-variant system for fine-scale genetic classification of PRRSV-2 ORF5 sequences in the United States, which has been tailored to meet the needs of animal health professionals.[9] Specifically, we outline criteria for defining PRRSV-2 genetic variants and the process for updating the nomenclature to accommodate expanding genetic diversity. We also introduce a webtool that can be used to assign newly generated sequences to the appropriate variant ID. Finally, we demonstrate how lineage-variant classification can enhance the utility of sequence data for the purposes of epidemiological monitoring and surveillance.

Materials and methods

Data source and pre-processing

Sequence data were obtained from the Morrison Swine Health Monitoring Project (MSHMP), which collects sequence data from both sow and grow-finish farms belonging to 38 production systems, accounting for > 55% of the US population.[10] 25,403 sequences were available from October 1, 2015 to December 31, 2023. We initiated the classification system in 2018 with 36 months of data (2015-18), then added new data every three months up until December 2023 to establish procedures for updating the classification system on a rolling basis. We have continued to update the system since then. For each 36-month dataset, sequences are aligned and a maximum likelihood tree is built using IQ-TREE2.

Classifying variants

Each quarter, groups of closely related sequences are identified in phylogenetic trees using a validated workflow involving custom-built code and clustering algorithms, with a maximum threshold set for the allowable average pairwise distance amongst sequences belonging to the same variant. These methods were established after a rigorous comparison of classification approaches to ensure robust and reproducible results.[7] Additional criteria for defining a variant were that the group must have a) five of more sequences, b) have high phylogenetic support for their shared ancestry in the tree (bootstrap > 85) to ensure reproducibility, and c) that the genetic distance to the nearest named variant must be > 2%. Any sequences belonging to clades that did not meet these requirements were labeled as "unclassified."

Naming convention

Nomenclature for variants was the sub-lineage to which the variant belonged, followed by an integer (ie, 1A.3 and 1H.3 are the third variants identified within sub-lineage 1A and 1H, respectively).

Assignment of new sequences

We created a web-tool for end-users to assign their sequences to the appropriate variant ID (**https://stemma.shinyapps.io/PRRSLoom-variants/**). This web-tool is updated quarterly, and does not retain any user data. Rather, it uses a machine learning algorithm to determine the first and second most likely variant ID for a given sequence. A sequence is considered unclassified if the assignment probability is < 0.25 (or where the highest probability was < 2x more than the second highest probability).

The updated assignment algorithms are also available as R and Python scripts so that they can be ported to external applications maintained by diagnostic laboratories or other groups. This ensures that all potential end-users will obtain the same variant classifications, regardless of the platform.

Results

Based on > 25,000 US open-reading-frame 5 (ORF5) sequences, sub-lineages were divided into genetic variants using a clustering algorithm (Figure 1). Through classifying new sequences every three months and systematically identifying new variants across eight years, we demonstrated that the lineage-variant classification system produced robust, reproducible results across time and can dynamically accommodate new genetic diversity arising from virus evolution. From 2015 to 2024, 168 variants were identified. On average, there were 48 active variants per year, of which 26 were common (detected at least 50 times).[9] Mean within-variant genetic distance was 2.4% (max: 4.8%). The mean distance to the closest related variant was 4.9%.

Figure 1: Example tree showing ORF5 sequences (sub-lineage 1H), colored according to A) RFLP-type and B) variant ID. The tree contains two major RFLP types (1-8-4 and 1-4-4), which are inter-mixed on the tree, demonstrating that RFLP-types do not convey accurate information on genetic relatedness. In contrast, variants are well-delineated in the tree (limited inter-mixing of colors), demonstrating that variant IDs accurately group together closely related sequences.

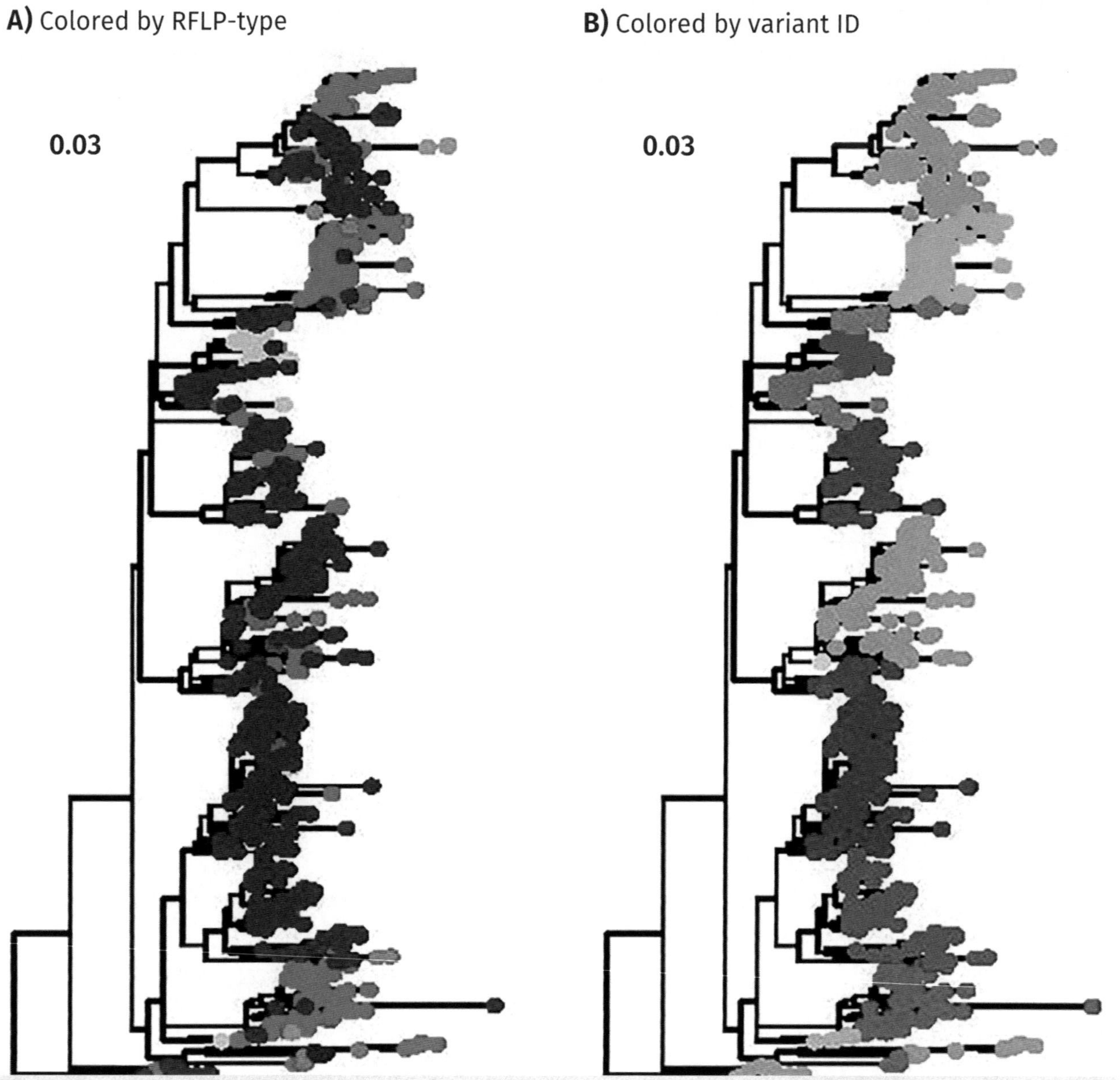

Figure 2: Dashboard for the publicly available webtool that can be used to assign sequences to the appropriate variant ID. Sequences can be copy-pasted or uploaded as a *fasta* file (no data is saved by the webtool). Outputs include the column "assign.final," which shows the variant IDs that best match the uploaded sequences. The second-best match is also shown.

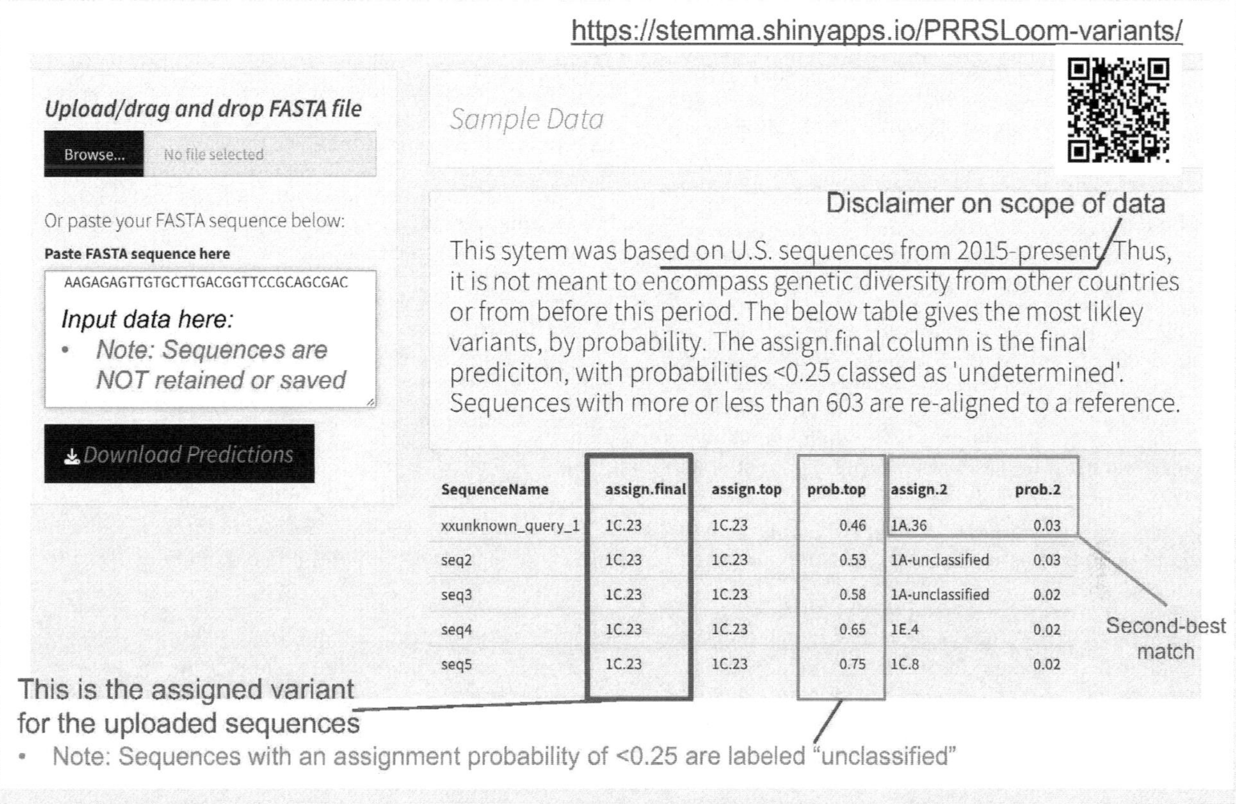

Initially, ~36% of sequences could not be reliably grouped into a well-supported variant and were considered "unclassified," but the unclassified rate reduced and stabilized to ~11% in 2020 and 2021, and ~7% in 2022 and 2023. Lineage 1A accounted for 94.6% of unclassified sequences. Lineage 1A has lower genetic diversity than other sub-lineages due to its more recent emergence approximately 10 years ago,[3] and thus divergence of this group into distinct genetic variants is less pronounced.

A routinely updated webtool (**https://stemma.shinyapps.io/PRRSLoom-variants/**) was developed and is publicly available for end-users to assign newly generated sequences to a variant ID (Figure 2). This classification system relies on US sequences from 2015 onwards; further efforts are required to extend this system to older or international sequences. The webtool also contains tabs listing all variants identified thus far, including the IDs of those associated with modified live vaccines. Other reported data include the total number of sequences assigned to a variant in the MSHMP dataset since 2015, year when the variant was first and last detected, and the US occurrence trend. The US occurrence trend is based on a variant's relative occurrence (number of sequences) two years ago compared the past year. Relative occurrences close to 1 indicate that the incidence of sequences across the past two years is relatively constant - or 'stable incidence'. If the occurrence of sequences has doubled year-over-year, then the occurrence trend is classed as 'accelerating'. If the occurrence of sequences has halved year-over-year, then the occurrence trend is classed as 'decelerating'. Variants that have been detected < 5 times in the past few years are classed as 'rare or extinct' in this. dataset.

Discriminating between new and previous PRRSV-2 viruses on the same farm

Lineage-variant classification had superior performance in discriminating between new and previous PRRSV-2 on the same farm. Using the MSHMP database, we identified 73 sow farms for which there were at least 4 sequences available within the same year. We confirmed whether these sequences belonged to the same apparent on-farm circulation event through phylogenetic analysis (ie, whether sequences clustered together in the tree, suggesting the circulation of single strain). The RFLP-type was not stable in approximately 43% of these on-farm circulation events. The use of variant IDs minimized this issue, with only ~6.5% of on-farm circulation events exhibiting instability in the variant ID.

Tracking between-farm spread and possible sources of introduction

Lineage-variant classification provide more accurate information for tracking between-farm spread than RFLP-types. As an example, we zoomed into the sequences collected during 2020 from 15 farms located in a small geographic area (Figure 3). Focusing on the circled farm (which was infected by variant 1H.26), we can trace six nearby farms with the same variant ID (Figure 3a). All these viruses cluster together in the tree, and a homology table shows that they are genetically very similar (< 2.5% different in the ORF5 gene). In contrast, we can trace six nearby farms with the same RFLP-type (RFLP-type 1-4-4, sub-lineage 1H, Figure 3b), but these are not the same farms that had 1H.26. An examination of the tree shows that these 1-4-4 viruses are not closely related (Figure 3c). Thus, relying on RFLP-types would result in identifying the wrong set of farms part of this outbreak, and thus lead to potentially erroneous conclusions on the likely modes of between-farm transmission.

Discussion

While the shortcomings of existing PRRSV-2 ORF5-based classification systems have long been apparent, recent advances in computational power and the creation of national sequence databases have created the opportunity to finally address these issues. Here, we outline the lineage-variant system for fine-scale classification of PRRSV-2 in the US, which is expandable to new genetic diversity as it evolves. We demonstrate that, as compared to RFLP-typing, lineage-variant classification more reliably groups viruses based on relatedness in the ORF5 gene, and provides better discrimination between unrelated viruses.

We lay out a validated workflow to update this classification system on a quarterly basis and develop procedures for the assignment of newly generated sequences via a centrally maintained assignment tool. This facilitates a unified naming scheme across the US that can dynamically adapt to evolving PRRSV-2 diversity across time and space. As of 2024, these variant IDs have now been reported on a routine basis by several veterinary diagnostic labs, including the University of Minnesota.

At present, the variant classifications are based on data from the US from 2015 to present, and sequences that are older or from other countries are likely to belong to distinct genetic variants that were not present in our dataset (and thus would be unclassified in this system). We continue to explore the expansion of lineage-variant classification to other countries where representative sequence data are available, which is often not the case. Another caveat is that variant IDs no not provide information on virulence or clinical picture. No classification system currently provides such information, likely because apparent virulence is influenced by co-infections and other external factors. Variant IDs also do not directly predict immunological cross-protection, although viruses labeled as the same variant are more genetically homologous.

Overall, a unified and more reliable classification system will facilitate communication about outbreaks amongst diagnostic labs and practitioners, improve epidemiological monitoring of emerging and endemic variants, and provide a framework for better integration of field insights with research across the epidemiological, virological, and immunological realms. Finally, this work

Figure 3: Data from 15 farms in a small geographic area in 2020. A. Map of farm locations, colored by the variant IDs detected at each farm. B. Map of farm locations, colored by the RFLP-type detected at each farm. The focal farm is circled in red, and green arrows are used indicate possible sources of infection based on recover of the same variant ID or RFLP-type. C. Genetic homology table (deeper red colors indicating higher relatedness) and phylogenetic trees (colored by variant ID or RFLP-type).

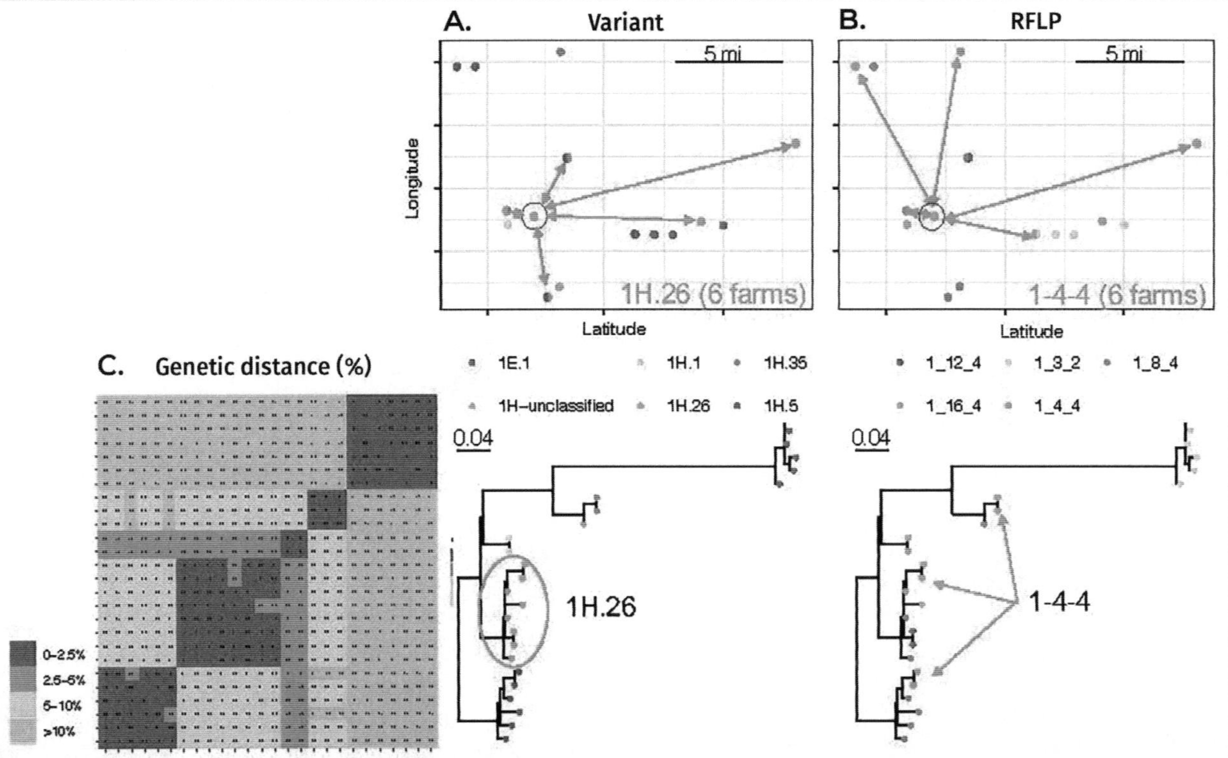

was conducted with iterative feedback from a working group of veterinarians, researchers, and diagnosticians. This close engagement with stakeholders and end-users has been crucial for the operationalization and adoption of the variant classification, ensuring that it is tailored to the needs of animal health professionals utilizing sequence data for disease management in the field.

References

1. VanderWaal K. 2024. Surprising but true: PRRSV one of the most sequenced viruses in the world, on National Hog Farmer. **https://www.nationalhogfarmer.com/livestock-management/surprising-but-true-prrsv-one-of-the-most-sequenced-viruses-in-the-world**

2. Wesley RD, Mengeling WL, Lager KM, Clouser DF, Landgraf JG, Frey ML. 1998. Differentiation of a porcine reproductive and respiratory syndrome virus vaccine strain from North American field strains by restriction fragment length polymorphism analysis of ORF 5. *J Vet Diagn Invest* 10:140-4.

3. Paploski IAD, Pamornchainavakul N, Makau DN, Rovira A, Corzo CA, Schroeder DC, Cheeran MCJ, Doeschl-Wilson A, Kao RR, Lycett S, VanderWaal K. 2021. Phylogenetic Structure and Sequential Dominance of Sub-Lineages of PRRSV Type-2 Lineage 1 in the United States. *Vaccines* 9:608.

4. Yim-im W, Anderson TK, Paploski IAD, Vanderwaal K, Gauger P, Krueger K, Shi M, Main R, Zhang JQ, Chao DY. 2023. Refining PRRSV-2 genetic classification based on global ORF5 sequences and investigation of their geographic distributions and temporal changes. *Microbiology Spectrum* 11:e02916-23.

5. Murtaugh MP, Stadejek T, Abrahante JE, Lam TT, Leung FC. 2010. The ever-expanding diversity of porcine reproductive and respiratory syndrome virus. *Virus Res* 154:18-30.

6. Cha SH, Chang CC, Yoon KJ. 2004. Instability of the restriction fragment length polymorphism pattern of open reading frame 5 of porcine reproductive and respiratory syndrome virus during sequential pig-to-pig passages. *J Clin Microbiol* 42:4462-4467.

7. VanderWaal K, Pamornchainavakul N, Kikuti M, Linhares D, Trevisan G, Zhang J, Anderson TK, Zeller M, Rossow S, Holtkamp DJ, Makau DN, Corzo CA, Paploski IAD. 2024. Phylogenetic-based methods for fine-scale classification of PRRSV-2 ORF5 sequences: a comparison of their robustness and reproducibility. *Frontiers in Virology* 4:1433931.

8. SDRS. 2024. PRRSV genotyping dashboard: RFLP and Lineages (accessed April 2024). *Swine Disease Reporting System* May, 2024. **https://fieldepi.org/domestic-swine-disease-monitoring-program/**

9. VanderWaal K, Pamornchainavakul N, Kikuti M, Zhang J, Zeller M, Trevisan G, Rossow S, Schwartz M, Linhares DCL, Holtkamp DJ, da Silva JPH, Corzo CA, Baker JP, Anderson TK, Makau DN, Paploski IAD. 2024. PRRSV-2 variant classification: a dynamic nomenclature for enhanced monitoring and surveillance. *bioRxiv* **doi:10.1101/2024.08.20.608841:2024.08.20.608841**

10. MSHMP. 2024. PRRS cumulative incidence (last updated April 2024). *Morrison Swine Health Monitoring Project* April, 2024. **https://mshmp.umn.edu/reports#Charts**

https://doi.org/10.54846/am2025/13

Estimating the effects of negative and positive pressure systems on PRRS occurrence

X. Yue, BSc, MSc, PhD; M. Kikuti, DVM, MPH, PhD; C. M. Melini, DVM, MA; C. A Corzo, DVM, MS, PhD
Department of Veterinary Population Medicine, University of Minnesota, Saint Paul, Minnesota

Introduction

Porcine reproductive and respiratory syndrome virus (PRRSV) continues to pose significant challenges to the US swine industry. While air filtration systems combined with biosecurity measures have demonstrated to help prevent PRRSV introductions,[1-3] the impact of spatial clustering, regional pig density, and ventilation pressure type (positive vs negative) have not been addressed. This project aims to estimate and compare the effects of different ventilation pressure air filtration systems on PRRS occurrence in US sow farms.

Materials and methods

Breeding herd data from the Morrison Swine Health Monitoring Project (MSHMP)[4] were collected spanning from 2009 to 2024. A database including Farm ID, geospatial coordinates, herd size, air filtration status (ie, not filtered, positive or negative pressure), and PRRSV outbreak data was assembled. We calculated the total number of new PRRSV outbreaks and the total number of weeks at risk for each farm based on their air filtration status. The number of farms within 35 km (21.75 miles) was also calculated. The total number of pigs in the county was obtained from the USDA 2022 census. Unfiltered farms within a 150 km (93 miles) radius of any filtered farm were included in the study. We used generalized additive mixed models (GAMM) with a negative binomial distribution to assess the effects of air filtration status on the total number of PRRSV outbreaks per farm, incorporating Gaussian process smooths for spatial autocorrelation, with the natural logarithm of the total number of weeks at risk as an offset. Herd size, number of farms within 35 km, and number of pigs in the county were included as covariates while farm ID was used as a random effect. A similar model was applied to a subset of farms with only positive and negative pressure filtration.

Results and conclusion

The study involved 413 sow farms, comprising 238 unfiltered, 128 with negative pressure, and 47 with positive pressure filtration. The filtered farms represented nearly one million sows. GAMM results showed that the rate of PRRS outbreaks was 46% lower ($P < .001$) in negative pressure-filtered farms and 59% lower ($P < .001$) in positive pressure-filtered farms when compared to unfiltered farms. Despite the relatively low number and shorter at-risk period that positive-pressure-filtered represented in the dataset, our analysis was able to detect a significantly lower PRRS outbreak rate compared to unfiltered farms. Furthermore, no difference was detected when comparing the rate of outbreaks between positive and negative pressure filtered farms. The non-significant difference in PRRS occurrence between farms with positive and negative pressure may be partly attributed to the unbalanced nature of the dataset, as negative pressure systems are more commonly used in current industrial practices.

This study offers valuable evidence on the effects of different air filtration systems on PRRS occurrence while controlling for factors such as regional pig density and comparing filtered vs unfiltered farms within the same region. Altogether, this study enhanced existing knowledge through a comprehensive longitudinal retrospective analysis.

References

1. Spronk G, Otake S, Dee S. Prevention of PRRSV infection in large breeding herds using air filtration. The Veterinary record. 2010;166:758. **https://doi.org/10.1136/vr.b4848**

2. Dee S, Cano JP, Spronk G, Reicks D, Ruen P, Pitkin A, et al. Evaluation of the long-term effect of air filtration on the occurrence of new PRRSV infections in large breeding herds in swine-dense regions. Viruses. 2012;4:654–62. **https://doi.org/10.3390/v4050654**

3. Alonso C, Murtaugh MP, Dee SA, Davies PR. Epidemiological study of air filtration systems for preventing PRRSV infection in large sow herds. Preventive Veterinary Medicine. 2013;112:109–17. **https://doi.org/10.1016/j.prevetmed.2013.06.001**

4. MSHMP. Morrison Swine Health Monitoring Program. October 5, 2024, Accessed October 31, 2024. **https://mshmp.umn.edu/**

https://doi.org/10.54846/am2025/14

Impact of production and health management strategies on PRRSV outbreak recovery in breeding herds

A. P. Poeta Silva, DVM, PhD; C. Rademacher, DVM; E. De Conti, DVM; T. Peterson, MS; J. Swaminathan, BS; K. Rupasinghe, BS; G. Silva, DVM, PhD; G. Trevisan, DVM, PhD; D. Linhares, DVM, PhD
Field Epi Team, Veterinary Diagnostic and Production Animal Medicine, Iowa State University, Ames, Iowa

Introduction

Recent research and field reports have highlighted the ongoing emergence of porcine reproductive and respiratory syndrome virus (PRRSV) strains and variants[1-3] alongside the increased clinical impact of PRRSV outbreaks on breeding herds. This study described the association of PRRSV outbreaks recovery metrics and health and production management strategies using data from recent (2017-2024) outbreaks.

Materials and methods

A total of 161 herd-outbreaks between 2017-2024 that included lineage information (derived from sequencing of outbreak-associated strains) were retrieved from the PRRS Outbreak Management Program (POMP) database. Outbreak recovery metrics were time-to-stability (TTS, weeks to achieve 13 consecutive weeks of PRRSV PCR-negative results in due-to-wean piglets samples), time-to-baseline-production (TTBP, weeks to recover the volume of weaned piglets comparing 21 weeks prior vs 52 weeks after outbreak), and total piglet losses per 1,000 sows (TL, number of piglets not weaned between the outbreak and achieving TTBP). Health and production management strategies were collected through a survey and used as explanatory variables in the analyses. Mixed-effect cox and zero-inflated poisson regressions were used to associate TTS and TL with selected health interventions. The production system was used as a random effect and lineages as a covariate term in both models.

Results

Figure 1 and 2 shows TTS and TL by systems, respectively. Analyses showed that batch farrowing was associated with 11 times faster (25 weeks) TTS and 32% lower TL compared to weekly farrowing ($P < .01$). Herd closure was associated with 5.3 times faster (40 weeks) TTS compared to no closure ($P < .01$). As exposures for sows, live virus inoculation (LVI) only and LVI+modified live vaccine (MLV) combination were associated with 2.3 times faster (10 weeks) TTS compared to MLV only ($P < .01$), while LVI only was associated with 40% higher TL (average of 2000 piglets not weaned) compared to MLV only and LVI+MLV ($P < .01$). Comparing post-outbreak gilt exposures, LVI+MLV were associated with 5.4 times faster (10 weeks) TTS compared to MLV or LVI only ($P = .07$), while LVI+MLV was associated with 17% lower TL (700 piglets not weaned) than LVI only ($P < .01$). Likewise, for gilts before outbreaks, any immunologic exposure (MLV only, LVI only, or LVI+MLV) was associated with 47% (2200 piglets not weaned) lower TL than no intervention ($P < .01$).

Discussion

The variation of TTS and TL across different systems revealed opportunities to improve PRRSV control management programs. These differences could be associated with different production and health management strategies used during outbreak control. For instance, when comparing health management strategies, faster TTS recovery was associated with herd closure and combinations of LVI and MLV in sows and gilts (during acclimation and as exposure during the outbreak control phase). Farms utilizing batch farrowing methods showed faster TTS recovery alongside lower TTL. As an ongoing project, additional outbreak information will be added to the analysis, as it becomes available, and results will be shared with the industry.

References

1. Martín-Valls et al (2023). **Doi: 10.1186/s40813-022-00298-3**
2. Xia et al (2023). **Doi: 10.1155/2023/9929573**
3. Kikuti et al (2021). **Doi: 10.1371/journal.pone.0259531**

Figure 1: Time-to-stability (TTS) by swine system (A to L) enrolled in the PRRS outbreak management program (POMP).

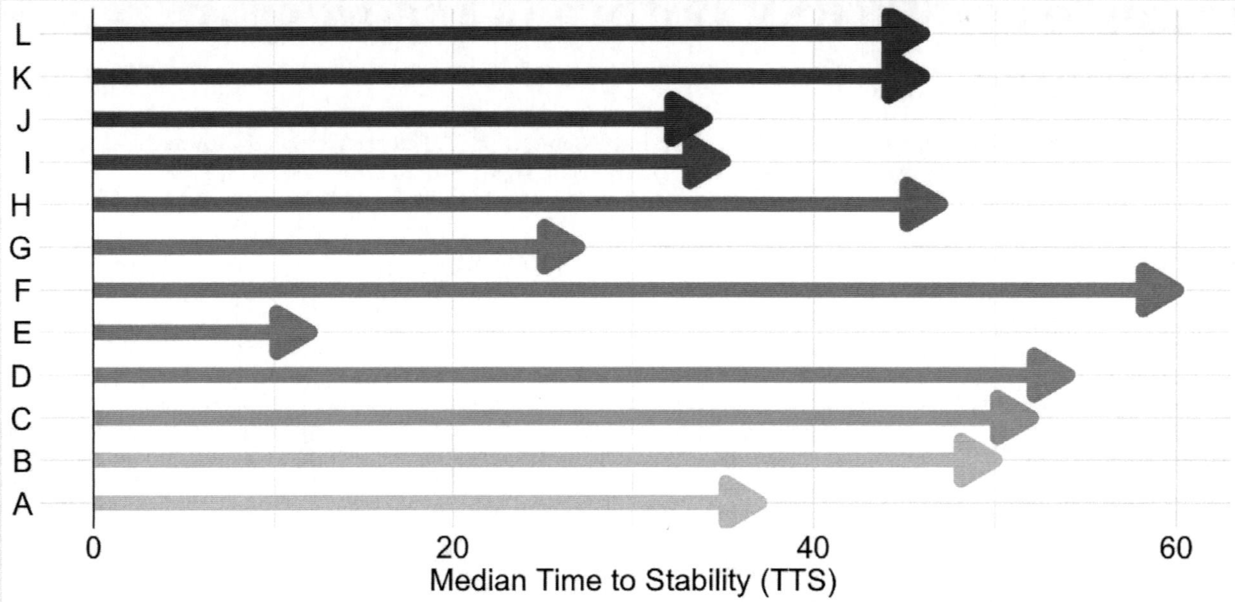

Figure 2: Total losses (TL) per 1,000 sows by swine system (A to L) enrolled in the PRRS outbreak management program (POMP).

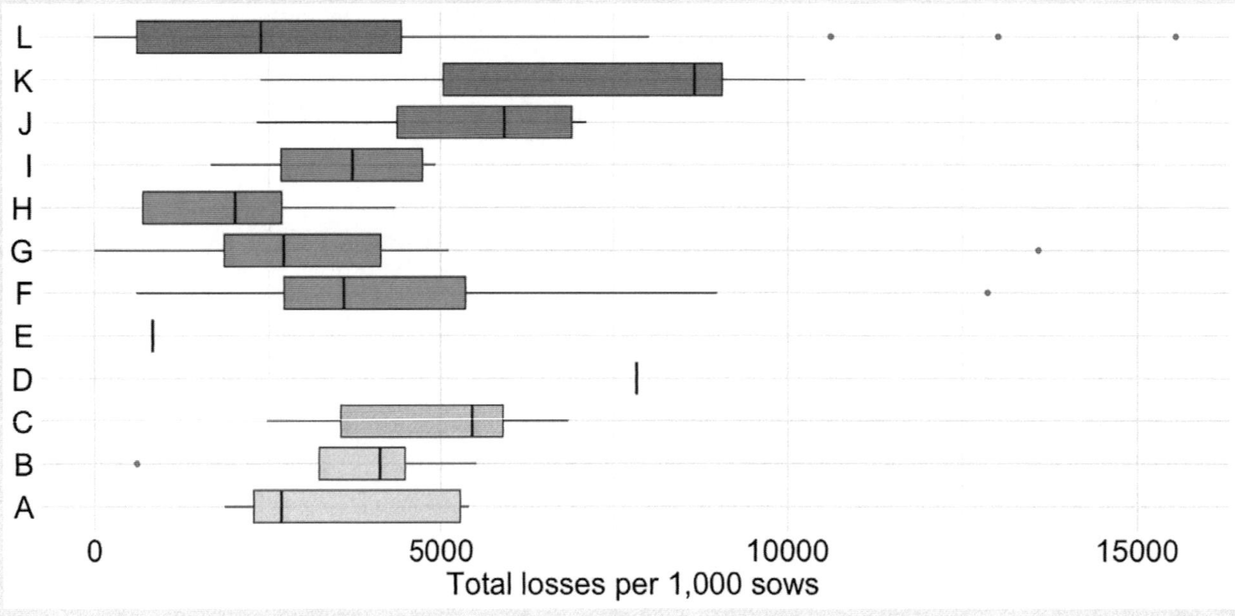

https://doi.org/10.54846/am2025/15

Productivity losses due to PRRSV, PEDV, and *Mycoplasma hyopneumoniae* in US breed-to-wean herds from 2016 to 2020

Onyekachukwu Osemeke, DVM, MS, PhD; Gustavo Silva, DVM, MS, PhD;
Daniel Linhares, DVM, MBA, PhD; Derald Holtkamp, DVM, MS
Department of Veterinary Diagnostic and Production Animal Medicine, Iowa State University, Ames, Iowa

Introduction

Porcine reproductive and respiratory syndrome virus (PRRSV), porcine epidemic diarrhea virus (PEDV), and *Mycoplasma hyopneumoniae* (Mhp) are three pathogens that significantly impact productivity.[1-3] However, there is limited information on the productivity impact of each of these pathogens when the effects of the other two are concurrently controlled for within the same dataset. This study aimed to assess the impact of PRRSV, PEDV, and Mhp statuses on the number of pigs weaned per mated female per year (PWSY) in US breeding herds.

Methods

Sow performance data, outbreak history, vaccination history, and pathogen (PRRSV, PEDV, Mhp) status by week were collected from 12 companies, encompassing 297 farms and over 68,000 herd-weeks between 2016 and 2020. A linear mixed regression model was used to estimate the least squares or adjusted mean of each productivity parameter given the pathogen status. Herd weeks were classified for Mhp as A (negative), B (controlled/not clinical), and C (uncontrolled/clinical). PEDV was classified as A (positive-stable, provisionally negative, or negative) and B (positive-unstable). PRRSV was classified as A (negative), B (positive-stable), C (positive unstable after 16 weeks post-outbreak), D (first 16 weeks following an outbreak in a previously positive herd), and E (first 16 weeks following an outbreak in a previously negative herd). The model outcome was the pigs weaned per mated female per year. The model used the PRRSV, PEDV, and Mhp statuses and the interaction between PRRSV and Mhp statuses as fixed effects. The production system, farm within the production system, and season within each year were included as random effects. The interaction between PED and PRRSV statuses was not assessed due to the absence of either one of the PED statuses in one or more levels of PRRSV statuses within the data.

A Tukey posthoc test evaluated the statistical significance ($\alpha = 0.05$) of PWSY differences between groups.

Results

The adjusted mean (and 95% confidence intervals) PWSY for PRRSV statuses are presented in Table 1. For PRRSV statuses, all pairwise comparisons showed statistically significant differences ($P < .05$). The adjusted mean PWSY for Mhp status group C was statistically lower than the other two groups ($P < .05$). The adjusted mean PWSY for PEDV statuses A and B were statistically different ($P < .05$). The interaction between Mhp and PRRSV statuses was statistically significant (Figure 1) with the effect of uncontrolled Mhp on productivity being most remarkable in herd weeks with the highest pressure of PRRSV infection (PRRSV statuses D and E).

Discussion

The PWSY is the throughput measure of breed-to-wean enterprises and the most economically translatable measure of sow productivity performance. PRRSV and PEDV statuses had the most profound effect on PWSY. The numbers provided here will aid swine practitioners in understanding how much more PWSY they could gain or lose by attaining better or worse disease categories and could justify investments in improving disease categories or in preventing attaining a worse status. The provided estimates were obtained from 2016 to 2020 data and may not represent the productivity realities of the assessed pathogens as of today.

Table 1: Least squares or adjusted means (with 95% confidence intervals) of pigs weaned per sow per year (PWSY) given disease pathogen (porcine reproductive and respiratory syndrome virus [PRRSV], porcine epidemic diarrhea virus [PEDV], and *Mycoplasma hyopneumoniae* (Mhp) statuses

PRRSV		Mhp		PEDV	
Status	PWSY	Status	PWSY	Status	PWSY
A	25.08 (23.52, 26.64) [a]	A	21.15 (19.54, 22.76) [a]	A	22.94 (21.38, 24.50) [a]
B	23.73 (22.17, 25.29) [b]	B	20.89 (19.33, 22.45) [a]	B	18.15 (16.58, 19.73) [b]
C	21.49 (19.93, 23.05) [c]	C	19.60 (17.99, 21.21) [b]		
D	18.79 (17.14, 20.44) [d]				
E	13.65 (11.96, 15.34) [e]				

[a-e] Levels within each column with different superscripts are statistically different ($P < .05$).

Figure 1: The pigs weaned per sow per year for the *Mycoplasma hyopneumoniae* (Mhp) statuses (A [negative], B [controlled/not clinical], and C [uncontrolled/clinical]) over the statuses of porcine reproductive and respiratory syndrome virus (PRRSV) (A [negative], B [positive-stable], C [positive unstable after 16 weeks post-outbreak], D [first 16 weeks following an outbreak in a previously positive herd], and E [first 16 weeks following an outbreak in a previously negative herd]).

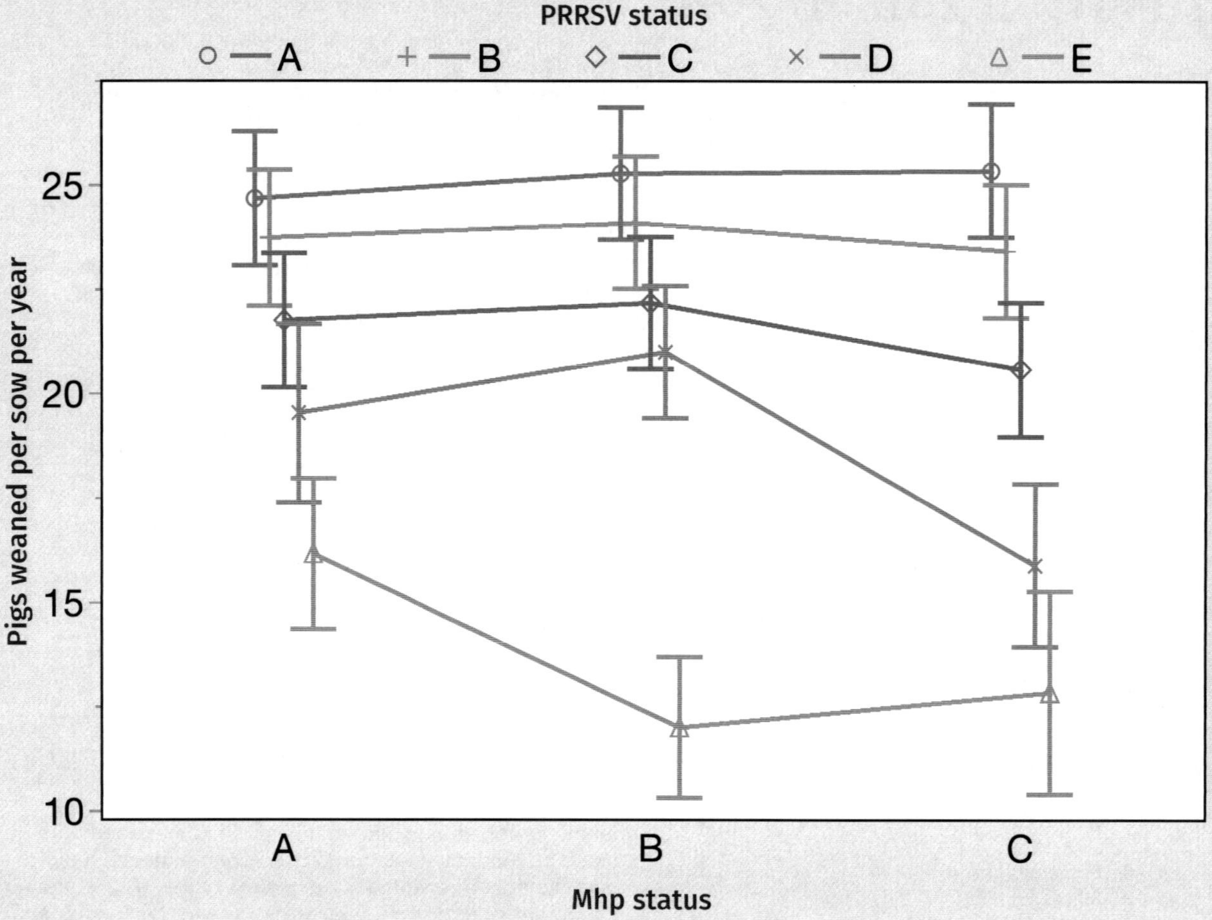

References

1. Schulz, L. L. & Tonsor, G. T. Assessment of the economic impacts of porcine epidemic diarrhea virus in the United States. *J Anim Sci* 93, 5111 (2015).

2. Boeters, M. *et al.* The economic impact of endemic respiratory disease in pigs and related interventions - a systematic review. *Porcine Health Manag* 9, (2023).

3. Holtkamp, D. J. *et al.* Assessment of the Economic Impact of Porcine Reproductive and Respiratory Syndrome Virus on United States Pork Producers. Journal of Swine Health and Production vol. 21 (2013).

https://doi.org/10.54846/am2025/16

The impact of the timing of PRRSV and swine enteric coronaviruses introduction on wean-to-market productivity

Kate Dion; Daniel C. L. Linhares; Gustavo S. Silva; Derald Holtkamp
Veterinary Diagnostic and Production Animal Medicine, Iowa State University, Ames, Iowa

Introduction and objective

Porcine reproductive and respiratory syndrome virus (PRRSV) and swine enteric coronavirus (SeCoV) can significantly decrease the productivity of growing pigs by increasing mortality rates and decreasing growth, but the impact of those pathogens may vary with age at the first introduction and the number of pathogens introduced. The objective of this study was to determine the impact of the timing of introductions of PRRSV and SeCoV on wean-to-market productivity and economic parameters.

Materials and methods

The pathogen status and timing of the introduction of porcine epidemic diarrhea virus, porcine deltacoronavirus, and PRRSV for 74 groups of pigs was determined by collecting eight oral fluid samples every two weeks per group of pigs and tested by RT-PCR. The timing of introduction for those pathogens was categorized as nursery (3-10 weeks of age), early finisher (11-21 weeks of age), and late finisher (22-33 weeks of age.) Generalized linear mixed regression models (SAS 9.4) were used to compare mortality, average daily gain (ADG), and feed conversion (FC) between pathogen combinations and time of pathogen introductions. A production economic model was used to determine differences in net profit per pig placed using productivity differences from baseline if the significance was lower than 0.25 in the regression model.

Results

Results are listed in Table 1.

Discussion and conclusions

The results of this study show that the impact of introducing multiple pathogens increases mortality compared to single pathogen introductions. Pathogen introduction timing in the nursery phase resulted in higher mortality than later introductions, and nursery introductions of SeCoV impacted ADG and FC. Net profit compared to negative groups is decreased due to increased mortality and reduced ADG and FC, with the most significant differences resulting from pigs with a nursery introduction of SeCoV. These results show the importance of implementing measures to prevent or delay introductions of pathogens to minimize the impact on wean-to-market productivity.

Table 1: Results of time and pathogen combinations on productivity and economic parameters

Time/pathogen combination	# of groups	Mortality	Average daily gain	Feed conversion	Difference from baseline, $/pig
Negative (baseline)	2	7.39%[a]	1.672[a]	2.664[ab]	–
Nursery PRRS	5	9.28%[b]	1.613[a]	2.645[a]	-$1.92
Nursery PRRS & SeCoV	3	13.21%[d]	1.548[a]	2.950[b]	-$20.27
Early finisher PRRS	36	9.74%[b]	1.634[a]	2.688[ab]	-$2.39
Early finisher PRRS & nursery SeCoV	18	12.45%[d]	1.577[a]	2.703[ab]	-$9.73
Early finisher PRRS & late finisher SeCoV	6	11.50%[c]	1.609[a]	2.769[ab]	-$4.18
Late finisher PRRS	4	7.51%[a]	1.646[a]	2.657[a]	$0.00

[a-d] Superscript letters indicate statistical differences, with a significance level of .05
PRRS = porcine reproductive and respiratory syndrome; SeCoV = swine enteric coronavirus

https://doi.org/10.54846/am2025/17

Evaluating the efficacy of alternative livestock trailer wash methods on reducing the risk of PEDV introduction to farm site areas when loading market pigs

E. S. Magalhaes[1]; E. Aljets[2]; M. P. Mil-Homens[2]; G. S. Silva[2]; P. Gauger[2]; D. C. L. Linhares[2]; M. Almeida[2]; J. Zhang[2]; R. Main[2]
[1]Department of Animal Science, Iowa State University, Ames, Iowa;
[2]Veterinary Diagnostic and Production Animal Medicine Department, Iowa State University, Ames, Iowa

Introduction

Effective trailer cleaning and disinfection protocols are crucial for controlling the spread of porcine epidemic diarrhea virus (PEDV) and other diseases of high consequence.[1] Livestock trailers returning from terminal points of concentration (slaughter facilities, buying stations, or cull markets) present as a primary risk factor for indirectly recirculating, amplifying, and broadly distributing disease-causing agents in US swine. The current lack of infrastructure and inability to clean and disinfect the greater expanse of livestock trailers returning from terminal points of concentration is a well-understood industry-level vulnerability of national importance. The principal objective of this study was to evaluate the effects of alternative livestock trailer cleaning methodologies on the relative risks of introducing diseases to farm sites when loading-out hogs to terminal points of concentration.

Materials and methods

Five cleaning treatments were tested under experimental conditions: Scrape & bake (S&B, dry clean & heat to 160°F), volume hose & disinfect (VH&D), power wash & disinfect (PW&D), positive control (PC, no cleaning), and a negative control (NC, no PEDV challenge and no cleaning). Model size trailers were inoculated with a standardized dose of PEDV laden fecal material. Each treatment had five replicates. The inoculum used to challenge the model sized trailers contained 100 mL of PEDV cell culture media (107TCID50/mL) mixed with 900 grams of PEDV-negative feces. For the NC treatment, 100 mL of PBS was used. Thirty minutes after inoculating the trailers, the treatments were applied, and foot traffic was simulated between the trailer and the farm-site area. Samples were collected from the floor of the trailer and farm-site area for qPCR analysis to estimate viral load pre- and post-treatment, and after foot traffic. Two 10-day-old PEDV naïve pigs per replicate were inoculated with the farm-site area samples to determine evidence of disease transmission from the trailer to the farm site. Statistical analyses included ANOVA and Fisher's Exact test.

Results

Pre-treatment swabs indicated a high level of PEDV (~ 9.5×10^6 genomic copies/mL) contained in the inoculum used to challenge the trailers. Post-treatment samples taken from the farm site area (chute) yielded an average of 6,148,127 genomic copies/mL of PEDV for the PC treatment, 94,417 genomic copies/mL for the S&B, 10,517 genomic copies/mL for the VH&D, and 68 genomic copies/mL for the PW&D treatments. Evidence of disease transmission to the farm site area was evident in all 5 replicates of the PC and in 4 of 5 replicates of the S&B treatment. No disease transmission to the farm site area was evident in the VH&D or the PW&D treatments. The NC treatment remained free of PEDV.

Discussion and conclusion

Under the conditions of this study, both water-based treatments evaluated were effective in mitigating disease transmission from the trailer to the farm site area. The S&B treatment was not effective. When compared to the PC, the VH&D and the PW&D treatments reduced the amount of PEDV nucleic acid tracked back to the farm site area by 99.83% and 99.99%, respectively. If broadly implemented, these findings suggest routinely cleaning and disinfecting all market haul trailers leaving terminal points of concentration by either of the water-based trailer cleaning treatments evaluated present as an opportunity to create a step-change in the magnitude of inter-premises disease transmission associated with market haul transport and elevate the state of preparedness across the US pork industry.

Reference

1. Lowe, J., Gauger, P., Harmon, K., Zhang, J., Connor, J., Yeske, P., Loula, T., Levis, I., Dufresne, L., Main, R., 2014. Role of transportation in spread of porcine epidemic diarrhea virus infection, United States. Emerging Infect. Dis. 20 (5), 872–874.

https://doi.org/10.54846/am2025/18

Retention rates of females through parity 3 in herds with high and low circulating *Mycoplasma hyopneumoniae*

L. Britton[1], MS; J. S. Pittman[2], DVM, MS, DABVP; G. Machado[1], DVM, MVSc, PhD; M. T. See[3], MS, PhD; G. W. Almond[1], DVM, MSc, PhD; W. L. Flowers[3], MS, PhD
[1]College of Veterinary Medicine, North Carolina State University, Raleigh, North Carolina; [2]Smithfield Hog Production, Warsaw, North Carolina; [3]Department of Animal Science, North Carolina State University, Raleigh, North Carolina

Introduction

Mycoplasma hyopneumoniae (MHP) is the causative agent of enzootic pneumonia in swine and leads to decreased growth, average daily gain, and reduced feed efficiency.[1] It has been estimated that MHP costs the pork industry 375 to 400 million dollars each year.[2] Economic analyses associated with the grow-finish phase[3] support that MHP elimination or stabilization strategies of the breeding herd would result in a high cost-to-benefit ratio. However, very little has been published or presented on the economic and production benefits of a MHP negative breeding herd. In theory, MHP elimination in breeding herds would involve fewer animals and number of farms and could have a sustained, longer-lasting impact compared with grow-finish facilities. The main objective of this study was to estimate the effects of MHP on reproductive performance and retention of sows by comparing production data of gilts entering MHP positive unstable farms with gilts entering farms undergoing MHP elimination.

Methods

Cohorts of ten gilts each were randomly selected from 8 farms of similar health status at four distinct farm entry time points: Two years prior to MHP elimination; one year prior; one week; and three months after MHP elimination began. Reproductive performance of these gilts was followed from farm entry at breeding through rebreeding after their third parity (n = 160 gilts per MHP status). All farms from which production records were evaluated went through a herd closure MHP elimination program. The farms received all their replacement gilts prior to their closure and subsequent exposure and did not receive any new animals until after the herd tested negative for MHP. Exposure was accomplished through either aerosolization via a fogging method or via an intranasal inoculation with MHP homogenates.[4] For the study, high MHP was defined as production without any MHP elimination or stabilization practices being used. Gilts that entered production two years and one year prior to MHP elimination were classified as high MHP cohorts. Low MHP was defined as production after a MHP elimination program was initiated. Gilts that entered production one week and three months after MHP elimination began were classified at low MHP cohorts. Survival analyses for retention rates were completed using proc LIFETEST and analyses of variance procedures for sow productivity data were completed with proc GLM in SAS®. The statistical model included farm, *Mycoplasma* group, entry replicate, entry date, production phase and appropriate interactions with age at first mating as a covariate.

Results

Sow productivity over three parities in low MHP produced more total pigs, pigs born alive, and pigs weaned ($P < .0001$) compared with high MHP. They also exhibited superior reproductive performance for productive days in the herd (397 ± 9.6 vs 316.5 ± 13.9; $P = .0007$); retention rate (65.4 ± 4.5 vs 48.3 ± 4.5; $P = .0499$); and litters farrowed (2.4 ± 0.1 vs 1.8 ± 0.1; $P = .0005$). Sows exposed to low MHP farrowed 1.3 and 1.4 more total and liveborn pigs, respectively, compared with those exposed to high levels of MHP. Both total number born and number born alive increased approximately one pig per litter from parity 1 through 3. Please see Table 1 for the effect of *Mycoplasma* status on sow productivity over 3 parities. Conception and farrowing rates both increased ($P \leq .0147$) for low MHP sows. Percentage of stillborn piglets, percentage of mummified fetuses, number of pigs weaned per litter, and the percentage of sows rebred after weaning were not affected by MHP status ($P > .1772$); parity ($P > .1404$); or their interaction ($P > .2898$).

Table 1: Effect of *Mycoplasma* status on sow productivity over 3 parities

	***Mycoplasma* status**		
Productivity variable	**High**	**Low**	***P*-value**
Retention rate (%)[1]	48.3 ± 4.5	65.4 ± 4.5	.0499
Productive days in herd	316.5 ± 13.9	397.6 ± 9.6	.0007
Litters farrowed	1.8 ± 0.1	2.4 ± 0.1	.0005
Total pigs born	23.7 ± 1.3	34.2 ± 1.2	< .0001
Number of pigs born alive	21.8 ± 1.2	32.1 ± 1.1	< .0001
Number of pigs weaned	18.2 ± 1.1	26.1 ± 1.1	< .0001

[1] (Sows rebred after parity 3 / gilts entering production) × 100

Discussion

These results indicate that a low MHP environment leads to better gilt retention within the herd and increased reproductive performance when compared to production in a high MHP environment. This study indicates the need for future work to be conducted to characterize and better understand the impact of MHP elimination on reproductive performance and retention within breeding herds. Eliminating MHP from a breeding herd will allow producers to retain animals for a longer period, produce more pigs, and have a potential additional profit of getting more weaned pigs out the door. The additional health benefits of downstream animals would be seen in decreased medication usage and cost, decreased time to market, and increased feed efficiency.[5]

References

1. Sibila M, Pieters M, Molitor T, Maes D, Haesebrouck F, Segalés J. Current perspectives on the diagnosis and epidemiology of *Mycoplasma hyopneumoniae* infection. Vet J. 2009 Sep; 181(3): 221–31.

2. Holtkamp, D. Economic impact of *Mycoplasma hyopneumoniae* on pig farms. Pig333.com. September 19, 2014. Accessed August 1, 2024. **https://www.pig333.com/articles/economic-impact-of-mycoplasma-hyopneumoniae-on-pig-farms_8936**

3. Marois-Créhan C, Segalés J, Holtkamp D, Chae C, Deblanc C, Opriessnig T, Fablet T. Interactions of *Mycoplasma hyopneumoniae* with other pathogens and economic impact. In Book: Mycoplasmas in swine. Chapter 7. Editors: Maes D, Sibila M, Pieters M. 2021 March; p 127-145.

4. Silva APSP, Marostica TP, McDaniel A, Arruda BL, Alonso C, Derscheid R, Yeske P, Linhares DCL, Giménez-Lirola L, Karriker L, Fano E, Zimmerman JJ, Clavijo MJ. Comparison of *Mycoplasma hyopneumoniae* response to infection by route of exposure. Veterinary Microbiology.Volume 258. 2021. 109118. ISSN 0378-1135. **https://doi.org/10.1016/j.vetmic.2021.109118**

5. Maes D, Deluyker H, Verdonck M, Castryck F, Miry C, Lein A, Vrijens B, de Kruif A. Effect of vaccination against *Mycoplasma hyopneumoniae* in pig herds with a continuous production system. Zentralblatt für Veterinärmedizin Reihe B. 1998; 45:495–505.

https://doi.org/10.54846/am2025/19

Determining the optimal infective dose of *Mycoplasma hyopneumoniae* lung homogenate when administered by intratracheal inoculation

D. Cagle[1], BS; E. Noblett[2], DVM, MS; J. Cagle[3], DVM; B. O'Brien[3], VMD, MS; D. Boykin[4], DVM; J, Pittman[4], DVM
[1]North Carolina State University, College of Veterinary Medicine, Raleigh, North Carolina; [2]Smithfield Hog Production, South Region, Laurinburg, North Carolina; [3]Merck Animal Health, Lenexa, Kansas; [4]Smithfield Hog Production, North Region, Tarboro, North Carolina

Introduction

Mycoplasma hyopneumoniae (MHP) affects 34.3% of US finishing herds, costing $2.58 USD per pig.[1] One disease control strategy utilized is intentional exposure of replacement gilts to MHP to create homogenous immunity, thereby reducing shedding to susceptible offspring.[2] This strategy requires that all gilts receive a sufficient infective dose of MHP at a young age due to its long infection duration. The objective of this study was to determine the optimal infective dose of a MHP lung homogenate administered intratracheally (IT) to replacement gilts. This information will help maximize the use of inoculum, reducing time, cost, and resources needed for creating the inoculum while ensuring consistently high infectivity rates.

Methods

MHP negative pigs (n = 115) were enrolled at approximately 7 weeks of age, tagged, and randomly assigned to a treatment group (T01: Non-inoculated controls [n = 24]; T02: Undiluted inoculum [Ct = 21.2] [n = 23]; T03: Dilution 1 [Ct = 28.1] [n = 21]; T04: Dilution 2 [Ct = 30.5] [n = 24]; T05: Dilution 3 [Ct = 33.7] [n = 23]). All pigs were housed in the same barn, but in different pens by treatment group, with T01 pigs located at the entrance of the barn, furthest from the exhaust fans, to minimize accidental exposure to MHP.

Challenge inoculum was created by mixing 10 tubes of MHP positive lung homogenate (40 mL each) together to create a uniform homogenate. This homogenate (40 mL) was added to ½ gal saline (1 gal distilled water + 7 tsp table salt), creating dilution 1. Dilution 1 (100 mL) was added to 300 mL saline to create dilution 2. Dilution 2 (100 mL) was added to 300 mL saline to create dilution 3. T02-T05 pigs received 10 mL of corresponding lung homogenate intratracheally.[3]

Pigs were sampled at enrollment and 14 days post-inoculation (dpi). Tracheal samplings were performed using a PCAI inner catheter, the end of which was cut into a tube containing 2mL PBS, and tested for MHP by PCR. Blood collected at enrollment was tested by MHP IDEXX ELISA to confirm pigs were of negative MHP status. The target MHP percent positive was 80% at 14 dpi, correlating with the production system's target. Mean Ct values of tracheal samples at 14 dpi were analyzed using One-way ANOVA. Pearson correlation coefficient was calculated for inoculum dose and 14 dpi sample Ct values.

Results

All pigs were confirmed MHP negative at enrollment with 100% negative on PCR and ELISA. One control pig tested positive on MHP PCR (Ct = 36.4) at 14 dpi, which may have been caused by movement of non-study pigs within the barn. There was a strong correlation between inoculum and tracheal Ct values collected at 14 dpi (Pearson correlation coefficient = 0.62) (Table 1). Using mean tracheal Ct values, dilution 2 had statistically similar infectivity rates as the undiluted inoculum and dilution 1 was statistically similar to dilution 2. While dilution 3 had a mean Ct similar to dilution 1, dilution 3 had a positivity rate below target (73%).

Table 1: Intratracheal *Mycoplasma hyopneumoniae* inoculation results

Treatment group	14 dpi positive PCR	Percent positive (%)	Mean Ct of tracheal samples
Control	1/24	4	37.1[a]
Undiluted (Ct = 21.2)	22/22	100	25[d]
Dilution 1 (Ct = 28.1)	19/20	95	29.7[b,c]
Dilution 2 (Ct = 30.5)	24/24	100	26.7[c,d]
Dilution 3 (Ct = 33.7)	16/22	73	31.9[b]

[a-d] Different superscripts within a column denote statistical difference at $P < .05$

Discussion

These results demonstrate that MHP inoculum with Ct ≤ 30.5 will successfully infect gilts when administered intratracheally. Further studies should be performed to refine positivity rates in swine given MHP homogenates with Ct values between 30.5 and 33.7, as well as studies testing these dilutions using other common inoculation techniques (eg, fogging/nebulization). It is likely had sampling timing been extended; an increased number of pigs would have tested positive at 21 dpi. This was not performed due to the production systems goal of 80% positivity 14 dpi, which was achieved in 3 of the 4 treatment groups.

Lung homogenate creation is a tedious process, requiring a veterinarian and assistants over multiple days, along with many materials, and is estimated to cost $14 per 40 mL tube of inoculum for this production system (D. Boykin, personal communication). Extending the inoculum from Ct = 28 to Ct = 30 will nearly double the number of pigs the collected inoculum can be used to infect, which in turn saves the production system time, cost, and overall resources needed in their *Mycoplasma hyopneumoniae* stabilization program.

References

1. Holtkamp D. Economic impact of *Mycoplasma hyopneumoniae* on pig farms. Pig 333. September 19, 2014. Accessed August 14, 2024. **https://www.pig333.com/articles/economic-impact-of-mycoplasma-hyopneumoniae-on-pig-farms_8936**

2. Pieters M, Fano E. *Mycoplasma hyopneumoniae* management in gilts. *The Veterinary Record*. 2016;178:122. **https://doi.org/10.1136/vr.i481**

3. Poeta Silva A, Marostica T, McDaniel A, Arruda B, Alonso C, Derscheid R, Yeske P, Linhares D, Giménez-Lirola L, Karriker L, Fano E, Zimmerman J, Clavijo M. Comparison of *Mycoplasma hyopneumoniae* response to infection by route of exposure. *Veterinary Microbiology*. 2021;258:109118. **https://doi.org/10.1016/j.vetmic.2021.109118**

https://doi.org/10.54846/am2025/20

Determining the ideal pH target to inhibit *Escherichia coli* growth in swine drinking water

R. Retterath[1], BS; R. Stika-Jensen[2], DVM; E. McDowell[2], DVM; E. Cleveringa, BS[2]
[1]University of Minnesota College of Veterinary Medicine, St. Paul, Minnesota;
[2]Pipestone Veterinary Services, Pipestone, Minnesota

Introduction

Escherichia coli (E coli) infection is a leading cause of morbidity and mortality in weaned pigs. *E coli* adheres to the microvilli of enterocytes where they colonize, proliferate, release enterotoxins, and cause excessive fluid secretion resulting in diarrhea. The enterotoxins cause extensive enterocyte damage which reduces water and electrolyte absorption. Intestinal epithelial damage can lead to septicemia, dehydration, and death.[1] The increased prevalence of antimicrobial-resistant *E coli* is becoming a concern as multi-drug treatments are not yielding successful results. One alternative treatment is to acidify the drinking water in swine facilities to decrease and prevent *E coli* intestinal colonization. Previous research shows that *E coli* is least likely to colonize when pH is between 4 and 5.[2] The objective of this study was to determine the ideal pH between a pH of 4 and 5 at which *E coli* growth is inhibited and thus provide a pH target for producers who are acidifying the water to minimize *E coli* infections in their pigs.

Methods and materials

This study was completed in two phases. Phase one determined the target pH levels of media while preventing contamination. Phase two quantified the *E coli* growth at different pH levels of the media.

Phase one

Eighty-eight percent lactic acid and sterile water were mixed in a 1:10 dilution as a stock solution. This stock was added to 100 mL of tryptic soy broth (TSB) in 0.025-1mL increments until pH levels of 4.0, 4.2, 4.4, 4.6, 4.8, and 5.0 were met and confirmed using a pH meter with an accuracy requirement of ± 0.2 pH increments. This phase was repeated twice for accuracy.

Phase two

Once pH levels had been produced and replicated, one colony of an F18 *E coli* strain from a clinical case was obtained on an inoculating loop, placed in TSB, and swirled. TSB without added stock solution was inoculated to act as a positive control. The TSB test containers were incubated for 22 hours at 37 degrees Celsius. After incubation, the samples were removed and diluted to 1:1,000, 1:10,000, and 1:100,000. Then 0.5 mL of each dilution was spread evenly over 5% bovine blood agar plates. Plates were incubated and evaluated after 12 hours. Colony counts and pictures were taken to quantify and visualize growth.

Results

Plates with pH of 4.20, 4.40, 4.60, 4.80, and 5.00 had quantifiable and visual surface growth at all dilutions. The plate with pH 4.00 had no visual or quantifiable surface growth at all dilutions (Table 1 & Figure 1).

Discussion

This in vitro work provides evidence that *E coli's* growth-inhibiting point is at a pH of 4.0, in other words, if pH is > 4.0, growth may occur. When administering an acidifier in swine drinking water, it is recommended to test with a pH meter with an accuracy requirement of ± 0.2 pH increments rather than using pH test strips. Further in vivo research is needed to understand the pigs' enteric luminal pH when provided water of pH 4.0 and whether this may affect water and feed intake.

Table 1: *E coli* colony counts at various pHs and dilutions

		Dilution ratio		
		1: 1,000	1: 10,000	1: 100,000
pH level	4.00	No growth	No growth	No growth
	4.20	14	2	3
	4.40	Too numerous to count (TNTC)	242	61
	4.60	TNTC	TNTC	545
	4.80	TNTC	TNTC	TNTC
	5.00	TNTC	TNTC	TNTC
	Control	TNTC	TNTC	TNTC

Figure 1: Visual observation of *E coli* colonies at various pHs and dilutions.

References

1. Zimmerman, Jeffrey J., et al. *Diseases of Swine*. 11th ed., John Wiley & Sons, Inc, 2019.

2. Bates, Carly, and Rachel Stika-Jensen. *How Low Should the pH Go?* Leman Conference, St. Paul, Minnesota, 2023.

https://doi.org/10.54846/am2025/21

Evaluation of population and individual-based sampling methods for surveillance of porcine astrovirus type 4 in a cohort of pigs during farrowing and nursery phases

E. Zwart[1], BS; E. Kettelkamp[2], DVM; A. M. Betlach[2], DVM, PhD
[1]Iowa State University, Ames, Iowa; [2]Swine Vet Center, St. Peter, Minnesota

Introduction

Porcine astrovirus type 4 (PAstV4) has recently emerged as a notable pathogen affecting the respiratory health of pigs.[1] The impact of PAstV4 infection on pig health and production is largely unknown but has been associated with respiratory disease, such as subacute tracheitis and bronchitis, lesions of epitheliotropic viral infections.[2] Moreover, there is a limited understanding of sampling strategies for the detection and surveillance of PAstV4 based on pig age or production phase. Increasing knowledge of sampling strategies allows for enhanced detection of disease, assessment of PAstV4 prevalence in the population, and may improve therapeutic interventions and control of disease. This pilot study aimed to assess the prevalence of PAstV4 and associated clinical signs in a cohort of pigs during the farrowing and nursery production phases and to compare population and individual-based sampling methods for PAstV4 surveillance.

Materials and methods

A prospective cohort study was conducted at a 2,500-head sow farm located in southern Minnesota with a clinical history of PAstV4 infection and diagnosis. In farrowing, 36 litters were randomly selected across three rooms. Sample size was determined using a 90% sensitivity, 100% specificity, and estimated prevalence of 2%, assuming 5% precision and 95% confidence (Epitools). Within each litter, 5 piglets were randomly selected, ear notched and sampled at 4, 11, and 18 days of age (doa). Piglet nasal swabs (NS), family oral fluids (FOF), and udder wipes (UW) were collected from each litter.[3] An UW was collected by wetting a Swiffer® Dry Sweeping Cloth in 50 mL of phosphate buffer saline and wiping the underline of the sow, focusing on the teats, followed by liquid extraction into a sterile conical tube.[4] Family oral fluids were collected by hanging a rope for 60 minutes at the level of the piglets and sow, followed by liquid extraction into a sterile conical tube.[5] At 21 doa, the cohort was weaned to a nursery where pigs were relocated to eight pens within one room. During the nursery phase, the cohort was sampled weekly. Nasal swabs were serially collected from a subset of individually tagged pigs (n = 10/pen). Moreover, oral fluids were collected from each pen (n = 8). A cough scoring system 0-6 (0: No cough, 1: Mild cough after encouraged movement, 2: Mild coughing while leaving the piglets undisturbed, 3: Moderate coughing after encouraged move, 4: Moderate coughing while leaving the piglets undisturbed, 5: Severe cough after encourage move and 6: Severe coughing while leaving the piglets undisturbed) was used throughout the study to assess and quantify respiratory signs.[6] Nasal swabs were pooled by litter or pen and submitted with the remainder of the samples to an accredited veterinary diagnostic laboratory for PAstV4 RT rt-PCR testing. Differences in results by sampling method were compared using the Cochran Q test followed by paired McNemar for multiple comparisons. The probability of resulting PCR positive for PAstV4 based on sampling method and time was evaluated using multivariate logistic mixed models. Multivariate linear mixed models assessed differences in Ct values by sampling method. Statistical significance was deemed at $P \leq .05$. Statistical analysis was conducted in R (version 4.2.3)

Results

In farrowing, a lower positivity rate ($P < .05$) was observed in NS (6.7-54.5%) compared to UW (80.0-92.9%) and FOF (53.3-100%) across all sampling events. At 11 and 18 doa, the positivity rate between FOF (100%) and UW (90.9 and 92.9%, respectively) was similar ($P = 1.00$). Higher mean Ct values were detected in NS (34.0) compared to FOF (30.5) and UW (30.7; $P < .001$). The probability of obtaining a positive PCR test for PAstV4 in FOF and UW samples was 18 times higher than NS ($P < .001$). Moreover, sampling at 4 doa showed a 90% reduction in the probability of detecting PAstV4 compared to 11 and 18 doa ($P = .04$). In the nursery, all samples were PAstV4 PCR positive. Higher Ct values were detected in NS (25.6) compared to OF (17.4; $P < .001$). Throughout the duration of the study, no cough was observed.

Discussion

Based on the results of this study, FOF or UW should be collected from litters ≥ 11 doa to optimize the detection of PAstV4 in farrowing for diagnostic and/or surveillance purposes. For the nursery phase, OF should be considered for PAstV4 detection due to a high positivity rate and ease of collection. This study was limited to studying only one farm, in which PAstV4 prevalence and infection dynamics may differ across farms. Further research of additional farms during various times of the year should be conducted to validate the findings of this study for differences in prevalence, clinical signs, and seasonality. Moreover, additional research would provide insight into how sampling strategies can improve the early detection and control measures of PAstV4. Research should be conducted to assess the isolation success rate using a sampling method to aid in possible vaccination development.

References

1. Rahe, Michael C., Michael, Alyona, Piñeyro, Pablo Enrique et al. Porcine Astrovirus 4 Detection in Lesions of Epitheliotropic Viral Infection in the Porcine Respiratory Tract, Transboundary and Emerging Diseases, 2023; 9113355. **https://doi.org/10.1155/2023/9113355**

2. Padmanabhan, A., Hause, B.M. Detection and characterization of a novel genotype of porcine astrovirus 4 from nasal swabs from pigs with acute respiratory disease. Arch Virol 2016;161, 2575–2579. **https://doi.org/10.1007/s00705-016-2937-1**

3. Garrido-Mantilla, J., Alvarez, J., Culhane, M. et al. Comparison of individual, group and environmental sampling strategies to conduct influenza surveillance in pigs. BMC Vet Res 15, 61 (2019). **https://doi.org/10.1186/s12917-019-1805-0**

4. Wheeler, M., Kettelkamp, E., Okones J., et al. Comparison of different sample types and assays for detecting Cystisospora suis at the litter level. AASV Annual Meeting 2024;(p. 60). **https://doi.org/10.54846/am2024/21**

5. Lebret, A., Normand, V., Berton, P. et al. Alternative Samples for Porcine Reproductive and Respiratory Syndrome Surveillance in an Endemic PRRSV-1-Infected Breeding Herd: A Descriptive Study. Veterinary Sciences. 2023;10(9), 558. **https://doi.org/10.3390/vetsci10090558**

6. Vicca, J., Stakenborg, T., Maes, D., Butaye, et al. Evaluation of virulence of *Mycoplasma hyopneumoniae* field isolates. Veterinary Microbiology, 2003;97(3-4), 177–190. **https://doi.org/10.1016/j.vetmic.2003.08.008**

https://doi.org/10.54846/am2025/22

Could we reduce vaccine reactogenicity in pigs?

S. Murray[1]; R. Tenbergen[2], DVM, MSc, BSc
[1]Ontario Veterinary College, University of Guelph, Ontario, Canada;
[2]Demeter Veterinary Services P.C., Woodstock, Ontario, Canada

Introduction

Reactogenicity is a term used in human medicine to describe reactions that follow vaccination.[1] In swine, vaccines stimulate the immune system both locally at the injection site and systemically, causing symptoms such as pyrexia, pain, fatigue, and myalgia, leading to the redirection of energy and nutrients intended for growth.[1] In children, non-steroidal anti-inflammatory drugs (NSAIDs) are commonly used to treat similar symptoms of pain and fatigue.[2] The objective of this study was to evaluate whether the application of an NSAID during vaccination of weaned pigs would reduce reactogenicity leading to improved welfare and performance.

Materials and methods

The study was initiated on a 750-head commercial farrow-to-grow farm, weaning at an average of 21 days. Piglets were assigned at 5-14 days of age to one of two groups. **1)** Vaccinated and NSAID (Treatment); **2)** Vaccinated (Control) and were balanced by weight, sex, and parity of the dam. Forty (40) piglets per group were assigned weekly for five consecutive weeks, for 200 pigs per treatment group. Piglets were weaned into a dedicated nursery room and sorted into one pen per treatment group. Five days post-weaning, all piglets were weighed and vaccinated with an intramuscular (IM) injection of a commercial porcine circovirus type 2, *Mycoplasma*, and *Lawsonia* combination vaccine (Circumvent CML®, Merck Animal Health). The treatment group also received an IM injection of an NSAID on the opposite side of the neck (off-label Banamine®, Merck Animal Health) at the same time. Behavioral observations were recorded 20 minutes before vaccination, 10 minutes post-vaccination, and hourly up to and including 6 hours post-vaccination. All pigs were weighed one week before leaving the nursery around 9 weeks-of-age. A subset of pigs (n = 76) was also weighed 24-hours post-vaccination. Vaccine-induced antibody response (titers) was compared using a quantitative immune-fluorescence antibody test for lawsonia and PCV2 and IDEXX ELISA for *Mycoplasma hypopneumoniae*. Descriptive statistics and a GLM were used to analyze the data.

Results

Twenty-four hours post-vaccination, 68% of Control group pigs lost weight compared to 37% of Treatment group pigs that lost weight ($P > .8$) (Table 1). The final nursery weights at 9-weeks of age were not significantly different ($P > .6$). Pigs displaying eating behavior at 10-min, 60-min and 120-min post-vaccination were numerically higher in the Treatment group compared to the Control group (6%, 20% and 14% versus 1%, 9% and 0%, respectively). In addition, Treatment pigs displayed higher drinking behavior at all time points recorded compared to the Control pigs. Furthermore, expected fighting behavior was observed in the Treatment pigs at 60-min and 120-min post-vaccination compared to none of this behavior in the Control pigs post-vaccination and Treatment pigs showed more activity (standing and walking) at all times points compared to the Control pigs from 120-min through 6-hours post-vaccination. Mortality in the Control group was numerically higher than the Treatment group (3.0% versus 1.1%, respectively). There was no negative impact on titer response in NSAID-treated pigs; there was no difference between treatment groups.

Discussion

Adverse reactions and decreased activity are expected following post-weaning vaccination in pigs. This leads to reduced welfare and may cause decreased feed intake and reduced performance. Although pigs may show compensatory growth through the nursery period, reducing the adverse effects of vaccination with the application of NSAIDs at the time of vaccination results in improved welfare of the treated pigs with no negative impact on the vaccine-induced antibody response.

References

1. Hervé C, Laupèze B, Del Giudice G, Didierlaurent AM, Tavares Da Silva F. et al. The hows and whats of vaccine reactogenicity. npj Vaccines 4, 39 (2019). **https://doi.org/10.1038/s41541-019-0132-6**

2. Zempsky W, Bell J, Mossali VM, Katchroo P, Siddiqui K. Common Self-care Indications of Pain Medications in Children. Paediatr Drugs. 2023; 25(3): 321–341. **doi: 10.1007/s40272-023-00562-1**

Table 1: Descriptive analysis of weight difference from vaccination to 24-hours post-vaccination

	Treatment (n = 38)	Control (n = 38)
Number of pigs that gained weight	15	8
Percentage of pigs that gained weight	39	21
Average weight gained (kg)	0.247	0.187
Number of pigs that lost weight	14	26
Percentage of pigs that lost weight	37	68
Average weight lost (kg)	0.220	0.208
Number of pigs with no weight change	9	4
Percentage of pigs with no weight change	24	11

https://doi.org/10.54846/am2025/23

Comparison of neonatal umbilical therapeutic interventions for the reduction of navel defects in downstream pigs

M. Wheeler[1], BS; M. Schleper[2], DVM; E. Kettelkamp[3], DVM
[1]Iowa State University, College of Veterinary Medicine, Ames, Iowa; [2]Chirstensen Farms, Sleep Eye, Minnesota; [3]Swine Vet Center, St. Peter, Minnesota

Introduction

Navel defects, often complicated by secondary infections and subsequent umbilical hernias, are a growing concern in swine production.[1] Earlier research indicates a prevalence of umbilical hernias in commercial wean-to-finish populations ranging from 0.4% to 1.5%.[2] A risk factor analysis concluded a strong association between hernias and infection exists. Umbilical hernias can result in million-dollar losses due to decreased primary marketable animals, condemnation of carcasses, and reduced growth potential.[1] Moreover, when disinfectant is not used when washing farrowing crates, the odds of navel defects were 16.9 times greater than when no disinfectant was used.[3] Even with different treatments, such as antibiotics or topical antiseptics, there has been minimal improvement and no standardization of umbilical cord care.[4] This study aimed to compare therapeutic interventions of litters at the time of birth to reduce the rate of navel defects in downstream pigs.

Materials and methods

A 5,000-head multiparous sow farm with a history of elevated grade outs due to navel defects in downstream pigs was utilized for this study. Enrolled litters were allotted to treatment and balanced by sow parity before farrowing. The litters received one of four umbilical treatments during farrowing: A 9" emasculator, Tech Dry Plus as a hygienic drying agent, Vetericyn Super 7 navel care solution applied topically, or Citrus Dry without directly applying the powder to the cord (control). Piglets were individually ear-tagged by the treatment group. Only actively farrowing litters were enrolled, excluding any piglets with dried umbilical cords, though wet piglets from the same litter were included. Tags were removed post-mortem to track pre-weaning mortality. Labor requirements for each treatment were timed for comparison. A previously described navel scoring system of 0-5 (0–no signs of swelling or hernia, 1–no hernia with scab, 2–no hernia with rosy, red appearance, 3–no hernia or scab being hard to the touch, 4-hard with infection when squeezed, 5-hernia present) was used at weaning on all tagged pigs.[5] Tagged pigs were followed throughout the nursery phase to determine if hernias developed.

Results

A total of 600 litters were enrolled: 147 emasculator, 154 Tech Dry Plus, 147 Vetericyn Super 7, and 154 controls. Tech Dry Plus required the least labor investment, averaging 10.2 seconds per piglet compared to 24.05 seconds for the emasculator, 15.7 seconds for Vetericyn Super 7, and 12.6 seconds for the control. Prewean mortality was numerically lower from the emasculator (9%, 144/1555) and control groups (9%, 166/1810) compared to Vetericyn Super 7 (12%, 197/1682), Tech Dry Plus (13%, 186/1482). Gilts had a navel defect score of 1 or higher in 31% of pigs (999/3144), while barrows scored 1 or higher in 24% of pigs (783/3221). Only one Vetericyn Super 7 gilt had a hernia present at navel scoring. Scores of 1 or higher were most frequent in the Vetericyn Super 7 treatment (30%, 509/1682), followed by the emasculator (30%, 470/1555), the control (24%, 452/1810). The Tech Dry Plus treatment observed the lowest quantity of navel defects (23%, 351/1482). Overall, 28% (1784/6365) of pigs enrolled exhibited a form of navel defect. Nursery mortality was numerically similar from the control group (2.2%, 37/1644), and Vetericyn Super 7 (2.3%, 35/1485) compared to the emasculator (2.5%, 36/1411), Tech Dry Plus (2.7%, 36/1296). Only one gilt from treatment group Vetericyn Super 7 was euthanized due to a documented umbilical hernia during the nursery phase.

Discussion

Navel defects and subsequent umbilical hernias continue to pose substantial economic and welfare issues within the swine industry. The results of this study suggest that using Tech Dry Plus as a drying agent numerically reduces labor requirements and tended to observe the lowest incidence of navel defects at weaning. This study was limited by enrolling only attended litters farrowed, cross-fostering within treatment, and the inability to keep enrolled litters intact through weaning for assessment. Further studies are necessary to validate these findings and explore the impact of day-one umbilical cord care on navel defects in downstream pigs.

References

1. Fordyce A. et al. Measuring birth weight and umbilical cord diameter at birth to predict subsequent performance in swine. Transl Anim Sci. 2020 Nov 19;5(1):txaa214. **doi: 10.1093/tas/txaa214**

2. Straw B, Bates R, May G. Anatomical abnormalities in a group of finishing pigs: prevalence and pig performance. *Journal Swine Health Prod* (2009); 17(1): 28-31.

3. Bracy M, et al. Risk factors for umbilical defects in commercial pigs. *American Association of Swine Veterinarians Annual Meeting.* Virtual (2021).

4. Hansen, ML., Larsen, I., Jensen, T.B. *et al.* Prevention of umbilical outpouchings and mortality in pigs: Meloxicam, tying, cutting, and chlorhexidine versus amoxicillin or no treatment? A clinical field trial. *Porc Health Manag* 10, 10 (2024). **https://doi.org/10.1186/s40813-024-00358-w**

5. Umbilical scoring system established by PIC and Bradley Wolter at The Maschhoffs, Carlyle, IL **https://www.nationalhogfarmer.com/hog-health/understanding-umbilical-hernias**

https://doi.org/10.54846/am2025/24

A comparison of two methods: Quantitative PCR (qPCR) vs autofluorescence technique in the diagnosis of *Cystoisospora suis* in a clinically affected farm treated with a new injectable combination of toltrazuril and iron for parenteral application (Forceris®)

Cesar Caballero[1], DVM, DVSc; Nicole Seebach[2], BSc; Ryan Tenbergen[3], DVM, MSc; Enrique Moncada-Fernandez[4], DVM, PhD; Daniel Sperling[5], DVM, PhD
[1]Ceva Animal Health Canada, Guelph, Ontario, Canada; [2]University of Guelph, Guelph, Ontario, Canada; [3]Demeter Veterinary Services PC, Woodstock, Ontario, Canada; [4]Ceva Animal Health, Lenexa, Kansas; [5]Ceva Santé Animale, Libourne, France

Background

Cystoisosporosis, caused by the parasite *Cystoisospora suis* (*C suis*), affects suckling piglets in farrowing rooms globally.[1,2] *C suis* is ingested orally from the environment and invades the intestinal lining resulting in diarrhea (Figure 1) and reduced weight gain.[3] Methods such autofluorescence (AF) microscopy, a sampling technique that works by the detection of ultraviolet excitation emitted from the oocytes from fecal smears, has been available for swine veterinarians during the last years.[4,5] Recently, a new sampling method (qPCR diagnostic test) that detects all stages of *C suis* has been developed.[4] Both detection methods have a higher sensitivity than flotation method in the detection and quantification *C suis*.[4,6,7]

Objective

The aim of this study was to compare both diagnostic techniques: qPCR test vs direct autofluorescence microscopy in piglets treated with a new injectable combination of toltrazuril and iron for parenteral application (Forceris®) (Group F) vs piglets with only injectable iron (Group C). Recommended dosage was 1.5 mL Forceris® suspension per animal (45 mg of toltrazuril and 200 mg of iron) as a single intramuscular injection behind the ear of newborn piglets (between 1 and 3 days of age). Each mL contains 30 mg of toltrazuril and 133.4 mg of iron (as gleptoferron, an iron dextran complex).

The outcomes measured in this study include a direct visualization of parasite final stage- oocysts (Autofluorescence technique) and detection of parasite DNA via qPCR technique.

Methods

In a clinically affected farm (400 sows), two diagnostic techniques were used to detect *Cystoisospora suis* infection in pre-weaning piglets. Litters were randomly selected and assigned to one of two groups: 12 litters (141 piglets) were injected with 1.5mL of Forceris® at 1-3 day of age (doa) (Group F); and 12 litters (141 piglets) received only iron at 1-3 doa (Group C).

Fecal samples were collected at 10-12 doa and 18-21 doa. Pooled fecal samples to detect *C suis* were then sent to two different laboratories for qPCR and AF testing (Figure 2). Cohen's Kappa (K) was used as the statistical method to quantify the level of agreement between qPCR technique and AF testing (Table 1).

Figure 1: Creamy yellow-white diarrhea in the corner of a stall in the farrowing room.

Sampling procedure qPCR- fecal and autofluorescence:

1. Cross-sectional sampling should be done on 12 litters per treatment group.
2. Samples will be taken individually from randomly selected piglets within litter by perineal stimulation/anal evacuation. Take samples of all different consistencies.
3. Collect 5-6 grams of pooled feces per litter from 4-5 piglets per litter (5 grams is approximately 1 teaspoon of feces. Do not sample litters or pigs with watery diarrhea.
4. Use screw on type vials, refrigerate after collection and ship on ice packs or some type of chilling.

Figure 2: Orange cap vials that contained the collected feces and were shipped to the lab for qPCR and autofluorescence detection.

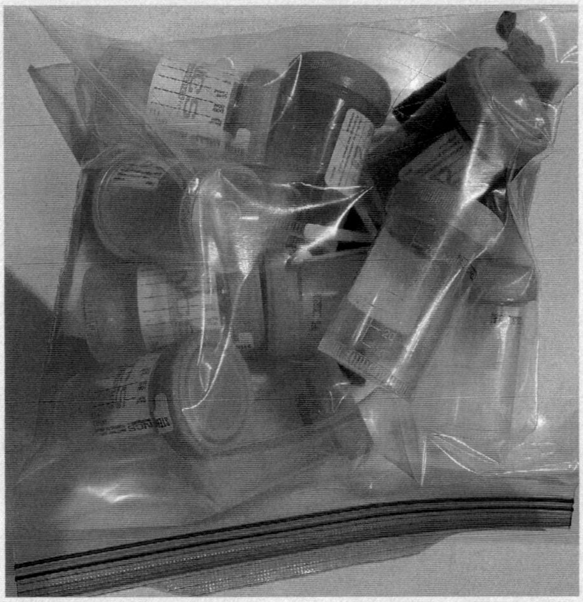

Table 1: Standards for strength of agreement for Cohen's Kappa (K) test

Poor	Slight	Fair	Moderate	Substantial	Almost perfect
≤ 0	0.01–0.20	0.21–0.40	0.41–0.60	0.61–0.80	0.81–1.00

Results

Under the conditions of this trial, efficacy of Forceris® was verified with two different diagnostic tests in the group F vs group C (Table 2). Concurrently, Cohen's Kappa (K) test confirmed a substantial agreement (0.61-0.80) between both tests at selected ages: 10-12 doa (Kappa"0.6667") and 18-21 doa (Kappa "0.7313") (Table 3).

Discussion

In infected piglets, oocysts excretion by *C suis* is usually short.[4] For that reason, sensitivity is a key for veterinary practitioners in the detection, diagnostics and control of cystoisosporosis in affected farms.[6,7] qPCR detection of *C. suis* DNA and correlation with AF technique and oocysts counts is challenging due to the *C suis* development cycle. Under the conditions of this trial, after confirming the efficacy of Forceris® with both qPCR technique and AF testing, the level of agreement between both tests was estimated with Cohen's Kappa (K). In conclusion, based on the actual results, there is a substantial agreement between both a direct visualization of parasite final stage- oocysts with the autofluorescence technique and detection of parasite DNA via qPCR method.

References

1. Dubey J.P. Coccidiosis in Livestock, Poultry, Companion Animals, and Humans. *CRC Press,* 1st edition 2020;125-145. **https://doi.org/10.1201/9780429294105**

2. Hinney B., Cvjetkovic V., Espigares D., Vanhara J., Waehner C., Ruttkowski B., Selista R., Sperling D. Cystoisospora suis control in Europe is not always effective. 2020. *Front. Vet. Sci.Sec.* Parasitology Volume 7. **https://doi.org/10.3389/fvets.2020.00113**

3. Worliczek H., Buggelsheim M., Saalmüller A., Joachim A. Porcine isosporosis: Infection dynamics, pathophysiology and immunology of experimental infections. *Wien Klin Wochenschr* 2007;119(3), 33–39. **https://doi.org/10.1007/s00508-007-0859-3**

4. Joachim A., Ruttkowski B., & Sperling D. Detection of Cystoisospora suis in faeces of suckling piglets - when and how? A comparison of methods. Porcine health management, 2018; 4, 20. **https://doi.org/10.1186/s40813-018-0097-2**

Table 2: Results autofluorescence (AF) vs qPCR between both groups at 10-12 days of age (doa) & 18-21 doa

	Results	AF 10-12 doa	qPCR 10-12 doa	AF 18-21 doa	qPCR 18-21 doa
Group C	Positive	11/12	8/12	9/12	7/12
Group C	Negative	1/12	4/12	3/12	5/12
Group F	Positive	1/12	0/12	1/12	0/12
Group F	Negative	11/12	12/12	11/12	12/12

Table 3: Kappa agreement results

10-12 days of age (doa)	Test + ve	Test -ve	Test 1 total	18-21 doa	Test + ve	Test -ve	Test 1 total
Test 2 +ve	8	0	8	Test 2 + ve	7	0	7
Test 2-ve	4	12	16	Test 2-ve	3	14	17
Test 2 total	12	12	24	Test 2 total	10	14	24

Kappa (95%CI) = 0.6667 (0.3855 - 0.9478)

Kappa (95%CI)= 0.7313 (0.4574 - 1.0053)

5. Wheeler M., Kettelkamp E., Okones J., Lauterbach., et al. Comparison of different sample types and assays for detecting Cystoisispora suis at the litter level. In proceedings of the 54th AASV Annual Meeting, Nashville, Tennessee.

6. Ruttkowski B., Joachim A., & Daugschies A. PCR-based differentiation of three porcine Eimeria species and Isospora suis. *Veterinary parasitology*, 2001; *95*(1), 17–23. **https://doi.org/10.1016/s0304-4017(00)00408-8**

7. Lalonde L. F., & Gajadhar A. A. (2011). Detection and differentiation of coccidian oocysts by real-time PCR and melting curve analysis. *The Journal of parasitology, 97*(4), 725–730. **https://doi.org/10.1645/GE-2706.1**

https://doi.org/10.54846/am2025/25

Lameness in conventional compared to organic swine finishing facilities

Elizabeth Taylor; Meghann Pierdon, VMD
Clinical Studies New Bolton Center University of Pennsylvania School of Veterinary Medicine

Statement of the problem

Lameness represents a significant health risk to growing swine. Lameness negatively impacts animal welfare as it decreases mobility, resulting in decreased feed intake and increased medication costs, which contribute to economic loss.[1] The etiology of lameness in growing pigs is not well studied. However, it is a multifactorial problem with links to husbandry, housing, disease, and nutrition.[2] Conventional and organic farms use different production practices, so quantifying the prevalence of lameness on these farms could generate hypotheses about causes of lameness and options for treatment. The organic farms in the study all met the organic standard which mandates the use of bedding, solid flooring, and a space allowance of roughly 25 sq ft per pig.[3] The commercial farms all met PQA plus guidelines for conventional production. The objective of this study was to compare the prevalence of lameness in grow-finish pigs on conventional and organic farms and identify differences that may contribute to lameness.

Materials and methods

A total of 16 grow-finish swine farms, all using commercial swine genetics, were enrolled in the study (n = 8 conventional farms, n = 8 organic farms). Farms were chosen from three production companies, 2 conventional and 1 organic, all in the same geographic region. Lameness scores were recorded at each farm by a single scorer. The number of pigs to score was determined by using the PQA plus site assessment Table 1, identifying the total number of pigs on the site, and assessing the recommended number of pigs. Lameness was scored on a 4-point scale where 0 is not lame, 1 is mild lameness, 2 is moderate lameness, and 3 is not able to rise. Prevalence of lameness on each farm was calculated by adding together the number of pigs that scored > 0 and dividing by the number of pigs scored. Oral fluids (OF) were pooled for all pens on conventional farms or from multiple ropes from the one large pen on each organic farm. Oral fluids were tested for porcine reproductive and respiratory syndrome (PRRS) virus by OF ELISA and *Mycoplasma hyosynoviae* by PCR. Feed samples were collected from each farm for nutritional analysis by ICP-MS. Zinc levels and the percentage of lame pigs were not normally distributed, so a non-parametric Mann Whitney u-test was used to compare median zinc levels and lameness prevalence on conventional compared to organic farms. Two by two tables were created, and a Fisher's exact test was used to look for differences in *Mycoplasma hyosynoviae* positivity and PRRS status in the conventional versus organic farms.

Results

The average size of conventional farms was larger than the organic farms and the pigs on the organic farms had been on the farm for longer at the time of the visit. There was a higher prevalence of lameness on conventional farms compared to organic farms. There was no difference in the *Mycoplasma hyosynoviae* positivity for conventional farms compared to organic farms. PRRS status was also not significantly different in conventional farms compared to organic farms. There were higher levels of zinc in organic feed compared to conventional feed (Table 1).

Discussion

Lameness in growing pigs could have different etiologies including infectious causes, such as *Mycoplasma hyosynoviae*, and non-infectious causes, such as trauma, osteochondrosis, or nutritional causes. PRRS virus infection may compromise the immune system in growing pigs and may lead to an increased risk of infectious arthritis. PRRS virus ELISA was used to determine if exposure to PRRS virus occurred at any point prior to this study. PRRS infection could be a risk factor for development of secondary bacterial arthritis that could contribute to lameness prevalence.[4] Infectious arthritis is typically of bacterial etiology, with specific links to agents such as *Mycoplasma hyosynoviae* and other *Mycoplasma* spp.[5] However, *Mycoplasma hyosynoviae* positivity and PRRS antibody status was not different between conventional

Table 1: Difference in characteristics on organic versus conventional finishing farms where lameness was scored. *Mycoplasma* status was determined by PCR and porcine reproductive and respiratory syndrome (PRRS) status was determined by ELISA on oral fluids.

	Organic farms (n = 8)	Conventional farms (n = 8)	*P*-value
Farm size	433 (319)	2689 (454)	NA
Days placed	78 (26)	49 (18)	NA
Lameness prevalence (%)	6.1 (2)	18.8 (8.8)	< .001
Proportion *Mycoplasma* positive	2/8	2/8	< .05
Proportion PRRS positive	1/8	3/8	< .05
Zinc (ppm)	1258 (250)	123 (14)	< .001

and organic farms, suggesting that the difference in lameness prevalence is likely due to other farm level factors, such as solid flooring and access to bedding. Flooring is an especially promising area to investigate as flooring type has been linked to lameness prevalence. Some studies have shown multi-surface systems with outdoor runs had higher risks of lameness compared to fully slatted pens,[6] whereas other studies have shown clinical leg problems such as leg weakness and stiff movement were worse on slatted flooring compared to solid floors with or without straw bedding.[7] Deep versus sparse bedding has also been shown to play a protective role on the incidence of lameness for varying floor types.[8] Variability among housing conditions in both conventional and organic systems poses a challenge to studying the effects of things like flooring type and bedding, so further studies are needed to evaluate specific links between housing conditions and lameness prevalence. Additionally, other farm factors such as stocking density, moisture levels, and hygiene can contribute to disease prevalence and lameness.[2] No population level antibiotics were used on organic farms in this study, so lower lameness prevalence on the organic farms cannot be attributed to antibiotic use. Nutritional links to lameness are also of interest, and deficiencies or excessive dietary intake of different vitamins and minerals can play a role in abnormalities contributing to lameness. Dietary zinc is an important factor in bone growth, strength, and bone mass; however, excessive zinc can increase the occurrence of articular cartilage disorders and increase incidence of osteochondrosis.[9] In this study, organic farms had higher levels of zinc in their feed compared to conventional farms, so potential nutritional links between dietary levels of vitamins and minerals and lameness should be further investigated. Further studies with larger sample sizes should evaluate specific husbandry practices on conventional and organic farms and their effect on lameness.

References

1. Jensen TB, Kristensen HH, Toft N. Quantifying the impact of lameness on welfare and profitability of finisher pigs using expert opinions. *Livestock Science*. 2012;149(3), 209–214. **doi.org/10.1016/j.livsci.2012.07.013**

2. Heinonen M, Pluym L, Maes D, Olstad K, Zoric M. Lameness in Pigs. In: Gross JJ, eds. Production Diseases in Farm Animals. Springer, Cham. 2024. **doi.org/10.1007/978-3-031-51788-4_18**

3. United States Department of Agriculture Agricultural Marketing Service National Organic Program (NOP); Organic Livestock and Poultry Standards. 2023. Accessed April 12, 2024. **https://www.federalregister.gov/documents/2023/11/02/2023-23726/national-organic-program-nop-organic-livestock-and-poultry-standards**

4. Brockmeier SL, Loving CL, Palmer MV, Spear A, Nicholson TL, Faaberg KS, Lager KM. Comparison of Asian porcine high fever disease isolates of porcine reproductive and respiratory syndrome virus to United States isolates for their ability to cause disease and secondary bacterial infection in swine. *Veterinary Microbiology*. 2017;203, 6–17. **doi.org/10.1016/j.vetmic.2017.02.003**

5. Pillman D, Surendran Nair M, Schwartz J, Pieters M. Detection of mycoplasma Hyorhinis and Mycoplasma hyosynoviae in oral fluids and correlation with pig lameness scores. *Veterinary Microbiology*. 2019;239, 108448. **doi.org/10.1016/j.vetmic.2019.108448**

6. Krieter J, Schnider R, Tölle KH. Health conditions of growing-finishing pigs in fully-slatted pens and multi-surface systems. DTW. Deutsche Tierarztliche Wochenschrift. 2004;111(12),462-466.

7. Jørgensen B. Influence of floor type and stocking density on leg weakness, osteochondrosis and claw disorders in slaughter pigs. *Animal Science*. 2003;77(3), 439–449. **doi.org/10.1017/s1357729800054382**

8. KilBride A, Gillman C, Green L. A cross-sectional study of the prevalence of lameness in finishing pigs, gilts and pregnant sows and associations with limb lesions and floor types on commercial farms in England. *Animal Welfare*. 2009;18(3), 215–224. **doi:10.1017/S0962728600000464**

9. van Riet MMJ, Millet S, Aluwé M, Janssens GPJ. Impact of nutrition on lameness and Claw Health in sows. *Livestock Science*. 2013;156(1–3), 24–35. **doi.org/10.1016/j.livsci.2013.06.005**

https://doi.org/10.54846/am2025/26

The before, during, and after of emergency swine depopulation with foam

G. Grant[1], BS; A. Bowman[2], DVM, PhD; E. Cortus[3], PhD; M. Leonard[4], BA; M. Culhane[1], DVM, PhD
[1]Department of Veterinary Population Medicine, University of Minnesota, Saint Paul, Minnesota; [2]Department of Veterinary Preventive Medicine, The College of Veterinary Medicine, The Ohio State University, Columbus, Ohio; [3]Department of Bioproducts and Biosystems Engineering, University of Minnesota, Saint Paul, Minnesota; [4]Department of Veterinary and Biological Sciences, University of Minnesota, Saint Paul, Minnesota

Introduction

Disease outbreaks in swine populations are significant risks to both animal health and industry productivity. Livestock depopulation is a critical measure to prevent further spread of infection within the industry.[1] During high-consequence disease outbreaks, efficient and humane depopulation methods are essential, emphasizing the need for a replicable, reliable, and safe approach for emergency depopulation.

Two well-known methods for depopulating groups of pigs are carbon dioxide gas (CO_2) and captive bolt (CB), though each has limitations.[2] High-concentration CO_2 stunning is used due to its efficiency and perceived benefits for animal welfare, such as minimal physical restraint, which reduces direct animal-operator interaction and stress.[3] Additionally, CO_2 is cost-effective and easier to administer without extensive training.[4] Currently, CO_2 is the only depopulation method for swine preferred by the American Veterinary Medical Association (AVMA).[5] However, CO_2 causes considerable discomfort[6,7] before animals lose consciousness; its acidic nature at high concentrations can irritate nasal mucosa when inhaled,[6] as well as cause air hunger and breathlessness, indicative of distress.[3] In contrast, CB stunning is less practical for mass depopulation as it requires individually killing animals and is fatiguing and impractical for use in large groups within a short time.[3] Stunning by CB requires workers to handle pigs individually and with a greater degree of contact, making it more labor-intensive and less scalable for mass depopulation efforts.[7] According to the AVMA preferred methods for swine depopulation, penetrating CB for nursery, growing, and mature pigs is contraindicated when bolt length and cartridge combination are not appropriate to the size of the animal, if pigs cannot be properly restrained, or if human safety cannot be assured.[8] Additionally, a penetrating CB may render an animal insensible, however a second CB application, pithing, or exsanguination may need to occur subsequently to ensure the irreversibility of death.[9]

The AVMA provides guidelines on humane depopulation criteria, emphasizing the need for minimal pain, worker safety, irreversibility, and environmental compatibility.[8] In recent years, high-expansion foam, particularly nitrogen-based foam, has shown promise as a humane and effective depopulation method with approvals for use for food animals both in the United Kingdom[10] and in the European Union (EU).[11] Using high-expansion nitrogen-based foam (N_2-foam) in a closed container rapidly displaces oxygen (O_2), creating an anoxic environment, while maintaining animal welfare practices.[12] As N_2-foam fills the container, it quickly pushes out ambient air; when the foam bursts, O_2 levels drop sharply to anoxic ($< 2\%$ O_2) conditions, thus avoiding prolonged exposure to hypoxia that can occur when pure gases mix gradually with air.[12] Although the foam itself may initially cause some stress and visual observation becomes difficult once the container fills, N_2-foam enables efficient group stunning with minimal human-animal contact, and acceptable animal welfare considerations.[12] Additionally, nitrogen gas provides an appropriate alternative when CO_2 supplies are limited.

For several years, researchers in the United States (US) and EU have been exploring the feasibility and effectiveness of N_2-foam as a humane euthanasia method for individual pigs and emergency depopulation of groups. Collaborative studies in the Midwestern US and the EU covered multiple food animal species. For swine, the N_2-foam humane killing process is approached from beginning to end in three stages: [1] assessing the efficiency of creating high-expansion N_2-foam for on-site humane killing of individuals (euthanasia) or groups (depopulation) ("Before"); [2] quantitatively assessing the process to note critical time points, durations, and pigs' physiological responses until reaching irreversible unconsciousness and death ("During"); and [3] evaluating the impact of foam on carcass decomposition during composting ("After").

Methods

Nitrogen gas production and foam creation (before)

An air-cooled, single-stage, electric, oil-injected screw air compressor (AC, Model G-7, Atlas Copco) and a nitrogen membrane generator (N_2-gen, Model 2700-09B, Terra Universal) were purchased to produce compressed air and extract nitrogen from ambient air, respectively. The AC delivered compressed air at 103 PSI to the N_2-gen, which separated nitrogen gas from the air at a pressure of 120 pounds per square inch (PSI) and a flow rate of 120 cubic feet per minute (CFM) to produce nitrogen of 98% purity. The nitrogen was either stored in a tank or directly used to generate foam in a specialized container (Model H1, HEFT International). With the gas source connected directly from the N_2-gen to the H1 container, the pressure was set between 116-145 PSI. Nitrogen purity, controlled by AC and N_2-gen pressure settings, varied but was maintained at levels targeted to produce 98% purity nitrogen gas. Gas production times were recorded, with purity measured by O_2 sensors (NeuLog® O_2 Sensor NUL-205 [item #369634] with a Neulog® Digital Display Unit [item #369730] and the NeuLog® Battery Pack Module [item #369732]) at the nitrogen outlet or within the H1 container. As nitrogen displaced O_2, the sensors measured residual O_2 levels and served to approximate nitrogen concentration ($100\% - O_2\ \% = \text{nitrogen}\ \%$). To create N_2-foam, nitrogen gas flowed from the source at a rate of 64 CFM with a pressure of 200 PSI into a pump in the container that allowed gas to mix with a 5% plant-based foam and water solution. The gas/foam mixture was then forced through stainless steel nozzles covered with a screen that produced large diameter bubbles (> 15 mm) filled with high-purity nitrogen gas ($> 98\%$).

Humane killing process (during)

All animals included in this study were handled according to University of Minnesota IACUC protocol 2403-41911A.

Several trials were conducted to evaluate the behavior, welfare, and efficacy of N_2-foam for pig euthanasia. In one small-scale trial, 14 pigs weighing approximately 20 kg each were placed under a surgical plane of anesthesia to minimize stress, then individually euthanized by anoxic asphyxiation in a closed N_2-foam-filled container. Nitrogen gas, with a purity above 98%, was supplied from a pressurized cylinder connected to the control unit of the container (Model H1, HEFT International). A 5% foam solution was prepared by mixing 0.5L of rapeseed foam concentrate and 10L of tap water. Each pig was placed individually into the container, which measured 1150×750×510 mm (L × W × H) and had non-slip flooring and a plexiglass lid for continuous visual monitoring. Once the pig was inside, the lid was closed, and the air valve was opened to allow the O_2 displaced by the N_2-foam to escape. The foam mixture was pumped in, and once it covered the screen, nitrogen flow started. Once N_2-filled bubbles filled the container, the air valve was closed. A final gas pulse bursts the foam, releasing nitrogen from all the bubbles to rapidly reduce O_2 levels to < 2%. The timer started when the nitrogen gas began to flow, and durations/occurrences of observed behaviors were recorded from that starting time point until the anoxic dwell time of 7 minutes 30 seconds was completed. It is important to note that few observations could be made of their behaviors for the brief period when the foam covered the pigs and before it burst. Table 1 lists the observations of pig behaviors (eg, vocalizations, escape attempts, exploring, blinking, shaking) recorded. Key metrics such as time to loss of posture (LOP), duration of convulsions, and time to cessation of movement (COM) were measured to assess euthanasia's effectiveness at causing irreversible unconsciousness/death while maintaining humane conditions. Post-mortem computed tomography (CT) scans were performed to determine if the mechanism of asphyxiation for N_2-foam treatment pigs was due to the removal of all O_2 (anoxia) or occlusion of the airways with foam/fluid. The CT results of N_2-foam pigs were compared to gaseous (CO_2 only) treatment pigs and liquid treatment pigs, which served as negative and positive controls, respectively.

In a separate large-scale field study assessing the feasibility of humanely killing groups of pigs using N_2-foam in the C3 HEFT unit, 87 growing pigs (approximately 40 kg each) were divided into three groups of 29 for depopulation with N_2-foam. The foam solution was prepared at a 4.7% concentration by combining 20L rapeseed concentrate with approximately 400L tap water, was mixed with nitrogen gas flowing into the container at a rate of 380 cfm at a pressure of 210 PSI dispensed through 24 nozzles (12 on each side) into a retrofitted 20-yard dumpster (HEFT C3 unit) with non-slip flooring and a white top roll cover. The pigs walked into the slip-proof flooring-covered container via a chute extending from the farm load out into the container and were not sedated. When the pigs were inside the container, the tarp was rolled across the top, and the N_2-foam delivery was started. Compressed nitrogen gas with a purity of 99.5% was supplied via slip tank/tube trailer, delivering up to 30,000 cubic feet (CF) of nitrogen at 3,200 PSI, regulated at the trailer down to 1,600 PSI for use with a T1227 regulator. Following the N_2-foam delivery process and holding the anoxic levels at < 2% O_2 for 12 minutes 30 seconds, the container was opened. Veterinarians confirmed death as described below. Upper respiratory tract gross examinations were conducted on 10% of all pigs as part of a partial necropsy to determine if foam was present in the mouth or nostrils.

In all trials, O_2 sensors were placed inside the container to verify that anoxic conditions (< 2% O_2) were rapidly achieved and maintained throughout the procedure; dwell times in the anoxic conditions were 7 minutes 30 seconds when using the H1 HEFT unit (small-scale) and 12:30 when using the C3 HEFT unit (large scale). Death was confirmed based on the absence of heartbeat,

Table 1: Pig behaviors and their definitions and recorded measurements

Behavior	Definition	Recorded measurement
Duration of convulsions	Uncontrolled full body shaking	Duration of the behavior in seconds
Escape attempt	Jumps up or pushes lid, door, or wall	Number of times observed within each treatment
Exploring floor	Snout touches the floor or the N¬2 vent	Number of times observed within each treatment
Exploring foam	Snout touches foam, pig is actively seeking contact with foam	Number of times observed within each treatment
Gasping	Deep breath through a wide-open mouth, which may involve stretching of the neck	Number of times observed within each treatment
Large movements	Full body shaking, all four legs paddling, convulsions, muscle excitation	Number of times observed within each treatment
Lateral recumbency	Lying down with side of body or belly in contact with the floor	Number of times observed within each treatment
Time to loss of posture	The inability of the pig to remain in upright position, strongly suggestive of unconsciousness	Length of time in seconds from the burst to onset of behavior
Small movements	One or two leg kicks, ear twitch, blinks	Number of times observed within each treatment
Time to cessation of movement	The moment after the last movement has been observed	Length of time in seconds from the burst to onset of behavior
Vocalization	Squeals or screams	Number of times observed within each treatment

respiration, corneal reflex, and pinch reflex. Key metrics, including time to LOP, duration of convulsions, and time to COM, were descriptively analyzed to evaluate the reproducibility and effectiveness of humane killing by N_2-foam.

Compost evaluation (after)

In January 2024, as part of a separate terminal study, three compost piles were constructed to evaluate the effect of N_2-foam on composting efficacy. Carcasses were placed in three separate compost piles (A, B, and C), with piles A and B containing N_2-foam-treated carcasses and pile C containing CB-treated carcasses. Piles were constructed by placing a 0.3-meter base layer of wood chips, a 0.1-meter layer of turkey litter, the carcasses (n = 3), another 0.4-meter layer of turkey litter, and then topped with wood chips until it reached a total pile height of 1.2 meters. Temperature sensors (HOBO data loggers [MX2303], Onset Inc., Bourne, MA) were placed at 15-cm (top) and 30-cm (core) depths in each pile, recording data every six hours for 20 days (n = 80). Thermophilic activity (104-149°F) was monitored as an indicator of composting efficacy.[13,14] Temperatures exceeding 149°F may harm beneficial microorganisms and impede carcass decomposition.[13] On day 20, subjective carcass decay scores were assigned to piles based on the degree of decomposition, ranging from 1 (minimal decomposition with a fully discernable carcass) to 5 (advanced degradation, with few to no brittle bones remaining). The piles were turned at 21 days and combined into a single large pile to reduce the amount of land used on the farm site; no further recordings or carcass scores were conducted after day 21.

Results

Nitrogen gas production and foam creation (before)

The AC/N_2-gen system successfully produced high-purity nitrogen (> 98%) that could be stored in a gas cylinder or used directly for foam creation. With an optimal inlet flow pressure of 101 PSI, the 40 CF cylinder achieved a maximum pressure of 250 PSI in 21 minutes, while the 300 CF cylinder reached a maximum pressure of 550 PSI in 60 minutes. These results demonstrate the system's ability to meet the demand for sufficiently high-volume N_2-foam production. However, some limitations in the N_2-gen ability to produce gas at high enough pressure to deliver the needed gas pulse into the H1 container directly and the capacity to produce large volumes consistently suggest that future improvements could enhance the system performance. Due to these limitations, only pressurized cylinders were used in the euthanasia procedures as the pressure and flow rate from a directly connected system were insufficient for the necessary gas pulse. Overall, the nitrogen purity consistently exceeded 98%, meeting the study's target, but further optimization of system components and linkages may be needed for greater efficiency in large-scale operations.

Humane killing observations (during)

In the small-scale trial involving 14 unconscious pigs, all pigs were humanely euthanized through anoxic asphyxiation. Death was confirmed after 7 minutes 30 seconds of immersion in anoxic conditions in an N_2-foam-filled H1 unit. The N_2-gen foam delivery system effectively created large-diameter bubbles filled with nitrogen. The bubbles pushed the air from the container, and upon bursting the foam bubbles, we rapidly created an anoxic environment, achieving < 2% O_2 within 11.25 seconds ± 1.25 seconds. The mean time to LOP was 28 ± 7.9 seconds, with subsequent convulsions lasting less than 24 seconds for all pigs. The mean time to COM was 3 minutes 53 seconds ± 38.78 seconds, with a maximum of 4 minutes 44 seconds. The CT scans revealed minimal foam/fluid aspiration, with foam rarely penetrating past the larynx during euthanasia (Table 2). Minimal foam/fluid was found distal to the pharynx. If any lungs contained fluid, those lungs were, on average, 0-50% filled (scores 1 or 2), with more fluid present in the caudal lung lobes. Foam/fluid findings in the N_2-foam pigs were comparable to those in the gaseous (negative) controls (data not shown). This is significant as it reduces the risk of physical distress from airway obstruction, further supporting the method's humaneness.

In the large-scale trial, 87 pigs were humanely killed in three runs of 29 pigs each in the C3 unit, with death confirmed after 12 minutes 30 seconds of immersion in an N_2-foam-filled, anoxic environment. The mean time to LOP and onset of convulsions was 27 seconds (range 20 to 30 seconds), and the average duration of convulsions was 150 seconds (range 100-170 seconds). COM occurred on average at 196 seconds (range 130 to 230 seconds). The most common behaviors observed included walking away from the foam, jumping, and head shaking. Escape behaviors were rare and typically stopped once the foam covered the pigs' heads. Vocalizations occurred when foam covered the pigs' heads and became more intense during convulsions, decreasing as convulsions lessened. Upper respiratory tract gross exams indicated that no pigs had foam occluding their nostrils or mouth, although two pigs exhibited scant foamy discharge from the snout.

Composting results (after)

Temperature profiles from the internal sensor (30-cm depth) of compost piles containing N_2-foam-treated and CB-treated carcasses are shown in Figure 1. Both piles displayed higher temperatures at the core compared to the top, with temperatures rapidly rising in the first 36-48 hours of pile construction. The N_2-foam-treated pile reached peak temperatures earlier than the CB-treated pile. No significant differences were observed in the thermophilic composting durations between the two piles ($\chi^2 = 2.0662$ $P > .5$) (Figure 1). However, pile A, which contained N_2-treated carcasses, exhibited higher carcass decay scores, indicating a more complete breakdown of the carcass. This suggests that the use of foam did not impede the composting process.

Discussion

Delivery of nitrogen gas via high-expansion foam is resource-efficient, safer, and more effective than traditional gas-only delivery systems, such as CO_2, and is more "hands-off" than CB. By using the foam as a vehicle for nitrogen gas, N_2-foam conserves resources, quickly filling the animal holding area and preventing gas leakage through imperfect seals. Additionally, nitrogen gas, derived from atmospheric air, is an abundant and sustainable alternative to CO_2, which is in high demand across various industries in the US and experiences seasonal supply fluctuations.[15] Anoxia, induced by lowering O_2 levels below < 2%, rapidly leads to death due to widespread cellular failure, particularly and acutely in the brain, where neurons are highly O_2-dependent.[16] Pigs exposed to N_2-foam transition from breathing air with 21% O_2 to an anoxic environment (O_2 < 2%) as the foam bursts, delivering nitrogen gas in a single, efficient pulse. Potential challenges with gas-based methods include sourcing, delivery, storage, vaporization rate, and risk of operator exposure to both the gas and potential pressurized equipment malfunctions.[17] Optimized pressure settings could enhance the N_2-gen system, improving the overall efficiency of N_2-foam production. The combination of plant-based

Table 2: Computed tomography result scores of the respiratory tract of 14 pigs humanely killed by N2-foam. Scores were assigned based on the presence of fluid in each section of the respiratory tract: 0 for 1-25% volume, 2 for 26-50% volume, 3 for 51-75% volume, and 4 for 76-100% volume.

Respiratory tract	Average	Minimum	Maximum
Left nasal cavity	1	1	2
Right nasal cavity	1	1	2
Left frontal sinus	2	0	3
Right frontal sinus	1	0	3
Left maxillary	2	0	4
Right maxillary	2	0	4
Pharynx glottis	1	0	2
Trachea	1	0	1
Right cranial bronchus	0	0	1
Right middle bronchus	0	0	1
Accessory bronchus	0	0	0
Right caudal bronchus	0	0	0
Left cranial bronchus	0	0	1
Left caudal bronchus	0	0	1
Right cranial lung	1	0	2
Right middle lung	2	1	3
Accessory lung	1	0	2
Right caudal lung	2	1	4
Left cranial lung	1	0	2
Left caudal lung	2	1	4

foam solutions and on-site nitrogen generation would enable on-demand supplies and facilitate depopulation in emergencies. Large-scale pork producers could benefit from utilizing mobile rendering trailers, which would facilitate worker safety and carcass disposal.[17] Animals can walk into the dump trailer on their own, be humanely killed, and be brought in the same dump trailer to the disposal area.[17] Performing humane killing on a vehicle like a dump trailer would also facilitate easier decontamination. The USDA approves of wet cleaning and disinfection of all exterior vehicle surfaces, where a mobile vehicle would allow for thorough physical removal of disease agents associated with organic matter at a separate location without risk of spread to the herd.[18,19] Future research conducted under varying farm and environmental conditions that also assesses the N2-foam method's full applicability in emergency animal management are needed.[20]

Swine industry stakeholders have expressed positive perceptions of foam-based depopulation methods, recognizing their ability to minimize pig distress, protect the emotional or psychological health of personnel, improve public reception to depopulation methods, enhance equipment accessibility, and foster a favorable overall impression.[5] This study confirms that high-expansion N_2-foam is a humane, feasible, and effective method for emergency swine depopulation. N_2-foam satisfies AVMA criteria by inducing a rapid, painless loss of consciousness and minimizing physical distress. In our studies, loss of consciousness was determined to occur when pigs had a LOP. Equating LOP to unconsciousness was based on prior work by Benson et al, who determined, using EEG measurements of brain activity, that there was no statistically significant difference between time to unconsciousness as measured by EEG or LOP for other food animals.[21] The quick creation of an anoxic environment, with rapid LOP and time to death in under four minutes yet with the consistency of physiological responses, indicates that N_2-foam can serve as a valuable alternative to traditional methods, such as CO_2 and CB. N_2-foam anoxia as a method of humane killing of pigs is comparable to well-studied water-based foam (WBF) methods. Pigs humanely killed with WBF are unconscious within two minutes, and brain death occurs in just over three minutes on average; both measurements are determined by EEG in research settings.[22] However, with WBF, there is a concern about foam aspiration. For N_2-foam with large bubble sizes, the absence of foam aspiration is important and further supports the humane nature of the method, as a lack of aspirated foam likely reduces distress and physical harm. While the use of anesthesia in the group of 14 pigs may have impacted the study results, as observations based on their natural behaviors could not be made, it was still possible to identify when the pigs lost posture, convulsed, or twitched. For example, even in the sedated pigs, the pigs' ears appear to be highly sensitive, with even a drop of water on their ear causing a twitch, demonstrating that their spinal cord reflexes remained functional. Although the

Figure 1: Temperature data of N2-foam-treated piles (piles A and B, green line) and CB-treated pile (pile C, blue line) and the daily ambient air temperature (black line). Thermophilic minimum and maximum (gray lines) represent the optimal conditions for the biologically active phases of mortality composting.[14]

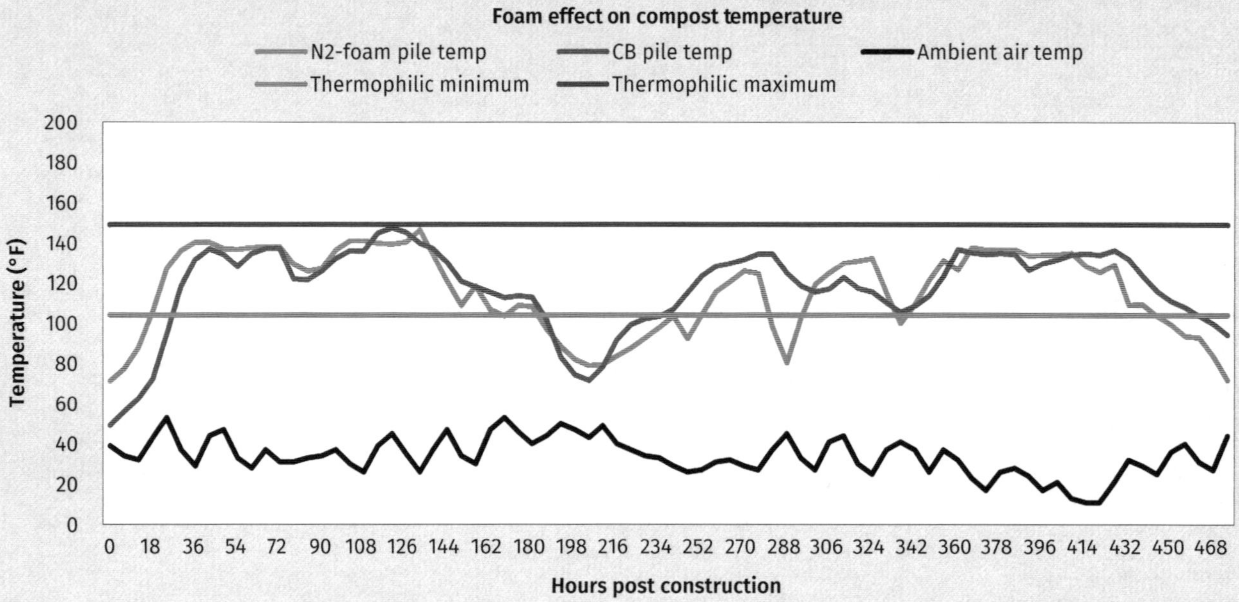

time to COM was, on average, under four minutes, the range and maximum time to COM cause us to recommend a dwell time of 7 minutes 30 seconds to ensure complete irreversibility and death. Overall, this N_2-foam is a hands-off approach that reduces risks to personnel and offers not only individual pig euthanasia but scalability for humanely killing groups of pigs, addressing key limitations of CB for large-scale depopulations.

Composting is a cost-effective and efficient method for carcass management, as the organisms involved in the biologically active process generate metabolic heat.[23] The results of the composting study demonstrate that N_2-foam treatment does not hinder decomposition, as N_2-foam-treated carcasses decomposed at a rate comparable to those euthanized by CB. Further research is needed to evaluate the method's applicability across various pig weights, environmental conditions, and disposal methods. The composting study solely provides preliminary findings, as it was a pilot investigation, with sensors removed after 21 days, just before the first pile turn. A more comprehensive composting study would involve extended monitoring to better understand long-term decomposition dynamics with foam.

Altogether, this work provides valuable information that can be used to improve humane killing methods, from resource procurement to carcass disposal. Generation of nitrogen gas from atmospheric air using an air compressor and a nitrogen generator is possible on a small scale and, with optimization and investment by industry, can be enhanced for large-scale, continuous nitrogen gas production. Causing irreversible unconsciousness and death repeatably and humanely is achievable when using N_2-foam on both individuals and groups of pigs. The pigs avoid pain from mucosal irritation caused by CO_2 gas; pigs exposed to N-gas-filled foam do not cough or appear to choke. This is because the N_2-foam bubbles are large and carry almost pure nitrogen gas. Therefore, the foam does not penetrate their upper respiratory tract while the pigs are still conscious, thus avoiding increased stress. The N_2-foam method of humane killing is a hands-off alternative to CB, which requires the handler to be in close contact with a loud and bloody, albeit rapid, instrument that causes death. Finally, carcasses of pigs humanely killed by N_2-foam can be disposed of in compost piles. Composting has worked well for US pig producers for routine mortality, and both state and federal agencies have invested time and training in ensuring composting can be used in emergencies,[24] and research has shown composting can kill pathogens.[25,26] Planning for a swine disease emergency needs to be thorough and studied before, during, and after to be of most benefit to the many involved.

Acknowledgments

We thank HEFT International for leasing the equipment used in the studies to the University of Minnesota and for their collaborations. We are especially grateful for the hard work of Cassidy Cordon, Kortney Dziedzic, Dr Timothy Goldsmith, Margret Tuisalo'o, Dr Declan Schroeder, and David Schmidt on these projects. We are also thankful for Lee Johnston's advice on pig carcass composting. This work was funded by the United States Department of Agriculture's National Animal Disease Preparedness and Response Program (NADPRP) awards AP21VSSP0000C030, AP21VSSP0000C021, and AP23VSSP0000C038; by the University of Minnesota's Swine Disease Eradication Center 2024 Research Project 1000002052; and with support from NPB Project Award No: PR-005723.

References

1. Berg C. The need for monitoring farm animal welfare during mass killing for disease eradication purposes. *Animal Welfare*. 2012;21(3):357-361. **doi:10.7120/09627286.21.3.357**

2. Arruda AG, Beyene TJ, Kieffer J, Lorbach JN, Moeller S, Bowman AS. (2020). A Systematic Literature Review on Depopulation Methods for Swine. *Animals: an open access journal from MDPI*, 10(11), 2161. **https://doi.org/10.3390/ani10112161**

3. Steiner AR, Flammer SA, Beausoleil NJ, et al. Humanely Ending the Life of Animals: Research Priorities to Identify Alternatives to Carbon Dioxide. *Animals: an open access journal from MDPI*. 2019;9(11):911. **doi:10.3390/ani9110911**

4. Boivin GP, Hickman DL, Creamer-Hente MA, Pritchett-Corning KR, Bratcher NA. Review of CO2 as a Euthanasia Agent for Laboratory Rats and Mice. *Journal of the American Association for Laboratory Animal Science*. 2017;56(5):491-499(9).

5. Park J, Campler MR, Cheng TY, et al. Perceptions of swine industry stakeholders on the use of water-based foam, high-expansion nitrogen foam, and carbon dioxide gas as methods of swine depopulation. *Preventive Veterinary Medicine*. 2024; 233:106356. **https//doi:10.1016/j.prevetmed.2024.106356**

6. Raj A, Gregory N. Welfare implications of the gas stunning of pigs 1. Determination of aversion to the initial inhalation of carbon dioxide or argon. *Animal Welfare*. 1995;4(1):273-280. **doi:10.1017/S096272860001798X**

7. Mullins C, Pairis-Garcia M, George K, et al. Determination of swine euthanasia criteria and analysis of barriers to euthanasia in the United States using expert opinion. *Animal Welfare*. 2017;26(4):449-459. **doi:10.7120/09627286.26.4.449**

8. AVMA American Veterinary Medical Association. AVMA Guidelines for the Depopulation of Animals: 2019 Edition. Published online 2019. Accessed November 10, 2024. **https://www.avma.org/sites/default/files/resources/AVMA-Guidelines-for-the-Depopulation-of-Animals.pdf**

9. Anderson KN, Deen J, Karczewski J, Zhitnitskiy PE, Vogel KD. History and best practices of captive bolt euthanasia for swine. *Translational Animal Science*. 2022;6(2):1-4. **doi:10.1093/tas/txac065**

10. Campbell M, Downes J, Gibson T, Lister S, McKeegen D, Rashbass P. AWC Opinion on the use of high expansion nitrogen foam for culling poultry. *Department for Environment Food & Rural Affairs*. Published online May 24, 2024. **https://www.gov.uk/government/publications/awc-opinion-on-high-expansion-nitrogen-foam-for-culling-poultry/awc-opinion-on-the-use-of-high-expansion-nitrogen-foam-for-culling-poultry#conclusions**

11. EFSA Panel on Animal Health and Welfare (AHAW), Nielson SS, Alvarez J, et al. The use of high expansion foam for stunning and killing pigs and poultry. *EFSA Journal*. 2024;22(7):e8855. **doi:10.2903/j.efsa.2024.8855**

12. Lindahl C, Sindhøj E, Brattlund Hellgren R, Berg C, Wallenbeck A. Responses of Pigs to Stunning with Nitrogen Filled High-Expansion Foam. *Animals*. 2020;10(12):2210. **https//doi:10.3390/ani10122210**

13. Lochner HL, Hutchinson ML, Wilson ML, et al. Characteristics and Sodium Pentobarbital Concentrations of Equine Mortality Compost Piles in the Upper Midwest. *Journal of Equine Veterinary Science*. 2022;114. **doi:10.1016/j.jevs.2022.104000**

14. Trautmann N, Olynciw A. Compost microorganisms. *Cornell Compost*; 1996. Accessed November 2, 2024. **http://compost.css.cornell.edu/microorg.html**

15. Naims H. Economics of carbon dioxide capture and utilization—a supply and demand perspective. *Environ Sci Pollut Res*. 2016;23:22226-22241. **doi:10.1007/s11356-016-6810-2**

16. Doyle KP, Simon RP, Stenzel-Poore MP. Mechanisms of ischemic brain damage. *Neuropharmacology*. 2008;55(3):310-318. **doi:10.1016/j.neuropharm.2008.01.005**

17. Lorbach JN, Campler MR, Youngblood B, et al. Comparison of Gaseous and Water-Based Medium-Expansion Foam Depopulation Methods in Cull Sows. *Animals*. 2021;11(11):3179. **doi:10.3390/ani11113179**

18. Guan J, Chan M, Brooks BW, Rohonczy E, Miller LP. Vehicle and Equipment Decontamination During Outbreaks of Notifiable Animal Diseases in Cold Weather. *Applied Biosafety*. 2017;22(3):90-137. **doi:10.1177/1535676017719846**

19. USDA. ASF Response: Depopulation, Disposal, and Decontamination Guidance—Option Matrices and Considerations. Accessed November 12, 2024. **https://www.aphis.usda.gov/sites/default/files/asf-3d-summary-guidance.pdf**

20. Kieffer JD, Campler MR., Cheng TY, et al. Evaluation of a Water-Based Medium-Expansion Foam Depopulation Method in Suckling and Finisher Pigs. *Animals: an open access journal from MDPI*. 2022;12(8):1041. **doi:10.3390/ani12081041**

21. Benson ER, Alphin RL, Rankin MK, Caputo MP, Kinney CA, Johnson AL. (2012). Evaluation of EEG based determination of unconsciousness vs. loss of posture in broilers. *Research in veterinary science*, 93(2), 960–964. **https://doi.org/10.1016/j.rvsc.2011.12.008**

22. Korenyi-Both J, Vidaurre J, Held T, et al. (2022). Description of electroencephalographic data gathered using water-based medium-expansion foam as a depopulation method for nursery pigs. *Scientific reports*, 12(1), 16798. **https://doi.org/10.1038/s41598-022-21353-7**

23. Hutchinson M, Duc HM, Flory GA, et al. Static Aerated Composting of African Swine Fever Virus-Infected Swine Carcasses with Rice Hulls and Sawdust. *Pathogens*. 2023; 12(5):721. **https://doi.org/10.3390/pathogens12050721**

24. USDA. Composting Livestock 2017 Livestock Mortality Composting Protocol. Published online 2017. Accessed November 12, 2024. **https://www.aphis.usda.gov/sites/default/files/livestock-mortality-compost-sop.pdf**

25. Duc HM, Ngan PH, Son HM, et al. (2022). The use of composting for the disposal of African swine fever virus-infected swine carcasses. Transboundary and emerging diseases, 69(5), e3036–e3044. **https://doi.org/10.1111/tbed.14659**

26. Carrau T, Malakauskas A, Masiulis M, et al. (2023). Composting of Wild Boar Carcasses in Lithuania Leads to Inactivation of African Swine Fever Virus in Wintertime. Pathogens (Basel, Switzerland), 12(2), 285. **https://doi.org/10.3390/pathogens12020285**

https://doi.org/10.54846/am2025/27

Computationally predicted T-cell epitope trends for 30 years of wild-type PRRSV-2 strains from the USA

J. P. Baker[1]; A. H. Gutierrez[2]; N. Pamornchainavakul[1]; A. S. De Groot[2]; K. VanderWaal[1]
[1]Department of Veterinary Population Medicine, University of Minnesota, Saint Paul, Minnesota;
[2]EpiVax Inc, Providence, Rhode Island

Introduction

Despite tremendous work on immune responses to porcine reproductive and respiratory syndrome virus (PRRSV)-2 infection, the correlates of protection are not completely understood. In particular, the paucity of information available on the relationship between cell-mediated immunity, and protective efficacy of existing PRRSV vaccines, has left a knowledge gap in our understanding of PRRSV.[1] Here, we begin to bridge this gap by mapping the T-cell epitopes contained in PRRSV viral structural proteomes (N = 163) and comparing the epitopes between viruses. T- cell epitope content comparison provides a unique perspective on viral diversity and evolution, and a basis for exploring cross-protection from the perspective of cell-mediated immunity. For this project, we utilized a high throughput bioinformatic pipeline (PigMatrix & EpiCC[2]) developed by EpiVax to computationally predict and compare MHC class I and class II epitopes in PRRSV-2 whole genomes from multiple ORF5 based phylogenetic lineages/years and to assess the extent to which predicted T-cell epitopes are shared between wild-type viruses and modified live vaccine (MLVs).

Methods

All structural proteins (GP2-GP5, M and nucleocapsid) of a phylogenetically diverse set of 161 wildtype PRRSV-2 viruses from the USA spanning the past ~30 years, as well as the five common MLVs, were obtained from GenBank. T-cell epitope content was identified for each virus using the PigMatrix tool, which screens potential epitopes and estimates whether they are likely to bind MHC class I (n = 11) and class II (n = 5) swine leukocyte antigens (SLA) based on comparisons with 'matrices' generated for each of these SLA.[3] For each protein T-cell epitope density was calculated as # of epitopes per 100 amino acid residues. Pairwise T-cell epitope content comparisons (EpiCC) were used to generate metrics of epitope cross-conservation (% epitopes shared) between viral structural proteomes.[4] These epitope cross-conservation metrics were used to group viruses into T-cell epitope "immune clusters" via classical multi-dimensional scaling (MDS) and k-means clustering.[5] T-cell epitopes in each viral genome/protein were also compared to the T-cell epitopes identified in the sequences of four PRRSV vaccines, and the change in coverage over time since vaccine introduction was calculated.

Results

The density of predicted T-cell epitopes varied across the structural proteins, with GP2 and M having the highest median epitope density, with 133 and 128 respectively, followed by GP4, GP5, and GP3 (114, 111, and 109, respectively) (Figure 1). The nucleocapsid protein had the lowest epitope density, at ~68 (Figure 1). This epitope density was generally stable over time, for the 30 years of data collection. However, the trends of pairwise comparisons of T-cell epitope content, between wild-type viruses and PRRSV vaccines, demonstrated that the epitopes in circulating strains have generally shifted away from the VR-2332 prototype PRRSV virus and from vaccines that have been on the market for over 10 years (Figure 2). More T-cell epitopes were shared within lineage compared to between lineages, and the clustering analysis correlated well with phylogenetic tree structure of wild-type viruses. Interestingly, older lineages tended to be clustered, regardless of genetic relatedness, whereas recent sub-lineages that have subsequently emerged have greater diversity in their T-cell epitope content despite having a comparable level of genetic relatedness.

Discussion

This is the first report of T-cell epitope content and cross-conservation in an extensive set of PRRSV-2 field-strains from the past three decades. One observation is that the structural protein with the highest relative abundance in the virion, the nucleocapsid protein, had the lowest predicted epitope density, which warrants further investigation. The EpiCC comparison supports the hypothesis that lineages are immunologically distinct and provides new evidence that existing MLVs may not be well matched to current wild-type strains, at least from the perspective of adaptive immunity (T-cell epitopes). Furthermore, there appears to be no single vaccine that has better coverage of the T-cell epitopes predicted in wild-type viruses. There is general expansion of the T-cell epitope landscape over time, though without obvious directional pressure within clusters through time. This underpins the need for updated vaccines that are better matched to the wild-type strains currently circulating in the field. Finally, this work lays the foundation for further investigation of PRRSV-2 T-cell epitopes, including which proteins are likely to be immunogenic from the T-cell perspective, and highlights the role that host T-cell responses may play in viral evolution and adaptation.

Figure 1: Calculated T-cell epitope density (class I and II) for each structural protein. Epitope density was calculated as # of predicted epitopes per 100 amino acid (AA) residues of the respective protein.

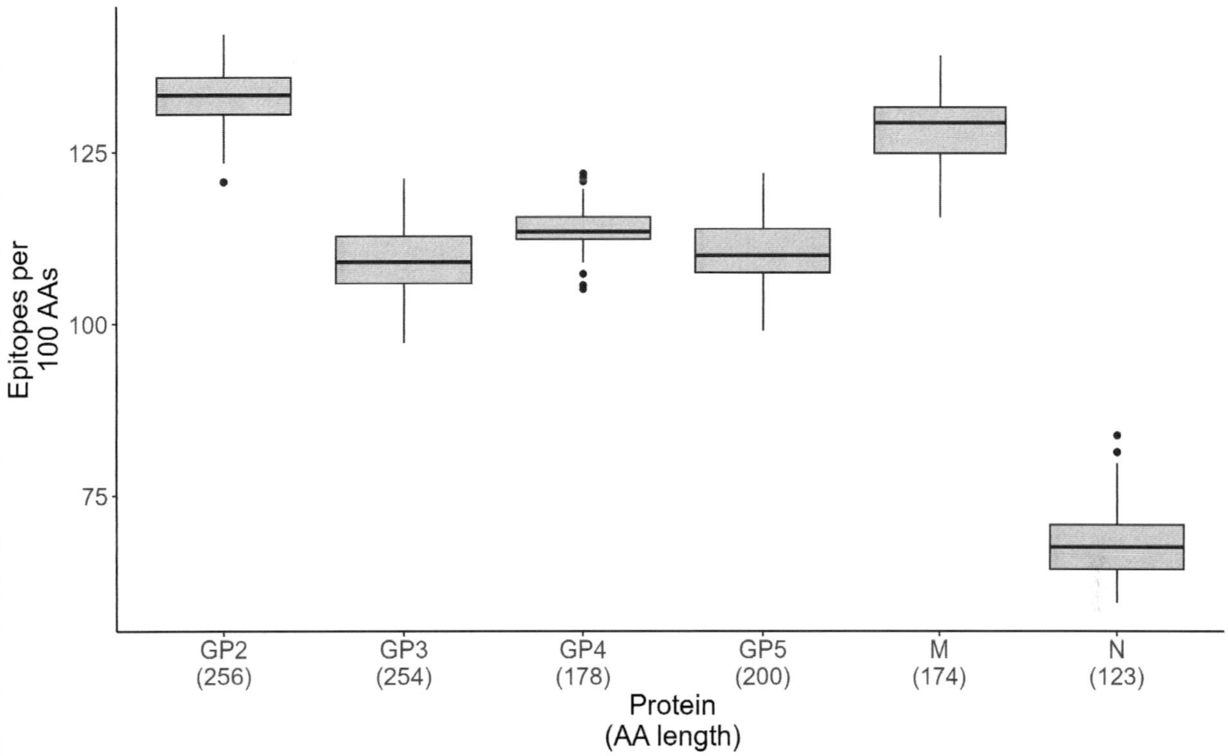

Figure 2: Pairwise percent shared T-cell epitopes for all wild-type viruses compared to VR-2332 and commercial MLVs. ORF5-based lineage is listed for each vaccine. Comparisons were shown from approximate date of vaccine introduction to the market through the present (Ingelvac MLV: 1994; Ingelvac ATP: 2004; Fostera: 2012).

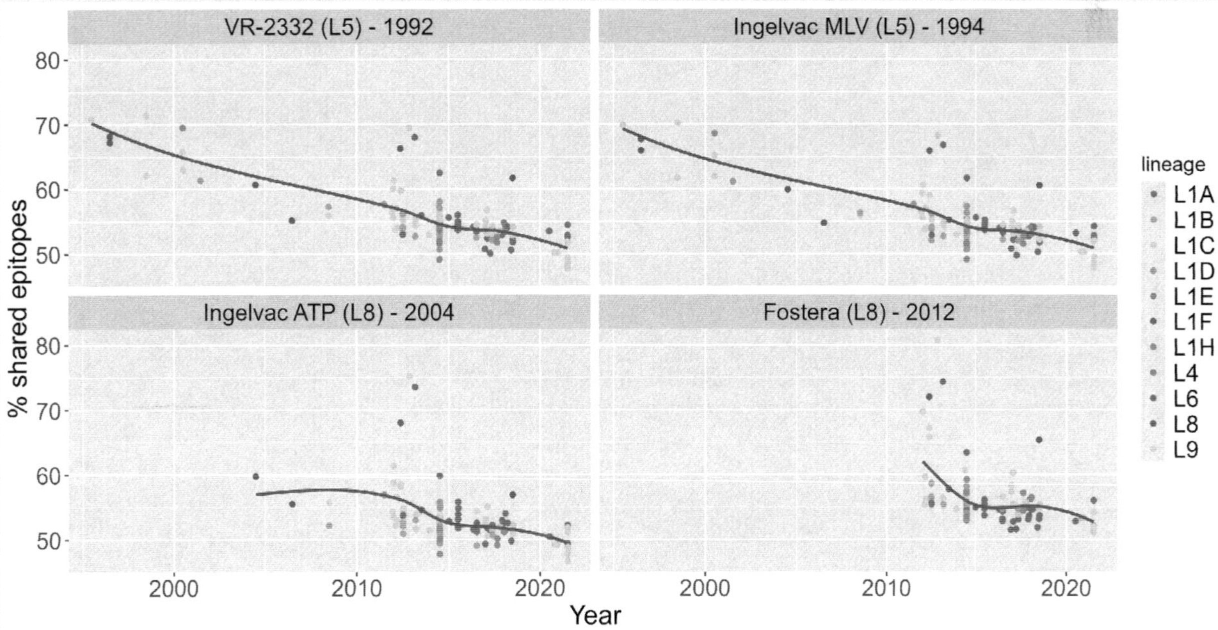

References

1. Fiers J, Cay AB, Maes D, Tignon M. A Comprehensive Review on Porcine Reproductive and Respiratory Syndrome Virus with Emphasis on Immunity. Vaccines (Basel). 2024 Aug 22;12(8):942. **doi: 10.3390/vaccines12080942**

2. Gutiérrez AH, Rapp-Gabrielson VJ, Terry FE, Loving CL, Moise L, Martin WD, De Groot AS. T-cell epitope content comparison (EpiCC) of swine H1 influenza A virus hemagglutinin. Influenza Other Respir Viruses. 2017 Nov;11(6):531-542. **doi: 10.1111/irv.12513**

3. Gutiérrez AH, Martin WD, Bailey-Kellogg C, Terry F, Moise L, De Groot AS. Development and validation of an epitope prediction tool for swine (PigMatrix) based on the pocket profile method. BMC Bioinformatics. 2015 Sep 15;16:290. **doi: 10.1186/s12859-015-0724-8**

4. Bandrick M, Gutiérrez AH, Desai P, Rincon G, Martin WD, Terry FE, De Groot AS, Foss DL. T cell epitope content comparison (EpiCC) analysis demonstrates a bivalent PCV2 vaccine has greater T cell epitope overlap with field strains than monovalent PCV2 vaccines. Vet Immunol Immunopathol. 2020 May;223:110034. Epub 2020 Mar 6. **doi: 10.1016/j.vetimm.2020.110034**

5. Chen J, Tan S, Avadhanula V, Moise L, Piedra PA, De Groot AS, Bahl J. Diversity and evolution of computationally predicted T cell epitopes against human respiratory syncytial virus. PLoS Comput Biol. 2023 Jan 10;19(1):e1010360. **doi: 10.1371/journal.pcbi.1010360**

https://doi.org/10.54846/am2025/28

Impact of cross-fostering on PRRSV transmission among piglets

Kristen Cleaver[1]; Elisa De Conti[1], DVM; Onyekachukwu Henry Osemeke[1], DVM, PhD; Daniel Linhares[1], DVM, MBA, PhD; Claire LeFevre[2], DVM; John Waddell[3], DVM
[1]College of Veterinary Medicine, Iowa State University, Ames, Iowa; [2]Carthage Veterinary Service Ltd., Carthage, Illinois; [3]Boehringer Ingelheim Animal Health USA Inc., Duluth, Georgia

Introduction

Cross-fostering (CF) is a common swine production practice used to increase litter uniformity, decrease pre-wean mortality, and reduce variability in wean weights.[1] Porcine reproductive and respiratory syndrome virus (PRRSV) is a major problem for the swine industry.[2] Efforts to reduce PRRSV transmission often include minimizing or eliminating CF.[3,4] However, recent studies have demonstrated that PRRSV transmission within litters was low at a time period when herds were approaching stability after a PRRSV infection.[5] We hypothesized that CF would not pose a significant risk of PRRSV transmission within litters. This study aimed to assess how CF piglets with a known PRRSV status influenced the probability of PRRSV infection in their new littermates.

Methods

An AASV PRRSV classification I-B, positive unstable low prevalence, breeding herd in the Midwest with a virulent, wild-type strain of 1-3-4 L1C.5 PRRSV was selected.[6] The study occurred 15 weeks after initial introduction and 11 weeks post-vaccination with a commercial PRRSV MLV vaccine. Routine CF was still being conducted. A total of 56 litters were enrolled over 3 days across 3 farrowing rooms. Blood swabs (BS) were collected from each piglet using an ear or tail vein within 24 hours after farrowing, and piglets were ear-tagged with individual identification numbers. Blood swabs were tested the same day at Carthage Diagnostic Laboratory (CDL) for PRRSV by reverse transcription quantitative polymerase chain reaction (RT-qPCR).[7] The RT-qPCR results determined piglet and litter status (PRRSV-positive or PRRSV-negative). Litters with no PRRSV-positive piglets were classified as PRRSV-negative litters. Litters with 1 PRRSV-positive piglets were classified as PRRSV-positive litters. Strategic CF into study groups was completed less than 48 hours after farrowing based on individual piglet and litter PRRSV status and sow teat count. A maximum of 1 piglet was entered into each litter. Piglet movements out of the litter were allowed at this time to reach functional teat count. The 5 study groups were: PRRSV-positive piglet to PRRSV-negative litter (P to N) (n = 15), PRRSV-negative piglet to PRRSV-positive litter (N to P) (n = 3), PRRSV-negative piglet to PRRSV-negative litter (N to N) (n = 13), PRRSV-positive litter with no piglets entered (P no CF) (n = 5), and PRRSV-negative litter with no piglets entered (N no CF) (n = 14). No piglet movements were allowed after the initial CF event, and litter counts were verified daily throughout the study. Tonsil-oral-scraping (TOSc) was collected from all sows on the final day of enrollment, after CF, and submitted to CDL for PRRSV RT-qPCR.[8] During the study, injections were administered with needles being changed between litters and castration instruments were disinfected between litters at processing. Piglet BS and sow TOSc were collected again, before any wean vaccinations were administered (piglets 19-21 days of age), and submitted to CDL for PRRSV RT-qPCR. Tongue tips (TT) were individually collected from all piglet mortalities, frozen, and taken to Iowa State University Veterinary Diagnostic Laboratory (ISU VDL) for PRRSV-RT-qPCR at the end of the study.[9] ORF5 sequencing was conducted at ISU VDL on 4 BS and 7 TOSc samples.[10] Regression models were constructed using R Statistical Software to measure the probability of PRRSV-negative pigs becoming PRRSV-positive and mortality given their study groups with sow PRRSV status from the initial TOSc sample as a fixed effect and farrowing room as a random effect.[11]

Results

Across all study groups, 10.7% (75/704) of the piglets became PRRSV-positive on final BS or TT after they initially tested negative. The percentages of piglets becoming PRRSV-positive were 4.2% (8/189) N to N, 7.3% (15/206) P to N, 7.1% (15/211) N no CF, 17.1% (6/35) N to P, and 49.2% (31/63) P no CF. The probability of a negative piglet becoming positive after CF was statistically higher ($P < .05$) in the P no CF group compared to the other 4 groups (Figure 1). All other comparisons showed no statistical difference ($P > .05$). There was no statistical difference in the mortality rates of piglets across the study groups (Figure 2). Sow PRRSV status at enrollment was not significantly associated with the probability of PRRSV-negative pigs becoming PRRSV-positive ($P = .29$). ORF5 sequencing was successfully completed on 2 BS and 4 TOSc samples. Homology to the wild-type 1-3-4 L1C.5 PRRSV strain ranged from 98.67-99.83%; indicating no evidence of a new PRRSV introduction or detection of vaccine strain.

Discussion

Given the conditions in this study, CF alone did not impact PRRSV transmission between piglets in a litter or piglet mortality rates prior to weaning. Practitioners can continue to take advantage of CF after a PRRSV introduction to improve productivity and animal welfare whilst ensuring that other biomanagement practices are properly followed.

Figure 1: Probability of piglets porcine reproductive and respiratory syndrome virus (PRRSV)-negative at birth becoming PRRSV-positive by weaning for each study group (N no CF [negative-no cross-fostered]: PRRSV-negative litter with no piglets entered; N to N: 1 PRRSV-negative piglet cross-fostered into PRRSV-negative litter; N to P: 1 PRRSV-negative piglet cross-fostered into PRRSV-positive litter; P to N: 1 PRRSV-positive piglet cross-fostered into PRRSV-negative litter; P no CF: PRRSV-positive litter with no piglets entered).

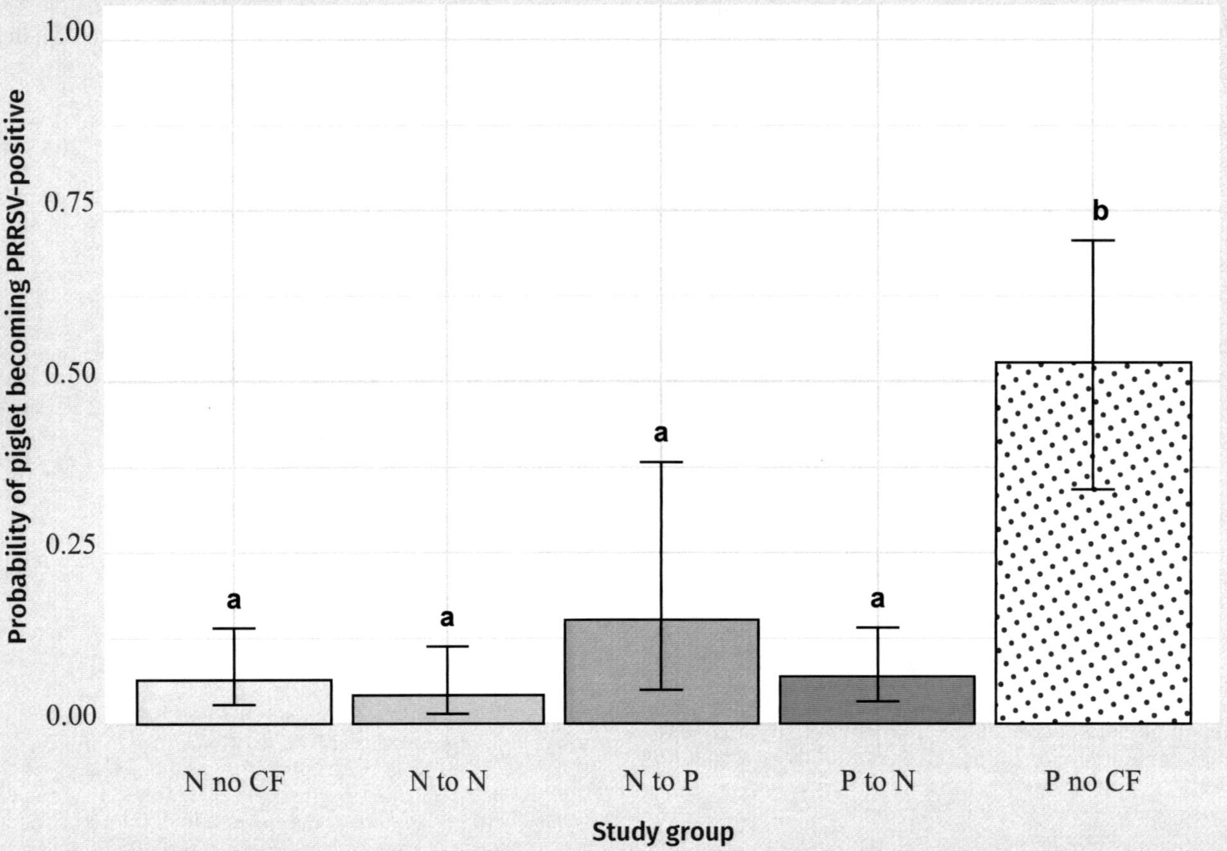

[a,b] Different superscripts are statistically different ($P < .05$)

Figure 2: Probability of piglet mortality for each study group (N no CF [negative-no cross-fostered]: PRRSV-negative litter with no piglets entered; N to N: 1 PRRSV-negative piglet cross-fostered into PRRSV-negative litter; N to P: 1 PRRSV-negative piglet cross-fostered into PRRSV-positive litter; P to N: 1 PRRSV-positive piglet cross-fostered into PRRSV-negative litter; P no CF: PRRSV-positive litter with no piglets entered).

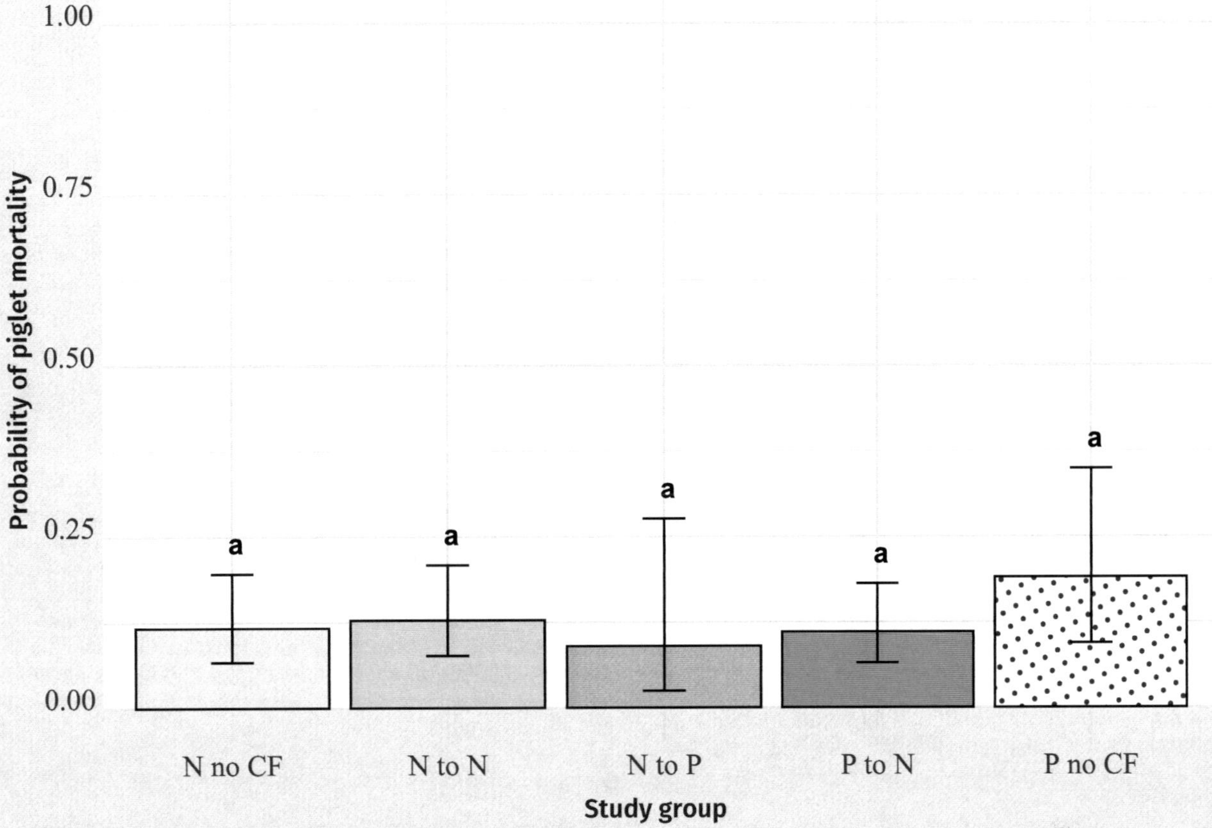

[a] Similar subscript indicates no statistical difference in study groups ($P > .05$)

References

1. Wattanaphansak S, Luengyosluechakul S, Larriestra A, Deen J. The impact of cross-fostering on swine production. *Thai J Vet Med*. 2002; 32:101–6.

2. Zimmerman J, Dee S, Holtkamp D, Murtaugh M, Stadejek T, Stevenson G, Torremorell M, Yang H, Zhang J. Porcine reproductive and respiratory syndrome viruses (porcine arteriviruses). *Diseases of Swine*. 2019; 685–708.

3. Dee S, Collins J, Halbur P, et al. Control of porcine reproductive and respiratory syndrome (PRRS) virus. Swine Health Production. 1996;4(2):95-98.

4. McCaw M. Effect of reducing crossfostering at birth on piglet mortality and performance during an acute outbreak of porcine reproductive and respiratory syndrome. *J Swine Health Prod*. 2000; 8:15–21.

5. Almeida M, Zhang M, Lopez W, Vilalta C, Sanhueza J, Corzo C, Zimmerman J, Linhares DCL. A comparison of three sampling approaches for detecting PRRSV in suckling piglets. *Prev Vet Med*. 2021; 194:105427. **doi:10.1016/j.prevetmed.2021.105427**

6. Holtkamp D, Torremorell M, Corzo CA, Linhares DCL, Almeida MN, Yeske P, Polson DD, Becton L, Snelson H, Donovan T, Pittman J, Johnson C, Vilalta C, Silva GS, Sanhueza J. Proposed modifications to porcine reproductive and respiratory syndrome virus herd classification. *J Swine Health Prod*. 2021; 29(5):261-270. **https://doi.org/10.54846/jshap/1218**

7. Osemeke OH, Cezar GA, Paiva RC, Moraes DCA, Machado IF, Magalhaes ES, Poeta Silva APS, Mil-Homens M, Peng L, Jayaraman S, Trevisan G, Silva GS, Gauger PC, Linhares DCL. A cross-sectional assessment of PRRSV nucleic acid detection by RT-qPCR in serum, ear-vein blood swabs, nasal swabs, and oral swabs from weaning-age pigs under field conditions. *Front Vet Sci*. 2023; 10:1200376. **doi:10.3389/fvets.2023.1200376**

8. Peng L, Poeta Silva APS, Moraes DCA, Yeske P, Osemeke OH, Magalhães ES, Silva Gustavo GS, Linhares DCL. Comparison of a novel rapid sampling method to serum and tonsil scraping to detect PRRSV in acutely infected sows. *Prev Vet Med*. 2024; 223, 106082. **https://doi.org/10.1016/j.prevetmed.2023.106082**

9. Machado IF, Magalhães ES, Poeta Silva APS, Moraes DCA, Cezar GA, Mil-Homens M, Osemeke OH, et al. Porcine reproductive and respiratory syndrome virus RNA detection in tongue tips from dead animals. *Front Vet Sci*. 2022; 9:993442. **doi: 10.3389/fvets.2022.993442**

10. Altschul S, Madden T, Schäffer A, Zhang J, Zhang Z, Miller W, Lipman D. Gapped BLAST and PSI-BLAST: a new generation of protein database search programs. *Nucleic Acids Res*. 1997; 25(17):3389-402. **doi: 10.1093/nar/25.17.3389**

11. R Core Team. R: A language and environment for statistical computing. R Foundation for Statistical Computing. Vienna, Austria. 2024. **https://www.R-project.org/**

https://doi.org/10.54846/am2025/29

Investigation of a compliance marker and cost of *Salmonella* vaccination by gel and oral drench

G. Duhe[1]; F. Leite[2], DVM, MS, PhD; J. Morgan[2], DVM, MVS
[1]College of Veterinary Medicine, Louisiana State University, Baton Rouge, Louisiana;
[2]Boehringer Ingelheim Animal Health USA Inc.

Introduction

Salmonella is a zoonotic pathogen that can pose significant risks to swine and human health. This pathogen can cause severe disease in pigs and persist in the environment, potentially leading to major economic losses.[1-3] Effective modified-live vaccines are available for *Salmonella,* which traditionally have been administered orally either via drench or drinking water.[4] While oral drenching can be done on the sow farm, it is labor-intensive, potentially reducing farm efficiency. Oral administration via gel is a novel delivery method in which pigs are vaccinated by consuming a gel combined with vaccine.[5] This method does not require individual pig handling since pigs consume the gel on their own after it is placed in the farrowing crates. This study aimed to evaluate *Salmonella* shedding following vaccination as a potential compliance marker, as well as measure the cost and labor efficiency when vaccinating by gel versus drench.

Methods

Part one of the study was performed to determine if shedding following vaccination with Enterisol Salmonella T/C® could serve as a compliance marker, along with the optimal timing of sampling. Rectal swabs (cotton swabs in 10 mL of saline) were collected from one randomly selected pig out of 24, 14–15-day old litters prior to vaccination and at 2-, 24-, 48-, and 72-hours post-gel vaccination. Longitudinal collection was used. Swabs were submitted for *Salmonella* PCR.

Part two compared administration time for oral drench versus gel in three different farrowing rooms with 24 crates/room and 768 pigs' total. Oral drench was administered via a two mL oral drench gun and a Matabi Doser Evolution 5-gallon backpack sprayer was used to apply 150 mL of gel -vaccine mixture per farrowing crate mat (12.5 mL /piglet). Administration time was recorded for both methods and a worker salary of $18 per hour was assumed, as well as including $0.03 per pig for the cost of gel in the cost analysis.

Results and discussion

Summarized in Table 1, results from part one revealed 0% (0/24) samples detected by *Salmonella* PCR prior to vaccination, 50% (12/24) at two hours, 12.5% (3/24) at 24 hours, and 0% (0/24) at 48- and 72-hours post-vaccination, indicating detection of vaccine shedding by PCR was highest at two hours post-vaccination. In part two of the study, the oral drench vaccination method took an average of one hour while the gel method took an average of two and a half minutes to vaccinate an average of 256 pigs over 24 crates. The average cost to vaccinate the three farrowing rooms was $0.14 per pig for oral drench versus $0.035 per pig for gel. Two people were required to vaccinate by oral drench while only one person was required to vaccinate by gel using the backpack sprayer.

Table 1: PCR results summarizing which pigs tested positive for the vaccine and at what time

Time of swab	PCR results
0 hours	0/24+
2 hours post vax	12/24+
24 hours post vax	3/24+
48 hours post vax	0/24+
72 hours post vax	0/24+

As seen in Figure 1, the cost-analysis revealed that gel vaccination was significantly more economical and less labor-intensive compared to oral drenching, assisting with bottom line economics from a production standpoint. Results from this study indicate that measuring *Salmonella* shedding by PCR in fecal swabs collected prior to and at two hours post vaccination via gel may be used as a compliance marker for the confirmation of Enterisol Salmonella T/C® vaccination status and that administration of oral vaccine via gel requires less labor and results in a lower cost than administration by oral drench.

References

1. "Salmonellosis." *Iowa State University,* Accessed 8 Sept 2024 **vetmed.iastate.edu/vdpam/FSVD/swine/index-diseases/salmonellosis**

2. Board, National Pork. "Salmonellosis and Salmonella Infections." *Pork Information Gateway,* 14 Apr. 2022, **porkgateway.org/resource/salmonellosis-and-salmonella-infections/**

3. Griffith, Ronald W., et al. "Salmonellosis." *Diseases of Swine,* no. 11, 29 Mar. 2019, pp. 912–925, **https://doi.org/10.1002/9781119350927.ch59**

4. "Enterisol Salmonella T/C®." *Boehringer Ingelheim Public Health,* Accessed 8 Sept. 2024. **bi-animalhealth.com/swine/products/enterisol/enterisol-salmonella-tc#:~:text=Modified%2Dlive%20vaccine%20provides%20rapid,tract%20to%20induce%20mucosal%20immunity**

5. Evaluation of the efficacy of oral vaccination against Salmonella by gel. Nerem J. et al., IPVS 2024.

Figure 1: Graph comparing the cost per pig for both methods and associated labor and materials.

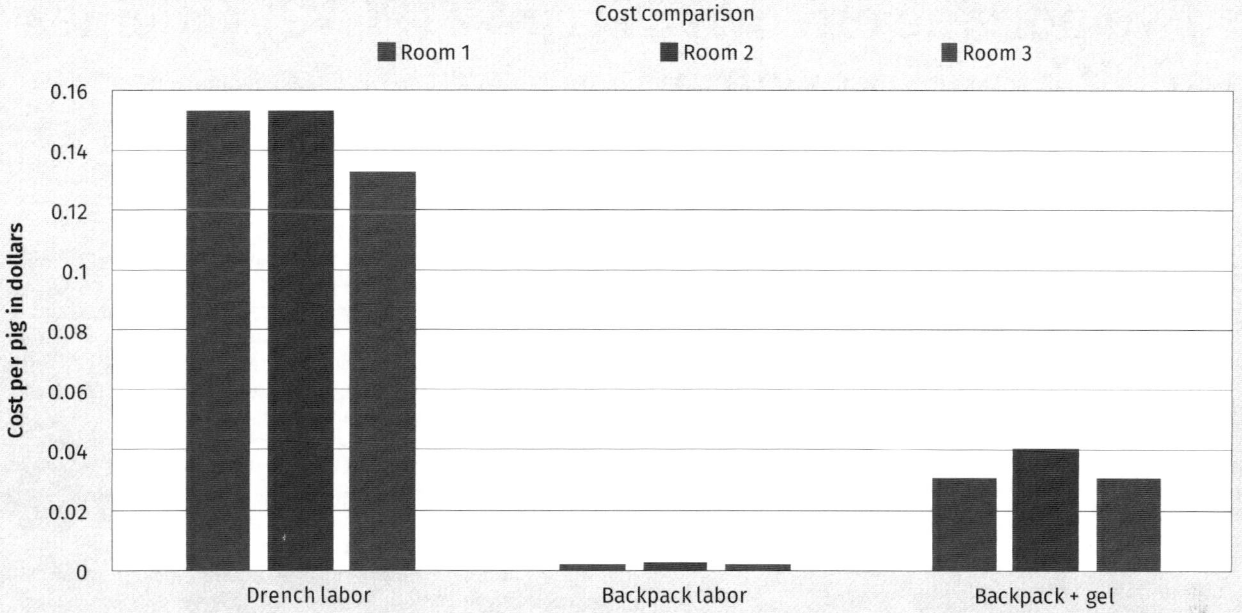

https://doi.org/10.54846/am2025/30

Comparison of cell mediated immunity induction by two porcine circovirus type 2 (PCV2) vaccines in adult breeding swine

J. Springer[1], BS; L. Glowzenski[2], VMD; N. Garbes[3], DVM; M. Jansen[3], DVM; D. Foss[3], PhD
[1]University of Illinois College of Veterinary Medicine, Urbana-Champaign, Illinois; [2]JBS Live Pork, Oakville, Iowa; [3]Zoetis, Parsippany, New Jersey

Introduction

Porcine circovirus type 2 (PCV2) causes multisystemic and reproductive disease in swine. While humoral immunity is important for protection against PCV2, cell mediated immunity (CMI) may be more important for virus clearance and shedding reduction.[1] It has been previously shown that CMI can be passively transferred from dams and can provide protection against PCV2 in newborn piglets.[2] Additionally, stimulation of CMI has been shown to be less impacted than humoral immunity stimulation in the face of maternally derived antibodies.[3] However, CMI is rarely evaluated in the field as it is expensive and time-consuming to measure. In the instances in which CMI is measured, it is typically done by utilizing one of two assays, either enzyme-linked ImmunoSpot (ELISpot) or flow cytometry.[4] CMI is driven by T cells, and each of these assays can be used to quantify T cells that are secreting interferon gamma (IFNγ) in response to antigen stimulation. The ELISpot assay measures how many T cells are secreting IFNγ. Flow cytometry works similarly as it can detect the cells that are secreting IFNγ, and it can also differentiate between T cell populations such as T helper cells (cells with both CD4 and CD8 receptors) or cytotoxic T cells (cells with only CD8 receptors).

Previously published studies in growing pigs evaluating PCV2-specific CMI through ELISpot testing demonstrated a greater CMI response in pigs vaccinated with Fostera® Gold PCV MH (Zoetis, Parsippany, NJ) than in pigs vaccinated with Circumvent® PCV-M G2 (Merck Animal Health, Duluth, GA).[5] The mechanism of this advantage is not fully understood, and may be attributed to the bivalent nature of Fostera Gold, the immunological activity of squalene emulsion adjuvants such as MetaStim® (the adjuvant in Fostera Gold), or a combination of both. PCV2-specific CMI in adult animals has not been widely evaluated. With the growing use of sow herd vaccination, it is beneficial to understand if CMI is induced when adult animals are vaccinated against PCV2, and if differences exist between vaccines. The objective of this study was to compare, through the use of flow cytometry, the CMI responses of sows vaccinated with two commercial vaccines, Fostera Gold PCV MH and Circumvent PCV-M G2 (Merck Animal Health, Duluth, GA).

Materials and methods

At a commercial 5,000 head sow farm in southeast Iowa, forty (40) parity one sows of similar genetics were randomly enrolled. All sows received a 2 mL dose of FLEXCombo® (Boehringer Ingelheim, Ridgefield, CT) as gilts at 27 weeks of age, and were approximately 57 weeks of age at enrollment. On study day zero (D0), sows were injected intramuscularly with 2 mL of either Fostera Gold PCV MH or Circumvent PCV-M G2. Whole blood and serum samples were taken on D0 and day eighteen (D18). Peripheral blood mononuclear cells (PBMC) were isolated from anticoagulated whole blood for flow cytometry through centrifugation and separation by density. Isolated PBMC were cultured with PCV2 antigen and stained to identify T cells and intracellular interferon gamma (IFNγ). The IFNγ-secreting T cells were identified and differentiated into two groups. The T cells with both CD4 receptors and CD8 receptors were labeled as CD4+/CD8+, and the cells with only CD8 receptors were labeled as CD4-/CD8+. The PCV2-specific response was calculated as the increase in IFNγ positive T cells from D0 to D18 for each of the cell populations. The Wilcoxon signed-rank test was performed to evaluate the differences between pre-treatment and post-treatment measurements. The test was conducted at a 90% confidence level using GraphPad Prism version 8.1.1 for Windows (GraphPad Software, San Diego, CA, USA). The null hypothesis of no difference was tested against the alternative hypothesis of a significant difference. A *P*-value less than 0.10 was considered indicative of statistical significance. Polymerase chain reaction (PCR) for PCV2 was run on serum samples to monitor for viremia.

Results

PCV2 PCR results were all negative, indicating no PCV2 viremia in sows. Flow cytometry results showed that both vaccines increased the number of PCV2-specific IFNγ secreting T cells on D18 ($P < .1$). The number of PCV2-capsid specific CD4+/CD8+ IFNγ secreting cells increased by an average of 3.57% for Fostera Gold PCV MH and 3.60% for Circumvent PCV-M G2, and the number of CD4-/CD8+ IFNγ secreting cells increased by an average of 2.36% for Fostera Gold PCV MH and 1.41% for Circumvent PCV-M G2 (Figure 1). There was no significant difference between the two vaccine groups in either cell population ($P > .75$).

Discussion

Cell mediated immunity to PCV2 is important to prevent PCV2-associated disease (PCVAD), clear the virus, and reduce shedding. Vaccine-induced, PCV2-specific T cells are expected to be present in two cell populations: T helper cells (CD4+/CD8+ in pigs) and cytotoxic T cells (CD4-/CD8+). Both of these cell types play a key role in PCV2 immunity. T helper cells provide immune help to both humoral and cell mediated immunity (in part by secreting IFNγ), and cytotoxic T cells function to eliminate virus-infected cells. The number of vaccine-induced PCV2-specific CD4+/CD8+ IFNγ cells observed in this study is similar to that seen with efficacious PCV2 vaccines administered to PCV2-negative growing pigs in a recent internal study (unpublished Zoetis data). Both groups of sows responded to the vaccines, but results did not show similar trends as observed previously in published growing pig studies. Earlier studies used an ELISpot assay to quantify the CMI response; this assay evaluates different populations of cells than the flow cytometry that was utilized in this study, which could contribute to the unexpected outcome. Additionally, all previous studies were conducted in PCV2-negative growing pigs. As such,

Figure 1: Percent change in IFNγ positive T cells.

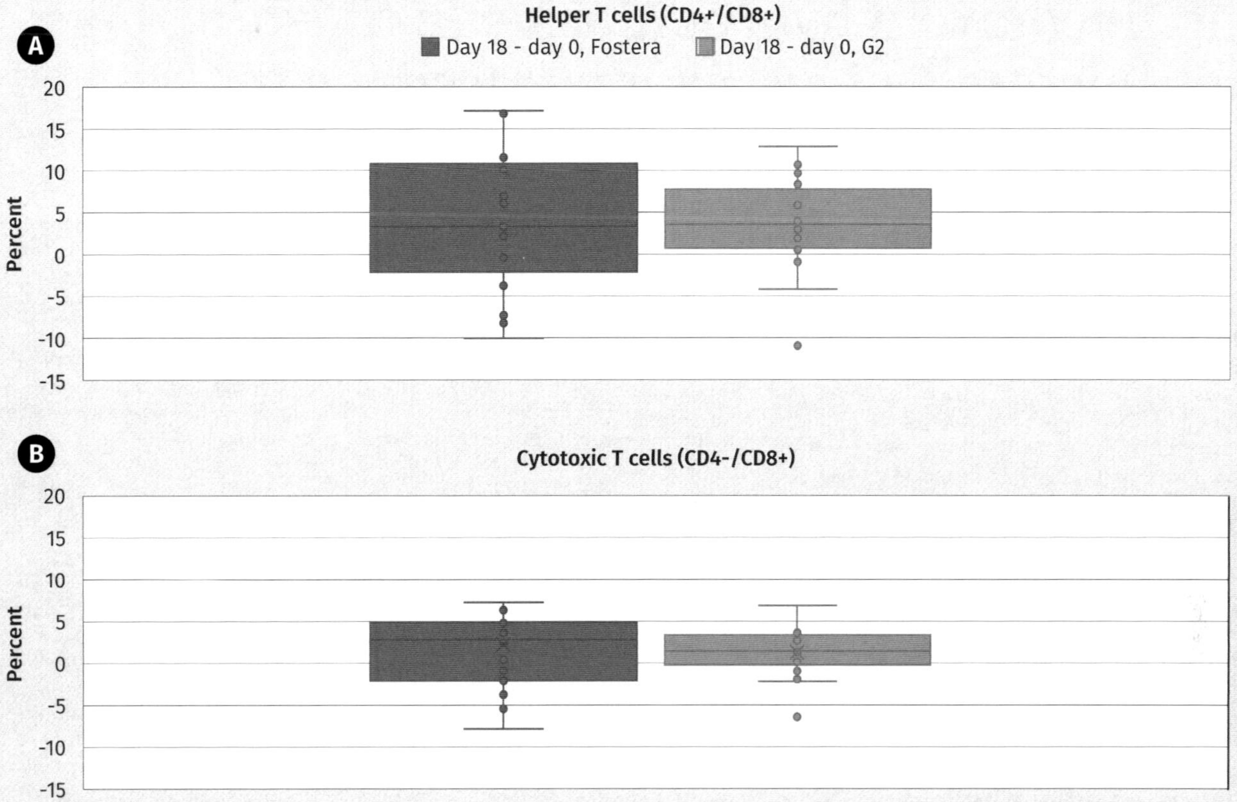

Porcine circovirus type 2 capsid-specific IFN-γ+ cells were determined in **A)** CD4+/CD8+ and **B)** CD4-/CD8+ T cells by flow cytometry. Graphs represent the difference in the number of each cell population between day 0 (pre-vaccination) and day 18 (post-vaccination), shown as a box plot with mean and quartile distributions.

G2 = Circumvent® PCV-M G2

pigs would not have had previous exposure or vaccination, so their T cells would be naïve to PCV2, potentially making the response more apparent. Sows in this study were likely exposed and definitely vaccinated, making them "primed" for PCV2 before the study began. Lastly, the farm experienced a porcine reproductive and respiratory syndrome virus break on D12 of the study. This active infection may have impacted the immune response by increasing the number of T cells actively secreting IFNγ without PCV2-specific activation. Due to the importance of passive CMI in protecting newborn piglets against PCVAD and the potential for CMI stimulation even if MDAs are present, further studies are warranted to further quantify and qualify the CMI response to PCV2 vaccination in adult breeding swine.

References

1. Fort M, et al. Development of cell-mediated immunity to porcine circovirus type 2 (PCV2) in caesarean-derived, colostrum-dervived piglets. *Veterinary Immunology Immunopathology.* 2009;129:101-107. **https://doi.org/10.1016/j.vetimm.2008.12.024**

2. Oh Y, et al. Protective effect of the maternally derived porcine circovirus type 2 (PCV2)-specific cellular immune response in piglets by dam vaccination against PCV2 challenge. *Journal of General Virology.* 2012;93:1556-1562. **https://doi.org/10.1099/vir.0.041749-0**

3. Martelli P, et al. Impact of maternally derived immunity of piglets' immune response and protection against porcine circovirus type 2 (PCV2) after vaccination against PCV2 at different age. *BMC Veterinary Research.* 2016;12:77. **https://doi.org/10.1186/s12917-016-0700-1**

4. Freen-van Heeren J, et al. Assessing Antigen-Specific T Cell Responses Through IFN-γ Enzyme-Linked Immune Absorbent Spot (ELISpot). *Kumar, V. (eds) Immune Homeostasis. Methods in Molecular Biology.* 2024;2782:209-226. **https://doi.org/10.1007/978-1-0716-3754-8_17**

5. Venegas-Vargas C, et al. Cellular and humoral immunity following vaccination with two different PCV2 vaccines (containing PCV2a or PCV2a/PCV2b) and challenge with virulent PCV2d. *Science Direct.* 2021;39:5615-5625. **https://doi.org/10.1016/j.vaccine.2021.08.013**

https://doi.org/10.54846/am2025/31

Comparative efficacy of water soluble tylvalosin and lincomycin hydrochloride in finishing pigs administered a *Lawsonia intracellularis* (Li or ileitis) oral challenge

H. Jung[1], BS; N. Winkelman[2], DVM
[1]College of Veterinary Medicine, The Ohio State University, Columbus, Ohio;
[2]Swine Services Unlimited Inc. (SSUI), Rice, Minnesota

Introduction

Ileitis is a prevalent enteric disease in modern swine production systems, where its conditions are particularly significant in grow to finish pigs. The disease adversely affects performance metrics such as average daily gain (ADG) and feed conversion ratio (FCR), and is rapidly transmitted via fecal to oral route.[1] The estimated economic loss due to an ileitis outbreak can range from \$5.98 to \$17.34[2] per pig, which highlights the need for identifying the most effective treatment. Antibiotics in either feed or water are considered primary treatment options, which includes both lincomycin[3] and tylvalosin,[4] for which there previously has been no direct comparison between the two. The objective of this study was to evaluate clinical, pathological, and performance differences in finishing pigs medicated with either water soluble tylvalosin or lincomycin hydrochloride after being challenged with a high dose of Li mucosal homogenate.

Materials and methods

Animals

A total of 170 Li-free pigs were obtained at 10 weeks of age, which consisted of 85 barrows and 85 gilts. Seventy-four pigs were selected for Li IPMA serum screening pre-selection for litter randomization. The pigs were all vaccinated for porcine circovirus 2 and *Mycoplasma hyopneumoniae (*Mhyo*)* at weaning, and were given 50g/ton carbadox in the nursery to ensure negative Li status[5] on arrival. The source farm is also free of porcine reproductive and respiratory syndrome (PRRS), swine influenza A virus (IAV-S), and Mhyo.

Randomization and allocation

On day -5 (D-5), all pigs were randomized by weight, gender, and litter, and allocated to three groups: a challenged, non-medicated, control group (T1; n = 50), and two challenged, medicated groups, receiving either water-soluble tylvalosin (T2; 50 ppm; n = 60), or water-soluble lincomycin hydrochloride (T3; 3.8 mg/lb bodyweight; n = 60). Five pigs were housed per pen. All animals were housed in two rooms (1 and 3) of the Swine Services Unlimited Inc (SSUI) research facility. Both rooms were ventilated with a negative pressure system with adequate heat supply. Temperatures and humidities were recorded daily. All pigs were given non-medicated feed and water *ad libitum* for the duration of the study. Feed was given as an SSUI manufactured diet.

Challenge

On D0, each pig was orally challenged with Li mucosal homogenate (1.85×10^{10} Li per pig). The concentration was precisely calculated using real-time PCR, and was a high dose that is necessary to induce clinical signs[6] in finishing pigs.

Treatment

Water treatments began on D13, when more than 10% of pigs showed clinical signs of ileitis. Untreated water and both tylvalosin and lincomycin hydrochloride were prepared fresh daily in separate 75-gallon tanks. Water for each group was then weighed out in a pail, hung on the ceiling and connected to each pen's water line for continuous gravitational flow into a nipple/cup drinker. After each 24-hour period, remaining water was weighed out and discarded to calculate precise average water consumption per pen if needed. Water medications were administered for 5 consecutive days.

Parameters measured

Clinical observations were recorded for individual pigs on D-1, 2, 6, 8,10, 12, 13, 15, 17, 19, 21, 22, 23, 24, 26, 27, 30, and 32. Four clinical parameters were scored from 0 (normal) to 3 (severe): Demeanor, abdominal appearance score (AAS), fecal diarrhea score (FDS), and fecal color score (FCS). Bodyweights (BW) were measured on D-5, 0, 13, 21, and 35, and were used to calculate ADG (lbs/day). Feed from each pen was weighed in every day, and weighed out on D -5, 13, 21, and 35 for average daily feed intake (ADFI; lbs/pig/day) and FCR (feed/gain) calculations. Intestinal porcine proliferative enteropathy (PPE) gross lesion severity index (SI) scores were calculated by summation of multiplied PPE lesion score by lesion length, in centimeters, for each section of the jejunum, ileum, cecum, and colon on D35, when all animals were necropsied upon termination of the study. The experimental unit for clinical observations, ADG, and SI was individual pigs. FCR was calculated at the pen level. Raw data regarding performance parameters have been used to calculate the least-squared means, and Tukey's method was used to assess the differences between each group.

Results

Mortality

Prior to D35, a total of six pigs died, three were from control (T1) and three from llincomycin hydrochloride (T3) group. In T1, two cases had a confirmed diagnosis of ileitis, with one presenting acute proliferative hemorrhagic enteropathy (PHE), and other in the form of chronic porcine intestinal adenomatosis (PIA). The third case remains unconfirmed due to severe hemorrhagic bowel syndrome (HBS), though there is a possibility that ileitis could have been present. T3 had one confirmed case with the disease in a form of PHE, one with HBS but no ileitis present, and one euthanized due to severe rectal prolapse. The euthanized pig was not confirmed with ileitis. The Fisher's Exact test was performed and there was no difference in mortality between the groups. No mortality was observed in tylvalosin-treated group (T2).

Clinical observations (Table 1) (Figure 1)

The proportion of animals with abnormal clinical observations (score > 0) was calculated between days 13 to 35. Demeanor score was excluded. FDS: T1–0.32, T2-0.076, T3-0.19, FCS: T1-0.18, T2-0.029, T3-0.10, AAS: T1-0.23, T2-0.056, T3-0.16. Statistical analysis is pending.

Performance parameters and PPE gross lesion SI

No significant differences were found in ADG (Table 2), FCR (Table 3), or ADFI (Table 4) before D13. However, during D13-35, T2 showed significantly better ADG than T1 ($P = .0007$), and T3 also had significantly higher ADG than T1 ($P = .0301$). There was no significance noted in ADG between T2 and T3 ($P = .2115$). For FCR during D13-21, T2 was significantly more efficient than T1 ($P = .0040$). T3's FCR was better than T1 but did not reach statistical significance ($P = .0730$), and there was no difference between T2 and T3 ($P = .3380$). In terms of ADFI during D13-21, both T2 and T3 had higher feed intake than T1, with P-values of 0.02 and 0.04, respectively. No significant difference in ADFI was noted between T2 and T3 ($P = .93$). Mean SI (Table 5) was highest in T1, followed by T3, and lowest in T2. Statistically significant differences were observed between T1 and T2 ($P = .006$), with no significant differences between T1 and T3 ($P = .129$) or between T2 and T3 ($P = .083$). Figure 2 refers to the percentage distribution of SI in terms of its severity per treatment.

Table 1: Clinical observations summary - percent abnormal

	FDS (fecal diarrhea score)			FCS (fecal color score)			AAS (abdominal appearance score)		
	D13- D22	**D23-D35**	**Overall**	**D13- D22**	**D23-D35**	**Overall**	**D13- D22**	**D23-D35**	**Overall**
T1 - Control	33.3%	31.1%	32.2%	18.7%	16.7%	17.7%	19.0%	27.4%	23.2%
T2 - Tylvalosin	6.1%	9.2%	7.6%	4.2%	1.7%	2.9%	6.9%	4.2%	5.6%
T3 - Lincomycin	20.1%	18.7%	19.4%	12.7%	6.9%	9.8%	14.7%	17.6%	16.1%

Figure 1: Clinical observations summary - percent abnormal.

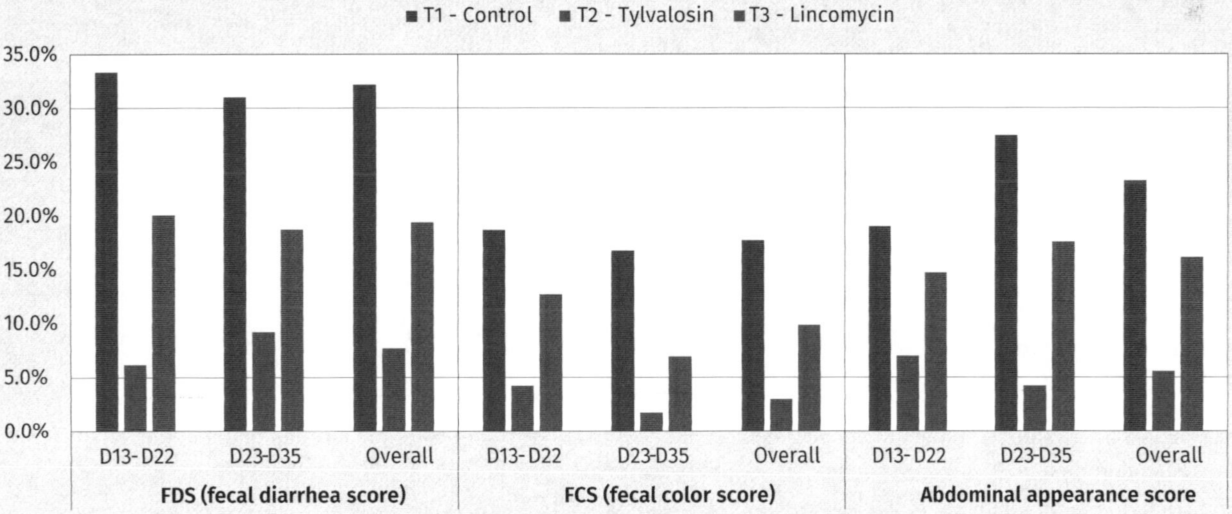

Table 2: Average daily gain summary

	D -5 to 13	D 13 to 21	D 21 to 35	D 13 to 35	D 0 to 35
T1 - Control	1.76[b]	1.44[b]	1.91[b]	**1.75[b]**	1.80[b]
T2 - Tylvalosin	1.70[b]	2.12[a]	2.34[a]	**2.26[a]**	2.11[a]
T3 - Lincomycin	1.80[b]	1.83[ab]	2.22[ab]	**2.07[a]**	2.00[ab]

[ab] Statistical significance ($P < .05$) is noted by different superscripts within a column

Table 3: Feed conversion ratio summary

	D -5 to 13	D 13 to 21	D 21 to 35	D -5 to 35
T1 - Control	1.63[b]	**2.69[b]**	2.31[b]	1.98[b]
T2 - Tylvalosin	1.69[b]	**1.81[a]**	2.17[b]	1.90[b]
T3 - Lincomycin	1.67[b]	**2.16[ab]**	2.20[b]	1.94[b]

[ab] Statistical significance ($P < .05$) is noted by different superscripts within a column

Table 4: Average daily feed intake summary

	D -5 to 13	D 13 to 21	D 21 to 35	D -5 to 35
T1 - Control	2.97[b]	**3.27[b]**	4.18[b]	3.45[b]
T2 - Tylvalosin	3.00[b]	**3.82[a]**	5.07[a]	3.89[a]
T3 - Lincomycin	3.09[b]	**3.77[a]**	4.71[ab]	3.78[ab]

[ab] Statistical significance ($P < .05$) is noted by different superscripts within a column

Table 5: Porcine proliferative enteropathy gross lesion severity index (SI) summary

	Mean SI of all pigs
T1 - Control	75.26[b]
T2 - Tylvalosin	25.53[a]
T3 - Lincomycin	33.41[ab]

[ab] Statistical significance ($P < .05$) is noted by different superscripts within a column

Figure 2: Porcine proliferative enteropathy (PPE) gross lesion severity index (SI) distribution in percentage per treatment.

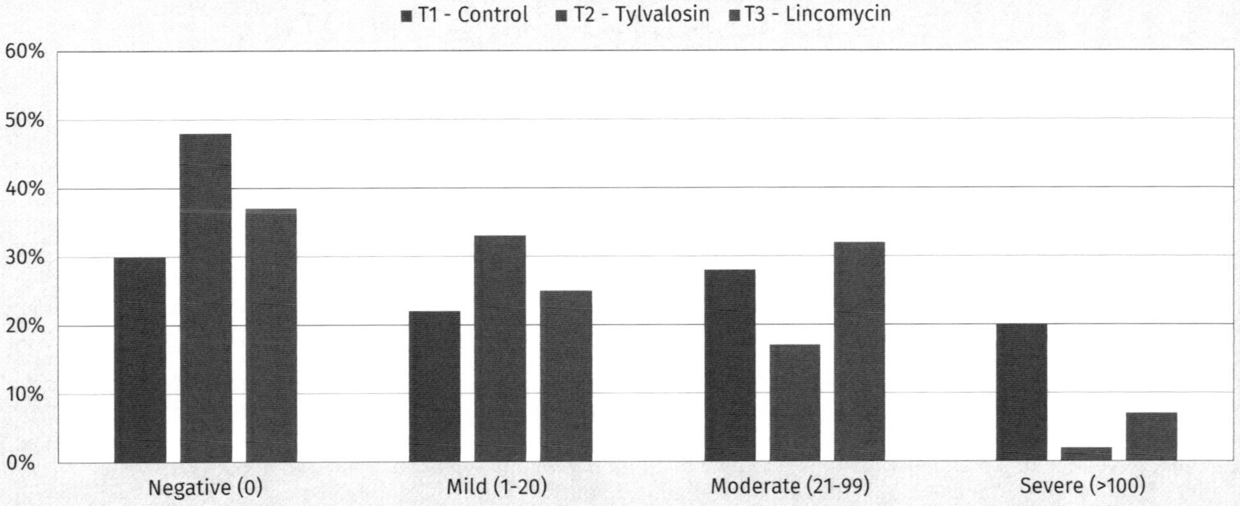

Discussion

Although chronic ileitis typically has a relatively low mortality rate, around 2%,[7] the challenge was effective, as shown by observed mortality rates across treatment groups. This provided a suitable context for evaluating the impact of each treatment. The trial demonstrated that pigs medicated with tylvalosin or lincomycin hydrochloride experienced improvements in mortality, diarrhea, body condition, and growth performance in the face of an experimental Li challenge. Tylvalosin-treated pigs showed significant improvements in ADG, FCR, ADFI, and SI compared to non-medicated pigs, and a numerical advantage over lincomycin hydrochloride-treated pigs. Lincomycin hydrochloride treated pigs also showed significant improvement in ADG and ADFI during D13-21 and a numerical advantage in other parameters compared to the control group. In contrast, the non-medicated control group also exhibited significantly more severe lesions in the GI tract, likely due to ileitis-related hyperplasia of immature crypt cells,[1] which compromised nutrient absorption and negatively impacted growth performance as observed. Notably, although both treatment groups exhibited significantly higher ADFI compared to controls, only the tylvalosin group showed a significant improvement in FCR. This may be attributed to tylvalosin's potential to reduce lesion severity and improve nutrient utilization, as supported by the significantly lower mean SI. The SI distribution further revealed that tylvalosin-treated pigs had the highest percentage of animals with low or negative SI scores, suggesting reduced disease severity and enhanced nutrient absorption. Given the prevalence and high economic cost of ileitis outbreaks, identifying and treating ileitis quickly is of extreme importance. Practitioners can apply the results and integrate either of these treatments, depending on specific treatment goals and available resources of each facility.

Acknowledgment

The authors thank Dr Gustavo Silva, Iowa State University, for statistical analysis.

References

1. Lawson G. H., Gebhart C. J. Proliferative enteropathy. 2000. *J Comp Pathol*, 122 (2–3), pp. 77-100.

2. Holtkamp D. 2019. Economic losses associated with ileitis. *MSD Animal Health.*

3. Winkelman, N. L., Crane, J. P., Elfring, G. D., Dal Kratzer, D., Meeuwse, D. M., Dame, K. J., Buckham, S. L., & Gebhart, C. J. 2002. Lincomycin-medicated feed for the control of porcine proliferative enteropathy (ileitis) in swine. *Journal of Swine Health and Production*, *10*(3), 107-111.

4. Guedes, R. M., França, S. A., Machado, G. S., Blumer, M. A., & da Costa Cruz, E. C., Jr. 2009. Use of tylvalosin-medicated feed to control porcine proliferative enteropathy. *The Veterinary record*, *165*(12), 342–345. **https://doi.org/10.1136/vr.165.12.342**

5. Wattanaphansak S., Singer R.S., Gebhart C.J. In vitro antimicrobial activity against 10 North American and European *Lawsonia intracellularis* isolates. *Vet Microbiol.* 2009;134(3–4):305–310. **doi: 10.1016/j.vetmic.2008.08.007**

6. Guedes, RMC, Winkelman, NL, & Gebhart, CJ. 2003. Relationship between the severity of porcine proliferative enteropathy and the infectious dose of *Lawsonia intracellularis*. *Veterinary Record* 153, 432-433.

7. Campillo, M., Smith, S. H., Gally, D. L., & Opriessnig, T. 2021. Review of methods for the detection of *Lawsonia intracellularis* infection in pigs. *Journal of veterinary diagnostic investigation : official publication of the American Association of Veterinary Laboratory Diagnosticians, Inc*, 33(4), 621–631. **https://doi.org/10.1177/10406387211003551**

https://doi.org/10.54846/am2025/32

Comparison of piglet vaccination timing protocols on early nursery stress and performance

M. Moore[1]; A. Reever[2], DVM, MS; B. Burton[2], DVM; J. Risser[3], DVM; R. Evelsizer[3], DVM
[1]College of Veterinary Medicine, Iowa State University, Ames, Iowa; [2]Suidae Health & Production, Algona, Iowa; [3]Elanco Animal Health, Greenfield, Indiana

Introduction

The morning of weaning is the current commercial standard for vaccination timing of wean-age piglets; however, vaccination timing varies from a few days pre-wean to several weeks post-placement, depending on labor capabilities and preferences. Understanding how the combination of weaning and vaccination events impact nursery performance is crucial for producers to make informed decisions on vaccination timing protocols. Previous research has shown that elevated stress and cortisol levels from events such as weaning, transport, and social integration can negatively influence vaccine efficacy but the effect on overall pig performance is not well understood.[1]

In the midst of other stressful events, vaccination on the day of weaning could have a detrimental impact on early nursery gain and survivability, which are important indicators of finishing performance. There may be an opportunity to reduce stress and improve pig performance by utilizing a different vaccination timepoint for wean-age pigs than the current commercial standard.

The objective of this study was to evaluate the differences in weight gain and survivability for several vaccination timing protocols.

Materials and methods

Piglets from a porcine epidemic diarrhea virus (PEDV) negative, *M hyopneumoniae* (Mhp) category IV (negative), and porcine reproductive and respiratory syndrome virus (PRRSV) category II-vx (stable vaccinated) commercial batch sow farm in northwest Iowa were enrolled in the study 3 days prior to weaning.[2] Two wean groups of piglets were utilized and weaned one week apart. Piglets ranged in age from 19 to 27 days at the time of weaning, with an average age of 23.2 days. Each pig was RFID tagged, weighed, and randomly allotted using a complete block design into one of three treatment groups at enrollment (n = ~350). Birth date, dam parity, sex, crate location, and sow ID were recorded.

Pigs were vaccinated with a commercially available porcine circovirus type II and Mhp vaccine and Prevacent PRRS according to their treatment group timing: 3 days pre-wean (A - pre-wean), on the morning of weaning prior to leaving the sow farm (B - wean), or 2 weeks post-placement in the nursery (C - 2 weeks post). Pigs from all treatment groups were comingled in designated pens in a commercial nursery. All pigs were weighed individually at 1-, 2-, and 4-weeks post-placement.

Hospitalizations and mortality dates and reasons were recorded for the duration of the study. Full value pigs were described as pigs that had not died, been hospitalized, or weighed less than 25 lbs at the conclusion of the study. Pigs were fed, raised, treated, and hospitalized in accordance with the farm's standard operating procedures. JMP software was used to analyze the data, and statistical significance was determined at $P < .05$ by ANOVA testing with post-hoc Tukey HSD, including enrollment weight as a covariate.

Results

No significant differences were identified between the treatment groups for pig weight or average daily gain, overall (Figures 1 and 2). However, when pigs were divided into weight categories, a statistically significant average daily gain (ADG) advantage was observed for the pigs vaccinated 2 weeks post-placement compared to the pigs vaccinated at or before weaning in the lightest 25% of pigs (< 12.19 lbs) at enrollment ($P = .006$). This difference was not observed for pigs weighing in the middle 50% (12.19-16.87 lbs) or the heaviest 25% at enrollment (> 16.87lbs) (Figure 3). Table 1 shows the least square means and *P*-values for metrics at 4-weeks post-weaning. Table 2 shows the least square means and performance data between treatment groups throughout the trial.

There were no significant differences in mortality or hospitalization rates, but it is important to note that mortality was very low for this nursery (~0.5%). Given the low mortality and although not statistically significant, there were numerical differences in the percentage of full value pigs at the end of the nursery (Figure 4).

Discussion

For this particular scenario, there was minimal advantage to adjusting vaccination timing and adding an additional handling event. However, for unhealthy flows, such as those experiencing a PRRS, PED or Mhp challenge, or farms weaning younger, lighter, or more variable weight pigs, there may be value in delaying vaccination timing to 2 weeks post-weaning.

This study did not assess differences in immune response or vaccine efficacy based on timing, which should also be taken into consideration. The results of this study can be immediately applied to help producers and veterinarians make informed protocol decisions that maximize efficiency and workflow without sacrificing performance in the nursery.

Figure 1: Average daily gain pre-wean to 4 week post-placement.

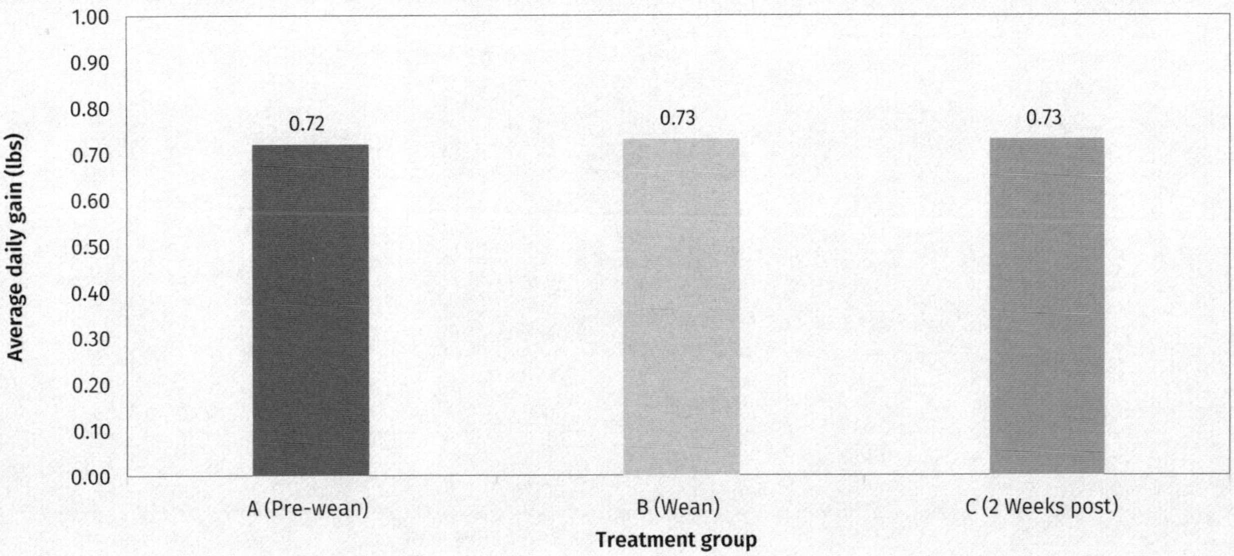

Figure 2: Weight over time by treatment group.

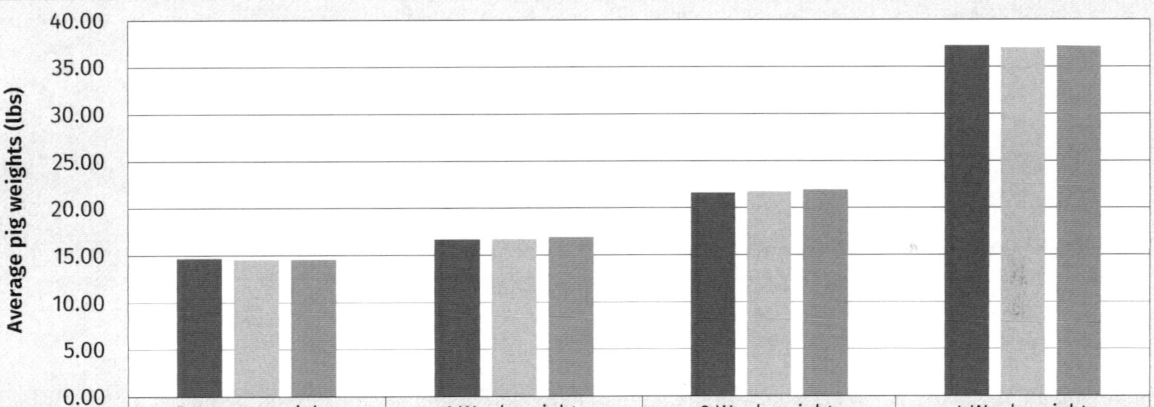

	Pre-wean weight	1 Week weight	2 Week weight	4 Week weight
A (Pre-wean)	14.70	16.67	21.53	37.19
B (Wean)	14.59	16.69	21.69	36.98
C (2 weeks post)	14.60	16.89	21.94	37.22

Figure 3: Average daily gain by treatment group weight subgroups.

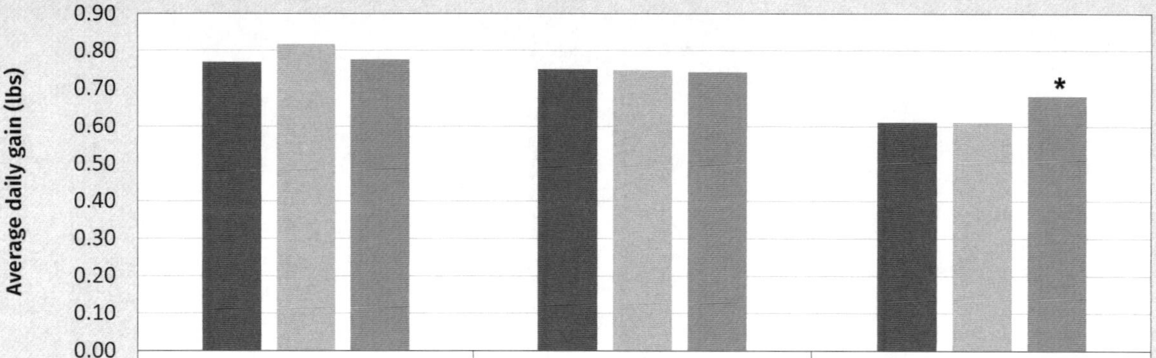

	Large	Medium	Small
A (Pre-wean)	0.77	0.75	0.61
B (Wean)	0.82	0.75	0.61
C (2 Weeks Post)	0.78	0.74	0.68 *

Treatment group subgroups

* Denotes statistical difference *P* = 0.006.

Table 1: Least squares means and *P*-values for metrics at 4-weeks post-weaning

Metric	Pre-wean	Wean	Post-wean	*P*-value
Pig weight (overall)	36.95 lbs	37.27 lbs	37.36 lbs	.613
Average daily gain (overall)	0.72 lbs/day	0.73 lbs/day	0.73 lbs/day	.613
Pig weight (light 25%, < 12.19)	29.20 lbs[b]	29.25b	31.27 lbs[a]	.006
Average daily gain (light 25%, < 12.19 lbs)	0.61 lbs/day[b]	0.61 lbs/dayb	0.68 lbs/day[a]	.006

Table 2: Least square means and performance data between treatment groups

Metric	Pre-wean	Wean	Post-wean
Number of pigs	345	355	365
Enrollment weight	14.70 lbs	14.59 lbs	14.60 lbs
Weight week 1 post-placement	16.67 lbs	16.69 lbs	16.89 lbs
Weight week 2 post-placement	21.53 lbs	21.69 lbs	21.94 lbs
Weight week 4 post-placement	36.95 lbs	37.27 lbs	37.36 lbs
Average daily gain (overall)	0.72 lbs/day	0.73 lbs/day	0.73 lbs/day
Full value pig %	88.04%	89.76%	91.79%

Figure 4: Percentage of full value pigs by treatment group.

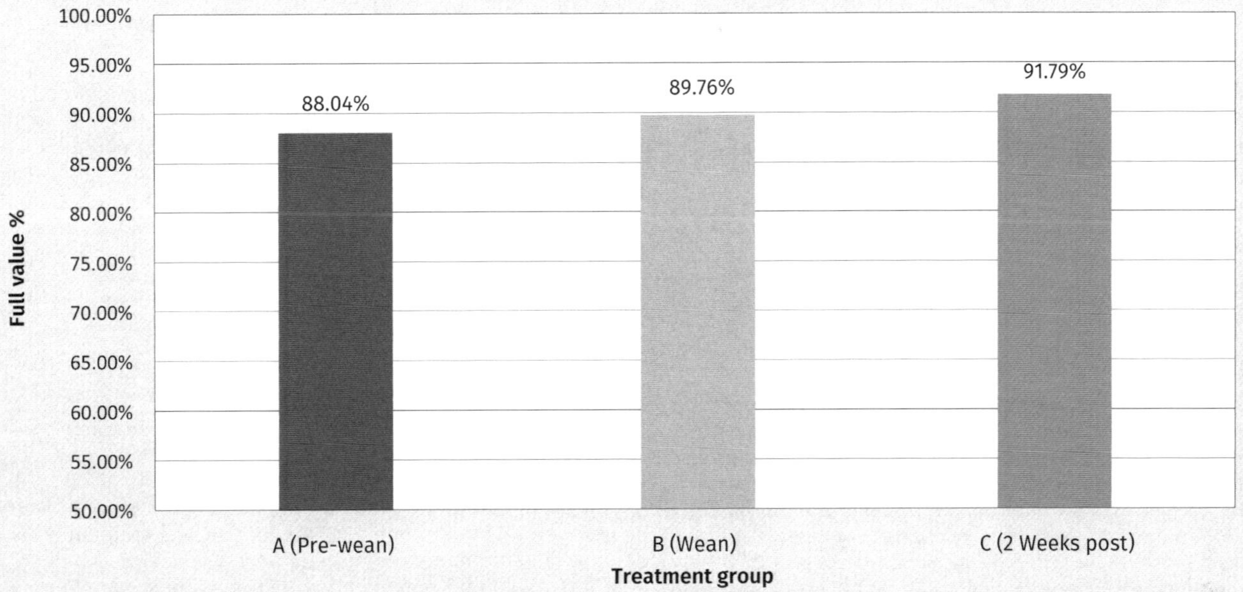

References

1. Augustyniak A, Pomorska-Mól M. Vaccination Failures in Pigs-The Impact of Chosen Factors on the Immunisation Efficacy. Vaccines (Basel). 2023 Jan 19;11(2):230. **doi: 10.3390/vaccines11020230. PMID: 36851108**

2. Holtkamp D, Clavijo M. New AASV Herd Classification Guidelines for PRRSV and *Mycoplasma hyopneumoniae*. ISU James D. McKean Swine Conference. November 4, 2021.

https://doi.org/10.54846/am2025/33

Antimicrobial susceptibility of *Actinobacillus pleuropneumoniae, Bordetella bronchiseptica, Pasteurella multocida* and *Streptococcus suis* isolated from diseased pigs in the United States and Canada, 2013 to 2023

Michael T. Sweeney
Zoetis, Veterinary Medicine Research and Development, Kalamazoo, Michigan

Introduction

Antimicrobial agents are critical to treat, control, and prevent disease in swine and other food animals. Responsible and timely antibiotic intervention is vital in controlling and mitigating disease incidence and spread, such as in swine respiratory disease (SRD) complex, which can endanger herd health and a sustainable food supply resulting in economic and commercial loss.[1] Of all the diseases that affect growing and finishing pigs, SRD is the most economically important as it is highly prevalent among indoor production facilities and can be difficult to treat and control. The treatment and control of swine respiratory disease requires an understanding of the complexities and interaction between the organisms that are present as well as management of the environment in which the pigs are raised in.[2] Primary pathogens for SRD complex may include *Mycoplasma hyopneumoniae, Actinobacillus pleuropneumoniae,* and *Bordetella bronchiseptica,* as well as viral agents. Common secondary pathogens include *Pasteurella multocida, Streptococcus suis, Glaesserella parasuis, Actinobacillus suis,* and *Salmonella choleraesuis.* These primary and secondary multi-etiologic pathogens act together to increase the severity and duration of SRD.[3]

Antimicrobial surveillance among veterinary bacterial pathogens obtained from clinical specimens provides a platform from which to detect emergence of resistance in animal populations. For the past 25 years, Zoetis has maintained a continuous surveillance program that monitors the antimicrobial susceptibility of antimicrobial agents against pathogens that cause SRD. In this presentation we report the percentages of *A pleuropneumoniae, B bronchiseptica, P multocida,* and *S suis* pathogens isolated from swine in the United States and Canada that were susceptible to the veterinary antimicrobial agents ampicillin, ceftiofur, danofloxacin, enrofloxacin, florfenicol, penicillin, tetracycline, tilmicosin, trimethoprim-sulfamethoxazole, and tulathromycin for a ten-year period from 2013 to 2023.

Methods

Veterinary diagnostic laboratories from the United States and Canada participated in this surveillance study and all *A pleuropneumoniae, B bronchiseptica, P multocida* and *S suis* isolates were recovered from diseased or dead pigs. Laboratories selected isolates based on their own protocols and were requested not to use antimicrobial susceptibility as a criterion for selection. Laboratories were also requested to submit no more than eight isolates per quarter year to prevent over-representation from any one geographic area. Each participating laboratory was also requested to send no more than one isolate of each bacterial species from a herd each quarter year to prevent the over-representation of bacterial clones from one region.[4-6]

In vitro susceptibility data were generated by performing minimal inhibitory concentration (MIC) testing at a central laboratory (Microbial Research Inc, Fort Collins, Colorado) which followed Clinical and Laboratory Standards Institute (CLSI) standardized methods and quality control guidelines during susceptibility testing.[7] The MIC values for all isolates were determined using a dehydrated broth microdilution system (Sensititre System; Thermo Fisher Scientific, Waltham, Massachusetts) which conforms to CLSI standards for testing of veterinary pathogens.[7]

Results

A pleuropneumoniae susceptibility to ampicillin was high to moderate over the study period from 2013 to 2023 (78.9% to 97.9%) and there was low susceptibility to tetracycline (8.6% to 29.2%). For tilmicosin, susceptibility from 2013 to 2023 ranged from 89.9% to 100% and was similar for enrofloxacin (94.7% to 100%).

A pleuropneumoniae isolates were 100% susceptible to ceftiofur, tulathromycin and florfenicol from 2013 to 2023.

For *B bronchiseptica* isolates in this study, clinical breakpoints are only available for ampicillin, tulathromycin and florfenicol. From 2013 to 2023, *B bronchiseptica* isolates were 100% resistant to ampicillin, while susceptibility to tulathromycin was high (99.1% to 100%) and was high to moderate for florfenicol (83.5% to 100%).

P multocida susceptibility to ampicillin from 2013 to 2023 was very high (97.5% to 100%) while the percentage of susceptibility to ceftiofur was 100%. *P multocida* susceptibility to enrofloxacin was very high (99.4% to 100%) and to florfenicol as well (98.9% to 100%). Susceptibility to penicillin (97.6% to 100%), tilmicosin (98.4% to 100%), and tulathromycin (98.6% to 100%) were also very high.

S suis susceptibility to ampicillin was very high from 2013 to 2023 (97.6% to 100%) while the percentage of susceptibility to florfenicol was also high (98.4% to 100%). Additionally, susceptibility was high for ceftiofur (92.8% to 98.6%) as well as enrofloxacin (95.3% to 98.9%). Penicillin susceptibility for *S suis* was moderate (78.7% to 88.2%) and was low for tetracycline (0% to 3.6%).

Discussion

Antimicrobial surveillance programs support antibiotic stewardship principles which require that all antibiotic prescribers (animal and human) assure good prescribing decisions that mitigate the emergence of resistance to preserve the effectiveness of antibiotics for veterinary and human medicine. Additionally, selecting the proper course of antimicrobial treatment for an animal, whether it's over the counter, prescribed, or through the Veterinary Feed Directive, should correlate with the Animal Medicinal Drug Use Clarification Act (AMDUCA).

The Zoetis SRD surveillance program has continuously obtained swine pathogens for over 25 years from veterinary diagnostic laboratories in North America that have then been tested for antimicrobial susceptibility. The purpose for this ongoing surveillance study is to summarize the antimicrobial susceptibility profiles of isolates from four different pathogenic bacterial species associated with SRD collected from laboratories in the United States and Canada over a 10-year period from 2013 to 2023.

The results of this surveillance study when using standardized susceptibility testing methods show high percentages of antimicrobial susceptibility among the major respiratory tract pathogens isolated from swine across the United States and Canada. This surveillance study continues to be useful in identifying the development of antimicrobial resistance among SRD target pathogens which is crucial for the prudent use of antimicrobial agents in veterinary medicine. Additionally, understanding the in vitro susceptibility of SRD pathogens isolated in the United States and Canada continues to be an important component of antimicrobial stewardship and "One Health".

Management practices used in modern pig farming such as manure management, age-segregation of pigs, and nutritional and metabolic awareness have profound influences on microbial interactions which may result in decreased disease among swine.[8] The high levels of antimicrobial susceptibility observed in this study and others may be attributed to specific health management practices within swine herds such as the "all-in, all-out" management practice system. Another management practice that may be contributing to overall high antimicrobial susceptibility rates is multi-site production where contained groups of pigs spend their production life in different facilities appropriately designed for each age group (site I: Breeding herd; site II: Nursery; site III: Finishing, all of which are located at separate geographical locations to minimize disease transmission). Future studies may be able to determine if these management practices influence antibiotic resistance changes over time, and if resistance reduction can be achieved through alternations in further enhanced housing and cleaning practices.

Conclusions

- The major SRD pathogens *A pleuropneumoniae, P multocida, S suis* and *B bronchiseptica* continue to show high rates of susceptibility towards key antimicrobial agents approved for the treatment of SRD in the United States and Canada.
- The lowest rates of susceptibility were seen for tetracycline against *A pleuropneumoniae, P multocida,* and *S suis,* and for ampicillin and florfenicol against *B bronchiseptica.*
- Continuous monitoring of antimicrobial susceptibility among swine pathogens provides up-to-date information about susceptibility trends for commonly used antimicrobial agents and is an important component of responsible use and antimicrobial stewardship for "One Health".

References

1. The National Pork Board. Pork Checkoff. Accessed June 25, 2021. **https://porkcheckoff.org/pork-production-management/swine-health/antibiotics/**

2. Global Ag Media. The Pig Site. Disease Guide Pneumonia. Accessed June 26, 2021. **https://www.thepigsite.com/disease-guide/pneumonia**

3. MacInnes JI, Gottschalk M, Lone AG, Metcalf DS, Ojha S, Rosendal T, Watson SB, Friendship RM. Prevalence of *Actinobacillus pleuropneumoniae, Actinobacillus suis, Haemophilus parasuis, Pasteurella multocida* and *Streptococcus suis* in representative Ontario swine herds. *Can J Vet Res.* 2008;72:242–248.

4. Portis E, Lindeman C, Johansen L, Stoltman G. Antimicrobial susceptibility of porcine Pasteurella multocida, Streptococcus suis, and Actinobacillus pleuropneumoniae from the United States and Canada, 2001 to 2010. *J Swine Health and Prod.* 2013;21:30–41.

5. Sweeney MT, Lindeman C, Johansen L, Mullins L, Murray R, Senn M, Bade, D, Machin C, Kotarski S, Tiwari R, Watts J. Antimicrobial susceptibility of *Actinobacillus pleuropneumoniae, Pasteurella multocida, Streptococcus suis,* and *Bordetella bronchiseptica* isolated from pigs in the United States and Canada, 2011 to 2015. *J Swine Health Prod.* 2017;25(3):106-120.

6. Sweeney MT, Gunnett LA, Kumar DM, Lunt BL, Galina Pantoja L, Bade D, Machin C. Antimicrobial susceptibility of *Actinobacillus pleuropneumoniae, Bordetella bronchiseptica, Pasteurella multocida,* and *Streptococcus suis* isolated from diseased pigs in the United States and Canada, 2016 to 2020. *J Swine Health Prod.* 2022;30(3):130-144.

7. Clinical and Laboratory Standards Institute (CLSI). CLSI Standard VET01-A5. Performance Standards for Antimicrobial Disk and Dilution Susceptibility Tests for Bacteria Isolated from Animals. Approved Standard. 5th ed. Wayne, Pennsylvania: Clinical and Laboratory Standards Institute; 2018.

8. Noyes N, Davies P, Nerem J, Dee S. Microbes, pigs and antimicrobial resistance. National Hog Farmer. September 8, 2020. Accessed June 22, 2021. **https://www.nationalhogfarmer.com/animal-health/microbes-pigs-and-antimicrobial-resistance**

https://doi.org/10.54846/am2025/34

CircoMatch: An immunoinformatics tool to predict and compare T cell epitopes across field viruses and vaccine strains

D. L. Foss[1], DVM, PhD; J. Angulo[2], DVM, MS; A. Aldaz[2], DVM, MBA; M. Bandrick[1], DVM, PhD; A. H. Gutiérrez[3], PhD; B. Pavao[3]; S. Monitz[3]; R. Nolan[3]; M. Ardito[3], F. E. Terry[3], MPH; A. S. De Groot[3], MD; W. D. Martin[3]
[1]Zoetis Inc., Kalamazoo, Michigan; [2]Zoetis Inc., Parsippany, New Jersey; [3]EpiVax, Inc., Providence, Rhode Island

Introduction

The mammalian immune system recognizes pathogens and vaccines in three general ways. These are innate recognition of pathogen associated molecular patterns (PAMPs), and adaptive immune recognition by B cells or T cells. Innate immune recognition is via pathogen recognition receptors (PRR) that have evolved to recognize PAMPs, initiate innate immunity and to provide the immunological context for the development of the adaptive immune response. The adaptive immune response includes both B cells and T cell that develop their antigen-specific receptors (BCR and TCR) via genomic rearrangement, cellular selection, and education. This process can be thought of as evolution of the immune system in real time. Signals from the innate immune system (cytokines and cell surface molecules) play a central role in driving the appropriate development of adaptive immunity, both B and T cell immunity.[1]

Immune recognition and effector mechanisms of B and T cells are different. B cells (and the immunoglobulins they produce) recognize antigens as three-dimensional shapes and these B cell-produced antibodies function to neutralize and opsonize pathogens. In contrast, T cells recognize viral antigens as small peptides (T cell epitopes) presented to them in the context of MHC molecules and are involved in recognizing and killing viral infected cells (cytotoxic T cells or CTL), as well as providing help to antibody-mediated immunity (helper T cells or Th cells).[1]

The appropriate or optimal protective immunity is pathogen specific. For porcine circovirus 2 (PCV2), T cell immunity is of particular importance for several reasons. PCV2 is a relatively slow, latent virus that disseminates within monocytes and replicates in activated lymphocytes, thus CTL are important for removing PCV2-infected cells. PCV2 has a high rate of mutation but is considered to have only one serotype.[2] This suggests that immune pressure on PCV2 may select for viruses with variation in T cell epitopes, rather than antibody-mediated selection. PCV2 also has specific immune suppression mechanisms,[3,4] several of which specifically suppress T cell-mediated response, suggesting an important role of T cells in PCV2 immunity. Expansion of T cells also provides immune memory. T cell responses to conserved T cell epitopes can provide protection across strains, and the T cell responses is less sensitive to MDA interference.[5] While anti-PCV2 antibodies likely play a role in PCV2 immunity, PCV2 vaccines can be efficacious, even in the absence of seroconversion.[6,7]

The rational design of vaccines must account for all three types of immune system recognition. The innate immune system interacts with non-antigen-specific components of vaccines including various PAMPs. Furthermore, many adjuvants interact with the innate immune system resulting in various cellular signals shaping the adaptive immune response. Vaccine design with respect to the B cell adaptive immunity is readily addressed by considering pathogen serotypes and by conducting neutralization assays and serological assays. Serology assays have been the standard experimental marker for vaccine efficacy throughout the history of vaccinology. In contrast, measuring T cell immunity is a relative challenge for vaccinologists. Recent decades have witnessed tremendous advances in measuring T cell immunity, including ex vivo methods such as ELISpot, flow cytometry, cellular killing assays, and others. None-the-less, application of these assays to an out bred population of animals like swine, and in larger studies as needed to demonstrate vaccine efficacy remains a significant challenge. Since T cells recognize viral peptides in the context of MHC molecules (known as SLA in swine), that presentation of antigens to T cells is said to be "restricted" by SLA type. SLA type (or a set of SLA alleles known as a haplotype) is a genomically encoded set of molecules that can be used to predict the specific viral peptides that can be presented to T cells. The ability to predict this T cell epitope (peptide) presentation can thus be used to predict the specific T cells that a given vaccine will generate and how these specific T cells are relevant to circulating PCV2 strains.

Recent advances in immunoinformatics have enabled the development of tools to predict T cell epitopes and compare them between vaccines and field strains.[8] Rapid and inexpensive sequencing technologies have facilitated knowledge of SLA alleles present in the commercial swine herd, and a large number of PCV2 viruses circulating in global swine herd. Computational prediction of peptide binding to SLA alleles has been described for several swine viruses including influenza A virus-swine (IAV-s) and PCV2. One of the best characterized methods for T cell epitope prediction in swine (PigMatrix) was developed by EpiVax and is used for CircoMatch.[9] CircoMatch uses PigMatrix based on 11 SLA class I and 5 SLA class II alleles to predict and compare T cell epitopes between vaccines and field strains. This system allows the comparison of large numbers of virus strains with several vaccines and was first validated experimentally with IAV-s.[10]

CircoMatch is a web-based tool that can be used to compare vaccine T cell epitopes with those of field strains. Results are expressed as an Epitope Content Comparison score (EpiCC) score. The comparison of many field strains with several vaccines can then be visualized in a radar plot of either EpiCC scores or as a T cell epitope coverage. Coverage refers to the percent of putative T cell epitopes that match between the vaccine and field strain. This work presents a set of 537 PCV2 sequences from North America, in relation to a bivalent PCV2a/PCV2b vaccine and two monovalent PCV2a vaccines.

Methods

A total of 1272 PCV2 capsid sequences were submitted to CircoMatch for analysis (as of October 2024) and included sequences from 2017 to 2024. Sequences were received as either full length PCV2 sequence or orf2 (capsid) sequence. The orf 2 sequence was identified and translated for the CircoMatch analysis. Sequences with ambiguous or missing nucleotides or with premature stop codons were edited to obtain full length capsid sequence. All editing of sequences is documented in the CircoMatch database. PCV2 sequences came from a total of 21 countries, including the Americas (Canada, United States, Mexico, Colombia, Brazil), Europe (Italy, Latvia, The Netherlands, Poland, Portugal, Spain, Belgium, Germany, Denmark, France), and Asia (Japan, Australia, Philippines, Taiwan, Thailand, Vietnam). Vaccine sequences were current commercial PCV2 vaccines. Results summarized in Figures 1 and 2 and Table 1 represent 537 sequences from North America.

The T cell epitope analysis of the PCV2 sequences was done as previously described.[11] Briefly, the identification of cell epitopes, based on predicted binding to SLA class I and class II alleles using the EpiVax PigMatrix, and included 11 class I alleles (SLA-1*01:01, 1*04:01, 1*08:01, 1*12:01, 1*13:01, 2*01:01, 2*04:01, 2*05:01, 2*12:01, 3*04:01, 3*05:01, 3*06:01, 3*07:01) and 5 SLA class II alleles (SLA-DRB1*02:01, 04:02, 06:02, 07:01, and 10:01). Epitope Content Comparison (EpiCC) scores were then determined for each of the sequences compared to each of the vaccines. Epitopes that were duplicated in the two components of the bivalent vaccine were not given extra score. The EpiCC analysis did not account for antigen dose in the vaccines. A baseline score was calculated for each field sequence by comparing it to itself. EpiCC scores were converted to percent T cell epitope coverage by dividing the EpiCC score of a sequence to its baseline score. Coverage represents the percent of the field isolate T cell epitope content for which the vaccine will likely induce memory T cells.

Results

The distribution of PCV2 genotypes is shown in Table 2 and Figure 3. Globally, 67.5% of the PCV2 sequences were PCV2d, with PCV2a, PCV2b and PCV2e making up 19.8%, 11.6% and 1.1% of the total, respectively. Among 537 sequence from North America (United States, 448; Canada, 83; Mexico, 6) the percent of PCV2a was higher (38.8% versus 19.9%) and the percent of PCV2d was lower (53.4% versus 67.5%) than the global distribution. While there are geographical and temporal differences in the ratio of PCV2 genotypes, PCV2d continues to be the most prevalent genotype.

The EpiCC scores for the North American PCV2 sequences are summarized in Table 2. The EpiCC baseline score represents the relative number of predicted T cell epitopes present in each of the four genotypes. The baseline EpiCC score for PCV2e is higher (12.16) than the other genotypes (10.41-10.79). The EpiCC scores for the bivalent PCV2a/PCV2b vaccine is higher for all genotypes than the monovalent PCV2a vaccines. The EpiCC scores can also be expressed as percent T cell epitope coverage. The largest advantage of the bivalent vaccine over the monovalent vaccines are for PCV2b and PCV2d strains with increases of 39.5-47.9% and 27.1-33.7% for PCV2b and PCV2d, respectively. All vaccines had low EpiCC scores for PCV2e, with less than 50% T cell epitope coverage. The individual percent T cell epitope coverages for three vaccines and 537 PCV2 sequences are shown as a radar plot (Figure 1). The highest variation in T cell epitope coverage is seen with PCV2a where all current vaccines have a T cell epitope coverage of 80-90% for most all PCV2a strains. The bivalent vaccine (containing PCV2b) has the highest T cell epitope coverage for PCV2b. The PCV2b component also provides greater T cell epitope coverage of PCV2d strains (average of 80.39%) compared with the monovalent PCV2a vaccines (60.15-63.24%) (Table 1 and Figure 1).

The overall T cell epitope coverage of the bivalent PCV2a/PCV2b vaccine for PCV2d field strains (80.39 ± 1.68) is similar to the T cell epitope coverage of the monovalent PCV2a vaccines for PCV2a field strains (83.20 ± 1.68), while the T cell epitope coverage for PCV2a monovalent vaccines for PCV2d field strains is markedly lower (61.70 ± 2.20) (Figure 2).

Discussion

PCV2d is the predominant PCV2 genotype among CircoMatch sequences. However, there continues to be significant number of PCV2a and PCV2b strains. The relative prevalence of PCV2 genotypes does vary between geographical regions, for example with 38.0% of strains being PCV2a in North America, while globally 19.8% of strains are PCV2a. In contrast, PCV2b sequences are more prevalent globally (11.6%) compared to North America (6.3%). The prevalence of PCV2e remains quite low, 1.1% globally. The collection of sequences for CircoMatch depends on the voluntary submission of sequences from veterinarians and diagnostic labs, so there may be bias toward selecting sequences that are associated with clinical disease, or that are not well controlled by current vaccines. Could the lower T cell epitope coverage of PCV2a vaccines for PCV2d field strains provide an opportunity for the increasing prevalence of PCV2d in the global swine population? The integration of vaccination history, clinical outcomes and strain sequences enabled by CircoMatch may help to address this question.

The matching of vaccine T cell epitopes to field strains may be particularly relevant for pathogens such as PCV2 where T cells play a key role in vaccine efficacy, however there is not a specific EpiCC score cutoff that defines vaccine efficacy. Determinants of vaccine efficacy are complex, however better matched vaccines have been shown to improve multiple parameters of vaccine efficacy. The average percent coverage of T cell epitopes of all vaccines for PCV2a sequences was 83.20%. Since these vaccines are generally considered efficacious it is expected that this level of T cell epitope coverage is adequate for vaccine efficacy. The bivalent vaccine (PCV2a/PCV2b) has an epitope coverage for PCV2d strains of 80.39% compared to only 61.70% for monovalent PCV2a vaccines. The bivalent vaccine containing PCV2b, has a much better coverage of PCV2d due to the closer sequence relationship of PCV2b to PCV2d and to the additional T cell epitopes provided by the PCV2a fraction. In a study of heterologous versus homologous PCV2 vaccine efficacy, a 36.46% increase in T cell epitope coverage gave biologically superior protection.[12] Considering the North American sequences discussed here, the PCV2a/PCV2b vaccine improved T cell epitope coverage by 30.4% compared to the monovalent PCV2a vaccines. An increase expected to be relevant to vaccine efficacy.

CircoMatch was originally conceived to compare T cell epitope content of vaccines with field strains as an aid to vaccine selection. In addition, the CircoMatch data base contains much data beyond sequences, including vaccination history, clinical signs, location and date of collection. We expect this data base to be useful to study PCV2 epidemiology and evolution, early detection of new variant strains and for diagnostic investigations.

Acknowledgments

The authors acknowledge and thank the many farmers, veterinarians and diagnostic laboratories for the collection, submission, and sequencing of PCV2 field strains.

Figure 1: T cell epitope coverage by three vaccines of 537 North American PCV2 capsid sequences arranged by genotype.

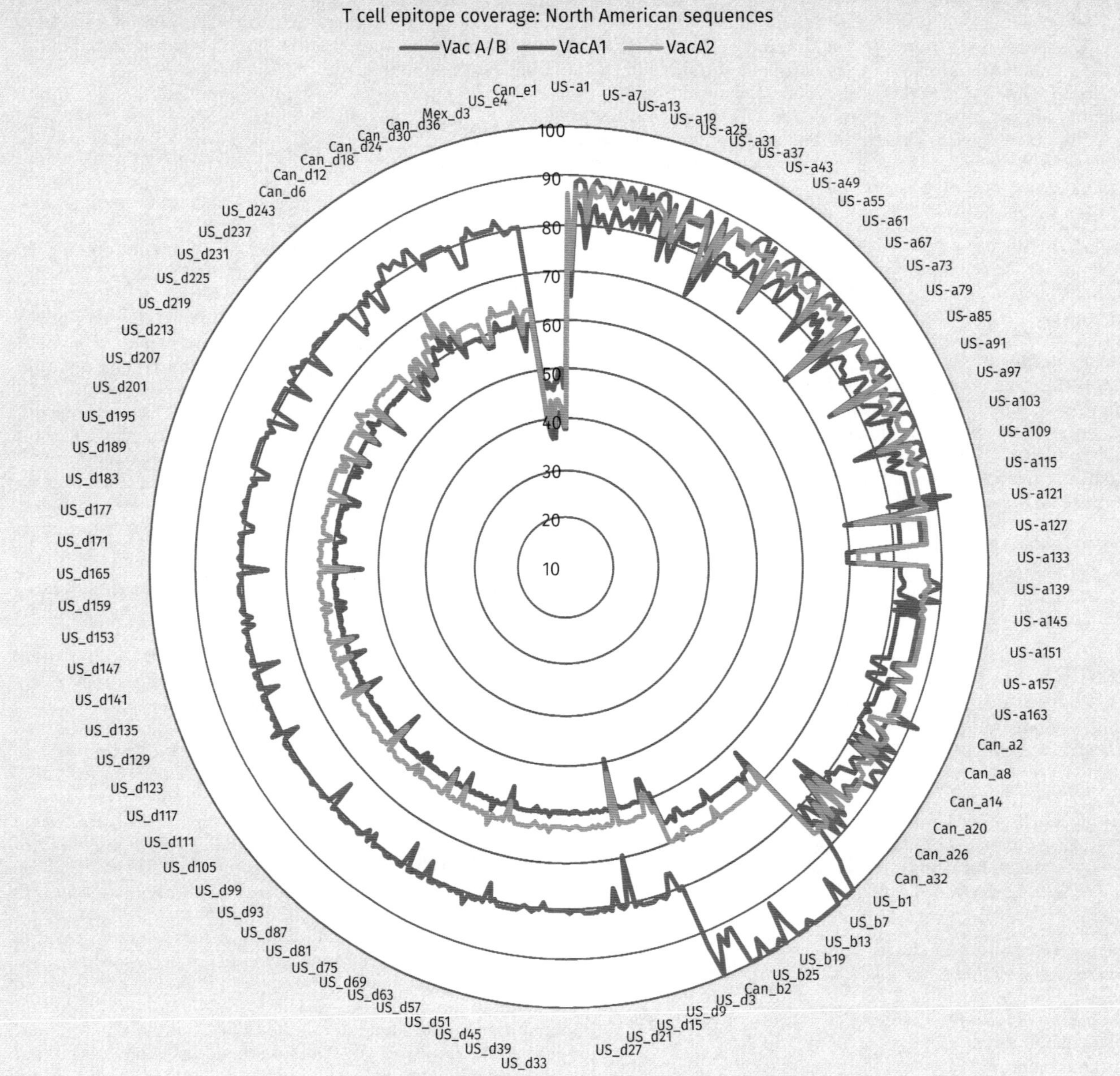

Figure 2: T cell epitope coverage of PCV2a/PCV2b bivalant vaccine and monovalent PCV2a vaccines relative to homologous and heterologous genotype sequences.

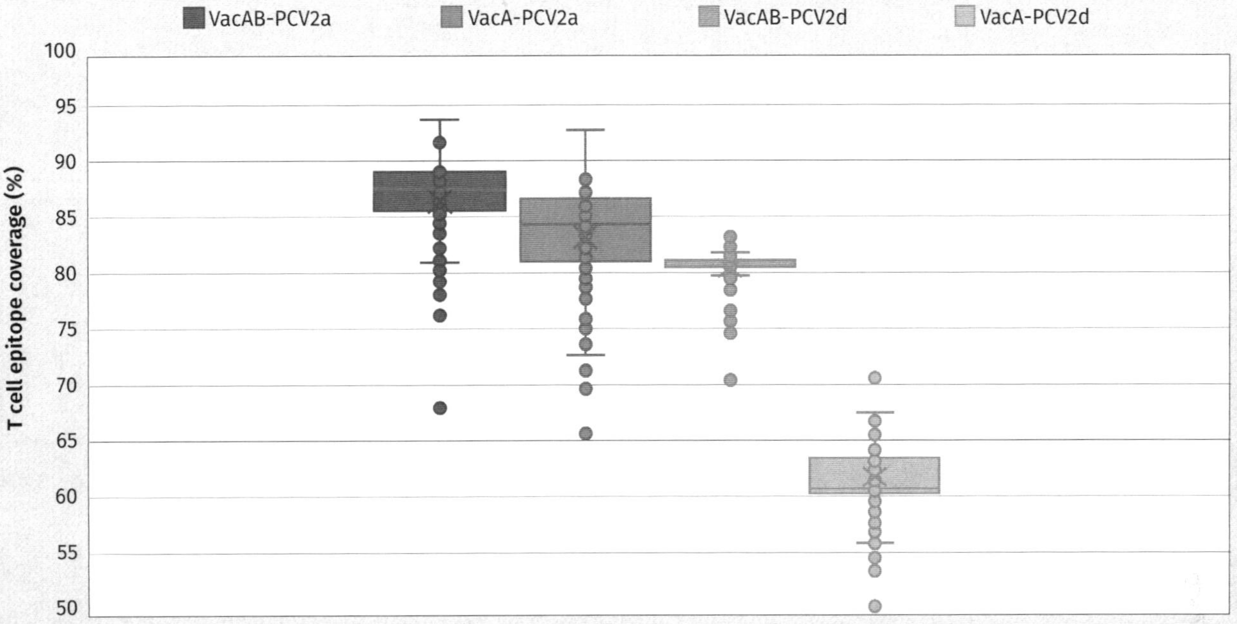

Table 1: Summary of EpiCC scores, percent T cell epitope coverage of three vaccines relative to 537 PCV2 sequences

Genotype	Number of isolates	EpiCC baseline	EpiCC score			T cell epitope coverage (%)			Vac AB increase in coverage (%)	
			VacAB	VacA1	VacA2	VacAB	VacA1	VacA2	VacA1	VacA2
All	537	10.61	8.81	7.23	7.60	83.07	68.06	71.56	24.0	17.9
PCV2a	204	10.77	9.31	8.74	9.18	86.35	81.16	85.24	6.5	1.4
PCV2b	34	10.79	10.58	7.15	7.58	98.04	66.28	70.26	47.9	39.5
PCV2d	287	10.41	8.37	6.26	6.58	80.39	60.15	63.24	33.7	27.1
PCV2e	12	12.16	5.86	4.79	5.07	48.18	39.41	41.70	22.4	15.8

Table 2: Distribution of PCV2 genotypes in CircoMatch (October 2024)

Source	Number of PCV2 sequences (percent)				
	PCV2a	PCV2b	PCV2d	PCV2e	Total
USA	167 (37.3)	29 (6.5)	243 (54.2)	9 (2.0)	448 (100)
Canada	36 (43.4)	5 (6.2)	39 (47.0)	3 (3.6)	83 (100)
Mexico	1 (16.7)	0 (0)	5 (83.3)	0 (0)	6 (100)
North America	204 (38.0)	34 (6.3)	287 (53.4)	12 (2.2)	537 (100)
Global (All)	252 (19.8)	148 (11.6)	858 (67.5)	14 (1.1)	1272 (100)

Figure 3: Distribution of PCV2 genotypes in CircoMatch. A) All sequences in CircoMatch; B) North American sequences (USA, Canada, Mexico).

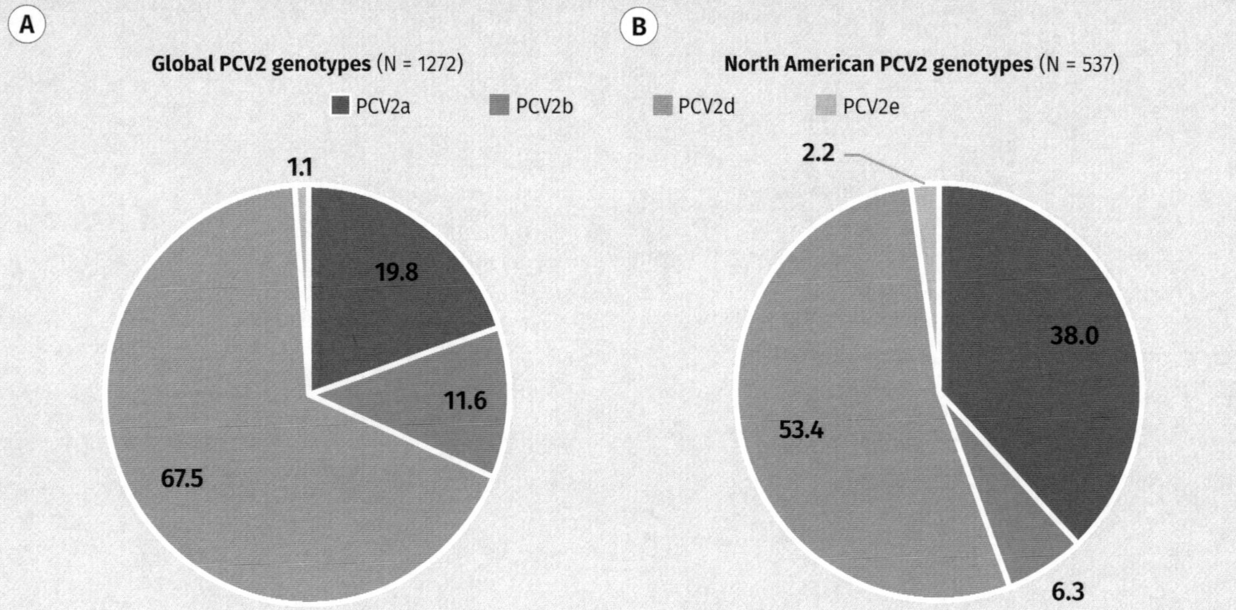

References

1. Punt J, Strandford S, Jones P, Owen OJ. *Kuby Immunology.* 2019; New York, W. H. Freeman and Company.

2. Franzo G, Segalés J. Porcine Circovirus 2 genotypes, immunity and vaccines: multiple genotypes but one single serotype. *Pathogens* 2020;9(12).

3. Shi R, Hou L, Liu J. Host immune response to infection with porcine circoviruses. *Animal Diseases* 2021;1(1):1-10.

4. Qian D, XingChen W, TongTong W, et al. Porcine circovirus type 2 suppresses IL-12p40 induction via capsid/gC1qR-mediated microRNAs and signalings. *Journal of Immunology* 2018;201(2):533-547.

5. Orije MRP, Maertens K, Corbière V, et al. The effect of maternal antibodies on the cellular immune response after infant vaccination: A review. *Vaccine* 2020;38(1):20-28.

6. Fort M, Sibila M, Pérez-Martín E, et al. One dose of a porcine circovirus 2 (PCV2) sub-unit vaccine administered to 3-week-old conventional piglets elicits cell-mediated immunity and significantly reduces PCV2 viremia in an experimental model. *Vaccine*; 2009 27(30):4031-4037.

7. Fenaux M, Opriessnig, Halbur PG, et al. A chimeric porcine circovirus (PCV) with the immunogenic capsid gene of the pathogenic PCV type 2 (PCV2) cloned into the genomic backbone of the nonpathogenic PCV1 induces protective immunity against PCV2 infection in pigs. *Journal of Virology* 2004;78(12):6297-6303.

8. Moise L, Gutiérrez A, Kibria F, et al. iVAX: An integrated toolkit for the selection and optimization of antigens and the design of epitope-driven vaccines. *Human vaccines & Immunotherapeutics* 2015;11(9):2312-2321.

9. Gutiérrez AH, Martin WD, Bailey-Kellogg C, et al. Development and validation of an epitope prediction tool for swine (PigMatrix) based on the pocket profile method. *BMC Bioinformatics* 2015;16(1)1-13.

10. Gutiérrez AH, Loving C, Moise L, et al. In vivo validation of predicted and conserved T cell epitopes in a swine influenza model. *PloS one* 2016;11(7).

11. Foss DL, Gutiérrez AH, M. Bandrick M, et al. Comparison of predicted T cell epitopes in porcine circovirus type 2 isolates from 2017 to 2021 and selected vaccines (EpiCC analysis) confirms the global relevance of a bivalent vaccine approach. *Veterinary Vaccine* 2023;2(1-10).

12. Bandrick M, Balasch M, Heinz A, et al. A bivalent porcine circovirus type 2 (PCV2), PCV2a-PCV2b, vaccine offers biologically superior protection compared to monovalent PCV2 vaccines. *BMC veterinary research* 2022;53(12).

https://doi.org/10.54846/am2025/35

Application of health-focused nursery feeding strategies to support pigs facing natural porcine reproductive and respiratory syndrome virus infection

B. N. Smith, DVM, PhD; M. McGhee, MS, PhD; W. Schweer, MS, PhD; A. Elsbernd, MS; C. González-Vega, MS, PhD
Cargill Animal Nutrition, Minneapolis, Minnesota

Introduction

Pathogenic disease continues to pose strong production challenges to the North American pork industry, contributing significantly to reduced growth performance and increased financial losses due to mortality. Among common pathogens, porcine reproductive and respiratory syndrome virus (PRRSV) remains one of the most negatively impactful, costing the US swine industry a recently estimated $1.2 billion in lost production annually.[1] Its wide distribution throughout the US herd and unique characteristics as a viral pathogen have made its prevention and treatment difficult in commercial production.[2] For that reason, there is interest in approaches to provide additional support to pigs facing PRRSV infection, including nutrition. Nutritional support of PRRSV infection has been investigated by numerous researchers, with a wide range of focuses including nutrient and ingredient composition of the diet, utilization of bioactive compounds, and other nutrition-related factors, that have demonstrated varying degrees of success.[3-7] Aside from PRRSV-specific nutritional interventions, there is also the potential for nutritional interventions developed to address the support of other disease presentations such as systemic bacterial infections like those caused by *Streptococcus suis* (*S suis*) or gastrointestinal disease to be beneficial. For that reason, the objective of this trial was to determine the effects of feeding four nursery feeding programs designed for different nursery health statuses on the growth performance and clinical response of pigs to a naturally occurring PRRSV infection.

Methods

At weaning, 2,304 pigs (initial body weight = 12.3 ± 0.08 lb; DNA genetics) were randomly allotted to one of four experimental treatments represented by 18 replicate pens each (32 pigs/pen, mixed sex) across two rooms. Each treatment received one of four nursery diet strategies: 1) Control, a standard nursery dietary program suited for a moderate enteric health challenge; 2) Cost strategy, a diet program aimed to reduce feed cost per gain and not specific to a disease presentation; 3) PRRSV strategy, an experimental program aimed to support pigs during active PRRSV infection; and 4) Systemic (SYS) strategy, a diet program aimed to support pigs during primary systemic *S suis* bacterial challenge. All diets were formulated to meet or exceed NRC dietary recommendations, excluding the SYS strategy which incorporated a restriction of standardized ileal digestible (SID) lysine of approximately 20% relative to the control treatment. Treatments were provided as a mash diet and were applied in phases 1 and 2 only (approximately 21 days), with all pigs receiving a common phase 3 diet through the end of the nursery period. The source population for pigs utilized in this trial were from a PRRSV+/*Mycoplasma hyopneumoniae*+ sow farm. Positive PRRSV status was confirmed on pen-based oral fluids; in addition to PRRSV-infection, there was observational evidence of mild to moderate enteric disease during the early nursery period. The use of in-feed and water-soluble medications were not restricted during this trial and left to the discretion of the herd veterinarian. Pig and feed weights were recorded by pen at the beginning and end of each phase for the calculation of growth performance metrics (average daily gain [ADG], average daily feed intake [ADFI], feed conversion ratio [FCR]). Mortality events were recorded by pen. Incidence of scouring was assessed and recorded on a pen basis on day 3, 7, 10, 14, and 20 using a binomial scoring system (no or yes for scouring present within a pen). Data were analyzed using the lme4, ImerTest, and forestinventory packages of R 4.1.2, and the statistical models included the fixed effect of treatment, as well as the random effects of barn and location within barn. Mortality and scouring incidence were analyzed as a probabilistic generalized mixed model with a binomial distribution. Tendencies were detected at $.05 < P < .10$ and significance was detected at $P < .05$. Least square means separation was performed using pairwise comparisons with Tukey's adjustment.

Results

Pigs were confirmed PRRSV positive post-weaning by pen oral fluids; however, the severity of challenge was mild with overall mortality at 1.74%. Recorded reasons for deaths included respiratory, intestinal, bacterial/systemic, chronic/small, and unknown. Although mortality was low overall, the SYS strategy resulted in numerically greater mortality compared to the other non-control dietary interventions (SYS strategy = 2.43% vs cost strategy and PRRSV strategy = 1.56%) indicating the SYS diet strategy was not well suited for this clinical presentation (Table 1).

Differences in average body weight were not detected ($P > 0.10$) on day 21 (conclusion of phase 2) or day 42 (conclusion of trial), but the control strategy and SYS strategy resulted in final average body weights of 58.45 and 57.41 lbs., respectively, representing the high and low range of average body weights among treatments. Growth performance variables (ADG, ADFI, and FCR) did not differ among treatments for the overall period, however, phase 1 and 2 growth performance did reflect differences among dietary interventions. The SYS strategy incorporated a 20% SID lysine restriction relative to the control treatment; thus, as expected, ADG during the first 2 phases was reduced ($P < .05$) for this treatment compared to the control. Average daily feed intake in the first 2 combined phases was greater ($P < .05$) for the control strategy compared to the cost strategy, with the PRRSV and SYS strategies intermediate. The PRRSV strategy increased ($P < .05$) ADFI in phase 1 compared to the other treatments, but the effect was not sustained in phase 2. The control strategy resulted in greater ($P < .05$) ADFI than the cost strategy and the SYS strategy in phase 1, and greater intake ($P < .05$) was also observed in phase 2 for

Table 1: Effects of dietary treatment applied during the first two dietary phases on growth performance and mortality of pigs facing naturally established porcine reproductive and respiratory syndrome virus (PRRSV) infection

		1 Control	2 Cost	3 PRRSV	4 SYS	SEM	*P*-value
Body weight (lbs)	Day 1	12.26	12.27	12.26	12.27	0.18	1.000
	Day 21	28.01	27.53	27.46	26.81	1.05	.166
	Day 42	58.45	57.84	58.05	57.41	1.20	.648
ADG (lbs)	Phase 1+2	0.76[b]	0.73[ab]	0.73[ab]	0.70[a]	0.042	.042
	Overall	1.11	1.09	1.10	1.08	0.022	.501
ADFI (lbs)	Phase 1+2	0.98[b]	0.90[a]	0.94[ab]	0.93[ab]	0.035	.003
	Overall	1.53	1.48	1.51	1.50	0.031	.195
Feed/gain	Phase 1+2	1.29[ab]	1.23[b]	1.29[ab]	1.34[a]	–	.014
	Overall	1.37	1.36	1.37	1.38	–	.533
Cost/gain ($/lb.)	Phase 1+2	0.47[b]	0.43[a]	0.50[c]	0.50[c]	0.010	< .001
	Overall	0.32[b]	0.31[a]	0.33[b]	0.33[b]	0.004	< .001
Scouring (probability, %)	Day 3	7.95[a]	56.49[b]	56.49[b]	42.78[ab]	17.84	.033
	Day 7	69.87	81.57	91.64	56.92	20.69	.128
	Day 10	4.30	19.33	43.45	25.00	18.85	.089
	Day 14	5.56	22.22	22.22	11.11	9.80	.459
Mortality (%)	Overall	1.39	1.56	1.56	2.43	0.642	.533

Abbreviations: SYS, systemic; SEM, standard error of mean; ADG, average daily gain; ADFI, average daily feed intake

that treatment compared to others. The cost strategy resulted in the lowest ($P < .05$) FCR in phase 1 and 2, whereas the SYS strategy resulted in the poorest FCR. By design, the cost strategy had the lowest ($P < .05$) cost per unit of gain in both phase 1 and 2 and overall, which was driven by superior FCR and reduced diet cost per ton. In contrast, the PRRSV strategy and SYS strategy had greater ($P < .05$) cost per unit of gain compared with the control and cost strategies in the first two phases.

The incidence of diarrhea was greatest at day 7 for all treatments (on average across treatments, 75% of pens had observed scouring), but by day 20, there was no longer any scouring observed. The control strategy resulted in the lowest incidence of scouring (day 3, $P < .05$; day 10, $P < .10$), while the cost strategy and PRRSV strategy had the greatest incidence of diarrhea at most time points.

Discussion

Although the level of health challenge was well-controlled in this population of pigs as evidenced by the 1.74% closeout mortality, pigs were indeed facing a combination of low-level pathogens: pigs were confirmed PRRSV+ at weaning and 3 weeks post-wean; moderate enteric disease was noted by veterinary observation and supported by the record of 75% of pens with diarrhea on day 7; and underlying *S suis* was a known challenge for this flow and required administration of Amoxicillin and Equisul via the water during the study period. Cumulatively when assessing numeric and statistically relevant outcomes, the control strategy resulted in the most favorable clinical outcomes among treatments, including having the lowest mortality and probability of scouring combined with the greatest phase 1 + 2 ADFI and greatest final body weight.

The outcomes for the cost strategy were in line with expectations. The treatment was designed to reduce diet cost per ton and thereby incorporated fewer feed additive technologies and a lower investment in nutritional adjustments for health support. The treatment resulted in improved FCR in phase 2, which contributed to the lowest cost per unit of gain observed in the first two phase and overall; however, the diet did not result in the highest feed consumption immediately post-weaning, which should be a primary goal of early pig nutrition. The diet resulted in favorable economic outcomes in the trial conditions but had there been stronger enteric disease pressure or more active PRRSV infection, it is hypothesized that the dietary treatment may result in less favorable health and growth outcomes with the absence of some nutritional health-supporting technologies.

The PRRSV strategy incorporated dietary modifications or utilization of ingredients that have been demonstrated to be beneficial under PRRSV-infection including increased utilization of soybean meal[3,8] and bovine spray-dried plasma,[9] increased structural fiber inclusion, increased inclusion of vitamins, and increased arginine[10] in the diet. Phase 1 feed intake was greatest for this treatment, which was a goal of the intervention given the reduction in intake expected during an acute PRRSV break. The treatment did not negatively impact growth performance overall; however, it did result in the greatest cost per gain. The nutritional adjustments may have elicited greater benefit during an acute PRRSV infection compared to the mild challenge present in the current study, but more research is needed to support this conjecture. Because of the challenge to deploy reactionary nutritional interventions in the face of a new PRRSV break, more research should be conducted under both acute and mild PRRSV infection to further elucidate economical nutritional interventions to support pigs during active PRRSV and not greatly increase diet cost if deployed during low-challenge periods as well.

The nutritional intervention applied in the SYS strategy has resulted in reduced mortality on average by 1% in five nursery research trials conducted in commercial conditions with pigs undergoing primary *S suis* infection; however, adverse impacts on pig survival and growth have been observed when the intervention was applied under non-primary *S suis* conditions.[11] Although herd health history and observational evidence of *S suis* was observed in the present study, it was not the only pathogen in play. A hypothesis for why this strategy is beneficial under primary *S suis* challenge is the impact this pathogen can have on total inflammation and oxidative stress the animal experiences[12-14]; by decreasing the contributions from metabolic processes to this cumulative burden through restriction of an essential nutrient, the total stress the animal experiences is less.[15] Restricting lysine in the present study did not improve livability, which further reinforces prior data that this nutritional adjustment is likely only beneficial under specific conditions. Since the degree of pathogenic challenge in this study was low and not driven by primary *S suis* challenge, the restriction had opportunity to more negatively impact growth in this population of pigs.

In summary, during a mild PRRSV challenge with concurrent moderate enteric challenge, the control diet resulted in the lowest incidence of scouring, reflecting the benefits of the enteric-driven formulation. As expected, the cost strategy resulted in the lowest cost of gain. This study highlights the importance of matching health-focused nutritional programs to the appropriate challenge severity.

References

1. Holtkamp, D., and Osemeke, O. 2024. "PRRS Now Costs Pork Producers $1.2 Billion per Year." National Hog Farmer, July 30, 2024.

2. Zimmerman, J. W, Dee, S. A., Holtkamp, D. J., Murtaugh, M. P., Stadejek, T., Stevenson, G. W., Torremorell, M., Yang, H., ad Zhang, J. (2019) Porcine Reproductive and Respiratory Syndrome Virus (Porcine Arteriviruses). In Zimmerman, J. Y., Karriker, L. A., Ramirez, A., Schwartz, K. J., Stevenson, G. W., and Zhang, J. (Eds.), *Diseases of Swine* (11th ed., pp 685-708). John Wiley & Sons.

3. Boyd, R. D., Johnston, M. E., and Zier-rush, C. 2010. Soybean meal level modulates the adverse effect of high immune stress on growth and feed efficiency in growing pigs. Proc. MN Nutr. Conf. 71: 167-174.

4. Bromm, J. J., Tokach, M. D., Woodworth, J. C., Goodband, R. D., DeRouchey, J. M., Flohr, J. R., Schmitt, R. A. M., Gebhardt, J. T., and Zarate, F. 2023. Effects of Increasing Alpha-Linolenic Acid on Growth Performance and Mortality Rate in PRRS-Virus Challenged Nursery Pigs. Kansas Agricultural Experiment Station Research Reports: Vol. 9: Iss. 7. **https://doi.org/10.4148/2378-5977.8510**

5. Smith, B. N., Morris, A., Oelschlager, M.L., Connor, J., and Dilger, R.N. (2019). Effects of dietary soy isoflavones and soy protein source on response of weanling pigs to porcine reproductive and respiratory syndrome viral infection. Journal of Animal Science. **doi: 10.1093/jas/skz135**

6. Smith, B.N., Oelschlager, M.L., Abdul Rasheed, M.S., and Dilger, R.D. (2020). Dietary soy isoflavones reduce pathogen-related mortality in growing pigs under porcine reproductive and respiratory syndrome viral challenge. Journal of Animal Science. **doi: 10.1093/jas/skaa024**

7. Little, E. A., Dunkelberger, J., Hanson, D., Eggert, J., Gonda, M. G., MacNeil, M. D., and Dee, S. 2024. Comparison of differences in performance whose sires were identified using different selection strategies after experimental infection with PRRSV. Trans. Anim. Sci. 8: txae128. **doi: 10.1093/tas/txae128**

8. Rochell, S. J., Alexander, L. S., Rocha, G. C., Van Alstine, W. G., Boyd, R. D., Pettigrew, J. E., and Dilger, R. N. 2015. Effects of dietary soybean meal concentration on growth and immune response of pigs infected with porcine reproductive and respiratory syndrome virus. Journal of Animal Science. 93(6): 2987-2997. **doi: 10.2527/jas.2014-8462**

9. Campbell, J. M., and Crenshaw, J. P. 2016. Orally fed spray dried plasma modulated immune response during respiratory challenges: A review. Journal of Animal Science. 94(3): 45-47. **doi: 10.2527/jas.2015-9844**

10. Che, L., Yang, P., fang, Z., Lin, Y., and Wu, D. 2013. Effects of dietary arginine supplementation on reproductive performance and immunity of sows. Czech Journal of Animal Science. 58(4): 167-175. **doi: 10.17221/6711-cjas**

11. Smith, B.N., Gonzalez-Vega, C., McGhee, M., Oeschlager, M., and Price, K. 2024. Nutritional approach to mitigation of clinical Streptococcus suis in the nursery: Learnings and considerations for application. AASV Annual Meeting 2024. [Oral Presentation].

12. Neila-Ibanez, C., Brogaard, L., Pailler-Garcia, L., Martinez, J., Segalas, J., Segura, M., Heegard, P.M., Aragon, V. Piglet innate immune response to *Streptococcus suis* colonization is modulated by the virulence of the strain. *Vet. Res.* 2021;52(145): 1-15. **doi: 10.1186/s13567-021-01013-w**

13. Segura, M., Aragon, V., Brockmeier, S. L., Gebhart, C., de Greeff, A., Kerdsin, A., O'Dea, M. A., Okura, M., Salery, M., Schultz, C., Valentin-Weigard, P., Weinert, L. A., Wells, J. M., Gottschalk, M. Update on *Streptococcus suis* Research and Prevention in the Era of Antimicrobial Resistance: 4th International Workshop on *S. suis* [Conference Report]. *Pathogens.* 2020;9(374): 1-39. **doi: 10.3390/pathogens9050374**

14. Segura, M., Calzas, C., Grenier, D., Gottschalk, M. Initial steps of the pathogenesis of the infection caused by *Streptococcus suis*: fighting against nonspecific defenses. *FEBS.* 2016;590: 3772-3799. **doi: 10.1002/1873-3468.12364**

15. Prunier, A., Heinonen, M., Quesnel, H. High physiological demands in intensively raised pigs: impact on health and welfare. *Animal,* 2010;4(6): 886-898. **doi: 10.1017/S175173111000008X**

https://doi.org/10.54846/am2025/36

Beneficial effects of using a water-dispersible organic acid / monoglyceride blend product on water quality and pig performance in a commercial nursery research barn

M. Whitney, PhD; Y. Chen, PhD; R. Wolfenden, PhD
Animal Nutrition, Eastman Chemical Company, Kingsport, Tennessee

This study was conducted to evaluate the effect of providing a water-dispersible organic acid and monoglyceride blend (Eastman Protaq™ Bond 50) in the drinking water on water quality and pig performance in a commercial nursery research barn. A total of 2,317 pigs were visually sorted upon entry by size, and even numbers of small, medium and large pigs were placed in pens. Pigs visually estimated to be < 4.5 kg were placed in "small" pens (n = 4). All other pens (n = 58) were randomly assigned to one of two treatment groups: (1) CON, standard water, pH = 7.6 and (2) PRO, same as CON except Protaq™ Bond 50 was added to stock solution and run through the water medicator to achieve a final dosage in water of 0.5 mL/L, targeting a water pH of 3.5. Pen-level growth performance (by phase) and removal rates were measured, and economics calculated for the 35d trial period. Pen scour scores were evaluated on a scale of 0-4 on days 4, 10, and 18. Growth performance and economics data were analyzed in R statistical software using a linear mixed model with treatment group and size as fixed effects and room as a random effect. Pen served as the experimental unit. Water samples were collected from CON and PRO on d22 (4 each) and analyzed for microbial levels.

Pig health was excellent throughout this study, regardless of treatment, with low mortality (< 1%) and very little diarrhea observed, resulting in no differences between groups ($P > .10$). Providing Protaq™ Bond 50 in the water numerically reduced bacteria (< 40 vs 6853 CFU/mL), coliform bacteria (59 vs 2420 CFU/100 mL), and *E Coli* (< 1 vs 156 CFU/100 mL) levels compared to untreated water, respectively (Table 1). Water consumption averaged 2.04 l/p/d for CON and 1.89 l/p/d for PRO. Growth performance data and economics are presented in Table 2. PRO pigs had a greater average daily feed intake (ADFI) ($P = .05$) and average daily gain (ADG) ($P = .01$) during phase 1 (d 0-10) compared to CON pigs. This difference continued in phase 2 (d 10-18), with PRO pigs having greater ADFI ($P = .01$) and ADG ($P < .05$). During the final phase 3 period (d 18-35), ADG, ADFI and feed to gain were similar between treatments ($P > .10$). Overall, for the entire 35d feeding period, ADFI was increased ($P = .05$) and ADG tended to be greater ($P < .10$) in PRO vs CON pigs, resulting in a significant increase in body weight (BW) of 0.21 kg/p at d 18 ($P < .05$) that was maintained at d 35 (0.24 kg/p BW, $P < .10$). Economics were estimated, with revenue of gain ($/p) tending to be greater in PRO vs CON pigs ($0.43, $P < .10$). Overall, income over feed cost was similar between treatments ($P > .10$).

Although animal health was excellent throughout this study, providing organic acids and short- and medium-chain monoglycerides in the water increased feed intake and weight gain the first half of the study, resulting in an increase in body weight at the end of the nursery period of 0.24 kg/p. Protaq™ Bond 50 is an effective and economic water acidification tool that enhances health and performance in nursery pigs during the stressful post-weaning period.

Table 1: Microbial analysis of water samples (collected d 22)

Treatment	Description	Bacteria (CFU/mL)	Fungi (CFU/mL)	Coliform bacteria (CFU/100 mL)	*E coli* (CFU/100 mL)
Control (CON)	Room 1a	8860	50	> 2420	411
CON	Room 1b	7400	40	> 2420	161
CON	Room 2a	4210	30	> 2420	20
CON	Room 2b	6940	20	> 2420	31
CON	**Average**	**6853**	**35**	**> 2420**	**156**
Protaq (PRO)	Room 1a	80	70	65	1
PRO	Room 1b	< 10	10	19	< 1
PRO	Room 2a	20	10	81	1
PRO	Room 2b	50	10	70	1
PRO	**Average**	**< 40**	**25**	**59**	**< 1**

Table 2: Pen-level comparison of water acidifier in the nursery (n = 2,317 pigs)

	Control (CON) (n = 28 pens)	Protaq (PRO) (n = 30 pens)	SEM	*P*-value
Bodyweight, kg				
d 0	4.77	4.78	0.02	.80
d 10	6.59	6.67	0.06	.08
d 18	9.62	9.84	0.13	.02
d 35	18.48	18.71	0.22	.08
Phase 1 (d 0-10)				
ADG, kg/d	0.17	0.19	0.005	.01
ADFI, kg/d	0.17	0.18	0.005	.05
F/G	0.97	0.95	0.02	.16
Phase 2 (d 10-18)				
ADG, kg/d	0.37	0.39	0.014	.04
ADFI, kg/d	0.42	0.44	0.009	.01
F/G	1.14	1.13	0.02	.90
Phase 3 (d 18-35)				
ADG, kg/d	0.51	0.51	0.005	.93
ADFI, kg/d	0.74	0.75	0.018	.31
F/G	1.45	1.47	0.01	.26
Overall (d 0-35)				
ADG, kg/d	0.38	0.39	0.005	.08
ADFI, kg/d	0.50	0.51	0.009	.05
F/G	1.31	1.31	0.01	.59
Treatments, %	2.38	2.80	0.44	.42
Mortality, %	0.64	0.67	0.23	.88
Fallbacks, %	1.57	1.54	0.35	.93
Economics				
Protaq cost, $/p	0	0.19	–	–
Feed cost, $/p	7.31	7.47	0.15	.02
Feed cost, $/kg gain	0.114	0.114	0.001	.72
Gain revenue, $/p[1]	21.81	22.24	0.34	.08
Income over cost, $/p	14.50	14.58	0.22	.69

[1] Value of gain = $1.65/kg to calculate revenue from weight gain.

ADG = average daily gain; ADFI = average daily feed intake; F/G = feed to gain

https://doi.org/10.54846/am2025/37

ProVent® ECL: A comprehensive probiotic to support swine challenged with multidrug-resistant *E coli*

K. R. Leistikow[1], PhD; K. Pederson[1], BS; J. Lee[2], MS; E. A. Galbraith[1], MS
[1]Microbial Discovery Group, Oak Creek, Wisconsin;
[2]United Animal Health, Sheridan, Indiana

Introduction

Multidrug-resistant (MDR) *Escherichia coli* is a growing concern for the swine industry, threatening animal health, food security, and economic sustainability. Antibiotic resistance limits treatment options for diseases like colibacillosis, characterized by diarrhea and increased mortality, while resistance mechanisms continue to evolve faster than new drugs can be developed. This challenge underscores the need for alternative, sustainable approaches to help manage MDR pathogens. Probiotics, beneficial microorganisms capable of modulating gut health, have emerged as promising tools for controlling MDR *E coli*. Beyond traditional antimicrobial production, probiotics employ diverse mechanisms to inhibit pathogen proliferation and mitigate virulence. These include direct pathogen inhibition, biofilm disruption, quorum sensing interference, and immune system modulation. In this study, we evaluated ProVent® ECL, a swine-derived probiotic formulation, for its ability to inhibit MDR *E coli* virulence using a multifaceted approach.

Methods

ProVent® ECL, a probiotic formulation comprised of strains isolated from of the gastrointestinal tract of high-performing swine, was tested against MDR *E coli* isolates obtained from veterinary diagnostic laboratories across the United States. ProVent® ECL strains were assessed for antagonistic activity against MDR *E coli* strains using cross streak and broth microdilution assays.[1] Biofilm inhibition and disruption were assessed using microtiter plate assays and crystal violet staining methods.[2] Immunomodulatory activity was evaluated by exposing intestinal epithelial cells to *E coli* lipopolysaccharide (LPS) with or without probiotics and measuring immune marker expression using RT-qPCR. Statistical analyses were performed using GraphPad Prism 10.1.2. Differences where $P < .05$ were considered significant. Differences in biofilm inhibition and disruption were analyzed using a 2-tailed unpaired t-test (2 groups), or one-way ANOVA, followed by Tukey's multiple comparisons (≥ 3 groups). Differential gene expression was analyzed using one-way ANOVA, followed by Tukey's multiple comparisons using log2 transformed data.

Results

ProVent® ECL effectively inhibited the growth of 79 MDR *E coli* isolates collected from diagnostic labs across the US. It also reduced biofilm formation (up to 90%) and disrupted pre-existing biofilms (up to 25%) in *E coli* strains with high biofilm-forming potential. Additionally, cell culture data demonstrated that ProVent® ECL reduced pro-inflammatory cytokine expression, particularly TNFα, in a dose-dependent manner in epithelial cells exposed to an *E coli* lipopolysaccharide (LPS) challenge. These findings suggest that specific probiotic combinations, applied at optimal concentrations, may prime the pig's innate immune system and mitigate the severity of *E coli*-associated diseases.

Discussion

These findings highlight the potential of probiotic cocktails like ProVent® ECL to mitigate MDR *E coli*-associated disease through a variety of direct and indirect mechanisms. By directly inhibiting bacterial growth and disrupting biofilm formation, ProVent® ECL targets multiple facets of *E coli* pathogenesis. Additionally, its ability to stimulate the host immune response represents an indirect yet critical mechanism of disease mitigation. These results underscore the potential of probiotics to be used as part of a comprehensive disease management strategy, offering a sustainable, non-antibiotic approach to reduce the severity of enteric bacterial infections. Importantly, leveraging probiotics strategically can also support antibiotic stewardship efforts in swine production by reducing reliance on conventional antimicrobials. Future work should explore how probiotics can complement other interventions, including antibiotic, vaccination, and improved husbandry practices. Together, these approaches can help safeguard swine production while addressing the growing challenge of antimicrobial resistance.

References

1. Balouiri, M., Sadiki, M. & Ibnsouda, S. K. Methods for in vitro evaluating antimicrobial activity: A review. J. Pharm. Anal. 6, 71–79 (2016).

2. O'Toole, G. A. Microtiter Dish Biofilm Formation Assay. JoVE J. Vis. Exp. e2437 (2011) **doi:10.3791/2437**

https://doi.org/10.54846/am2025/38

A *Bacillus*-based product, ProVent® ECL, improved piglet recovery after a porcine reproductive and respiratory syndrome (PRRS) wild-type break

C. L. Puls[1], PhD; M. McCallum[1], BS; K. McCormick[1], MS; J. D. Spencer[1], PhD; J. Lee[1], MS; E. Galbraith[2], MS; B. Arnold[1], BS
[1]United Animal Health, Sheridan, Indiana; [2]Microbial Discovery Group, Oak Creek, Wisconsin

Introduction

Recent estimates report that porcine reproductive and respiratory syndrome virus (PRRS) costs the US swine industry in excess of 1.2B USD.[1] Clinical signs often manifest with increased abortions, low piglet viability, increased pre-wean mortality, and poor reproductive performance in sow herds. Downstream, increased mortality, reduced weight gain, and poorer feed conversion is often observed. Strain virulence and pathogenicity varies greatly and impacts system and herd severity and profitability. Furthermore, many PRRS infections occur in swine with one or more bacterial pathogens, a state known as co-infection. Common industry practices for mitigating the effects of PRRS often include multiple interventions, including herd closures, use of killed or modified-live vaccines, increased use of antibiotics, and reduced pig movements. While the use of nutritional products is not often first of mind, there is evidence that feeding microbial based products may help mitigate PRRS symptoms and reduce the time to reach pre-PRRS introduction baseline production values. One such direct fed microbial is ProVent® ECL (United Animal Health, Sheridan, IN), a *Bacillus*-based product containing multiple strains shown to be immunostimulatory. Zuckerman et al[2] reported feeding of ProVent® ECL during a PRRS co-infection challenge model reduced rate of lung colonization, eliminated and reduced presence of PRRS in the lung, and reduced severity of lung pathology. Use of ProVent® ECL may prove beneficial in improving sow herd and downstream nursery performance during a PRRS wild-type break and subsequent recovery period.

Methods

A 1,000 head research sow farm (Sheridan, IN) experienced a Lineage 1A 1-7-4 wild-type PRRS introduction in December 2023. Reduced sow lactation feed intakes and increased abortions were observed prior to diagnostic confirmation. Onset of symptoms were observed approximately 14 days post-farrowing. Gilts on farm are developed onsite from weaning. A total of 216 sows are batch farrowed every 5 weeks. Farrowing rooms are equipped with a computerized feed system that allows for individual and daily sow feed intakes. Gestation rooms are all-in-all-out by farrowing group, allowing for cleaning and disinfection between groups. All sows and piglets are tagged with radio-frequency identification (RFID) tags corresponding to the LeeO individual animal identification system. A timeline of key events associated with the PRRS wild-type introduction are presented in Table 1. Tilmicosin was started in gestation feed at 363 g/ton and in lactation feed at 181 g/ton within 2 weeks of clinical sign onset. Concurrently, all piglets received a tulathromycin injection at processing and at weaning. At the discretion of the herd veterinarian, whole herd disease homogenization was accomplished with administration of live virus inoculation approximately 3-4 weeks later.

Table 1: Schedule of PRRS break and interventions

Event	Date
Onset of clinical signs	December 22, 2023
Tilmicosin started in gestation/ lactation feed	December 28, 2023
Live virus inoculation	January 15, 2024
ProVent® ECL started in gestation feed	February 2, 2024
Farrowed	March 25, 2024
Weaned	April 15, 2024

Due to the unique opportunity of a wild-type PRRS introduction on a research sow farm, a controlled research study using a randomized complete block design was then conducted to evaluate the effectiveness of ProVent® ECL at mitigating PRRS symptoms in sows and progeny. Sows were randomly allocated on the basis of parity and body weight to one of two dietary treatments: 1) Antibiotic Control containing 363 and 181 g/ton Tilmicosin in gestation and lactation feed, respectively – no direct fed microbial vs 2) Antibiotic Control + ProVent® ECL administered in feed at 2 lb/ton. There were 112 sows allocated to each dietary treatment. Experimental diet feeding began three weeks after herd live virus inoculation at approximately 65 days gestation and was fed through lactation. Sows were weighed at entry into the farrowing room, at weaning, and at subsequent breeding. Piglets were weighed individually within 24 hours of birth and at weaning. Cross-fostering was allowed within dietary treatment and was completed within 24 to 48 hours of birth to match teat count. Two samples of processing fluids were collected from piglets and showed PRRS Ct values of 26.2 and 26.9, respectively. Lactation length was approximately 20 days. Approximately 4 days prior to weaning, 30 piglets were randomly selected from litters and bled to determine piglet viremia. Samples (6 pools of 5 pigs) were submitted to Iowa State University Veterinary Diagnostic Laboratory for PRRS viral load and showed PRRS circulation with Ct ranges of 24.4 to 33.4 with one of the pools testing negative for PRRS (> 37).

At weaning, piglets (n = 1,392) were given an injection of tulathromycin and moved to a research nursery and placed into pens of 12 on the basis of sow treatment and body weight. Nursery dietary treatments mirrored sow dietary treatments and there were 58 pen replicates per dietary treatment during the nursery growth performance evaluation. Pigs were weighed at weaning and at nursery exit 42 days later. All feed additions and feed remaining in the feeder at the time of pig weighing was recorded and used to calculate daily feed intake and feed conversion. All morbid animals or mortalities were weighed and recorded during the nursery period.

Results

Of the 112 sows/treatment originally allocated to study, 111 controls and 108 ProVent® ECL-fed sows farrowed. Largely, piglet birth weight, stillborns, and pre-wean mortality had already returned to pre-PRRS break baseline values. The percentage of mummified piglets was still 3-4x greater than pre-PRRS wild-type introduction baseline values.

There was no difference ($P > .05$) between Control and ProVent® ECL-fed treatments for most sow farrowing and lactation metrics. Interestingly, pigs from ProVent® ECL-fed sows tended to be lighter at birth (0.14 lb difference; $P = .06$) and were lighter at weaning compared to controls (15.2 vs 14.5 lb, respectively; $P = .01$).

Nursery growth performance data are presented in Table 2. Despite lighter wean weight, there was no difference in final nursery exit weight (47.3 vs 46.9 lbs, respectively; $P = .42$). ProVent® ECL-fed pigs had improved feed conversion in the nursery period (1.47 vs 1.45, respectively; $P = .01$). Additionally, mortality was reduced 3 percentage points for ProVent® ECL-fed pigs compared to Control-fed pigs (5.3 vs 2.3%; $P = .002$). Due to the dramatic difference in mortality, further analysis was conducted to evaluate mortality adjusted growth performance (ie, closeout performance). ProVent® ECL-fed pigs had greater pen BW gain (+14.6 lb/pen; $P = .04$), tended to have greater ADG (1.3%; $P = .10$), and had improved feed conversion (3.3%; $P < .001$) compared to Control-fed pigs.

Discussion

The lack of responses in sows fed ProVent® ECL was somewhat surprising and not as expected. One realistic explanation to consider is the inherent variability commonly observed in PRRS wild-type infections. There were multiple sows that farrowed few born alive pigs in both treatment groups which could have influenced results. Nonetheless, while ProVent® ECL has been documented to improve sow performance metrics in commercial sow environments,[3-5] results from this study differed which could, in part, be due to the health changes and challenges associated with a wild-type introduction with a virulent strain of PRRS.

However, progeny nursery performance and incidence of mortality was improved for ProVent® ECL-fed pigs compared to controls. These results validate previous research showing ProVent® ECL can improve performance of nursery pigs during a PRRS challenge.[2] The difference in mortality is particularly interesting as downstream mortality following a wild-type PRRS introduction is often elevated for long periods of time during the recovery period. These results show that ProVent® ECL can significantly reduce offspring nursery mortality and improve growth performance even when other common interventions to mitigate effects of PRRS are already being used.

References

1. Onyekachukwu, H. O., C. Corzo, M. Kikuti, X. Yue, S. Vadnais, G. Silva, D. Linhares, and D. Holtkamp. 2024. Updates on the economic impact of PRRSV to US pork producers. Page 174 in Proceedings 2024 Allen D. Leman Swine Conference. St. Paul. MN.

2. Zuckerman, F. A., R. Husmann, W. Chen, P. Roady, J. Pfeiff, K. R. Leistikow, M. Duersteler, S. Son. M. R.King, and N. R. Augspurger. 2022. Bacillus-based direct-fed microbial reduces the pathogenic synergy of a coinfection with Salmonella enterica serovar choleraesuis and porcine reproductive and respiratory syndrome virus. Infection and Immunity. 90(4):e00574-21. **https://doi.org/10.1128/iai.00574-21**

3. Hensch, M., J. Miller, J. Spencer, and J. Lee. 2021. Optimizing gut health in the face of enteric endemics. Pages 140-145 in Proceedings 2022 AASV Annual Meeting.

4. Mason, B. 2022. Using ProVent® ECL to help solve an enteric health challenge within a sow unit. Pages 173-176 in Proceedings 2022 AASV Annual Meeting.

5. Specht, T., M. Heimerl, and B. Heimerl. 2020. ProVent® ECL, a unique bacillus-based direct-fed microbial (DFM), as a tool to improve wean-to-first service interval distribution and sow productivity. Pages 119-123 in Proceedings 2022 AASV Annual Meeting.

Table 2: Effects of ProVent® ECL on nursery pig performance

Item	Treatments		SEM	*P*-value
	Control	ProVent® ECL		
BW, lb				
Day 0 (weaning)	15.3	14.5	0.30	< .001
Day 42	47.3	46.9	0.71	.42
Overall ADG, lb	0.76	0.77	0.012	.37
Overall ADFI, lb	1.11	1.11	0.016	.92
Overall F:G	1.47	1.45	0.007	.01
Nursery mortality, %	5.3	2.3	-	.002
Closeout performance				
Pen BW gain, lb/pen	368.9	383.5	6.38	.04
ADG, lb	0.76	0.77	0.011	.10
F:G	1.51	1.46	0.010	< .001

BW = body weight; ADG = average daily gain; ADFI = average daily feed intake; F:G = feed to gain

https://doi.org/10.54846/am2025/39

Breeding better mothers for a free farrowing system

C. Lipori[1], MSc; I. Reimert[1], PhD; J. R. Dunkelberger[2,3], PhD; E. F. Knol[2], PhD; I. M. van As[4], PhD; N. M. Soede[1], PhD
[1]Wageningen University & Research, The Netherlands; [2]Topigs Norsvin, Den Bosch, The Netherlands;
[3]Topigs Norsvin USA, Bloomington, Minnesota; [4]Denkavit Netherlands B.V., The Netherlands

Introduction

Society insists on welfare reform in livestock production. Providing housing that is more consistent with an animal's "natural" environment is perceived as better animal welfare. Therefore, producers are exploring alternative means of housing pigs, such as replacing farrowing cates with open farrowing pens. Providing sows with extra space during farrowing and lactation is intended to enhance sow welfare, but may compromise piglet welfare. In fact, results from a previous study show that the risk of piglet crushing was 14% higher for piglets farrowed in a free farrowing vs crated system.[1] Therefore, novel management strategies are needed to optimize production and sow vs piglet welfare in these types of housing systems. The objective of this study was to evaluate genetic selection for mothering ability as a potential strategy to improve welfare and production in free farrowing systems.

Materials and methods

Animals

Topigs Norsvin Large White x Landrace sows (N = 56) were used for this study. Sows were group-housed during gestation and transported to the farrowing unit at ~109 days of gestation. In the farrowing unit, half of the sows were assigned to the free farrowing (FF) (N = 28) group, and the remaining half to the crated (C) group (N = 28) where parity structure was balanced across groups. Sows in the FF group were housed in pens with an area of 8 m^2 and sows in the C group were housed in crates from 2 days prior to their due date through day 5 post-farrowing. The area of each crate was 1.8 m^2. Cross-fostering was permitted within the first 24 hours post-farrowing, but restricted to moving piglets between litters of the same housing type. Sows and piglets were monitored throughout the duration of the study using cameras mounted above each pen or crate.

Breeding values

Estimated breeding values (EBVs) for mothering ability (EBV_m) were obtained via Topigs Norsvin's breeding value estimation pipeline, which are currently estimated using data collected from sows housed in crated systems. EBV_m represents an individual's genetic merit for mothering ability, estimated as the maternal effect of piglet survival from birth through weaning. EBVs for this trait are expressed as a percentage, where sows with a positive value have better-than-average genetic merit for piglet survival, and sows with a negative value have poorer genetic merit for piglet survival, compared to the population average. EBV_m for sows included in this study ranged from -1.92% to +3.08%.

Traits

Mortality was recorded at the piglet level from day 0 through day 5 post-farrowing. Reasons for mortality included: Stillborn, nonviable, unknown, and crushed. Piglets were classified as "nonviable" if they had either a low birth weight (below 1.65 pounds), a non-vital appearance, or a congenital defect. Piglets were classified as "crushed" if they showed signs of bruising or imprints. All instances of crushing were validated using video footage.

Statistical analyses

All statistical analyses were performed using SAS software. Each mortality reason (stillborn, non-viable, unknown, crushed) and overall mortality (due to any reason) were analyzed as separate (binary) traits (1 = died, 0 = survived) using PROC GENMOD. Fixed effects fitted in the model included housing type (2 levels: FF vs C), parity class (of the foster sow; 2 levels: parity 1 or 2 vs parity 3+), and EBV_m (of the foster sow) as a covariate. A compound symmetry structure for the residuals was used to account for repeated measurements, with "sow" specified as the repeated subject. For all analyses presented, all pairwise interactions were fitted and removed if not significant ($P > .05$).

Results and discussion

Overall piglet mortality was significantly higher for FF (14.2%) vs C (5.9%) litters ($P = .002$), as expected. The main reason for piglet mortality was crushing. Consistent with this finding, mortality due to crushing was also significantly higher for FF (11.8%) vs C litters (1.1%) ($P < .001$). No evidence of a difference was detected between FF and C litters for mortality due to stillbirth ($P = .44$), non-viability ($P = .15$), or unknown deaths ($P = .98$). Results from this study also showed that EBV_m was significantly associated with overall piglet mortality, where a higher/more favorable EBV_m (of the foster sow) was associated with lower piglet mortality ($P = .04$). A higher EBV_m was also associated with significantly lower mortality due to stillbirth ($P = .02$) or crushing ($P = .002$), but had no significant effect on mortality due to non-viability ($P = .83$) or unknown causes ($P = .18$).

Results from this study show that, during the first five days post-farrowing, overall piglet mortality was significantly higher for piglets housed in a free-farrowing vs conventional (crated) pen. This finding is consistent with results from previous studies, suggesting a potential trade-off between sow and piglet welfare. The main cause of piglet mortality in this study was piglet crushing, which was significantly higher for FF vs C litters. Piglet crushing occurs when sows fail to exercise care when moving or laying. Mothers who actively communicate with their piglets, respond to piglet vocalization, and use caution when laying or changing posture, are needed to maximize production and piglet welfare in FF systems. Therefore, the objective of this study was to evaluate genetic selection for mothering ability as a potential strategy to improve production and welfare in FF systems.

Selection for enhanced mothering ability is already part of Topigs Norsvin's breeding goal. Genetic merit for mothering ability is estimated as the maternal effect on piglet survival, using data collected from sows housed in conventional/crated systems. Therefore, an additional objective of this study was to evaluate whether breeding values for mothering ability, derived using data collected from one type of housing environment, are also predictive of mothering ability in a different housing environment. Results show that a better EBV_m was associated with lower stillborn, crushing, and piglet mortality from 0 to 5 days post-farrowing. These findings indicate that selection for better mothering ability is expected to reduce piglet losses during lactation, regardless of housing type.

In conclusion, results from this study indicate that selection for better mothering ability, using data collected from sows housed in a crated system are predictive of mothering ability for sows housed in a free farrowing system. Therefore, genetic selection for enhanced mothering ability should be part of a strategy to maximize production and welfare when shifting from a conventional to free farrowing system.

Reference

1. Glencorse D, Plush K, Hazel S, et al. Impact of non-confinement accommodation on farrowing performance: a systematic review and meta-analysis of farrowing crates versus pens. *Animals*. 2019; 9. **doi:10.3390/ani9110957**

https://doi.org/10.54846/am2025/40

Inhibiting viral transmission in animal husbandry: Impact of sodium diformate alone or in combination with medium chain fatty acids: An overview

Christian Lückstädt; Christoph Hutter; Stevan Petrović
ADDCON GmbH, Bitterfeld, Germany

Introduction

The role of animal feed as a vehicle for the transport and transmission of viral diseases has only been known since a relatively short time. It was only encountered in 2014 during the porcine epidemic diarrhea virus epidemic in North America.[1] To understand this newly identified risk factor, numerous studies have recently been carried out. It was found that the risk of viral transmission in feed cannot be underestimated. Therefore, there is currently growing attention to the role of feed and feed ingredients, which are interpreted as a moderate-risk transmission vehicle for the African swine fever virus (ASFV) and other swine viral pathogens.[2,3] Since it was furthermore demonstrated that certain viruses can also be transmitted orally in liquids (eg, drinking water), research was also initiated in this regard.[4]

This knowledge motivated the scientific community to develop mitigation strategies to reduce the feed- and water mediated infection probability with the aid of additives. There is growing evidence that certain feed additives can mitigate the risk substantially. Since formaldehyde is long known to be an effective antibacterial disinfectant, it was among the first substances to be tested on its anti-viral properties. It proved to be successful in reducing the porcine epidemic diarrhea virus in feed.[5] However, the substance is no longer globally approved and furthermore seen critically by regulating authorities and consumers due to being a trigger for certain cancers. As such further additives with potential anti-viral properties had to be checked. Early work on biosecurity aspects and pathogen inactivation in high-risk animal by-products identified formic acid as a possible additive.[6] This short-chain organic acid is used in animal nutrition since almost 60 years and is in Europe also registered as a disinfectant. As such this group of Scandinavian researchers used the substance in their mitigation strategy for mashed animal by-products which are stored at low pH-levels (4.5) over a period of 130 days. However, a high dosage (3% inclusion rate) was needed to achieve this stable pH and the subsequent impacts on various viruses. Such dosages might be feasible for large scale storage of said material, like fish silage. However, they cannot be applied to finished feed – which is transported globally and directly consumed by the different animal species. Another promising class of approved feed additives with a documented impact against enveloped viruses, consists of medium chain fatty acids or their glycerides.[7] The authors tested the impact of a commercial mixture of C8:C10:C12 (MCFA) as well as glycerol monolaurate (GML) on the ASFV infectivity in feed. While the MCFA-mixture showed no impact at any dosage, GML exerted an 88% reduction in viral infectivity, but again with a 2% inclusion rate only. Since it was known that the virucidal activity of monolaurin becomes more effective if the pH is lowered to around 4.2, it was interesting to follow that approach.[8] The current study therefore reports the available information (published and unpublished) on the impact of various medium chain fatty acids, including monolaurate, under an acidic environment supplied by sodium diformate (patented; produced and registered by ADDCON) – a double salt of formic acid.

Impact against ASF virus

One of the first trials was designed to evaluate the viability of ASFV (p72, genotype II) over time (0-, 1-, 3- and 7 days post-inoculation) in commercial swine feed containing either 0% or 0.3% of an agglomerate of sodium diformate and monolaurate (FORMI® 3G, ADDCON, hereafter abbreviated to 3G). The feed bags were incubated at room temperature (25°C) with a viral concentration of 10^8 HAD_{50}/mL. After the appropriate post-inoculation incubation period, the surviving virus was eluted from the samples using RPMI 1640 medium with 5% fetal bovine serum. Each treatment used a set of triplicate samples that were combined and used for a single titration and inoculation into cells. Virus titers (HAD_{50}/mL) were calculated by the Karber method. The quantity of ASFV was determined by real-time PCR to measure C_t-value. A significance level of .05 was used in all tests.[9]

Mean abundance rates of ASFV in the positive control as well as FORMI 3G-feed are shown in Table 1. The ASFV titration assay on cell cultures showed that the feed acidifier had a significant reduction activity against ASFV throughout the whole trial period, beginning only a few hours after the initiation of the trial. The 0.3% 3G inclusion into the diet was able to inhibit the virus within less than one hour significantly ($P = .013$), from 4.72 to 3.99 $Log_{10}HAD_{50}$. From day 1 onwards, the reduction was highly significant ($P < .001$). On day 7, the ASFV was inhibited completely. It was commented that the additive can reduce ASFV infectivity in commercial feed at low dosages and can be consequently an economical and sustainable approach to curb the disease transmission while offering a strongly reduced infection probability for pigs that might consume virus-contaminated feed. A subsequent trial, using the same methodology, found for an agglomerate of sodium diformate and medium chain fatty acids a similar pattern in the reduction of the ASFV in swine feed under tropical conditions (Table 2).[10]

Impact against porcine reproductive and respiratory virus (PRRSV)

Based on the evidence that monolaurate (GML) is more effective in inhibiting pathogens in an acidic environment, a further study aimed to investigate the effects of GML under acidic conditions on the expression of PRRS virus of nursery pigs.[11] 1520 nursery pigs (42 days old) were randomly assigned to two trial groups with 20 replicates (pens) per group. In the control group, 0.1% GML was added

Table 1: Relative abundance (Log_{10} HAD_{50}) of African swine fever virus in positive control (PC) and 0.3% Formi 3G in swine diets over time

Time	PC	3G	Difference (%)	*P*-value
Day 0	4.72[a]	3.99[b]	-81.4	.013
Day 1	4.60[a]	3.52[b]	-91.7	.0001
Day 3	4.07[a]	2.15[b]	-98.8	.0002
Day 7	3.59[a]	0b	-100	.0000

Table 2: Relative abundance (Log_{10} HAD_{50}) of African swine fever virus in positive control (PC) and 0.3% Diformate-MCFA in swine diets over time

Time	PC	DF-MCFA	Difference (%)	*P*-value
Day 0	4.72[a]	4.10[b]	-76.0	.013
Day 1	4.60[a]	3.35[b]	-94.4	.0004
Day 3	4.07[a]	2.27[b]	-98.4	.0006
Day 7	3.59[a]	0[b]	-100	.0000

to the basal diet, while the treatment group received 0.5% Formi 3G in the basal diet. Throughout the experimental period, blood samples were collected from nursery pigs every two weeks, and PRRSV Ct values in their blood were assessed using real-time PCR. Significance of differences between the treatment group and the control group was assessed using the *P*-value ($P < .05$).

During the first four weeks (42-70 day-old), the PRRSV antibody levels in the 3G-group were significantly higher than those in the control group. Moreover, the PRRS Elisa S/P ratio in 3G-group remained stable within the range of 1.737 - 2.128, whereas 0.710 - 2.006 were measured in the control group. Regarding the virus transcription levels (Ct value), the PRRSV exhibited lower Ct values in the control group compared to the 3G-group, indicating a higher viral load in the former. The PRRSV expression decreased continuously in the 3G-group and after six weeks became undetectable (Figure 1).

The authors commented that the co-addition of sodium diformate and GML (3G) is associated with reduced PRRSV expression. Its mechanism of action may involve reducing the production of pro-inflammatory cytokines and enhancing interferon-γ release. The results also demonstrated that under the acidic environment, GML has a better effect on virus inhibition. Therefore, employing GML in an acidic environment - thus Formi 3G, appears to be a more reasonable strategy for the inhibition of PRRSV in nursery pigs.

Figure 1: The effect of adding Formi 3G on the expression of porcine reproductive and respiratory syndrome virus (PRRSV) in nursery pigs.

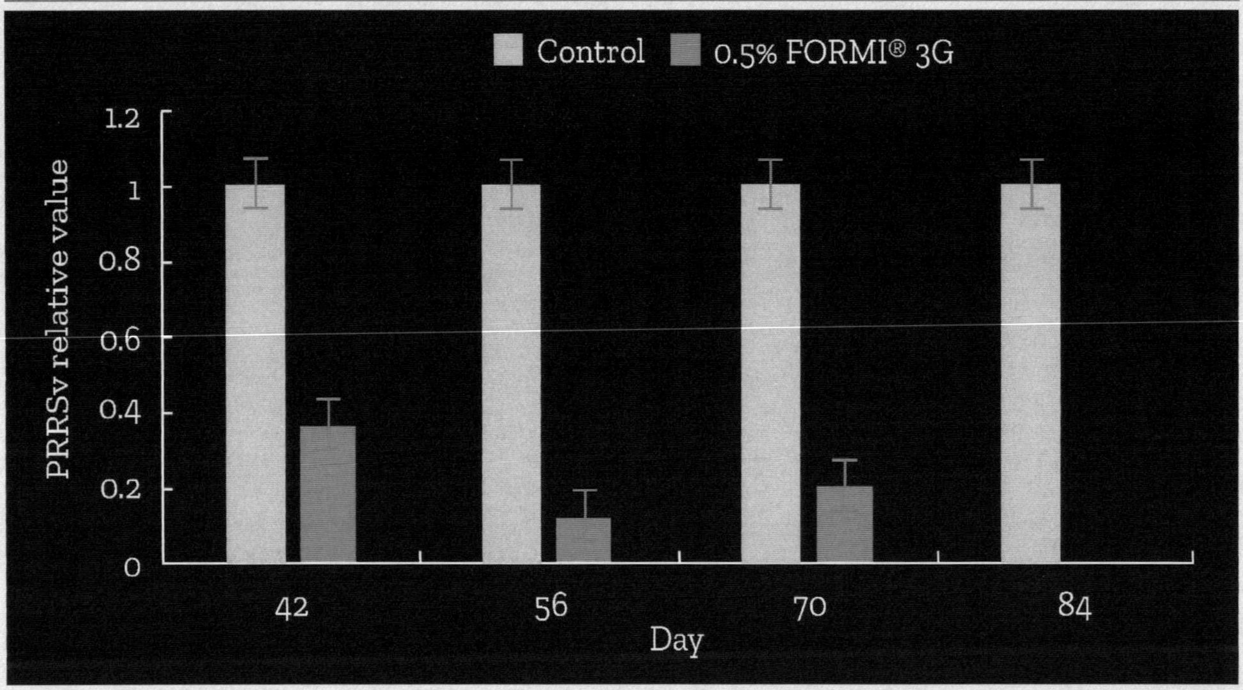

Anti-viral impact in water

Several viruses may be transmitted via drinking water.[4] This is especially true for the different influenza viruses, as indicated by recent research.[12] To validate the impact of a formate/propionate-MCFA mixture (ADDCON XL forte), applied to the drinking water of pig and poultry, on various water-transmittable viruses, a research collaboration with the Scientific institute of Veterinary Medicine, Belgrade was started. Researchers found that the application of XL forte at 0.5 L to 1000 L of drinking water was able to reduce the virus titer of the PRV-virus (Pseudorabies) within 30 minutes from 4 Log_{10}/mL to zero. Similar results were achieved with 1 L of the product in 1000 L drinking water on the impact against Influenza virus type A/H3N2. Also here was the virus titer zero after 30 minutes. The lower dosage of 0.5 L/1000 L needed for the same complete inhibition of the virus titer 24 hours.

It can be concluded that mixtures of organic acids and medium chain fatty acids may have the chance to inhibit the transmission of enveloped viruses via drinking water. However, due to the scarcity of data, more research might be needed.

Further reports

Referring to the virus reduction in animal by-products by means of formic acid application to reach a low pH (4.5),[6] the authors tested the application of sodium diformate (Formi NDF, ADDCON, hereafter abbreviated as NDF) on a commercial pig feed, challenged with the classical swine fever virus (CSFV). Again, these tests were carried out in collaboration with the Scientific Institute of Veterinary Medicine in Serbia. According to the British Standards Institution, a product is considered effective in its anti-viral action, if a reduction of the viral titer of at least 4 log_{10} is achieved.[13] Based on this assumption, the trial was carried out with triplicate measurements (Table 3).

The authors could prove that also with the addition of 0.6% sodium diformate, reaching a pH of only around 5.4 in the pig feed, a complete reduction of the virus load is possible. The CSFV is considered as a rather weak virus. The achieved results warrant further investigation - especially since Formi NDF is a widely used additive in pig and poultry nutrition, which is focused today mainly on its anti-bacterial and digestive enhancing properties, thus ensuring gut health and not yet for its potential anti-viral activity.

It is the strong belief of the authors that continuous development of appropriate and innovative formulations to boost the anti-viral performance and commercially utilize above mentioned additives for water and feed delivery could help these additives becoming vital tools in sustainably combating various viral infections in animal husbandry.

Table 3: Relative abundance of classical swine fever virus titer (Log_{10}/mL) in positive control (PC) and 0.6% sodium diformate (Formi NDF) treated commercial pig feed over time

Time	PC	NDF	Difference (%)	P-value
Day 0	4.33[a]	4.17[a]	-3.7	.12
Day 1	4.17[a]	0[b]	-100	.0000005
Day 3	4.08[a]	0[b]	-100	.000001
Day 7	4.08[a]	0[b]	-100	.000001

References

1. Dee S, Niederwerder MC, Patterson G, Cochrane R, Jones C, Diel D, Brockhoff E, Nelson E, Spronk G, Sundberg P. 2020. The risk of viral transmission in feed: What do we know, what do we do? Transbound Emerg Dis. 67:2365-2371.

2. Jones CK, Woodworth J, Dritz SS, Paulk CB. 2019. Reviewing the risk of feed as a vehicle for swine pathogen transmission. Vet Med Sci. **https://doi.org/10.1002/vms3.227**

3. U.S. Department of Agriculture. 2020. The foreign animal disease preparedness and response plan (FAD PReP)-African swine fever response plan: The red book. Maryland: Department of Agriculture.

4. Niederwerder MC, Stoian AM, Rowland RR, Dritz SS, Petrovan V, Constance LA, et al. 2019. Infectious dose of African swine fever virus when consumed naturally in liquid or feed. Emerg Infect Dis. 25:891–7.

5. Dee S, Neill C, Clement T, Singrey A, Christopher-Hennings J, Nelson E. 2015. An evaluation of porcine epidemic diarrhea virus survival in individual feed ingredients in the presence or absence of a liquid antimicrobial. Porc Health Manag;1:9.

6. Vinnerås B, Samuelson A, Emmoth E, Nyberg KA, Albihn A. 2012. Biosecurity aspects and pathogen inactivation in acidified high risk animal by-products. J Environ Sci and Health, Part A: Toxic/Hazardous Subst and Environ Eng, 47:8, 1166-1172.

7. Jackman JA, Hakobyan A, Zakaryan H, Elrod CC. 2020. Inhibition of African swine fever virus in liquid and feed by medium-chain fatty acids and glycerol monolaurate. J. Anim Sci and Biotech. 11:114.

8. Hilmarsson H, Traustason BS, Kristmundsdottir T, Thormar H. 2007. Virucidal activities of medium- and long-chain fatty alcohols and lipids against respiratory syncytial virus and parainfluenza virus type 2: Comparison at different pH levels. Arch of Virol, 152: 2225-2236.

9. Lückstädt C. 2022. Impact of an agglomerate of sodium diformate and monolaurate on the reduction of African Swine Fever virus in commercial pig feed. Proceedings IPVS 2022, p. 692.

10. Lückstädt C. 2022. Acidifier reduce African swine fever virus in commercial pig feed under tropical conditions. Book of abstracts, Tropentag 2022, p. 288.

11. Chuang WY, Lückstädt C. 2024. Impact of an agglomerate of sodium diformate and monolaurate on the reduction of PRRS virus load in nursery pigs. Proceeding IPVS & ESPHM 2024, p. 341.

12. Kenmoe S, Takuissu GR, Ebogo-Belobo JT, Kengne-Nde C, Mbaga DS, Bowo-Ngandji, et al. 2024. A systematic review of influenza virus in water environments across human, poultry, and wild bird habitats. Water Research X, 22: 100210.

13. British Standards Institution. 2013. Chemical Disinfectants and Antiseptics. Quantitative Suspension Test for the Evaluation of Virucidal Activity in the Medical Area. Test Method and Requirements. BSEN 14476:2013.

https://doi.org/10.54846/am2025/41

Effect of dietary vitamin D on PRRSV disease severity and immune response in nursery pigs

K. G. Keen[1]; E. van Heugten[2]; S. Hough[3]; P. Sitthicharoenchai[1]; S. Eguiluz-Hernandez[1]; V. Bamrung[1]; M. Rahe[1]
[1]Department of Population Health and Pathobiology, North Carolina State University, Raleigh, North Carolina;
[2]Department of Animal Science, North Carolina State University, Raleigh, North Carolina;
[3]dsm-firmenich, Plainsboro, New Jersey

Introduction

Vitamin D_3 is a fat-soluble micronutrient vital for skeletal health across mammalian species, with increasing interest in characterizing its immunomodulatory capabilities. Prior research reported significant increases in circulating white blood cells and macrophage antimicrobial capacity in weaning pigs fed diets containing higher vitamin D levels compared to control.[1] As such, we hypothesized vitamin D can bolster the immune system in the face of pathogenic challenge and improve disease outcomes. Our objective was to determine the effect of dietary vitamin D_3 on porcine reproductive and respiratory syndrome virus (PRRSV)-challenged nursery pigs.

Methods

Forty-two weaning-aged pigs were designated to one of three experimental diets, which differed in vitamin D_3 international units per kilogram (IU/kg): Marginal (200 IU/kg), standard (1,500 IU/kg), or supplemented (1,500 IU/kg + 2,000 IU/kg 25-hydroxyvitamin D_3). Following a diet acclimation period of twenty-seven days, pigs were assigned to their experimental treatment groups, which differed in diet and challenge status: Negative (marginal diet + no challenge; n = 6), marginal (marginal diet + PRRSV challenge; n = 12), standard (standard diet + PRRSV challenge; n = 12), or supplemented (supplemented diet + PRRSV challenge; n = 12). On day 0 post-challenge (DPC 0), all pigs in the marginal, standard, and supplemented treatment groups were challenged intranasally and intramuscularly with a lineage 1 PRRSV-2 1-4-4 L1C (L1C.5) variant at $5x10^5$ $TCID_{50}$/pig. Body weight and temperature data were collected throughout the trial, as were nasal swab and blood samples. At DPC 14, a final necropsy was conducted on all remaining animals. Rib and femur bone collection was performed for later analysis. Gross pathology on lungs were noted and lung tissue sections were collected for histopathology, RNA isolation, and single-cell suspensions. PCRs for 1-4-4 L1C were conducted on blood, nasal swab, and lung samples. Isolated peripheral blood mononuclear cells (PBMCs) and lung cells were used for IFN-y ELISpot analysis. Serum was analyzed at the ISU veterinary diagnostic laboratory for PRRSV-specific IgG and IgM ELISA, and fluorescent focus neutralization. Clinical signs were assessed through body weight, feed intake, feed conversion, and temperature data.

Results

No significant differences in clinical signs were noted between challenged treatment groups, though all challenged groups had significantly poorer production compared to the unchallenged control. Similarly, no prominent differences in gross pathology, viremia, or nasal shedding were observed between the challenged groups. When stimulated *ex vivo* with the L1C.5 variant, PBMC IFN-y production was significantly higher at DPC 14 in the marginal group compared to control. Interstitial pneumonia was significantly higher in the marginal and supplemented groups compared to both the control and standard groups. While not significantly different, both PRRSV-specific antibodies and neutralizing antibodies were lower in the marginal group compared to standard or supplemented groups (Figures 1 and 2).

Discussion

Dietary vitamin D_3 does not appear to affect the acute stage of PRRSV infection or disease severity. While ELISpot analysis demonstrated the marginal group having the highest IFN-y production, there was not a statistical difference when compared to the supplemented group. Ongoing IFN-y ELISpot analysis of lung cells will provide further insight into differences in cytokine production between groups and uncover if there is a positive correlation between interstitial pneumonia scores and IFN-y producing cells. The antibody data suggests that decreased vitamin D_3 can negatively impact the humoral immune response. Whether this is due to a direct effect on plasma cells, a more complex interaction involving innate immunity, or something else is unclear. A future study, with an increased post-challenge period, will provide a clearer picture of antibody response in addition to the recovery phase between the dietary groups.

Figure 1: Porcine reproductive and respiratory syndrome virus (PRRSV)-specific serum IgG and IgM titers (reported as ratio of sample/known positive) at DPC 14. Serum was analyzed at the ISU veterinary diagnostic laboratory using a commercial IDEXX PRRS ELISA kit. Black = negative control; red = marginal + PRRSV challenge; green = standard + PRRSV challenge; blue = supplemented + PRRSV challenge. Data was analyzed using a one-way ANOVA with Tukey's post-hoc and a significance level of *P* < .05. "***" indicates a significance level of ≤ .001.

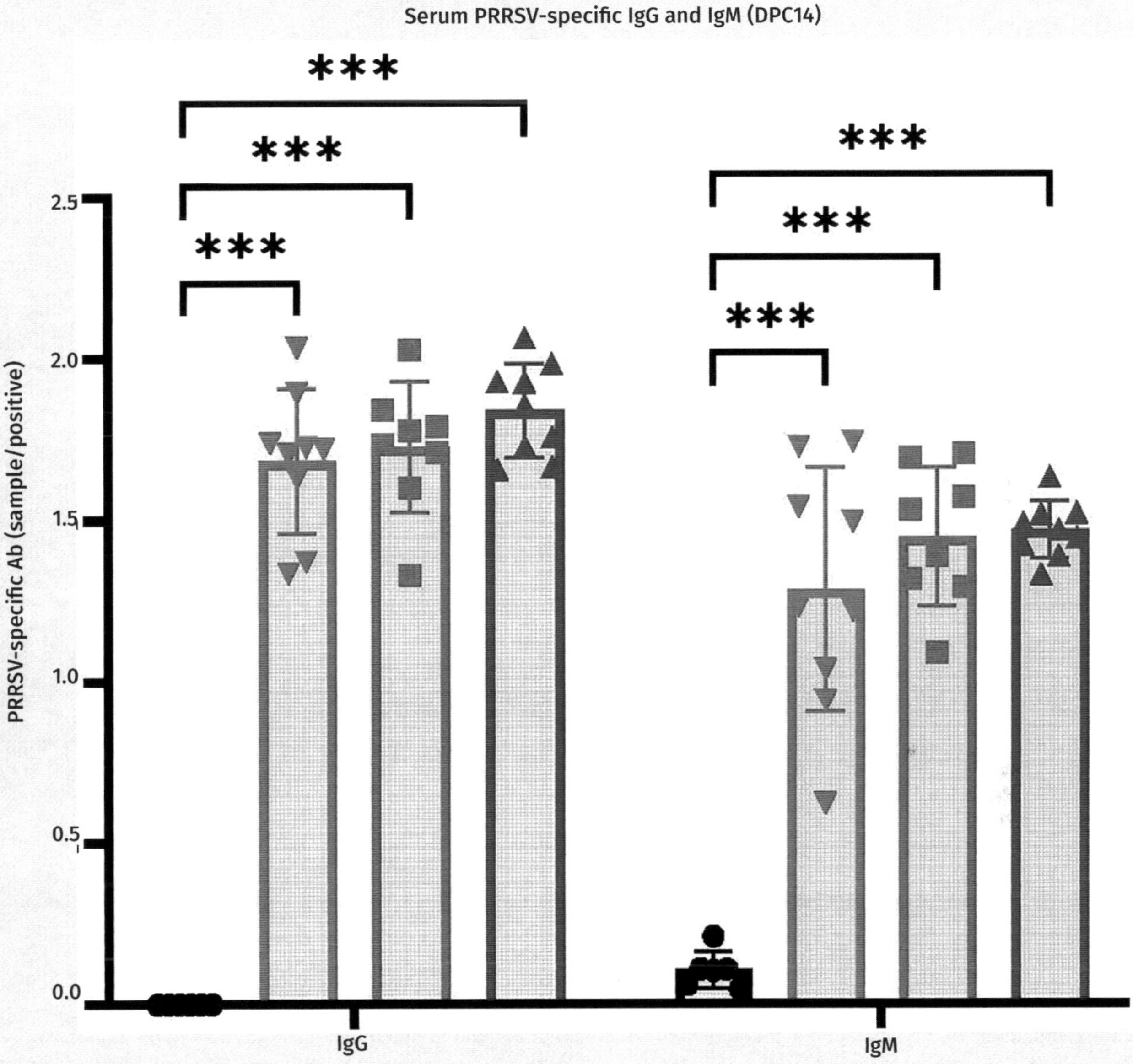

DPC = day post challege

Figure 2: Porcine reproductive and respiratory syndrome virus (PRRSV)-specific serum neutralizing antibody titers (reported as Log 2) at DPC 14. Serum was analyzed at the ISU veterinary diagnostic laboratory using in-house fluorescent focus neutralization assay. Data was analyzed using a Kruskal-Wallis non-parametric test with a Dunn's post-hoc and a significance level of $P \leq .05$.

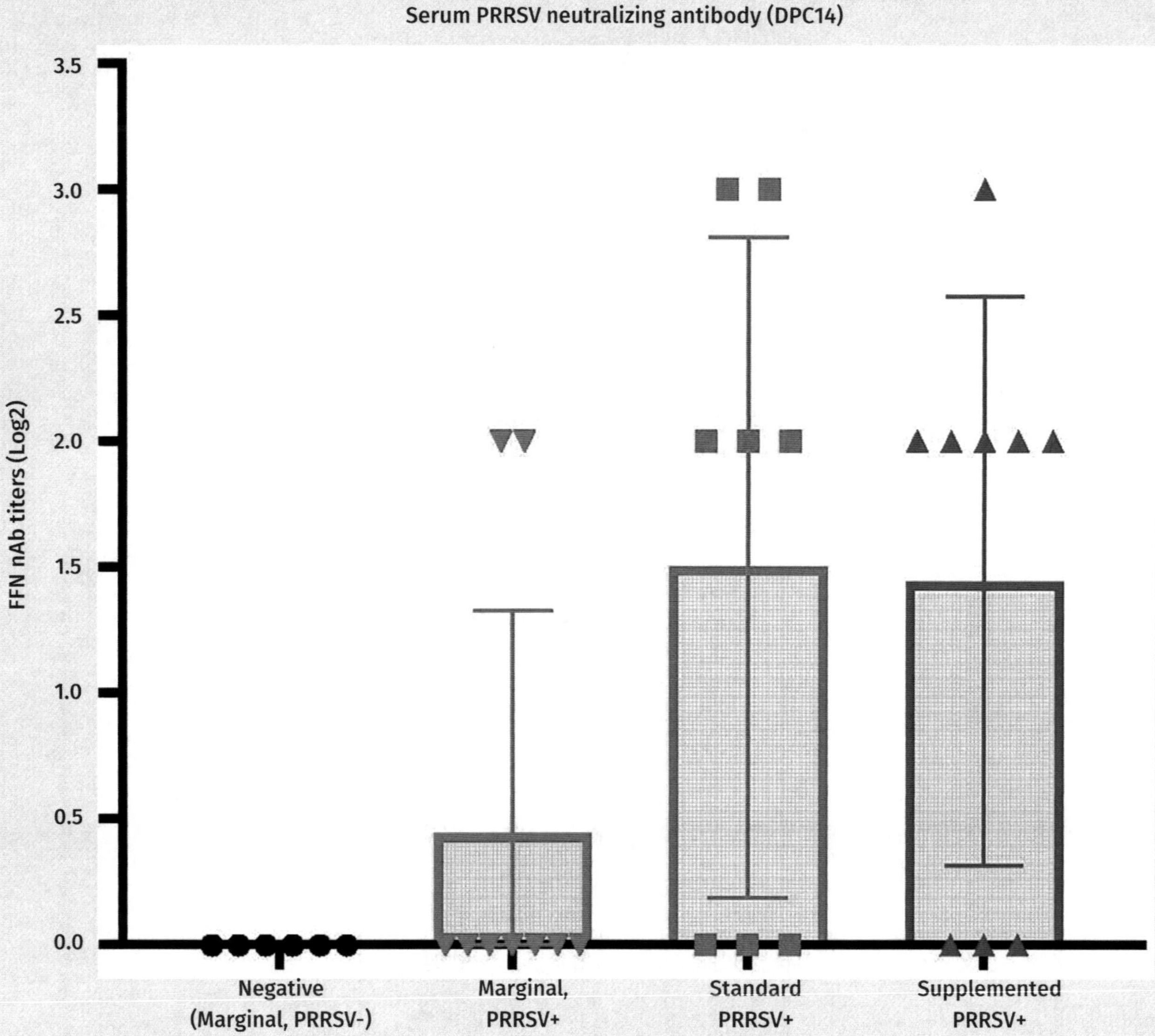

DPC = day post challege

Reference

1. Konowalchuk, Jeffrey D., et al. Modulation of Weanling Pig Cellular Immunity in Response to Diet Supplementation with 25-Hydroxyvitamin D3. *Veterinary Immunology and Immunopathology.* 2013.

https://doi.org/10.54846/am2025/42

Serum 25-OH-D_3 as an indicator of vitamin D status and bone health in swine, and a novel testing technique to verify status

S. Hough, DVM; J. Bergstrom, PhD; L. Zheng, PhD; P. Gott, PhD
dsm-firmenich, Plainsboro, New Jersey

Introduction

Lameness and metabolic bone disease are issues impacting the lifetime performance of sows and the robustness and livability of finishing pigs. Between 2012 and 2021, sow mortality in the US increased from 8.16 to 13.56% with lameness identified as one of the leading reasons for humane euthanasia and mortality.[1] Lameness in finishing pigs has also increased due to the demands of higher growth rates of modern breeds. There is a greater allocation of nutrients towards muscle deposition and immunity relative to bone structure. Lame or unhealthy pigs have been shown to have reduced serum levels of the form of vitamin D_3 known as 25-hydroxycholecalciferol (25-OH-D_3).[2] Serum 25-OH-D_3 is considered the best biomarker of vitamin D status.[3] Therefore, antemortem testing of serum 25-OH-D_3 could be used as a good indicator of bone health, to support interventions to avoid lameness, metabolic bone disease can be a major contributor to lameness in pigs. It is characterized by a disturbance in the growth, modelling and remodeling of bone and leads to diseases such as rickets, osteomalacia, fibrous osteodystrophy and osteoporosis/osteopenia.[4] The major causes of metabolic bone disease include deficiencies or imbalances in important vitamins and minerals, such as inadequate absorption of phosphorus (P) or low amounts of digestible P in the diet. There can also be inadequate or excess calcium (Ca) in the diet or an imbalance in the Ca:P ratio. As mentioned, low levels of circulating 25-OH-D_3 in the blood can also contribute to deficiencies in Ca or P.

Diagnostics

Bone ash, bone density, and bone mineral measurements are also considered essential in diagnostic investigations of lameness. However, bone ash results can be different depending on which laboratory method used (eg, whether bones are defatted or not prior to ashing) and the 'normal range' that is referenced.[5] The reference range that is frequently used to characterize 'normal' bone ash is 58-62%, which is based on information using defatted bone from healthy finishing and adult pigs.[6] Defatting of bone is a time-consuming process that is not often used in diagnostic laboratories. If bone samples are not defatted before determining their ash content, the bone ash results will be considerably lower. Therefore, for bone that has not been defatted, bone ash results < 45% are concerning.[7]

Microscopic bone lesions are also useful to confirm a diagnosis of metabolic bone disease, and the 10th rib should be considered for submission, as it can be the most sensitive for identifying abnormalities due to an inadequate dietary supply of P and vitamin D.[2]

Serum 25-OH-D_3 reference ranges that are commonly used by diagnostic labs are extrapolated from a publication from 1994.[8] Recent findings to characterize the vitamin D status in varying ages of pigs were shared at the 2021 AASV meeting and showed that most vitamin D concentrations were found to be below published reference ranges.[9] This information confirmed what was found in previous work completed by Darin Madson, which showed that vitamin D_3 supplementation levels in the diet were, in most cases, well above recommended NRC levels during all growth stages and pigs were still considered deficient.[10] The problem with measuring serum vitamin D in swine is that normal age-based reference intervals for serum 25(OH)D_3 have not been clearly defined in recent years. We must establish what "normal" levels should be and that can only be determined from properly executed prospective studies. An even more recent survey showed that piglets are deficient for various vitamin and mineral concentrations in serum and tissues when compared to historical reference ranges.[11] Consideration should also be made for the sow and how her levels of serum 25-OH-D_3 could impact piglet levels during the critical weaning period.[12]

As well as vitamin D status, the incidence of lameness is also associated with health status (eg, with porcine reproductive and respiratory syndrome virus [PRRSV] and Seneca Valley virus [SVA] status.[13] Health status is correlated with serum 25-OH-D_3, bone ash, and bone density.[2,14] Importantly, a low serum 25-OH-D_3 status is not common in healthy pigs that are housed outdoors, particularly during summer.[15]

Both infectious and toxic agents that may be causing lameness issues need to be ruled out, so joints, spinal cord and skeletal muscle should also be considered as part of the full examination. The causes for lameness are multifactorial, with infections, nutrition, genetics, conformation, facility, environment, and trauma all having an impact, and many clinical expressions of lameness appear to be the result of cumulative insults over time. Astute, accurate and specific case definitions are invaluable in a diagnostic investigation.

Supplementation

One way to overcome the difference in bioavailability of vitamin D supplemented as cholecalciferol compared with that produced via sunlight is to supplement directly with 25-OH-D_3 in the diet. Vitamin D can be acquired through injection, the diet, or endogenously by exposure to UVB light. All routes are typically biologically inert and require multiple phases of hydroxylation in the body for activation. Hy-D®, a source of 25-OH-D_3 for feed or water supplementation, is a product that provides an efficient form of the vitamin that does not require the initial liver hydroxylation step. Also, 25-OH-D_3 is more bioavailable than Vitamin D_3, due to its higher polarity and much stronger affinity for the vitamin D-binding protein.[16] At the 2021 AASV meeting, Dr Hough shared some graphs that illustrated the difference in serum levels in gilts, sows and growing pigs fed equivalent levels (IU) of D_3 and Hy-D®. Even at equivalent levels for vitamin D_3 and Hy-D®, serum levels were different, indicating improved bioavailability with Hy-D®.[9] Research over the past 13 years has generated substantial evidence for this. It has been shown that maternal supplementation of 50 μg/kg 25-OH-D_3 as Hy-D® on top of cholecalciferol vitamin D_3 (2,000 IU/kg) throughout gestation and lactation significantly increased sow serum 25-OH-D_3 at farrowing (from 34.6 to 65.4 ng/mL) as well as bone ash, femur strength and bone density in the newborn piglets, compared with supplementing cholecalciferol alone.[17] It has also been shown that maternal supplementation of 50 μg/kg 25-OH-D_3 (as Hy-D®) on top of 500 IU/kg of cholecalciferol increased serum 25-OH-D_3 in gilts and sows, as well as their progeny's average daily gain during the grower-finisher phase and final bodyweight at slaughter (+ 4 kg), and tended to improve gain:feed from weaning until harvest, compared with supplementing cholecalciferol alone.[18] Even direct supplementation of 50 μg/kg 25-OH-D_3 (as Hy-D®) to pigs from weaning until harvest (6 to 110 kg BW) increased serum 25-OH-D_3 levels and reduced the incidence of osteochondrosis lesions (associated with lameness) when compared with supplementing cholecalciferol alone at 1,800–1500 IU/kg.[19] Collectively, these and other studies have enabled dsm-firmenich to establish robust reference ranges for serum 25-OH-D_3 levels in pigs. The recommended ranges can be found in Table 1.

When choosing the source of vitamin D, consider the quality of research that has been conducted using those products in swine. There are vitamin D metabolites that are not particularly beneficial or safe for improving vitamin D nutrition, and some products promoted for their vitamin D content may contain metabolites or impurities that can be harmful. For instance, products that may contain the hormonal 1,25-OH_2-D_3 metabolite can increase the risk for hypercalcemia in pigs.[20] Some products that do not provide pure 25-OH-D_3 can cause a decrease in feed intake and provide no benefit to the animal.[21] Hy-D® is made using a proprietary process to ensure the purity of 25-OH-D_3 and the absence of other metabolites, resulting in a vitamin D supplement with a very high safety margin.[22] Other sources or supplements for exogenous vitamin D are not the same.

Table 1: Recommended serum reference ranges

25-OH-D3 serum level in swine (ng/mL)	Indication
30-60	Ca and P metabolism and bone health
> 60	Muscle growth, fetal development, improved lifetime performance, reduce farrowing complications

Figure 1: Dried blood spot (DBS) sampling.

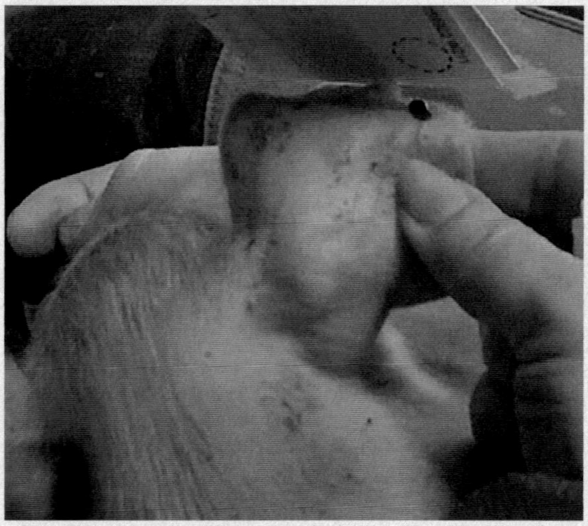

New diagnostic offering

dsm-firmenich has developed a new way to collect whole blood samples, where just one drop of blood is needed to determine the vitamin D status. This innovation is called dried blood spot testing or DBS. An example of the technique can be found in Figure 1. This allows practitioners to sample multiple animals easily. The transfer of the sample to the diagnostic lab is easier than preparing and shipping serum as a biological sample. There is a third-party lab involved that has validated this testing in humans and swine, with poultry and ruminants to follow. There is a user-friendly digital interface that allows customers to order sampling kits and view test results in a timely manner. dsm-firmenich will offer interpretation and support after the results are available.

Conclusion

Lameness is a significant welfare and cost issue in pig production. A significant amount of lameness cases are inherently linked to Ca, P and vitamin D status, and these can be affected by health status, management and nutrition. Available data indicate that 25-OH-D_3 is one of the reliable biomarkers for assessing bone health and the risk for lameness, such as diagnosing metabolic bone disease. Pigs reared in modern farms without access to sunlight generally have a lower vitamin D status (serum 25-OH-D_3 < 30 ng/mL), even when there are high dietary levels of cholecalciferol vitamin D_3 supplementation in the diet. In these situations, dietary supplementation with 50 μg/kg 25-OH-D_3 in both the maternal and offspring diets can significantly improve vitamin D status, which can support better bone and joint health and growth performance. Reach out to dsm-firmenich for status testing and resources.

References

1. National Hog Farmer. 2021 sow mortality analysis. Accessed November 8, 2024. **https://www.nationalhogfarmer.com/hog-health/2021-sow-mortality-analysis**

2. Williams, H., M. Tokach, J. Woodworth, R. Goodband, J. DeRouchey, J. Bergstrom, M. Rahe, C. Siepker, P. Sitthicharoenchai, S. Ensley, S. Radke and J. Gebhardt. Diagnostic survey of analytical methods used to determine bone mineralization in pigs. *Journal of Animal Science*. 2024: 102, 1-15.

3. Jones, G. Metabolism and biomarkers of vitamin D. *Scandinavian Journal of Clinical and Laboratory Investigation*. 2012; 243:7-13.

4. Merck Veterinary Manual Musculoskeletal System, Lameness in Pigs. Topic: Overview of Lameness in Pigs. Accessed October 29, 2024. **https://www.merckvetmanual.com/musculoskeletal-system/lameness-in-pigs/overview-of-lameness-in-pigs**

5. Wensley, M. R., C. M. Vier, J. T. Gebhardt, M. D. Tokach, J. C. Woodworth, R. D. Goodband, and J. M. DeRouchey. Technical note: assessment of two methods for estimating bone ash in pigs. *Journal of Animal Science*. 2020: 98:1–8.

6. Madson, D. What we are seeing at the diagnostic lab. 2013 Proceedings from the Allen D. Leman Swine Conference.

7. Darin Madson, Iowa State University Veterinary Diagnostic Lab, USA, personal communication.

8. Puls, R. Vitamin Levels in Animal Health: Diagnostic Data and Bibliographies. Sherpa International. 1994.

9. Hough, S, J. Bergstrom and J. Hahn. Recent trends in vitamin D serum status in swine. AASV 2021 proceedings.

10. Arnold, J., D. Madson, S. Ensley, J. Goff, C. Sparks, G. Stevenson, T. Crenshaw, C. Wang and R. Horst. Survey of serum vitamin D status across stages of swine production and evaluation of supplemental bulk vitamin D premixes used in swine diets. *Journal of Swine Health and Production.* 2015; 23 (1), 28-34.

11. Greiner, L., S, Elefson, S. Radke, C. Hagen, D. Humphrey, S. Becker. A survey of vitamin and trace mineral ranges for diagnostic lab reporting from conventionally raised swine. *Journal of Swine Health and Production.* 2022: 30(5), 282-291.

12. Goff, J.P., R.L Horst and E.T. Littledike. Effect of sow vitamin D status at parturition on the vitamin D status of neonatal pigs. *Journal of Nutrition.*1984; 114, 163-169.

13. Paiva, R., C. Moura, P. Thomas, B. Haberl, L. Greiner, C. Rademacher, A. Paula Silva, G. Trevisan, D. Linhares and G. Silva. Risk Factors associated with sow mortality in breeding herds under one production system in the Midwestern United States. *Preventative Veterinary Medicine.* 2023: 213, 105883.

14. Baker, J., M. Wagner, and D. Baumert. The influence of a porcine reproductive and respiratory syndrome virus infection on vitamin and mineral levels in serum, liver, and bone of recently weaned pigs. Proceedings of the 2019 Annual Meeting of the AASV.

15. Jakobsen, S. S., J. Jakobsen, and J. P. Nielsen. Vitamin D levels in sows from five Danish outdoor herds. *Animals,* 2022; 12:299.

16. Burild, A., C. Lauridsen, N. Faqir, H. M. Sommer and J. Jakobsen. Vitamin D3 and 25-hydroxyvitamin D3 in pork and their relationship to vitamin D status in pigs. *Journal of Nutritional Science.* 2016; 5(3), 1-6.

17. Zhou, H., Y. Chen, Y. Zhuo, G. Lv, Y. Lin, B. Feng, Z. Fang, L. Che, J. Li, S. Xu, and D. Wu. 2016. Effects of 25-hydroxycholecalciferol supplementation in maternal diets on milk quality and serum bone status markers of sows and bone quality of piglets. *Journal of Animal Science,* 88:476–483.

18. Tsai, T.C., B. A. Sam, S. Dritz, C. Vier, J. Bergstrom, S. Hough, U. Orlando, K, Blake, N. Davis, H. Grogan, K. Coble and C. Maxwell. 2023 PSIV-1 Nutrient Intervention During Gilt Development Leads to Effects on Progeny Growth Responses. *Journal of Animal Science.* 2023: 101, supp 2, 307-308.

19. Sugiyama, T., T.K Chung, H. Yonekura, E. Azem and T. Hayakawa. Effects of 25-hydroxycholecalciferol on the development of osteochondrosis in swine. *Journal of Animal Science.* 2013; 84, 341-349.

20. Schlegel, P., D. Guggisberg and A. Gutzwiller. Tolerance to 1,25 dihydroxyvitamin D3 glycosides from Solanum glaucophyllum by the growing pig. *Research in Veterinary Science.* 2017. 112. 119-124.

21. Okafor, P.C.J. and N. Homwong. Dietary Hydroxyvitamin D3 Improved Serum Concentration Level and Alkaline Phosphatase Activity during lactation but Had Meager Impact on Post-Farrowing Reproductive Performance in Sows. *Animals.* 2024: 14: 419-432.

22. Rosenberg, S., G. Weber, A. Erhardt, U. Holler, U, Wehr and w. Rambeck. Tolerance Evaluation of overdosed dietary levels of 25-hydroxyvitamin D3 in growing piglets. *Journal of Animal Physiology and Animal Nutrition,* 2015; 100, 371-380.

https://doi.org/10.54846/am2025/43

Assessment of houseflies (*Musca domestica*) as a potential reservoir of porcine reproductive and respiratory syndrome virus in swine compost facilities under field conditions

Grant Allison[1], DVM; Phil Gauger[2], DVM, MS, PhD; Onyekachukwu Osemeke[2], DVM, MS; Gene Spellman[3], BS
[1]Walcott Veterinary Clinic, Walcott Iowa; [2]Vet Diagnostic and Production Animal Medicine, Veterinary Diagnostic Laboratory, Iowa State University, Ames, Iowa; [3]Central Life Sciences, Schaumburg, Illinois

Introduction

The use of a compost facility is a common, cost-effective way to dispose of mortalities in swine facilities. Decomposition of the carcass is typically achieved through utilization of a primary and secondary heating cycle of the pile. If functioning properly, the primary heating cycle will reach 120°F – 150°F within 1 to 2 days. After the primary bin is full, the primary heating cycle will commonly run for 60 to 90 days. However, the length of the primary cycle is dependent on the size of carcasses placed in the bin with 30 days being a typical cycle for farrowing house losses but up to 6 months for breeding stock. Once the primary bin temperature decreases to 110°F the composting material in the primary bin is removed and placed in the secondary bin. The process of moving material from the primary to secondary bin results in redistribution of water and provides for the additional oxygen needed to maintain bacterial growth. The secondary heating cycle, typically 60- to 90-days, is usually sufficient to reduce the carcass of large breeding stock to a few bones devoid of soft tissue. Temperature within the pile is not uniformly distributed and little to no active bacterial growth will occur when the pile reaches 60°F. If the pile fails to heat up, the material is too dry or wet. In some cases, failure to reach the proper temperature requires unloading the bin and mixing hot (active) compost material to restart the process.[1,2] Porcine reproductive and respiratory syndrome virus (PRRSV) can remain viable for 1 to 6 days at 68°F (20°C).[3] PRRSV can be inactivated at a temperature of 122°F (50°C) for more than 60 minutes.[4] The infectivity half-life of PRRSV is prolonged 14.6 hours at a temperature of 68°F (20°C) in the presence of manure.[5] Isolation of PRRSV in lung tissue and serum has been reported in samples stored at 77°F (25°C) seventy-two hours after necropsy.[6]

Materials and methods

Fly traps (n = 6) were hung around the compost facility of a 2,400 head tunnel ventilated wean-to-market site (Barn 1) in Eastern IA. The compost facility is about 60 ft from Barn 1. Similar traps (n = 8) were placed around the compost facility of the sow herd from which Barn 1 pigs were sourced, located 20 miles away. The sow farm was PRRSV-positive unstable and had wildtype PRRSV (1-2-4 L1C.2). Flies were trapped via Captivator® jug liquid traps (Starbar®, Central Life Sciences) using distilled water to solubilize trap attractant. Flies were allowed to enter the trap by their own volition. Negative control traps were hung that prevented flies from entering the device but allowed airborne matter to circulate into the trap. Flies were removed with a fish net and processed as a 10% homogenate solution using phosphate buffered saline. Homogenates were clarified by centrifugation and the supernatants analyzed at the Iowa State University veterinary diagnostic laboratory (ISUVDL) using the rRT-PCR test for PRRSV. In addition, all trap liquid, maggot, and leachate samples were likewise submitted for rRT-PCR testing at ISUVDL.

Results

Lung tissues collected from pigs the day of weaning at the sow farm before they were moved to Barn 1 were positive for PRRSV (1-2-4 L1C.2). At both 22- and 60-days post placement, oral fluid from Barn 1 pigs remained positive for PRRSV (1-2-4 L1C.2). There was 100% homology between all sample collections that were positive with the PRRSV 1-2-4 L1C.2 strain. In contrast, flies collected 22 days post-placement found all traps placed at the Barn 1 compost facility was positive for a different strain of PRRSV (1-4-4 L1C.5). As the Barn 1 compost facility served several other barns within the system, their PRRS status was reviewed. Oral fluid collected from Barn 2, 22 days prior to placement of pigs at Barn 1 was positive for PRRSV (1-33-4 L1C.5). Oral fluid collected from both Barn 3 and Barn 4 at day 8 and 14 respectively post-placement of pigs in Barn 1 were positive for PRRSV (1-4-4 L1C.5). There was 99.1% - 99.8% homology between the PRRS strain found in the flies from Barn 1 compost and the PRRSV strain found in oral fluid samples collected from Barns 2, 3 and 4. There was 90.2% - 90.3% homology between the PRRSV L1C.5 strains and the PRRSV L1C.2 strains (Table 1). Samples collected from the Barn 1 compost facility (60 days post placement) as well as samples collected from the sow farm compost (95 – 96 days post weaning) were all negative for PRRSV. The trap and control liquids were PRRSV rRT-PCR negative. Flies' positive for PRRSV were found only after hanging traps for seventy-two hours at one point in time. No PRRSV was found in flies or trap liquid hung for 2 hours – forty-eight hours, as well as maggots or leachate (Table 2).

Table 1: Sequence homology and distance from porcine reproductive and respiratory syndrome positive flies

Site	Barn 2 (oral fluid)	Sow farm (lung)	Barn 3 (oral fluid)	Barn 4 (oral fluid)	Barn 1 (flies)	Barn 1 (oral fluid)	Distance from barn 1 flies
Barn 2 (oral fluid)	xxxxx	90.2%	99.0%	99.1%	99.2%	90.2%	1.4 miles
Sow farm (lung)	90.2%	xxxxx	90.2%	90.3%	90.4%	100%	20 miles
Barn 3 (oral fluid)	99.0%	90.2%	xxxxx	99.8%	99.8%	90.2%	3.8 miles
Barn 4 (oral fluid)	99.1%	90.3%	99.8%	xxxxx	99.1%	90.3%	4.1 miles
Barn 1 (flies)	99.2%	90.4%	99.8%	99.8%	xxxxx	90.3%	
Barn 1 (oral fluid)	90.2%	100%	90.2%	90.3%	90.4%	xxxxx	60 feet

Table 2: Sequence and rRT-PCR results

Site	Date	Sample	No. positive	Strain	Comment
Barn 2	4/23/24	Oral fluid	1	1-33-4 L1C.5	
Sow farm	5/15/24	Lung	1	1-2-4 L1C.2	Day of weaning
Barn 3	5/23/24	Oral fluid	1	1-4-4 L1C.5	
Barn 4	5/29/24	Oral fluid	1	1-4-4 L1C.5	
Barn 1 (72 hours)	6/7/24	Flies	6 of 6	1-4-4 L1C.5	Ct 29.9 – 33.8 4466 – 67479 copies/mL
	6/7/24	Trap liquid	0 of 6		
	6/7/24	Control liquid	0 of 2		
	6/7/24	Maggots	0 of 2		
	6/7/24	Oral fluid	3 of 3	1-2-4 L1C.2	Ct 29.2 – 30.7
Barn 1 (2 hours)	7/15/24	Oral fluid	3 of 3	1-2-4 L1C.2	Ct 31.0 – 36.2
	7/15/24	Trap liquid	0 of 9		
	7/15/24	Control liquid	0 of 3		
Sow farm compost (2 hours)	8/20/24	Flies	0 of 1		
	8/20/24	Trap liquid	0 of 1		
	8/20/24	Control liquid	0 of 1		
Sow farm compost (48 hours)	8/21/24	Flies	0 of 8		
	8/21/24	Trap liquid	0 of 8		
	8/21/24	Control liquid	0 of 2		
	8/21/24	Maggots	0 of 1		
	8/21/24	Leachate	0 of 1		

Discussion

It has been reported that houseflies experimentally infected with PRRSV can infect a naïve population of swine over 0.07 miles.[7] In this study, greater than 99.0% homology was found in the PRRSV sequence from flies (Barn 1 compost) to PRRSV strains 1.4, 3.8 and 4.1 miles (Barns 2, 3 and 4, respectively) distant from the Barn 1 compost. It is unknown if the PRRSV-carrying flies from Barns 2 flew to the compost facility (Barn 1) or were transferred by taking mortalities from Barn 2 to the compost facility or if they picked PRRV up from materials in the compost. It is possible that PRRS carrying flies flew from the compost facility (Barn 1) and infected pigs in Barn 3 and/or Barn 4. However, it is clear that the PRRSV strain identified in the oral fluid samples from Barn 1 did not match the strain found in flies, exhibiting only 90.4% homology. Flies typically stay within a radius of one-half to 2 miles of their origin. However, reports of traveling up to twenty miles is reported as they travel for food or for oviposition.[8] This study reveals that flies found in a compost facility can harbor a strain of PRRSV found 1.4 to 4.1 miles distant from the facility and different from the strain in the nearest barn. Flies found in compost facilities may serve as a potential conduit of PRRSV that can be transmitted to naïve populations of swine distant from the compost facility. It is therefore important for swine practitioners to further fortify their biosecurity programs involving insect control around their farms and carcass disposal sites.

References

1. Extension, Iowa State Univ. Composting Swine Mortalities in Iowa. PM 1917 Reviewed November 2008.

2. Kohl, Kris; Stender, Dave. Field Tips for Successful Swine Mortality Composting. ISU James D. McKean Swine Disease Conf., Nov. 2020.pg 45-47.

3. Dee, Scott; Deen, John; Rossow, K; Weise, C; Eliason, R; Otake, S; Han Soo Joo; Pijoan, C. Mechanical transmission of porcine reproductive and respiratory syndrome virus throughout a coordinated sequence of events during warm weather. Can J Vet Res. 2003 Jan; 67(1): 12–19.

4. Jill van Kessel, MSc; Stacy Storm, MSc; Hans Deason, MEng; Elaine Van Moorlehem; Nathalie Berube, MSC; Shirley Hauta; Champika Fernando; Janet Hill, PhD; Terry Fonstad, PhD, PEng, Pag, Fec; Volker Gerdts, DVM. Time and termperature requirements for heat inactivation of pathogens to be applied to swine transport trailers. Journ. Sw. Hlth and Prod. Vol 29, No 1 pg 19-28.

5. Daniel C.L.; Linhares; Montserrat;Torremorell; Han Soo Joo; Robert B. Morrison, Infectivity of PRRS virus in pig manure at different temperatures. Vet, Micro. Vol 160, Issue 1-2, Nov 2012. Pg23-28.

6. W. G. Van Alstine; C. L. Kanitz, G; W. Stevenson. Time and temperature survivability of PRRS virus in serum and tissues. J Vet Diag. Invest 5:621-622 (1993).

7. Pitkin AN, Otake S, Deen J, Moon RD, Dee SA. Further assessment of houseflies (Musca domestica) as vectors for the mechanical transport and transmission of porcine reproductive and respiratory syndrome virus under field conditions. *Can J Vet Res* 2009; 73:91-96.

8. Extension, Penn State Univ. House Flies. Updated Jun 19,2023

https://doi.org/10.54846/am2025/44

A comparison of two different quantitative PCR (qPCR) sampling methods in the detection of coccidiosis in Canadian farms

Cesar Caballero[1], DVM, DVSc; Nicole Seebach[2], BSc; Ryan Tenbergen[3], DVM, MSc; Enrique Moncada-Fernandez[4], DVM, PhD; Daniel Sperling[5], DVM, PhD
[1]Ceva Animal Health Canada, Guelph, Ontario, Canada; [2]University of Guelph, Guelph, Ontario, Canada; [3]Demeter Veterinary Services PC, Woodstock, Ontario, Canada; [4]Ceva Animal Health, Lenexa, Kansas, USA; [5]Ceva Santé Animale, Libourne, France

Introduction

Porcine coccidiosis, a protozoal disease caused by the organism *Cystoisospora suis,* can affect preweaning pigs with watery or creamy white to yellow diarrhea, mortality and stunted growth,[1,2] which causes significant economic losses worldwide with high prevalence rates.[3,4] Swine veterinarians have used clinical observations of scours starting at the second and third week of age and laboratory confirmation to assess coccidiosis.[2] Techniques such as flotation and autofluorescence (AF) microscopy can detect oocytes obtained from fecal smears.[5] Recently, a new sampling method with a higher sensitivity (qPCR diagnostic test) that detects all stages of *C suis,* which helps with the short timeframe of oocysts excretions, was developed.[5-7]

Objective

The aim of this study was to compare two qPCR diagnostic protocols: Pooled fecal vs environmental samples to confirm the presence of *Cystoisospora suis* in the Canadian farms included in this study.

Methods

Over 11 months (Nov 2023 - Sept 2024), 17 clinically affected Canadian farms (4 different provinces), experiencing pre-weaning diarrhea with no toltrazuril treatment were involved in this trial. In each farm, sampling procedure included two protocols in parallel in the same litters at the second (2nd) and third (3rd) week of age (woa). (Figure 1).

Protocol #1. Sampling procedure (qPCR- fecal)

1. Cross-sectional sampling should be done on 10 litters.
2. Samples will be taken individually from randomly selected piglets within litter by perineal stimulation/anal evacuation. Take samples of all different consistencies (Figure 2).
3. Collect 5-6 grams of pooled feces per litter from 4-5 piglets per litter (5 grams is approximately 1 teaspoon of feces).
4. Do not sample litters or pigs with watery diarrhea.
5. Use screw on type vials, refrigerate after collection and ship on ice packs or some type of chilling.

Protocol #2. Sampling procedure (qPCR- environmental)

1. Sampling should be done on the same 10 litters.
2. Use either Swiffer or dish towels/cloths or scrub sponges.
3. Pre-soak the Swiffer or dish towel/cloth or scrub sponge in PBS, distilled water or tap water about 20-30 mL.
4. From each litter, collect environmental fecal samples from 5 different areas, then squeeze into a conical tube. Best to use a 50 mL conical tube for ease of transfer to the tube.
5. Refrigerate after collection and ship on ice packs or some type of chilling.
6. Change gloves between litters.

Cohen's Kappa (K) was used as the statistical method to quantify the level of agreement between fecal and environmental qPCR techniques (Table 1).

Figure 1: Fecal and environmental kit for farm sampling.

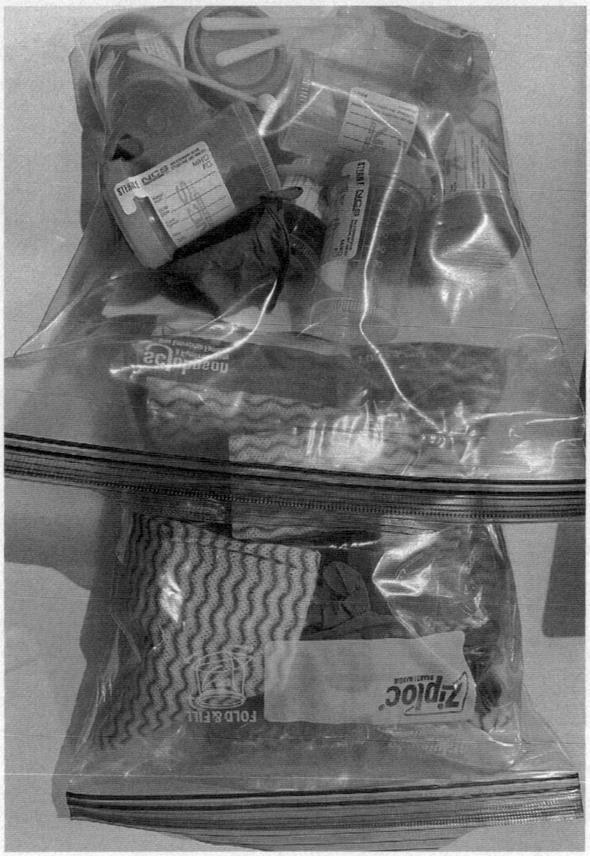

Figure 2: Fecal samples different consistency.

1 = Hard/firm; 2 = Low moisture; 3 = High moisture; 4 = Watery

Results

In total, 343 pooled fecal and environmental samples from 17 different farms were collected in parallel. The PCR cutoff value was Ct-value of 35. Under the conditions of this trial, Cohen's Kappa (K) test confirmed a substantial agreement (0.61-0.80) between both tests at both second and third woa (Kappa"0.645"). Also, with a total of 343 samples in parallel, the number of observed agreements where: 300 (87.46% of the observations, 16.61% [+ve]; 70.84% [-ve]) (Table 2).

Discussion

In infected piglets, oocysts excretion by *C suis* is usually short. For that reason, sensitivity is a key in the detection, diagnostics and control of coccidiosis in affected farms. Two qPCR diagnostic protocols (fecal and environmental) for the detection of *Cystoisospora suis* DNA were implemented. Under the conditions of this trial, after testing 17 clinically affected Canadian farms (4 different provinces) with 343 samples in parallel, all of them experiencing pre-weaning diarrhea with no toltrazuril treatment, the results concluded that there is a substantial agreement between both techniques: a direct detection of parasite DNA with qPCR assay with pooled fecal samples and environmental samples. The actual environmental protocol shows a significant advantage for field veterinarians when sampling collection is implemented, making it simpler to use, less stressful and easier to handle in each litter.

References

1. Lindsay, D.S., Dubey, J.P., Santin Durán, M. Chapter 66: Coccidia and Other Protozoa. In *"Diseases of Swine"*, Editors: Straw B.E., Zimmerman, J.J., D'Allaire, S., Taylor, D.J. Blackwell Publishing, Ames, Iowa, USA, 11th Edition. 2019. p.1015-1027.

2. Joachim, A., Ruttkowski, B., & Sperling, D. Detection of *Cystoisospora suis* in faeces of suckling piglets - when and how? A comparison of methods. *Porcine health management*, 2018; 4, 20.

3. Hinney B., Cvjetkovic V., Espigares D., Vanhara J., Waehner C., Ruttkowski B., Selista R., Sperling D. *Cystoisospora suis* control in Europe is not always effective. 2020. *Front. Vet. Sci.Sec.* ParasitologyVolume 7. **https://doi.org/10.3389/fvets.2020.00113**

4. Aliaga-Leyton, A., Webster, E., Friendship, R., Dewey, C., Vilaça, K., & Peregrine, A. S. An observational study on the prevalence and impact of Isospora suis in suckling piglets in southwestern Ontario, and risk factors for shedding oocysts. *The Canadian veterinary journal* 2011; 52(2), 184–188.

5. Hommonay Z, Vanhara J, Kars-Hendriksen S, Smits H, Sperling D, Kiss I. Specific real-time PCR assay for detection of *Cystoisospora suis* in piglet fecal samples. 2021. *Proceedings of the European Symposium of Porcine Health Management*, Bern, 438.

6. Mundt H.C., Joaquim A., Becka M., Daugschies A., Betlach A.M., Kettlkamp E., Rossow S., Payne B., Comparative field evaluation of a new *Cystoisospora suis* PCR in an endemic sow herd. In *proceedings of the 54th AASV Annual Meeting*, Nashville, Tennessee.

7. Wheeler M., Kettelkamp E., Okones J., Lauterbach., et al. Comparison of different sample types and assays for detecting *Cystoisispora suis* at the litter level. In *proceedings of the 54th AASV Annual Meeting*, Nashville, Tennessee.

Table 1: Standards for strength of agreement for Cohen's Kappa (K) test

Poor	Slight	Fair	Moderate	Substantial	Almost perfect
≤ 0	0.01 – 0.20	0.21 – 0.40	0.41 – 0.60	0.61 – 0.80	0.81 – 1.00

Table 2: Comparison of fecal vs environmental sampling methods

All farms	q PCR fecal (+ve)	q PCR fecal (-ve)	q PCR fecal total
q PCR Environmental (+ve)	57 (16.61%)	27	84
q PCR Environmental (-ve)	16	243 (70.84%)	259
q PCR Environmental (-ve)	73	270	343 (100%)

https://doi.org/10.54846/am2025/45

Estimating PRRSV immunogenicity to four challenge strains with an in-silico T cell model

J. M. Hammer[1]; J. Risser[1]; A. H. Gutierrez[2]; W. Martin[2]; A. S. De Groot[2]
[1]Elanco Animal Health, Greenfield, Indiana; [2]EpiVax, Inc., Providence, Rhode Island

Introduction

Porcine reproductive and respiratory syndrome virus (PRRSV) continues to evolve presenting a significant challenge for producers, veterinarians and vaccine developers. PRRSV vaccines must provide cross-protection against highly diverse strains. Vaccines that contain T cell epitopes well-matched to circulating strains are more likely to induce strong cell-mediated memory responses upon challenge or field exposure. Presented here is a preliminary study to assess immunogenicity through PigMatrix.[1] The technology uses published research,[2] propriety PigMatrix[1] and the epitope content comparison (EpiCC) algorithm[3,4] to predict and quantify T cell epitopes of Prevacent PRRS to four PRRSV field challenge strains' epitopes of structural proteins (GP2b [E], GP2, GP3, GP4, GP5, GP6 [M], and GP7 [N]).[5] These strains represent current circulating strains,[6,7] and the structural proteins evaluated all have been reported as having antigenic regions.[8] Vaccine to field strain relatedness defined by EpiCC scores quantifies vaccine "coverage" of these field strain T cell epitope content. This multiphase project is expected to develop a PRRSV T cell immunogenicity predictive tool to help elucidate immunological fit of PRRS vaccines to PRRSV field strains.

Method

T cell epitope prediction software (PigMatrix[1] and the EpiCC analysis[9]) were being applied and will be validated to estimate PRRSV immunogenicity of Prevacent® PRRS to the challenge strains facilitating further application to field strain.

T cell epitope prediction

PigMatrix prediction matrices uses the pocket profile method modified with a well-defined EpiMatrix binding preferences for human major histocompatibility complex (MHC) pockets.[1,9] PigMatrix[1] assessment scores range from approximately -3 to +3 and are normally distributed. PigMatrix[1] assessment scores above 1.64 are defined as high potential affinity binding to the MHC (hits). These higher affinity peptides are potentially immunogenic and worthy of further consideration in the model. It is expected about 5% of all assessments to score above 1.64.

Seven PRRSV structural proteins (E, GP2, GP3, GP4, GP5, M, and N)[4] amino acid sequences from Prevacent PRRS and four wildtype PRRSV-2 challenge strains[2] (NADC20, NADC30, VR2332, NC174) were obtained as input data. These input sequences were parsed into overlapping frames. Each 9-mer frame (9 amino acid peptide) was evaluated with respect to the previously selected class I (13) and class II (8) swine leukocyte antigen (SLA) to predict SLA binding affinity.[1,10]

EpiCC analysis

EpiCC[9] uses SLA class I and II PigMatrix[1] predictions to estimate the T cell epitope content shared between Prevacent PRRS vaccine sequence and a challenge strain and calculates a relatedness score (EpiCC score).[5] The 9-mer epitopes are considered shared if they are identical to or if they are "matched" by JanusMatrix,[11] which predicts binding to the same SLA and exactly matched at the T cell receptor (TCR) face.

To quantify vaccine T cell epitope coverage, the EpiCC score of each Prevacent PRRS-challenge strain comparison was divided by the challenge strain T cell epitope content (baseline EpiCC score) and expressed as a percentage. An example is shown in Figure 1. The T cell epitope content of a challenge strain is well matched or "covered" by Prevacent PRRS if the challenge strain baseline and the EpiCC score are similar. High coverage means the predicted T cell epitopes are similar and therefore should be cross protective. High T cell epitope coverage values can help to identify vaccines that may confer the broadest cross-reactive cell-mediated immunity and protection.

The EpiCC scores, baseline EpiCC scores, and T cell epitope coverages of individual proteins of each strain were combined to assess the overall shared epitope content between Prevacent PRRS vaccine and challenge strains.

Results

Putative T cell epitopes

Seven PRRSV structural proteins from Prevacent PRRS and 4 wild-type PRRSV-2 challenge strains or 35 input sequences were parsed into 6,010 individual 9-mer frames. Each frame was assessed for SLA binding potential with respect to 13 SLA class I alleles and 8 SLA class II alleles. In total, 126,210 frame-by-allele assessments were performed. Considering all the analyzed sequences, 5,607 putative SLA class I T cell epitopes and 3,799 putative SLA class II T cell epitopes were identified. The numbers of putative T cell epitopes identified in our analysis are shown in Table 1. From Table 1 in total, the four PRRSV challenge strains contained slightly fewer SLA class I (1,100 – 1,130) predicted T cell epitopes than the vaccine (1,146). As shown in Table 1 for the individual proteins E and N had the smallest number of putative epitopes and GP2 the largest. For class II, only NC174 contained more T cell epitopes than the vaccine. NC174's GP2 had the largest difference in number of predicted class II epitopes compared to the vaccine.

EpiCC scores

Table 2 shows the summary of Baseline EpiCC scores, EpiCC scores, and T cell epitope coverage. Figure 2 depicts Baseline EpiCC scores, EpiCC scores and T cell Epitope Coverage for a visual comparison. The baseline EpiCC scores ranged from 8.14 (N, VR2332) to 18.06 (E, NADC30), which suggests that T cell epitope density differed among these proteins (Table 2). E proteins had on average, the highest baseline scores. N proteins had the lowest baseline scores, indicating the lowest total T cell epitope content. However, within each protein, baseline scores were similar among strains.

Figure 1: An example of the T cell epitope content comparison (EpiCC) analysis.

Vaccine A and field strains (strains C and D) are screened for binding likelihood to a set of SLA alleles (one shown); 9-mers (rows) predicted to bind to a specific SLA allele (binding probability column) are shown in light (top 5%) or dark (top 1%) blue. (A) The baseline EpiCC score is defined by comparing the T cell epitope content of each sequence to itself. (B) For the comparison between VacA and field strains, the EpiCC score is based on shared T cell epitope content. Note that cross-conserved T cell epitopes, as defined by JanusMatrix, are considered shared (*3A). (C) T cell epitope coverage of each strain is calculated by dividing the EpiCC score of each vaccine-strain pair by the field strain's Baseline EpiCC score.

The combined Baseline EpiCC scores (Table 2) were also similar, ranging from 13.94 (VR2332) to 14.67 (NC174). The average combined EpiCC score was 8.3, ranging from 7.24 (NC174) to 8.9 (NADC20). Thus, the vaccine shares more T cell epitope content with NADC20. For these four PRRSV strains, the EpiCC scores in descending order were: NADC20, VR2332, NADC30 and NC174. However, the differences in EpiCC scores among these strains are small. For individual proteins, the EpiCC score ranged from 5.72 (N, NADC30) to 11.33 (M, NADC20). M had the highest average combined EpiCC score (10.52) and N had the lowest (6.25). The low EpiCC scores for N proteins are explained by their low Baseline EpiCC scores (ie, low T cell epitope content density). For proteins with low Baseline EpiCC scores, EpiCC scores are low even if the vaccine and the strain share a large percentage of their T cell epitopes.

T cell epitope coverage

The average combined coverage was 58.09%, with individual coverage scores ranging from 52.76% (NC174) to 62.72% (NADC20) (Figure 2) indicating the vaccine covers more of the T cell epitope content predicted in NADC20. For these four PRRSV strains, the vaccine T cell epitope coverage in descending order was: NADC20, VR2332, NADC30 and NC174. However, the differences in combined T cell epitope coverage among these strains are small, in particular, between VR2332 and NADC20 and between NADC30 and NC174.

For individual proteins, T cell epitope coverage ranged from 39.56% (GP2, NC174) to 82.02% (N, NADC20). N had the highest average coverage (74.1%) and GP5 had the lowest (49.98%). The vaccine covered at least 65% of the T cell epitope content identified in N proteins; however, the N sequences contained the lowest T cell epitope content. For NC174, the lowest coverage observed for GP2 was explained by an elevated Baseline EpiCC score (NC174 T cell epitope content) and a low EpiCC score (shared T cell epitope content). With exception of GP4 and GP5, vaccine coverage of NC174 T cell epitopes was the lowest. GP5 was the only protein where NC174 T cell epitope coverage was the highest.

Conclusions

Based on the combined T cell epitope content of PRRSV seven structural proteins, Prevacent® may confer broad cross-reactive cell-mediated immune response against the challenge strains. The model predicts the T cell immune response in descending order: NADC20 (62.72%), VR2332 (61.34%), NADC30 (55.55%), and NC174 (52.76%). These predictions rank the challenge strains in the same order as the challenge study rank. While cross-reactive responses are expected at this level of coverage, they might be limited. Although a specific coverage threshold that may confer protection or partial protection are unknown, it is hypothesized that higher T cell epitope coverage would likely result in stronger cross-reactive cell-mediated immune responses. However, the small differences in T cell epitope coverage between NADC20 and VR2332 and NADC30 and NC174 might not produce measurable differences experimentally. This evaluation is the initial phase of development of a PRRSV model to predict the immunogenicity of emerged or emerging PRRSV strains to Prevacent PRRS. The model when fully developed and validated will help swine

Table 1: Number of putative T cell epitopes. Class I and II T cell epitopes predicted for seven proteins from one vaccine and four PRRSV strains are shown. Differences in (1) the number of alleles used for prediction (13 alleles for class I and eight alleles for class II), and (2) the length of the input sequences, has an impact on the number of epitopes predicted

Alleles	Strain	Total[a]	E (73 aa)[b]	GP2 (256 aa)	GP3 (254 aa)	GP4 (178 aa)	GP5 (200 aa)	M (174 aa)	N (123 aa)
Class I (13 alleles)	Vaccin	1,146	65	293	211	167	181	166	63
	VR2332	1,100	75	270	218	156	161	157	63
	NADC20	1,125	65	28	216	156	177	164	63
	NADC30	1,130	68	262	222	152	172	185	69
	NC174	1,106	63	28	207	151	160	176	69
	Sum	**5,607**							
Class II (8 alleles)	Vaccin	763	47	171	145	109	146	10	39
	VR2332	731	48	178	130	106	134	95	40
	NADC20	741	51	163	135	10	140	102	42
	NADC30	762	57	172	141	103	148	102	39
	NC174	80	57	20	152	104	136	107	44
	Sum	**3,799**							

[a] Total number of putative T cell epitopes identified in the analyzed proteins per strain;

[b] Length of the input protein sequence.

Number of putative T cell epitope is colored from low (blue) to high (orange) by SLA allele class.

Table 2: Summary of EpiCC scores and T cell epitope coverage

Assessment	Strain	Combined[a]	E (73 aa)	GP2 (256 aa)	GP3 (254 aa)	GP4 (178 aa)	GP5 (200 aa)	M (174 aa)	N (123 aa)
Baseline EpiCC score	VR2332	13.94	17.18	16.57	12.76	14.13	14.45	13.71	8.14
	NADC20	14.19	16.64	16.25	12.94	14.27	15.42	14.52	8.35
	NADC30	14.43	18.06	16.02	13.40	13.75	15.76	15.47	8.42
	NC174	14.67	17.45	18.00	13.50	13.77	14.50	15.44	8.94
	Average	14.31	17.33	16.71	13.15	13.98	15.03	14.79	8.46
EpiCC score	VR2332	8.55	10.47	9.99	8.18	8.50	6.20	10.32	6.60
	NADC20	8.90	11.07	10.09	7.61	9.33	7.02	11.33	6.85
	NADC30	8.02	9.90	7.36	6.30	9.11	8.36	10.87	5.72
	NC174	7.74	9.43	7.12	6.16	8.97	8.48	9.55	5.85
	Average	8.30	10.22	8.64	7.06	8.98	7.51	10.52	6.25
T cell epitope coverage, %	VR2332	61.34	60.97	60.28	64.12	60.16	42.89	75.25	81.04
	NADC20	62.72	66.51	62.07	58.77	65.40	45.51	78.06	82.02
	NADC30	55.55	54.83	45.92	47.02	66.27	53.03	70.29	67.93
	NC174	52.76	54.02	39.56	45.64	65.13	58.48	61.85	65.41
	Average	58.09	59.08	51.96	53.89	64.24	49.98	71.36	74.10

[a] Combined scores calculated by combining the class I- and II-restricted T cell epitope content of the seven analyzed proteins. Baseline EpiCC scores, EpiCC scores, and T cell epitope coverage are colored from low (blue) to high (orange).

Figure 2: EpiCC scores and T cell epitope coverage. The combined Baseline EpiCC scores, EpiCC scores and T cell epitope coverage percentages are in the first box on the left. Baseline EpiCC scores, EpiCC scores and T cell epitope coverage percentages for seven individual proteins from four PRRSV strains are represented graphically in the following boxes.

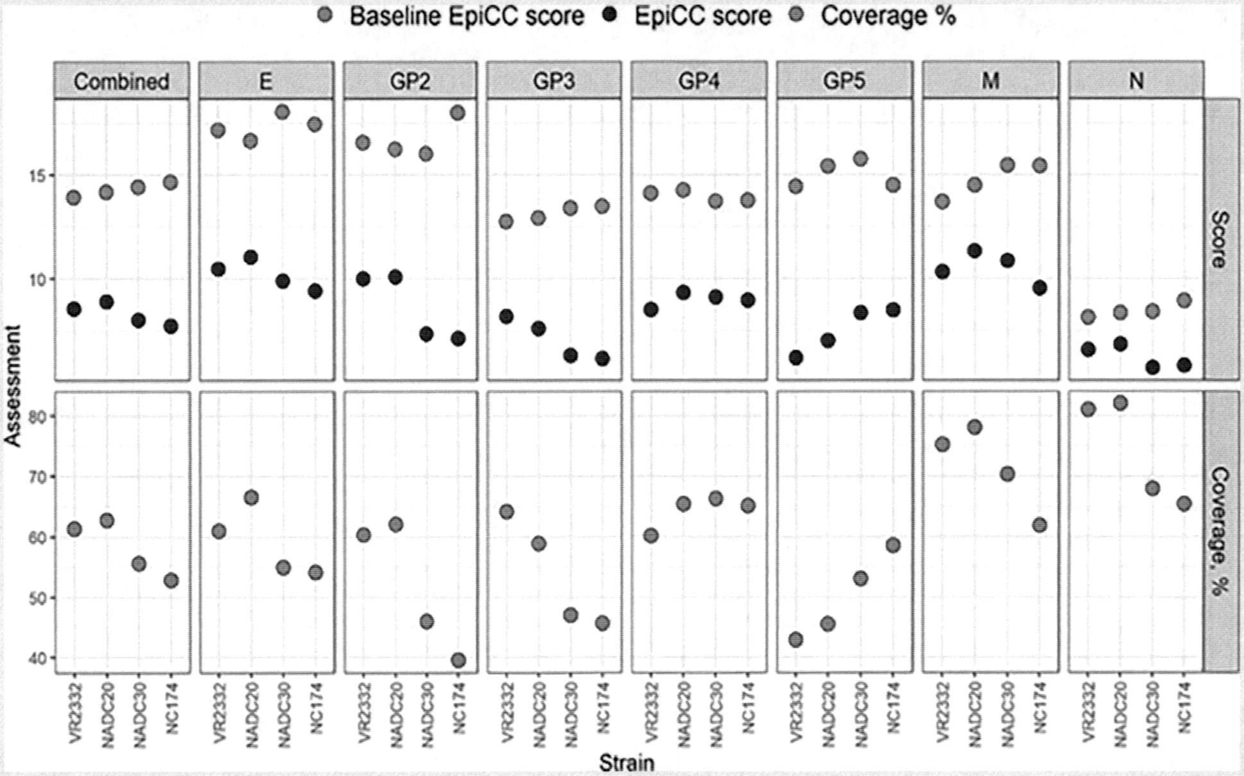

veterinarians predict the immunological fit Prevacent PRRS and potentially other vaccines currently on the market. It may also delineate the T cell epitopes important in building a broad immunological response to PRRSV.

References

1. Gutiérrez, A. H., W. D. Martin, C. Bailey-Kellogg, F. Terry, L. Moise and A. S. D. Groot (2015). "Development and validation of an epitope prediction tool for swine (PigMatrix) based on the pocket profile method." *BMC Bioinformatics* 16(1): 290.

2. Proctor, J., I. Wolf, D. Brodsky, L. M. Cortes, A. Frias-De-Diego, G. W. Almond, E. Crisci, T. T. N. Watanabe, J. M. Hammer and T. Käser (2022). "Heterologous vaccine immunogenicity, efficacy, and immune correlates of protection of a modified-live virus porcine reproductive and respiratory syndrome virus vaccine." *Frontiers in Microbiology* 13: 977796.

3. De Groot, A. S. D., L. Moise, F. Terry, A. H. Gutierrez, P. Hindocha, G. Richard, D. F. Hoft, T. M. Ross, A. R. Noe, Y. Takahashi, V. Kotraiah, S. E. Silk, C. M. Nielsen, A. M. Minassian, R. Ashfield, M. Ardito, S. J. Draper and W. D. Martin (2020). "Better Epitope Discovery, Precision Immune Engineering, and Accelerated Vaccine Design Using Immunoinformatics Tools." *Frontiers in Immunology* 11: 442.

4. Moise, L., A. H. Gutiérrez, S. Khan, S. Tan, M. Ardito, W. D. Martin and A. S. D. Groot (2020). "New Immunoinformatics Tools for Swine: Designing Epitope-Driven Vaccines, Predicting Vaccine Efficacy, and Making Vaccines on Demand." *Frontiers in Immunology* 11: 563362.

5. Lunney, J. K., Y. Fang, A. Ladinig, N. Chen, Y. Li, B. Rowland and G. J. Renukaradhya (2014). "Porcine Reproductive and Respiratory Syndrome Virus (PRRSV): Pathogenesis and Interaction with the Immune System." *Annual Review of Animal Biosciences* 4(1): 1-26.

6. Guo, J., Z. Liu, X. Tong, Z. Wang, S. Xu, Q. Chen, J. Zhou, L. Fang, D. Wang and S. Xiao (2021). "Evolutionary Dynamics of Type 2 Porcine Reproductive and Respiratory Syndrome Virus by Whole-Genome Analysis." *Viruses* 13(12): 2469.

7. Paploski, I. A. D., N. Pamornchainavakul, D. N. Makau, A. Rovira, C. A. Corzo, D. C. Schroeder, M. C. Cheeran, A. Doeschl-Wilson, R. R. Kao, S. Lycett and K. VanderWaal (2021). "Phylogenetic Structure and Sequential Dominance of Sub-Lineages of PRRSV Type-2 Lineage 1 in the United States." *Vaccines (Basel)* 9(6).

8. Vanhee, M., W. Breedam, S. Costers, M. Geldhof, Y. Noppe and H. Nauwynck (2011). "Characterization of antigenic regions in the porcine reproductive and respiratory syndrome virus by the use of peptide-specific serum antibodies." *Vaccine* 29(29-30): 4794-4804.

9. Gutiérrez, A. H., C. Loving, L. Moise, F. E. Terry, S. L. Brockmeier, H. R. Hughes, W. D. Martin and A. S. D. Groot (2016). "In Vivo Validation of Predicted and Conserved T Cell Epitopes in a Swine Influenza Model." PLoS ONE 11(7): e0159237.

10. Gutiérrez, A. H., V. J. Rapp-Gabrielson, F. E. Terry, C. L. Loving, L. Moise, W. D. Martin and A. S. D. Groot (2017). "T-cell epitope content comparison (EpiCC) of swine H1 influenza A virus hemagglutinin." *Influenza and Other Respiratory Viruses* 11(6): 531-542.

11. Moise, L., A. H. Gutierrez, C. Bailey-Kellogg, F. Terry, Q. Leng, K. M. A. Hady, N. C. VerBerkmoes, M. B. Sztein, P. T. Losikoff, W. D. Martin, A. L. Rothman and A. S. D. Groot (2013). "The two-faced T cell epitope: Examining the host-microbe interface with JanusMatrix." *Human Vaccines & Immunotherapeutics* 9(7): 1577-1586

https://doi.org/10.54846/am2025/46

Pradofloxacin: A novel dual-targeting antibiotic with a low propensity for resistance selection

J. M. Blondeau, MSc, PhD, RSM(CCM), SM(AAM), SM(ASCP), FCCP; Provincial Lead, Clinical Microbiology, Saskatchewan Health Authority; Head, Clinical Microbiology, Royal University Hospital & University of Saskatchewan; Clinical Associate Professor of Biochemistry, Microbiology and Immunology, Clinical Associate Professor of Pathology, Clinical Associate Professor of Ophthalmology, University of Saskatchewan, Saskatoon, Saskatchewan, Canada

Introduction

Swine respiratory disease (SRD) is a complex disease involving an interplay between viruses and management and environmental factors such as overcrowding, mixing different sources of animals, continuous flow ventilation and temperature fluctuations predisposing animals to secondary bacterial infections.[1,2] In particular, *Pasteurella multocida* is the most frequent opportunist bacterial pathogen and it's an organism with multiple virulence factors making it a successful pathogen causing considerable morbidity and mortality.[3] A number of other bacteria are also associated with SRD – some of which are listed below. SRD has an enormous impact on the swine industry in the North America and globally costing in excess of a billion dollars annually in the USA alone. Antimicrobial agents remain an important therapy for SRD and investigations with such agents are necessary to understand the strengths and weaknesses of each drug and to help determine which agent or agents are appropriate for optimizing therapy while at the same time being considerate of concerns regarding antimicrobial resistance.

Pradofloxacin (Pradalex™) is the newest of the veterinary fluoroquinolones to be approved (2024) for the treatment of swine respiratory disease caused by *Bordetella bronchiseptica, Glasserella (Haemophilus) parasuis, Pasteurella multocida, Streptococcus suis* and *Mycoplasma hyopneumoniae* (**www.elanco.com**). This drug has broad spectrum activity against Gram-positive and Gram-negative bacteria, some anaerobic bacteria and against atypical bacteria such as *Mycoplasma* species and others. The drug is an advanced generation fluoroquinolone and is dual targeting meaning that it simultaneously targets two enzymes critical for DNA replication giving it a low propensity for selection of antimicrobial resistance during exposure of bacterial populations to the drug.[4]

Fluoroquinolones are concentration dependent bactericidal antibiotics that kill bacteria by inhibiting enzymes in the bacterial cell (DNA gyrase and topoisomerase IV) critical for DNA replication.[5] Older generation quinolones such as enrofloxacin, danofloxacin, marbofloxacin and orbifloxacin preferentially target DNA gyrase in Gram-negative bacteria or topoisomerase IV in Gram-positive bacteria with some exceptions. Pradofloxacin, a third-generation quinolone is a novel dual targeting agent, simultaneously inhibiting both DNA gyrase and topoisomerase IV – a lethal effect. Dual targeting antibiotics are less likely to select for mutant subpopulations as it would require an organism to simultaneously possess 2 mutations in the genes encoding for the primary targets thereby reducing the likelihood for resistance selection.

Pharmacologically, pradofloxacin was investigated in pigs (n = 6) at a dose of 7.5mg/kg BW (Elanco data on file). The maximum serum concentration (C_{max}) was 2.5µg/mL with the time to maximum concentration being 0.75 hour and the volume of distribution was 2.53 L/kg. The area under the drug concentration curve was 17.56mg*h/L and the ½ life elimination time was 8.5 hours. The mean residence time was 6.39 hours. Pradofloxacin is approximately 33% protein bound and when considering the free (f) drug concentrations, target attainment for efficacy against Gram-negative bacteria (AUC/MIC ≥ 72) or Gram-positive bacteria (AUC/MIC ≥ 40) was exceeded for all relevant pathogens (ie, *P multocida* 1164, *A pleuropneumoniae* 390 and *S suis* 94). High target attainment (low MIC values and favorable PK/PD) and short drug residence time further reduces the likelihood for selection of antimicrobial resistance.

In vitro measurements to determine susceptibility or resistance of a bacteria to an antimicrobial agent is by the minimum inhibitory concentration (MIC) – a standardized measurement utilizing 10^5 (100,000) colony forming units per milliliter (CFU/mL) and controlled for numerous variables including media, drug concentration tested, pH and incubation temperature/atmosphere/duration. The determined MIC is compared to the established breakpoint whereby a value at or below the susceptibility breakpoint is considered susceptible and above the breakpoint is considered resistant. The mutant prevention concentration (MPC), described more than 20 years ago,[6] is an *in vitro* measurement utilizing $\geq 10^9$ (1 billion) CFUs tested on a series of agar plates containing a range of drug concentrations and following incubation (24 and 48 hours), the lowest drug concentration preventing any growth (100% inhibition) is the MPC.[5] This higher inoculum for MPC testing was selected as spontaneous mutant bacterial cells may occur in bacterial populations when the organism density is between 10^7-10^9 CFUs/mL. Such high bacterial densities have been reported in acute infections (many studies from humans and some from animals) of the central nervous system, lungs, urinary tract and others. The Mutant Selection Window (MSW) is a drug concentration range which is determined from the measured MIC and MPC drug concentrations and the difference in those 2 values determines the width of the MSW. A narrow MSW is better than a wider window. For drugs with therapeutic concentrations above the MSW, both fully susceptible and mutant bacterial cells (if present) would be inhibited by the drug. Drug concentrations below the MSW and as such, below the MIC, would not inhibit bacterial cells. However, for therapeutic drug concentrations that fall within the MSW (above the MIC but below the MPC), inhibition of the susceptible bacterial cells would occur, but mutant subpopulations would not be inhibited and selectively enriched in the presence of the drug. In this model, ideally, drugs would have MPC values at or below the susceptibility breakpoint.

Another *in vitro* measurement applied to antimicrobial agents and bacteria is time kill experiments.[7] In these measurements, varying densities of bacteria are exposed to clinically relevant drug concentrations (MIC, MPC, C_{max}, Maximum tissue drug concentration) and the $\log_{10}$ reduction and percent kill of viable cells recorded at predetermined timed intervals and compared to the starting cell density at time zero. This allows determination of bactericidal versus bacteriostatic activity along with the speed and extent of bacterial killing. Such data can be compared with the therapeutic drug concentrations to determine the likelihood for organism eradication over the dose or dosing interval. For kill experiments conducted in our research laboratory a minimum of 3 unique strains are used with triplicate measurements at each time point and the values averaged such that each time point there are 9 independent measurements (3 strains and measured in triplicate).

As concentration dependent agents, quinolone activity is defined by 2 pharmacodynamic/pharmacokinetic (PK/PD) parameters namely the maximum serum (C_{max}) to MIC ratio and the area under the drug concentration curve (AUC) to MIC ratio. The precise values of these ratios have been debated but a C_{max}/MIC ratio of 8-12 and an AUC/MIC >100 have been associated with a favorable clinical outcome and minimization of resistance.[8]

Investigations with pradofloxacin and swine bacterial pathogens

Actinobacillus pleuropneumoniae (n = 29) is not a label indication for pradofloxacin in swine but investigations – with this pathogen were conducted in our research laboratory.[9] Pradofloxacin was tested to determine MIC and MPC values and the concentrations of pradofloxacin inhibiting 50% (MIC_{50}) 90% (MIC_{90}) and 100% (MIC_{100}) were 0.031μg/mL, 0.031μg/mL and 0.125μg/mL. The corresponding MPC_{50}, MPC_{90} and MPC_{100} values were as follows respectively: 0.031μg/mL, 0.063μg/mL and 0.125μg/mL. The C_{max}/MIC_{90} ratio was 51.6 and the C_{max}/MPC_{90} ratio was 41.9. By comparison, the AUC/MIC_{90} ratio was 1075 and the AUC/MPC_{90} ratio was 273. Pradofloxacin therapeutic drug concentrations would be expected to exceed the MSW (considering the MIC_{90} and MPC_{90} values) for > 24 hours of the dose. In kill measurements conducted using 10^5 CFU/mL and over 180 minutes with sampling at 5, 10, 15, 20, 25, 30, 60, 120 and 180 minutes after drug exposure, pradofloxacin at the MPC drug concentration killed 1 $\log_{10}$ (72.1%) of viable cells at 60 minutes, 2.6 $\log_{10}$ (99.7%) at 120 minutes and 4.3 $\log_{10}$ (99.9%) at 180 minutes. Considering the C_{max} drug concentration, pradofloxacin killed 2 $\log_{10}$ (98.9%) of cells following 10 minutes of drug exposure and 2.5-5.7 $\log_{10}$ (99.4-100%) following 15 minutes to 180 minutes of drug exposure. Finally at the maximum tissue (lung) concentration (based on the value determined in cattle), pradofloxacin killed 1.9 $\log_{10}$ (94.4%) of cells following 15 minutes and 2.3-5.7 $\log_{10}$ (97.4-100%) following 20-180 minutes of drug exposure.

Pasteurella multocida (n = 41): The MIC_{50}, MIC_{90} and MIC_{100} values were ≤ 0.016μg/mL, ≤ 0.016μg/mL, 0.031μg/mL and the corresponding MPC values respectively were ≤ 0.016μg/mL, 0.031μg/mL and 0.063μg/mL. The C_{max}/MIC_{90} ratio was 165 and the AUC/MIC_{90} ratio was 1075 and the C_{max}/MPC_{90} ratio was 85.2 and the AUC/MPC_{90} ratio was 273. Pradofloxacin therapeutic drug concentrations would exceed the MSW (considering MIC_{90} and MPC_{90} values) for > 24 hours of the dose. At the MPC drug concentration pradofloxacin killed 1.3 $\log_{10}$ (56.6%) of cells following 120 minutes and 1.6 $\log_{10}$ (61.1%) following 180 minutes. At the C_{max} drug concentration, pradofloxacin killed 1.8 $\log_{10}$ (81.7%) of cells following 10 minutes of drug exposure and 4.2 $\log_{10}$ (98.8%) following 180 minutes. At the maximum tissue (lung) drug concentration, pradofloxacin killed 1.4 $\log_{10}$ (91.7%) of cells following 5 minutes of drug exposure and killing increased to 2.7-3.4 $\log_{10}$ (99.1-99.8%) following 20-60 minutes of drug exposure. When higher densities of bacteria (10^6-10^9 CFU/mL) were used in kill assays, pradofloxacin following 2 hours of drug exposure at the C_{max} drug concentration, killed 2.2 $\log_{10}$ (99.2%) of 10^6 CFU/mL, 1.9 $\log_{10}$ (96.1%) of 10^7 CFU/mL, 2.5 $\log_{10}$ (99.6%) of 10^8 CFU/mL and 2.1 $\log_{10}$ (98.9%) of 10^9 CFU/mL. Killing increased to 99.9-100% over all densities following longer drug exposures (ie, 4-24 hours). Nearly identical kill rates were seen following exposure to the maximum tissue (lung) drug concentration with 91.3-99.8% of cells killed following 2 hours of drug exposure and increasing to 99-100% kill with longer drug exposure (4-24 hours). The MIC values reported here for swine isolates of *P multocida* are consistent with MIC values for *P. multocida* strains from cattle and reported by us in 2024.[7] Figure 1 is a schematic representation of the pradofloxacin drug concentration curve in swine and the MSW for *P multocida*.

Streptococcus suis (n = 29): The MIC_{50}, MIC_{90} and MIC_{100} values were 0.031μg/mL, 0.25μg/mL and 0.25μg/mL and the corresponding MPC values were 2μg/mL, 2μg/mL and 2μg/mL. The C_{max}/MIC_{90} ratio was 21.1 and the AUC/MIC_{90} ratio was 137.6; the C_{max}/MPC_{90} ratio was 1.3 and the AUC/MPC_{90} ratio was 8.6. Kill experiments conducted with a bacterial density of 10^5 CFU/mL and over 180 minutes showed that pradofloxacin killed 0.5-0.7 $\log_{10}$ (66.8-68.2%) of bacterial cells following 120-180 minutes of drug exposure to the MIC drug concentration. By comparison, using the MPC drug concentration, pradofloxacin killed 0.9-2.1 $\log_{10}$ (87.7-99.2%) of bacterial cells following 60-180 minutes of drug exposure. At the C_{max} drug concentration, pradofloxacin killed 0.6-2.4 $\log_{10}$ (68.3-99.4%) of bacterial cells following 30-180 minutes of drug exposure. Finally at the maximum tissue (lung) drug concentration, pradofloxacin killed 0.6-1.7 $\log_{10}$ (71.6-97.1%) of cells following 60-180 minutes of drug exposure.

Figure 1: Pradofloxacin drug curve in swine considering the mutant selection window for *Pasteurella multocida*.

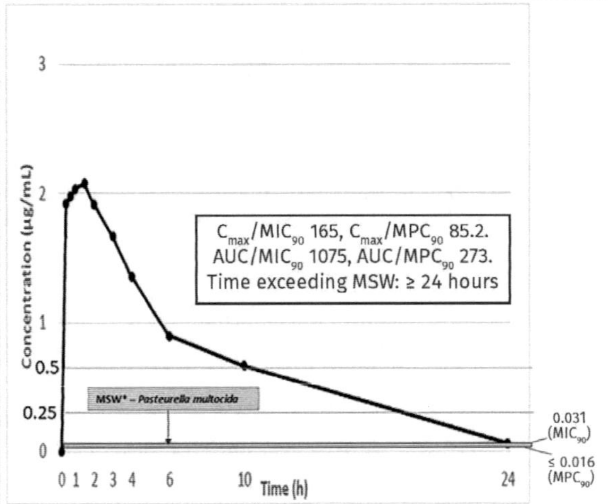

* MSW = mutant selection window.

Conclusions

Pradofloxacin is a novel dual targeting veterinary fluoroquinolone approved for swine respiratory disease. This dual targeting mode of action results in lower MICs increasing potency while also reducing the likelihood of resistance selection. Investigations in our laboratory with key swine respiratory pathogens showed low MIC and MPC values and when considering C_{max}/MIC_{90} and AUC/MIC_{90} ratios, pradofloxacin greatly exceeds preestablished values shown to be associated with favorable clinical outcomes and minimization of resistance. Low MPC values in the context of therapeutic drug concentrations show pradofloxacin will exceed the MSW for > 24 hours of the dose. In kill studies with varying densities of bacteria and with clinically relevant drug concentrations, pradofloxacin was rapidly bactericidal with kill occurring shortly after drug exposure and reaching > 95-100% kill with drug exposures within the 24 hours of the dose. Rapid bactericidal activity and 100% reduction in bacterial populations further reduce the likelihood for resistance. The late Dr Charles Nightingale is credited with coining the phrase "Dead bugs don't mutate!" Pradofloxacin appears to be an important addition to antimicrobials currently available for the treatment of swine respiratory disease.

References

1. Brockmeier, S.L.; Halbur, P.G.; Thacker, E.L. Porcine Respiratory Disease Complex. In *Polymicrobial Diseases*, Brogden, K.A., Guthmiller, J.M., Eds.; ASM Press: Washington, DC, 2002.

2. Done, S.H. Environmental factors affecting the severity of pneumonia in pigs. *Vet. Rec.* 1991, *128*, 582-586, **doi:10.1136/vr.128.25.582**

3. Vu-Khac, H.; Trinh, T.T.H.; Nguyen, T.T.G.; Nguyen, X.T.; Nguyen, T.T. Prevalence of virulence factor, antibiotic resistance, and serotype genes of Pasteurella multocida strains isolated from pigs in Vietnam. *Veterinary world* 2020, 13, 896-904, **doi:10.14202/vetworld.2020.896-904**

4. Sykes, J.E.; Blondeau, J.M. Pradofloxacin: a novel veterinary fluoroquinolone for treatment of bacterial infections in cats. *Vet. J.* 2014, 201, 207-214, **doi:10.1016/j.tvjl.2014.06.008**

5. Blondeau, J.M.; Hansen, G.T.; Metzler, K.L.; Hedlin, P. The role of PK/PD parameters to avoid selection and increase of resistance: mutant prevention concentration. *J. Chemother.* 2004, 16, 1-19.

6. Dong, Y.; Zhao, X.; Domagala, J.; Drlica, K. Effect of fluoroquinolone concentration on selection of resistant mutants of *Mycobacterium bovis* BCG and *Staphylococcus aureus*. *Antimicrob. Agents Chemother.* 1999, 43, 1756-1758.

7. Blondeau, J.M.; Fitch, S.D. Comparative *In Vitro* killing by Pradofloxacin in Comparison to Ceftiofur, Enrofloxacin, Florfenicol, Marbofloxacin, Tildipirosin, Tilmicosin and Tulathromycin against Bovine Respiratory Bacterial Pathogens. *Microorganisms* 2024, 12, 996.

8. Schentag, J.J.; Gilliland, K.K.; Paladino, J.A. What have we learned from pharmacokinetic and pharmacodynamic theories? *Clin. Infect. Dis.* 2001, 32, S39-46.

9. Blondeau, J.M.; Fitch, S.D. Comparison of the minimum inhibitory and mutant prevention drug concentrations for pradofloxacin and 7 other antimicrobial agents tested against swine isolates of *Actinobacillus pleuropneumoniae* and *Pasteurella multocida*. *Molecules* 2024, In Press.

https://doi.org/10.54846/am2025/47

Creating a more robust pig

John Sonderman[1], PhD; Brady McNeil[1], MS; Amanda Cross[1], PhD, MBA;
Christine Mainquist-Whigham[1,2], DVM, MS; Tom Rathje[1], PhD, MBA
[1]DNA Swine Genetics, Columbus Nebraska; [2]Pillen Family Farms, Columbus, Nebraska

Introduction

One metric that is most closely associated with throughput and total system profitability is the number of full value pigs. However, pig mortality has increased from 2017 to 2023 by 3.43% resulting in 23.5% of pigs born alive not making it to market in 2023.[1,2] Genetic companies have been able to select on traits correlated with survival such as birth weight, wean weight, feed intake, and against defect traits such as umbilical and scrotal hernias. However much of the focus on decreasing mortality fell on the producers. Recently, genetic companies have been able to implement new strategies to select for improved pig robustness through selection on allele frequency and large-scale commercial test data with pedigreed market pigs.

E coli

F18 *Escherichia coli* (*E coli*) is a leading cause of nursery death loss and culls due to diarrhea. A specific genotype has been shown to be resistant to F18 *E coli*. To prove this within the DNA full program (L246) population, three genotypes were evaluated, Genotype A was homozygous for the susceptible allele, Genotype B was heterozygous for F18 *E coli* resistant alleles, and Genotype C was homozygous for F18 *E coli* resistant alleles. A total of 179 pigs were exposed to an F18 *E coli* isolate 4 days post weaning at a geometric mean concentration of 9.8 x 10^9 colony forming units per milliliter over a 3-day period.[3] Pens of 12 pigs separated by genotype were tracked for growth, feed conversion, and mortality for 34 days post-weaning. Pigs in Genotype C had significantly ($P < .05$) higher ADG at 381 grams per day compared to 281 and 299 grams per day for Genotype A and B respectively. Genotype C had a lower FCR at 1.35 compared to Genotype A at 1.67 ($P < .01$). Mortality rates were 26.7%, 18.3% and 0% for Genotypes A, B, and C respectively ($P < .01$). By incorporating selection for the homozygous resistant genotype across maternal and terminal lines, genetic companies can increase the prevalence of pigs resistant to F18 *E coli* and create more full value pigs at marketing.

Full program test

Nucleus and commercial environments produce different stressors on pigs due particularly to the difference in disease pressure between the two environments. In order to select animals that will survive in a commercial environment, genetic lines need to be evaluated within a commercial environment. Evaluating offspring of nucleus level sires in a commercial environment causes the sires to re-rank. DNA has followed over 250,000 pedigreed commercial pigs representing over 350 nucleus sires and recorded mortality from birth through the finish phase. In a low health setting, sire's offspring have mortality ranging between 12-28% from birth to market, with post-weaning mortalities accounting for 2-11%. While survivability is a lowly heritable trait, since 2021 we have seen a reduction of 2.6% of the mean birth to market mortality.

Maternal selection

While direct selection for a reduction in mortality has resulted in genetic progress. Indirect selection on traits correlated with survival also has had an effect. Specifically, selection for birth weight and weaning weight. Over the last 5 years, the number of DNA full program pigs weighing less than 900 grams at birth have decreased by 50%. Based on survivability curves, pigs that weigh less than 900 grams have below a 60% chance of survival from birth to weaning. Within the DNA L200 farms, pre-wean mortality has dropped 5% due to the increase in pig size at birth. Weaning weights have also improved over 225 grams over a similar timeframe, leading to an easier to start pig in the nursery.

Conclusion

A total system approach is required to create more full value pigs. In addition to management and health factors, producers need to work with and understand what their genetic supplier can provide. By utilizing genomics and large commercial level data, genetic companies can make progress in traits not previously possible.

References

1. MetaFarms. (2021). Production Analysis Summary for U.S. Pork Industry: 2017-2021. **https://porkcheckoff.org/wp-content/uploads/2022/05/MetaFarms-2021-Production-Analysis-for-NPB.pdf**

2. MetaFarms. (2023). Production Analysis Summary for U.S. Pork Industry: 2019-2023. **https://porkcheckoff.org/wp-content/uploads/2024/07/2024-Production-Analysis-Full-Report-1.pdf**

3. Welch MW, Cross AJ, Solar Diaz IDP, Johnson DC, Parr E, Rathje TA, Borg RC, Boler DD. Genetic selection reduces mortality and improves growth performance of nursery pigs after an experimental virulent F18 enterotoxigenic Escherichia coli (ETEC) challenge. *Leman Swine Conference.* St. Paul, MN, 2024. 144.

https://doi.org/10.54846/am2025/48

Efficacy evaluation of INGELVAC PRRS® MLV against a PRRSV 1-4-4 L1C.5 variant challenge

R. Philips, DVM; J. Fiechtner, DVM; G. Haiwick, BS
Boehringer Ingelheim Animal Health USA Inc., Duluth, Georgia

Introduction

Porcine reproductive and respiratory syndrome (PRRS) is a complex and costly disease, due in part to the high degree of genetic variation among PRRS field isolates. Throughout its 30-year history, INGELVAC PRRS MLV has demonstrated its ability to provide heterologous protection against numerous wild-type PRRSV strains that differ from the vaccine strain in terms of nucleotide sequence and lineage.[1-3] The use of INGELVAC PRRS vaccines can significantly reduce lung lesions following challenge with heterologous PRRSV isolates (86-94% ORF5 nucleotide similarity) in the three-week-old pig respiratory challenge model.[1-3] In October 2020, a new PRRSV 1-4-4 L1C.5 variant strain emerged in the United States that has been associated with high mortality and morbidity of pigs.[4,5] A previous study has demonstrated that modified live PRRS vaccines offer a level of protective efficacy against the PRRSV L1C variant strain.[6] The objective of this study was to evaluate the efficacy of INGELVAC PRRS MLV (Boehringer Ingelheim Animal Health USA Inc., Duluth, Georgia) in a three-week-old pig respiratory challenge model, against a PRRSV RFLP 1-4-4 L1C.5 variant field isolate that was isolated in 2021.

Materials and methods

A randomized, blinded vaccination-challenge study was performed in 98, three-week-old pigs from a PRRS naïve source. Pigs were confirmed to be PCR negative for PRRS prior to the initiation of the study. The study consisted of 3 treatment groups of pigs. Group 1; n = 46, was vaccinated with INGELVAC PRRS MLV (2 mL intramuscular (IM) per label recommendations), on day 0. Group 2; n = 46, was a non-vaccinated challenge control treatment group that received a 2 mL placebo IM injection on day 0. Group 3; n = 6, was a non-vaccinated and non-challenged strict control group that served as an environmental control. Pigs were housed in rooms by group during the vaccination period. Group 3 pigs were removed and euthanized prior to challenge and examined for lung lesions and confounding bacterial pathogens. At day 28 of the study (D28), all pigs were comingled and challenged with 2.0 mL IM and 2.0 mL intranasally (1 mL per nostril) of inoculum containing $10^{4.99}$ Log_{10} $TCID_{50}$/dose of virulent PRRSV RFLP 1-4-4 L1C.5 variant. INGELVAC PRRS MLV has 86.4% similarity to the PRRSV L1C.5 variant challenge isolate based on ORF 5 sequence. Blood was collected from all pigs on days 0, 7, 14, 21, 27, 29, 31, 33, 35 and 42. Serum samples were tested by RT-PCR for the presence of viremia and by ELISA for the presence of anti-PRRSV antibody. Viremia testing for PRRSV was done by the Iowa State University Veterinary Diagnostic Lab, Ames, Iowa, using quantitative PCR (qPCR). Results of the qPCR were reported as "positive or negative" and also as PRRSV quantity in units of genomic copies per mL serum. Pigs were weighed at day 0, day of challenge (D28), and termination of the study (D42) to assess average daily weight gain (ADWG). On D42 (14 days post-challenge), all pigs were necropsied and lungs were scored for the presence of macroscopic lesions. Data were analyzed using SAS® Generalized Linear Mixed Models. Pairwise comparisons between groups were conducted as appropriate using a level of confidence of 0.05 with multiplicity-adjusted *P*-values to indicate statistical significance. The multiplicity adjustment was accomplished using ADJUST=SIMULATE in the LSMESTIMATE option in SAS® GLIMMIX.

The study design is summarized in Table 1.

Results

Lung lesions: The day 42 (14 days post-challenge) lung lesion assessment found the vaccinated treatment group to have significantly lower median gross lung lesion scores compared to the non-vaccinated (Placebo) challenged treatment group. The median percent lung lesion score for INGELVAC PRRS MLV were 14.13%, compared to a median lung lesion score of 28.10% for the non-vaccinated challenged treatment group.

Post-challenge lung lesion results are summarized in Table 2.

Average daily weight gain: In the post-challenge phase of the study (day 28-day 42) the vaccinated treatment group demonstrated a significant increase in ADWG compared to the non-vaccinated (Placebo) challenged treatment group. ADWG for INGELVAC PRRS MLV was 0.35 lbs per day, compared to a 0.07 lbs per day ADWG for the non-vaccinated challenged group.

The post-challenge ADWG results are summarized in Table 3.

Viremia: Following challenge, the non-vaccinated challenge control pigs became viremic within 24 hrs of infection (D29), and all animals remained viremic during the post-challenge period. In comparison to the challenge control group, the INGELVAC PRRS MLV vaccinated group had significantly lower levels of viremia as measured and analyzed by Log_{10} genomic copies at D29 and D42 of the study - D29 and D42; ($P < .0001$ and $P < .0002$ respectively).

Post-challenge viremia results are summarized in Figure 1.

Table 1: Study design

Group	N	Treatment	Day 0	Day 28	Day 42
1	46	INGELVAC PRRS® MLV	Vaccinated	PRRSV 1-4-4 L1C.5 challenge	Necropsy – lung lesion evaluation
2	46	Placebo	Sterile diluent non-vaccinated		

Table 2: Day 42 lung lesions (median %) in vaccinated and non-vaccinated pigs challenged with PRRSV L1C.5 variant

Group	Treatment	Median percent lung score
1	INGELVAC PRRS® MLV	14.13[a]
2	Placebo - challenge control (non-vaccinated)	28.10[b]

[a,b] Different letters indicate significant ($P \leq .05$) differences

Table 3: Post-challenge average daily weight gain (ADWG) (day 28-42)

Group	Treatment	ADWG in lbs (95% CI)
1	INGELVAC PRRS® MLV	0.35[a] (0.26, 0.45)
4	Placebo - challenge control (non-vaccinated)	0.07[b] (-0.03, 0.16)

[a,b] Different letters indicate significant ($P \leq .05$) differences

Discussion and conclusions

The primary objective of this study was to evaluate the efficacy of INGELVAC PRRS MLV against a current and virulent heterologous PRRSV 1-4-4 L1C.5 variant isolate in a respiratory challenge model. While the PRRSV L1C.5 variant challenge proved to be highly virulent, INGELVAC PRRS MLV demonstrated protection compared to the non-vaccinated challenge control group. Mitigating this negative biological impact can have a significant economic benefit. In this study, the INGELVAC PRRS MLV group demonstrated a significant improvement in ADWG compared to the challenge control group during the post-challenge period. INGELVAC PRRS MLV vaccinated pigs had a 5-fold increase in ADWG when compared to the non-vaccinated challenge control group. In addition, INGELVAC PRRS MLV demonstrated a significant reduction in the level of post-challenge viremia compared to non-vaccinated challenge controls.

This study is another example demonstrating the ability of INGELVAC PRRS MLV to protect against a relevant PRRSV challenge. Implementing vaccine in a systematic methodology for PRRS control can mitigate the consequences of infection and subsequently improve the health and performance of pigs.

INGELVAC PRRS® is a registered trademark of Boehringer Ingelheim Vetmedica GmbH, used under license.

Figure 1: Post-challenge viremia – least square means of genomic copies on Log_{10} scale by group & day.

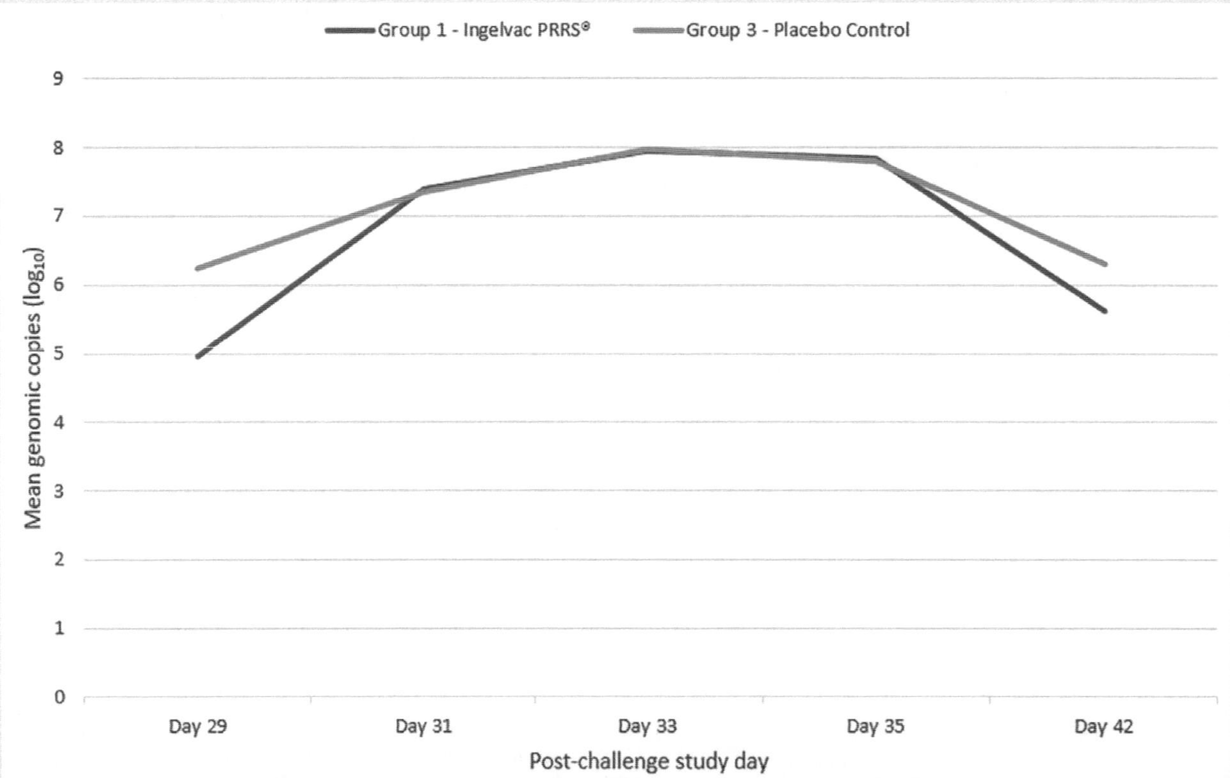

References

1. Patterson, A, et al. 2019. Efficacy of Ingelvac PRRS® modified live virus vaccine against heterologous Porcine Reproductive and Respiratory Syndrome virus challenges. *J Vaccines Vaccin 10:407.*

2. Patterson, A., et al. 2013. Modified live PRRSV vaccination is efficacious following challenge with eight genetically diverse PRRSV isolates. *A.D. Leman Swine Conference Proceedings.*

3. Philips, R, et al. 2021. Comparative efficacy evaluation of two modified-live PRRS vaccine against a heterologous PRRSV 1-7-4 challenge. *AASV Annual Meeting Conference Proceedings.*

4. Trevisan, G, et al. 2021. Complete Coding Genome Sequence of a Novel Porcine Reproductive and Respiratory Syndrome Virus 2 Restriction Fragment Length Polymorphism 1-4-4 Lineage 1C Variant Identified in Iowa, USA. *Microbiol Resour Announc.* 10:e0044821.

5. Kikuti, M, et al. 2021. Emergence of a new lineage 1C variant of porcine reproductive and respiratory syndrome virus 2 in the United States. *Front Vet Sci.* 8:752938.

6. Rawal, G., Zhang, J, et al. 2023. Comparative evaluation of protective efficacy of six commercial PRRSV-2 MLV vaccines against the emergent PRRSV 1-4-4 L1C variant strain in weaned pigs. *ISU James McKean Swine Disease Conference Proceedings.*

https://doi.org/10.54846/am2025/49

Evaluation of the effect of Enterisol Salmonella T/C® vaccination on the reduction of *Salmonella* on farm and at slaughter

L. F. Hernig[1]; F. Leite[2]; F. de Quadros[1]; J. D. Kich[2]; A. Coldebella[3]; R. T. Lippke[1]; D. P. Fagundes[1]; T. R. M. Ce[4]
[1]Boehringer Ingelheim Animal Health Brazil; [2]Boehringer Ingelheim Animal Health USA Inc.; [3]EMBRAPA Suínos e Aves, Concórdia, Brazil; [4]Seara Alimentos LTDA, Brazil

Introduction

Salmonella is a significant pathogen for human and animal health and can impact food safety. In this scenario, the swine industry targets *Salmonella* reduction in animals and food products since pork has been recognized as one of the major sources for human salmonellosis around the world.[1] Pigs may harbor *Salmonella* even when they are asymptomatic, persisting in the host with intermittent fecal shedding.[2] Interventions at the farm level such as biosecurity, cleaning, disinfection, and vaccination are important tools for swine health, which improve performance, and reduce the chances of carcasses contamination at the slaughterhouse.[3] Studies have demonstrated that vaccination can be a viable approach to reduce *Salmonella* in lymph nodes at slaughter under field conditions.[4] Moreover, vaccination has been recognized as one of the main tools to prevent salmonellosis.[2,3] The aim of this study was to evaluate the effect of vaccination with Enterisol Salmonella T/C® on *Salmonella* excretion in feces and colonization of lymph nodes at slaughter.

Methods

Farrowing phase

The study was conducted in a commercial farm in Santa Catarina state, south of Brazil, between August 2023 and January 2024. A total of eighty litters with an average of 14 days of age, from the same farm and farrowing batch, were selected and divided into two treatments: 1) vaccinated group and 2) non-vaccinated control group. Piglets were individually weighed, and ear tagged for identification. The piglets were assigned to treatment group randomly after blocking of litters by sow parity and weight of litters, with 40 litters being in the vaccinated group and another 40 litters assigned to the non-vaccinated control group. A total of 2,110 piglets: 1,048 piglets (533 males and 515 females) were applied to the vaccinated group and 1,062 piglets (546 males and 516 females) to the control group. The vaccinated group received a full 2 mL dose of Enterisol Salmonella T/C® vaccine, which contains a *S.* Choleraesuis and a *S.* Typhimurium vaccine strain (Boehringer Ingelheim Animal Health USA Inc.), orally by drench at 14 days of age. The control group received 2 mL of saline orally at the same age. The vaccination process was done one day after the weighing of animals. All piglets from this study were monitored from the lactation phase to the slaughterhouse.

Nursery phase

Piglets from this study were all weaned with an average of 21 days of age and moved together to the same nursery farm. In the nursery, vaccinated and control groups were allocated into the same pens in a proportion of 50% of each treatment with males and females housed separately. In the second week of the nursery, a subgroup of 120 piglets were selected from 20 pens (3 piglets/treatment/pen) to take blood and fecal samples. These same animals were sampled again (blood and fecal) in the last week of the nursery to evaluate seroconversion for *Salmonella* by ELISA (Svanovir, IDEXX) and excretion of *Salmonella*, respectively. At the end of the nursery phase all piglets were individually weighed and taken to a finishing barn.

Finishing phase

Similar to the nursery phase, vaccinated and control groups were allocated into the same pens in a proportion of 50% of each treatment with males and females housed separately. The same subgroup was sampled again in two moments, including in the last week of finishing phase when all animals were individually weighed.

Transport and slaughterhouse

All animals were transported to the same slaughterhouse. First, the vaccinated group was loaded onto the truck onto the second floor, and then the control group was loaded onto the first floor. The floor surface of the truck was sampled with overshoes on both floors before the animals were loaded for transport to the slaughterhouse. Likewise, samples were collected from the floor surface and drinkers of the lairage pens where the animals were housed at the slaughterhouse. These samples were collected and sent to the laboratory to check for *Salmonella* contamination, through bacteriological analysis. At the slaughterhouse, ileocecal lymph nodes of 80 animals from the same subgroup (40/treatment) were collected and sent to the laboratory for bacteriological analyses to investigate the presence of *Salmonella*.

Statistical analyses

For the weight data, analysis of variance was applied, considering the effects of pen, sex and treatment and the random effect of pen within treatment. The MIXED procedure of SAS (2012) was used to perform this analysis. The evaluation of the OD% data of the ELISA serological test for *Salmonella* was performed using PROC MIXED of the SAS (2012), according to XAVIER (2000). The structure used in the analysis was chosen based on the lowest value of the Akaike Information Criterion (AIC). The estimation method used was restricted maximum likelihood. In addition to the previous analysis, the OD% data were also transformed into a qualitative response in positive and negative, considering the cutoff points of 10, 20 and 40%. The analysis of these data was performed using the generalized estimating equations (GEE) proposed by Zeger and Liang (1986) and Liang and Zeger (1986). In this case, the probability distribution was considered binomial. The GENMOD procedure of SAS (2012) was used to test the fixed effects of pen, treatment, time after vaccination and the interaction between the last two factors. The working correlation matrix used was the Exchangeable. The chi-square ($\chi2$) test was used to evaluate the effect of the treatments on the presence of *Salmonella* in feces on the farm before slaughter and at the slaughterhouse and in ileocecal lymph nodes. In addition, the McNemar test was applied to evaluate the effect of the sample type on the isolation of *Salmonella*.

Results

The growth performance data is shown in Table 1. As expected, both treatments had similar ($P > .05$) body weight at the beginning of the study (at 13 days of age). However, even though no statistically significant difference between treatments was observed, the weight of the vaccinated treatment group was 0.4 lb higher at the end of the nursery phase (66.18 vs 65.79 lbs, $P = .292$) and 2.64 lbs higher at the end of the finishing phase (292.99 vs 290.35 lbs, $P = .076$) than the non-vaccinated animals.

Considering the serological results, more pigs were positive for antibodies against *Salmonella* in the vaccinated group than the control group in the four timepoints evaluated (Figure 1). There was also an effect of time after vaccination on the result obtained, however the interaction between treatment and time after vaccination was not significant, indicating that the treatment effect was independent of time. The seroconversion in the control group validated the natural *Salmonella* infection that occurred during the trial.

When evaluating *Salmonella* culture results, the proportion of pigs shedding *Salmonella* increased more than two times before loading on the farm and after transport to the slaughterhouse in the control group (20.0% to 42.5%, respectively) ($P < .05$, Figure 2). In the vaccinated group, the number of pigs shedding *Salmonella* remained similar on the farm and at the slaughterhouse (22.5% to 27.5%, respectively) (Figure 2). This indicates that transport did not increase shedding of *Salmonella* in the vaccinated group while an increase did occur in the control group. When evaluating culture results of the ileocecal lymph nodes measured at slaughter, a significantly lower percentage ($P < .02$) of lymph nodes were positive for *Salmonella* in the vaccinated animals (2.5%) compared to the control non-vaccinated animals (17.5%) (Figure 3). All samples collected from the floor surface of the truck (before loading the animals to the slaughterhouse), lairage pen floor and lairage pen drinkers tested negative for *Salmonella*.

Discussion

Salmonella can lead to foodborne illness. In Brazil, the Ministry of Agriculture, Livestock and Supply, established the normative instruction N° 60, of December 20, 2018, for microbiological control of pig carcasses, with the objective of evaluating the hygiene of the process and reducing the prevalence of pathogenic agents.[5] With this, several companies in the country have made investments to improve manufacturing practices both at the

Table 1: Body weight (BW) (mean ± SE) measured at beginning of the study, end of the nursery and finishing phases

	Treatment		
	Control	Vaccinated	*P*-value
Initial BW, lb - 13 days of age	9.05 ± 0.07	8.88 ± 0.07	.098
Nursery BW, lb - 48 days of age	65.79 ± 0.26	66.18 ± 0.26	.292
Finishing BW, lb - 146 days of age	290.35 ± 1.32	292.99 ± 1.32	.076

Figure 1: Percentage of pigs tested positive for *Salmonella* at different timepoints post vaccination measured by ELISA of serum samples.

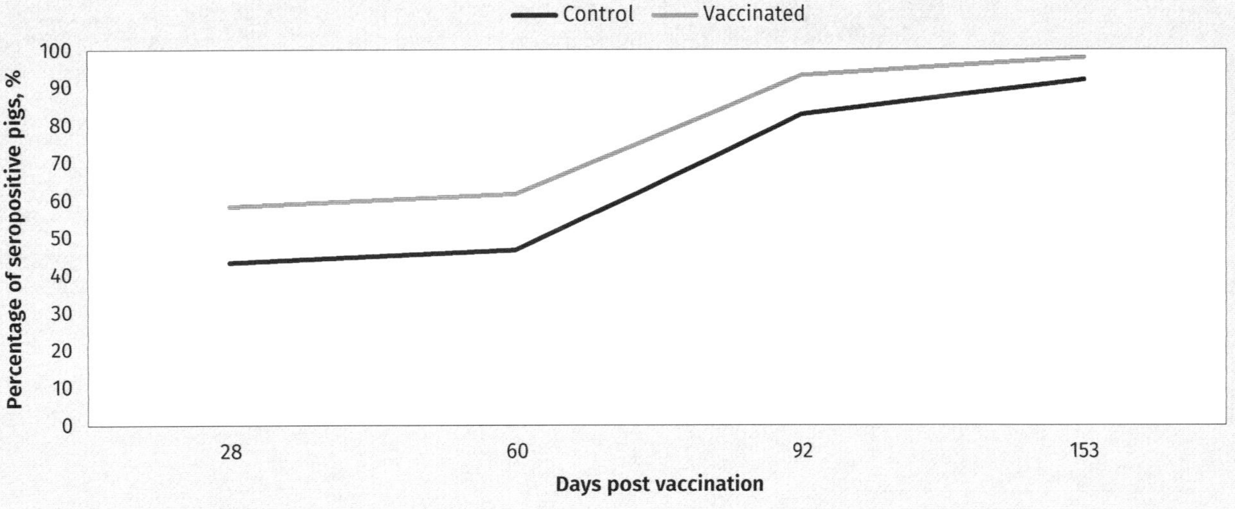

Figure 2: *Salmonella* isolated in fecal samples on the farm and at the slaughterhouse in the control and vaccinated groups before and after the transport to slaughterhouse.

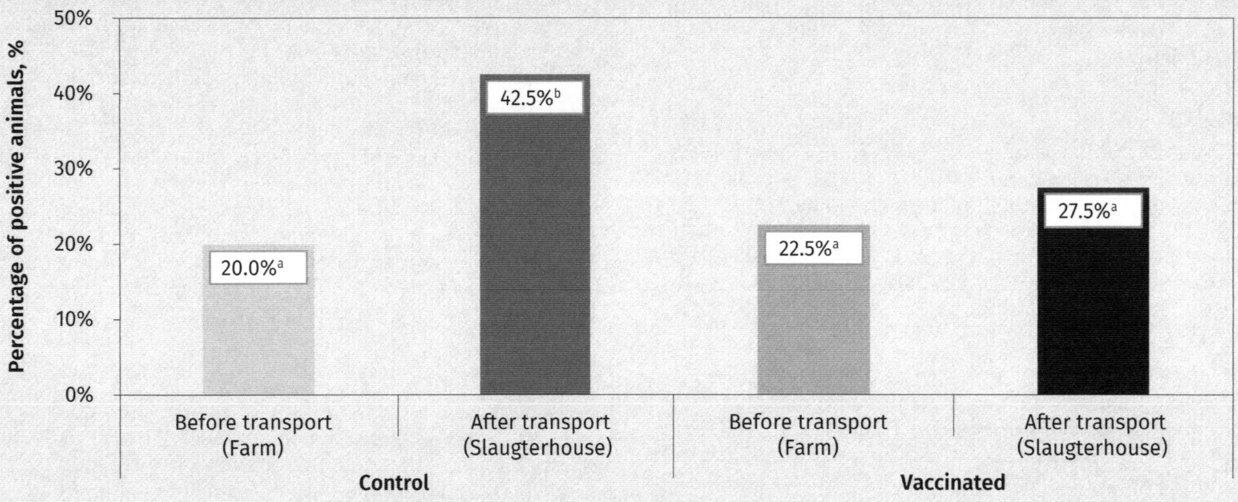

a, b Different superscript letters indicate statistical significance among each group ($P \leq .05$)

Figure 3: Percent of positive ileocecal lymph nodes for *Salmonella* in the vaccinated and control treatments collected at slaughter.

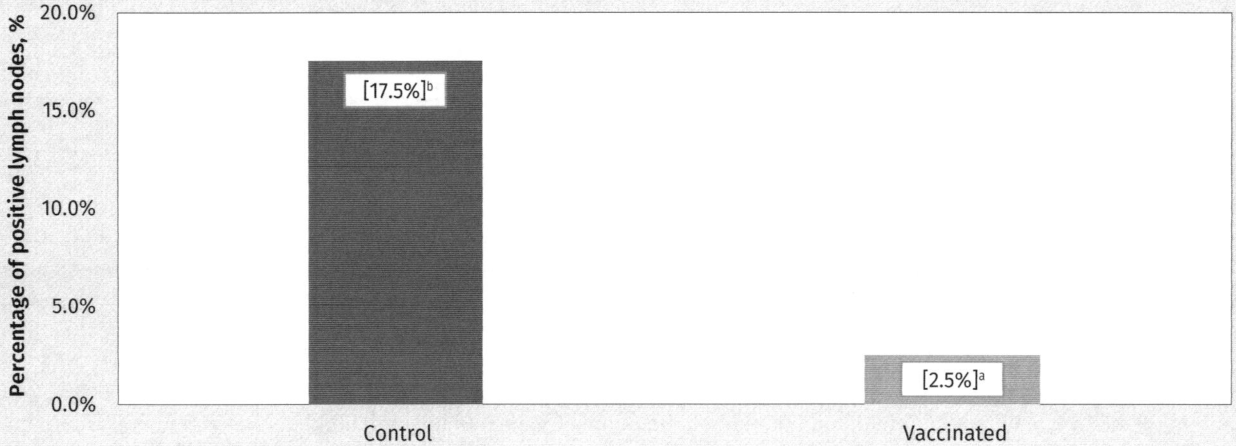

a, b Different superscript letters indicate statistical significance ($P = .025$)

slaughterhouse and on farm. In this scenario, this study evaluated the effect of vaccination on reducing *Salmonella* levels using an avirulent live vaccine which contains attenuated strains of *S* Choleraesuis and *S* Typhimurium.

Pigs challenged with *Salmonella* can have a negative impact on growth performance.[6] In this study, a tendency ($P = .07$) was found for vaccinated pigs to be 2.64 lbs heavier than non-vaccinated pigs at market age. Knowing that the vaccine has proven to protect pigs from negative impacts on growth performance,[7,8] it is likely that vaccinated animals were protected and, therefore, presented better performance when compared to unvaccinated animals, since the only difference between the groups was the use of Enterisol Salmonella T/C® vaccine.

Asymptomatic pigs infected with *Salmonella* can harbor and shed the pathogen in the feces for a long time increasing the contamination of the farm environment and lairage at the slaughterhouse.[2] This can happen especially in stressful moments.[3] In this study, samples were collected mainly after stressful moments that happen routinely in a farm, such as weaning, vaccination, and transport. In this study, a significant increase of 112% of pigs shedding *Salmonella* was observed in unvaccinated animals before and after being transported to slaughter. This increase in shedding due to transport was not observed in vaccinated animals, suggesting that the vaccine was effective in decreasing transport or stressed induced shedding of *Salmonella*.

A protective effect was also observed when evaluating lymph nodes colonized with *Salmonella* at slaughter. The vaccinated group had significantly fewer pigs with positive ileocecal lymph nodes (2.5%) than the control group (17.5%) ($P < .05$). The ileocecal lymph node was selected for culture since it is the most consistent tissue colonized by *Salmonella* in an infected animal, it is easier to harvest from the mesenteric viscera without contaminating either the sample or the carcass,[9] and because the ileocecal lymph node drains the cecum, a section of gut favorable to *Salmonella* colonization.[10] The effect of Enterisol Salmonella T/C® in reducing shedding and lymph node colonization has also been observed in previous studies.[8,11,12]

In Brazil, the seroprevalence of *Salmonella* in the nursery and finishing phases range from 3% to 84%.[4] In this study, the percentage of animals seropositive for *Salmonella* was around 90% in control animals at the end of the finishing phase demonstrating a high level of natural exposure to the pathogen. Moreover, different than other studies that found the lairage area as an important source of *Salmonella* in slaughterhouse and that carcass contamination originates from the environment rather than from the pig, in this study *Salmonella* was not found in lairage pens and truck floors prior to loading of pigs.[13,14] This difference could be due to the study animals being the first batches received at the slaughter plant on the day of sample collection.

In conclusion, these results demonstrate that vaccination with Enterisol Salmonella T/C® can be a useful strategy to reduce the level of *Salmonella* shedding at the slaughterhouse, which would help to control the spread of this pathogen. This vaccine was also effective in reducing the colonization of ileocecal lymph nodes at slaughter, reducing the possibility of meat contamination.

References

1. EFSA. Scientific report on the European Union one health 2018. zoonoses report. *EFSA J.* 2019, 17, 5926.

2. Griffith, R.W. et al. Salmonellosis. In Disease of Swine. Zimmerman, J. J.; Karriker, L.A.; Ramirez, A. *et al.* 2019, Ed. 11, p. 912 -925.

3. De Busser, E. V., De Zutter, L., Dewulf, J. et al. Salmonella control in live pigs and at slaughter. *The Veterinary Journal* 196.1 (2013): 20-27.

4. Schwarz P, Kich JD, Kolb J, Cardoso M. Use of an avirulent live Salmonella Choleraesuis vaccine to reduce the prevalence of Salmonella carrier pigs at slaughter. *Veterinary Record.* 2011;169(21):553.

5. Ministério da Agricultura, Pecuária e Abastecimento. Accessed in: **Normative instruction SDA/MAPA N° 60/2018.pdf**

6. Li M M, Seelenbinder K. M, Ponder M A, et al. Effects of dirty housing and a Salmonella Typhimurium DT104 challenge on pig growth performance, diet utilization efficiency, and gas emissions from stored manure. *Journal of Animal Science.* 2017; 95.3:1264-1276.

7. Leite F, Arruda P, Jordan M, Bearson S., Efficacy of Enterisol® Salmonella T/C in protecting against monophasic Salmonella I 4,[5],12:i:-. AASV. 2022. **https://doi.org/10.54846/am2022/52**

8. Cline G, Kaiser TJ, Claxton KT, Jordan DM, Bryson WL. Efficacy of a Salmonella typhimurium and Salmonella choleraesuis combination vaccine against S. typhimurium challenge in growing pigs. Leman Swine Conference p.61, 2014.

9. Carlson, Allan R., and Thomas Blaha. "In-herd prevalence of Salmonella in 25 selected Minnesota swine farms." *Journal of Swine Health and Production* 9.1 (2001): 7-10.

10. Bahnson, P. B., Snyder, C., Omran, L. M. Salmonella enterica in Superficial Cervical (Prescapular) and Ileocecal Lymph Nodes of Slaughtered Pigs, *Journal of Food Protection*, V. 69, (4), 2006, p. 925-927, **https://doi.org/10.4315/0362-028X-69.4.925**

11. Leite F, Arruda P, Jordan M, Bearson S., Efficacy of Enterisol Salmonella T/C® in reducing the shedding and colonization of multidrug resistant monophasic Salmonella enterica serovar I 4,[5],12:i:-. AASV 2023. **https://doi.org/10.54846/am2023/74**

12. Seate J, Kolb J., Sun T. et al., Vaccination With Enterisol® Salmonella T/C Reduces Salmonella Enterica Colonization Of Ileocecal Lymph Nodes In Growing Pigs. Safepork Conference p. 127, 2017.

13.De Busser, E. V., Maes, D., Houf, K., et al. Detection and characterization of Salmonella in lairage, on pig carcasses and intestines in five slaughterhouses. International Journal of Food Microbiology, (2011). 145(1), 279–286.

14. Botteldoorn, N., Heyndrickx, M., Rijpens, N. et al. L. Salmonella on pig carcasses: positive pigs and cross contamination in the slaughterhouse. Journal of Applied Microbiology, (2003) 95(5), 891–903.

https://doi.org/10.54846/am2025/50

Enhancing PCV2 control: Homologous vs heterologous vaccination in coinfection scenarios

M. Kroeger[1]; E. Fano.[2,3]; A. Sponheim[2]; K. J. Schwartz[1]; F. L. Leite[2]; O. Gomez-Duran[4]; L. Lecznieski[4]; P. Piñeyro[1]
[1]Department of Veterinary Diagnostic and Production Animal Medicine, Iowa State University, Iowa;
[2]Boehringer Ingelheim Animal Health USA Inc, Duluth, Georgia;
[3]Pipestone System, Pipestone, Minnesota; [4]Boehringer Ingelheim Vetmedica GmbH, Ingelheim am Rhein, Germany

Introduction

Porcine circovirus 2 (PCV2) is a single-stranded DNA virus that causes a wide range of clinical diseases in pigs, collectively known as porcine circovirus-associated disease (PCVAD). These diseases include systemic, lung, enteric, reproductive, and skin conditions, often accompanied by growth retardation, weight loss, and respiratory distress. PCV2 primarily infects lymphoid tissues, leading to granulomatous inflammation and immune cell depletion. Diagnosis of PCVAD involves clinical signs, histological lesions, and detection of PCV2 through in situ hybridization or immunohistochemistry.

The PCV2 genome consists of two open reading frames, ORF1 and ORF2, encoding the replicase and capsid proteins, respectively. Due to a high mutation rate, several PCV2 subtypes have emerged, with PCV2a, PCV2b, and PCV2d being the most commonly associated with outbreaks. The prevalence of these subtypes has shifted over time, with PCV2a dominant in the 1990s, followed by PCV2b and, more recently, PCV2d. However, the clinical significance of these subtypes and their virulence remains unclear. Vaccines for PCV2 have been available since 2006 in the United States and have proven effective in preventing clinical PCVAD, although they do not completely prevent viral replication.

PCV2 vaccines induce both humoral and cell-mediated immune responses that aid in viral clearance, reduce clinical disease, and prevent lesions. While neutralizing antibodies are crucial for protection, mutations in the capsid protein can affect the vaccine's ability to cross-protect against different subtypes.[ref] Additionally, co-infections, particularly with porcine respiratory and reproductive syndrome virus (PRRSV), can worsen PCV2 disease by enhancing viral replication and immune suppression, leading to increased disease severity. Studies suggest that co-infection with PRRSV and different PCV2 subtypes may exacerbate disease outcomes.

This study aims to evaluate the efficacy of homologous and heterologous PCV2 vaccines in a PCV2d/PRRSV co-infection model. By comparing clinical outcomes, pathological changes, viral load, and immunological responses, this research provides insights into vaccine efficacy in co-infection scenarios and contributes to the development of more effective PCV2 vaccination strategies.

Materials and methods

In the experimental design, sixty 21-day-old crossbred pigs were used, sourced from a farm with a low prevalence of PCV2. On day 0, all pigs were vaccinated with Ingelvac PRRSV and a commercial PCV2a vaccine (PCV2a, n = 20), commercial PCV2d vaccine (PCV2d, n = 20), or left unvaccinated for PCV2 (NVX, n = 20). All pigs were challenged with PCV2d (1mL intramuscular and 1mL intranasal, 5 log10/2mL dose) and PRRSV 1-7-4 (2mL intramuscular, 4.0 TCID50/mL) on day 28. The pigs were monitored for clinical signs, and samples were collected throughout the study. Body weight and daily gain were measured, and pigs were treated with antibiotics based on clinical scoring for signs of illness. Upon euthanasia at day 56, necropsies were performed to assess gross lung lesions. Fresh and fixed lung, lymph nodes, and tonsil were collected for further analysis.

Histological and immunohistochemical analyses were carried out to evaluate the severity of lesions and the presence of PCV2 antigen in tissues. Gross lesions, including pneumonia and lymphoid depletion, were scored, and tissues were examined for viral presence using PCV2-specific antibodies. Quantitative PCR was also used to assess PCV2 viral loads in serum and tissues, categorizing the infection status based on viral load.

Serum samples were tested for neutralizing antibodies against the PCV1-2 chimeric viruses expressing the PCV2d challenge, PCV2a vaccine, and PCV2d vaccine Cap sequences using a serum virus neutralization assay. Additionally, PRRSV viremia was quantified using RT-PCR. Statistical analyses were performed using ANOVA to compare treatment groups, with significance set at an alpha value of 0.05. This comprehensive approach aims to provide insights into the effectiveness of PCV2 vaccines and the role of PCV2 in respiratory disease in pigs.

Results

Clinical evaluation

In this study, the impact of co-challenging pigs with PCV2d and PRRSV 1-7-4 was assessed. While a significant reduction in average daily weight gain (ADG) was observed in the NVX group, no significant differences were observed between vaccine treatment groups, although ADG was numerically higher in the PCV2d group. There was a notable difference in treatment frequency, with PCV2a vaccinates requiring nearly twice the number of treatments compared to the PCV2d vaccinates. Mortality rates varied across groups, with the NVX group experiencing significantly higher mortality (60%) compared to the vaccinated groups ($P < .0001$). The PCV2a vaccinates had a mortality rate of 15%, and the PCV2d vaccinates had the lowest mortality at 5.5%. Mortality occurred between 15 and 26 days post-challenge.

Macroscopic and microscopic lesion evaluation

Gross lung lesions were not significantly different between the treatment groups, although interstitial pneumonia was significantly more frequent in the NVX group compared to the PCV2d vaccinates. Histopathological analysis revealed that the NVX group exhibited significantly more severe lesions, including moderate to severe lymphoid depletion and inflammation in various tissues, such as lymph nodes, tonsils, and lungs. Conversely, both vaccinated groups (PCV2a and PCV2d) had fewer and less severe lesions, with the PCV2a vaccinates showing slightly more severe lesions than the PCV2d vaccinates.

Quantification of PCV2 viral load and antigen presence by IHC

Viral load measurements revealed that the NVX group had significantly higher PCV2 viral load in the lymph nodes, tonsils, and lungs compared to both vaccinated groups (Figure 1A).

In addition, PCV2d significantly reduces the viral load in tissues in the face of a homologous challenge. Immunohistochemistry (IHC) further confirmed that PCV2 antigen detection was significantly higher in the NVX group, with more extensive staining across tissues. In contrast, the PCV2a and PCV2d vaccinates showed minimal or no antigen detection, indicating a strong protective response (Figure 1B). Furthermore, PCV2d vaccinates had significantly lower antigen detection in the iliac (ILN) and mesenteric lymph nodes (Figure 1B).

Figure 1: Porcine circovirus type 2 (PCV2) tissue viral load and antigen detection in the lymph nodes, tonsil, and lung. PCV2 viral load assessed by qPCR (A). Each section of lymph node (LN), iliac (ILN), mesenteric (MLN), tuberculosis (TBLN), tonsil, and lung was assessed blinded for the treatment and histological score was based on amount of PCV2 antigen detected within lesions by IHC: Score 0 = no positive cells stained, score 1 = few scattered cells (< 10% of cells) stained positive for PCV2 antigen, score 2 = multifocal clusters and multiple foci (10-50% of cells) stained positive for PCV2 antigen, score 3 = severe and diffuse (> 50% of cells) stained positive for PCV2 antigen. Average IHC score by treatment group (B).

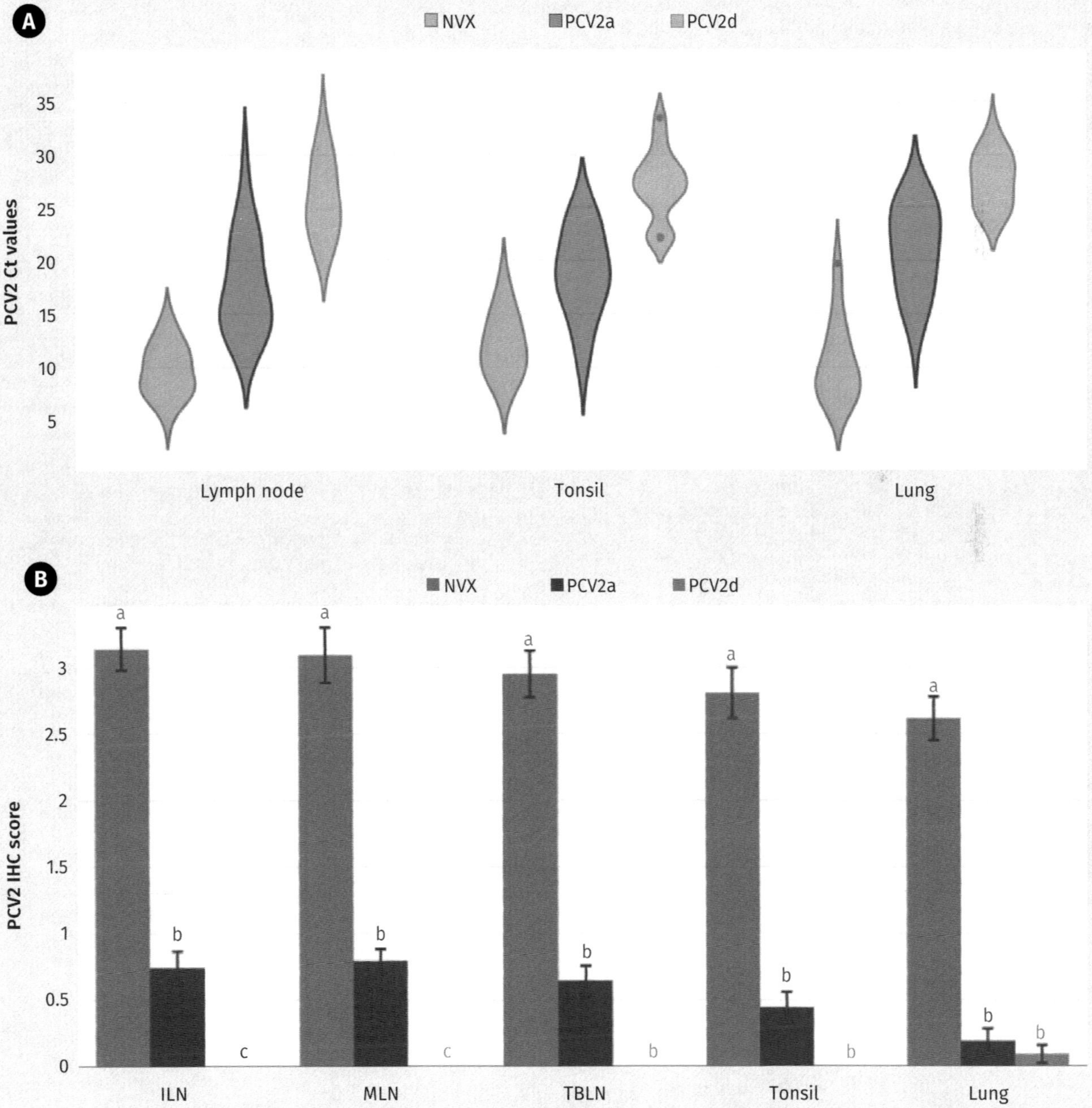

Analysis of PCV2 and PRRSV viremia dynamics

Viremia levels were significantly higher in the NVX and PCV2a groups compared to the PCV2d vaccinates at D42 and D56. At D42, 85% of the NVX group had viremia levels exceeding 10^7 gcn/mL, whereas the PCV2d group had undetectable or subclinical viremia. By D56, the majority of PCV2a vaccinates had viremia levels lower than those observed in the NVX group (Figure 2).

Total and neutralizing antibody response to PCV2 following challenge and vaccination

Total antibody responses measured by ELISA were higher in both vaccinated groups prior to challenge compared to the NVX group. By D56, the NVX group had caught up, showing comparable total antibody levels to the vaccinated groups (Figure 3A). Regarding neutralizing antibodies (NA), the PCV2d vaccinates had significantly higher titers anti-PCV2d to both vaccine and challenge strain prior to challenge on D28 (Figure 3B, 3D). Following the challenge, the NVX group exhibited the highest NA titers by D56, (Figure 3B). While both vaccinated groups had high NA titers against their respective PCV1-2 vaccine chimeric virus prior to challenge, the NVX group demonstrated strong cross-reactivity and NA responses against the PCV1-2a vaccine and PCV1-2d vaccine chimeric viruses following challenge (Figure 3C, 3D).

Discussion

PCV2 has become a significant and economically impactful pathogen in the global swine industry. Since the introduction of PCV2 vaccines in the United States in 2006, they have successfully prevented clinical PCVAD. These vaccines, primarily based on the PCV2a subtype, have evolved to include formulations for other subtypes, such as PCV2b and PCV2d. Despite this progress, there have been cases of vaccine failures, particularly with heterologous vaccines, raising concerns about the cross-protection provided against different PCV2 subtypes under field conditions. Studies suggest that while PCV2a-based vaccines offer protection against PCV2d challenges, clinical disease has been challenging to replicate experimentally, requiring co-infection with other pathogens, such as PRRSV, to trigger PCVAD. Co-infection with PRRSV exacerbates the severity of PCVAD, increasing the presence of PCV2 antigen in tissues and prolonging viral shedding. However, the effectiveness of commercial vaccines against co-infection with PCV2d and contemporary PRRSV strains has not been fully assessed.

This study aimed to evaluate the efficacy of commercial PCV2a and PCV2d-based vaccines following co-challenge with PCV2d and a contemporary PRRSV strain (1-7-4). Both vaccines reduced mortality compared to non-vaccinated controls, but PCV2a vaccinates displayed higher viral loads, more severe histological lesions in addition to requiring a greater number of treatments compared to PCV2d vaccinates. Some of the PCV2a vaccinates had moderate lymphoid depletion and moderate levels of viral antigen detection in lymphoid tissues, suggesting a minority developed clinical PCVAD, whereas the PCV2d vaccinates exhibited none to slight lymphoid depletion with no detectable antigen, indicating subclinical infection and a greater level of protection. Viremia levels were significantly lower in PCV2d vaccinates, supporting homologous vaccination provides superior efficacy under co-infection conditions. Additionally, both vaccines induced similar total antibody responses, but the PCV2d vaccinates elicited higher NA titers against the PCV1-2d challenge chimeric virus. Vaccination favored the earlier development of subtype specific NAs, however, cross-neutralizing antibody responses were observed in both vaccinated groups following challenge. These findings suggest that while both vaccines offer protection, subtype-specific differences in immune responses may affect the severity of disease, underscoring the need for tailored vaccination strategies based on regional epidemiology and emerging PCV2 strains.

Figure 2: Viral distribution of porcine circovirus type 2 (PCV2) and porcine reproductive and respiratory syndrome virus (PRRSV). PCV2 viremia expressed as genomic copy numbers per mL (gcn/mL-1) on D28-56 (0-28 days post-challenge) Differences in PCV2 viremia was calculated by 2-way ANOVA. Bars represent the SE of the mean and significance was established at $P < .05$.

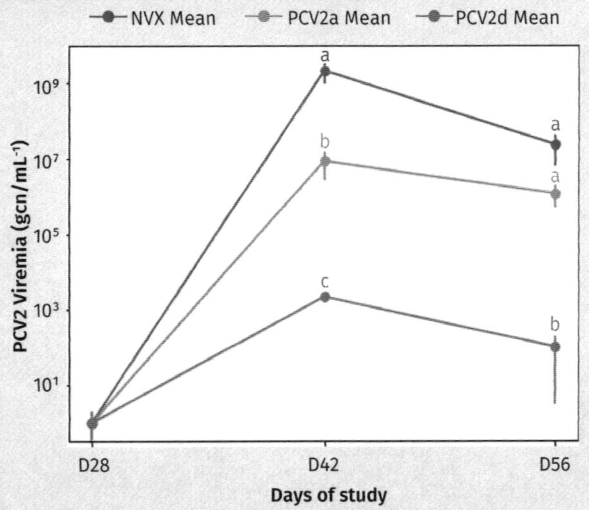

References

1. Franzo G, Segalés J. 2020. Porcine Circovirus 2 Genotypes, Immunity and Vaccines: Multiple Genotypes but One Single Serotype. 9:1049.

2. Guo L, Fu Y, Wang Y, Lu Y, Wei Y, Tang Q, Fan P, Liu J, Zhang L, Zhang FJPo. 2012. A porcine circovirus type 2 (PCV2) mutant with 234 amino acids in capsid protein showed more virulence in vivo, compared with classical PCV2a/b strain. 7:e41463.

3. Opriessnig T, Xiao C-T, Gerber PF, Halbur PG, Matzinger SR, Meng X-J. JTJogv. 2014. Mutant USA strain of porcine circovirus type 2 (mPCV2) exhibits similar virulence to the classical PCV2a and PCV2b strains in caesarean-derived, colostrum-deprived pigs. 95:2495.

4. Segalés J. JErov. 2015. Best practice and future challenges for vaccination against porcine circovirus type 2. 14:473-487.

5. Fort M, Olvera A, Sibila M, Segalés J, Mateu EJVm. 2007. Detection of neutralizing antibodies in postweaning multisystemic wasting syndrome (PMWS)-affected and non-PMWS-affected pigs. 125:244-255.

6. Meerts P, Misinzo G, Lefebvre D, Nielsen J, Bøtner A, Kristensen CS, Nauwynck HJ. JBVR. 2006. Correlation between the presence of neutralizing antibodies against porcine circovirus 2 (PCV2) and protection against replication of the virus and development of PCV2-associated disease. 2:1-11.

Figure 3: Porcine circovirus type 2 (PCV2) total antibody levels detected by ELISA in PCV2a and PCV2d vaccinates increased post-challenge on D28 but remained statistically similar throughout the study (A). The not vaccinated (NVX) group had lower titers until D56, when levels matched those of vaccinates (A). PCV2d vaccinates had significantly higher NA titers against the PCV1-2d challenge and vaccine chimeric virus before challenge (D28) compared to PCV2a vaccinates and the NVX group (B, D). PCV2a vaccinates showed higher NA titers against the PCV1-2a vaccine chimeric virus pre-challenge compared to the PCV2d vaccinates and NVX group (C). NA titers were expressed as the highest dilution with ≥ 90% reduction in fluorescent foci. ELISA and neutralization assays were analyzed using two-way ANOVA with Tukey's correction (*P* < .05). Bars indicate SE of the mean.

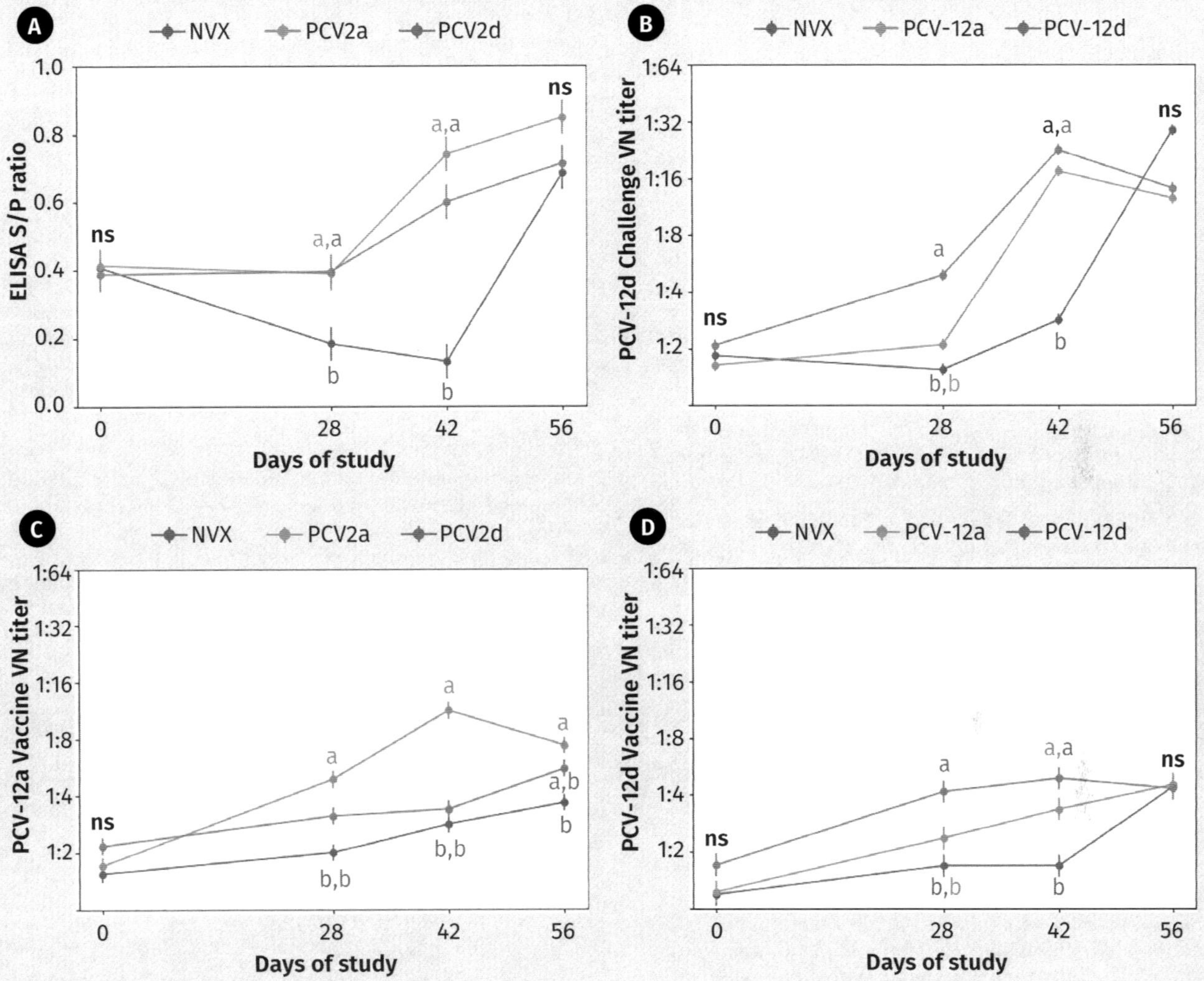

https://doi.org/10.54846/am2025/51

Improving transport biosecurity and reducing risk of disease spread with biosecurity management software

C. Mainquist-Whigham, DVM; K. Andersen, BSc, AGR; A. Boakes, DAGR

Introduction

Research confirms that people and vehicle movements play a significant role in the spread of swine diseases like porcine epidemic diarrhea (PEDV) and porcine reproductive and respiratory syndrome (PRRSV). Dr Derald Holtkamp and a team of researchers at Iowa State University reviewed infections of PRRSV, PEDV and porcine deltacoronovirus (PDCoV) at growing-pig sites to investigate biosecurity gaps. They followed 75 groups of pigs from nursery to finish, all of which were negative for PRRSV at placement. Of these, only two stayed negative after being placed. The researchers identified one pattern that stood out from the data. It was how closely the frequency of events (such as movements, deliveries, maintenance, etc) corresponds to infection of PRRSV, PEDV, and PDCoV, in both the nursery and finishing phases.[1]

Additionally, farm connections such as shared personnel and animal movements have been confirmed by the USDA in the spread of highly pathogenic avian influenza (HPAI) among dairy herds.[2]

Pillen Family Farms is a 3rd generation, family owned and operated swine farm based out of Nebraska. The company piloted Farm Health Guardian biosecurity management software to test its ability to improve their transport biosecurity program and reduce the risk of disease spread between farm properties.

Methods

The pilot project involved 184 Pillen farm properties, 197 vehicles, and 71 users of the app from February – May 2024. Vehicles were linked to the software through GPS link, or by installing a GPS device to record visits to geo-fenced properties. Of the 197 vehicles, 25 were installed with a GPS taillight device, with the balance connected to the software through GPS link to existing navigation system. Vehicle types included feed trucks, market livestock trucks, nursey livestock trucks, sales livestock trucks, and trailers. Farm properties were geofenced in the Farm Health Guardian software system, and included farrow to wean barns, finishing barns, gilt development units, gene centres, nucleus farms, nurseries, and nursery finishers. Planned visits to the properties were to be validated by the Farm Health Guardian software system. Following the initial pilot project period, the software use continued, and three transport biosecurity protocols were configured for ongoing compliance monitoring between April and October 2024.

Results

A total of 22,744 property visits were recorded by the software during this period, with 17,789 (78%) of those visits being trucks/vehicles recorded through GPS link or devices. The balance were people visits recorded by the app. The software was configured to send breach alerts for trucks if they violated any one of three biosecurity rules for visits between farm properties. For example, visits to two distinct truck washes (a high-speed wash and then an audited truck wash) were required for livestock trailers travelling from any production site with a health status of 4 or higher, before going back to a DNA site. A second rule was for dedicated trailers for a specific flow of sow to finishing barns, which includes its own truck wash. A third rule was feed truck visit order, which was to go from high health to lower health status properties for feed deliveries during the course of the week.

Biosecurity rules for people movements between farms were verified by configuring 18 health statuses for downtime requirements. The lower the number the higher the health status, with DNA sites having the highest health status and Finisher PEDV+ having the lowest health status. For example, going from a finisher (levels 9 & 10) to a DNA genetics site (level 0), requires three nights downtime. When the sites and their corresponding health statuses were configured in the Farm Health Guardian software, a notification was sent to health staff if a user attempted to visit a site without meeting the required downtime (Figure 1).

For most of the weeks (16 out of 25), there were no recorded transport breaches (Figure 2). A minimal number of breaches (three or fewer per week) occurred in 9 of the 25-week period. The ability to visualize the breach data and receive real-time notifications helped ensure a very low number of breaches were maintained by personnel. This demonstrates the impact of real-time feedback and visualizing data to increase awareness and influence biosecurity implementation.

Figure 1: Example of a breach notification alert sent through the Farm Health Guardian software when a user attempts to visit a site without meeting the required downtime.

Figure 2: Breaches per week shown visually.

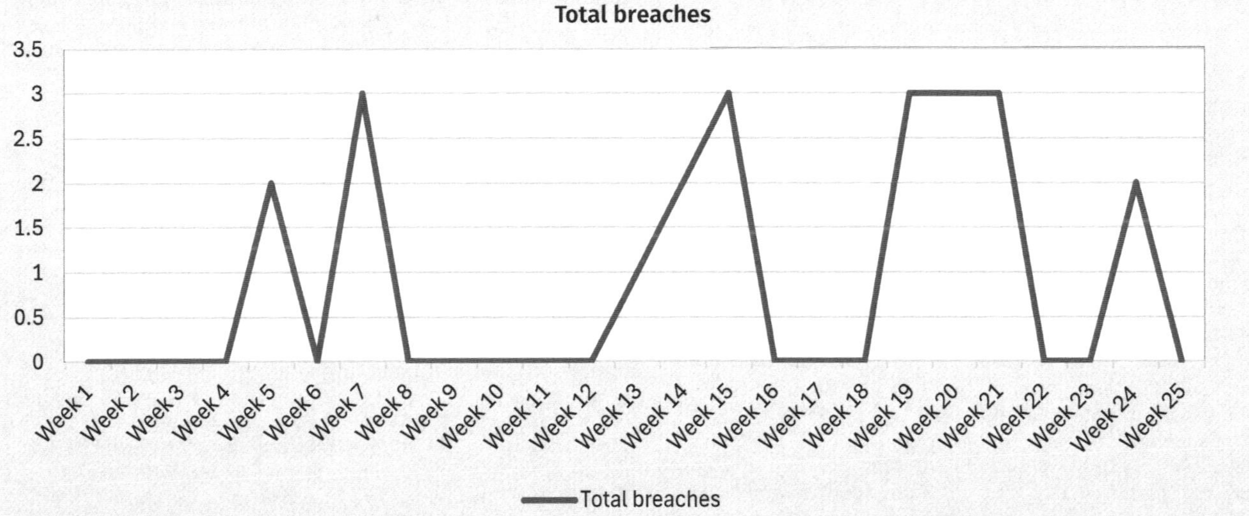

Figure 3: A map generated by Farm Health Guardian software showing movements between farm properties using demonstration data. Green lines indicate a clean truck; red lines indicate truck movement between properties within the incubation period; orange dots are properties that had indirect contact with the index farm; and grey dots represent properties that had no contact.

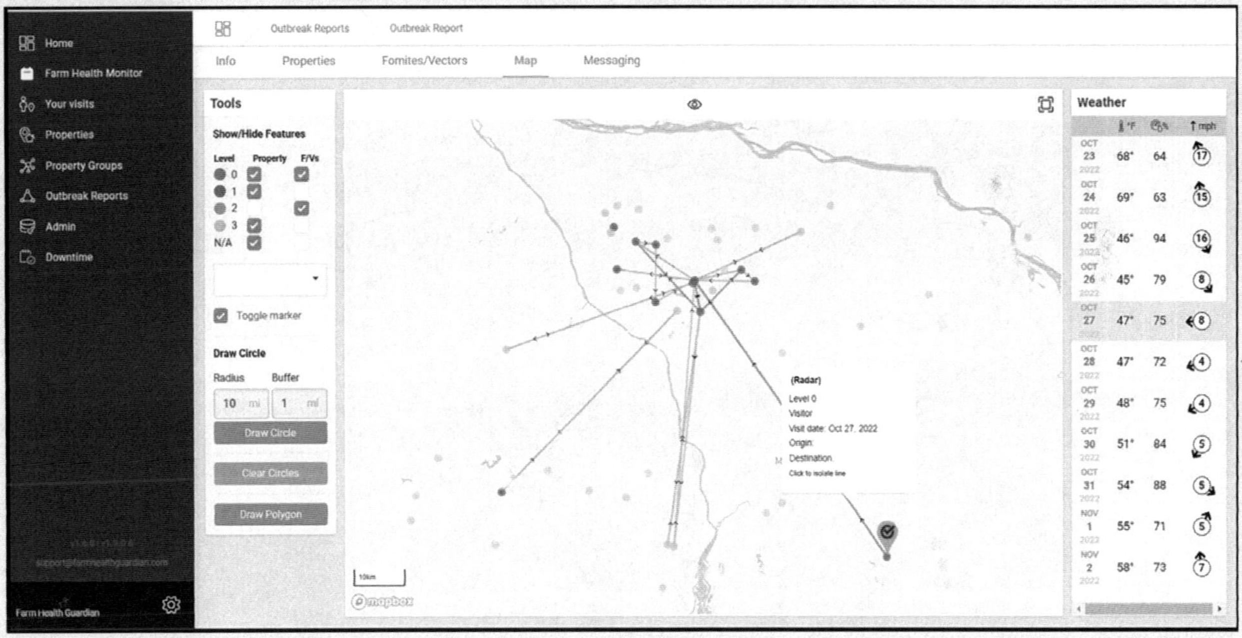

In addition, the software recorded examples where equipment normally dedicated to a site, was used on another site or, other trailers visited sites that typically have dedicated equipment. These examples of non-dedicated equipment use occurred in 14 of 25 weeks. These changes are intentional and approved decisions based on business needs at the time and may not have increased biosecurity risks. The FHG software and visualization of data provide an accurate reflection of what is actually happening, proving to be a valuable tool for a large system.

Discussion

Research shows that real-time notifications about biosecurity breaches greatly improves compliance by personnel, reducing the risk of disease spread.[3] When disease hits or is suspected, the speed of response is critical in preventing spread which reduces risk of additional financial losses. (In the US, PRRS is estimated to cause annual losses of around $1.2 billion[4] from 2016 to 2020; the 2014 PED outbreak is estimated to have resulted in a loss of approximately $1 billion).[5] One of the lengthiest and most onerous parts of a disease investigation is finding out who travelled to what farm and where they went to next. In a disease simulation exercise using Farm Health Guardian technology, Pillen had complete access to this data in minutes, via a trace out report generated by the software (Figure 3) versus days to manually gather the same information through phone calls, emails, texts, and paper logbooks. Additionally, using the software to conduct disease investigations greatly improved the confidence of the information. Digital, unbiased, automated data guarantees the accuracy of the trace out report, which in turn helps ensure that disease response is more efficient and effective.

The company says it's an efficient way for their team to keep everyone informed and to receive communication about the health status of different farms. Ryan Strain, director of transportation and environmental at Pillen testifies to the impact the technology has had. "Biosecurity isn't a simple subject," he says. "Every day things change. We also rely on human interpretation. This technology helps eliminate varied responses to what's happening in the health pyramid."

References

1. Swine Health Information Center, SHIC-funed Project Examines Growing Pig Site Biosecurity Gaps. June 16, 2022. **https://www.swinehealth.org/shic-funded-project-examines-growing-pig-site-biosecurity-gaps/**

2. US Department of Agriculture Animal and Plant Health Inspection Service. Highly Pathogenic Avian Influenza H5N1 Genotype B3.13 in Dairy Cattle: National Epidemiologic Brief. June 8, 2024. **https://www.aphis.usda.gov/sites/default/files/hpai-dairy-national-epi-brief.pdf**

3. Diaz, I., Quintana, E., Abad, G., Digital biosecurity on farms and accelerating learning. Pig 333. 24 April 2023. **https://www.pig333.com/articles/digital-biosecurity-on-farms-and-accelerating-learning_19109/**

4. Roepke, D., Growing Losses from PRRS Cost Pork Producers $1.2 Billion Per Year, New Study Shows. Iowa State University. July 30, 2024. **https://research.iastate.edu/2024/07/30/growing-losses-from-prrs-cost-pork-producers-1-2-billion-per-year-new-study-shows/**

5. Weng, L., Weersink, A., Poljak, Z., de Lange, K., von Massow, M., An economic evaluation of intervention strategies for Porcine Epidemic Diarrhea (PED). Preventive Veterinary Medicine. 1 November 2016. **https://www.sciencedirect.com/science/article/abs/pii/S0167587716303890**

https://doi.org/10.54846/am2025/52

Real-time feed delivery data: What might we learn about disease biosecurity/transmission from Feed Allocation Systems (FAS)?

Andrea Dixon[1], PhD; Jordan Gebhardt[1], DVM, PhD; Mike Sanderson[1], DVM, MS;
Joel Stave[2], BS; Tyler Holck[2], DVM, MS, MBA
[1] College of Veterinary Medicine, Kansas State University; [2] Prairie Systems by Ever.Ag

Introduction

Understanding of feed biosecurity has rapidly improved in recent years. Much has been learned about the role that feed ingredients and/or complete feed may play in the dissemination of bacterial and viral pathogens. More recently, however, efforts have been directed towards improving our understanding of other aspects of feed biosecurity, namely transportation logistics for complete feed delivery to swine farms. Feed delivery is one of the most frequent events occurring as swine farm, only second in frequency to employee entrance. On most farms, feed trucks deliver feed on a weekly, if not more frequent, basis. Understanding the potential role that feed trucks and feed truck drivers may play when considering feed delivery biosecurity is an important consideration. To date, there is limited information objectively documenting the contact network for feed deliveries in the US swine industry. Thus, a significant limitation in our ability to properly design and calibrate disease transmission models which would include feed delivery logistics exists.

Feed Allocation System (FAS) is a grow finish management solution focused on feed budgeting, ordering, and tracking software developed and marketed by Prairie Systems Inc., a subsidiary of Ever.Ag. FAS was initially offered to the US swine industry in 2006 and has grown rapidly. By leveraging mobile technology, producers and growers have an easy, effective, and empowering process to manage their biggest expense, feed. In 2020, MyFeedWatch was added as a tool for FAS users to allow mills and drivers to have an electronic means to scan and confirm what feed has been delivered to a specific bin in real time. FAS coupled with MyFeedWatch captures both feed ordering and delivery data including the feed mill of origin, delivery truck and driver, and farm site and specific bin for each batch of feed ordered. Utilized by approximately 40% of the US swine industry and still growing, the FAS database represents over 450,000 loads of feed delivered annually representing approximately 12 million tons of feed. Data for this analysis was provided with approval of FAS customers participating in the use of MyFeedWatch.

Through this analysis using data of digital feed delivery records, our objective was to perform a preliminary exploratory analysis to evaluate key questions regarding feed delivery information using the Feed Allocation System along with MyFeedWatch to serve as a foundation for future work to further characterize the contact network structure of feed deliveries within the US swine industry to be used in disease modelling and response efforts.

Methods

Data were extracted from the FAS system into a .csv file. From there, data handling, cleaning, and summary statistics were performed using the R language version 4.4.1 (R Core Team, 2024).[1] Summary statistics reported include mean, median, range, and standard error of the mean.

Data from Prairie Systems spanning 2019 to 2024 was used representing a total of 560,686 observations with each observation representing a delivery to a specific feed bin on farm. A total of 17,226 observations were excluded as duplicates. Sixty-four observations relating to nine feed mills were removed as they had fewer than 14 days of data in the entire dataset. Data for 2019 and 2020 were very sparse comprising data across only five and 17 weeks, respectively, and all observations in these years were removed (n = 2,814).

After data management, the data set (n = 540,582 deliveries to a feed bin) represented 57 unique feed mills located in 10 states, which provided feed to 1,908 farms located in 12 states from January 1st, 2021 to September 30th, 2024. The total number of drivers in the dataset was 967.

Results

Through this effort, we aimed to answer several specific questions regarding some of the key attributes of the contact network in the feed delivery supply chain. Some important questions we aimed to answer through this exploratory analysis are described in greater detail in the following sections.

How many farms are represented within the dataset with at least one feed delivery each week?

To begin, an understanding of the base dataset used within this analysis is important. Figure 1 demonstrates the number of farms within at least 1 feed delivery recorded into MyFeedWatch by week. The data clearly demonstrates an increase in the number farms capturing data using the system.

How many feed deliveries occur on a farm during a week?

Based on the data a core question wanting to be answered was how many feed deliveries occur on a farm in a given 1 week period. Results demonstrate that the median number of deliveries to each farm was between 3 and 4 deliveries per week (Table 1). The range of deliveries per farm per week varies greatly. Ranging from as few as 1 delivery per week to 48 deliveries per week. When further evaluating the farms with a very high number of deliveries per week, these represented very large farms with many bins and different diets thus it is believed these data points are not simply an artifact, but a small percentage of farms with a very high delivery frequency.

How many different feed truck drivers deliver to a farm during a week?

Further building on how many total deliveries to a farm occurred within a week, a logical follow-up question would be how many different drivers were performing these deliveries. Based on the

Figure 1: Number of farms with a feed delivery recorded in MyFeedWatch by week and year.

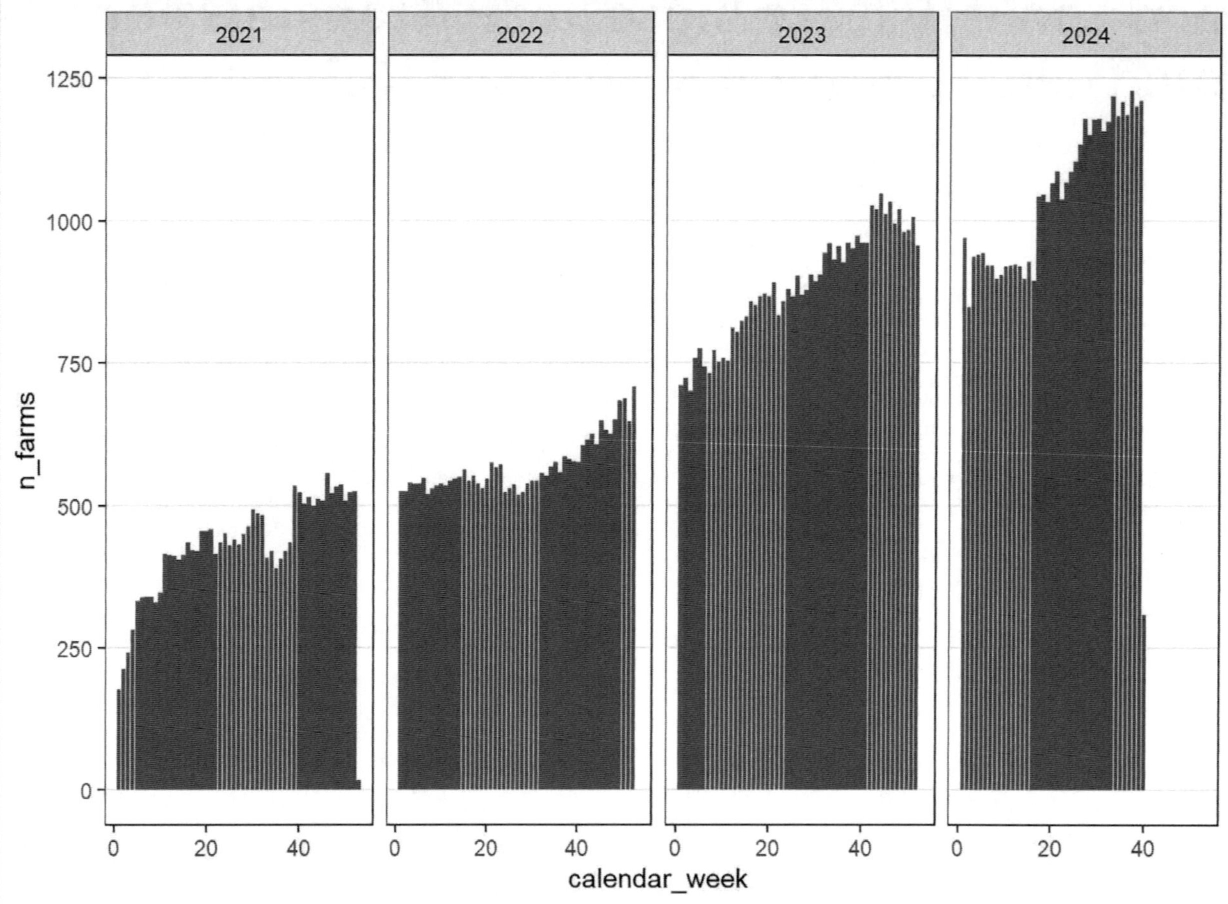

Table 1: Descriptive statistics for number of deliveries per farm per week by year

Year	Weeks of data included[1]	Mean	Median	Range
2021	53	3.6	3	1-38
2022	52	3.8	3	1-48
2023	52	3.9	4	1-47
2024	40	4.0	4	1-48

1 Data was extracted in September 2024 with only 40 weeks of data generated in 2024 to that point.

data capture process for MyFeedWatch, data is summarized by driver, not by truck. A logical question would be how many different trucks does this represent, but currently we are only able to determine how many unique drivers performed these deliveries. The number of trucks making deliveries could be either greater than the number of drivers if a driver uses multiple trucks throughout the week, or number of trucks could be fewer than number of drivers if multiple drivers are using the same truck throughout the week.

On any given week, the median number of feed truck drivers delivering feed to an individual farm was between 1 and 2 (Table 2; Figure 2). This indicates that within a farm, there is not a large number of different drivers (and assuming feed trucks) that would be arriving and present biosecurity risk for visiting other farms. The range in the number of drivers per farm does vary substantially, however, with some farms having up to 10 different drivers delivering feed to them within a single week.

Within a day, how many different farms does a single feed truck driver deliver to?

The next question aimed to address how many unique farms does a single driver deliver feed to within a single day. The median number of farms visited per day was 2, with a wide range with the greatest number of farms visited per day in the dataset being 24 (Table 3; Figure 3). This was an extreme observation, but there were several observations that ranged from 10-15 farms per day. Further exploring these data did not indicate any specific artifact in the data that would indicate this interpretation would be invalid. However, there are limitations in the use of the technology such as ensuring that each driver using only their own login information on a single device to ensure data accuracy for future data exploration efforts.

Within a week, how many different farms does a single feed truck driver deliver to?

Extending beyond a single day, we then focused on understanding how many different farms would a single feed truck driver deliver to within a week. Values represent the number of unique farms visited per week. If a single farm was visited on multiple days within the same week, it was only counted once. Within the 4 years of data, a truck driver visited approximately 4 unique farms per week (Table 4; Figure 4). However, like other responses, there is a significant degree of variation with some drivers visiting 30+ farms per week.

Table 2: Descriptive statistics for number of feed truck drivers delivering to a farm per week by year

Year	Weeks of data included[1]	Farms with deliveries	Mean	Median	Range
2021	53	737	1.6	1	1-7
2022	52	946	1.7	1	1-8
2023	52	1372	1.7	2	1-9
2024	40	1665	1.8	2	1-10

[1] Data was extracted in September 2024 with only 40 weeks of data generated in 2024 to that point.

Figure 2: Boxplot of number of feed truck drivers delivering to a farm by year.

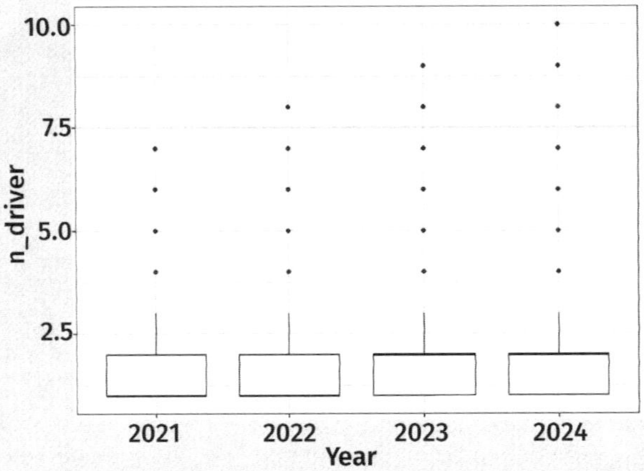

Table 3: Descriptive statistics for number of farms a feed truck driver delivers to per day by year

Year	Days	Drivers	Mean	Median	Range
2021	352	362	2.3	2	1-12
2022	361	452	2.2	2	1-11
2023	364	551	2.4	2	1-24
2024	274	567	2.4	2	1-16

Figure 3: Boxplot of number of farms a driver visits per day by year.

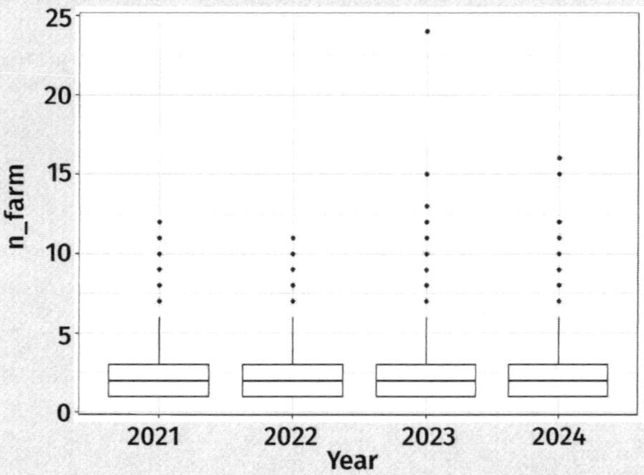

Figure 4: Boxplot of number of unique farms a driver visits per day by year.

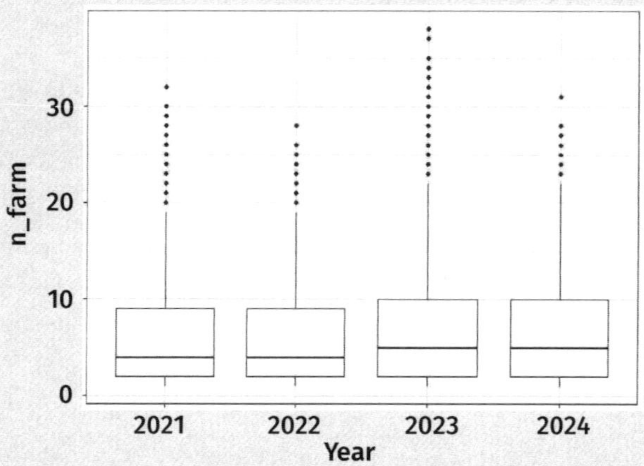

Table 4: Descriptive statistics for number of unique farms a feed truck driver delivers to per week by year

Year	Weeks	Drivers	Mean	Median	Range
2021	53	362	6.0	4	1-32
2022	52	452	5.8	4	1-28
2023	52	551	6.5	5	1-38
2024	40	567	6.4	5	1-31

Additional opportunities for further exploration

The data generated through the FAS system using MyFeedWatch is the most robust set of data currently available regarding the contact network of feed deliveries in the US swine industry across multiple production systems. Additional questions regarding the feed mill itself are of high value, such as how many feed deliveries does a feed mill generate per day? How many unique farms does that represent? What about per week? Likewise, questions could be generated about the degree of connectivity between two feed mills on a single farm. Exploration of these questions is warranted.

Using the data available from the farm-level, we explored the number of farms within a week that received feed from more than one feedmill. In 2021, there were 20 instances of a farm receiving feed from more than one feed mill within a single week, with 37, 21, and 36 in 2022, 2023, and 2024, respectively. Based on this dataset having multiple feed mills deliver to the same farm within the same week is not a frequent event, but does occur. Such situations could present as nursery feed being made by one feed mill that has pelleting capabilities, and the later dietary phases being fed in mash form where another feed mill would manufacture those diets. Based on these data, this event of deliveries from multiple mills within the same week does occur and presents a biosecurity challenge with now multiple delivery contact networks across different feed mills now being connected.

Herein we do not have a detailed analysis of feed deliveries by feed mill because in nearly all of the feed mills included in the dataset, a subset of feed manufactured and delivered goes to customers that are not currently using the MyFeedWatch technology. Therefore, summarizing feed mill data using only those customers that are using the MyFeedWatch system likely would under report the degree of connectivity between farms within and between mills. Nonetheless, it is an important area to understand to fully characterize the contact networks of feed deliveries within the US swine industry.

Discussion

The preliminary findings from this study provide a baseline understanding of the contact network of feed deliveries in the US swine industry. As the results demonstrate, feed trucks are likely the most frequent event where drivers and equipment are visiting multiple farms per day and per week. Therefore, understanding feed delivery as a potential biosecurity risk is critical. Using data such as generated documenting feed deliveries is important for disease modelling efforts to understand where and how biosecurity practices could be implement to have the greatest efficacy while minimizing labor and resource allocation. Access to this information in a real-time manner has tremendous potential for disease traceability. Results of this report can be further explored to further characterize feed contact networks with information being used for future disease modelling efforts.

Reference

1. R Core Team (2024). R: A Language and Environment for Statistical Computing. R Foundation for Statistical Computing, Vienna, Austria. **https://www.R-project.org/**

https://doi.org/10.54846/am2025/53

Sustainable use of antibiotics in boar semen production

K. Simmet[1], DVM; C. Simmet[1], DVM; J. Quackenbush[2]
[1]Minitüb GmbH, Tiefenbach, Germany
[2]Minitube USA Inc., Verona, Wisconsin

Introduction

In modern animal production, responsible antibiotic use is crucial to meet societal expectations, combat development of resistance, and ensure sustainability. Minitube leads the way in promoting responsible antibiotic use in assisted reproduction technologies. Our sustainable initiatives focus on product transparency, selection and appropriate use of antibiotics as well as innovation. The intent of this white paper is to outline our commitment to this important initiative.

Our ethos

We prioritize responsible antibiotic use in boar semen extenders, guided by:

- Transparency: Our transparency ensures that our customers have a complete understanding of the antibiotic components in our semen extender products.
- Selective antibiotic use: We source our ingredients only from GMP approved manufacturers. We use scientifically based evidence to select the appropriate antibiotic(s), and purposely exclude WHO-classified highest priority critically important antimicrobials (HPCIA) from our products.
- Appropriate use: We offer a comprehensive array of standard operating procedures (SOPs) for hygienic boar semen production and precise antibiotic use, featuring novel ways to apply antibiotics at the efficient in vitro concentration.
- Innovative concepts: We invest in development of alternatives to antibiotic reliance and pioneer sustainable alternatives to traditional antibiotics.

Transparency

Our commitment to transparency supports our customers' sustainability management, regulatory compliance and personnel safety goals. We provide certificates of analysis (CofAs) and material safety data sheets (MSDS) for each product and batch we produce and market, which always state the antibiotics used and their precise concentration. Our GMP-certified facility is open for scheduled customer visits.

Selective antibiotic use

We only use antibiotic compounds that have been scientifically proven to work effectively and safe in animal semen extenders. We then source our antibiotics from carefully selected suppliers complying with European (Ph Eur), British (BP) or US (USP) Pharmacopoeia. Lastly, we purposely exclude any HPCIA antibiotics (eg, quinolones, colistin, polymyxins, cephalosporins, antibiotics for human-use only) to aid in the global effort of preventing resistances against essential human and veterinary antibiotics. With this, Minitube contributes to the WHO's One Health approach.

Standard operating procedures (SOPs)

Hygienic conditions during semen collection and processing minimize antibiotic use and resistance. Our equipment is designed for thorough cleaning, and our SOPs ensure prudent antibiotic use. Our concept of accurate dosage of antibiotics (ADA) optimizes antibiotic dosing and enhances efficacy while reducing resistance potential. Laboratory automation and software modules help to achieve precise dosing, ensuring standardized semen dose quality and avoiding human error.

Innovative concepts

We pioneer sustainable alternatives to traditional antibiotics:

- Hypothermic storage: Storing semen at 5°C inhibits bacterial growth without antibiotics. Over the last decade, we developed Androstar® Premium, a semen extender that supports hypothermic long-term storage, and the necessary procedures to preserve boar semen at 5°C while maintaining fertility and fecundity in the field.
- Organic bactericidal substance (OBS): OBS in semen extenders controls bacterial contamination naturally. OBS effectively inhibits bacterial growth at 17°C, reducing the required concentration of antibiotics or even eliminating the need for antibiotics for certain contaminants.

Conclusion

Our commitment to sensible and sustainable antibiotic use in assisted reproductive technologies remains an underlying principle of Minitube's culture as we strive to support global health and food security. This commitment is demonstrated through our prioritization in transparency, prudent antibiotic selection, comprehensive SOPs, and ongoing development of alternatives to antibiotic reliance. We invite our customers to join us in this mission for the highest safety and efficacy standards.

https://doi.org/10.54846/am2025/54

SEQUIVITY® ID: A newly licensed prescription vaccine for intradermal vaccination

Susan Knetter[1]; Mark Mogler[1]; Chandra Morgan[1]; Allison Hoehn[1]; Ruizhong Shen[1]; Pravina Kitikoon[2]; Supraja Puttamreddy[1]; Faris Jirjis[1]; Ruud Segers[3]
[1]Merck Animal Health, Rahway, New Jersey; [2]Previously with Merck Animal Health, Rahway, New Jersey; [3]MSD Animal Health, The Netherlands

Introduction

Skin-based vaccination has long been a focus of human and animal vaccine development, due to the ease of administration, reduction of pain, and the diversity of skin immunology.[1] The layers of the skin are rich with both innate and adaptive immune cells, capable of stimulating a variety of immunological responses. Multiple antigen presenting cell phenotypes are present in the epidermis and dermis, such as Langerhans cells (LC), dermal dendritic cells (DDC) and dermal macrophages, specially equipped to drive humoral and cell-mediated immune responses through antigen presentation.[2] The resident lymphocyte population is also diverse, and includes CD4+, CD8+ and γδ T cells, as well as memory and plasma B cells.[3,4] In both humans and pigs, intradermal (ID) vaccination has been explored to combat multiple diseases, including respiratory disease caused by influenza.[1,5] The continued antigenic drift and shift of influenza A virus in swine (IAV-S) results in genetic diversity that poses a major challenge for vaccine development, and as such, a variety of IAV-S vaccines have been introduced.[6,7] Vaccines are based on whole inactivated virus, recombinant protein, or replicon particle vaccine platforms, and available as autogenous, commercial, or veterinary prescription products.[7] The veterinary prescription platform technology allows customized immunogens of interest against a variety of pathogens to be included in non-replicating, nonviable vaccines, produced using a standardized production platform.[6] Until now, there were no veterinary prescription vaccines, against IAV-S or any other pathogen, available for intradermal vaccination in pigs. Here, we conducted three studies to evaluate intradermal administration of an HA RNA Particle (RNA-P) vaccine with Microsol Diluvac Forte® (MDF) adjuvant (SEQUIVITY® ID) for safety in pigs and pregnant sows, and efficacy against an H1N1 challenge as an enhanced tool against IAV-S in the swine industry.

Methods

Three studies were conducted to evaluate an adjuvanted, intradermal H1 RNA-P vaccine (SEQUIVITY ID, test vaccine). The vaccine antigen was based on the HA gene of a currently circulating US H1N1 IAV-S isolate. In all studies, the frozen vaccine was thawed at point of use and mixed with Microsol Diluvac Forte adjuvant. All vaccinations were administered as a 0.2 mL dose using the hand-held IntraDermal Application of Liquids (IDAL®) device, which is a battery powered, intradermal injector. In Study 1, the H1 RNA-P vaccine was formulated to target a minimum effective potency level, and the seroresponse and efficacy of the vaccine were evaluated. Sixty seronegative pigs were enrolled in the study at 20-21 days of age (post-weaning). On the day of vaccination, 30 pigs were intradermally administered test vaccine, and 30 litter-matched pigs were intradermally administered placebo vaccine, formulated with green fluorescent protein (GFP) RNA-P antigen, in place of the H1 RNA-P. All study pigs received a booster dose three weeks later at six weeks of age. All pigs were observed every other day for local or systemic reactions to the vaccine for 15 days following each vaccination. Blood was collected on the day of each vaccination and weekly following the booster dose to evaluate the serological response to vaccination by hemagglutinin inhibition (HI) titer. At the time of challenge, twenty pigs per treatment group were randomly assigned to a necropsy subgroup, and the remaining ten pigs per group were assigned to a shedding subgroup. At 31 days post-booster vaccination, all study pigs were intratracheally challenged with 6.1 log_{10} $TCID_{50}$ of A/swine/Iowa/A01778877/2016 (H1N1), homologous to the vaccine H1 antigen. The necropsy subgroups were euthanized at 5 days post challenge (DPC), to evaluate the lungs for percent macroscopic lesions and viral load in bronchoalveolar lavage fluids. The shedding subgroups were swabbed daily for nasal shedding evaluation by live virus titration of nasal swab fluids until shedding had resolved for two consecutive days, at 8 DPC.

To evaluate the safety of the vaccine in studies 2 and 3, two vaccine serials were formulated to contain > 100-fold more RNA-P antigen than the serial used in study 1. The two serials were used in approximately equal numbers of animals within each study. The safety of the vaccine in male and female pigs at 21-24 days of age was evaluated in study 2. Seven hundred and thirty-three pigs at three separate farms were vaccinated with one of two serials of the vaccine, receiving a booster dose three weeks later of the same serial. All study pigs were monitored daily for three weeks following each vaccination, or until any observed adverse events resolved. In study 3, the safety of the vaccine in pregnant sows was assessed by vaccinating 479 sows twice three weeks apart with one of two serials. Study sows were either in early (30-40 days), mid (50-60 days), or late (70-80 days) gestation at the time of the first vaccination. Non-vaccinated sows (n = 237) in each phase of gestation were also enrolled as controls. All sows were observed for three weeks after each vaccination timepoint, and sows and their progeny were observed at farrowing and for seven days after farrowing.

Results

In study 1, all pigs vaccinated with the test vaccine seroconverted and remained seropositive at time of challenge, with peak HI titers (geomean titer 1733.5) at 7 days post-booster. Following challenge, the vaccine prevented nasal shedding, as no vaccinated pigs shed live H1N1, as compared to 92% of placebo pigs positive for nasal shedding in the challenge period (Figure 1). Peak shedding occurred on 3 DPC and was resolved by 7 DPC. At the time of necropsy, the severity of lung lesions was reduced in the vaccinated pigs by 95%, compared to the lung lesions observed in the placebo pigs (Figure 2). There was no detectable virus in the lungs of vaccinated pigs, while 45% of placebo pigs were positive for H1N1 viral load. Vaccinated pigs had no systemic reactions and limited local injection site reactions following vaccination with either the test vaccine or the placebo vaccine (majority of reactions < 2 cm in diameter, resolving on average within five days).

Figure 1: Study 1, mean nasal shedding titers as measured by live virus titration of nasal swabs collected daily following challenge with H1N1 virus.

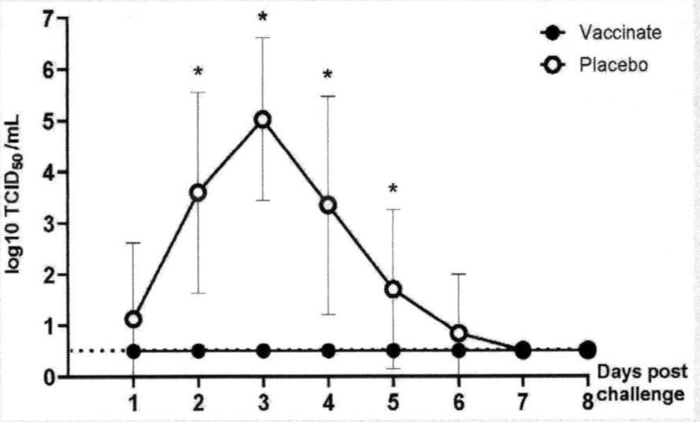

* Statistically significant difference within day post-challenge between the vaccinated and placebo group ($P < .05$).

Figure 2: Study 1, mean percent macroscopic lung lesion scores of vaccinated and placebo pigs at five days post-challenge with H1N1 virus.

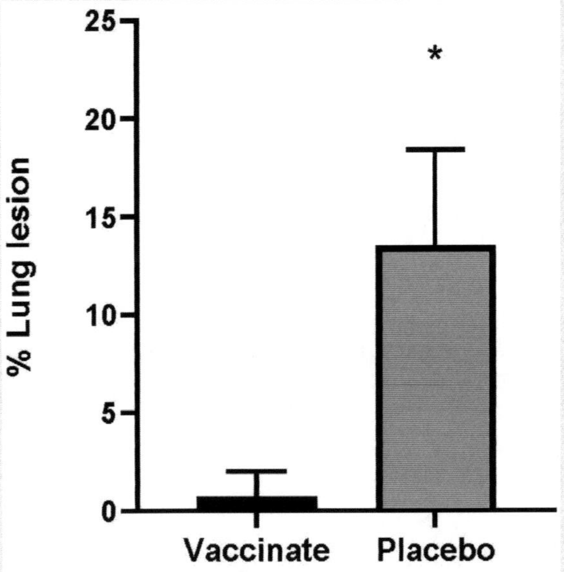

* Statistically significant difference between the vaccinated and placebo group ($P < .05$).

In study 2, no anaphylactic reactions or systemic adverse events attributable to the vaccine were observed. Local injection site reactions occurred in 5% of pigs following the first vaccination and in 30% of pigs following the second vaccination. No abscesses were observed at the injection sites. The majority of these reactions were mild, measuring less than 1.5 cm, and the mean duration of these reactions was 1.4 and 3.9 days following first and second vaccination, respectively.

In study 3, none of the systemic adverse events observed were attributable to the vaccine. The incidence of adverse events (ie, anorexia, return to estrus, etc) in vaccinated sows was not different across gestation phases, or when compared to the non-vaccinated controls. The incidence of abortion was the same whether sows were vaccinated or controls, occurring in 1.3% of each group. Local injection site reactions were observed in 35% sows, of which 71% were less than 1.5 cm in diameter. Reactions to the intradermal vaccination typically appeared within one day of vaccination and resolved in an average of 8 days after the first vaccination, and 10 days after the second vaccination. Vaccination and phase of gestation did not impact the number of live born, mummified, stillborn, or weak born pigs, nor did it impact the number of live pigs born to vaccinated sows observed one-week post-farrow, when compared to the controls.

Discussion

The three studies demonstrated the safety and efficacy of this intradermal, adjuvanted RNA-P vaccine (SEQUIVITY ID) administered in a two-dose regimen. Intradermal vaccination with SEQUIVITY ID stimulated robust HI antibody titers to the vaccine fraction, with peak titers as early as 7 days post-booster. This immune response was efficacious, limiting disease in pigs from the IAV-S challenge. Vaccination eliminated viral load in the lungs and nasal shedding of the live virus, and significantly reduced macroscopic lung lesions when compared to controls. The vaccine was demonstrated to be safe in weaned pigs, as vaccination resulted in mild injection site reactions with no systemic adverse reactions. No anaphylaxis was observed following vaccination. Further, intradermal vaccination with SEQUIVITY ID was safe in pregnant sows in all phases of gestation. No systemic adverse reactions were observed, and there were no significant differences in vaccinated sows when compared to non-vaccinated sows for farrow metrics, incidence of abortion and return to estrus, or in the health status of their progeny.

Taken together, the three studies demonstrated that ID administration of this RNA-P vaccine in MDF adjuvant was efficacious against IAV-S H1N1, and safe for use in pigs and pregnant sows at all gestation phases. Prevention of viral shedding provides evidence for the ability of the vaccine to reduce IAV-S disease spread on farm, with the potential to reduce IAV-S circulation and reassortment. The pairing of SEQUIVITY ID and IDAL technologies presents veterinarians and swine producers a new set of innovative solutions to combat swine diseases.

References

1. Có-Rives I, Chen AY, Moore AC. Skin-Based Vaccination: A Systematic Mapping Review of the Types of Vaccines and Methods Used and Immunity and Protection Elicited in Pigs. Vaccines (Basel). 2023 Feb 16;11(2):450. **doi: 10.3390/vaccines11020450**

2. Behazine Combadiere & Christelle Liard (2011) Transcutaneous and intradermal vaccination, Human Vaccines, 7:8, 811-827, **DOI: 10.4161/hv.7.8.16274**

3. Mueller S.N., Zaid A., Carbone F.R. Tissue-Resident T Cells: Dynamic Players in Skin Immunity. Front. Immunol. 2014;5:332. **doi: 10.3389/fimmu.2014.00332**

4. Debes G.F., McGettigan S.E. Skin-associated B cells in health and inflammation. J. Immunol. 2019;202:1659–1666. **doi: 10.4049/jimmunol.1801211**

5. Hernandez-Franco JF, Yadagiri G, Patil V, Bugybayeva D, Dolatyabi S, Dumkliang E, Singh M, Suresh R, Akter F, Schrock J, Renukaradhya GJ, HogenEsch H. Intradermal Vaccination against Influenza with a STING-Targeted Nanoparticle Combination Adjuvant Induces Superior Cross-Protective Humoral Immunity in Swine Compared with Intranasal and Intramuscular Immunization. Vaccines (Basel). 2023 Nov 7;11(11):1699. **doi: 10.3390/vaccines11111699**

6. Petro-Turnquist E, Pekarek M, Weaver, E. Swine influenza A virus: challenges and novel vaccine strategies. Front Cell Infect Microbiol. 2024; 14:1336013.

7. Kitikoon P, Knetter SM, Mogler MA, Morgan CL, Hoehn A, Puttamreddy S, Strait EL, Segers RPAM. Quadrivalent neuraminidase RNA particle vaccine protects pigs against homologous and heterologous strains of swine influenza virus infection. Vaccine. 2023 Nov 13;41(47):6941-6951.

https://doi.org/10.54846/am2025/55

Sequivity® IAV NA keeps pigs growing in the face of an early nursery IAV-S challenge

Brett O'Brien[1], VMD, MS; Chandra Morgan[1]; Ashley Leblond[1]; Pravina Kitikoon[2]
[1]Merck Animal Health, Lenexa, Kansas; [2]Formerly with Merck Animal Health, Lenexa, Kansas

Introduction

Influenza A virus in swine (IAV-S) has been reported to cost $3-10 per growing pig.[1,2] The Sequivity® IAV-S NA vaccine offered by Merck Animal Health can be administered to pigs as young as three days of age (DOA) to protect growing pigs against clinical IAV-S. This vaccine is labelled to be given to piglets as two 1 mL doses staggered three weeks apart. This study's objective was to compare the nursery growth performance between pigs vaccinated with Sequivity IAV-S NA versus unvaccinated controls in the face of an early nursery IAV-S challenge.

Materials and methods

Two piglets of similar body weight from 48 litters of IAV-S naïve dams were enrolled and randomly assigned to treatment group (T01: Positive controls; T02: Sequivity IAV-S NA vaccinated) at approximately three DOA (study day 0). Vaccinated pigs received either a 1 mL intramuscular dose of Sequivity IAV-S NA at approximately 3 and 21 DOA while control pigs received 1 mL of sterile saline. On study day 19, pigs were weaned into a contract research nursery while maintaining litter integrity within pens. All pigs were vaccinated with PRRSV MLV at 5 weeks of age (WOA) and challenged intratracheally at 6 WOA (study day 40) with an H1N1 gamma strain. All pigs were weighed on study days 0, 18, 39, 50, and 68. A subset of 30 animals per treatment group had nasal swabs collected on study days 39, 41, 43, 45, 47, and 50 and were tested by IAV-S viability and standard qPCR at the University of Minnesota. IAV-S sequencing was performed from a subset of positive PCR samples from study days 39, 41, 43, and 50. All animals were bled on study days 0, 18, 39, 47 and 68 and tested by IAV-S serological assays (hemagglutinin inhibition [HI], neuraminidase inhibition [NI], and NP ELISA) at Merck Animal Health. General health observations were recorded daily throughout the study and clinical counts of respiratory signs (cough, increased respiration rate, nasal discharge, and lethargy were recorded daily from study day 41 to 55 and then weekly thereafter. The average daily gain (ADG), body weight, and serology data were analyzed with general linear mixed models with litter as a random effect using SAS version 9.4. Nasal shedding data was analyzed using a fitted repeated measures model. A linear Pearson correlation coefficient was calculated for standard versus viability PCR results.

Results

IAV-S PCR, sequencing, and serology results 1 day prior to challenge (study day 39) demonstrated that an H1 alpha N1 classic virus had infected the study population roughly 1-2 weeks prior to administration of the gamma challenge strain. On study day 39, three control pigs were NP ELISA and NI positive whereas all control pigs were serologically negative on study days 0 and 18, indicating IAV-S infection within the previous 14 days. All Sequivity IAV-S NA vaccinated pigs were NP ELISA negative on study day 39. Since the Sequivity IAV-S NA vaccine does not contain the IAV-S nucleoprotein, it is a 'differentiating infected from vaccinated animals' (DIVA) capable vaccine when sera from vaccinated pigs exposed to wild type virus are tested with the NP ELISA assay. Nasal swabs from study day 39 had PCR positivity rates of 90% in the control animals and 77% in the Sequivity IAV-S NA vaccinated pigs indicating a high infectivity rate in the IAV-S strain that infected the pigs on study. This H1N1 strain was the only virus sequenced from all time points tested. As often seen with IAV-S infections, there was a low level of respiratory clinical signs in pigs (2.7% cough, 0.6% lethargy, 3% increased respiration rate, and no nasal discharge). Likewise, general health observations recorded prior to study day 41 did not include any clinical symptoms associated with an IAV-S infection. With early nursery IAV-S infection, Sequivity IAV-S NA vaccinated pigs had a 2.9-pound advantage over controls during the last 4 weeks of the nursery period ($P = .03$, see Table 1). Sequivity IAV-S NA vaccinated pigs had a 2-pound numerical advantage over the unvaccinated controls during the entire nursery period ($P = .2$). Figure 1 shows control pigs had 1-2 logs higher viral shedding levels than vaccinated pigs during the late infection period monitored before study day 45 ($P < .05$). There is a strong correlation between the standard and viability PCR test (Pearson correlation coefficient = 0.66). During the late infection period, the viability PCR reported approximately 4 logs of virus less than the standard PCR test which indicates that the standard PCR detected a high level of dead virus in the nasal passages towards the end of the infection period. Prior to the gamma H1 challenge, all Sequivity IAV-S NA vaccinated pigs had high NI titers on the N2 2002a assay which is a vaccine specific NA that does not cross-react with the H1N1 virus that infected the pigs (backtransformed logarithmic mean titer of 43 on study day 18, and mean titer of 1316 on study day 39) as shown in Table 2. On study days 47 and 68, the NP ELISA positivity rate was reduced in vaccinates compared to controls (see Table 3) and mean S/P titers were significantly lower in controls than Sequivity NA vaccinated pigs ($P < .01$).

Table 1: Mean average daily gains of Sequivity NA vaccinated pigs compared to controls

Treatment group	Farrowing	Early nursery (3-6 WOA)	Late nursery (6-10 WOA)	Entire nursery period (3-10 WOA)
Positive controls	0.585	0.600	1.264[b]	0.972
Sequivity IAV-S NA	0.586	0.591	1.368[a]	1.011

[a,b] Different superscripts within a column denote statistical significance $P \leq .05$

WOA = weeks of age

Figure 1: Nasal shedding (viability PCR).

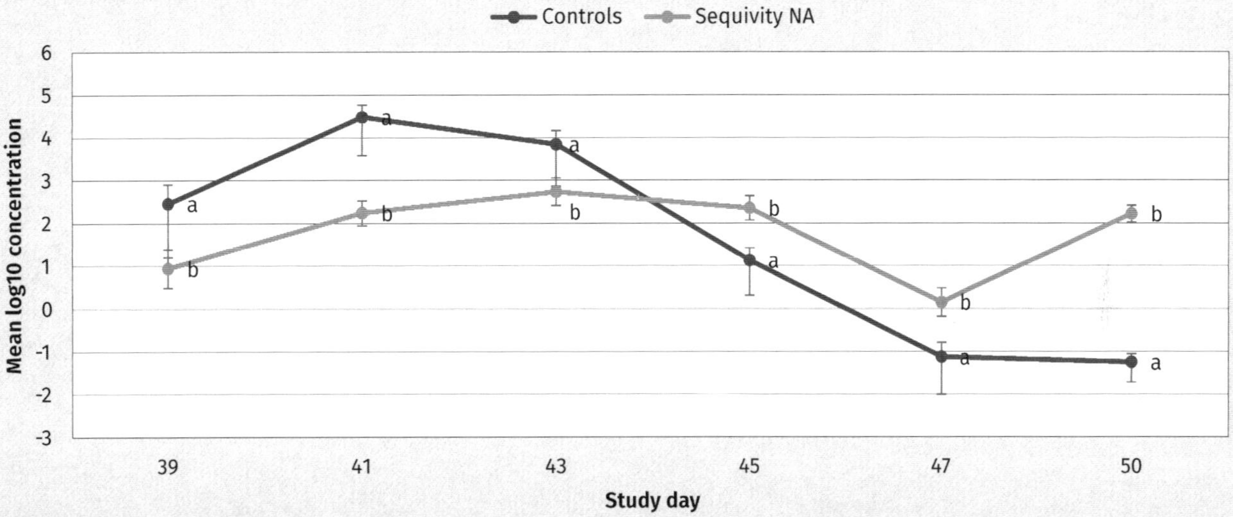

[a,b] Different superscripts within study day denote statistical significance at $P < .05$

Table 2: N2 2002a IAV-S backtransformed logarithmic mean NI titers (vaccine specific NA)

Study day	Controls	Sequivity IAV-S NA	*P*-value
0 (enrollment)	9	9	.4
18 (2nd vaccination)	9	43	< .01
39	9	1316	< .01
47	9	651	< .01
68	9	295	< .01

Titer ≥ 20 is positive

Table 3: Mean IAV-S NP ELISA titers and positivity rates

Study day	Mean NP S/P	Number positive	Mean NP S/P	Number positive	*P*-value
0 (enrollment)	0.993	0/30	1.029	0/30	.24
18 (2nd vaccination)	1.177	0/30	1.144	0/30	.3
39	0.85	3/30	0.893	0/30	.16
47	0.564	19/30	0.757	5/30	< .01
68	0.411	25/29	0.723	9/29	< .01

Titer ≤ 0.60 is positive

Discussion

While the clinical picture of IAV-S infection can be mild in growing pigs, the negative impact to their growth can be substantial as witnessed in this study. Sequivity IAV-S NA is an effective tool to improve growing pigs' performance in the face of a heterologous IAV-S early nursery challenge. This DIVA capable vaccine also allowed researchers to identify the introduction of a different IAV-S strain prior to challenge via serological testing.

References

1. Haden et al. Assessing production parameters and economic impact of swine influenza, PRRS and Mycoplasma hyopneumoniae on finishing pigs in a large production system. *AASV annual meeting proceedings*. 2012; 75-76.

2. Donovan et al. How we deal with Swine Influenza Virus in a multisite production system. *Allen D. Leman Swine Conference proceedings*. 2011; 75.

https://doi.org/10.54846/am2025/56

Critical components of a successful IAV-S stabilization program for breeding herds

Cameron Schmitt, DVM, MS
Pipestone Veterinary Services, Independence, Iowa

Introduction

Many breeding herds in the US struggle with influenza A virus in swine (IAV-S). Multiple strains of H3N2, H1N1 and H1N2 are presently endemic in the US swine herd. Infection with IAV-S can result in high fever, coughing, sneezing, poor feed intake and damage to the epithelia of the airways, which results in increased susceptibility to bacterial co-infections and pneumonia. IAV-S is an important cofactor in the porcine respiratory disease complex (PRDC), where it causes the greatest economic impact. Multiple studies in production systems have demonstrated this negative economic impact in reduced average daily gain (ADG), feed conversion, and full value pigs as well as increased mortality, cull rates, and treatment costs. In one study the overall economic impact of IAV-S was determined to be $3.23 loss per pig.[1] Given the high costs incurred by this disease, there has been an increased effort to eliminate IAV-S from breeding herds with the goal of weaning IAV-S negative pigs. Prevention of respiratory disease and decreasing the impact of PRDC will reduce the use of antibiotics, improve livability, sustainability, public health, and animal welfare. Production of IAV-S negative weaned piglets is attainable for many sow farms provided that the multiple complexities of IAV-S transmission, vaccine program, pig flow, and biosecurity at the farm and/or production system level are taken into consideration when designing an effective stabilization or elimination protocol.[2]

Goal setting and planning

Most pork production systems considering stabilizing their breeding herds for IAV-S are currently unstable. An unstable herd is one in which the virus can be detected via PCR (sample types include lung tissue, nasal swabs, nasal wipes, oral fluids, udder wipes, or other appropriate sample types) in the farrowing house (dams and/or piglets). Virus can often be detected from other areas of the sow farm including the gilt developer unit (GDU), gestation barns, weaned pig holding areas, and/or any onsite nurseries or finishers. See Table 1 for a review of IAV-S breeding herd IAV-S classifications based on diagnostic history and clinical signs. Unstable herds can be further subdivided into categories based on the timeframe since infection. During the acute phase (epidemic) of the herd's infection following a new IAV-S strain introduction, clinical signs may be present and there is typically a high prevalence of the population shedding virus. Farms that continue to circulate virus in a subpopulation of susceptible animals after the initial infection are classified as endemically infected (chronic phase) and may or may not have respiratory clinical signs due to existing immunity and decreased viral circulation. Additionally, some IAV-S strains (ie, some pandemic H1N1 strains) may not elicit pronounced clinical signs in individual pigs or in different populations of pigs. The stabilization program recommended is the same regardless of the phase of the infection.

Understanding the farm's IAV-S clinical and vaccination history as well as the potential risks for IAV-S introduction and circulation is key in developing overarching goals for IAV-S control in the farm. A herd's IAV-S history and current IAV-S status (which IAV-S strains are present and where virus shedding is occurring) help determine the appropriate IAV-S control goal and pathway for the herd. Make sure to collect diagnostic samples from every air space within the farm. Furthermore, sequences derived from diagnostic sampling will assist in selecting the appropriate vaccine to use in the breeding herd. With regards to vaccine selection, we have seen the best success utilizing vaccines that are good matches (high homology and key sites match) with the farm specific strains. Farm layout as well as animal and people movements are critical to determine if enhanced biosecurity steps are needed for a successful stabilization program.

Table 1: Summary of breeding herd influenza A virus in swine (IAV-S) classifications based on diagnostic monitoring history and clinical signs of the herd

Herd category	PCR positive samples-farrowing	PCR positive samples-whole herd	Clinical signs-farrowing	Clinical signs-whole herd	HA positive sera	NA positive sera	NP positive sera
I. Unstable	+	+	+/–	+/–	+	+	+
II. Stable (wean pig negative)	–	+	–	+/–	+	+	+
III. Provisionally negative	–	–	–	–	+/–	+/–	+/–
IV-A. Negative vaccinated	–	–	–	–	+/–*	+/–*	+/–*
IV-B. Negative unvaccinated/naive	–	–	–	–	–	–	–

* Negative when DIVA vaccine used. May be positive if whole inactivated virus vaccine is used and will require negative serology from sentinel pigs to confirm.

Once the plan for the farm has been established, it is critical to communicate the plan and obtain buy in from all key constituents (personnel, farm owner, manager, veterinarian). Assign an IAV-S project leader, an on-farm project foreperson, and a primary diagnostic contact who will coordinate diagnostic sample collections (supplies, schedule, shipping, dissemination of results. Generate a one-year timeline and review with everyone. It is critical that everyone knows what will be happening and when. There are three phases to an IAV-S stabilization program:

- Phase 1 is week 1-6 after the start of mass vaccination of the sow herd when IAV-S circulation will still be present and can be detected on the farm.
- Phase 2 is 6 to 10 weeks following mass vaccination of the herd and marks the time period when the farrowing house is expected to be populated with IAV-S negative piglets.
- Phase 3 begins after farrowing house diagnostics confirm active IAV-S circulation is no longer detectable (3 consecutive weekly negative PCR results) and involves ongoing surveillance to confirm the status of the farm (expected stable).

Regular project communication is key to its success and recommend setting up weekly or biweekly calls with all key members (veterinarian, farm manager, etc) to help ensure the plan is being followed. More frequent calls are usually necessary during the initial 4 months of the program and will allow the team to talk through any questions or issues that arise along the way and allow for a rapid response.

Biosecurity

Biosecurity is a critical component in the execution of a successful IAV-S stabilization program on breeding herds.[3] First, it is important to educate all farm personnel and visitors the role they could serve in either bringing in new IAV-S strains within the herd or transmission of IAV-S within the herd to uninfected animals. It is recommended to educate everyone working in or visiting the farm to understand what the clinical signs of influenza are (fever, cough, sore throat, runny or stuffy nose, muscle or body aches, headaches, or fatigue) and if they exhibit these signs, they should not enter the farm. Some farms will require anyone planning to enter the farm to take their temperature and not permit anyone exhibiting a fever (> 100°F) to enter. Farms may take the extra precaution of requiring personnel and visitors wear an appropriately fitted N95 mask to help reduce the infection risk of IAV-S between people and pigs within the farm. Vaccination of farm personnel against influenza is another optional layer of protection to help limit the introduction of new IAV-S strains into the herd. New employees should monitor themselves closely for clinical signs of infection as they are at risk of contracting IAV-S in unstable farms given it is a zoonotic disease. Influenza can survive on surfaces for up to 48 hours, thus supplies entering the farm should be disinfected and allowed to dry in a segregated area (disinfection and drying [D&D room]) before entering the farm. Anyone disinfecting supplies should wear gloves and protective clothing (not clothes worn into the barns) and immediately remove and dispose of these when leaving the supply area.

It is critical to communicate what parts of the farm are actively circulating IAV-S throughout the whole stabilization timeline. Ideally employees from positive areas of the farm should not enter negative areas. If that is not possible, it is preferable for farm personnel to start in negative areas before entering positive ones. If one must move from positive to negative areas, it is recommended to shower and change clothes. An example of this would be weaning days in the farrowing house when most farm personnel must first handle weaned pigs which would be suspected IAV-S positive until 10 weeks after the stabilization program begins. After weaning is completed, all personnel should shower and change into new coveralls and clean boots before working with the pigs in younger rooms in farrowing. Likewise, hallways where the weaned pigs travelled should be cleaned and disinfected before it is trafficked again by pigs or people. If this cannot be done in other areas of the farm (ie, GDU), a change of boots and coveralls with a hand wash/glove change station can be implemented. It is also recommended to put boot wash stations outside of positive areas to minimize the spread of IAV-S in the hallways. While water and soap help remove organic debris and possible viable virus from the skin, it is recommended to follow-up with an alcohol based hand sanitizer to ensure complete kill of the virus. A good understanding of farm layout as well as animal and people movements is critical to create an effective biosecurity protocol during the stabilization program.

It is also important to recognize that common areas in the farm where personnel cross-over (ie, office, break room, supply or medication rooms, boot area) are a potential area where IAV-S might reside on surfaces. It is necessary to keep these areas clean and disinfected every day. Wipe down or disinfect commonly touched surfaces daily on the farm (ie, tables, counters, chairs, door handles). Additionally, farm personnel should thoroughly wash their hands and arms with soap and water followed by an alcohol based hand sanitizer after working with the animals. Boots should be removed and soiled clothing changed before entering common spaces with other farm personnel.

Map all traffic patterns between positive and negative animals and ensure that they do not overlap. If areas overlap, it is important during phase 2 to make sure they are disrupted or discontinued. Given the high amount of traffic in farm hallways, it is important to clean and disinfect hallways immediately after animal movements. Always double check that the hallway is clean before bringing new groups of animals through the farm. It is possible that an adult sow could become infected in a dirty hallway in which IAV-S positive weaned pigs recently travelled. Likewise, equipment can transfer influenza and facilitate entry of the virus within/around farms. Therefore, equipment should be cleaned and disinfected before moving between different locations, rooms and barns. To create traffic pattern rules, ask all farm personnel about their regular daily traffic patterns on the farm. A map of this information should be included in the farm biosecurity plan.

There are enhanced biosecurity practices important during phase 2 of the project that are centered around the two highest IAV-S circulation centers within the breeding herd: the farrowing house and GDU. Cross fostering should be restricted to the first 24 hours of life. Fallback piglets should only be moved within the same farrowing room. Nurse sows should have their udders wiped down with chlorhexidine solution before receiving a new litter. Personnel should wear gloves at all times in farrowing and change gloves between litters. Personnel should also avoid stepping in and out of farrowing crates. Processing equipment should be disinfected between each litter and needles changed between litters. All suckling piglets must be weaned when they reach 21 days of age regardless of weight. Wean entire rooms and make sure to clean, disinfect, and dry the room before reloading with sows. Do not hold any piglets back in farrowing crates or hot nurseries. Perform chores in farrowing house rooms from youngest to oldest. If possible, segregate farrowing house personnel into 3 groups (1. Front end [assisted farrowing and in-room cross-fostering]; 2. Processors [work with piglets from processing until 10 days of age]; 3. Late rooms [work with piglets from day 10 until weaning]). If it is necessary to visit younger IAV-S negative suckling pigs after working with older suckling pigs which are presumed IAV-S positive, shower and change coveralls and boots beforehand. When entering rooms with piglets from dams that have been double mass vaccinated for IAV-S, use only clean boots and coveralls and gloves. Do not allow traffic patterns of negative animals to contact or cross those of positive animals. In the farrowing house, it is expected that IAV-S circulation will continue through the first 9-10 weeks of the elimination program. After that, the goal is to stepwise remove or "walk" influenza out of the farm with each room of pigs weaned. It is critical to follow the sanitation and traffic guidelines to ensure IAV-S from older pigs is not carried back to infect rooms with younger piglets. Piglets are born negative to influenza and the goal is to keep them that way. In the GDU, notify employees of each room's IAV-S status (door signage may help). Like the farrowing house, employees should chore IAV-S negative pigs before positive ones (usually oldest to youngest). A boot wash, change of clothes and hand wash/ sanitization or glove change should be implemented after leaving positive rooms/ages. In IAV-S positive rooms, change gloves between pens. Heat check boars should be dedicated to working either the GDU or the sows (NOT BOTH). Ideally GDU heat check boars are contained in their own area and in a location that is either within or nearest the GDU. GDU heat check boars travelling outside the GDU should not be given time to interact with mature sows on farm. Heat check boars should not be allowed to interact with positive rooms or ages. If they are put in a positive room, they should remain there until phase 2 is concluded. Shut GDU down to all new arrivals until at least one month after the second mass vaccination or when IAV-S diagnostics confirm no IAV-S circulation in the GDU. Do not allow traffic patterns of negative animals to contact or cross those of positive animals.

Vaccination program

Mass vaccination is used to create consistent protective immunity to circulating strains for all animals at the same time. Sows and gilts with high antibody titers to the circulating farm strain will help ensure suckling piglets are adequately protected through maternally derived antibodies thereby reducing or eliminating viral circulation in the farrowing house. On day 1 of the IAV-S stabilization program, all adult animals (gilts, sows, boars) and all animals housed on the farm's site (gilts in GDU, nursery or finishing pigs) should receive 1 full dose of IAV-S vaccine. These same animals should be mass-vaccinated again 3-4 weeks later. Please remember that following a mass vaccination of any swine must comply with the labelled vaccine withdrawal period. It is advised to coordinate the timing of cull loads before planned vaccination events. In order to maintain high maternal antibodies in weaned pigs, after mass vaccination, it is recommended to administer a pre-farrow IAV-S vaccine booster during phase 2. After which, a quarterly mass vaccination program to mature adult animals in conjunction with routine replacement gilt and boar vaccination is recommended for at least the following year. One year of negative virus detection though a surveillance program should be achieved before reducing the frequency of IAV-S vaccination in the sow herd.

If replacement gilts are retained from the farm (internal multiplication), it is advised to vaccinate them with IAV-S vaccine as early as possible. This can be done with either a Sequivity® IAV-S NA vaccine at 3 days and 3 weeks or a vaccine containing HA at 10 or more weeks of age when maternal antibodies are expected to wane.[4] Serological testing of replacement gilts can help guide replacement gilt vaccination timing. To pass from phase 2 to phase 3, three consecutive weeks of udder wipes or nasal swabs from late farrowing litters should be negative. If there is an onsite holding room, pigs housed within it should be sampled weekly with nasal swabs. All samples should be collected using a low prevalence strategy (assume 5% prevalence). The GDU requires three negative monthly oral fluids from every three weeks of age using a low prevalence sampling strategy. All age groups of replacement gilts and boars should be tested with oral fluids 1 and 3 weeks before arrival to the sow farm to ensure no new introductions of virus. It would also be recommended to test the new gilt group within the first three days after arrival. Any clinical signs of respiratory disease anywhere on the farm should be investigated with appropriate diagnostics using the high prevalence strategy. After reaching phase 3, monitoring for virus must continue on a monthly basis. Phase 3 concludes after 3 consecutive months of negative PCR testing across all sectors of the farm. The farm is now classified as provisionally negative. All age groups of replacement gilts and boars should be tested with oral fluids 1-2 weeks before arrival to the sow farm to ensure no new introductions of virus and any clinical signs of respiratory disease anywhere on the farm should be worked up. After one year of successfully remaining virus free on all parts of the farm, the farm may decide to proceed to negative status. Because of the IAV-S introduction risks from both people and replacement gilts or boars, it is recommended to maintain IAV-S vaccination within the breeding herd.

Conclusion

The keys to a successful IAV-S stabilization program in a breeding herd relies on proper diagnostic surveillance and monitoring programs, planning, frequent and thorough communication, a high farm commitment level, enhanced biosecurity programs and the appropriate vaccination program. These program components have been incorporated based on IAV-S elimination experiences and research. When all these areas are addressed, there is a high likelihood of successfully eliminating all active circulating IAV-S from the breeding herd.

References

1. Haden et al. Assessing production parameters and economic impact of swine influenza, PRRS and *Mycoplasma hyopneumoniae* on finishing pigs in a large production system. *AASV annual meeting proceedings.* 2012; 75-76.
2. Schmitt et al. Putting all pieces together to eliminate influenza from sow farms. *Leman conference proceedings.* 2021; 41.
3. Lopez-Moreno et al. Evaluation of internal biosecurity practices combined with sow vaccination to wean influenza negative piglets. *Leman conference proceedings.* 2021; 59-60.
4. Koep et al. Use of a hemagglutinin (HA) and neuraminidase (NA) vaccine program to stabilize a sow farm for influenza. *AASV annual meeting proceedings.* 2024; 127-129.

https://doi.org/10.54846/am2025/57

Florfenicol in swine drinking water: The advantages of flexible dosing for veterinary prescriptions

P. Smith, PharmD, MBA; B. Payne, DVM; K. Bretey, DVM
Veterinary Pharmaceutical Solutions (VPS), St. Peter, Minnesota

Introduction

Respiratory, enteric and systemic diseases in swine occur in modern swine production involving groups of animals that require fast-acting therapy that can be implemented across populations with minimal handling and stress. Timely delivery of therapeutic agents via the drinking water reduces the need to individually inject pigs when disease outbreaks occur and veterinarians conclude that population therapy becomes the prudent choice in terms of efficacy, welfare, positive health outcomes and resource management. There are FDA-approved florfenicol oral drugs for swine intended for use via drinking water; however, they have not been consistently manufactured and available to veterinarians for administration to swine. In their absence, compounded florfenicol has been used by the swine industry to combat disease outbreaks. UltraFlor™, a compounded concentrated formulation with a wider breadth of dose dilution and miscibility compared to the FDA-approved drugs, is the latest innovation in florfenicol water soluble drugs available to producers via veterinary prescription.

Florfenicol as a viable water medication option

Florfenicol is in the class of amphenicol antimicrobial drugs, whose mode of action disrupts bacterial protein synthesis by blocking enzymes on the bacteria's 50S ribosome subunit. It is a broad-spectrum, time-dependent, bacteriostatic antibiotic used in swine for the treatment of respiratory, enteric and systemic infections. Its antibacterial activity includes gram-negative bacilli and gram-positive cocci.[1]

A susceptibility summary of *Streptococcus suis* (*S suis*) isolates from the University of MN from 2014-2022 identified florfenicol as having the least resistance (1.35%), behind ampicillin, on its panel of antibiotics.[2] In one study evaluating percent resistance of *Pasteurella multocida* (P mult) to florfenicol, it found that there were no resistant isolates out of the 72 tested.[3] A summary of F18 *E coli* from 2010-2022 at the Iowa State University Veterinary Diagnostic Lab, showed that florfenicol was only one of two antibiotics increasing in F18 susceptibility in the previous five years.[4] The other drug with increasing sensitivity for F18 was ceftiofur which is not legally available for oral administration in livestock.[5] Oral florfenicol has been found to achieve concentrations above the MIC_{50} and MIC_{90} for P mult, *Glaesserella (Hemophilus) parasuis* (GPS) and *Actinobacillus pleuropneumoniae* (APP).[6] Oral florfenicol is highly bioavailable and rapidly absorbed.[7] Florfenicol's lipophilic attribute has been shown to moderately (46% available) cross the blood-brain barrier of calves.[8] Florfenicol serum concentrations have been maintained above swine pathogen MIC's (1μg/mL) when swine drank 100ppm florfenicol-treated drinking water for a five day period.[9]

Response to an industry need and request from the field

While there are FDA-approved and labeled florfenicol products for use in swine drinking water, such as Florcon® (Aspen Veterinary Resources, Liberty, MO), none are currently available in the marketplace. Compounding a drug that is essentially equivalent to an FDA-approved drug is illegal. However, compounding the essentially equivalent drug when it is not available in the marketplace is legal. Such is the case with compounded florfenicol for use in swine drinking water. This makes it a viable option for veterinarians if there are no other suitable alternatives higher on the AMDUCA (Animal Medicinal Drug Use Clarification Act) decision-tree.[10]

In the absence of a labeled florfenicol product for use in swine while disease outbreaks continued, veterinarians requested Veterinary Pharmaceutical Solutions (VPS), an FDA and Boards of Pharmacy inspected pharmacy that adheres to the FDA Code of Federal Regulations Title 21 (21CFR) and US Pharmacopeia (USP) Chapter 795 for non-sterile preparations, formulate a 1.347% solution of florfenicol that could be compounded as a ready-to-use (FF-RTU) stock solution antimicrobial for use in a medicator at a 1:128 dilution with drinking water. FF-RTU was made using active ingredients from FDA-approved final form products with either an assigned Abbreviated New Animal Drug Application (ANADA) or a New Animal Drug Application (NADA) number which is a requirement for livestock and poultry compounding with this drug.[5] The veterinarian must prescribe compounded products to specific animals and supply the indication(s), directions for use and withdrawal period.

FF-RTU instructions for use were to add the product to a stock solution bucket without further dilution of water and proportion (1:128) into the drinking water. This provided the same targeted drinking water concentration, 400mg/gallon, as the FDA-approved products. The FF-RTU stock solution process differed from the previous labeled florfenicol options that were formulated as 2.3% solutions, requiring careful dilution with water before addition to the stock solution bucket on farm. This labeled drug dilution process was associated with significant heat production and precipitation easily occurred if it was diluted beyond the label instructions. Additionally, the label on the FDA-approved product warned not to use in proportioners if the water hardness was greater than 275 ppm. Another warning indicated that perianal inflammation could occur. The author's personal experience showed this to be true along with transient watery diarrhea. Slaughter withdrawal in swine on the labeled 2.3% solutions was 16 days. Residues have not been tested with the FF-RTU at this time. Both previous labeled florfenicol and FF-RTU were generally not miscible with other medications.

Continued innovation

To mitigate the multiple limitations of existing florfenicol options, VPS reformulated its florfenicol further into UltraFlor™. UltraFlor is a concentrated (4.25%) water soluble compounded florfenicol that can be diluted with on-farm water in stock solution buckets at varying ratios to provide flexible final dosing levels. To achieve the original 400mg/gallon concentration in the water delivered to the pig, the practitioner prescribes an UltraFlor gallon to be mixed with 2.2 gallons of water to make 3.2 gallons of stock solution (Table 1). The water-soluble properties of UltraFlor allow the product to be diluted in the stock solution bucket by up to 50% without sacrificing stability of the active ingredient (Table 1). This ability allows the prescribing veterinarian to adjust the dose of product being delivered to the pigs based on knowledge of and experience with the disease scenario being treated. A reasonable strategy in treating various infectious diseases may include dilution of the product to a level that efficaciously treats the disease, but that mitigates the adverse effects of perianal inflammation and diarrhea often seen with higher doses. This dilution strategy may also allow longer duration treatments without dehydration consequences. Practitioners have been able to achieve success in providing therapy by diluting UltraFlor down another 50% and administering the medication for 5 days, versus their original 2-day directions, to achieve longer therapy without the previous side effects of perianal inflammation.

Added benefits are the solubility and miscibility of UltraFlor. While warnings of previous labeled products not being used in water hardness > 275ppm, UltraFlor has successfully been utilized in farms with water hardness much greater, up to 3,556 ppm. UltraFlor is miscible with VPS's compounded 0.6% meloxicam soluble product. Practitioners have prescribed these products to treat respiratory, enteric and systemic bacteria and the inflammation of the tissues that goes along with those infections.

Additionally, the innovative concentration of this product results in a 3.2X reduction in product volume leading to substantially lower transport and inventory resource requirements.

Previous labeled florfenicol was indicated for the treatment of swine respiratory disease associated with APP, P mult, *Salmonella choleraesuis*, and *S suis*. Other uses are considered extra-label drug use (ELDU) and are permitted for veterinary prescription under AMDUCA and 21 CFR 530.13 when there is no medically appropriate drug higher on the decision-tree.[10] Example ELDU uses with veterinary prescriptions include treatment of virulent *E coli* infections that are resistant to current labeled products, systemic infections involving *Salmonella typhimurium*, and virulent, but susceptible, secondary bacteria to PRRSV and IAV-S infections.

Table 1: Dilution ratios recommended for UltraFlor™

To achieve	Volume of UltraFlor	Volume of water	Total volume of stock solution
400mg/gal (original concentrate)	1 gallon	2.2 gallons	= 3.2 gallons
300mg/gal (75% original rate)	1 gallon	3.8 gallons	= 4.8 gallons
200 mg/gal (50% original rate)	1 gallon	5.4 gallons	= 6.4 gallons

Best practices

Make and mix stock solution fresh daily. Warm water will improve the homogeneity and solubility of the stock solution. UltraFlor will be more viscous if it is cold and make it harder to mix; bring to room temperature if it has been shipped in cold weather. Poor water quality may make it more difficult and take longer to mix the stock solution. This has been found in water characterized with elevated sulfates and minerals that are described as levels that may result in impaired mineral absorption and osmotic diarrhea. Facility owners should evaluate how to provide better quality water to their livestock. Utilizing a stir-pump or water agitator that creates heat in the bucket will not degrade the potency of the florfenicol of UltraFlor. When the stock solution is first made, bubbles will arise to the top of the bucket. Let the bubbles dissipate before putting in the filtered medicator draw-off hose.

Discussion

Current practices with UltraFlor™ as an effective water medication for swine include both 1) use of the product at traditional rates for short periods of time (generally 2-3 days duration) as well as 2) use at lower dosage rates utilizing the dilution capability, but for longer treatment durations (up to 5 or more days). This lower-dose-with-longer-duration strategy appears to be providing efficacious treatment of both *E coli* enteric issues as well as reduction of mortality due to bacteria secondary to respiratory viral pathogens, while at the same time mitigating potential adverse side effects. The authors anticipate additional successful protocols utilizing the flexible dosing ability of UltraFlor will be developed and adopted as veterinary practitioners gain further experience with this innovative tool.

References

1. Papich M. Florfenicol. Saunders Handbook of Veterinary Drugs. 4th ed, 2016;327-329. **doi:10.1016/B978-0-323-24485-5.00264-3**

2. Chagas S, Antimicrobial resistance of *Streptococcus suis* from diseased pigs in the USA – a retrospective study from 2014-2022. Leman Conference. St Paul, MN. 2023. p137-8.

3. Zhou, C, Lara C, et al. Study of the susceptibility of *Pasteurella multocida* strains to florfenicol. IPVS. Copenhagen, Denmark. 2006. p31-8.

4. Paiva R, Burrough E, Piñeyro P, Almeida M. Characterization of hemolytic E. coli and antimicrobial susceptibility from ISU-VDL submitted cases from 2010 to 2022. AASV. 2024. P 46-8. **doi.org/10.54846/am2024/14**

5. National Archives and Records Administration. Code of Federal Regulation: Title 21. Ch 1 Subch E. Part 530. US National Archives. November 8, 2024. Assessed November 11, 2024. **https://www.ecfr.gov/current/title-21/chapter-I/subchapter-E/part-530**

6. Kelly, R, Fleck, R. Florfenicol susceptibility of porcine respiratory isolates compared with serum and plasma levels of florfenicol administered orally. IPVS. South Africa. 2008. P08.058.

7. Liu J, Fung KF, Chen Z, Zeng Z, Zhang J. Pharmacokinetics of florfenicol in healthy pigs and in pigs experimentally infected with Actinobacillus pleuropneumoniae. Antimicrob Agents Chemother. 2003 Feb;47(2):820-3. **doi: 10.1128/AAC.47.2.820-823.2003**

8. Craene BA, Deprez P, et al. Pharmacokinetics of florfenicol in cerebrospinal fluid and plasma of calves. Antimicrobio Agents and Chemoth. V41: n9. Sep 1997. p1991-5.

9. Zhou C, Varma KJ. Pharmacokinetics of florfenicol in swine following intravenous, oral gavage and drinking water administration. *Proceedings of the 19th IPVS Congress*, Copenhagen, Denmark, 2006, Vol 2: 433.

10. AASV. Extralabel drug use algorithm. Assessed November 13, 2024. **https://www.aasv.org/wp-content/uploads/2024/06/extralabel-use-algorithm.pdf**

https://doi.org/10.54846/am2025/58

The effect of feed and water consumption patterns on subsequent reproductive performance

Katie Holtz, PigEasy, LLC, Templeton, Iowa

Introduction

US sow mortality has been steadily rising, with a 6.9% increase reported between 2014 and 2023, according to Bradley Eckberg of MetaFarms. In 2023, the average sow death loss reached a record-high 15.8%, underscoring the urgent need to prioritize animal selection, husbandry practices, and strategies to improve sow longevity to reverse this concerning trend.[1]

Optimizing feed and water intake during lactation is crucial for ensuring reproductive success and sow longevity. Adequate nutrition also improves milk production and reduces piglet mortality. With today's prolific sow producing larger litter sizes and birthweights, meeting their heightened nutritional demands is more critical than ever. Without precise monitoring of individual sow intake, farm feeding programs cannot be effectively managed, potentially compromising productivity and animal welfare.

By monitoring feed and water intake amounts and frequency in real-time, farms can uncover intake patterns that help identify high-performing sows and those likely to underperform in future litters, signaling the need for culling. An economical precision feeding solution is vital for sow farms of all sizes, enabling more efficient daily management and long-term analysis of intake data. These insights improve decision-making, optimize sow performance, and enhance overall farm productivity.

Research and methods

Magdiel Lopez Soriano, Caleb J. Grohmann, and Timothy J. Safranski from the University of Missouri conducted a study[2] analyzing daily feed and water intake data collected via MealMeter 2.0 smart feeders on a sow farm in southwest Minnesota in March 2023. The research investigated how lactation feed and water intake patterns influence the time it takes for sows to return to heat after weaning, known as the wean-to-estrus interval (WTE).

The findings revealed that sows with more consistent feeding and drinking habits during lactation were less likely to have a WTE exceeding five days, making them ready for breeding sooner. Surprisingly, variations in feed and water intake patterns also appeared to reduce the likelihood of prolonged WTE, a result that warrants further exploration. These findings highlight the critical role of monitoring and analyzing sow feed and water intake patterns during lactation to enhance reproductive efficiency and overall herd performance.

Further analysis is underway to examine feed and water intake patterns based on sow parity, reproductive performance, and longevity. The objective is to provide producers with actionable insights to identify top-producing animals within the herd and make informed breeding decisions to enhance overall productivity and efficiency.

Results and discussion

To address the demand for a streamlined and cost-effective precision lactation feeding solution, PigEasy, LLC enhanced its already high-performing feeder by integrating sensors to accurately monitor feed and water intake and frequency. This data is reported in real-time to a cloud-based system known as MealMeter 2.0. Sows dispense feed and water manually, offering a reliable and high-performing feeder without electrical complications and tedious software management.

The system's straightforward mobile app enables daily management tasks such as responding to intake alerts and conducting sow health analyses. This innovation simplifies precision feeding while providing actionable insights to improve both individual sow care and overall herd productivity.

Further analysis of feed and water intake is necessary during other stages of the sow's reproductive cycle, including gilt development and pre-breeding phases. These insights would assist in improving gilt selection, individual sow care, and overall herd productivity.

PigEasy remains committed to creating innovative solutions that are simple, economical, and reliable, continually addressing the evolving needs of the modern sow.

References

1. Goyal, A. (2023, October 9). U.S. sow mortality trends continue to climb. *National Hog Farmer.* **https://www.nationalhogfarmer.com/livestock-management/u-s-sow-mortality-trends-continue-to-climb**

2. Lopez Soriano, M., Grohmann, C. J., & Safranski, T. J. (2024). Effect of feed and water consumption patterns during lactation on wean-to-estrus interval in sows. *Journal of Animal Science, 102*(Supplement_2), 346. **https://doi.org/10.1093/jas/skad401.625**

https://doi.org/10.54846/am2025/59

A comprehensive vaccine approach to protect piglets from IAV-S in the face of maternal antibodies

J. Griffel[1]; K. Crawford[2], DVM, MS; R. Saltzman[2], DVM; P. Winter[1], PhD
[1]Phibro Animal Health Corporation, Omaha, Nebraska; [2]VRI, Ames Iowa

Introduction

Influenza A of swine (IAV-S) virus is a primary cause of acute respiratory disease in pigs. Many production systems have faced increased respiratory disease due to IAV-S infections in pigs of all ages; nursery through finishing. Vaccination against influenza has long been a tool to prevent and control the disease in the pig population and is commonly used in sows.[1] Vaccination of the sows typically provides protection against clinical disease associated with IAV-S infection in adult sows and increases passive immunity in piglets as well. The role of maternal derived antibodies (MDA) is to protect young piglets from clinical disease.[1,2] Current commercial and autogenous IAV-S vaccines contain one, two, or up to five different influenza isolates. The goal of this study was to develop a multi-variant inactivated autogenous IAV-S vaccine that induces protection when administered to piglets in the presence of maternal antibodies.

Materials and methods

Study design

Sixteen (16) cross-bred pregnant sows were sourced from a commercial herd. Sows were confirmed to be IAV-S negative by IAV- NP ELISA test kit (IDEXX, Westbrook, ME) at Iowa State University Veterinary Diagnostic Laboratory. Sows were vaccinated at 6 weeks and 2 weeks pre-farrowing. At farrowing a total of 135 piglets were randomized into one of nine treatment groups (n = 15), consisting of four control groups and five treatment groups. Treatment groups are summarized in Table 1. Vaccinations were administered at 1-2 days of age (Day 0) followed by a booster at Day 21 (Groups 1, 2 and 5), only at Day 21 (Groups 6 and 7), or at Day 21 followed by a booster at Day 42 (Groups 3, 4, 8 and 9). Day of challenge (0 DPC) was Day 42 (for Groups 1, 2 and 5) and Day 63 (for Groups 3, 4 and 6-9). All pigs were euthanized 5 days post challenge (DPC). Blood was collected at various timepoints throughout the study in serum separator tubes and serum was stored at -80°C until testing.

Table 1: Summary of study design

Treatment group	Pigs	Antigen	Adjuvant	Route	Volume	IAV-S vaccination	IAV-S challenge
1	15	Control / PBS	No	IM	1cc / 2cc	Day 0 / Day 21	Day 42
2	15	Control / PBS	Emulsigen-D (12%)	IM	1cc / 2cc	Day 0 / Day 21	Day 42
3	15	Control / PBS	No	IM	2cc	Day 21 / Day 42	Day 63
4	15	Control / PBS	Emulsigen-D (12%)	IM	2cc	Day 21 / Day 42	Day 63
5	15	300 HA / 300 HA	Emulsigen-D (12%)	IM	1cc / 2cc	Day 0 / Day 21	Day 42
6	15	300 HA	Emulsigen-D (12%)	IM	2cc	Day 21	Day 63
7	15	600 HA	Emulsigen-D (12%)	IM	2cc	Day 21	Day 63
8	15	300 HA / 300 HA	Emulsigen-D (12%)	IM	2cc	Day 21 / Day 42	Day 63
9	15	600 HA / HA 600	Emulsigen-D (5%) Emulsigen-D (12%)	IM	2cc	Day 21 Day 42	Day 63

IAV-S = Influenza A virus swine; PBS = Phosphate buffered saline; HA = Hemagglutinin; IM = Intramuscular

Clinical assessment

Pigs were weighed at Days 0, 21, 42, 47, 63 and 68 of the study. Weight gain was calculated by the difference in weights recorded at 0 DPC and 5 DPC. Rectal temperatures were recorded from -3 DPC through 5 DPC. A numerical value was assigned to each pig for coughing, respiratory, and behavioral scores every day during challenge. The coughing score ranged from 0 = normal, 1 = mild cough, 2 = moderate cough and 3 = severe - loud barking cough. Respiratory score ranged from 0 = normal, 1 = mild, 2 = moderate and 3 = severe - respiratory distress. Behavioral score ranged from 0 = normal, 1 = depressed - slight decrease in movement, 2 = lethargic - requires aggressive physical stimulations before moving and 3 = recumbent.

Vaccines and vaccination

All IAV-S vaccines were formulated with the following five isolates: Pandemic-H1N1, Gama-H1N1, Delta 1-H1N2, Delta 2-H1N1 and Cluster 4-H3N2. The vaccines were adjuvanted with either 5% or 12% Emulsigen®-D (MVP Adjuvants, Omaha, Nebraska). Control vaccines were comprised of phosphate buffered saline (PBS) only (Groups 1 and 3) or in combination with 12% Emulsigen-D (Groups 2 and 4). Control (Groups 1 and 2) and treatment (Group 5) were vaccinated intramuscularly with 1cc of vaccine at Day 0 and 2cc at Day 21. Controls (Groups 3 and 4) and treatment (Groups 6, 7, 8 and 9) were vaccinated intramuscularly with 2cc of vaccine at Day 21 and or Day 42. Vaccination schedules are summarized in Table 1.

Challenge and necropsy

The challenge material (IAV-S strain Delta 1-H1N2) was grown on Madin-Darby canine kidney (MDCK) cells in Dulbecco's Modified Eagle Medium (D-MEM, Fisher Scientific). All pigs were challenged intranasally (2cc) with the Delta 1-H1N2 isolate at a dose of 2×10^6 $TCID_{50}$/mL, 1cc of challenge material was administered to each nostril using a 3.0cc syringe attached to a commercially available atomizer.

On Day 47, all pigs, in Groups 1, 2 and 5 were humanely euthanized in accordance with the 2020 AVMA's Panel on Euthanasia and necropsied along with remaining groups on Day 68 of the study. Total lung lesions for each pig were determined by a veterinary pathologist blinded to the study: Bronchoalveolar lavage fluids were collected from each pig using Dulbecco's Modified Eagle Medium (D-MEM, Fisher Scientific) supplemented with 1% penicillin/streptomycin (Gibco, Fisher Scientific).

Results

Clinical observations

Rectal temperature data showed no difference between treatment groups post challenge (103.7-104.6° F) when compared to baseline temperatures (103.6-104.6° F) prior to challenge administration. Coughing and respiratory scores (data not shown) showed an increase in both severity and incidence in groups that received a single vaccination (Day 21), compared to treatment groups receiving two vaccinations (Day 21 and Day 42).

Weights

Piglet weight data showed no significant difference between control groups and treatment groups through Day 42 of the study. Pigs vaccinated at Day 0 and Day 21, followed by intranasal challenge at Day 42 (Groups 1, 2 and 5) had the greatest individual group weight gain post challenge. Treatment Group 5 gained less overall weight post challenge when compared to control Groups 1 and 2. Treatment Groups 8 and 9 (vaccinated at Day 21 and Day 42 followed by intranasal challenge at Day 63) showed greater weight gain post challenge when compared to treatment groups receiving only a single vaccination at Day 21 and then challenged at Day 63 (Figure 1). Groups 8 and 9 average weight post challenge was higher when compared to the control groups. All groups that received two vaccinations outperformed those which only received a single vaccination.

Serology

The hemagglutination inhibition (HI) titers for all treatment groups showed high HI titers at Day 0 with all groups sustaining protective levels ($\geq$ 40) against all five vaccine fractions at Day 14. By Day 21 of the study, all groups still maintained HI titers above protective levels for 2 of the 5 vaccines fractions (Gama-H1N1, Delta 1-H1N2). At Day 42, HI titers dropped below 40 for groups 1, 2 and 5. On Day 63, treatment Groups 8 and 9 (vaccinated on Day 21 and Day 42) showed significantly higher HI titers in comparison to the control and other treatment groups, receiving only one dose of vaccine The piglets in Groups 8 and 9 seroconverted after vaccination and before challenge, showing protective HI titers against all 5 vaccine strains.

Lung assessment

Lungs scores for all groups vaccinated at Day 0 and Day 21 then challenged at Day 42 showed no significant differences (Figure 2). Treatment Group 5 showed elevated lung scores (11.25%) when compared to control Groups 1 and 2 (8.48%, 8.88% respectively). Lungs scores for treatment Groups 6 and 7 (vaccinated on Day 21) had the highest lungs scores (16.52% and 12.07%) for any group in the study. Control Groups 3 and 4 vaccinated on Day 21 and Day 42 then challenged on Day 63 had lungs scores of (4.76% and 7.48%, respectively). Treatment Groups 8 and 9 had the lowest lung scores (0.74% and 1.05%) for the study and were statically significant ($P < .05$) when compared to control Groups 3 and 4 (Figure 3).

PCR results

Samples of bronchoalveolar lavage fluid (BALF) were collected at necropsy from all pigs in the study and tested by quantitative Polymerase Chain Reaction (qPCR) to determine the viral genomic load in the lungs. Testing was completed at Iowa State University Veterinary Diagnostic Laboratory. Cycle threshold (Ct) values show a statistically significant difference between Group 5 and control Groups 1 and 2. Group 5 animals had a higher Ct (23.8), indicating a lower viral load in the lungs, when compared to control Groups 1 and 2 (20.2 and 20.8). Control Groups 3 and 4 showed similar Ct results to the previous control groups with Ct values of 21.4 and 21.5. Treatment Group 6 indicated a slight reduction in viral genomic load with a Ct of 22.9 when compared to control Groups 3 and 4. For treatment Group 7, a greater and more significant reduction was seen with a Ct of 26.0. The greatest and most significant reduction in viral load was evident in treatment Groups 8 and 9 as these treatment groups had Ct values of 36.9 and 36.4 respectively (Figure 4). Quantitative analysis (data not shown) of the collected BALF samples mirrored the results that were seen with the Ct values.

Figure 1: Percent weight gained post challenge.

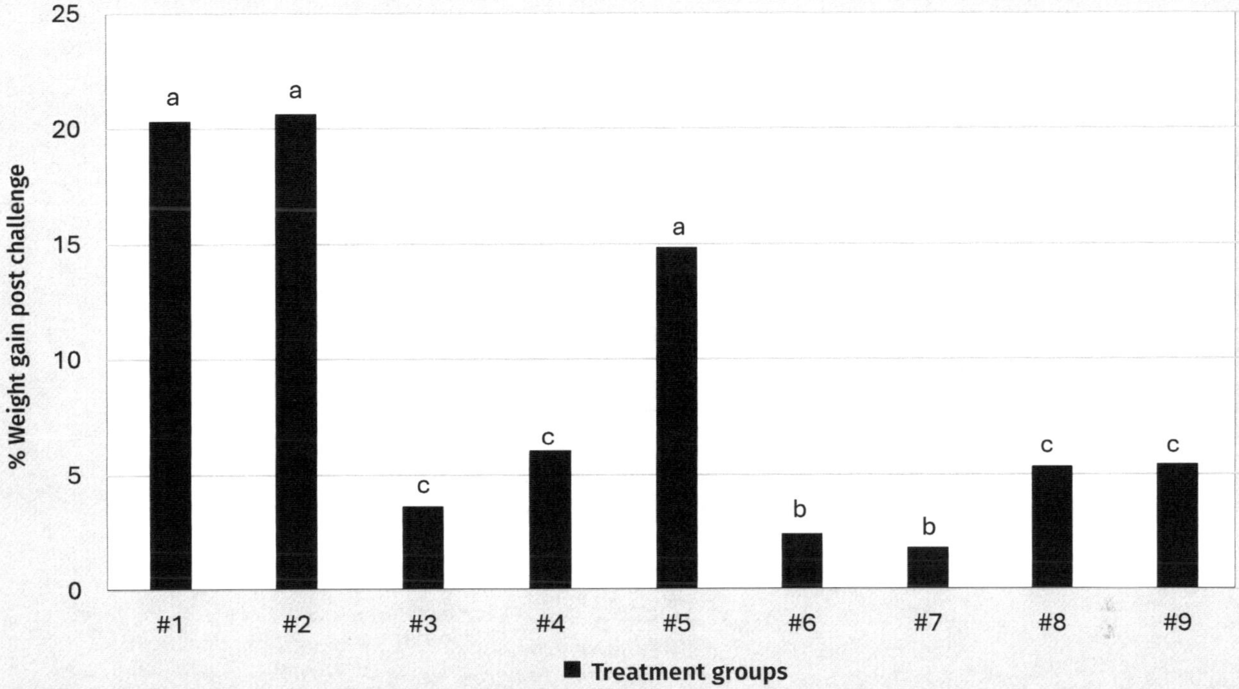

a = Groups vaccinated on Day 0 and Day 21, challenged on Day 42. b = Groups vaccinated on Day 21 and challenged on Day 63. c = Groups vaccinated on Day 21 and Day 42, challenged on Day 63.

Figure 2: Lung scores for groups (see Table 1) vaccinated on Day 0 and Day 21.

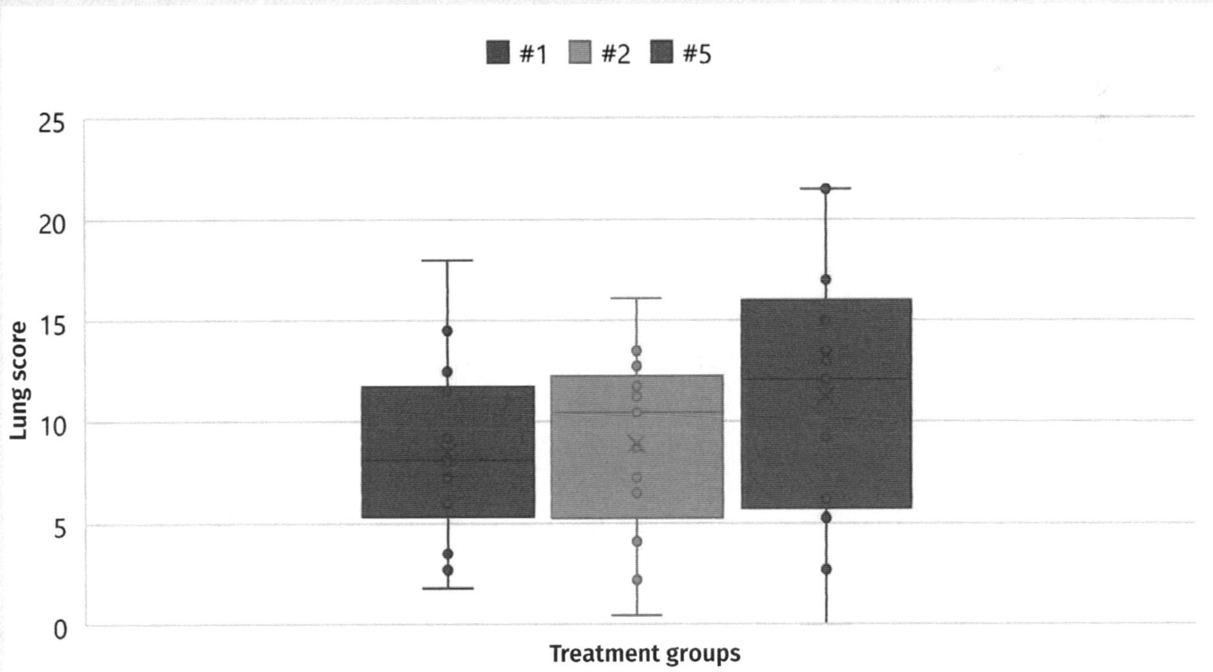

Figure 3: Lung scores for groups (see Table 1) vaccinated on Day 21 and/or Day 42.

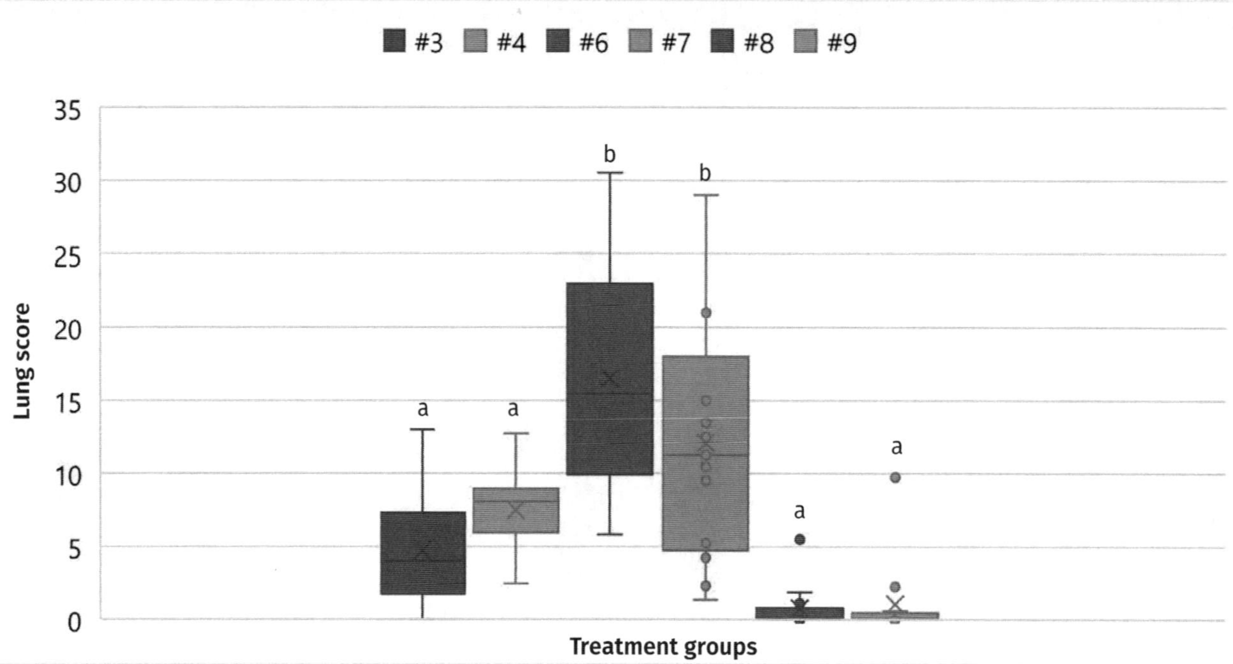

a = Groups vaccinated on Day 21 and Day 42. b = Groups vaccinated on Day 21 only.

Figure 4: Cycle threshold values for all treatment groups (see Table 1).

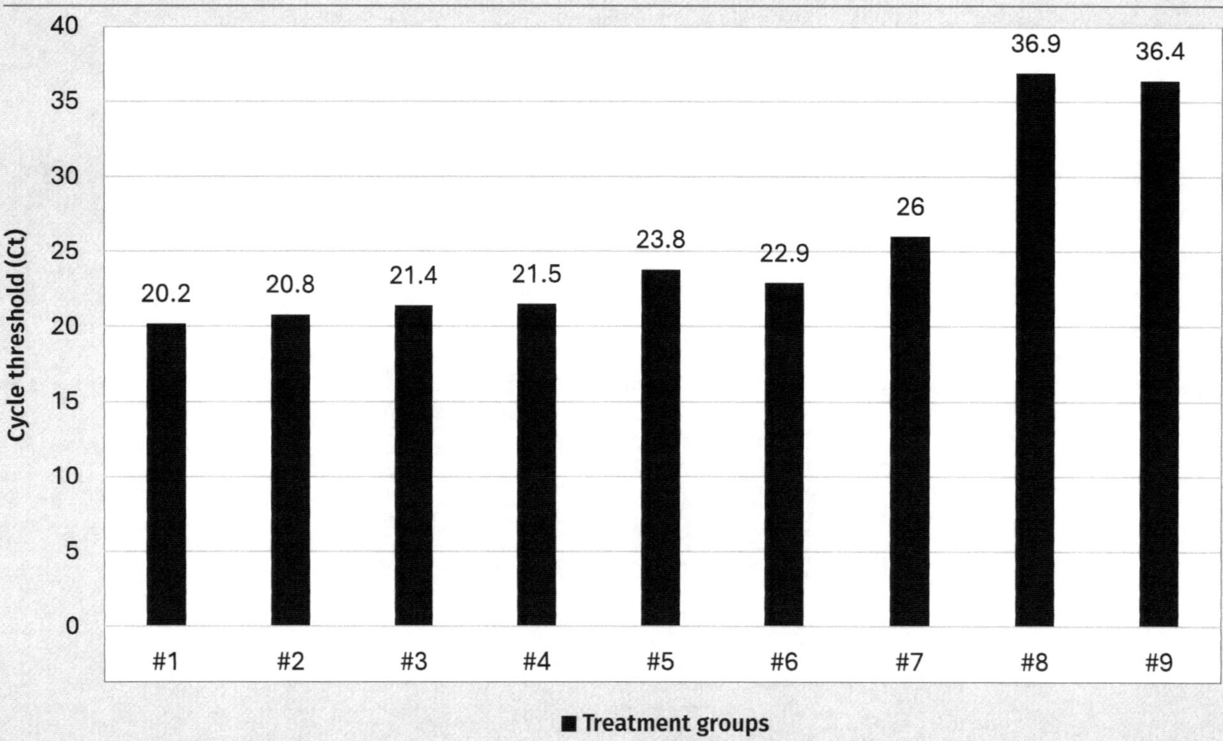

Conclusion

Recent modeling and empirical work on influenza A (IAV-S) suggest that piglets play an important role as an endemic reservoir.[3] Production systems have seen increased respiratory disease due to IAV-S infections, leading to significant economic impact on the US industry. Vaccinating against IAV-S has long been an effective tool to help control and prevent disease in sows. The goal for this study was to develop a multi-variant inactivated autogenous IAV-S vaccine that induces protection when administered to piglets in the presence of maternal antibodies. Our current data demonstrates that vaccination of commercial piglets at weaning (3 weeks of age) followed by a booster 3 weeks later (6 weeks of age) is effective when utilized as part of an IAV-S vaccination program.

References

1. Sandbulte, M.R.; Spickler, A.R.; Zaabel, P.K.; Roth, J.A. Optimal Use of Vaccines for Control of Influenza A Virus in Swine. *Vaccines* 2015, *3*, 22-73. **https://doi.org/10.3390/vaccines3010022**

2. Petro-Turnquist, E; Pekarek, MJ; Weaver, E; Swine influenza A virus: challenges and novel vaccine strategies. *Front. Cell. Infect. Micobiol.* 14:1336013 **https://doi.org/10.3389/fcimb.2024.1336013**

3. L.A. White, M. Torremorell, M.E. Craft. Influenza A virus in swine breeding herds: Combination of vaccination and biosecurity practices can reduce likelihood of endemic piglet reservoir, Pre Vet Med, Volume 138, 2017, Pages 55-69. **https://dx.doi.org/10.1016/j.prevetmed.2016.12.013**

https://doi.org/10.54846/am2025/60

From lab to field: Integrating CASA for optimal semen dose success

R. Ausejo; L. Dominguez
Department of Biotechnology R&D, Magapor S.L., Ejea de los Caballeros, Spain

The importance of precision in the adjustment of seminal doses

Artificial insemination in sows is a fundamental technique in pig reproduction that enables farmers to optimize productivity and enhance herd genetics. As the pork industry becomes increasingly competitive, the ability to efficiently produce healthy piglets with desirable traits is crucial. A key factor in achieving successful artificial insemination lies in the precise adjustment of the concentration of semen doses used.[1]

The semen dose, which contains sperm, must be carefully adjusted to ensure effective fertilization. Research has shown that the concentration of sperm can significantly influence conception rates. When sperm concentration is too low, the probability of successful fertilization decreases, resulting in lower conception rates and longer intervals between farrowing.[2] This inefficiency not only affects the immediate productivity of the sow but also has cascading effects on the overall profitability of the farm.

Conversely, if the concentration of sperm is excessively high, it can lead to a phenomenon known as "saturation" where sperm compete against one another for fertilization. This competition can hinder the sperm's ability to reach and fertilize the ovule, ultimately reducing the chances of conception.[3] Therefore, achieving the right balance in sperm concentration is essential to optimize reproductive results.

Moreover, the health and well-being of the sows are also impacted by improper dosing. An ideal sperm concentration can contribute to better overall reproductive health, resulting in more viable piglets and improved litter sizes.[4] In this presentation, we will delve into the implications of incorrect dosing practices and highlight the importance of meticulous management in the artificial insemination of sows.

To achieve the necessary precision in semen concentration, the implementation of a Computer Assisted Sperm Analysis (CASA) system is invaluable. CASA technology allows for accurate and objective assessment of sperm quality, including parameters such as motility, morphology, and concentration. These metrics are critical for determining the viability of semen and ensuring optimal dosing for artificial insemination.

CASA systems as a tool for optimal dosing

The integration of CASA systems, such as Magavision developed by Magapor S.L, represents a crucial step in the modernization of laboratories, as it allows for objective, reliable, and standardized analyses across different studs and labs.

Magavision uses machine learning-based algorithms to analyze various parameters related to semen quality. These algorithms have been trained on a large set of pre-labeled images and optimized to minimize prediction errors and achieve optimal accuracy. This enables a more detailed and precise assessment of sperm detection and classification.

One of the most relevant parameters in semen quality analysis is the total sperm concentration, which allows to determine the amount of extender to add to the ejaculate to achieve the target concentration for doses. To ensure the accuracy of the obtained data, the concentration of doses and ejaculates was validated with NucleoCounter (Figure 1), a fluorescence-based analysis device that is considered the gold standard for determining sperm concentration in semen. The correlation obtained was over 98%.

However, there may be certain discrepancies in concentration in samples with a high level of agglutination. The algorithm can detect and quantify sperm agglutinations, providing a percentage of the total area they occupy in the analyzed field. This allows for a more accurate concentration calculation by subtracting the agglutination area from the total analyzed area. Otherwise, the sperm concentration in the field would be underestimated.

Another key parameter is the identification of abnormal sperm forms, as sperm with abnormal morphology have a significantly reduced chance of fertilizing an ovule. By identifying morphology, we can adjust the concentration by subtracting abnormal sperm cells from the total count to calculate what is known as useful concentration. Thanks to the complex patterns identified by the machine learning-based algorithm, along with high-resolution video cameras and high-quality optics from the microscope, we have achieved recall and precision coefficients above 0.95 (Table 1) in predicting different types of abnormal forms (distal droplets, proximal droplets, detached heads, folded tails, and coiled tails).

Lastly, our CASA system calculates sperm kinetic parameters. Sperm motility, or their ability to move effectively through the female reproductive tract, is essential for fertilization. Even if sperm morphology is normal, if it is static or has very low progressive motility, its ability to advance and reach the ovule will be severely limited. Therefore, just as with sperm with abnormal morphology, it is interesting to count only motile or progressive sperm for calculating useful concentration. There is a lack of standardization in kinetic parameters among different CASA systems, which can make it difficult to compare results between laboratories or studies. Consequently, Magavision has been compared with other systems to minimize this variability through consensus on established criteria for calculating various parameters of total and progressive motility, DCL, DSL, DAP, VCL, VSL, VAP, STR, LIN, WOB, ALH, BCF, and MAD.

In conclusion, it is important that the CASA system adapts to the needs of each insemination stud, as each has its own criteria for adjusting semen doses. This is why Magavision is fully customizable, allowing each user to select parameters for dose concentration adjustment based on total or progressive motility, as well as each type of abnormal form, with the ability to determine the correction weight of each on the total. Adjustments for both motility and morphology can be applied simultaneously, deducting static and abnormal sperm cells only once. It is also possible to establish different dose calculation adjustment criteria based on boar breed.

Figure 1: Comparison between the concentration measured with Magavision and NucleoCounter.

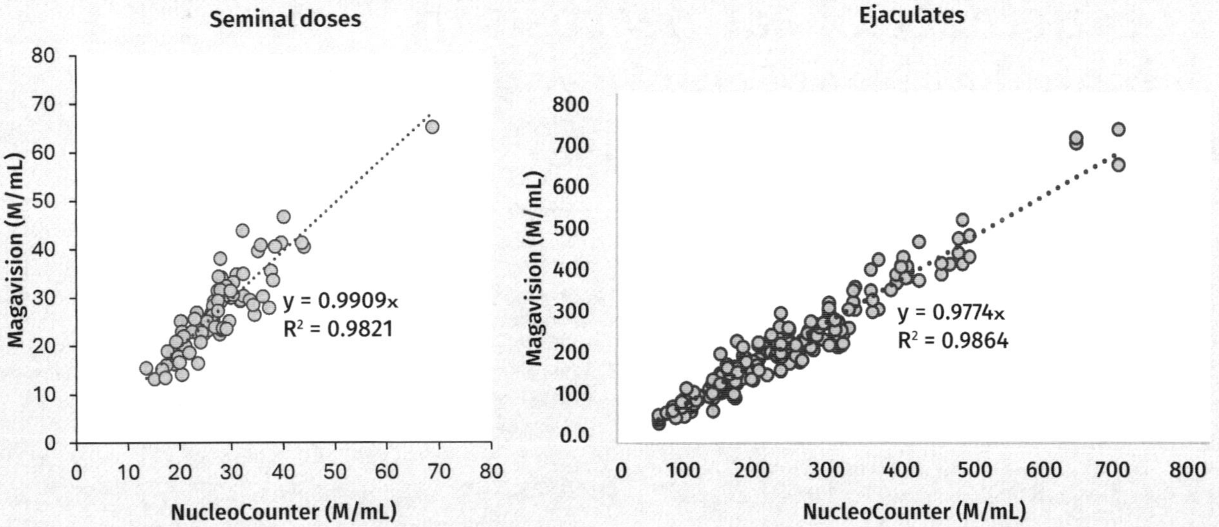

Table 1: Precision and recall of the abnormal forms

	Precision	Recall	N
Number of sperm	0.990888	0.95781	12518
Abnormal shapes	0.992536	0.959481	4597

Other factors to consider and automation of the analysis process for greater objectivity

To achieve optimal dosing of semen doses, in addition to a CASA system with precise software, we must consider other factors such as:

- Proper handling for sample homogenization before analysis, correct pipetting, and appropriate use of dilution factors.
- Using a counting chamber that meets the requirements to ensure accurate concentration and motility measurements.
- Using a high-quality extender that ensures the preservation of semen doses.
- Correct dilution and packaging process of the doses.
- Automated analysis with the CASA system for greater objectivity.

It is essential to emphasize the importance of conducting as objective an analysis as possible, minimizing human error to the maximum extent. Automated analysis, possible with the Magavision CASA system's SmartStage, allows for a more objective and faster analysis. With this motorized stage, the system can autofocus and move automatically to the same central coordinates in each well, eliminating potential errors in focus level that could lead to missed abnormal forms or sampling near the edge of the well, where sperm are more concentrated. In terms of speed, with a motorized stage accommodating up to 4 counting chambers, up to 20 ejaculates can be analyzed in under 6 minutes without human intervention (Table 2), which helps streamline workflow in the insemination center.

References

1. García CL, et al. Optimization of artificial insemination in swine: the role of semen quality and concentration. Theriogenology. 2016;85(5):903-909. **https://doi.org/10.1016/j.theriogenology.2015.10.005**

2. García LJ, et al. Effect of sperm concentration on the fertility of sows after artificial insemination. Reprod Domest Anim. 2010;45(2):242-246. **https://doi.org/10.1111/j.1439-0531.2008.01321.x**

3. Kastelic JP, et al. Semen handling and storage: effects on fertility. Anim Reprod Sci. 2014;148(1-2):59-66. **https://doi.org/10.1016/j.anireprosci.2014.05.008**

4. Márquez G, et al. Influence of sperm dose on reproductive performance of sows. Livest Sci. 2012;147(1-3):22-27. **https://doi.org/10.1016/j.livsci.2012.04.008**

Table 2: Time of analysis with the SmartStage according to the numbers of fields

		Number of analysis			
		5	10	15	20
Number of fields	1	0:44	1:30	2:18	3:04
	2	0:50	1:48	2:44	3:43
	5	1:24	2:55	4:27	5:58

https://doi.org/10.54846/am2025/61

Improvement in growth performance and the occurrence of ear necrosis in Vepured® vaccinated animals compared to an optimized diet against oedema disease

R. Jordà[1]; S. Van Poucke[2]; A. Puigredon[1]; B. Moreno[1]; L. Nodar[1]; M. Simon-Grifé[1]; N. Bernal[1]; I. Ballarà[1]
[1]HIPRA, Amer (Girona), Spain; [2] SYN+, Zulte, Belgium

Introduction

Oedema disease (OD) is caused by Verotoxin 2e (Vt2e) produced in the small intestine by pathogenic serotypes of *Escherichia coli* (*E coli*). The subclinical form of the disease is characterized by a decreased growth rate, a slight increase in mortality and a higher probability of having *Streptococcus suis* problems or post-weaning diarrhoea but in the absence of specific clinical signs.[1,2] Furthermore, the vasculitis caused by oedema disease[3] may be one possible cause of porcine ear necrosis.

Some feed management strategies such as reduction of crude protein have been reported as being effective in controlling OD, however these measures have obvious detrimental effects on the growth performance of the animals.[4] The aim of this study was to evaluate the effect on growth performance of Vepured® (Laboratorios Hipra SA), an inactivated vaccine indicated to prevent mortality and reduce clinical signs of oedema disease, compared to an optimized diet with a lower percentage of crude protein and a higher digestibility of the protein sources but with the same level of energy.

Methods

The present study was conducted under field conditions on a Vt2e positive farm without clinical oedema signs in Belgium. One farrowing batch of 192 sows arranged into two independent compartments (n = 96) was selected. Piglets born in compartment A were vaccinated with Vepured® at 2 days of age (VG). Piglets born in compartment B remained un-vaccinated against oedema disease (NVG).

At weaning (24 days of age), approximately 500 piglets from each group (VG = 509; NVG = 508) were randomly selected and moved to the nursery site. There, the piglets were allocated into two different compartments depending on their sex. Each compartment consisted of alternating pens for VG (N = 8) and NVG (N = 8). All piglets were vaccinated 4 days after weaning against PCV2 and *Mycoplasma hyopneumoniae* with the same protocol in both groups.

During the nursery period, piglets from NVG were fed with an isocaloric optimized diet with a lower percentage of crude protein (NVG 16.5% versus VG:18%). At 12 weeks of age, all piglets were moved to the fattening unit. During the fattening period both groups received the same standard-farm diet but from a different silo for each group so the feed consumption could be measurable.

The presence of Vt2e was confirmed during the nursery and fattening stages in oral fluids using VEROCHECK® (qPCR of the gene coding for the VT2e) (Tables 1 and 2).

The body weight of all piglets from both groups was measured at weaning, twice at the end of the nursery phase (at 10 and 12 weeks of age) and at the end of fattening. All animals were sent to slaughterhouse on the same day, five days after weighing. Mortality and clinical signs during nursery stage such as post-weaning diarrhoea and ear necrosis were also monitored. For ear necrosis evaluation, Malik et al[5] scoring was used during 7 consecutive days from the onset of clinical symptoms.

Results

No weight differences were found between groups at weaning. A statistically significantly higher growth was observed in the Vepured® group compared to the optimized diet group when animals where weight at 10 weeks of age (P = .001), but no weight differences were observed between groups at 12 weeks of age (Figure 1). Consequently, the average daily weight gain from weaning until 10 weeks of age was 23g/day higher in the VG compared to NVG (P = .001) (Figure 2).

Statistically significant differences were also observed at the end of the fattening phase, when the Vepured® vaccinated group had a statistically significantly higher growth (P = .002) resulting in a 3.05 kg of difference between groups (Figure 3). The average daily weight gain for the fattening period was 24g/day higher in the VG compared to the NVG (P = .014) (Figure 4), and the global feed conversion rate (FCR) was equivalent between groups.

Regarding the development of ear necrosis, the Vepured® vaccinated group had statistically significantly fewer observations of ear necrosis (15.74 %) compared to the optimized diet group (36.12%) (P < .001) (Figure 5).

No statistical differences were observed between groups regarding the occurrence of diarrhea (VG:37.75% and NVG:36.73%, P = .80) and mortality (VG:6.1% and NVG:5.3%, P = .59).

Discussion

Vaccination against OD with Vepured® significantly improved productive parameters at the end of nursery (10 weeks of age) and fattening stages by 3,05 kg more on average. Additionally, Vepured® vaccination reduced the occurrence of ear necrosis during the nursery phase in a farm with subclinical presentation of OD. This study highlights the importance of a good diagnostic on subclinical oedema disease, and the consequent beneficial effects of preventing it by vaccination.

On the other hand, further analysis should be done in order to deeply understand the mechanism of action of the ear necrosis, however it's clear how the effect of vaccination reduced significantly the incidence and the severity of the lesions, and improved the general performance of the pigs until the end of the fattening period.

Table 1: VEROCHECK® results on nursery stage. Verotoxin 2e (VT2e) (Ct): POS < 38.5.

Group	Sample	VT2e PCR results	Ct values
Vepured®	1	NEG	–
	2	NEG	–
	3	NEG	–
Control	1	NEG	–
	2	NEG	–
	3	POS	30.1

Table 2: VEROCHECK® results on fattening stage. Verotoxin 2e (VT2e) (Ct): POS < 38.5.

Group	Sample	VT2e PCR results (12 woa)	Ct values	VT2e PCR results (17 woa)	Ct values
Vepured®	1	NEG	–	POS	36.5
	2	NEG	–	NEG	–
	3	NEG	–	NEG	–
	4	NEG	–	NEG	–
	5	NEG	–	NEG	–
	6	NEG	–	NEG	–
Control	1	NEG	–	NEG	–
	2	NEG	–	NEG	–
	3	NEG	–	NEG	–
	4	NEG	–	NEG	–
	5	NEG	–	POS	37.5
	6	POS	37.1	POS	38.4

woa = weeks of age

Figure 1: Weights evolution from weaning to the end of nursery stage.

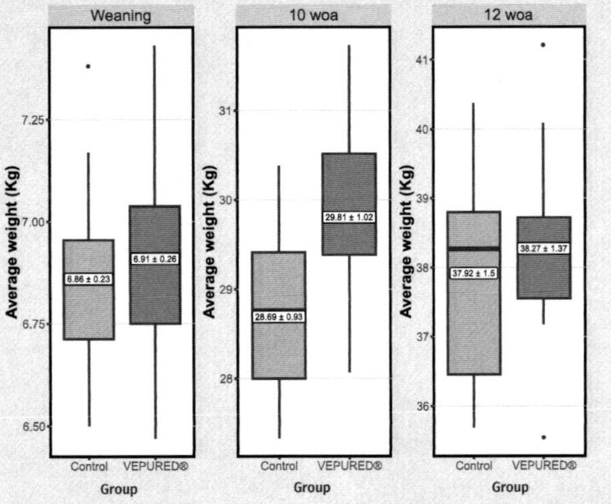

woa = weeks of age

Figure 2: Average daily weight gain from weaning to 10 weeks of age. Statistically significant differences between Vepured® and control groups (*P* = .001; linear regression test).

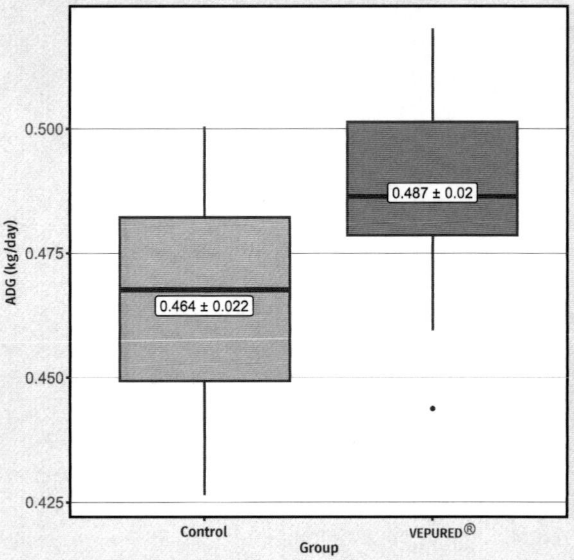

Figure 3: Weight at the start and end of fattening phase of Vepured® and control groups. Statistically significant differences between Vepured® and control groups (*P* = .002; Linear regression test).

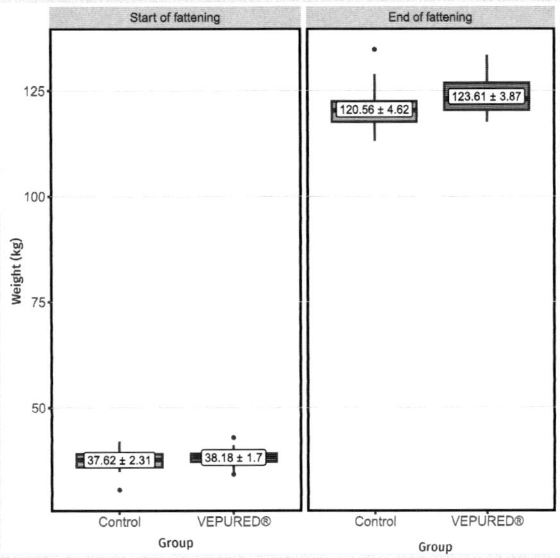

Figure 4: Average daily weight gain from all fattening period. Statistically significant differences between Vepured® and control groups (*P* = .014; linear regression test).

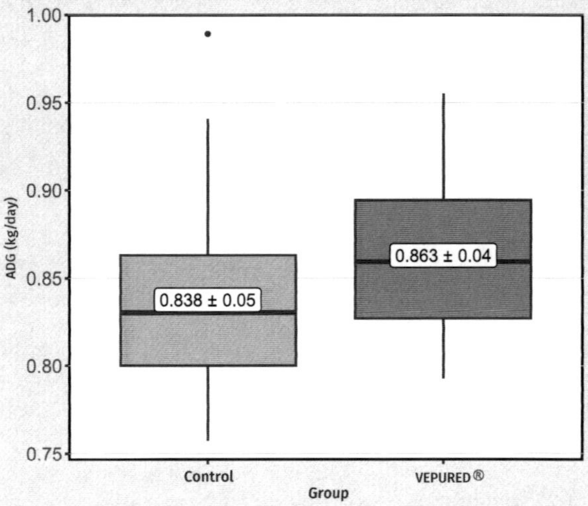

Figure 5: Percentage of ear necrosis observations in Vepured® and control groups, where 0 = absence and 1 = presence of ear necrosis. Statistically significant differences between Vepured® and control groups ($P < .001$; Linear regression test).

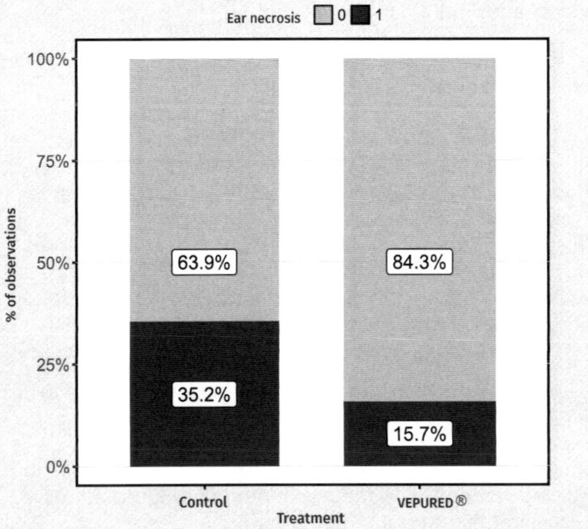

References

1. Zimmerman, J.J. et al. Diseases of Swine. 11th ed. Wiley-Blackwell, 2019. ISBN: 9781119350859.

2. De Jong, E. et al. 2022 Comparison of two *E. coli* vaccines and their effects on a farm with clinical problems of *S. suis*. Proceedings 13th ESPHM.

3. Miller, L.M., Gal, A. Cardiovascular System and Lymphatic Vessels. In: Zachary, J.F. *Pathologic Basis of Veterinary Disease* 6th ed. Elsevier, 2017. ISBN:978-0-323-35775-3.

4. Bertschinger, H.U. et.al. Evaluation of low nutrient, high fibre diets for the prevention of porcine Escherichia coli enterotoxaemia, Veterinary Microbiology, 1979, 3, 281-290.

5. Malik, M. et al. Porcine ear necrosis. *The Veterinary Journal* 2021 May;271.

https://doi.org/10.54846/am2025/62

Exploring biosecurity barriers – boot bath disinfection

Gisele Mouro Ravagnani, DVM, MSc, PhD
Technical Marketing Manager – Lanxess Biosecurity Solutions

Biosecurity barriers are crucial in the farming industry to help prevent the introduction and spread of diseases, thereby protecting the health of the animals and maintaining productivity. Biosecurity measures, which are defined and implemented, provide strategies to help protect animals, humans, as well as the environment. Some examples include: infrastructure (eg, Danish entry), technologies, techniques (eg, vermin control) and hygiene practices (eg, changing of boots and washing hands, terminal disinfection).

Best biosecurity practices also include strong communication management in the movement of animals, personnel and equipment. The potential of direct exposure to infected animals is a strong risk factor in introducing disease into a farm. People can act as a mechanical vector, introducing and spreading infectious diseases on farms. Preventive measures, such as changing boots and washing hands, are needed to mitigate the risk of exposure. Unfortunately, biosecurity compliance related to biosecurity measures in barn anterooms has been shown to be generally inadequate in the area of animal production systems.

The introduction and maintenance of an effective on-farm biosecurity program requires several elements, including: Farm and barn designs facilitating implementation of biosecurity measures; regular communication with all personnel and visitors educating them about potential threats and biosecurity risks; and training programs for all farm personnel, designed to reinforce the benefits of biosecurity in preventing infectious disease transmission.

Different studies have identified people-related biosecurity risk factors for multiple diseases. For example, the lack of a hygiene barrier in a broiler barn entrance increases the chances of *Campylobacter* contamination by more than three times. Van de Giessen et al[1] reported that chances of *Campylobacter* contamination are reduced by five times when people wash their hands, change farm boots and use footbaths. A study on infectious bronchitis in Quebec, Canada showed that failure to change boots between barns increased the risk of infection 10-fold.

According to Racicot et al,[2] boot and handwashing compliance were only at 53% and 36%, respectively. These failures to prevent the introduction of pathogens at the farm level pose a significant risk for the entire poultry production industry, and ultimately to trade and consumers.

Boot bath disinfection is a critical component of biosecurity measure on farms, particularly in poultry, livestock, and other animal operations. It helps to prevent the spread of pathogens by ensuring that all footwear entering and exiting sensitive areas is properly sanitized.

To extract the maximum benefit that boot disinfection offers, it is necessary to know how to manage it correctly: 1) The boot bath should be placed at all entry and exit points of animal housing areas, feed storage, and other critical zones to ensure that everyone passes through; 2) Choose an appropriate disinfectant that is effective against the pathogens of concern; 3) Make sure it is compatible with the material of the boot bath and follows the manufacturer's recommendations for dilution and contact time; 4) Prefer choosing liquid solutions for boot dips as it is often more effective at killing pathogens quickly due to the nature of liquid disinfectants, these, penetrate surfaces and provide thorough coverage and also ensure even application and coverage on boots, which helps to improve pathogen control; 5) Before disinfection, boots should be cleaned to remove any organic matter, dirt, or debris, this ensures the disinfectant can work effectively; 6) Mix the disinfectant product with water (if required) according to the manufacturer's instructions, ensure the concentration is adequate for effective disinfection and 7) Pour the disinfectant solution into the boot bath, filling it to the recommended level, it should be deep enough to fully submerge the soles of the boots.

All personnel must step into the boot bath before entering the animal areas. Ensure they walk through the solution, making sure the soles and sides of the boots come into contact with the disinfectant. Allow the boots to remain in the disinfectant solution for the recommended contact time. This is crucial for effective pathogen kill. The contact time can vary depending on the disinfectant used, so follow the product's guidelines.

Do not forget to regularly check the boot bath to ensure it has not become contaminated or diluted. Refill or replace the disinfectant solution as needed. Periodically clean the boot bath itself to remove any accumulated debris and residual disinfectant. This helps maintain its effectiveness and prolongs boot bath's lifespan. Ensure that everyone follows the boot bath protocol. Regularly inspect compliance and address any issues to maintain biosecurity standards.

By following these steps, you can significantly reduce the risk of introducing or spreading pathogens into sensitive animal areas, thus enhancing overall biosecurity on the farm.

Disinfectants should be selected based on efficacy, ease of use, cost of use. The introduction and maintenance of an effective on-farm biosecurity program mitigates the risk of pathogen spread and in the long-term also saves you valuable time and money.

References

1. Van de Giessen AW, Tilburg JJHC, Ritmeester WS, Plas JVD. Reduction of *Campylobacter* infections in broiler Flocks by application of hygiene measures. *Epidemiol Infect.* (1998) 121:57–66. **doi: 10.1017/S0950268898008899**

2. Racicot M, Cardinal AM, Tremblay D, Vaillancourt JP. Technologies monitoring and improving biosecurity compliance in barn anterooms. Frontiers in Veterinary Science. 9:1005144. **doi: 10.3389/fvets.2022.1005144**

https://doi.org/10.54846/am2025/63

Water biology impacts after water line cleaning and disinfection with peracetic acid

G. E. Doughan, DVM; B. K. Walthart, BS; K. J. Skoland, BS; M. B. Petersen, DVM, PhD, MPH; J. T. Brown, DVM, PhD; C. E. Schau, BS; G. R. Clark, BS; J. L. Bonnema, DVM, MS; L. A. Karriker, DVM, MS, DACVPM
Swine Medicine Education Center, Iowa State University College of Veterinary Medicine, Ames, Iowa

Introduction

Previous work has demonstrated that pigs can generally tolerate a wide range of water qualities with minimal impact on growth performance.[1-8] Health and production impacts of microbiological components of water, or the involvement of biofilms or biofilm shedding events in water distribution systems (WDS) (ie, swine water lines and drinking systems) were often not investigated in previous work. Outside of water quality itself, swine WDS are "open", ie, often non-circulating, low-pressure systems, and can serve as a pathogen introduction point.[9,10] Antimicrobial resistance genes (ARGs) were present in swine water line biofilm organisms conferring resistance to multiple drug classes and multidrug resistance.[10] Water should more thoroughly be considered as a biosecurity concern, and water line biofilms could be a reservoir for drug resistance. Currently, there is no clear guidance for swine water line sanitation practices to improve water quality or antimicrobial administration. Therefore, water line sanitation has not been widely performed across much of the United States' swine industry. Such guidance has been researched and developed in poultry[11-16] but needs further investigation and adaptation for swine WDS.

CID 2000 Pro 2% solution (CID LINES an Ecolab Company) is a commercially available peracetic acid (PAA), a stabilized hydrogen peroxide peroxyacetic acid mixture, water line cleaner and can be applied to water lines in-between groups of pigs when no pigs are in the farm due to Environmental Protection Agency label restrictions. PAA has been utilized in water line sanitation due to its strong oxidation and broad-spectrum disinfection properties to help eliminate biofilm and mineral scale.[17] In a water line cleaner comparison study, CID 2000 Pro 2% solution resulted in the quickest reduction of bacterial counts at four hours post treatment and continued reduction at 24 hours.[18] CID 2000 Pro is labeled for applications at 2% or 1:50 for terminal line cleaning, but most US swine producers operate medicator dosing systems that are at a fixed rate of 1:128, or 0.78%.

The objectives of the study were to comprehensively assess the following before and after application of CID 2000 Pro at a 1:128 or 0.78% rate for 24 hours: **1.** Quantitatively and qualitatively assess water line biofilm regrowth after water line cleaning; **2.** Characterize the composition of the water line biofilm (bacterial, viral, and antimicrobial resistance genes); **3.** Assess water quality parameters and their relationship with biofilm formation and composition, and; **4.** Test for phenotypic resistance to common water medications administered in swine WDS, specifically chlortetracycline (CTC) and lincomycin hydrochloride (L) via drug-infused culture plates.

Methods

An observational, longitudinal study assessing the impacts of a one-time water line cleaning and disinfection event on water quality, biofilms, genetic presence, and phenotypic expression of ARGs was performed. Using the site source of private well water, this study assessed the rate of biofilm regrowth on six wean-to finish sites in northwest and west-central Iowa, USA. Water and water line samples were collected across 10 time points over a 78-day period from new, replaceable polyvinyl chloride (PVC) pipe sections plumbed from the original main water line (ie, termed coupon side streams) and from water sources on the farm. Initial water and water line samples were collected prior to administering CID 2000 Pro at 1:128, which charged the water lines and was allowed to act for a 24-hour period between groups of pigs. Samples were collected at pre-treatment (0), at 24 hours after CID 2000 Pro had been applied and water lines were flushed (1), then 3, 5, 7, 14, 21, 42, 56, 77 days post-treatment. Water samples were collected from the farm's well and a flush valve from the coupon side stream in a two-roomed wean-to-finish site, while water line samples were collected from PVC in the coupon side stream at each time point.

Water and water line samples were collected aseptically in field conditions. Water samples were collected from a sanitized spigot (well sample) or water line flush valve (from coupon side streams), where water was flushed for at least 20 seconds prior to collection. For water line samples, the exterior pipe of a single coupon from each room was washed with soap and water to remove dust and debris. A 2% diluted bleach solution was applied to the exterior of the water pipe and extraction tools for at least 2 minutes to remove any external contamination.[19]

Objective 1: Quantitative and qualitative assessment of water line biofilm regrowth rate after CID 2000 Pro application

Aerobic and anaerobic standard plate counts were used to quantitatively estimate water line biofilm regrowth rates. Samples were taken from 2.54 cm (1-inch) water line sections aseptically collected in two rooms from each wean-to-finish site at each time point (n = 240). Biofilm was aseptically removed from the water line pipe by scraping the inside of the water line pipe with the plastic end of a sterile puritan swab and rinsed in 10 mL 0.9% saline in a Whirl Pak bag. Once the biofilm was removed, it was transferred into a 15 mL conical tube with 10 mL 0.9% saline. Conical tubes were centrifuged to create a pellet and sent to the Iowa State University Diagnostic Laboratory (ISU VDL) for testing. Photographs were taken of the interior of the water line with a borescope inspection camera at installation and for qualitative assessment of biofilm development in the coupon side stream at each time point.

Objective 2: Characterize swine water line biofilms

For each time point, one 7.62 cm (3-inch) water line pipe section from each coupon side stream from the six sites was submitted, minus one sample that could not be recovered due to contamination (n = 119). Water line sections were submitted to the ISU VDL for 16S rRNA analysis and metagenomic analysis.

Objective 3: Assess water quality parameters and their relationship with biofilm formation and composition

At each time point, 500 mL of water was collected from three locations (n = 180); the well, and the two coupon side streams from the flush valve. Each water sample was tested at the ISU VDL for trace minerals, water quality panel, coliforms, and pH assessment.

Objective 4: Assessing phenotypic resistance

Phenotypic resistance was assessed by obtaining water line sections prepared identical to objective 1. Organisms from the biofilm pellet were re-suspended and cultured on CTC or L drug-infused culture plates. Drug-infused sheep's blood agar culture plates were created for each drug (CTC or L) and the concentration of the drug was calculated at a rate to match what it would be in the water line at the time of treatment . Cultured organisms that fell under the spectrum of the drug were considered resistant, while organisms that did not fall under the spectrum of the drug were not included as resistant (ie, intrinsic resistance).

Results from this work for each objective are undergoing statistical analysis. Preliminary results on water line biofilm regrowth rate and water quality results are available but undergoing formal statistical analysis.

Results

Objective 1: Quantitative and qualitative assessment of water line biofilm regrowth rate after CID 2000 Pro application

Biofilm and biofilm regrowth rate in water lines were quantified using aerobic and anaerobic standard plate counts. Preliminary results are displayed in Figures 1 and 2. Biofilm quantities ranged from 0 colony forming units (CFUs)/ mL to 3.3 billion CFU's/ mL. In most cases, the off-label 0.78% administration of CID 2000 Pro completely reduced or greatly reduced biofilm quantities post-treatment (between time points 0 and 1). Biofilm quantities differed per site, and some sites displayed room differences. Biofilm regrowth appeared to grow back to original or surpassed original biofilm levels three to seven days after treatment with CID 2000 Pro.

Objective 2: Characterize swine water line biofilms

Identification of organisms and ARGs in swine WDS biofilms through 16S rRNA and metagenomic sequencing is pending.

Objective 3: Assess water quality parameters and their relationship with biofilm formation and composition

Preliminary water results demonstrated that the room water samples had higher average coliform counts than those present from samples taken at the well, as seen in Figure 3. Average *Escherichia coli* counts in water samples were more variable and in many cases were largely site dependent which can be noted in Figure 4. The highest *E coli* counts in the rooms at site 5 mimic the findings in the coliform room averages.

Objective 4: Phenotypic resistance

Primary and opportunistic pig pathogens were identified in preliminary phenotypic resistance to CTC and/or L drug- infused plates. Pig pathogens identified include a hemolytic F18 *E coli* with EAST1 toxin and PAA (attaching and effacing) adhesion genes and *Staphylococcus hyicus*. Opportunistic pathogens were numerous, including *Klebsiella pneumoniae, Aeromonas hydrophila, Aeromonas caviae, Citrobacter freundii, Enterobacter hormaechei* and others. Normal flora of the pig's respiratory, vaginal or intestinal tract were also isolated on L or CTC drug infused plates or both plate types.

The hemolytic F18 *E coli* was isolated from site 5. It was first isolated on day 5 sample collection point on a CTC drug-infused plate with EAST1 and PAA genes. It was isolated again on day 7 sample collection date from both a CTC and L drug-infused plate. *E coli* was not isolated again until day 21 sample collection in which a separate *E coli* was isolated and characterized as hemolytic with a PAA adhesion gene but no fimbrial or toxin genes (Table 1). In conclusion, an *E coli* capable of causing disease in pigs was isolated from water line biofilms and displayed phenotypic resistance to drugs commonly administered in water lines.

Discussion

Water is an essential part of daily pig production and there is a need for further characterization of its impact on swine health, production, and description of best practices for water management. Four objectives were assessed to aid in characterization of water line biofilms, water quality dynamics, estimation of genetic and phenotypic resistance to commonly administered water antimicrobials and tested the effect of applying 0.78% CID 2000 Pro to aid in further establishment of best practices in swine WDS management.

Preliminary results demonstrate that a one-time water line cleaning and disinfection event with CID 2000 Pro reduced water line biofilms through aerobic and anaerobic standard plate counts, but biofilms quickly re-established three to seven days post disinfection event. This suggests that CID 2000 Pro was successful at reducing or eliminating biofilms, but a continual cleaner and disinfectant is needed to manage biofilms long term.

Preliminary results suggest that water quality declines after the well samples compared to water samples from rooms in terms of increased coliforms and site dependent *E coli* concentrations. Coliforms indicate livability conditions for other pathogens in water. Average *E coli* counts in water samples were more variable and in many cases were largely site dependent on their presence which was noted in Figure 4. Water samples collected at the well may not be representative of what the pig is consuming due to contamination through the WDS. It is recommended to sample water from multiple sources (from the well or source water and in the rooms) in routine water testing to better represent and quantify true water quality on farms.

Preliminary phenotypic resistance assessment for CTC and L resistant organisms demonstrate that there are primary and opportunistic pig pathogens with phenotypic resistance to commonly administered water antimicrobials. These results provide further motivation to develop best practices for water line cleaning and disinfection mechanisms and the administration of water medication to livestock as concerns surrounding antimicrobial stewardship rise. These challenges expose critical knowledge gaps in the impact of water on the pig and best practices to improve WDS management.

Figure 1: Preliminary anaerobic and aerobic standard plate count results from water line samples from site 1 demonstrating water line quantitative biofilm counts in colony forming units/ mL. Data was log transformed and demonstrates rate of biofilm regrowth after off-label water line treatment with 0.78% CID 2000 for 24-hour dwelling time. Rooms are differentiated by their navigational orientation.

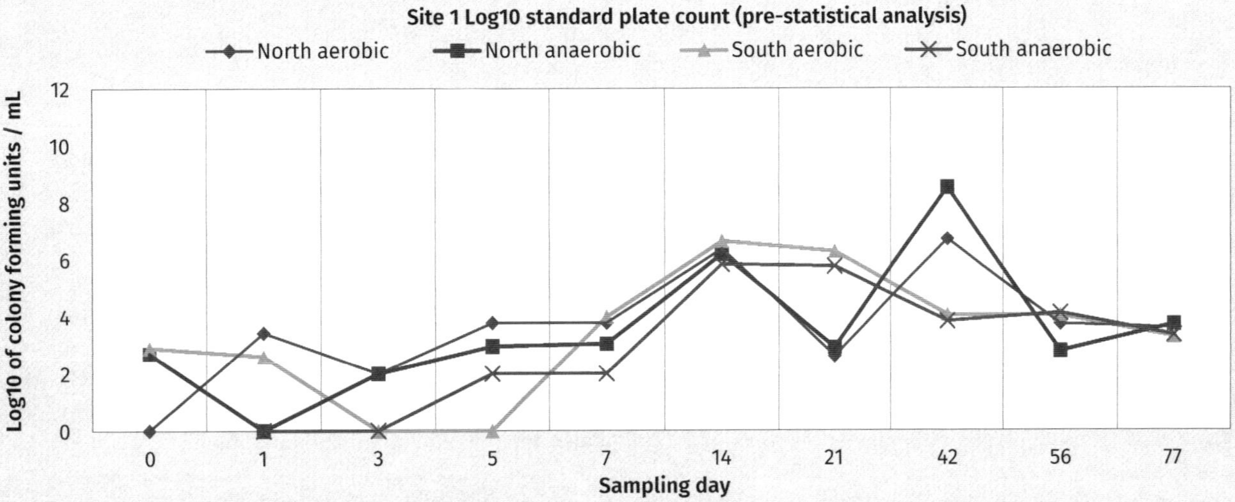

Figure 2: Preliminary anaerobic and aerobic standard plate count results from water line samples from site 2 demonstrating water line quantitative biofilm counts in colony forming units/ mL. Data was log transformed and demonstrates rate of biofilm regrowth after off-label water line treatment with 0.78% CID 2000 for 24-hour dwelling time. Rooms are differentiated by their navigational orientation.

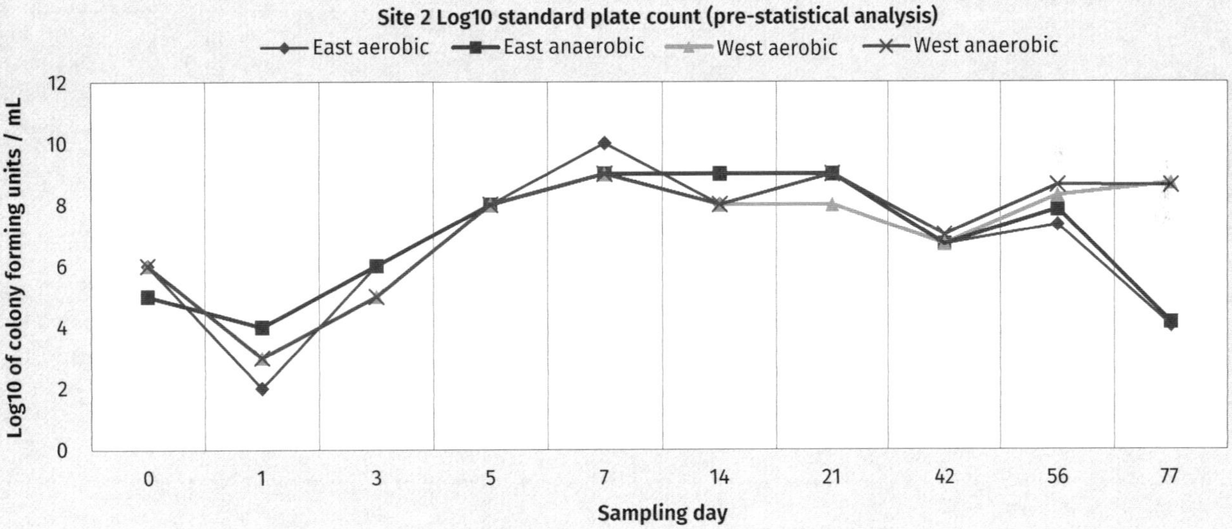

Figure 3: Preliminary coliform results from water samples aseptically taken from a well access point (spigot) and from flush valves located in the rooms at a site. Well coliform averaged across all six wean-to-finish sites and room counts were averaged together across the six sites.

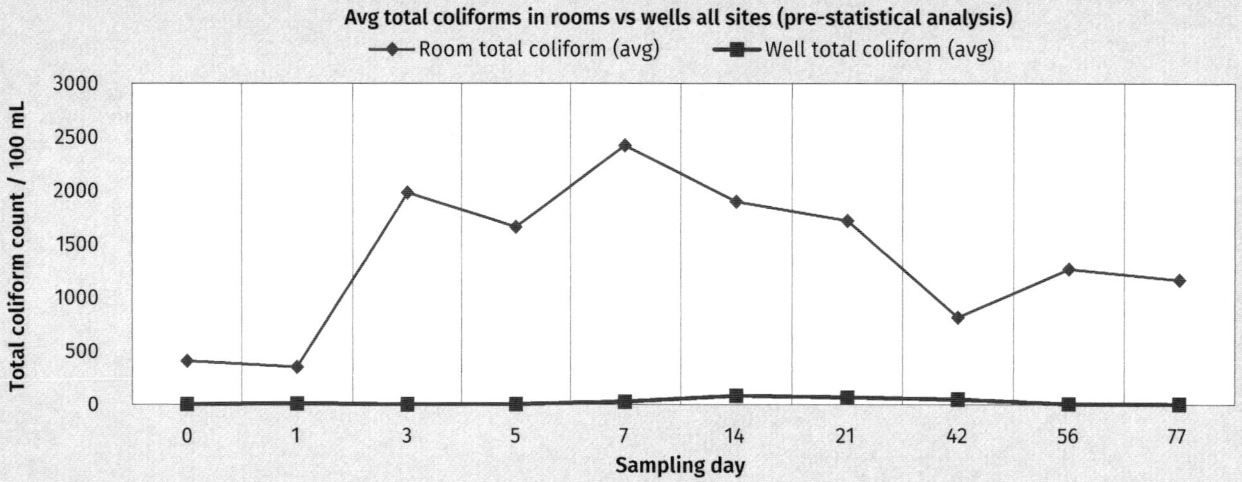

Figure 4: Preliminary average *E coli* counts from water samples aseptically taken from well access point (spigot) and from flush valves located in rooms at a site. *E coli* averages were broken out into highest and lowest sites to demonstrate site 5's elevation of the room average, while the lowest total *E coli* counts (sites 1 and 3) were more consistent with presence at the well.

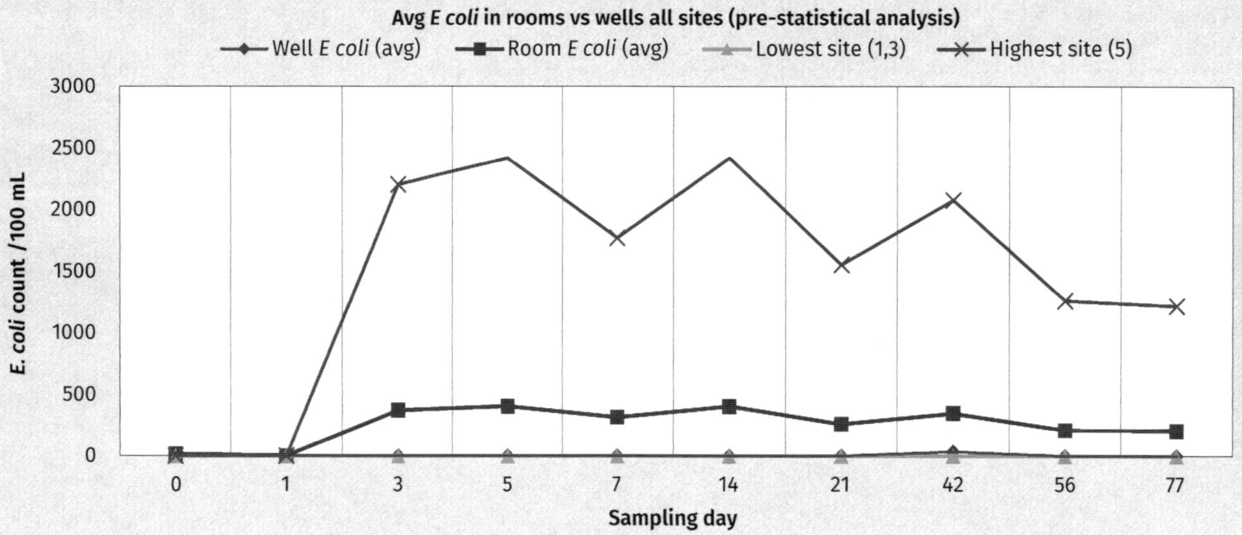

Table 1: Preliminary results from site 5 characterizing *E coli* isolated from water line samples plated on blood agar drug-infused plates

Sampling date	Day 5	Day 7	Day 21
Plate(s) isolated	CTC	CTC + L	CTC
E coli characteristics	• Hemolytic • F18 pillus • PAA adhesion • EAST1 toxin	• Hemolytic • F18 pillus • PAA adhesion • EAST1 toxin	• Hemolytic • PAA adhesion

CTC = chlortetracycline; L = lincomycin hydrochloride

References

1. Lozinski, B.M., et al., Effects of water quality on growth performance and health of nursery pigs. *Transl Anim Sci*, 2022. 6(1): p. txac002-txac002.

2. McLeese, J.M., Water intake patterns in the weanling pig: effect of water quality, antibiotics and probiotics. *Annals of Warsaw Agricultural University*, SGGW-AR., 1992. **54**(1): p. 135-142.

3. Patience, J., Water quality issues in pork production, in *Allen D. Leman Swine Conference*. 2011. p. 157-164.

4. Nyachoti, C.M., A. Regassa, and J.F. Patience, Water in Swine Nutrition, in *Sustainable Swine Nutrition*. 2022. p. 37-63.

5. Samuel, R., Total dissolved solids (TDS) less than 1000 ppm *in drinking water did not impact nursery pig performance. Veterinary Sciences*, 2022. **9**(11): p. 622.

6. Anderson, D. and S. Stothers, Effects of saline water high in sulfates, chlorides and nitrates on the performance of young weanling pigs. *Journal of Animal Science*, 1978. 47(4): p. 900-907.

7. Anderson, J.S., D.M. Anderson, and J.M. Murphy, The effect of water quality on nutrient availability for grower/finisher pigs. *Canadian journal of animal science*, 1994. 74(1): p. 141-148.

8. Nyachoti, C.M., J.F. Patience, and I.R. Seddon, Effect of water source (ground versus surface) and treatment on nursery pig performance. *Canadian Journal of Animal Science*, 2005. 85(3): p. 405-407.

9. Doughan, G., Karriker, L., 2022. Water biology: Implications for pig health, production, and biosecurity. In *proceedings of the ISU James D. McKean Swine Disease Conference, Ames, IA*. November 3-4, 2022. Pg. 66-71.

10. Doughan, G., Karriker, L., Mou, K. 2023. *Water biology:* The next frontier for biosecurity. In *proceedings of the 54th Annual meeting of the American Association of Swine Veterinarians*. March 4-7, 2023. Pg. 346-350.

11. Watkins, S. 2017. Clean Water Lines for Flock Health. *Avian Advice*, University of Arkansas Extension. September 2017. Pg. 1-5. **https://poultry-science.uark.edu/_resources/pdf/AvianAdvice_Sept2017.pdf**

12. Maharjan, P., Development of a Biofilm Model for Evaluating Poultry Drinking Water Sanitation Procedures, in *Poultry Science*. 2016, University of Arkansas. p. 113.

13. Watkins, S. 2021. Water Line Sanitation. *Aviagen Brief* Pg. 1-4. **https://tmea.aviagen.com/assets/Tech_Center/Broiler_Breeder_Tech_Articles/English/AviagenBrief-WaterLineSanitationUpdate-2021-EN.pdf**

14. Scantling, M. and S. Watkins. Identify Poultry Water System Contamination Challenges. *University of Arkansas Sytem Division of Agriculture Research and Extension*. Pg 1-4. **https://www.uaex.uada.edu/publications/pdf/FSA-8011.pdf**

15. Raut, R., et al., Impacts of on-farm water sanitation practices on microbial hygiene in poultry waterlines and efficacy of sodium hypochlorite-based product on foodborne pathogens. *Journal of Applied Poultry Research*, 2024. 33(3): p. 100425.

16. Galgano, S., et al., Impact of precursor-derived peracetic acid on post-weaning diarrhea, intestinal microbiota, and predicted microbial functional genes in weaned pigs. *Frontiers in Microbiology*, 2024. 15.

17. Kitis, M., Disinfection of wastewater with peracetic acid: a review. *Environment International*, 2004. 30(1): p. 47-55.

18. Hancock, A., J. Hughes, and S. Watkins, In Search of the Ideal Water Line Cleaner. *Avian Advice* 2007. 9(1): p. 1-4.

19. Fischer, M., et al., Efficacy Assessment of Nucleic Acid Decontamination Reagents Used in Molecular Diagnostic Laboratories. *PLoS One*, 2016. 11(7): p. e0159274.

https://doi.org/10.54846/am2025/64

Soybean meal functional compounds: Impact on pig health and physiology

Amy L. Petry; Brooke Bowen; Lisa Weaver; R. Dean Boyd

Introduction

While soybean meal (SBM) is well known as a quality source of protein, approximately 40% of processed SBM is something other than protein, moisture, and oil.[1] It has been hypothesized that this fraction contains extra-nutrient components that may mitigate respiratory disease effects and improve energy partitioning for growth in commercially reared pigs.[2,3] SBM is an abundant and diverse source of functional compounds[4] in the non-protein fraction and bioactive peptides that arise from parent proteins.[4] A functional or phytochemical compound is a naturally derived, dietary compound that is regularly consumed and impacts a particular physiological process. SBM's benefits have been documented for growth, viability, and medical intervention of pigs growing under high and low immune stress conditions. Production improvements with low immune stress were reported by Cemin et al,[5] who reported a linear improvement in the ability of nursery pigs to thrive as dietary SBM levels increased (6000 pigs, two systems). A noteworthy phenomenon has also emerged while validating SBM net energy (NE) in commercial settings. SBM NE values exceeded a recent calorimetric estimate for SBM by at least 20%.[3] These results do not necessarily conflict with classically derived NE estimates but may represent the conservation of maintenance energy that otherwise might have been used to counter pathogens or other stressors in the commercial environment. This paper provides a targeted review of the functional compounds in SBM that may be involved in its disease-mitigating effects in swine and potentially improving energy partitioning for growth under conditions of subclinical disease.

Emerging evidence for value of SBM functional compounds in pigs

Important findings are beginning to emerge, showing how functional compounds in SBM could act therapeutically in the commercial sector to reduce impaired growth, reduce medication needs, and improve viability under systemic disease challenges (Figure 1).[6]

Several studies have been published by the Ryan Dilger lab using young pigs infected with PRRSV to create the health disturbances commonly encountered in commercial practice (impaired growth, mortality). They investigated the recovery rate from infection with and without isoflavones (ISF) in their diet. Isoflavones have been shown to have anti-inflammatory, antiviral and anti-oxidative effects primarily within the GI tract.[7] These studies significantly advanced our understanding of soybean ISF value in effectively countering the ubiquitous respiratory pathogen PRRSV. Our research has evaluated the effect of isolated SBM functional compounds in nursery pig diets, relative to intact SBM,[8] to clarify the role of these compounds on nursery pig gastrointestinal physiology, systemic health, and inflammatory markers and aid in elucidating their mechanism of action in pigs. The results revealed a 30% decrease in white blood cells, particularly neutrophils, in pigs fed diets with isolated isoflavones and the SBM control, indicating an immune cell modulatory effect. Another interesting finding was improved systemic and gastrointestinal antioxidant capacity with diets including soy lecithin and phytosterols. The study's insights suggest that soy's functional components, particularly its lipids and polyphenol components, could confer health benefits in young pigs (Figure 2).

Functional compounds: Classification and potential mechanisms of action

'Functional foods' have emerged as a critical concept in human nutrition, with a growing body of clinical evidence supporting their role in preventing and managing chronic illnesses. Soybeans have been recognized as a functional food due to their dense concentration of functional compounds.[9] A functional or phytochemical compound is a naturally derived, dietary compound that is regularly consumed and impacts a particular physiological process.[10] Functional compounds are not considered nutrients, as their lack does not correlate with any specific deficiency. Many cell culture and population studies within the human sciences have identified soybeans as a source of at least 14 functional compounds,[4] including phytic acid, peptides, triterpenes, phenolics and flavonoids, fatty acids, carotenoids, coumarins, protease inhibitors, oligosaccharides, and dietary fibers. These compounds can act alone or synergistically with the gut microbiome to create metabolites that impact health and physiology. Processed SBM contains four critical categories of functional compounds that may contribute to health-oriented pork production objectives: 1) polyphenols and terpenoids, 2) bioactive peptides, 3) dietary fiber and oligosaccharides, and 4) functional lipids. These four categories represent approximately 70% of the functional compounds found in soybeans that have recognized effects within human nutrition and are most prevalent in soybean meal.

I. Polyphenols and terpenoids

Soybeans produce polyphenolic compounds through their secondary metabolism, which include phenolic acids, isoflavones, and saponins. Polyphenols within the bean are multifunctional compounds essential for defense against plant pathogens while contributing to the plant's overall health and environmental adaptation.[11] Eight phenolic acids (p-hydroxy benzoic, chlorogenic, coumaric, ferulic, gentisic, salicylic, syringic, and vanillic acids) are consistently present in soybeans.[4] These compounds can be found in an accessible form, a free conjugate form, or as a bound component and are all derived from phenylalanine.[12] The bioavailability of bound phenolic acids still needs to be established in swine. In vitro, monogastric digestion models suggest an 8% to 28% bioavailability.[13] Isoflavones, classified as flavonoids, resemble 17β-estradiol, and include three molecules—daidzein, genistein, and glycitein—each present in four structural motifs in soybeans. They are bioavailable in their aglycone state (called daidzein, genistein, glycitein). Conversely, their glycosidic form (daidzin, genistin, glycitin) is bound to glucose and is not bioavailable.[4] Isoflavones bound to a carbohydrate motif can be hydrolyzed by β-glycosidases produced by the host or microbiota. Once hydrolyzed, they are bioavailable and can be absorbed predominately in the small intestine or metabolized further by microbiota. Most absorbed isoflavones are metabolized by

Figure 1: Cumulative mortality of weaned pigs infected with PRRS virus over 161 d post-infection if fed diets with and without isoflavones. (Adapted with permission from Smith et al, 2020).[6]

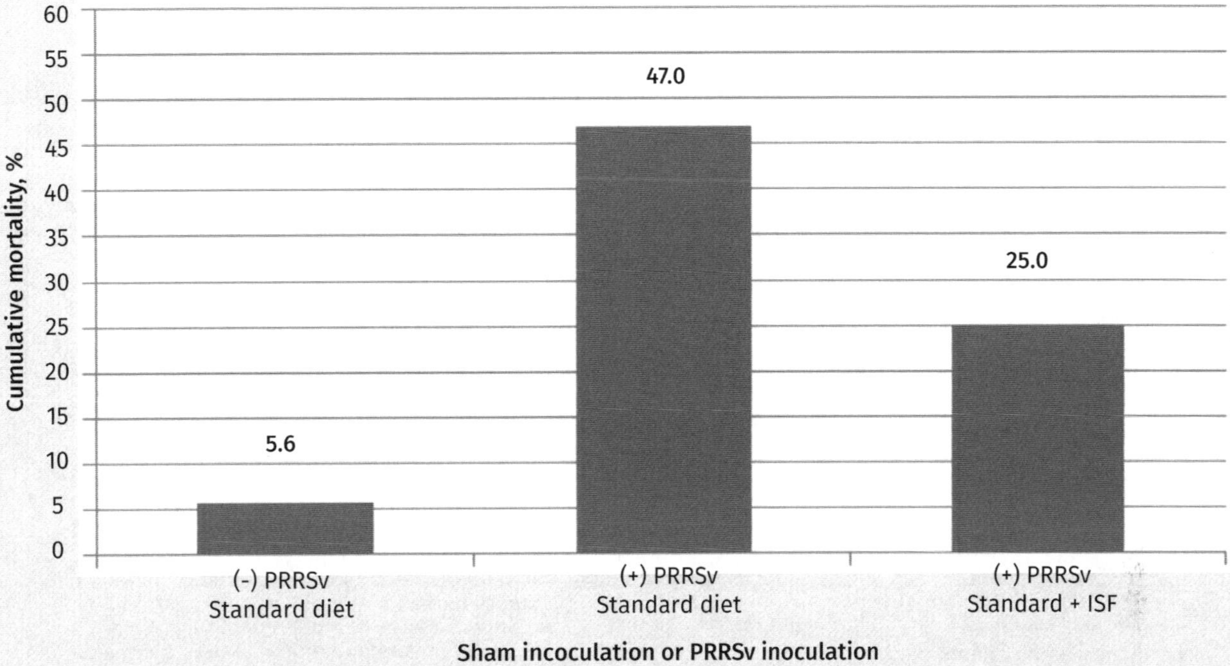

Weaned pigs were divided into 3 groups and housed as groups in a Biosafety Level-2 containment facility. Pigs were adjusted to the facility for 7 d and then given either sham (PBS control) or PRRSV intranasal inoculation. Pigs were fed either a (1) standard diet (sham, +PRRSV) or (2) standard diet + ISF (+PRRSV) for each diet phase.

PBS = phosphate buffered saline; PRRSV = porcine reproductive and respiratory syndrome virus.

enterocytes, with only a fraction entering circulation (< 5%).[14] Saponins are surfactant terpenoid compounds consisting of a sapogenin core attached to one or more carbohydrate moieties and found predominately in the germ. Five distinct sapogenins (A–E) have been characterized in soybeans, each potentially glycosylated with three monosaccharides: galactose, arabinose, rhamnose, glucose, xylose, or glucuronic acid.[15] Like isoflavones, their bioavailability depends on them being hydrolyzed from the glycone to the aglycone form.

The antioxidant, antimicrobial, and anti-inflammatory properties of polyphenols, flavonoids, and terpenoids have garnered significant attention in human and animal science research. Microbial metabolism can transform the more complex polyphenol structures into potent phenolic metabolites, such as equol and salicylic acid.[16] However, the bioavailability of polyphenols and terpenoids in pigs remains poorly understood. These compounds must first be hydrolyzed from their plant matrix through processing or digestion and then absorbed in the gut. Soy phenolic and flavonoid compounds act as antioxidants through electron donation and metal chelation, a process influenced by the special arrangement of their functional groups, protecting against oxidative damage to proteins, nucleic acids, and lipids.[17] Research has highlighted the effectiveness of antioxidant properties of phenolic acids and isoflavones in both cellular and enzymatic systems.[18,19] Specifically, isoflavones combat oxidation by neutralizing free radicals and inhibiting singlet oxygen without producing additional radicals.[20] This unique property allows it to recycle as an antioxidant more effectively than many other functional compounds. Dietary soy isoflavones in nursery pigs have improved jejunal antioxidant levels and regulated gut barrier integrity, potentially through a means of mitigating ROS.[21,8] Additionally, equol, a microbial metabolite produced from daidzein, has greater antioxidant activity than its parent compound and is more readily absorbed into the bloodstream.[22]

Soy isoflavones possess complementary anti-inflammatory and antioxidant properties. Their primary mechanism involves immunomodulatory actions through inhibiting tyrosine-specific protein kinases and NF-κβ transcription, which are key players in regulating inflammation and producers of cellular oxidants.[23,24] Saponins can also mediate NF-κB transcription factors that mitigate pro-inflammatory responses.[25] Likewise, phenolic compounds' anti-inflammatory properties suppress inflammatory mediators like prostaglandins and nitric oxide, potentially reducing inflammation.[26] Diet supplementation with phenolic-enriched soy isoflavones has been linked to enhanced immune function, marked by reduced diarrhea incidence and lower endotoxin levels in lipopolysaccharide (LPS)-induced, immune-challenged nursery pigs.[27] They also activate mononuclear cells, bolstering the phagocytic response through MAPK and NF-κB pathway modulation in mice models.[28] Furthermore, cell culture work suggests phenolic acids may contribute to gut health by inhibiting the COX-2 enzyme, implicated in inflammatory responses, and thus, decreasing pro-inflammatory prostaglandins in activated cells.[19]

Figure 2: Summary of soybean functional compounds and health-related and physiological roles. (Graphic created with BioRender.com).

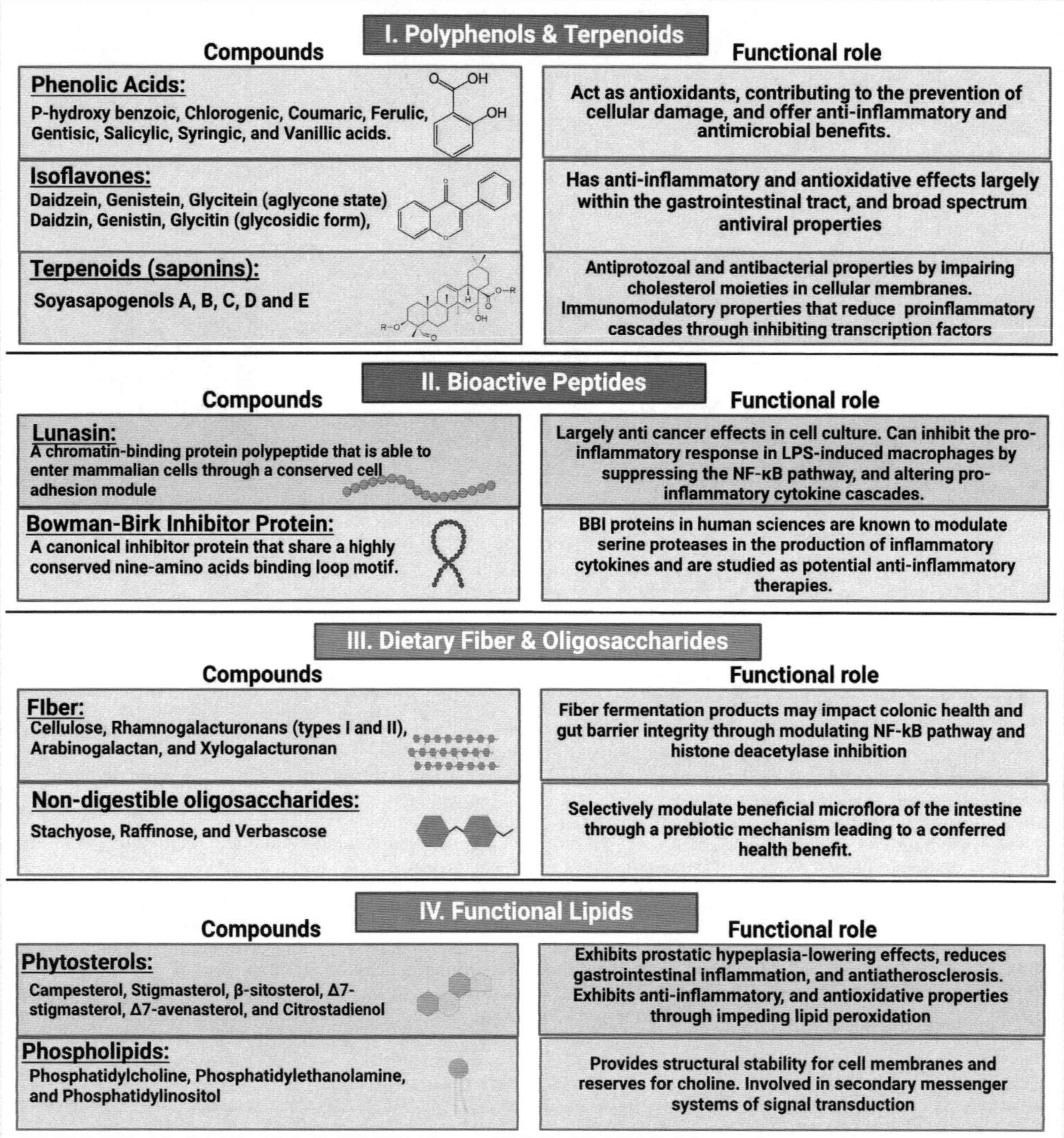

Isoflavones have demonstrated antiviral properties across various viruses by interfering with different stages of the virus life cycle, including viral entry, replication, protein synthesis, and virus envelope formation.[29] Genistein is particularly noted for its broad-spectrum antiviral efficacy against various viruses, including those with differing envelope structures and genetic materials.[30] The effectiveness of flavonoids and isoflavones in antiviral defense, however, depends on the dosage used, ranging from physiological to supraphysiological levels. An in-depth discussion of the potential antiviral and antibacterial mechanisms of soy isoflavones and saponins, a subclass of terpenoids, is found in the comprehensive analysis by Smith and Dilger.[7] Their findings suggest that administering soy-derived isoflavones or saponins to infected animals could potentially enhance recovery, lower mortality rates, and stabilize performance in specific conditions.

II. Bioactive peptides

Soybean meal protein is a composite of protein types or families, 80% to 90% of which are glycinin and beta-conglycinin.[31] These two major proteins undergo enzyme cleavage during digestion (by endo- and exo-peptidases) and through gut microbiome activity after ingestion. Protein cleavage can also occur if soybean meal is fermented before being added to the diet. Since proteins are chains of amino acids, when they are cleaved, the result is individual amino acids and a mixture of smaller peptides. Those peptides with amino-acid-to-amino-acid linkages for which an endopeptidase does not exist appear to represent a designed outcome.[32] The fact that some peptides are bioactive themselves, and have effects beyond just delivering amino acids, has been known for some time. Bioactive peptides are usually not active until released from the parent protein and are generally between two and 20 amino acids long.[33] Lunasin, and the Bowman-Birk inhibitors (BBi) are small proteins that are self-existent bioactive peptides (oligo-peptides), larger than other bioactive peptides identified from soybean.[33] Lunasin has long been known for anticancer activity and has anti-inflammatory activity,[34] while BBi has been associated with antiviral and anti-inflammatory activity.[35] Except lunasin and BBi, the activity of proteases is required to release the active peptide, so the ability to consistently produce the same mixtures of peptides through the digestive process will likely depend on species, development stage, and gut microbiome composition.[36] Bioactive peptides may have both local and systemic effects. With respect to the latter, dipeptides (two amino acids) are easily absorbed, but there is evidence that various mechanisms take up longer peptides.[37] Soy bioactive peptides target localized cell proliferation, so may be anticarcinogenic.[38] Soy lunasin has long been noted for its anti-cancer and anti-inflammatory activity. The anti-inflammatory activity may be due to a tripeptide, arginine-glycine-aspartate, found within lunasin, which could be released by digestion and then disrupts the inflammatory cascade.[34] Finally, the soybean Bowman-Birk trypsin inhibitor (BBi) is often considered a bioactive peptide. This small protein, almost twice as large as lunasin, has been shown to have anti-inflammatory and anti-viral activities.[35,39] BBi is believed to be quite stable, due to its multiple disulfide bonds. Unlike other bioactive peptides which are released by digestion, BBi's ability to inhibit proteases may be responsible for its activity. The inflammatory response involves serine proteases, and inhibiting these proteases reduces pro-inflammatory cytokine release.[35] Likewise, the inhibition of key enzymes involved in virus replication may be the source of BBi's antiviral activity.[35]

III. Dietary fiber and oligosaccharides

Carbohydrates constitute approximately 30% to 35% of soybean's dry matter content.[40] These carbohydrates are predominantly nonstructural, including a minor fraction of starch, sugars, oligosaccharides, and dietary fiber. Solvent-extracted SBM is particularly noted for its dietary fiber and prebiotic oligosaccharides content, averaging 12 ± 6% total dietary fiber and 5 ± 2% non-digestible oligosaccharides.[41] The fermentable fiber in SBM is characterized by a variety of sugars such as galactose, arabinose, and fructose, as well as uronic acids like galacturonic acid, with notable quantities of non-digestible galactooligosaccharides (NDOs) including predominantly stachyose, with smaller amounts of raffinose, and verbascose.[40] While dietary fiber can impede nutrient digestibility, its fermentable components are valued for their beneficial effects on gastrointestinal health.[42] Nevertheless, NDOs may exert antinutritive effects in young pigs, potentially causing gastrointestinal disturbances and resulting in loose fecal consistency.[1] Pigs cannot digest soy fiber or NDOs, and instead depend on microbial fermentation. These NDOs can exhibit a prebiotic effect in the gastrointestinal tract at low concentrations. Above 3%, NDOs may negatively influence the digestibility and energy value of SBM; however, within the concentrations typical in corn-SBM-based grower diets, they may serve as prebiotics, promoting the proliferation of beneficial bacteria such as *Lactobacillus* and *Bifidobacterium*.[40] Moreover, soybean NDOs may inhibit the production of putrefactive compounds from protein fermentation in the large intestine.[43] This modulation of gut microbial communities by soybean NDOs may leverage a prebiotic-like effect to reduce protein-derived catabolites in swine's intestinal luminal content.[44] Additionally, the fermentation of galactooligosaccharides is known to stimulate *Bifidobacteria* growth in humans, thereby altering colonic microbiota.[45] *Bifidobacteria* are known for improving gut barrier integrity, vitamin synthesis, and immune function enhancement.[46]

IV. Functional lipids

The functional lipid content of SBM, such as phytosterols and phospholipids, is subject to variation based on processing conditions and growing intensity.[47] These lipids are present in lower quantities and exhibit considerably more variability than other functional compounds in SBM, conceivably playing a lesser role in its functional attributes. However, soy lecithin, a phospholipid, can be reintroduced into SBM through 'addbacks' during the soy oil refining process, resulting in fluctuating concentrations in end products. Phospholipids, integral to cell membrane composition, include two fatty acids, a phosphate group, and a glycerol backbone. In soybeans, prevalent phospholipids include phosphatidylcholine, phosphatidylethanolamine, and phosphatidylinositol.[48] Additionally, phytosterols, which are plant sterols analogous in structure to cholesterol, are present in soybeans as six isomers (campesterol, stigmasterol, β-sitosterol, Δ7-stigmasterol, Δ7-avenasterol, and citrostadienol) comprising approximately 0.4% of the total weight.[49] In humans, phytosterols and phospholipids are valued for their contribution to cell membrane integrity, heart and liver health, and antioxidant effects.[50] However, when consumed by young animals and fish, they can be antinutritive.[51] Soybean phospholipids, including glycerophospholipids and phosphatidylcholine, can aid in forming cell membranes and the intestinal mucus layer. Phospholipids play a crucial role in intestinal mucus composition, existing as a continuous layer on the luminal side of the mucus gel. They are present within the mucus in the form of liposome-like aggregates, and form a monolayer on the surface of mucosal cells. Clinical applications of phosphatidylcholine in humans have shown anti-inflammatory properties, primarily driven by modifying the prostaglandin pathways and inhibiting TNF-α-induced inflammatory responses.[52] The phytosterols found in soybeans, such as β-sitosterol and campesterol, are beneficial for lowering cholesterol and may also influence immune responses related to colitis.[53] Currently, there is a dearth of knowledge on the direct effects of phytosterols and phospholipids found in SBM, as compared to those contained in soy oil.

Conclusion

While SBM is known in swine nutrition for its protein content, it is the non-protein fraction — rich in polyphenols, terpenoids, bioactive peptides, fiber, functional lipids and other functional compounds — that may play a role in modulating health outcomes in pigs and may be considered as complementary nutrition factors. Data indicate possible benefits of SBM in disease resilience and physiological responses, yet the mechanisms of action remain to be fully elucidated, and bioavailability will be a key factor in efficacy.

Acknowledgments

Brought to you by U.S. Soy. Fully funded by the National Soybean Checkoff.

References

1. Choct, M., Y. Dersjant-Li, J. Mcleish and M. Peisken, 2010. Soy oligosaccharides and soluble non-starch polysaccharides: A review of digestion, nutritive and antinutritive effects in pigs and poultry. Asian-Aust. *J. Anim. Sci.* Vol. 23. No. 10:1386- 1398.

2. Boyd, R.D., M.E. Johnston, J. Usry, P. Yeske and A. Gaines. 2023. Soybean meal mitigates respiratory disease-impaired growth in pigs. *Feedstuffs,* October digital edition, page 1. **https://informamarkets.turtl.co/story/feedstuffs-october-2023/page/2/1**

3. Boyd, R.D. and A.M. Gaines. 2023. Soybean meal NE value for growing pigs is greater in commercial environments. *Feedstuffs,* August digital edition, page 1. **https://informamarkets.turtl.co/story/feedstuffs-august-2023/page/2/1**

4. Basson AR, Ahmed S, Almutairi R, Seo B, Cominelli F. Regulation of Intestinal Inflammation by Soybean and Soy-Derived Compounds. *Foods.* 2021; 10(4):774. **https://doi.org/10.3390/foods10040774**

5. Cemin, H. S., H. E. Williams, M. D. Tokach, S. S. Dritz, J. C. Woodworth, J. M. Derouchey, K. F. Coble, B. A. Carrender, and M. J. Gerhart. 2020. Estimate of the energy value of soybean meal relative to corn based on growth performance of nursery pigs. *J. Anim. Sci.* Biotech 11:70-78. **https://doi.org/10.1186/s40104-020-00474-x**

6. Smith, N.S., M.L. Oelschlager, M.S.A. Rasheed, R.N. Dilger. 2020. Dietary soy isoflavones reduce pathogen-related mortality in growing pigs under porcine reproductive and respiratory syndrome viral challenge. *Journal of Animal Science,* 98(4), February 2020. **https://doi.org/10.1093/jas/skaa024**

7. Smith, B.N. and Dilger, R.N., 2018. Immunomodulatory potential of dietary soybean-derived isoflavones and saponins in pigs. *Journal of Animal Science,* 96(4), pp.1288-1304.

8. Bowen, B., 2023. Evaluating the relationship of bioactive functional compounds innate to soybean meal on nursery pig performance, gut physiology, nutrient digestibility, and systemic health. *Thesis.* Texas Tech University.

9. Messina, M., Duncan, A., Messina, V., Lynch, H., Kiel, J., & Erdman Jr, J. W. (2022). The health effects of soy: A reference guide for health professionals. *Frontiers in Nutrition,* 1837.

10. Dillard, C.J. and German, J.B., 2000. Phytochemicals: nutraceuticals and human health. *Journal of the Science of Food and Agriculture,* 80(12), pp.1744-1756.

11. Singh, S., Kaur, I. and Kariyat, R., 2021. The multifunctional roles of polyphenols in plant-herbivore interactions. *International Journal of Molecular Sciences,* 22(3), p.1442.

12. Xu, Minwei, Jiajia Rao, and Bingcan Chen. Phenolic Compounds in Germinated Cereal and Pulse Seeds: Classification, Transformation, and Metabolic Process. *Critical Reviews in Food Science and Nutrition* 60, no. 5 (March 8, 2020): 740–59. **https://doi.org/10.1080/10408398.2018.1550051**

13. Zhu, L., Li, W., Deng, Z., Li, H. and Zhang, B., 2020. The composition and antioxidant activity of bound phenolics in three legumes, and their metabolism and bioaccessibility of gastrointestinal tract. *Foods,* 9(12), p.1816.

14. Larkin, T., Price, W.E. and Astheimer, L., 2008. The key importance of soy isoflavone bioavailability to understanding health benefits. *Critical Reviews in Food Science and Nutrition,* 48(6), pp.538-552.

15. Zhao, G., Zhu, L., Yin, P., Liu, J., Pan, Y., Wang, S., Yang, L., Ma, T., Liu, H. and Liu, X., 2022. Mechanism of interactions between soyasaponins and soybean 7S/11S proteins. *Food Chemistry,* 368, p.130857.

16. Setchell, Kenneth DR, Nadine M. Brown, and Eva Lydeking-Olsen. "The clinical importance of the metabolite equol—a clue to the effectiveness of soy and its isoflavones." *The Journal of Nutrition* 132, no.12 (2002): 3577-3584.

17. Mahfuz, Shad, Qinghui Shang, and Xiangshu Piao. Phenolic Compounds as Natural Feed Additives in Poultry and Swine Diets: A Review. *Journal of Animal Science and Biotechnology* 12, no. 1 (April 7, 2021): 48. **https://doi.org/10.1186/s40104-021-00565-3**

18. Wiseman, H., J. D. O'Reilly, H. Adlercreutz, A. I. Mallet, E. A. Bowey, I. R. Rowland, and T. A. Sanders. Isoflavone Phytoestrogens Consumed in Soy Decrease F(2)-Isoprostane Concentrations and Increase Resistance of Low-Density Lipoprotein to Oxidation in Humans. *The American Journal of Clinical Nutrition* 72, no. 2 (August 2000): 395–400. **https://doi.org/10.1093/ajcn/72.2.395**

19. Russell, W., and Duthie, G., Plant Secondary Metabolites and Gut Health: The Case for Phenolic Acids. *Proceedings of the Nutrition Society,* 70(3), pp.389–396. **https://doi.org/10.1017/S0029665111000152**

20. Ruiz-Larrea, M.B., Mohan, A.R., Paganga, G., Miller, N.J., Bolwell, G.P. and Rice-Evans, C.A., 1997. Antioxidant activity of phytoestrogenic isoflavones. *Free Radical Research,* 26(1), pp.63-70.

21. Li, Y. P., Jiang, X. R., Wei, Z. X., Cai, L., Yin, J. D., & Li, X. L. (2020). Effects of soybean isoflavones on the growth performance, intestinal morphology and antioxidative properties in pigs. *Animals,* 14(11), 2262-2270.

22. Setchell, K.D. and Clerici, C., 2010. Equol: pharmacokinetics and biological actions. *The Journal of Nutrition,* 140(7), pp.1363S-1368S.

23. Dia, V.P., Berhow, M.A. and Gonzalez De Mejia, E., 2008. Bowman–Birk inhibitor and Genistein among Soy compounds that synergistically inhibit nitric oxide and prostaglandin E2 pathways in lipopolysaccharide-induced macrophages. *Journal of agricultural and food chemistry,* 56(24), pp.11707-11717.

24. Sanchez, A. and A. Vazquez. 2017. Bioactive peptides: A review. *Food Qual. Safety* 1:29-46. **Doi:10.1093/fqs/fyx00**

25. Francis, G., Z. Kerem, H.P.S. Makkar, and K. Becker. 2002. The biological action of saponins in animal systems: a review. Br. *J. Nutr.* 88:587–605. **doi:10.1079/BJN2002725**

26. Kang, J.H., M.K. Sung, T. Kawada, H. Yoo, Y.K. Kim, J.S. Kim, and R. Yu. 2005. Soybean saponins suppress the release of proinflammatory mediators by LPS-stimulated peritoneal macrophages. *Cancer Letters* 230, no. 2 (2005): 219-227.

27. Zhu C, Wu Y, Jiang Z, Zheng C, Wang L, Yang X, et al. Dietary soy isoflavone attenuated growth performance and intestinal barrier functions in weaned piglets challenged with lipopolysaccharide. *Int Immunopharma.* 2015;28(1):288–94.

28. Morimoto, M., Watanabe, T., Yamori, M., Takebe, M., & Wakatsuki, Y. (2009). Isoflavones regulate innate immunity and inhibit experimental colitis. *Journal of Gastroenterology and Hepatology,* 24(6), 1123-1129.

29. Andres, Aline, Sharon M. Donovan, and Mark S. Kuhlenschmidt. "Soy Isoflavones and Virus Infections." *The Journal of Nutritional Biochemistry 20,* no. 8 (August 1, 2009): 563–69. **https://doi.org/10.1016/j.jnutbio.2009.04.004**

30. Ninfali, P., Antonelli, A., Magnani, M. and Scarpa, E.S., 2020. Antiviral properties of flavonoids and delivery strategies. *Nutrients*, 12(9), p.2534.

31. Natarjan et al. Characterization of Storage Proteins in Wild (Glycine soja) and Cultivated (Glycine max) Soybean Seeds Using Proteomic Analysis J. Agric. *Food Chem.* 2006, 54, no 8, 3114–3120. **https://doi.org/10.1021/jf052954k**

32. Kitts, D.D. and Katie Weiler. Bioactive proteins and peptides from food sources. Applications of bioprocesses used in isolation and recovery. *Curr Pharm Des.* 2003;9(16):1309-23. **doi: 10.2174/1381612033454883**

33. Chatterjee, Cynthia, Stephen Gleddie, and Chao-Wu Xiao.. Soybean Bioactive Peptides and Their Functional Properties Nutrients 2018 10, no. 9: 1211. **https://doi.org/10.3390/nu10091211**

34. Lule, V.K., Garg, S., Pophaly, S.D., Hitesh, and Tomar, S.K. (2015), Potential Health Benefits of Lunasin: A Multifaceted Soy-Derived Bioactive Peptide. *Journal of Food Science*, 80: R485-R494. **https://doi.org/10.1111/1750-3841.12786**

35. Agata Gitlin-Domagalska , Aleksandra Maciejewska and Dawid Debowski, Bowman-Birk Inhibitors: Insights into a Family of Multifunctional Proteins and Peptides with Potential Therapeutical Applications (2020) *Pharmaceuticals* 13, 421; **doi:10.3390/ph13120421**

36. Chong Wang, Fan Zhao, Yun Bai, Chunbao Li, Xinglian Xu, Karsten Kristiansen, Guanghong Zhou, Effect of gastrointestinal alterations mimicking elderly conditions on in vitro digestion of meat and soy proteins, 2022, *Food Chemistry,* 383: 132465, **https://doi.org/10.1016/j.foodchem.2022.132465**

37. Amigo, Lourdes, and Blanca Hernández-Ledesma. 2020. *Current Evidence on the Bioavailability of Food Bioactive Peptides Molecules* 25, no. 19: 4479. **https://doi.org/10.3390/molecules25194479**

38. Ben O. de Lumen, Lunasin: A Cancer-Preventive Soy Peptide, *Nutrition Reviews,* Volume 63, Issue 1, January 2005, Pages 16–21. **https://doi.org/10.1111/j.1753-4887.2005.tb00106.x**

39. de Almeida Barros, B., da Silva, W.G., Moreira, M.A. et al. In silico characterization and expression analysis of the multigene family encoding the Bowman–Birk protease inhibitor in soybean. *Mol Biol Rep* 39, 327–334 (2012). **https://doi.org/10.1007/s11033-011-0742-1**

40. Karr-Lilienthal, L. K., C. T. Kadzere, C. M. Grieshop, and G. C. Fahey. "Chemical and Nutritional Properties of Soybean Carbohydrates as Related to Nonruminants: A Review." *Livestock Production Science* 97, no. 1 (October 1, 2005): 1–12. **https://doi.org/10.1016/j.livprodsci.2005.01.015**

41. *National Animal Nutrition Program*.2024. **https://animalnutrition.org/feed-composition-database-signup?destination=feed-composition-database**

42. Jha, R., Fouhse, J. M., Tiwari, U. P., Li, L., & Willing, B. P. (2019). Dietary fiber and intestinal health of monogastric animals. *Frontiers in Veterinary Science,* 6, 48.

43. Nakata, T., Kyoui, D., Takahashi, H., Kimura, B. and Kuda, T., 2017. Inhibitory effects of soybean oligosaccharides and water-soluble soybean fibre on formation of putrefactive compounds from soy protein by gut microbiota. *International Journal of Biological Macromolecules*, 97, pp.173-180.

44. Zhou, X.L., Kong, X.F., Lian, G.Q., Blachier, F., Geng, M.M. and Yin, Y.L., 2014. Dietary supplementation with soybean oligosaccharides increases short-chain fatty acids but decreases protein-derived catabolites in the intestinal luminal content of weaned Huanjiang mini-piglets. *Nutrition Research,* 34(9), pp.780-788.

45. Roberfroid, M.B., 1997. Health benefits of non-digestible oligosaccharides. *Dietary Fiber in Health and Disease,* pp.211-219.

46. Liao, S.F. and Nyachoti, M., 2017. Using probiotics to improve swine gut health and nutrient utilization. *Animal Nutrition,* 3(4).

47. Wang, Tong, and Lawrence A. Johnson. "Survey of Soybean Oil and Meal Qualities Produced by Different Processes." *Journal of the American Oil Chemists' Society* 78, no. 3 (March 1, 2001): 311–18. **https://doi.org/10.1007/s11746-001-0262-7**

48. Wu, Yingzi, and Tong Wang. "Soybean Lecithin Fractionation and Functionality." *Journal of the American Oil Chemists' Society* 80, no. 4 (April 2003): 319–26. **https://doi.org/10.1007/s11746-003-0697-x**

49. Zhu, Liuying, Wenting Li, Zeyuan Deng, Hongyan Li, and Bing Zhang. The Composition and Antioxidant Activity of Bound Phenolics in Three Legumes, and Their Metabolism and Bioaccessibility of Gastrointestinal Tract. *Foods* 9, no. 12 (December 2020): 1816. **https://doi.org/10.3390/foods9121816.**

50. Castro-Gómez, P., Garcia-Serrano, A., Visioli, F. and Fontecha, J., 2015. Relevance of dietary glycerophospholipids and sphingolipids to human health. *Prostaglandins, Leukotrienes and Essential Fatty Acids,* 101, pp.41-51.

51. Couto, A, T. M. Kortner, M. Penn, A. M. Bakke, Å. Krogdahl, and A. Oliva-Teles. "Effects of Dietary Phytosterols and Soy Saponins on Growth, Feed Utilization Efficiency and Intestinal Integrity of Gilthead Sea Bream (Sparus Aurata) Juveniles." *Aquaculture* 432 (August 20, 2014): 295–303. **https://doi.org/10.1016/j.aquaculture.2014.05.009**

52. Treede, I., Braun, A., Sparla, R., Kuhnel, M., Giese, T., Turner, J.R., Anes, E., Kulaksiz, H., Fullekrug, J., Stremmel, W. and Griffiths, G., 2007. Anti-inflammatory effects of phosphatidylcholine. *Journal of Biological Chemistry,* 282(37), pp.27155-27164.

53. Juritsch, A.F. and Moreau, R., 2018. Role of soybean-derived bioactive compounds in inflammatory bowel disease. *Nutrition Reviews,* 76(8), pp.618-638.

https://doi.org/10.54846/am2025/65

Efficacy of Tilmovet® for PRRS control by feed medication in weaned pigs

Ulrich Klein[1], DVM, MS; Wouter Depondt[1], DVM; Thomas Fangman[2], DVM; David Nolan[2], DVM
[1]Huvepharma® NV Belgium; [2]Huvepharma® Inc. Peachtree City, Georgia

Introduction

One of the important features of porcine reproductive and respiratory syndrome virus (PRRSV) is its high genetic and anti-genetic diversity which is constantly evolving. Emerging strains of PRRS arose in the last years in Spain (Rosalia PRRSV-1 strain) and in the US (PRRSV L1C variant of RFLP 1-4-4 & PRRSV L1C 1-2-4). These strains are highly virulent and disseminate quickly in corresponding pig sectors.

First cases of PRRS Rosalia infections were detected in Spain in 2020. The strains circulation in Spain has undergone several recombination events with other PRRSV-1 isolates, which resulted most likely in enhanced virulence.[1] Rosalia PRRSV-1 isolates are currently present in numerous Spanish pig farms. Characteristic of the highly virulent Rosalia strain is that it can affect adult animals and piglets equally.

Rosalia has considerable impact in affected farms, regardless of whether they were previously PRRSV positive and were implementing a PRRSV vaccination program.[1]

In the case of sows: Fever, inflammation and discomfort are caused by Rosalia followed by significant reproductive problems (fetal death, high abortion rates, stillbirths, sow mortality). In nurseries viremic piglets affected by Rosalia show increased respiratory infections, reduced performance, decreased feed conversion rates and increased mortality rates.

Commercial modified live or inactivated vaccines currently available do not guarantee complete protection due to cross-reactivity between vaccine strains and field strains.

It is well known that pulmonary alveolar macrophages (PAMs) are the primary target of PRRSV infection and location of virus replication. It has been demonstrated that tilmicosin exhibits PRRSV-antiviral effects when concentrated in the pulmonary alveolar macrophages.[2] Rapid uptake of tilmicosin throughout the respiratory tract and highest intracellular accumulation in comparison to other macrolide antibiotics in lung macrophages, is the basis for its blocking effect of PRRSV replication and the high antibacterial effect of Tilmovet®.[3-8]

A reduction in clinical signs and performance improvements were shown in PRRS-infected pigs when medicated with tilmicosin.[9] In most of these studies animals were experimentally infected with PRRSV and medicated with tilmicosin via feed.[10-12] Published study results of one field study conducted in Spain in which Tilmovet® AC was administered via drinking, show successful control of PRRS Rosalia infection in naturally infected nursery pigs.[13]

The purpose of this paper is to describe a clinical case of Rosalia PRRSV-1 infected weaned pigs where Tilmovet® Premix (tilmicosin phosphate) was administered via feed to control PRRSV infection and to present data on viremia, antibody titers, clinical and production parameters and severity of PRRS infection.

Materials and methods

The field study was conducted in a transition farm in Spain which was affected by a Rosalia PRRS infection outbreak two months before study onset. Beside PRRS infection the animals on the farm were also affected by *Glaeserella parasuis, Mycoplasma hyopneumoniae* and *Escherichia coli* post weaning. 800 animals divided into two groups at arrival. Group 1: Tilmovet® Premix treatment 21 consecutive days at 4, 5 and 6 weeks of age (dose 16mg/kg bw). The clinical effect of Tilmovet® Premix was compared to a non-medicated control group (Table 1).

Parameters assessed: blood samples for assessment of viremia (qPCR) and PRRSV antibodies (ELISA); PRRSV clinical signs (nasal discharge, cough, behavior, body condition), DWG, FCR, health status, mortality rate. ELISA cut-off value: S/P ratio ≥ 0.4 = positive for antibody against PRRSV. On the Spanish farm diverse bacterial respiratory pathogens (*G parasuis, M hyopneumoniae*) were present during the study.

Results and discussion

PCR blood testing at study start showed a high-level of PRRSV infection. The percentage of pigs PCR-positive at the start of the study (3 weeks of age) was 51% in the control group and 46% in the Tilmovet® treatment group.

For the evaluation of the effect of Tilmovet® Premix medication three subgroups were created according to the piglet's qPCR results (ct value) at the enrolment of the study. The three subgroups were composed of those pigs that were negative to qPCR (ct value > 40), those that had low viral load at the entrance (ct value between 30-40) and those animals with high viremia on the day of the enrolment (ct value < 30).

In PCR-negative pigs at the study start (3 weeks of age) a positive effect of Tilmovet® Premix medication was determined. After the infection peak at 6 weeks of age the PRRS infection was reduced by 26% (week 8) and by 23% (week 10) as shown in Figure 1. In non-medicated pigs the percentage of positive pigs increased by 8% on week 8 and by 5 % on week 10.

PCR low-level positive pigs at the beginning of the study showed in the Tilmovet® medicated group a higher infection reduction by 30% (6 weeks of age) and by 40% at 8 weeks of age vs the control group, in which the viremia decreased by 15% on age week 6 and by 8% on age week 8. The PCR results are shown in Figure 2.

In high-level PCR-positive pigs at the beginning of the study, a higher infection reduction by 18% at 6 weeks of age and 40% at 8 weeks of age in the Tilmovet® group vs the control group with 8% (week 6) and 29% (week 8) was determined (see Figure 3).

Rosalia PRRS infection medication improved the survival rate. As a consequence, as shown in Figure 4, a higher survival rate in the Tilmovet group (82.0%) in comparison to the control group (77.5%) over the whole study period was determined.

Beside PRRSV infection the pigs in the treatment and control group were affected in parallel by bacterial infections. Based on local veterinarian decisions, additional curative treatments were required for better infection control. Primarily based on respiratory infections a higher number of concomitant treatments were needed in non-medicated pigs in comparison to Tilmovet®-medicated pigs as shown in Table 2.

A higher daily weight gain was determined in pigs with Tilmovet® Premix medication in comparison to the control group (see Table 3).

Table 1: Description of treatment groups in the Rosalia study in Spain

Treatment groups	Dosage (mg/kg bw)	Treatment duration	Age medication period
Group 1 (control)	---	0 days	No medication
Group 2 (Tilmovet® Premix)	16mg/kg bw	21 days	4, 5, 6 weeks (start day 0 to day 21)

Figure 1: Evaluation of PRRS viral load of negative pigs at study start, from the beginning to the end of the experimental trial.

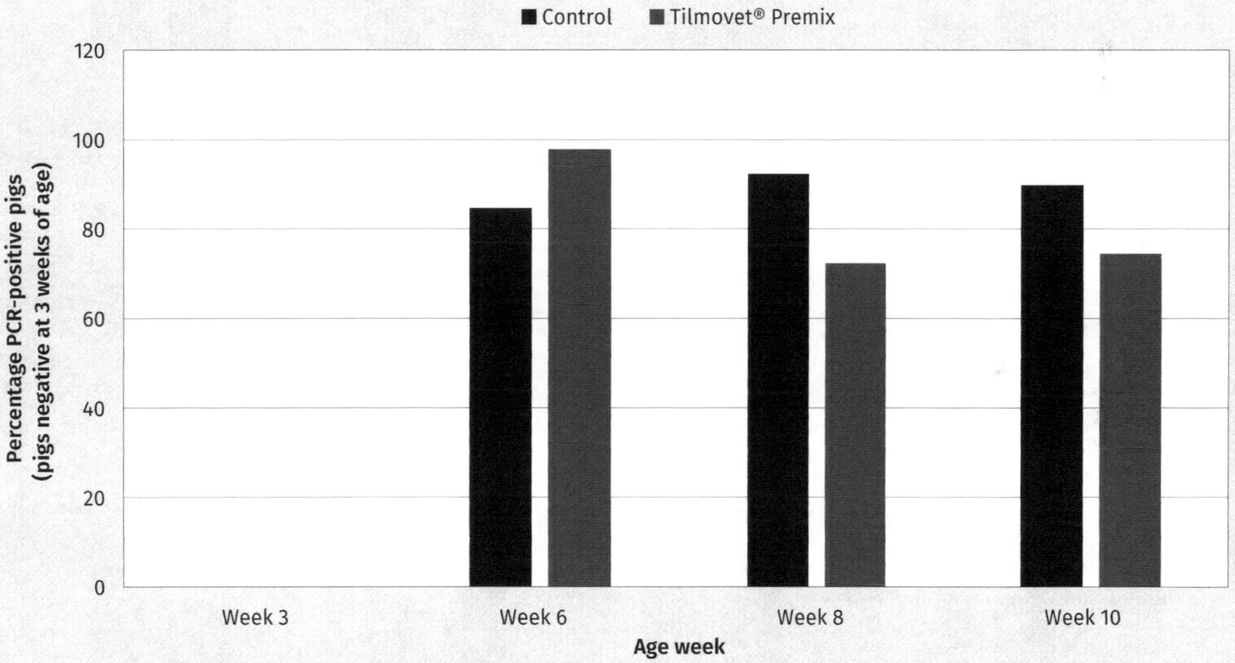

Figure 2: Evaluation of PRRS viral load of low-level PCR-positive pigs at study start, from the beginning to the end of the experimental trial.

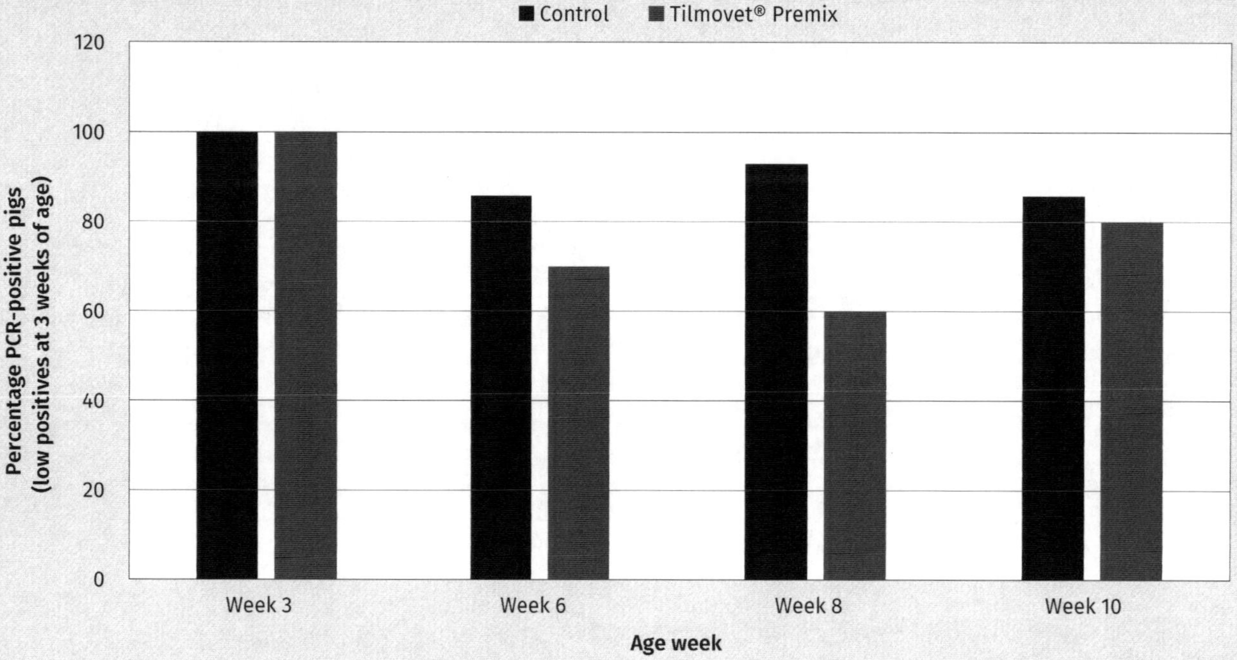

Figure 3: Evaluation of PRRS viral load of high-level PCR-positive pigs at study start, from the beginning to the end of the experimental trial.

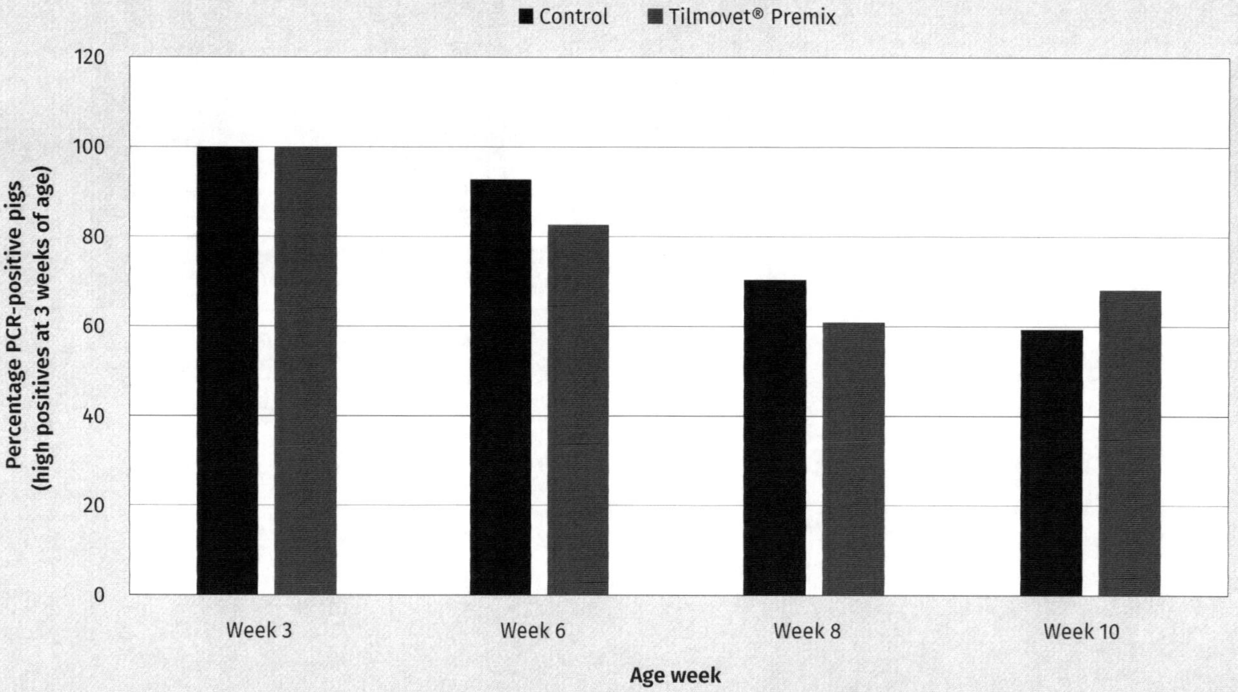

Figure 4: Survival rate in both treatment groups during the study.

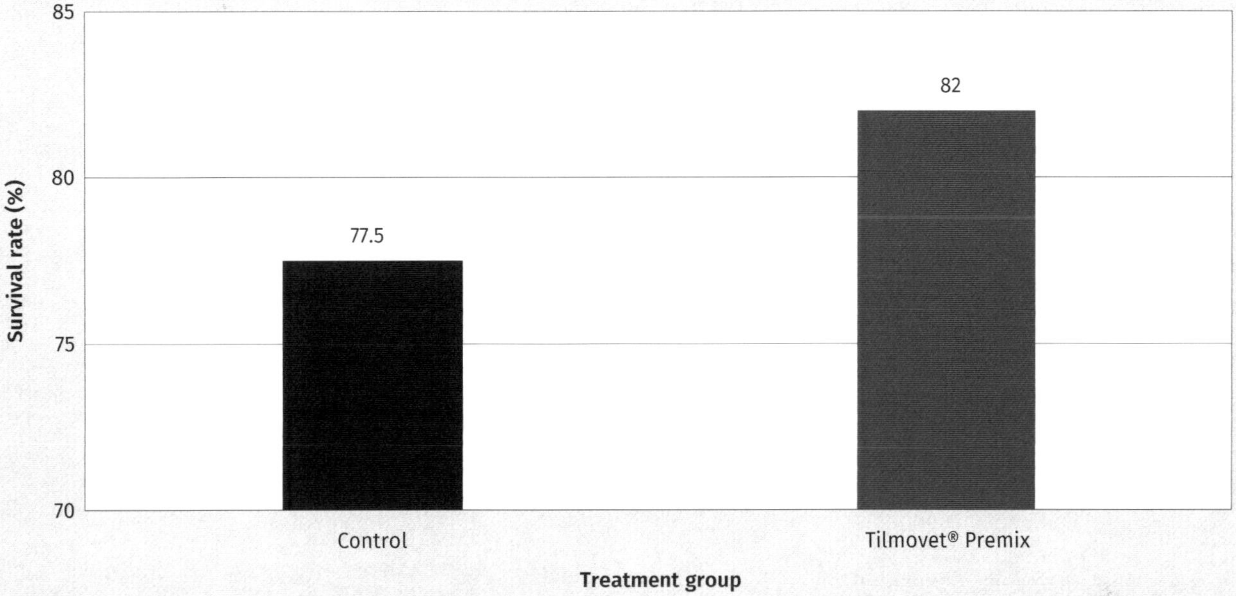

Table 2: Number of concomitant treatments during the study in the treatment groups

Treatment group	Concomitant treatments received
Control	115 times
Tilmovet® Premix	97 times

Table 3: Body weight and average daily gain (ADG) of pigs in the two treatment groups

Age (weeks)	Control (kg)	Tilmovet® Premix (kg)
3	5.30	5.35
12	17.97	18.73
ADG (kg)	0.270	0.280

Conclusion

Study results were generated in a Spanish farm being naturally affected by Rosalia PRRSV-1 infection. A high PRRS infection level was determined in the farm at study start. Secondary respiratory pathogens were present during the study.

The study results are showing the positive clinical effect of the Tilmovet® Premix medication program in pigs naturally infected by the highly pathogenic Rosalia PRRS variant at high PRRS infection levels at study start.

Following the viremia peak, the positive effect of Tilmovet® Premix medication led to a reduction of PRRSV infection, virus load and spreading and viremia duration.

As shown in different investigations, there is high correlation between qPCR results and the detection of actively replicating PRRS virus.[14,15] In the study verified by the blood qPCR results, a PRRS viremia decrease and depression of PRRS virus replication in the Tilmovet®-medicated group was determined. The PRRS viremia decrease and duration and the depression effect on PRRS virus replication at Tilmovet® Premix medication was identified to be higher in PCR-negative pigs and low-level PCR-positive and viremic pigs in comparison to high level PCR-positive pigs.

Based on these results it can be concluded that it is useful to start with Tilmovet® Premix medication in PRRSV viremic herds early before all pigs have become viremic.

The PRRS infection suppression and virus spread reduction in the Tilmovet® medicated group resulted in higher survival rates. The relatively low survival rates found in the study in both treatment groups reflect the aggressive potential of the Rosalia variant, which was affecting the farm the first time two months before the study started.

Lower level on concomitant treatments required in the Tilmovet® treatment group confirm the positive antibacterial effect provided by Tilmovet® against the bacterial pathogens (*G parasuis, M hyopneumoniae*) present during the study on the farm.

Better zootechnical results in the case of daily weight gain in medicated pigs at the end of the study are the result of the antiviral and antibacterial effect of Tilmovet® Premix during the study.

The use of Tilmovet® Premix based on the study results can be considered in infection situations caused by highly pathogenic PRRSV strains post-weaning as an effective strategy to stabilize and control PRRSV-viremia situations in the presence of porcine respiratory disease complex bacterial pathogens in weaned pigs.

References

1. Martin-Valls G.E. et al. (2023). Introduction of a PRRSV-1 strain of increased virulence in a pig production structure in Spain: virus evolution and impact on production. *Porcine Health Management*, 9:1; **https://doi.org/10.1186/s40813-022-00298-3**

2. Du, Y. *et al.* (2011). Antiviral activity of tilmicosin for Type 1 and Type 2 Porcine Reproductive and Respiratory Syndrome Virus in cultured Porcine Alveolar Macrophages. J. Antivir. *Antiretrovir.* Volume 3, pp.28-33. **doi :10.4172/jaa.1000031**

3. Blais, J. and S. Chamberland (1994). *Proc. 13th IPVS Congress Bangkok*, p.331.

4. Hawser, S. (2022). European Collection of Veterinary Pathogens (VetPath V) - minimal inhibitory concentration (MIC) testing, *IHMA Study Report* 3638 September 2022.

5. Pridmore, A. (2018). European Collection of *Mycoplasma* Pathogens (MycoPath II): collection, storage, identification and MIC determination for *Mycoplasma* strains isolated from cattle, pigs and poultry. *Don Whitley Scientific Study Report* 015/16.

6. Scorneaux, B., Shryock, T. (1998). Intracellular accumulation, subcellular distribution and efflux of tilmicosin in swine phagocytes. *J. Vet. Pharmacol. Ther.*21(4): pp.257-268.

7. Stoker, J. et al. (1996). Conc. Of tilmicosin in pig serum and respiratory tissue following administration via feed at level of 400g/tonne. *Proc. 14th IPVS Congress Bologna*, p.656.

8. Stuart, A.D. et al (2008). Tylvalosin a macrolide antibiotic inhibits the in-vitro replication of European and American PRRSV. *The Pig Journal*, 61, 42-48.

9. Depondt, W. et al. (2014). Stabilizing PRRSV with Tilmovet®. *Proc. 6th ESPHM Congress Sorrento*, p.163.

10. Benfield, D.A. *et al.* (2002). An evaluation of the effects of tilmicosin in feed of nursery pigs inoculated with porcine reproductive and respiratory syndrome virus. *Proc. American Association of Swine Veterinarians*, pp.87-91.

11. Lehe, K. *et al.* (2003). Case study: effect of Pulmotil® on mortality and growth in nursery pigs with clinical PRDC initiated by PRRSV. *Proc. American Association of Swine Veterinarians*, pp.129-130.

12. Molitor, T.W. *et al.* (2001). Tilmicosin affects Porcine Reproductive and Respiratory Syndrome Virus replication. *Proceedings Leman Conference*, 28, 31-32.

13. Klein, U. *et al.* (2023). Efficacy of Tilmovet® AC for PRRS control by water medication in weaned pigs. *Proc. 54th American Association of Swine Veterinarians*, pp.166-170.

14. Christopher-Hennings, J. *et al.* (1995). Detection of PRRSV by PCR. *J. Clin. Micro*; 33, 1730-1734.

15. Christopher-Hennings, J. *et al.* (2002). Porcine reproductive and respiratory syndrome (PRRS) diagnostics: Interpretation and limitations. *J. Swine Health Prod.*; Volume 10 (No.5), 213-218.

https://doi.org/10.54846/am2025/66

Vacasan®, Huvepharma's tulathromycin injection, administered to sows pre-farrowing to reduce sow and piglet mortality

P. Schneider[1], DVM; T. Fangman[2], DVM, MS, DABVP; D. Nolan[2] DVM; J. Harshman[2] PhD
[1]Innovative Agriculture Solutions LLC, Ames, Iowa; [2]Huvepharma, Peachtree City, Georgia

Introduction

This sow study was modeled after an NIH funded study[1] evaluating women over a 40-day postpartum period. In the NIH funded study it was determined that a single-dose of Azithromycin given during labor significantly reduced maternal sepsis and death in women from low- and middle-income countries. Tulathromycin (Vacasan®) is derived from azithromycin (Azalides a class of macrolide antibiotics).

Methods and results

This sow study was conducted on a PRRS-free, *Mycoplasma hyopneumoniae*-free farm with below industry average sow mortality. The study was designed to demonstrate a reduction in sow death rate and an expected improvement in suckling pig survivability when 1367 farrowing sows received Vacasan® (treatment) and 1377 farrowing sows did not receive Vacasan® at the time of farrowing (control). Parity was balanced between control and treatment 1.97 vs 1.96 (P = .959). Sows receiving Vacasan® were treated with 5mL IM (1.1 mL of 100mg/mL per CWT) of Vacasan® 1-day prefarrowing and monitored for 30 days after farrowing. Treatment groups were randomly assigned across farrowing groups for each week. However, weaning days across each farrowing group resulted in a statistically increased (Table 1) number of lactation days (P = .014) for the Vacasan® treated sows (21.83 days) compared to non-treated sows (21.36 days). Additional treatments were recorded for each group of sows as per normal farm protocol and were equal. Neither group of offspring received any treatment in the study. Treating sows with Vacasan® provided a significant reduction in the percentage of sow mortality in the 30 days post-partum 2.5% vs 1.4% (P = .03) and resulted in an improvement in pigs weaned per litter 11.8 vs 11.4 (P = .001). No differences were detected in total born, born alive, stillborns, or mummies at farrowing. The means and P-values are reported in Table 1. Sow mortality values were reported as percentages (humanely euthanized and dead) and does not include removals (culls).

Discussion

The longer lactation period observed in the treated sows did not increase the sow mortality in the Vacasan® treated group. In addition, Vacasan® treated sows weaned more pigs. This study suggest administration of Vacasan® at the time of farrowing is a viable solution to reduce sow mortality and improve pigs weaned per sow.

Reference

1. Tita ATN, Carlo WA, *et al.* Azithromycin to prevent sepsis or death in women planning a vaginal birth. *New England Journal of Medicine 2023.* **DOI: 10.1056/NEJMoa2212111**

Table 1: Summary of litter performance and sow mortality for sows and piglets receiving Vacasan® 1-day prior to farrowing compared to control sows not receiving Vacasan® at time of farrowing

Items	Control	Vacasan®[1]	P-value
Number of sows	1,377	1,367	-
Total born	14.78	14.66	.435
Born alive	14.02	13.88	.349
Stillborn	0.55	0.58	.465
Mummies	0.22	0.20	.492
Pigs weaned	11.41	11.83	.001
Lactation days	21.36	21.83	.014
Sow mortality, %	2.54	1.39	.030

[1] Vacasan® (Huvepharma, Peachtree City, Georgia)

https://doi.org/10.54846/am2025/67

A novel feed mitigant (FeedARMOR) in combination with hypochlorous water treatment shows additional benefit through reduced rates of sow prolapses

F. Sandberg, PhD; K. Soltwedel, MS, PhD; R. Fischer, MS, PhD; W. Kwon, MS, PhD
Furst-McNess Company, Rockford, Illinois

Introduction

The importance of whole farm biosecurity to reduce the probability of entry of pathogens onto a farm, and especially sow farms has been well established, and the significant role of feed mitigants well determined.[1] Ultimately, disease flows from the top of the production pyramid to the bottom, and therefore protecting boar studs and sow farms has become adopted across the industry, yet many producers still don't use mitigation. A key objection to implementation is the additional production cost of using what is 'insurance' without any direct observable benefit. The cost per sow per year ranges from lowest $7 to $8 to highest $20 per sow. Furthermore, pathogens can enter farms from animals, people, feed, water and air, and without an integrated approach to focus on all of those key risk areas farms are left exposed. As we discern between which feed mitigants to use, we need to consider three things: **1)** Efficacy as a mitigant **2)** Safety to mill staff and feed milling equipment **3)** Safety to animal and animal performance and **4)** Net cost per sow to the producer. For example, formaldehyde has been a lower cost option when considered as the product alone (cost per ton) but when correcting for impacts on milling processes, as well deterioration of amino acids and vitamins in the feed, and, documented negative impact on animal performance[2] the actual cost per sow per year is more than simply the cost of the product in the feed.

As we develop whole farm bio-security programs that address animals, people, feed, water and air as potential vectors for disease to enter farms we need to describe the OVERALL impact of any mitigant and/or intervention on the system to determine the cost of bio-security in an objective apples to apples comparison as opposed to simply looking at one dimension of the cost matrix (cost per ton). Overcoming this hurdle of communication and analysis of real economics could potentially support greater adoption of whole farm biosecurity programs.

A key indirect benefit from using enhanced biosecurity is reduced sow mortality driven by the normal function of organic acids, medium chain fatty acids, essential oils. Sow mortality occurs from many causes, but sow prolapses (POP; POP/1000 sows/week) have continued to be a significant issue in the industry. Dr Jason Ross as part an extensive survey conducted by Iowa State University across 104 farms, and 385,000 sows reviewed this critical industry issue, summarized sow mortality, and found that 21% of mortality was due to POP.[3] Furthermore, Dr Ross determined that there was no impact of sow parity on POP so any successful interventions can benefit the whole farm. It was determined that water treatment reduced POP from 0.52 to 0.28 ($P < 0.01$) with no difference between chlorine or peroxide-based water treatments. The worst POP was untreated well water (0.53 POP), followed by rural water (0.45 POP), then treated well (0.31 POP) and the lowest was treated pond (POP 0.22). Another key factor identified from this large analysis was that a sow body condition score (BCS) of 1 (thin sows) was 6x more likely to prolapse than BCS of 3 (optimal condition), with BCS of 2 (slightly thin) intermediate. Finally, it was developed that a perineal score of 3 increased prolapse rate from 1% for 1-2 perineal score, to 7.2% ($P < .01$). Other factors had slight impacts, but these three were marked. Thus, if we can a) address sow body condition b) eliminate sows from the herd with a perineal score of > 2 and c) deploy a water treatment system we may be able to significantly impact POP rate.

We report here the findings from six farms with a prolonged history of POP that pursued different levels of attention to addressing sow body condition, and water treatment as well as implementing a feed mitigant (FeedARMOR) and the impact on POP. This paper attempts to summarize the findings, explore causality, and present a systematic 4 step approach for implementing a whole farm biosecurity program with the additional benefit of potentially significant reductions in POP that builds on the extensive work by Ross.[3]

Methods

FeedARMOR has been implemented on numerous farms, and based on field observations and customer feedback 6 farms were identified from two systems with high quality sow records and a retrospective analysis was conducted. Data was evaluated for before and after implementing the feed mitigant. The farms were all of one genetic, with a mix of commercial and genetic multiplier farms. The farms ranged from low POP rate to high POP rate as defined by Ross.[3] There was also a notable difference in water type between the two systems. An overview of the characteristics of the 6 farms are shown in Table 1. Performance records were analyzed to determine impacts on POP as well as sow mortality in general.

Results

In Table 2 is a summary of the change of POP before and after implementation of the feed mitigant. There were different responses for the two systems. In System 1 that utilized hard well water modest reductions were observed (14% numerical reduction, P = non-significant). Whereas in System 2 with pond and river water, some large reductions in POP (70% reduction, $P < .05$). In System 2 commercial Farm 4 observed a reduction from 1.99 to 0.63 with this farm having routinely stayed at a POP level of around 2 for more than 2 years with many interventions attempted over that time frame. Figure 1A, B displays the time series change in POP and sow mortality for Farm 4, showing a reduction of 8% in sow mortality.

Discussion

The results presented here have shown different responses in sow POP with and without a feed mitigant (FeedARMOR). In several farms, large reductions in sow POP rate (> 50%) was observed and appeared independent of the base POP level. As shown previously by Ross[3] there were interactions with water source with the impact of FeedARMOR and hypochlorous acid (HOCL) water treatment on POP. The response being much more dramatic in softer types of water, as compared to hard well water. It is a key question to further resolve the issue of POP to understand why there is such a difference between hard and soft water on POP, and how external factors such as water treatment or feed ingredients may impact POP. Farm 4 pursued a long-term process to impact sow POP including using calipers to improve sow condition. The implementation of water treatment alone did not resolve POP, yet when combined with the implementation of a feed mitigant, POP significantly changed. This was then observed at other farms, including farms where water treatment was not yet in place.

Based on this approach, the steps followed that demonstrated the greatest improvement in POP across these 6 farms was: 1) strict management of sow condition; 2) implementation of a water sanitation program; 3) the use of soft sources of water and 4) implementation of a feed mitigant (FeedARMOR). A key question from these observations is whether implementing a water softener, prior to HOCL treatment, in conjunction with the feed mitigant, could generate the same level of response as that observed here on sow farms with naturally soft water with harder sources of water.

From a financial perspective, the cost of most mitigants is around $10 per sow per year which translates to $0.36 per pig weaned (assuming PSY = 28). It has been modelled extensively that 1% in sow mortality increases cost of production of weaned pigs by $0.40 (range $0.30-$0.50), or that 1 dead sow has a cash cost to the farm of $1000 ± $300 depending on when in its production life mortality occurs. So, to neutralize the cost of a mitigant, a reduction in sow mortality of 1% would need to be observed. Water treatment systems range in cost from $2 per sow per year up to as high as $15 per sow per year. Given the importance of bio-security for sow farms, and the subsequent benefits on downstream productivity (not factored here as a financial gain) investing $0.45 to $0.60 per weaned pig to implement feed mitigation and water sanitation only requires a 1 to 1.5% reduction in sow mortality (again without assigning any value to down stream improved performance). This is a different perspective to the cost of on farm biosecurity and greater attention should be put on direct secondary benefits from using feed mitigation and/or water sanitation, such as reduced POP shown here to support greater adoption of feed and water biosecurity measures.

Table 1: Characteristics of 6 farms evaluated for impact of feed mitigant (FeedARMOR) on before and after pelvic organ prolapse in sows, where all farms were of the same genetics

System	Sow farm name	Number of sows	Water source	Water treatment and duration	Feed mill
1	Farm 1 commercial	1,500			Feed Mill A
	Farm 2 commercial	1,800	Hard well water	Intermittent HOCL	Feed Mill B
	Farm 3 commercial	1,800			
2	Farm 4 commercial	3,200	Pond	HOCL	Feed Mill C
	Farm 5 genetic multiplier	1,200	Well	None	
	Farm 6 genetic multiplier	4,000	River	None, then HOCL	

HOCL = hypochlorous acid

Table 2: Sow pelvic organ prolapse (POP) rate (all types combined) across 6 farms before and after implementing FeedARMOR as the feed mitigant

		Average POP (POP/1000 sows/week)		
System	Sow farm name	Before using FeedARMOR	After using FeedARMOR*	Reduction (%)
1	Farm 1 commercial	2.17	1.88	14%
	Farm 2 commercial	1.64	1.42	14%
	Farm 3 commercial	1.64	1.42	14%
2	Farm 4 commercial	1.99	0.63	69%
	Farm 5 genetic multiplier	0.21	0.09	56%
	Farm 6 genetic multiplier	0.26	0.04	85%

* Average sow prolapse rates were compared at 10 weeks for System 1 and 18 weeks for System 2 after the introduction of FeedARMOR.

Figure 1: The historical incidence of **A)** sow prolapses (POP) on a commercial farm that had a prolonged historical issue with POP before and after implementation of the FeedARMOR feed mitigant and **B)** the associated annualized sow mortality.

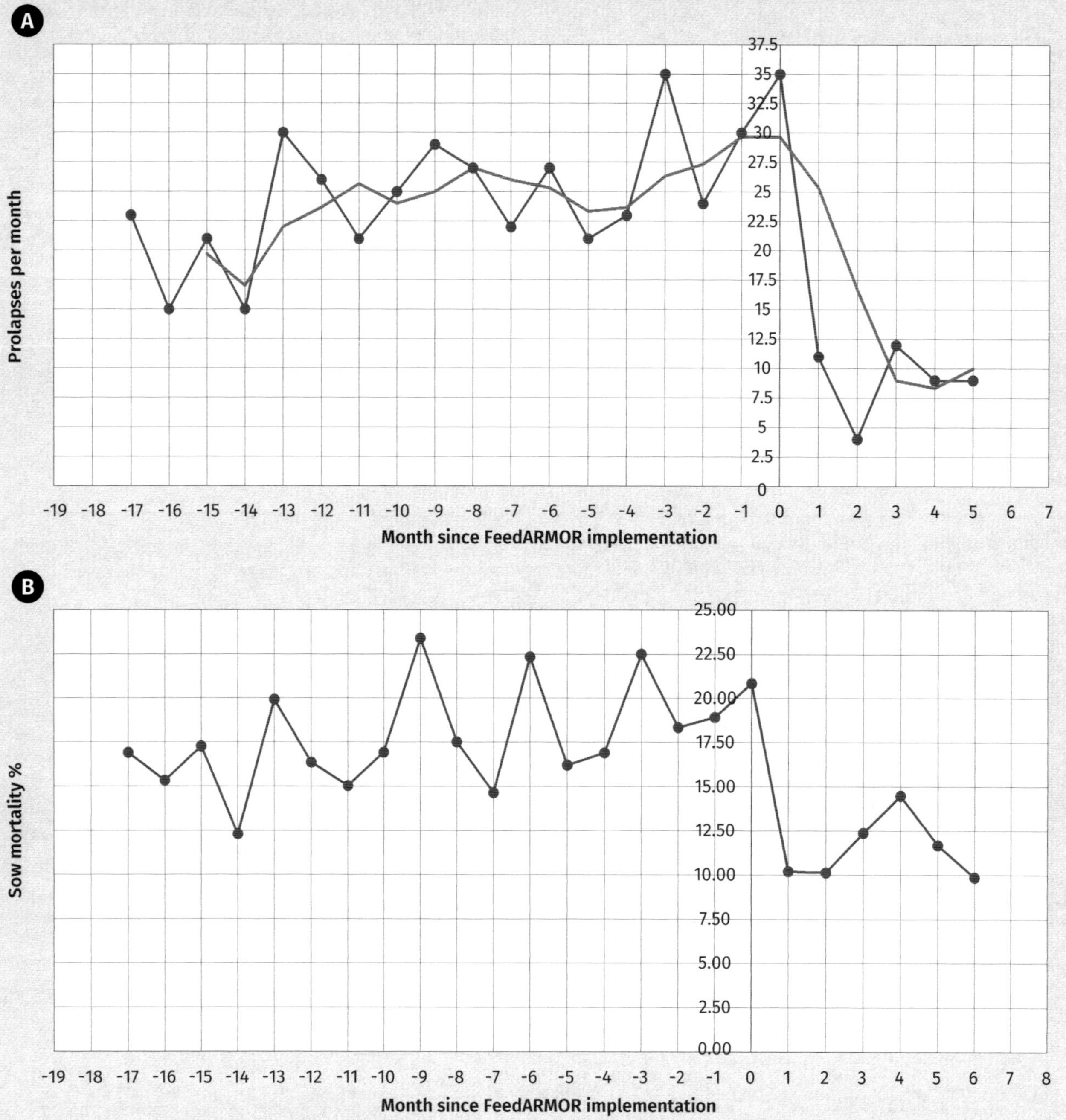

References

1. Dee *et al.* (2020) Transboundary and Emerging Disease: 00: 1-13.

2. Williams *et al.* (2018) Nature Scientific Reports: 8:8164.

3. Ross, J. (2019) Pork Checkoff Research Report: 'Identification of putative factors contributing to pelvic organ prolapse in sows'. NPB #17-224.

https://doi.org/10.54846/am2025/68

Correlation of intestinal lesions due to ileitis and fecal shedding of *Lawsonia intracellularis* in finishing pigs treated with either water-soluble Aivlosin or lincomycin hydrochloride after an ileitis challenge

N. Winkelman[1], DVM; H. Jung[2]; A. Mueller[1], DVM
[1]Swine Services Unlimited, Inc., Rice, Minnesota; [2]The Ohio State University, Colombus, Ohio

Introduction

Lawsonia intracellularis (Li) is an obligate intracellular bacterium that can cause porcine proliferative enteropathy (PPE), or ileitis.[1] Li is transmitted by the fecal-oral route, while the defining lesion for PPE is the thickening of the intestinal mucosa due to crypt epithelial cell proliferation[1]. Ileitis is widespread throughout commercial swine finishing operations and carries a significant economic burden, costing swine producers roughly $6 to $17 per head.[2] As intestinal lesions become more severe, so does the production impact on the pigs affected. This emphasizes timely diagnosis and intervention with an antibiotic to reduce the severity of intestinal lesions and fecal shedding throughout the population. The objective of this study was to correlate intestinal lesions due to ileitis and fecal shedding of Li in finishing pigs that were treated with either water-soluble Aivlosin or lincomycin hydrochloride after an experimental ileitis challenge.

Materials and methods

One hundred and seventy 10-week-old commercial pigs were placed into a finisher in central Minnesota on Day -5 (D-5). Pigs were randomized by weight, gender, and litter and allocated to one of three treatment groups, per pen. Pigs were confirmed negative for Li at placement into the finisher by IPMA antibody testing and for porcine reproductive and respiratory syndrome virus (PRRSV). On D0, all animals were orally inoculated with Li mucosal homogenate (1.85×10^{10}). Water treatments began on D13, when > 10% of the pigs were observed by the attending veterinarian to have clinical symptoms consistent with ileitis. Starting on D13, pens in the control group received no medication (n = 50), Aivlosin (ECO Animal Health; 62.5% w/w tylvalosin as tylvalosin tartrate) at 50 ppm for five consecutive days[3] (n = 60), or lincomycin hydrochloride soluble powder (Huvepharma) at 66 ppm for five consecutive days[4] (n = 60) (Table 1). The study was terminated on D35, when all animals were euthanized.

Fecal samples were collected from the same 30 pigs in the control group and 36 pigs in each of the treated groups at D7, D13, D21, and D35. These samples were submitted to University of MN VDL and tested individually for Li polymerase chain reaction (PCR). Gross intestinal PPE lesion severity index (SI) scores were calculated for each individual pig on D35. SI scores were the summation of multiplied PPE lesion score (0-3) by length in centimeters (cm) for each section of the jejunum, ileum, cecum, and colon. Pearson correlation and linear regression models were built to estimate the strength of the correlation and impact of the treatments on SI, fecal PCR ct, and average daily gain.

Results

PPE mortality

Four pigs died from ileitis prior to the end of the trial, indicating a significant Li challenge. Three pigs died in the control group, and 1 pig died from acute proliferative hemorrhagic enteropathy on Day 22 in the lincomycin group. No PPE mortalities occurred in the Aivlosin treated pigs.

Gross PPE lesion scores

Average SI scores from pigs within each treatment group were analyzed using the Wilcoxon rank sum test (Table 2). Pigs treated with Aivlosin, on average, had significantly less severe SI scores when compared to the pigs from control pens ($P = .006$) and tended to have significantly less severe SI scores when compared to the pigs treated with lincomycin ($P = .083$).

Figure 1 shows the PPE gross lesion severity index distribution on Day 35 post-challenge. Pigs were categorized by SI scores into negative (no lesions), mild (1-20cm), moderate (21-99 cm), and severe lesion (> 100 cm) SI scores. The Aivlosin treated pigs had the highest percentage of negative and mild category PPE pigs and the lowest percent moderate and severe pigs compared to both lincomycin treated pigs and the challenged controls.

Fecal PCR shedding

Figure 2 shows the mean PCR ct (cycle threshold) values (from a 40-cycle run) by treatment on Days 7, 13, 21 and 35 after the Li challenge. Interestingly, no shedding had occurred by D7 post-challenge, indicating a very long incubation period even in heavily challenged pigs. However, by D13, when over 10% of the pigs showed clinical diarrhea and/or gauntness, all the pigs were shedding high levels of Li with a mean PCR ct value of 26.0 to 26.6.

Although there were no statistically significant differences in fecal PCR shedding on D21, the Aivlosin treatment did reduce fecal shedding numerically with a 28.2 ct value, compared to 26.6 ct in the controls, and 25.7 in the lincomycin groups. Based on U of MN VDL Li quantitative standard curves, this is 8.3 times fewer Li bacteria per gram of feces (1.14×10^5 vs 9.4×10^5 qPCR) in the Aivlosin pigs vs lincomycin pigs. By D35, natural cell-mediated immunity reduced mean ct values to 36-37. However, all fecal samples tested (n = 102) were still shedding Li at the end of the study.

Table 1: Trial design – 3 treatments (170 pigs total)

Treatment	Treatment duration	# Pens	# Pigs/pen	*Lawsonia* challenge (10^9)	Trial duration
1-Challenge control	5 days	10	5 (50 pigs)	Yes	35 days
2-Aivlosin®-50 ppm	5 days	12	5 (60 pigs)	Yes	35 days
3-Lincomycin-3.8 mg per lb of bodyweight	5 days	12	5 (60 pigs)	Yes	35 days
		Total 34	Total 170		

Table 2: Mean PPE gross lesion severity index summary

Severity index scores	Mean
Control	75.26[a]
Aivlosin	25.53[b]
Lincomycin hydrochloride	33.41[ab]

Statistical significance ($P < .05$) is noted by different superscripts within a column

PPE = porcine proliferative enteropathy

Figure 1: PPE gross lesion SI distribution in percentage per treatment.

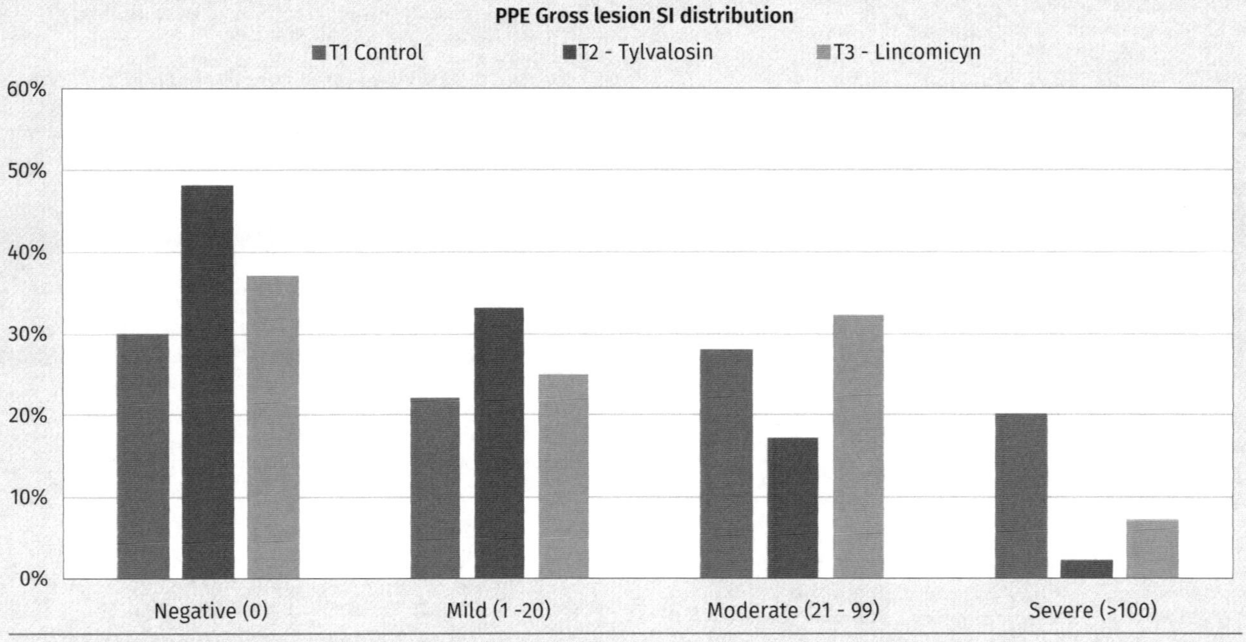

PPE = porcine proliferative enteritis; SI = severity index

Figure 2: Mean PCR ct values by day and treatment.

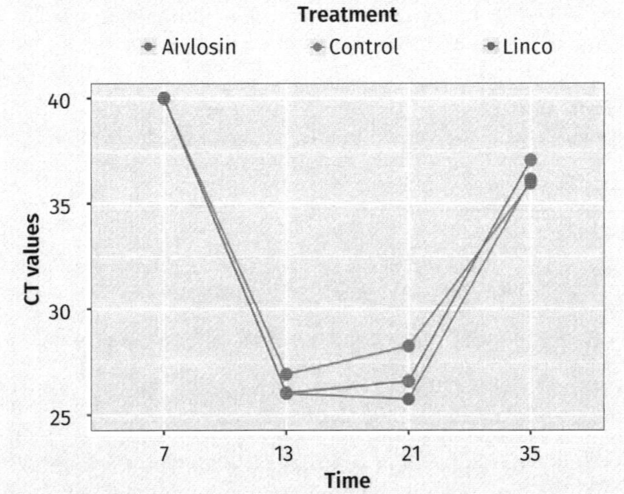

Figure 3: All treatments - fecal PCR on D13 vs PPE severity index.

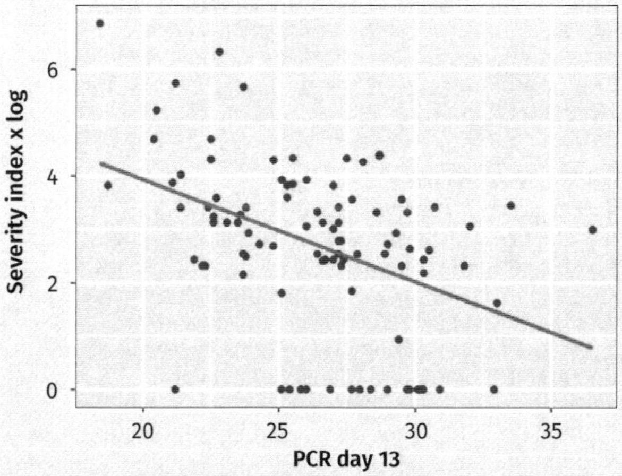

PPE = porcine proliferative enteritis

Correlations/scatter plots

Figure 3 shows a scatter plot with a regression line with SI log scale compared to fecal Li PCRs on Day 13 post-challenge with all treatments. There is a moderate correlation of -0.42 (P-value ≤ .001) between the PCR on Day 13 and the PPE lesion severity score. For each ct unit increase, there is a decrease on the log SI by -0.19.

Figure 4 shows all 3 treatment regression lines for the same analysis. There is no difference between treatments on D13 prior to treatment. When testing correlations of PCR shedding on D21 and D35 to PPE Severity index, there was no statistical significance. This is likely because SI scores peak around Day 21 post-challenge, so that by Day 35, both fecal shedding and SI had dramatically improved. The reason the study timeline went to D35 was to collect performance data beyond peak lesions.[5]

Figure 4: By treatment – fecal PCR on D13 vs PPE severity index.

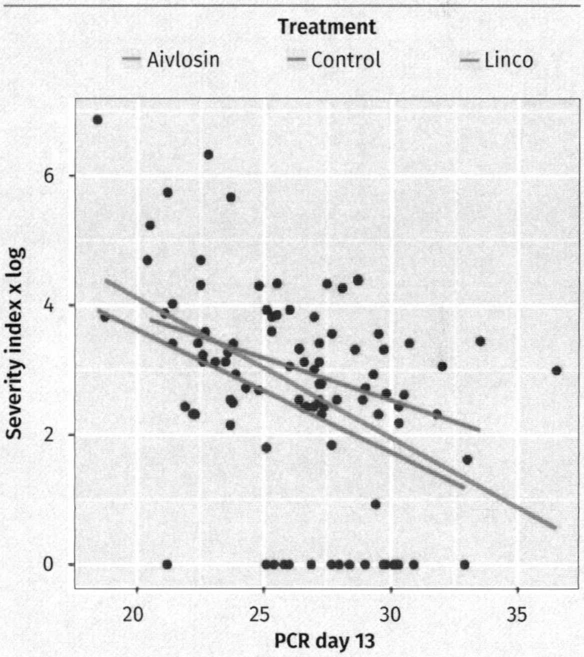

PPE = porcine proliferative enteritis

Discussion

When to water medicate

Lawsonia intracellularis PCR from fresh fecal samples and/or oral fluids are an important ante-mortem diagnostic tool both for diagnosing clinical ileitis and for routine monitoring for subclinical ileitis. The challenge for the swine practitioner is to be able to interpret Li PCR results and intervene accordingly with appropriate water medications to mitigate PPE intestinal lesions, performance (average daily gain [ADG], feed conversion ratio [FCR]) loss, and subsequent economic loss. Whenever known clinical signs of ileitis (diarrhea, fecal color, gauntness) occur, water medicating that group of pigs is warranted. Due to the long incubation time of ten or more days with Li, when clinical signs are present, much of the remaining population is also subclinically affected and should be treated with antibiotics. For example, in this study on Day 13 post-challenge, only about 10% of the pigs were affected clinically, yet the remaining 90% were subclinically affected with low ct values (high shedding) as shown in Figure 2.

In field outbreaks pigs are not infected on the same day as in challenge trials, so there may not be 100% clinically or subclinically affected pigs, but due to the rapid fecal to oral spread and the low infectious dose necessary to cause disease, there will be a high percentage of infected pigs that will warrant water medication. Therefore, water medication should be considered whenever clinical signs are present or PCR ct values are below a certain threshold, eg, < 30 ct.

Aivlosin vs lincomycin

This is the first study with direct comparison of two commonly used water medications for ileitis in a Li challenge study. The trial demonstrated that both Aivlosin and lincomycin treated pigs showed improvements in mortality, clinical scores, pathological PPE lesions, and performance parameters over nontreated controls. Aivlosin pigs showed significant improvements in ADG, FCR, and feed intake (Table 3), and a numerical advantage over the lincomycin pigs. Lincomycin treated pigs also showed significant advantage in ADG, and feed intake over the controls.[5]

Therefore, this trial shows Aivlosin to be the preferred water medication compared to lincomycin in Li orally challenge grow-finish pigs.

Table 3: Performance (ADG, ADFI, FCR) summary D13 to D35

	ADG	FCR	ADFI
Control	1.75^{b}	2.69^{b}	3.27^{b}
Aivlosin	2.26^{a}	1.81^{a}	3.82^{a}
Lincomycin	2.07^{a}	2.16^{ab}	3.77^{a}

Statistical significance ($P < .05$) is noted by different superscripts within a column

ADG = average daily gain; ADFI = average daily feed intake; FCR = feed conversion ratio

Acknowledgment

The authors would like to thank Dr Gustavo Silva, at the Iowa State University College of Veterinary Medicine Diagnostic Laboratory for assisting on statistical analysis for this project.

References

1. Zimmerman, J., Karriker, L.,Ramirez, A., et al. 2019. *Diseases of Swine: Eleventh Edition*. Chapter 58, Proliferative Enteropathy. P 898 – 911. Published 2019 by John Wiley & Sons, Inc.

2. Holtkamp, D. 2019 Economic losses associated with ileitis. MSD Animal Health

3. Aivlosin water soluble granules label information. Pharmgate Animal Health, Wilmington, NC. https://pharmgate.com/usa/aivlosin-water-soluble/. Accessed 11/01/2024.

4. Lincomycin hydrochloride soluble powder label information. Huvepharma, Peachtree City, GA. https://www.huvepharma.us/product/lincomycin-hcl/. Accessed 11/01/2024.

5. Jung, H., Winkelman, N., Comparative efficacy of water soluble tylvalosin and lincomycin hydrochloride in finishing pigs administered a Lawsonia intracellularis (Li or ileitis) oral challenge p ? AASV Annual Meeting 2025

https://doi.org/10.54846/am2025/69

Responding to a sow farm PRRSV outbreak: Aivlosin® treatment, planned exposure, and sampling timeline

E. Kettelkamp, DVM
Swine Vet Center, Saint Peter, Minnesota

Introduction

Porcine reproductive and respiratory syndrome virus (PRRSV) remains a significant challenge for sow farms, causing substantial economic losses due to reproductive failure in sows and respiratory disease in piglets.[1,2] PRRSV infection also disrupts the innate immune response, complicating disease control and vaccination efforts.[1] Effective response strategies involve a combination of treatments, controlled virus exposure, and meticulous sampling to monitor the infection dynamics and evaluate intervention efficacy.[3] Tylvalosin, an antimicrobial agent, has been shown to reduce local and systemic proinflammatory cytokines after challenges with respiratory pathogens in sows and piglets.[4] Additionally, Tylvalosin has demonstrated efficacy in reducing bacterial pneumonia recovery from the lungs and may reduce virus shedding in piglets following transplacental PRRSV infection in sows.[4] This case describes applying Aivlosin® (62.5% w/w Tylvalosin as Tylvalosin Tartrate, Pharmgate Animal Health) treatment, planned exposure to the homologous virus, and a structured sampling timeline in response to a PRRSV outbreak in a commercial sow herd.

Materials and methods

The project was conducted on a 2,500-head commercial sow farm located in central Minnesota. The barn was built in an H-style, with two farrowing barns and two gestation barns connected by a central hallway and office. The farm was PRRSV, *Mycoplasma hyopneumoniae* and porcine epidemic diarrhea virus naïve. PRRSV was initially detected on April 10, 2024 (week 15), as PRRS 1-2-3 L1C.5 via real-time polymerase chain reaction (rt-PCR) testing of serum samples from sows showing clinical signs including fever, off-feed, and abortion. Clinical signs and production data, including off-feed sows, abortions, sow mortality, and piglet mortality, were recorded daily. All samples were sent to an accredited veterinary diagnostic laboratory for testing and sequencing for this project.

Whole-herd natural planned exposure was conducted 4- and 10-weeks post-outbreak (weeks 18 and 25), and live virus inoculation (LVI) was conducted 11 weeks post-outbreak (week 26). Homologous PRRSV LVI was performed using serum collected from clinically affected animals to synchronize immunity across the herd and promote stable virus circulation. Sows received intramuscular serum injections at a standardized dilution.

Aivlosin® was administered in the water to sows for 5 days at one pack per 44,000 lbs in weeks 17, 18, and 26 to align with the timing of the outbreak, first natural planned exposure and the LVI event. In addition, feed medications included Pennchlor (chlortetracycline, Pharmgate Animal Health) at 1800 grams per ton in the gestation diet for 14 days and Pulmotil® (tilmicosin, Elanco Animal Health) at 181 grams per ton in the lactation diet for 21 days. Medications were used strategically during key periods of the outbreak and LVI events to mitigate bacterial infections and support recovery efforts. The farm staff applied injectable spot treatments as needed for individual sows and piglets exhibiting fever, anorexia, abortions, or signs of respiratory disease.

To assess the prevalence of PRRSV in the herd after initial detection, 30 sows were identified for serial serum collection at three time points: after the outbreak, before natural planned exposure, and one week after natural planned exposure (weeks 16, 18, 19). Samples were tested for PRRS q-PCR. Two weeks after LVI (week 28), groups of six sequentially housed animals were randomly selected from 20 regions, blocked by location, within the gestation and breeding barns. These groups were sampled and tested for PRRS rt-PCR. After the feed drop, one dry Swiffer® sweeping cloth (trough swab) was used to collect saliva from the feed trough of each group of dams. Additionally, another trough swab was collected from the full length of each row in the barn. A modified tonsil-oral-scrubbing (TOSc) tool was created using a post-cervical artificial insemination rod covered with a 4"x4" piece of square cotton gauze held in place with a rubber band.[5] TOSc and blood samples were collected from all animals in the study and pooled by group.

The farm underwent planned construction of the breeding and gestation barns, requiring each barn to be emptied sequentially. Half of the sow herd was moved three weeks after LVI to an off-site gilt development unit (GDU) location. These sows moved back to the farm by breed group weekly from week 32 to week 37. In week 40, the adjacent barn of the sows were moved off-site and returned in week 44 to the sow farm. During their time off-site, sows were monitored weekly using oral fluids for PRRS rt-PCR testing. All pens of sows scheduled to ship were tested for PRRS rt-PCR using oral fluid samples within 24 hours of shipment to the sow farm. Upon return to the sow farm, all sows were placed into an isolation barn and tested for PRRS PCR at 2- and 7-days post-arrival before being released into the sow herd. Beginning week 41, processing fluids were collected on the farm weekly for PRRS PCR testing.

Results

During the outbreak period, clinical signs attributed to the PRRSV introduction included off-feed sows (300-350 total), daily abortions (78 total), and sow mortalities (67 total), along with significant piglet losses reaching up to 352 piglets per day (8294 total). By the time of LVI (week 26), there were no recorded abortions or notable piglet losses, and clinical stability was achieved. Performance trends before, during, and after the PRRS outbreak reflected the progression of recovery. Before the outbreak (weeks 14-16), farrowing rates averaged 91–94%, with average piglets weaned per sow reaching 15.3. During the outbreak (weeks 17-25), farrowing rates declined to as low as 84.1%, and liveborn piglets per litter decreased significantly, reaching a low of 5.8 per litter. Mummified fetuses accounted for up to 47.2% of total births, and average piglets weaned per sow dropped to 3.1 in the worst weeks. Performance metrics gradually improved after the herd interventions (weeks 26-42). Farrowing rates rebounded above 89%, liveborn piglets increased to 14.8 per litter, and mummified fetuses dropped to under 6%. Average piglets weaned per sow per year increased to over 28.9 by week 42, reflecting a return to productivity.

After diagnosis of acute PRRSV infection (week 16), the first serial sow serum samples resulted in 15/30 serum samples were q-PCR-positive, with viral loads ranging from 23,082,700,000 to 494,398,000,000 ORF 6 RNA copies/mL and corresponding infectious virus particles (IVP) per mL ranging 9052-193,881 IVP/mL. Only 1/30 sows tested ELISA positive. Before the first natural planned exposure (week 18), 29/30 serum samples were PRRSV q-PCR positive with ORF 6 RNA copies/mL ranging from 8,839 to 268,829,000 copies/mL, with corresponding IVP ranging from 0.0035 to 105.4231 IVP/mL. One week after natural planned exposure (week 19), 22/30 sows tested q-PCR positive with viral loads ranging from 8,325 to 18,056,900 copies/mL, with corresponding IVP ranging from 0.0033 to 7.0811 IVP/mL. The average IVP/mL week 18 before exposure was 11.43 in the breeding barn and 5.67 in the gestation barn respectively, with a total average of 8.55. After natural planned exposure, the average IVP/mL decreased to 1.16 in the breeding barn and 0.05 in the gestation barn, with a total average of 0.60, highlighting the reduction in viral activity across both locations.

The groups of six sows sampled during week 28, including trough swabs, TOSc samples, and serum pooled by group, all tested PRRS rt-PCR negative. At all testing time points off-site, pre- and post-shipment, sows tested negative by PRRS rt-PCR (week 32-44). All processing fluid samples have been PRRS rt-PCR negative weeks 41-45.

Discussion

Serial sow serum sampling during the outbreak, as well as before and after natural planned exposure, provided insight into viral loads (ORF 6 RNA copies/mL and infectious particles per mL) over time, demonstrating the impact of whole herd natural planned exposure and outbreak infection dynamics. Sows and piglets showed improvement in clinical signs within the first week of Aivlosin® administration. By week 28, LVI and natural planned exposure resulted in a synchronized viral infection across the sow herd. Measured viremia peaked within the first three weeks after PRRSV detection and subsequently declined to baseline levels by week 28, indicating the effective establishment of herd exposure to the virus. Despite multiple movements of sows off-site and within the sow farm due to construction events, sow shedding of PRRSV remained undetectable. Subsequently, piglet health stabilized in the flow within 15 weeks of the LVI event.

Combining Aivlosin® treatment with planned exposure provided a practical and effective response to PRRSV in a commercial sow herd. Aivlosin® mitigated bacterial complications, while planned exposure promoted herd immunity. The structured sampling timeline allowed for precise monitoring of viral dynamics and intervention outcomes. Future cases should explore optimizing exposure protocols and herd treatments to mitigate production losses and optimize herd immunity.

References

1. Zimmerman JJ, Karriker LA, Ramirez A, Schwartz KJ, Stevenson GW. Diseases of Swine. 10th ed. Wiley-Blackwell; 2012.

2. Holtkamp DJ, Yeske PE, Polson DD, et al. Economic impact of PRRS: 2012 update. J Swine Health Prod. 2013;21(2):72-84.

3. Dee SA, Torremorell M, Thompson B, et al. Planned exposure to PRRS virus in sow herds. Vet Clin North Am Food Anim Pract. 2005;21(1):111-128.

4. Rodriguez, A.L., Fowler, V.L., Huether, M. et al. (2023). Effects of a water-soluble formulation of tylvalosin on disease caused by porcine reproductive and respiratory syndrome virus alone in sows or in combination with *Mycoplasma hyopneumoniae* in piglets. BMC Vet Res 19, 31. **https://doi.org/10.1186/s12917-023-03571-x**

5. Peng Li, Silva APSP, Moraes DCA, Yeske P, Osemeke OH, Magalhães ES, De Sousa E Silva G, Linhares DCL. Comparison of a novel rapid sampling method to serum and tonsil scraping to detect PRRSV in acutely infected sows. Prev Vet Med. 2024 Feb;223:106082.

https://doi.org/10.54846/am2025/70

Tax planning considerations for veterinarians and veterinary practice owners

Andrew Kleis, AIF®

Introduction

As they say, there are only two undisputable, unavoidable things in life: Death and taxes. And, sadly, taxes will be the single largest expense - both from a corporate and personal standpoint - in the life of any veterinarian. In this presentation, we will explore a few areas which may provide useful tools for both the practicing veterinarian and veterinary practice owners.

Charitable gifting strategies

Charitable giving can be an effective tool to reduce tax liabilities while also supporting the causes you are passionate about and which may provide value and benefit to your business.

Charitable gifting strategies can be effective in any year to reduce your taxes - but may be especially valuable in years which you have an abnormally high income. Federal law allows you to deduct up to 60% of your adjusted gross income for cash contributions. Further limitations apply to other types of donations.

One potential solution - especially in high income years - is the utilization of a Donor Advised Fund (DAF). A DAF allows you to:

- Make a contribution that is deductible in the year of the donation.
- Continue to provide direction on the use/investment management of the funds over time (up to and exceeding your lifetime).
- Provide direction on the eventual gifting of those assets over time to qualified charitable organizations.

This is particularly useful if you need to deduct a high level of income, but don't intend to provide gifts of that level immediately. You can continue to grow the value of the assets while directing their contributions at a later date.

Another solution - especially powerful for use with organizations like the AASV Foundation - is the gift of highly appreciated stock. For example, let's say you were an early investor in a company like Nvidia which has grown rapidly in value over recent years. In this scenario, let's imagine a $10,000 investment that is now worth $100,000. A sale of the position would result in a capital gain of $90,000 which could have a negative impact on your tax rate.

Instead, you could donate all or a portion of the shares directly to a qualified charity like the AASV Foundation. This allows you to 1) avoid the taxes due on the sale of the stock; 2) deduct the value of the appreciated stock that you donate in the current year.

Finally, for those attendees who have reached the Required Minimum Distribution age for their retirement accounts, a Qualified Charitable Deduction directly from an IRA can be especially powerful. By paying directly from your IRA, you can reduce your taxable income and potentially reduce your Medicare premiums without any need to itemize deductions in your tax filing.

Proper retirement plan structure

The dirty secret about the US Tax Code is that it is structured primarily to benefit business owners. Knowing how to utilize your business to benefit your tax bill is a fundamental task for any veterinary practice owner.

In many cases, this starts with the retirement plan structure a business owner utilizes. Not all plans are created equally. There are significant differences between the contribution amounts allowed and the funding requirements for employers, employees, and highly compensated individuals.

Things to consider include:

- How much, as a business owner, are you hoping to contribute to your retirement plan?
- Do you want to allow Roth and pre-tax contributions?
- Do you want to allow loans?
- Would you like to include a profit-sharing component?

The right plan structure - in the right situation - can allow a business owner to deduct the cost of the plan for the business and up to $70,000 in income for 2025. Working with a professional to review your options is vital to using your business to benefit you for retirement.

This is just a cursory overview of the options available to veterinarians today. A detailed review of your tax situation - and good, proactive tax planning - is crucial to your long-term financial goals.

https://doi.org/10.54846/am2025/71

Nerium indicum eliminates porcine reproductive and respiratory syndrome (PRRS) in sows with natural infection

Theresa W. Fossum[1]; Luke Strehle[2]; Joe Raemakers[3]; Rick Matos[4]; Sean Ford[5]
[1]Phoenix Animal Wellness, San Antonio, Texas; [2]Nebraska Veterinary Services, West Point, Nebraska; [3]Innovar LLC, Plano, Texas; [4]Raemakers Nutrition, Santa Cruz, California; [5]Dr Fossum's Pet Care, Carefree, Arizona

Introduction

Nerium indicum, commonly known as oleander, has gained attention for its potential therapeutic effects, particularly its antiviral properties. One significant area of concern in veterinary medicine is porcine reproductive and respiratory syndrome (PRRS), a viral infection that causes substantial economic losses in the swine industry. PRRS caused an estimated 1.2 B per year in lost production in the US pork industry from 2016 to 2020, which represents an 80% increase from a decade earlier.[1] "After 40 years of experience fighting this virus, veterinarians and producers are still losing the battle against PRRS," said Holtkamp, who led a 2012 study that estimated the virus caused $664 million in annual losses from 2006 to 2010.[2]

The PRRS virus (PRRSV) leads to reproductive failure in sows and respiratory disease in pigs, necessitating the exploration of alternative treatment options. Compounds derived from *Nerium indicum* may offer promising antiviral activity against PRRSV, providing a potential avenue for disease management.

Understanding PRRS and PRRSV

PRRS is caused by PRRSV, which belongs to the Arteriviridae family. During the past 20 years, there has been an enormous amount of research on PRRSV, and although we have learned a great deal, details on control of the disease for all types of swine-raising operations are far from complete. PRRS is the most economically important disease now affecting producers. Swine industry consolidation of the past 15 years has led to entire production systems being designed around strategies for controlling or eliminating this disease.[3]

PRRSV is known for its high mutation rate, leading to diverse strains that complicate vaccine effectiveness. PRRSV destroys up to 40% of the pig's macrophages, removing a significant part of the pig's defense mechanism which allows secondary infections by bacteria and other viruses, particularly those that affect the respiratory system such as swine influenza, to proliferate and cause damage. Infected pigs experience weakened immune responses, making them more susceptible to secondary infections, particularly pneumonia.

PRRSV is known to cross the placenta in late gestation (after approximately the 72nd day) and may reach a high titer in fetuses. It may kill all or some of the fetuses, most probably due to hypoxia associated with an arteritis that develops in umbilical vessels. Abortions are common (up to 40%),[2] the result of either the effects of acute disease and fever in sows or infection and death of fetuses.[1] There is no specific treatment for PRRS. Rapid elimination of PRRSV in sows could reap enormous benefits across the swine industry.

Antiviral action of *Nerium indicum* against PRRSV

Research suggests that *Nerium indicum* may exhibit antiviral effects against PRRSV through several specific mechanisms,[4] which have been found to occur when it is used against other enveloped viruses, such as SARS-CoV-2: [5]

1. **Inhibition of viral entry:** Compounds found in *Nerium indicum* can disrupt the binding of PRRSV to host cell receptors. By altering the lipid composition of the cell membrane or blocking specific receptor sites that the virus uses to enter cells, these compounds can prevent the initial infection.
2. **Interference with viral replication:** Once PRRSV enters a host cell, it begins replicating its RNA genome. *Nerium indicum* extracts have been shown to downregulate the production and translocation of intracellular GRP78, which is primarily responsible for the unfolded protein response of the innate immune system, and to inhibit the activity of enzymes essential for viral replication, such as RNA-dependent RNA polymerase. These activities can significantly reduce the viral load within the host and can cause the formation of progeny virus particles with extremely low infectivity, thereby ultimately leading to lower levels of infection in the animal.
3. **Induction of antiviral cytokines:** *Nerium indicum* enhances the production of interferons and other cytokines involved in the antiviral response. By boosting the immune response, these compounds help the host's body combat the virus more effectively, reducing the severity of PRRS symptoms.
4. **Inhibition of viral protein synthesis:** Research indicates that compounds from *Nerium indicum* can interfere with the synthesis of viral proteins necessary for assembling new viral particles. By blocking the translation of PRRSV proteins, these compounds can hinder the virus's ability to replicate and spread.
5. **Modulation of immune response:** Beyond direct antiviral activity, *Nerium indicum* can enhance the overall immune response. This modulation includes promoting the activity of T-cells and macrophages, which play critical roles in the immune defense against viral infections. A stronger immune response can lead to quicker clearance of PRRSV from the system.

Results

Depending upon the specific virus studied, the above-mentioned antiviral activity of extracts from *Nerium indicum* may not result in reduction of viral titer; however, the antiviral activity of *Nerium indicum* in enveloped viruses such as PRRSV, results in the production of progeny virus particles with substantially reduced infectivity.

Preliminary in vitro studies have demonstrated that extracts from *Nerium indicum* can significantly reduce PRRSV replication in cultured porcine alveolar macrophages (PAMs), which are key targets for the virus.[6] These studies have shown a marked decrease in viral RNA levels and infectious virus production when cells were treated with *Nerium indicum* extracts prior to and during PRRSV infection. An initial in-vivo trial in four PRRS-positive sows showed no fetal loss and all sows were PRRS PCR negative within 21 days. Additionally, some of the liters were PRRS negative, which was not expected because these sows were PRRS positive prior to the initiation of treatment. A larger study of 50 diseased sows is currently underway.

Safety and toxicity considerations

While the potential antiviral properties of *Nerium indicum* are promising, safety remains a critical concern. We have determined safe doses in both sows and piglets and have seen no adverse effects of treating either group when doses less than 15 ug/kg/day orally are administered in the feed. Importantly, Nerium indicum extract completely eliminated the occurrence of late-term abortion, stillbirth, and piglet mortality in a small study conducted on four pregnant PRRSV positive sows.

Current studies

To fully understand the antiviral potential of *Nerium indicum* against PRRSV future research should focus on:

- Clinical trials: A 50 sow study is currently underway to in PRRS-positive sows to determine efficacy, optimal dosages, and safety of *Nerium indicum* extracts in real-world conditions.

Conclusion

Nerium indicum shows potential as an antiviral agent against PRRSV, with mechanisms that include inhibiting viral entry, replication, and protein synthesis while modulating the immune response. As the swine industry seeks effective solutions to manage PRRS, the exploration of *Nerium indicum* may lead to innovative therapeutic strategies. However, rigorous research is necessary to ensure safety and efficacy, paving the way for new treatments in the fight against viral diseases in livestock.

References

1. Holtcamp D. Growing losses from PRRS cost pork producers $1.2 billion per year, new study shows. July 29, 2024. Accessed October 29, 2024. **https://www.news.iastate.edu/news/2024/07/29/prrscost**
2. Holtcamp D. Quantifying risk factors for PRRS virus introduction into swine herds through the use of the PRRS risk assessment – NPB #08-255. November7, 2012. Accessed October 12, 2024. **https://porkcheckoff.org/wp-content/uploads/2021/02/08-255-HOLTKAMP-ISU.pdf**
3. **vetmed.iastate.edu/vdpam/FSVD/swine/index-diseases/porcine-reproductive**
4. Newman RA, Sastry KJ, Arav-Boger R, Cai H, Matos R, Harrod R: Antiviral effects of oleandrin. J Exp Pharm. 2020:12,503-515.
5. Ha DP Shin W_J. Lui Z, Doche, ME, et al: Targeting stress induction of GRP78 by cardiac glycoside oleandrin dually suppresses cancer and COVID-19. Cell and Bioscience 2024:14,115-140.
6. Newman RA, Abdelsalam K, Buterbaugh R, Sastry KJ: Efficacy of oleandrin and PBI-05204 against viruses of importance to commercial pig health management. *Front. Anim. Sci* 2024:5, 1-9.

https://doi.org/10.54846/am2025/72

Using prescription platform vaccines for control of emerging swine diseases

Keith Erlandson, MS, MBA, DVM, DABVP (SHM)
Medgene Laboratories, Brookings, South Dakota

Introduction

Prescription platform vaccines are a new category of vaccines authorized by the United States Department of Agriculture. Released in 2015, USDA Veterinary Services Memorandum 800.213 outlines process for obtaining licensure of a platform vaccine, and USDA Veterinary Services Memorandum 800.214 outlines the procedures for manufacturing a prescription platform vaccine based on a licensed platform vaccine.

A platform vaccine uses a single vector or expression system (VES) with a gene of interest (GOI) inserted to make the product of interest. Once the GOI is inserted into the VES a recombinant master seed is created.[1,2]

The prescription platform vaccine approach allows for the creation of a library of ready-made vaccine targets that can be used to formulate a veterinarian prescribed vaccine to meet the needs of a particular farm, system, region or nation.[3] Based on surveillance in the field, genetic sequences from pathogens are obtained and compared.[4] From these sequences, GOIs are inserted into a VES which then allows the vector to produce an antigenic protein for incorporation into the prescription platform vaccine.

The prescription platform vaccine is administered by or on the orders of a licensed veterinarian under a valid Veterinary Client Patient Relationship (VCPR). The prescription platform vaccine must undergo safety and purity testing; however, the prescription platform vaccine does not need to undergo efficacy testing, and therefore no claims of efficacy are allowed.

For pathogens with low antigenic variation, there is the ability to manufacture a vaccine providing broad coverage, similar to the concept of commercially licensed vaccines. In the case where the variability of the antigen is large, it may be necessary to manufacture vaccines with specific antigens guided by the genetic sequence of pathogens on individual farms or systems, more in keeping with the autogenous vaccine concept. Unlike autogenous vaccine, the GOI need not originate from the farm where the vaccine will be used. This allows for the development of a GOI library. With a library already in place, the time to manufacture of vaccine can be reduced compared to a situation where a new master seed must be created for each unique situation.

Young[4] provides a more detailed explanation on selection and production of prescription platform vaccines.

Emerging pathogens

Emerging diseases are a fact of life in modern swine production. The ability of manufacture a vaccine in a relatively short timeframe makes them a valuable part of the swine practitioner's toolbox when dealing with these emerging diseases. At the time of this writing Medgene has manufactured and practitioners are implementing vaccination for several emerging pathogens. By the time of the presentation, observations should be available regarding the implementation of vaccines for Senecavirus A, porcine astrovirus 4, porcine sapovirus and porcine sapelovirus. Clinical observations from these situations will be shared.

References

1. USDA-APHIS, Veterinary Services Memorandum No. 800-213. USDA Animal and Plant Health Inspection Service, Washington, DC 2015.

2. USDA-APHIS, Veterinary Services Memorandum No. 800-214. USDA Animal and Plant Health Inspection Service, Washington, DC 2018.

3. Erdman, M.M., Clough, N.E., and Hauer, P.J., Review of updated regulations and product license categories for veterinary vaccines in the United States. *J Am Med Assoc 2020.* 257: 1142-1147.

4. Young, A., From farm to bottle -checkpoints in the selection and production of protein-bases prescription vaccines to meet the needs of evolving disease pressures. In: *Proc Am Assoc Swine Vet 2024.* 134-138.

https://doi.org/10.54846/am2025/73

Optimal iron dosing for improved growth. A review of three birth-to-market studies

C. Olsen, DVM, MS; W. Lyons, DVM
Pharmacosmos Inc

Introduction

Supplementation of iron via iron injections has been a routine practice in pig production for decades.[1,2] And as our knowledge of pig production has evolved over time, so too has our understanding of the iron demands of baby pigs. While there is much work still to be done, a growing body of research makes it increasingly clear that once-standard doses of injectable iron are no longer adequate to prevent the development of clinical or sub-clinical iron deficiency anemia (IDA) at weaning.[3-6] The consequences of IDA are vast yet subtle; while they may not be immediately noticeable, especially with sub-clinical IDA, the cumulative and long-term effects can have considerable economic impacts. Examples of such effects include differences in gene expression with iron-replete pigs having numerous gene related changes indicating improved glucose metabolism, improved gut wall integrity, and improved temperature regulation among other things.[7] In another study, iron replete pigs had reduced fecal scores, improved feed intake, and improved growth when faced with *Escherichia coli* and rotavirus related scours in the nursery.[8] Recently, a follow-up study was done after Pierce et al[7] reported that iron dose impacted the expression of a gene known to regulate body temperature. In the follow-up study by Jones et al,[9] it was reported that baby pigs given two, 200 mg doses of injectable iron dextran had improved thermoregulation over a six-day evaluation period during and after weaning. It is conceivable that each of these reported changes may not result in discernable clinical or economic changes by themselves alone. However, these changes do not occur in a vacuum. When IDA is present, each of these systems are likely impacted to a degree, and when taken together one can see how the cumulative impact over time can begin to manifest as measurable outcomes that impact the health, productivity, and economics of pig production. Altogether, this helps to explain why improvements in post-weaning growth are consistently observed in pigs given two, 200 mg doses of iron dextran early in the suckling period.[6,10-12] Until recently, much of the work studying the relationship between iron dose and pig growth has focused on the early post-weaning phase of growth[3,6,8,13,14] and it was not known if that early life growth improvement would carry through finishing. Now, two birth-market studies have been published since 2023[11,12] and a third has been completed which will be reported here. Therefore, the objective of this report is to summarize three independent trials evaluating the impact that two doses of Uniferon® have on birth to market pig growth.

Methods

Study 1

The first study was conducted at the University of Kentucky and consisted of 144 pigs split into two treatment groups. Complete details of the study have previously been published (Chevalier et al, 2023).[11] Briefly, all pigs were given a 200 mg intramuscular (IM) dose of Uniferon® at 24 hours of age. At 6-8 days of age littermate pairs of the same sex and similar weight were assigned to either the control or treatment group. All pigs in the treatment group received an additional 200 mg IM dose of Uniferon® at this time. Weaning occurred at 22-25 days of age and pigs were placed in a nursery facility with six pigs from the same treatment group per pen. Pigs were assigned to pens by weight within each treatment such that the heaviest pigs six pigs were penned together followed by the next six until all 12 pens were filled. After five weeks in the nursery pigs were moved to a finisher facility with larger pens. Pen integrity was maintained with the same six pigs from the nursery placed together in a pen in the finisher. Common diets were fed to all pigs in the nursery and finisher which included an added 100 mg/kg of iron as ferrous sulfate. Individual body weight and blood hemoglobin (Hb) concentration was measured in all pigs at birth, enrollment, weaning, end of nursery and at the end of the study. In the post-weaning phase body weight and feed disappearance was recorded every two weeks. Data were analyzed by least-squares analysis of variance using the generalized linear model as a complete block design in SAS 9.4. The individual pig was the experimental unit for body weight, daily gain and hemoglobin measurements. Pen was the experimental unit for feed intake and gain/feed ratio data.

Study 2

The second study was conducted on a commercial sow farm in Tennessee using a genetic cross of PIC C29™ sows bred to PIC 337™ boars. Complete details of the study have previously been published (Gomez Cruz et al, 2024).[12] Briefly, 590 pigs were enrolled at 24 hours of age and split into three treatment groups. Three gilts and three barrows of similar size were selected from each litter and randomly assigned to the three treatment groups by sex. Pigs were tagged with numbered and color-coded tags to identify individual pigs and treatment groups. Uniferon®, a 200mg/mL iron dextran product which is labeled for the prevention and treatment of iron deficiency anemia was used for all treatments. The control group received a single 200 mg IM dose of iron at 24 hours of age. The first treatment group (Single) received 400 mg of Uniferon® at 24 hours of age given as two separate 1 mL IM injections. The second treatment group (Split) received a 200 mg IM dose of iron at 24 hours of age followed by an additional 200 mg IM dose when pigs were processed at 6 days of age. Pigs were weaned at an average of 22 days of age and placed in a 4,800 head wean to finish site. Pens were double stocked for approximately six weeks and all study pigs were randomly distributed across pens in one barn on the site and remained on the site through finishing. Individual body weight and blood Hb was measured in all pigs at enrollment, six days of age and at weaning. Additionally, individual body weight was measured at nine weeks of age and at the end of the study which was prior to the first cut of hogs being marketed from the site at approximately 18 weeks post-weaning. Data were analyzed as a randomized block design with three treatments using the linear model in R version 4.2.3. At the final weight collection, some individual tag numbers were illegible.

Study 3

The third study was conducted on a commercial sow farm in Denmark using Danish genetics. In total 328 pigs were enrolled across two treatment groups. All pigs received a single 200 mg IM dose of Uniferon® at 24 hours of age. At six days of age pigs were assigned to treatment groups by selecting two similar sized pigs of the same sex from a litter and randomly assigning one to treatment and one to the control group. An equal number of pairs were enrolled into each group with remaining pigs from each litter left off-trial. Pigs in the treatment group received a second 200 mg IM dose of Uniferon® at this time. Weaning occurred at an average age of 28 days and pigs were placed in an onsite nursery with study pigs spread randomly across pens. At the end of the nursery period, pigs were moved to a finisher with treatment groups comingled randomly across pens. Blood Hb was collected from each pig at weaning and individual weights were taken at enrollment, weaning, end of nursery and before marketing. Data analysis was performed in R version 4.4.1 using a linear model to test the effect of treatment on live weight. Enrollment weight was included as a covariate in the model.

Results

Study 1

Data from this trial are shown in Table 1. Pigs which received two doses of iron dextran had a higher blood Hb concentration at weaning compared to controls (133 vs 107 g/L, $P < .01$). Over the 19 week wean-finish period, pigs in the 400 mg group had a .05 lb improvement in ADG ($P = .04$) resulting in a 7 lb heavier live weight ($P = .04$) at market compared to controls.

Table 1: Summary of data from Study 1. Effect of injectable iron dose on blood hemoglobin concentration, live weight and growth performance in pigs on a university research farm.[11]

	Iron dose		
Parameter	**200 mg**	**400 mg**	**P-value**
Hemoglobin, g/L			
Birth	99	99	.13
Weaning	107	133	< .01
Live weight, lb			
Birth	3.4	3.4	.53
Weaning (22-25 days)	16.4	16.3	.67
End of nursery (7 weeks of age)	42.0	43.9	.03
Market (19 weeks on feed)	248	255	.04
Wean-market growth performance			
Average daily gain, lb	1.78	1.83	.04
Feed intake, lb	4.14	4.18	.74
Gain/feed ratio	0.43	0.44	.33

Study 2

Data from this trial are shown in Table 2. Pigs in both treatment groups which received two doses of iron dextran had improved growth performance compared to controls. Hemoglobin at weaning was improved in both treatment groups (134 and 127 g/L) compared to control pigs (119 g/L, $P < .05$). Average daily gain over the 19 week wean-market period was improved by .05-.06 lb in treatment pigs ($P < .05$) resulting in 7-8 lb heavier live weights at market ($P < .05$) compared to control pigs receiving a single 200 mg dose. Even though some of the numbers on the tags were illegible, data shown in Table 2 includes all pigs which had color-coded tags representing their respective treatment group (n = 419 pigs).

Study 3

Data from this trial are shown in Table 3. In this study, pigs which received 400 mg of iron dextran had numerically higher blood Hb at weaning however it was not statistically significant (114 vs 112, $P = .15$). Growth performance was improved with weaning weight (14.0 vs 13.1, $P = .01$) and market weight (211 vs 203, $P = .05$) both higher in treatment pigs vs controls respectively. Average daily gain over the 15 week wean-market period had a tendency for improvement in treatment pigs (1.56 vs 1.50 lb, $P = .06$).

Table 2: Summary of data from Study 2. Effect of injectable iron dose and timing on blood hemoglobin concentration, live weight, and growth rate in pigs on a commercial US pig farm.[12]

	Treatment group*		
Parameter	**Control†**	**Single‡**	**Split§**
Hemoglobin, g/L			
Birth	94^{a}	100^{b}	92^{a}
Weaning	119^{b}	134^{b}	127^{b}
Live weight, lb			
Birth	3.4	3.5	3.5
Weaning (22 days)	14.7^{a}	14.8^{ab}	15.4^{b}
End of nursery (6 weeks on feed)	46.1^{a}	48.0^{ab}	48.4^{b}
Market (19 weeks on feed)	253^{a}	266^{b}	267^{b}
Wean-market average daily gain, lb	1.89^{a}	1.98^{b}	1.99^{b}

a,b Different superscripts within rows indicate P-values < .05

† Single 200 mg dose of IM iron dextran at 24 hours of age

‡ Two, 200 mg IM dose of iron dextran at 24 hours of age given as two separate 1 mL doses

§ 200 mg IM dose of iron dextran at 24 hours of age followed by an additional 200 mg IM dose at 6 days of age

Table 3: Summary of data from Study 3. Effect of injectable iron dose on blood hemoglobin concentration, live weight and growth rate in pigs on a Danish pig farm.

Parameter	Iron dose		*P*-value
	200 mg	400 mg	
Hemoglobin, g/L			
Weaning	112	114	.15
Live weight, lb			
Enrollment (6 days)	5.5	5.5	.78
Weaning (28 days)	13.1	14.0	.01
End of nursery (7 weeks on feed)	65.9	69.2	.07
Market (15 weeks on feed)	203	211	.05
Wean-market average daily gain, lb	1.50	1.56	.06

Discussion

In this report, three separate studies conducted over a two-year period from 2023 to 2024 are summarized. The studies were all performed in different geographic regions, in different herds with very different management, health and production conditions. Study 1 was done at the University of Kentucky swine farm which is a closed herd free of all major swine pathogens and experiences little mortality.[11] Study 2 was on a commercial swine farm in Kentucky, USA. This farm used standard production practices and stocking densities and was a relatively healthy herd with low wean-finish mortality compared to industry averages.[12] The third study was done on a commercial farm in Denmark. This farm used common Danish genetics, and health and production were typical for the Danish swine industry.

Each study compared a single 200 mg dose to two, 200 mg doses of iron dextran however each varied in the timing of administration of both doses. A farms piglet handling procedures often dictate convenient times for iron administration and the variety of conditions across these three studies provide significant flexibility for producers when implementing a two-dose protocol. In all three studies, the first iron dose was given at 24 hours of age which is important given that pigs are born with very low iron stores that are rapidly depleted without early supplementation.[14,15] The second dose was given by 8 days of age in all studies. Again, this is important as previous studies that have given a second 200 mg dose later in lactation, or at weaning have not found significant improvements in post-weaning growth performance.[16,17]

Remarkably, despite the many differences in each of these studies, administration of two, 200 mg doses of iron dextran by 8 days of age resulted in a 7-8 lb improvement in live weight by the end of each study. As previously stated, the consequences of IDA are vast yet subtle. Yes, it is true that pigs given sub- optimal doses of injectable iron will grow and will make it to market. And without intentional measurement, a 7-8 lb difference in live weight is likely to go unnoticed given the complexity of pig production and the range of factors that impact pig health and growth. However, what is known, is that nearly all cells in the pig's body have a requirement for iron.[18] And while growth may continue, when iron demand is not met, the consequences of early life IDA never fully recover.[19] As demonstrated in this report, the cumulative and incremental improvements associated with optimal iron dosing result in meaningful improvement in birth to market growth are not exclusive to specific production conditions. Implementation of two doses of iron dextran is simple and the timing of administration is flexible given the constraints detailed in this report are followed. The outcome of successful implementation of two doses of Uniferon® are proven and can improve the health and growth of pigs while to benefiting the profitability of pig production operations globally.

References

1. Ullrey D, Miller E, West D, Schmidt D, Seerley R, Hoefer J, & Luecke R. Oral and Parenteral Administration of Iron in the Prevention and Treatment of Baby Pig Anemia. *J Anim Sci.* 1959;18:256-263.

2. Kernkamp H, Clawson A, & Ferneyhough R. Preventing Iron-Deficiency Anemia in Baby Pigs. *J Anim Sci.* 1962;21:527-532.

3. Jolliff J & Mahan D. Effect of injected and dietary iron in young pigs on blood hematology and postnatal pig growth performance. *J Anim Sci,* 2011;89(12):4068-4080. **https://doi.org/10.2527/jas.2010-3736**

4. Bhattarai S & Nielsen J. Association between hematological status at weaning and weight gain post-weaning in piglets. *Livestock Science,* 2015;182:64-68. **https://doi.org/10.1016/j.livsci.2015.10.017**

5. Perri A, Friendship R, Harding J, & O'Sullivan T. An investigation of iron deficiency and anemia in piglets and the effect of iron status at weaning on post-weaning performance. *J Swine Health Prod.* 2016:24:10-20.

6. Olsen C. Field trial results: a second dose of injectable iron to maximize growth. Proceedings 51st annual meeting of the American Association of Swine Veterinarians. 2020.

7. Pierce J, Lyons J, Chevalier T, & Lindemann M. Effects of a second iron-dextran injection administered to piglets during lactation on differential gene expression in liver and duodenum at weaning. *J Anim Sci.* 2024;102:1-10. **https://doi.org/10.1093/jas/skae005**

8. Johnson A, Dittrich B, Cole A, Prodell M, Fritz S, Li W, Fregulia P, Lyons W, & Jang Y. Effects of second iron injection before weaning on growth performance, hematological parameters, and fecal score of pigs fed nursery diets with different dietary iron levels under natural disease challenge. Proceedings of the 27th International Pig Veterinary Society Congress. 2024.

9. Jones M. Improvements in piglet thermoregulation with higher iron doses. Proceedings of the 50th Allen D. Leman Swine Conference. 2024.

10. Olsen C & Fredericks L. Impact of iron dose and hemoglobin concentration on wean-finish weight gain. Proceedings of the 25th International Pig Veterinary Society Congress. 2018.

11. Chevalier T, Lyons W, Paczosa D, Rentfrow K, & Lindemann M. A second iron injection administered to piglets during lactation improves hemoglobin concentration, growth performance, and carcass characteristics at slaughter. *J Anim Sci.* 2023;101:1-8. **https://doi.org/10.1093/jas/skad270**

12. Gomez Cruz F, Krantz S, Lyons W, & Pierce J. Comparison of 200 mg injectable iron dextran vs 400 mg in either a single timepoint or split-dosed into two timepoints on growth and hemoglobin in pigs from birth to end of finishing. Proceedings of the 55th annual meeting of the American Association of Swine Veterinarians. 2024.

13. Haugegaard J, Wachmann H, & Kristensen P. Effect of supplementing fast-growing, late-weaned piglets twice with 200 mg iron dextran intramuscularly. *The Pig Journal.* 2008;61:69-73.

14. Chevalier T, Monegue J, & Lindemann M. Effects of iron dosage administered to newborn piglets on hematological measures, preweaning and postweaning growth performance, and postweaning tissue mineral content. *J Swine Health Prod.* 2021;29(4):189-199. **https://doi.org/10.54846/jshap/1207**

15. Ekman L & Iwanska S. Studies on iron metabolism in normal and anemic nursing pigs. *Zentralbl Veterinarmed A*. 1966;13(7):585-595. **https://doi.org/10.1111/j.1439-0442.1966.tb00397.x**

16. Kamphues J, Manner K, & Netzer C. Effects of a 2nd Iron Injection in Suckling Piglets on Iron Retention and Performance Before and After Weaning. Proceedings of the 12th annual International Pig Veterinary Society Congress. 1992.

17. Peters J & Mahan D. Effects of neonatal iron status, iron injections at birth, and weaning in young pigs from sows fed either organic or inorganic trace minerals. *J Anim Sci*. 2008;86(9):2261-2269. **https://doi.org/10.2527/jas.2007-0577**

18. Wang J & Pantopoulos K. Regulation of cellular iron metabolism. *Biochem J*. 2011;434(3):365-381. **https://doi.org/10.1042/bj20101825**

19. Knight L, & Dilger R. Longitudinal Effects of Iron Deficiency Anemia and Subsequent Repletion on Blood Parameters and the Rate and Composition of Growth in Pigs. *Nutrients*. 2018;10(5):632. **https://doi.org/10.3390/nu10050632**

https://doi.org/10.54846/am2025/74

Two iron dextran injections before weaning improved postweaning growth performance, hemoglobin levels, and fecal score of pigs under natural disease challenge

Alaina Johnson[1]; Barbara Dittrich[1]; Aleah Cole[1]; Marie Prodell[1]; Scott Fritz[2]; Wenli Li[3]; Priscila Fregulia[3]; Wesley Lyons[4]; Young Dal Jang[1,5]
[1]University of Wisconsin-River Falls; [2]Kansas State University; [3]US Dairy Forage Research Center, USDA-Agricultural Research Service, Madison, Wisconsin; [4]Pharmacosmos Inc.; [5]University of Georgia

Introduction

Iron injection at birth is a common practice to prevent piglets from developing iron deficiency anemia during the suckling period in pig production,[1] because newborn piglets are susceptible to iron deficiency due to insufficient iron reserves at birth and a low level of iron in sow milk.[2] However, with increasing weaning age up to 28 days and a fast growth rate of suckling pigs, a single iron injection may not be sufficient to meet the iron needs of pigs for the entire suckling period. Therefore, an additional iron injection during the first 12 days of life may be necessary to meet the iron demands for growth and subsequent blood volume expansion through the weaning phase. The NRC recommendation for dietary iron concentration in nursery diets is 100ppm, however, a Kansas State University industry survey showed that the majority of swine nutritionists formulating diets for nursery pigs provide significantly more iron in the first two phases of the nursery rations.[3,4] It has been assumed that due to the immaturity of the small intestine of newly weaned pigs, along with the bioavailability of dietary iron, more iron is needed to adequately supplement growing nursery pigs. Conversely, the large amount of iron in the intestine can lead to the proliferation of pathogenic bacteria such as *Escherichia coli* and *Salmonella* spp causing postweaning diarrhea. Research into the gene expression of pigs at weaning receiving either 1 or 2 doses of 200mg iron dextran showed that pigs receiving 2 doses had gene expression coding for improved storage of iron in the liver and an improvement in small intestinal health and resilience.[5] Because pigs do not have an efficient excretion mechanism for parenteral iron, the risk of small intestinal bacteria overgrowth and subsequent diarrhea is negligible when compared to dietary iron. The goal for this study was to compare the effects of a second iron injection and different dietary iron levels in the postweaning diets on growth performance, hematological parameters and fecal score of pigs undergoing a natural disease challenge postweaning.

Methods

All procedures used in this study were approved by the Institutional Animal Care and Use Committee of University of Wisconsin-River Falls (Protocol # 20-21-56350). The experiment was conducted in the nursery facility at Mann Valley Farm of University of Wisconsin-River Falls (WI, USA). A total of 70 newborn piglets [initial body weight (BW): 1.83 ± 0.31 kg; Yorkshire × Yorkshire, Yorkshire × Duroc, Yorkshire × Duroc × Duroc; 40 males and 30 females] from 7 lactating sows with 8-12 piglets per litter were allotted into 4 treatments within the litter in a 2 × 2 factorial design (main factors were the number of iron injection and dietary iron level in nursery diets) based on BW and sex. At d 2-3 of age (d 0 of experiment), all piglets received the first intramuscular (IM) injection of 200 mg iron and were processed based on the standard protocol of UWRF Mann Valley Farm in which male piglets were castrated on d 10 of age. Piglet sets were created by having 4 littermates within the same sex that had similar BW. Within each set, 2 piglets were assigned to the one iron injection treatment and the others to the second iron injection treatment. Treatments were: **1)** Negative control (NC): No additional iron injection + NC diets (100 mg/kg iron) in nursery period, **2)** IRON+NC: Second IM iron injection (200 mg) at 5 d after the first injection + NC diets in nursery period, **3)** Positive control (PC): No additional iron injection + PC diets (200 mg/kg iron) in nursery period, and **4)** IRON+PC: Second IM. iron injection (200 mg) at 5 d after the first injection + PC diets in nursery period. A common iron-dextran product (UNIFERON® 200, Pharmacosmos, Inc., Watchung, NJ) was used for both injections. The piglets in the second iron injection treatments (IRON+NC and IRON+PC) received an additional 200 mg of IM iron injection at 5 d after the first injection (d 5 of experiment) and all piglets were weaned at d 25 of experiment (d 27-30 of age). There was no cross-fostering among litters and no creep feed was provided to the piglets during the entire suckling period. At weaning, pigs were moved to the nursery facility, mixed, and sorted by BW and sex to avoid sow effects. Then, the pigs within iron injection treatments were divided into nursery diet treatments in 3 replicate pens per treatment balanced by BW and sex with 5 or 6 pigs per pen for a 27-d postweaning growth period. After weaning, piglets tested positive for hemolytic *E coli* strains and Rotavirus A and C. This was confirmed as fresh feces were collected at d 6 and 13 postweaning and analyzed at the Iowa State University Veterinary Diagnostic Laboratory (ISU-VDL, Ames, IA).

Data and sample collection

The BW of each pig was recorded at initial (d 0 of experiment), d 5, 11, 18, and 25 (weaning) of experiment. The BW and pen feed consumption was recorded at d 0, 6, 13, 20, and 27 postweaning for calculation of average daily gain (ADG), average daily feed intake (ADFI) and gain to feed ratio (G:F). Blood samples from all pigs were collected from the jugular vein into K_3 EDTA tubes each day when BW was recorded for hematocrit analysis. A drop of blood from the ear veins of the pigs was loaded into disposable microcuvettes via capillary action and analyzed for hemoglobin levels. During the nursery phase, the fecal score was recorded every day for the entire experimental period using a 4-scale fecal score system (1 = normal, 2 = soft, looser than normal feces, slight diarrhea, 3 = moderate diarrheic feces, and 4 = liquid, severe diarrhea) by observing individual piglets in each pen and assessing signs of stool consistency in the pen.

Statistical analysis

All data were analyzed with ANOVA for a randomized complete block design using PROC MIXED of SAS (version 9.2; SAS Inst. Inc., Cary, NC). The experimental unit was an individual pig for growth performance in the suckling period and blood data, and a pen for growth performance and fecal score data in the nursery period.

Results

The second iron injection significantly increased hemoglobin ($P < .05$) and hematocrit ($P < .05$) levels in the suckling period from d 11 to 25 of experiment (Table 1). In the postweaning period (Table 2), the pigs in the two-iron injection treatment tended to have greater final BW ($P < .10$, tendency) at d 27 postweaning and ADG in d 13-27 ($P < .10$ tendency) and 0-27 ($P < .10$, tendency) postweaning and had significantly greater ADFI in d 0-13 ($P < .05$), 13-27 ($P < .05$), and 0-27 ($P < .05$) postweaning than those in the one-iron injection treatment, while the pigs in the two-iron injection treatment had lower fecal score in d 13-27 ($P < .05$) and 0-27 ($P < .05$) postweaning than those in the one-iron injection treatment. The dietary iron levels had no significant effect on growth performance (Table 2) except for the pigs fed the PC diets showing lower fecal score compared with those fed the NC diets in d 13-27 postweaning ($P < .10$, tendency).

Discussion

The current study evaluated the effect of second iron injection given to suckling pigs during the suckling period on postweaning growth performance, hematological parameters, and fecal characteristics when pigs were fed nursery diets with or without additional iron supplementation. Due to limited iron supply through the placenta and milk that is associated with a decline in hemoglobin level near weaning, suckling pigs may need an additional iron injection to meet their iron needs in the late suckling and early nursery periods.[1,6,7] However, dietary iron content from calcium phosphate and limestone and additional iron supplementation using inorganic sources such as ferrous sulfate may exceed the iron requirement estimates for nursery pigs.[4] This may result in a high level of unabsorbed iron remaining in the intestine, causing pathogenic bacteria proliferation because *E coli* and *Salmonella* require iron for their abundance and virulence.[8,9] Reducing the free iron level in the gut while still maintaining sufficient body iron status with an additional iron injection may be beneficial for pigs to reduce postweaning gut health issues such as diarrhea and improve postweaning growth performance. In the current study, piglets were naturally challenged with rotavirus and *E coli* within the first week after weaning, as confirmed by the ISU-VDL. These enteric pathogens are common in weaning pigs, resulting in postweaning diarrhea, and therefore, the response of nursery pigs to treatments in the current study was under the condition of enteric disease challenge. In addition, since there were virtually no interactions observed, discussion focused on

Table 1: Preweaning growth performance, hemoglobin and hematocrit levels of pigs receiving one or two iron injections in the suckling period[a]

	Treatment[b]			
Item	1 injection	2 injections	SEM	P-value
Body weight, kg				
d 0	1.84	1.83	0.07	.79
d 5	3.19	3.18	0.12	.96
d 11	5.04	5.13	0.17	.60
d 18	7.21	7.33	0.20	.61
d 25 (weaning)	9.61	9.81	0.25	.52
Hemoglobin, g/dL				
d 5	8.8	8.6	0.40	.54
d 11	11.1	11.7	0.53	.05
d 18	10.3	12.2	0.34	< .0001
d 25 (weaning)	10.1	12.4	0.47	< .01
Hematocrit, %				
d 5	31.4	31.1	1.03	.68
d 11	35.3	37.6	1.28	< .01
d 18	32.8	38.2	0.96	< .01
d 25 (weaning)	31.7	37.9	1.06	< .0001

[a] n = 35 per treatment from all pigs used in the experiment.

[b] One-iron injection: 200 mg iron injection at 2-3 d of age (d 0 of experiment) and no additional iron injection in the suckling period; two-iron injection: 200 mg iron injections at 2-3 d of age and 5 d after the first injection (d 5 of experiment).

Table 2: Postweaning growth performance and fecal score of pigs receiving one or two iron injections in the suckling period and different dietary iron levels in nursery period[a]

	Injection[b]		Dietary Fe[c]			*P*-value		
Item	1 injection	2 injections	100 ppm	200 ppm	SEM	Injection (I)	Dietary Fe (D)	I×D
Body weight, kg								
d 0	9.71	9.91	9.77	9.84	0.67	.15	.57	.97
d 13	11.09	11.76	11.54	11.31	0.29	.16	.60	.66
d 27	18.84	20.75	20.02	19.58	0.60	.08	.62	.35
ADG, kg/d								
d 0-13 (Phase I)	0.099	0.150	0.133	0.116	0.022	.16	.61	.66
d 13-27 (Phase II)	0.554	0.642	0.606	0.590	0.028	.08	.71	.28
d 0-27 (overall)	0.335	0.405	0.378	0.362	0.022	.08	.62	.35
ADFI, kg/d								
d 0-13 (Phase I)	0.287	0.348	0.329	0.307	0.013	.02	.29	.84
d 13-27 (Phase II)	0.923	1.033	0.999	0.957	0.028	.04	.35	.18
d 0-27 (overall)	0.617	0.703	0.676	0.644	0.019	.03	.29	.27
G:F								
d 0-13 (Phase I)	0.334	0.426	0.394	0.365	0.066	.37	.77	.88
d 13-27 (Phase II)	0.601	0.623	0.609	0.615	0.024	.56	.86	.88
d 0-27 (overall)	0.542	0.576	0.559	0.559	0.028	.43	.99	.75
Fecal score								
d 0-13 (Phase I)	2.40	2.23	2.31	2.33	0.08	.19	.88	.55
d 13-27 (Phase II)	2.13	1.69	2.04	1.78	0.08	.01	.08	.94
d 0-27 (overall)	2.26	1.95	2.17	2.04	0.07	.03	.28	.78

[a] n = 3 replicate pens per treatment. Initial body weight was used as a covariate in statistical analysis for growth performance.

[b] One-iron injection: 200 mg iron injection at 2-3 d of age and no additional iron injection in the suckling period; two-iron injection: 200 mg iron injections at 2-3 d of age and 5 d after the first injection.

[c] NC: negative control with 100 mg/kg iron in nursery diets and PC: positive control with 200 mg/kg iron in nursery diets.

the main effects of iron injection and dietary treatments. In the preweaning period, although the second iron injection increased hemoglobin and hematocrit levels, there was no significant difference in the BW of piglets until weaning. This result agrees with previous studies that showed no difference in preweaning growth performance, although an additional iron injection to suckling pigs increased hemoglobin and hematocrit levels in the suckling period.[6,7,10] In addition, over 50 and 70% of piglets in the one-injection treatment had anemic (< 9.0 g/dL) and suboptimal hemoglobin levels (< 11.0 g/dL) at weaning, respectively, while no piglets had suboptimal hemoglobin levels when receiving the second iron injection. Chevalier et al, reported that it may take several days for injected iron to be incorporated into hemoglobin compounds.[1] Our previous studies also reported that the response of pigs to the second iron injection for growth performance may not appear until the nursery period.[6,10] Although the second iron injection in the suckling period did not show any growth performance response in the suckling period, the pigs receiving two iron injections during the suckling period tended to have greater BW at d 27 postweaning and growth rate in late and overall nursery periods with 10% difference in BW at d 27 postweaning. This resulted from increased feed intake throughout the nursery period. Chevalier et al, reported that the second iron injection to suckling pigs enhanced postweaning growth performance, which was maintained until the pigs reached the market weight.[7] Pierce et al, reported that a second iron injection upregulated the expression of Claudin 1, and genes associated with epithelial cell proliferation in the duodenum as well as genes involved in vitamin D metabolism in the liver.[5] It also downregulated the expression of genes related to gluconeogenesis and lipid synthesis in the duodenum tissue and urea synthesis in liver, which may lead to substantial energy savings that can contribute to improved growth rates in piglets receiving the second iron injection. Reduced fecal score indicating firmer feces in the two-injection treatment compared with the one-injection treatment in the current study indicates that an additional iron injection may benefit gut environment of weaning pigs. In conclusion, the second iron injection given to suckling pigs improved postweaning growth performance and the levels of hemoglobin and other hematological parameters, reduced fecal score indicating firmer feces. An additional 100 ppm of dietary iron over the requirement estimate increased hemoglobin levels, reduced fecal score in the late nursery period but did not affect postweaning growth. The current study also demonstrated that the effect of the second iron injection on hematological parameters could be maintained high until the first few weeks postweaning, while its effect on growth performance could be observed in the late nursery period, although the second iron injection is performed in the early suckling period.

References

1. Chevalier, T.B., H. J. Moneque, and M. D. Lindemann. 2021. Effects of iron dosage administered to newborn piglets on hematological measures, preweaning and postweaning growth performance, and postweaning tissue mineral content. *J. Swine Health Prod.* 29:189-199.

2. Matte, J. J. and I. Audet. 2019. Maternal perinatal transfer of vitamins and trace elements to piglets. *Animal.* 14:31-38. **doi: 10.1017/S175173111900140X**

3. Faccin, J. E. G., M. D. Tokach, R. D. Goodband, J. M. DeRouchey, J. C. Woodworth, and J. T. Gebhardt. 2023. Industry survey of added vitamins and trace minerals in U.S. swine diets. *Transl. Anim. Sci.* 7: txad035, **doi: 10.1093/tas/txad035**

4. NRC. 2012. *Nutrient requirements of swine.* 11th rev. ed. National Academic Press, Washington, DC.

5. Pierce, J. L., J. W. Lyons, T. B. Chevalier, M. D. Lindemann. 2024. Effects of a second iron-dextran injection administered to piglets during lactation on differential gene expression in liver and duodenum at weaning. *J. Anim. Sci.* 102:skae005. **doi: 10.1093/jas/skae005**

6. Albers, S. E., E. A. Pintens, P. K. Isensee, C. M. Lemanski, and Y. D. Jang. 2022. Effect of additional iron injection to suckling pigs on hematocrit level during the suckling period. *Animals.* 12:2980. **doi: 10.3390/ani12212980**

7. Chevalier, T. B., W. Lyons, D. B. Paczosa, G. K. Rentfrow, and M. D. Lindemann. 2023. A second iron injection administered to piglets during lactation improves hemoglobin concentration, growth performance, and carcass characteristics at slaughter. *J. Anim. Sci.* 101:1-8. **doi: 10.1093/jas/skad270**

8. Domínguez-Acuña, L. and F. G. Portillo. 2022. Ferrous iron uptake is required for *Salmonella* to persist within vacuoles of host cells. Infect. Immun. 90:e00149-22. **doi:10.1128/iai.00149-22**

9. Guo, B.X., Q. Q. Wang, J. H. Li, Z. S. Gan, X. F. Zhang, Y. Z. Wang, and H. H. Du. 2017. Lipocalin 2 regulates intestine bacterial survival by interplaying with siderophore in a weaned piglet model of *Escherichia coli* infection. *Oncotarget.* 8:65386-65396. **doi: 10.18632/oncotarget.18528**

10. Lima, G., C. M. Lemanski, A. N. Rank, and Y. D. Jang. 2024. Effect of iron injection strategies for suckling piglets on pre- and post-weaning growth performance and hematocrit levels. *Rev Colomb Cienc Pecu.* (in press). **doi: 10.17533/udea.rccp.v38n2a3**

https://doi.org/10.54846/am2025/75

Oral NSAID impact on production and additional treatments of grow-to-finish pigs challenged with respiratory viruses

Morgan Almeida[1]; Brian Payne[2], DVM; Gustavo de Sousa Silva[3], DVM, MS, PhD; Hans Rotto[4], DVM, MS
[1]College of Veterinary Medicine, Iowa State University, Ames, Iowa; [2]Veterinary Pharmaceutical Solutions; [3]Veterinary Diagnostic and Production Animal Medicine department, Iowa State University, Ames, Iowa; [4]Innovative Agriculture Solutions LLC, Ames, Iowa

Introduction

Porcine reproductive and respiratory syndrome virus (PRRSV) and influenza A virus (IAV) contribute to significant losses during the growing production phase, leading to inflammation and susceptibility to secondary challenges. Non-antibiotic tools are needed to mitigate these harmful sequelae. Non-steroidal anti-inflammatory drugs (NSAIDs) reduce inflammation, fever, and pain. This study aimed to evaluate the mortality and medication impact of oral meloxicam in naturally PRRSV- and IAV-challenged pigs. A cost-benefit analysis was performed to assist practitioners in making health and economic recommendations to their clients.

Methods

Thirty-four pairs of barns (n = 68) were enrolled in this case-control field trial based on respiratory clinical signs with PRRSV and IAV PCR detection. Paired barns were defined as the same site, flow, health status at placement, feeding, and age of pigs. Control barns maintained routine medication plans while case barns received an additional five days of 0.6% compounded meloxicam at a rate of 1.3 pints per 5 gallons of stock solution, targeting 0.13 mg/kg/day (assuming 1-gallon drinking water/100lbs of pig). Subjective analysis was performed to assess the timing of treatment (early, on-time, or late) and the accuracy of farm staff medication logging. Sites were analyzed by type of infection status (IAV only [1], PRRSV only [2], IAV and PRRSV [3]). Data on mortality, injectable treatments, and water treatments were recorded from day 1 (start of meloxicam administration) through the following ten weeks for a comparative cost analysis between the case and control barns. Generalized linear mixed models were performed in R to determine statistical significance in the cumulative mortality.

Results

Twenty-two barn pairs with adequate record-keeping, timing (early and on-time), and proper protocol adherence were included in the following results. Control barns averaged 1,596 pigs/barn (n = 17,551 pigs), and case barns averaged 1,630 pigs/barn (n = 17,928 pigs). Age at the start of treatment was 10.5 ± 2.5 SD weeks. Cumulative mortality rates at 4- and 10-weeks post meloxicam administration were 1.1% and 0.9% less in case barns compared to control barns, respectively (*P*-value < .05). By infection status, at week 4, case barns for status 1, 2, and 3 had a reduction in cumulative mortality (*P*-value < .05). The 10-week cumulative mortality had a reduction in status 2 and 3 as compared to the control barns (*P*-value < .05). On average, 17.7 more pigs/barn survived with meloxicam administration. For this production system, an annual 1% decrease in mortality is a savings of $0.50 per pig placed. The average retail cost of 5 days of meloxicam was $0.06/pig/day, including a calculated range of meloxicam disappearance per pig of 0.12-1.39mg/kg/day. In case barns, oral medication treatments had an average of $0.04/pig less. There was no statistical difference between treatment groups in the injectable or oral medication costs per pig. Barns receiving meloxicam in the water during viral respiratory challenges had an ROI of $0.48/pig in the barn compared to controls.

Discussion

Overall, reducing antibiotic use and mortality is an industry goal. Meloxicam case barns at both 4 and 10 weeks had a statistically reduced mortality; there was no difference in the cost of oral or injectable medications between case and control barns. Numerous sites fell out of the study due to incorrect implementation of treatment or poor record keeping, reducing the number of barns for comparison. These results reflect preliminary analysis of the first thirty-four pairs of barns enrolled; an additional thirty-four pairs have ongoing data collection for the 10-week data set.

Poster 02
https://doi.org/10.54846/am2025/76

Evaluation of the response to antibiotic treatment in newly placed finishing pigs

B. B. Anderson[1], BS; M. Bracy[2], DVM; J. Hall[2], BS
[1]University of MN CVM, Saint Paul, Minnesota; [2]Seaboard Foods, Guymon, Oklahoma

Introduction

Swine veterinarians are critical to the creation of antibiotic treatment protocols for specific health issues. To ensure responsible antibiotic use, veterinarians must monitor the impact of these protocols, but utilizing untreated control pigs to do so is difficult in practice. This study aims to evaluate the percent survivability in newly placed commercial finishing pigs treated with enrofloxacin per label vs untreated controls.

Materials and methods

Feeder pigs weighing approximately 65 lbs. were enrolled three days after placement into 12 barns across 6 finishing sites with no more than 75 pigs enrolled per barn. During enrollment, an observer trained in the production system's treatment protocol identified animals sick enough to be treated in the following disease categories; respiratory, fallback, gastrointestinal and septic. Pigs were tagged, gender determined, girth size measured, and rectal temperature taken with a GLA thermometer. Next, an assistant directed the caretakers to administer enrofloxacin intramuscularly (IM) at a dose of 3.4 mLs per 100 lbs to pigs bearing ID numbers that had been randomly assigned to treatment. Pigs randomly assigned to control were not treated or retreated during the study period. At 7- and 14-days post-enrollment the observer, who remained blinded as to treatment group, revisited the site to identify which pigs had died or were deemed sick enough to be retreated. Pigs that had died or required euthanasia were necropsied and cultures were submitted to verify the presence of labeled bacteria. A final count of surviving pigs was performed on day 42-56. Data were analyzed with a generalized linear mixed model (Proc Glimmix, SAS ver 9.4). Percent survived to day 42-56 was the primary outcome variable and percent sick enough for retreatment at 7 and 14 days was secondary. For analysis pigs with girth measurements below the average were designated as light and above were designated as heavy.

Results

A total of 750 pigs were enrolled; of these 4 were septic, 323 were fallbacks, 419 had respiratory disease and 4 had gastrointestinal disease. Labeled respiratory pathogens (either *Pasteurella multocida* or *Streptococcus suis)* were cultured from all sites, as was porcine reproductive and respiratory syndrome virus (PRRSV) found on PCR. There was no evidence of differences in girth size, gender or rectal temperature between the two treatment groups, indicating successful randomization. There were no significant differences between the fallbacks and respiratory pigs for percent survived or sick enough to retreat and they were analyzed together. Due to low numbers, gastrointestinal and septic pigs were excluded. There was no difference between treatment groups in survival rates (90.84% for controls, 93.21% for treated, $P = .223$). For the fallback cases, heavy/light was significantly associated with percent survived. For the respiratory cases and for the combined respiratory/fallback cases, girth size and heavy/light was significantly associated with the probability of being sick enough for retreatment.

Discussion

Pigs treated per label for respiratory disease with enrofloxacin had a numerically 3% higher survival rate compared to controls, with a number needed to treat (NNT) of 29.3. All sites were PRRSV positive and on a standard medicated feed ration, which may have dampened the treatment effect. This study also demonstrates that smaller pigs have a more chronic disease course and a higher chance of mortality.

https://doi.org/10.54846/am2025/77

Evaluation of intradermal vaccination in piglets to improve welfare and labor efficiency, reduce stress responses and pain, and minimize injuries to piglets and producers

M. Benedetti; S. Murray; F. De Grau, DVM, DVSc, MVZ, EPA

Introduction

Needle-free intradermal (NFID) vaccination of pigs is a growing practice, but the technology uptake in Canada has been slow despite the approval of NFID devices. Producers are not confident in the benefits or the available products, and many of the studies available are not realistic to the vaccination processes on-farm. Intradermal (ID) vaccination is believed to cause less pain and proven to reduce stress responses compared to intramuscular (IM) vaccination. It is also more efficient for producers and carries less risk of adverse reactions or injury to staff through contact with needles. This study aimed to prove these hypotheses in a research environment realistic to swine farms in Canada, to increase interest in intradermal vaccination and demonstrate how it can help counter labor shortages in the industry.

Methods

1123 three-week-old weaner piglets were randomly assigned into treatment groups by pen: An ID group (482 piglets), an IM group (464 piglets), and a control group (177 piglets). Piglets were vaccinated in a random order with an injector device (with a capped or uncapped needle) or Merck's IDAL® device. They were handled realistically, using a pig board to push the group to the front of the pen, and not picked up individually. During vaccination, time, audio, and video were recorded. 190 piglets (eighty from ID and IM groups, thirty from the control group) were randomly selected for bleeding one-hour post-vaccination. Two hours post-vaccination, three ID and three IM groups were recorded for thirty minutes after being given access to a novel object (a pig ball). The rest of the pens' behaviors were recorded via scan observations at two- and three-hours post-vaccination. A statistical analysis was done using descriptive statistics and mixed-effects ML regression.

Results

See Figure 1, where ID piglets had lower cortisol levels (mean = 94.37 nmol/L, max = 268 nmol/L) than IM piglets (mean = 185.16 nmol/L, max = 579 nmol/L) ($P < .001$). ID piglets had less vocalizations > 1000 Hz (ID mean screams per pen = 2.36, IM mean = 16.64) and less intense vocalizations (mean intensity = 61.14 dB, max intensity = 84.63 dB) than IM piglets (mean intensity = 68.20 dB, max intensity = 89.89 dB). ID vaccination was faster (mean time per pen of 01:30.26 min, compared to 01:54.51 min for IM groups), resulted in no adverse reactions (compared to eleven in the IM group), and no human injury (compared to two needle pokes while vaccinating the IM groups). ID piglets interacted with the ball in larger numbers (mean of 2.02 pigs touching the ball at any given one-minute interval, vs 0.24 pigs in the IM group) and for longer (ID mean = 23:29.67 min, IM mean = 07:51.68 min), while also moving the ball farther (ID mean = 139.33ft, IM mean = 6.67ft). ID piglets were also more active over the observation period.

Figure 1: Treatment group vs serum cortisol level. Piglets were chosen at random for bleeding one-hour post vaccination, in the order in which they were vaccinated. Piglets with undetectable serum cortisol levels (< 28 nmol/L) were excluded from this chart.

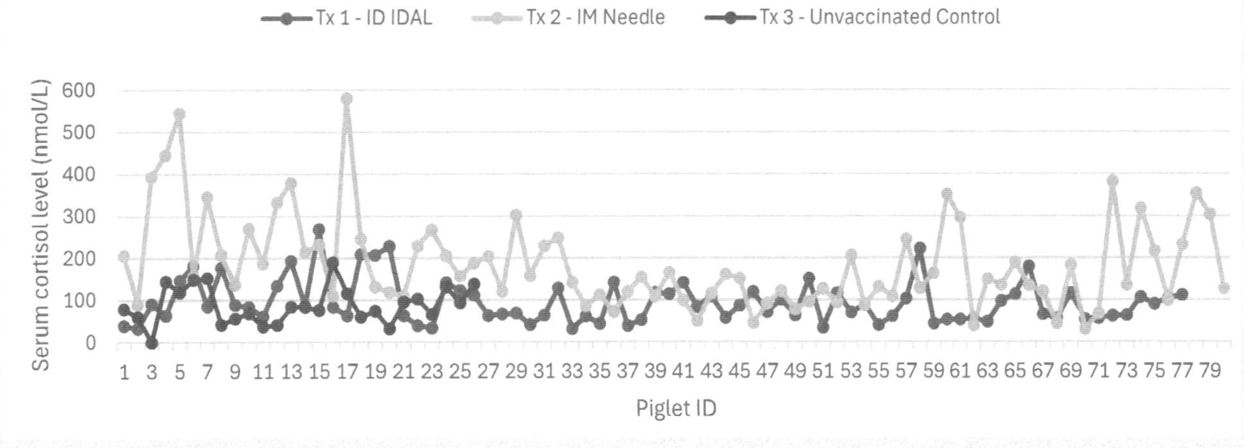

ID = intradermal; IM = intramuscular

Discussion

Our results suggest that intradermal vaccination can be an effective and safe method of protecting our swine, while at the same time enhancing animal welfare and promoting higher worker safety with better time management. It can help to relieve some of the stresses caused by nationwide labor shortages. To the best of the author's knowledge, this is the first study on the benefits of the IDAL® device done in Canada. It was also done on a scale much larger than similar studies, with new parameters and more realistic methods. Others are encouraged to research further for the education of producers nationwide.

https://doi.org/10.54846/am2025/78

Evaluation of lameness and lesion scores when a crude fiber concentrate is applied at time of pen mixing in gestating females

M. Wilhelm[1], BS; R. Stika-Jensen[2], DVM; R. Edler[2], MS; C. Haden[2], DVM

[1]Iowa State University, College of Veterinary Medicine, Ames, Iowa; [2]Pipestone Veterinary Services, Pipestone, Minnesota

Introduction

Sow lameness has been a growing problem within the industry. While the issue is multifactorial, part of the increase can be attributed to an increase in prevalence of group gestational housing. It has been shown that group gestation may cause an increase in lameness through increased social interaction and mobility.[1] With the increase in social interaction specifically around the time of pen mixing, often leading to increased inter-animal aggression and the development of a social hierarchy.[2] A crude fiber concentrate has been identified as a potential tool to help mitigate these issues when introduced at the time of pen mixing. This substance has three potential mechanisms of action. First, as a novel substance, which may provide distraction. Second, when consumed, the fiber may increase satiety. Third, the fiber could act as a floor drying agent, potentially reducing injuries as fighting occurs. The objective of this study was to determine if placing a crude fiber concentrate on the slats of pens during mixing would reduce fighting, subsequent gestational lameness and improve overall production.

Materials and methods

A 5000 head sow farm located in southern Wisconsin was selected for this study. Forty-two pens were observed over a 6-week time period. Females were sorted by parity (gilts, P1/P2, P3+) and placed within pens post pregnancy evaluation (35-40 days bred). Control pens were mixed without a crude fiber concentrate. Treatment pens received a crude fiber concentrate (Arbocel®). Pens designated to receive Arbocel® were randomly selected. At pen mixing, Arbocel® was placed on the slats, following a predetermined protocol. To evaluate a potential reduction in fighting, within 7 days of mixing, trained evaluators lesion scored each female in the pen using a fighting lesion scale with scores ranging from 0-6. Lesion scores were analyzed using ordinal logistic regression where treatment and parity were included as fixed effects and pen standard errors were included as a cluster variable in the model. To evaluate a potential reduction in lameness, lameness treatments were monitored including the number of treatment days and injections. Treatment information was analyzed using a generalized linear model with a binomial distribution and logit link function. Treatment and parity were included as fixed model effects.

Results

Regarding fighting, which was measured via fighting lesion scores, statistical analysis concluded that when controlling for parity and adjusting for pen error, Arbocel® had no impact on lesion scores. Interestingly, data revealed that multiparous sows have statistically greater odds of having a higher lesion score, or more significant fighting lesions when compared to gilts (Figure 1). Regarding treatments, pens that received Arbocel® on the slats had a higher percent of sows treated multiple times (30.29%) compared to control pens (22.39%). Similarly, gilt pens were more likely to have a higher percentage of the pen treated (7.91%) and have multiple treatments required for recovery (47.31%) compared to other pen types. Given the reduced likelihood of fighting lesions found in gilts compared to sows, this finding of increased number of animals treated and increased number of injections needed for recovery was unexpected. There are several limitations to this study including seasonality and the representation of only a single genetic line. Welfare and production impacts secondary to group housing continue to be a major concern for our industry and further studies are needed to determine appropriate mitigation strategies.

Figure 1: Average lesion score according to treatment group and pen type.

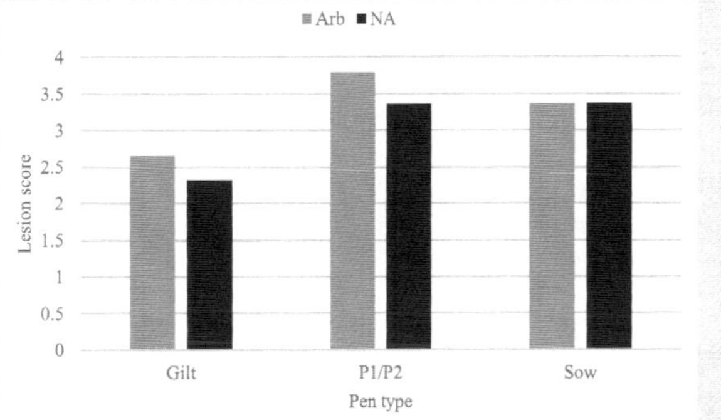

Arb = Arbocel®; NA = no Arbocel®; P1/P2 = parity 1/parity 2

References

1. A novel objective quantifiable measure of lameness in gestating sows. A. Shulman, T. D. Parsons.

2. Impact of parity on aggression between gestating sows upon entry into large-pen group housing. Nicole Johnson Vallagares, T.D. Parsons.

https://doi.org/10.54846/am2025/79

Factors influencing flank, tail, and ear lesions in conventionally housed finishing pigs

Y. Samar, MS; M. Pierdon, BA, VMD
The School of Veterinary Medicine at the University of Pennsylvania

Introduction

The etiology of flank bites, tail bites, and ear tip necrosis are all different but under-studied and may have multiple farm and pen level factors that influence their prevalence. Levels of sodium in the feed and mycotoxins, as well as space allowance, pen size, and pig age may be related to the occurrence of these lesions. The purpose of this study is to assess whether farm or pen level factors are associated with the prevalence of ear, flank, and tail lesions in conventional finishing pigs.

Methods

Seven farms from two production companies were included in this study with 669 pens and a total of 19,190 pigs. The average pig age was 17.2 ± 2.5 weeks of age. On each farm every pig in every pen was evaluated for the presence of flank, tail, and ear lesions and the pen size was measured and the pen location recorded. Feed samples from each farm were tested at a Veterinary Diagnostic Analytical Chemistry Lab and the minerals, sodium and zinc, were quantified by ICP-MS and the mycotoxin screening panel was done by LC-MS/MS. The mycotoxins fumonisin and deoxynivalenol (DON) were detectable on the panel and thus included in the statistical analysis. Mixed effect logistic regression was used for models at the farm and pen level. At the farm level, the fixed effects examined were sodium, zinc, DON, fumonisin, age, pen size, and total pigs per pen. The random effect included was production company. At the pen level, the fixed effects included pen size, number of pigs, and sq ft per pig. The random effect was farm.

Results

At the farm level, there was a significant negative relationship between fumonisin concentration in the feed and prevalence of flank lesions ($\beta = -.05$; $P < .001$). Fumonisin levels in the feed were positively correlated with the prevalence of ear lesions ($\beta = .06$; $P = .001$). Tail lesion prevalence was not influenced by any of the farm level variables. At the pen level, there was a positive significant relationship between the square footage per pig and tail bite prevalence ($\beta = .002$; $P = .001$). There was a significant negative relationship between the square footage per pig and ear lesion prevalence ($\beta = -.001$; $P = .02$). None of the pen level variables were significantly related to flank lesion prevalence.

Discussion

The relationship between fumonisin and increased ear lesions and decreased flank lesions may be influenced by the type, concentration, or presence of toxin binder used in the feed, which was not disclosed by the farms so more research on this relationship is warranted. We did not find any relationship between the farm level variables we recorded and tail lesions which could be a result of the similar environment and layout of the farms leading to limited variability in husbandry and farm type. The relationship between increased pen square footage and increased tail lesions may be related to farmers moving pigs into sick pens where they have higher space allowance or it could be a result of increased activity and ability to move around when stocking density is lower. Future randomized studies could focus on varying square footage and characterizing the severity of the tail lesions in relation to number of pigs in a pen and square footage per pig. Since ear tip necrosis may have infectious causes, the increase in ear necrosis as square footage decreased could be linked to pathogen spread between pen-mates. Future analysis of whether the location within the facility is related to the occurrence of these lesions is on-going.

https://doi.org/10.54846/am2025/80

Evaluation of PRRS modified live virus vaccination in combination with killed autogenous PRRS vaccination on performance of growing pigs challenged with 1-4-4 PRRS virus

S. Boss[1], BS; P. Thomas[2], DVM, MS; C. A. A. Moura[2], DVM, PhD; R. Evelsizer[3], DVM; J. Risser[3], DVM
[1]Iowa State University, College of Veterinary Medicine, Ames, Iowa; [2]Iowa Select Farms, Iowa Falls, Iowa;
[3]Elanco Animal Health, Greenfield, Indiana

Introduction

Porcine reproductive and respiratory syndrome (PRRS) virus has negatively impacted pork producers for decades due to its high genetic variation in field isolates, making it challenging to immunize against effectively. Economic losses occurring in the finishing stage typically result from decreased weight gain and increased mortality, resulting in significant costs to the industry. The objective of this study was to evaluate whether growing pigs vaccinated with a commercial PRRS modified-live virus vaccine (MLV) in combination with a killed PRRS autogenous vaccine (KV) would have improved performance in comparison to pigs vaccinated with a commercial MLV vaccine alone when challenged with 1-4-4 PRRS virus.

Methods

For the study, pigs were selected from a PRRS negative (category IV) commercial sow farm. Weaned pigs were placed at a PRRS clean wean-to-finish site and co-mingled in pens. The study started one week post weaning when 330 sampled pigs were individually weighed, ear tagged, and allocated to a treatment group on a three-pig rotation. On day 1, group 1 was given 1 mL of PRRS MLV, and group 2 and 3 were given 1 mL of PRRS MLV and 1 mL of PRRS KV. Vaccine reactions were monitored and recorded. On day 21, group 3 was given a second dose of 1 mL PRRS KV. Prior to the challenge, trial pigs were confirmed positive for PRRS MLV via PCR sequencing using a clamp for a specific vaccine strain to differentiate from a wild-type infection. On day 42, ear tags were identified, and initial weights were taken for each pig. All treatment groups were then injected with 1 mL of 1-4-4 L1C PRRS positive serum diluted with sterile water to achieve 680 viral particles per pig ($10^{2.8}$). The dosage used was 34.1x the median infectious dose of 50% exposed animals for 1-4-4 PRRS L1C variant found by a study done at the University of Minnesota.[1] Oral fluid samples were collected and sequenced 2 weeks post-injection to confirm successful inoculation. Mortalities, treatments, and pulls were recorded by a blinded farm manager for 4 weeks following live virus inoculation. On day 69, final individual pig weights were recorded using a market hog scale. Data analyses were completed with JMP 16. Individual pigs were the experimental units and analyses were conducted for final weight (AFW) and average daily gain (ADG) by running linear regression analyses. Weight on day 1 was included as a confounder in all the models. Significance level was .05 with a sample size designed to detect a 5-pound difference in weight among groups, with an 80% power and 95% confidence level.

Results

At initial vaccine administration, vaccine reactions were observed in 10.5% of pigs in treatment groups 2 and 3. No mortalities were observed from these reactions. Numerically, treatment group 1 had the highest AFW at 93.38 lbs and ADG at 1.17 lbs/day followed by treatment group 2 with AFW of 92.84 lbs and group 3 with AFW of 91.57 lbs. Although, after the challenge, treatment group 2 had the highest ADG of 1.60 lbs/day followed by group 3 with ADG of 1.56 lbs/day and group 1 with ADG of 1.55 lbs/day. Overall, no significant differences were observed among treatment groups for AFW or ADG.

Discussion

In this scenario, there was no statistical advantage to using a combination of PRRS MLV and PRRS KV for improving performance in PRRS challenged pigs. Although, failure to see a difference in outcomes between groups could have been due to each pig receiving a full dose of PRRS MLV vaccine 6 weeks prior to challenge. Clinical signs of PRRS were mild post-challenge, so an increased challenge dose may have been helpful in producing signs of disease. No positive controls (pigs unvaccinated when challenged) were used for comparison in this study.

Reference

1. Melini C.M, Kikuti M, Torremorell M. Do all PRRSV variants have the same level of infectivity? Swine Health Information Center. November 24, 2023. **https://mshmp.umn.edu/sites/mnshmp.umn.edu/files/2023-12/PARTICIPANT%20-%20SHMP%202023l24.21%20%5BPRRS%20variant%20infectivity%20level%5D.pdf**

https://doi.org/10.54846/am2025/81

A case study: Characterization of *Mycoplasma hyopneumoniae* replacement gilt exposure program in commercial sow systems

M. Perkins[1], BS; J. Pittman[2], DVM; J. Risser[3], DVM; R. Evelsizer[3], DVM
[1]Iowa State University College of Veterinary Medicine, Ames, Iowa; [2]Smithfield Hog Production, Smithfield, Virginia; [3]Elanco Animal Health, Greenfield, Indiana

Introduction

Mycoplasma hyopneumoniae (MHP) causes enzootic pneumonia in swine, presenting as a nonproductive cough that results in decreased growth rates and feed efficiency. Sows shedding MHP during lactation expose piglets who, if infected, later display clinical signs in grow-finish. Naïve gilts (P0s) entering a positive sow farm are most susceptible to MHP, so P0s and first parity sows (P1s) entering the farrowing units from multiplication sites with high health status have been identified as the highest-risk population to transfer MHP to nursing piglets. Exposure protocols inoculate P0s at an early age and control the period of MHP shedding before farrowing. This study aimed to identify diagnostic benchmarks that support effective MHP exposure and control protocols in replacement P0s entering commercial sow farms.

Materials and methods

A cross-sectional study was conducted from May through August 2024 to evaluate two swine production flows. Flow A included 2 nurseries, 2 gilt development units (GDU), and 12 commercial sow farms. Flow B included 2 nurseries, 2 GDUs, and 6 commercial sow farms. Flow A and B implemented MHP acclimation protocols in January 2023 prior to this study. The author analyzed data from 4 populations: nursery (9 weeks of age [WOA]), GDU (15 WOA), P0s recently placed on sow farms (27 WOA), and gestating P0s and P1s. Deep tracheal swab samples were collected from individual animals in each population using an inner post cervical artificial insemination catheter rod passed orally into the trachea to reach the tracheal bifurcation. The samples were evaluated for MHP by PCR and pooled in increments of 5. Deep tracheal swab samples from 30 P0s at two nurseries in both Flow A and B were collected before MHP exposure to confirm negative status across the populations; they were then challenged with MHP lung homogenate by aerosolization at 9 WOA.

Results

In Flows A and B, both nurseries tested PCR negative for MHP. Six weeks later, at the finisher, the same group of pigs were 100% positive on PCR for MHP with an average CT level of 25.6 (20.6-29.7) and 26.4 (19.6-34.3) in respective Flows. Each Flow had 2 farms used for individual sampling of recently placed P0s on sow farms: Flow A farms were 60/60 and 45/60 positive on PCR for MHP with an average CT level of 29.4 (21.5-36.9); while Flow B farms were both 60/60 positive on PCR for MHP with an average CT level of 25.9 (19.9-35.9).

In Flow A, P0s and P1s due-to-farrow in the next turn were individually sampled from 12 commercial sow farms: 24/1440 samples were positive on PCR for MHP with an average CT level of 33.6 (28.7-37.7). In Flow B, P0s and P1s due-to-farrow in the next turn were individually sampled from 6 commercial sow farms:12/713 samples were positive on PCR for MHP with an average CT level of 29.6 (23.0-34.9).

Discussion

This study focused on characterizing the efficacy of MHP exposure in replacement P0s and monitoring MHP control in commercial sow herds. Since P0s pose the highest risk of MHP exposure to their nursing piglets, an acclimation protocol aims to minimize MHP shedding at first farrowing. The results of this case study reveal that after a controlled MHP exposure in the nursery, shedding begins at GDUs and continues at the time of P0 placement on commercial sow farms. Controlled MHP exposure provides incoming P0s with an immune response upon subsequent exposure and allows shedding to end before interaction with the most naïve population: nursing piglets. Diagnostic testing at both GDUs and P0s recently placed on sow farms indicates that the MHP exposure protocol is effective. This study's low MHP prevalence detected among due-to-farrow P0s and P1s suggests minimal pathogen shedding and supports implementing MHP exposure protocols in commercial sow systems.

 https://doi.org/10.54846/am2025/82

Correlation between velocity patterns of pig spermatozoa and phagocytosis by neutrophils

Yaneli Hernandez-Flores[1], EMV; Adan Sepúlveda-Montes[2], DVM, MS; Gonzalo Rocha-Chávez[3], DVM, PhD
[1]Vet student, University of Guadalajara; [2]Centro Universitario del Sur, U de G;
[3]Departamento de Promoción Preservación y Desarrollo de la Salud, CUSUR, U de G

Introduction

Artificial insemination (AI) in pigs is currently the standard reproduction tool as it brings benefits like prevention of health risks and increasing genetic progress. However, sperm cells that underwent stressing manipulation such as freezing or sex sorting shows not only reduced viability, but also impaired ability to fertilize eggs.[1] Furthermore, it is a well-known fact that spermatozoa elicit an immediate immune response of the inseminated sow and phagocytosis of spermatozoa by polymorphonuclear neutrophils (PMNs) is the first line of defense and hence the main clearance mechanism of the reproductive tract. This response triggered when the first drop of semen arrives the sow's tract starts with slow or immotile sperm cells.[2] On the other hand, velocity is not a common parameter to consider in a regular semen evaluation test, but specific parameters such as straight-line velocity (VLS), curvilinear velocity (VCL) or average path velocity (VAP) are detailed parameters of the spermatozoa movement that can be accomplished with computer assisted semen analysis (CASA).[2.] The objective of this study was to correlate three velocity parameters of spermatozoa in a pig semen dose and its percentage of phagocytosis of blood derived PMNs. We hypothesize that the slower the sperm cell moves, the more susceptible is of being eaten by PMNs.

Methods

Thirty-seven pig semen doses were analyzed and VAP, VLS and VCL was determined by CASA system (Androvision, Minitube Germany). Simultaneously an aliquot of each semen dose was subjected to a phagocytosis assay using blood derived PMNs as described by Jing-Chun.[2] Results were expressed as a percentage of sperm cell phagocyted after incubation of 45 min. Finally, the results were subjected to a Pearson's correlation analysis and significance was set to $P < .05$.

Results

No correlation was found between VCL or VSL with phagocytosis ($r = -0.29$ and -0.53 respectively $P > .05$); however, a strong negative relationship $r = -0.7215$ was found between VAP and phagocytosis indicating that the slower the average path velocity of a sperm cell, the greater the phagocytosis is (Figure 1).

Discussion

Slow-moving spermatozoa is not a concern in normal artificial insemination. However, when special reproduction techniques are being used, velocity may play an important role in fertilization. Previous studies correlated specific velocity parameters with components of seminal plasma in fishes[3]; however, few studies in pigs correlate phagocytosis and sperm cell velocities. The findings in this study may serve as a basis in the quest for a method of improving velocity in weakened but "specialized" sperm cells.

Figure 1: Phagocytosis by polymorphonuclear neutrophils (PMN). (1) bacteria being eaten by a PMN. (2) intact sperm cell and (3) sperm cell inside a PMN.

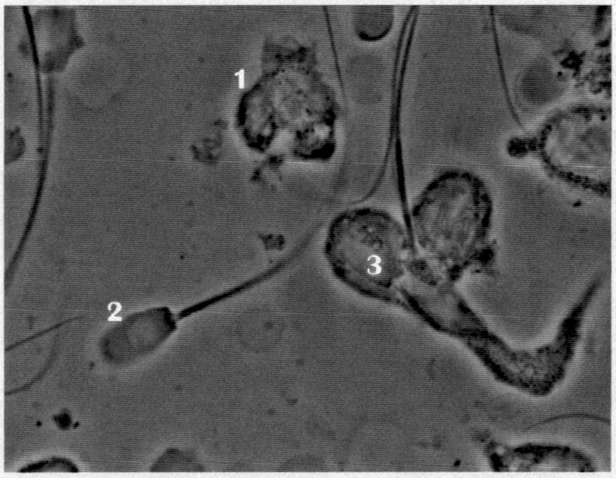

References

1. Dominguez FJ, Ngula J, Tejedor Sl, Alegre GB, Martinez PF, Alonso VM, Gonzalez MR, Gomez SC. (2017). Sperm stimulants in swine artificial insemination. Cría y Salud, Journal of Veterinary Medicine; 13 (66): 1-16.

2. Jing-Chun Li, Hechuan Wang, Yanbing Li, Minghui Guo, and Qun Zhang. (2023). Dexamethasone affects the chemotaxis and phagocytic activity of neutrophils for boar spermatozoa and the quality of liquid preserved boar semen in vitro. Theriogenology. 195 (2): 46-54.

3. Mustafa Özgür, Selim Erdoğan, Songül Aydemir, and Hatice Yumuşakbaş. (2023). Evaluation of relationship between sperm cell velocities and fatty acids contents of semen seminal fluid in the two trout fish species. Pakistan Journal of Zoology. 55. 863-870.

https://doi.org/10.54846/am2025/83

Use of a progesterone test strip for management of replacement gilts and weaned P1 sows

Maeve Davis[1], BS; Kayla Blake[2], DVM, MAg; Jessica Davenport[2], DVM; Kimberlee Baker[2], DVM, MS
[1]Virginia-Maryland CVM, Blacksburg, Virginia; [2]JBS Live Pork, Russellville, Arkansas

Introduction

Primiparous sows (P1s) are at risk for post-weaning anestrus, which is managed by injection with PG600 and review of management practices. PG600 acts as an FSH and LH analog in swine and promotes growth of follicles and ovulation in anestrous weaned sows. Due to limited supply, it must be targeted for gilts with inactive ovaries (prepubertal or post weaning anestrus). Another breeding source is replacement gilts who are often synchronized with altrenogest, which must be applied when animals are post pubertal. With appropriate use, standing heat is expected seven to nine days after removal of altrenogest. Understanding the progesterone status of animals is critical for the proper application of both PG600 and altrenogest. Recently a new pig-side progesterone test strip (PTS) has become available to approximate the level of progesterone in sows or gilts on site instead of sending it to a diagnostic lab, saving time and cost. The objective of this study was to evaluate the PTS for the management of anestrus P1s with proper PG600 use and for the evaluation of pubertal status of gilts before altrenogest treatment to ensure the gilts had reached puberty.

Materials and methods

The PTS is a lateral flow immunoassay that requires three drops of serum and is read after 15 minutes. High (>10 ng/mL) P4 levels are consistent with either pregnancy or diestrus (luteal phase) and will be positive on the test window. Low P4 levels indicate anestrous or follicular phase and will be negative on the test window. 38 commercial P1s not exhibiting behavioral estrus 12 to 19 days post weaning were enrolled, body condition scored (BCS) with MK calipers and tested with the PTS. P1s that tested negative were given PG600 and all enrolled sows were followed until 40 days post weaning during which time all breed or cull dates were recorded. For the second study objective, 225 replacement gilts at 27-weeks of age were enrolled and tested using the PTS on the day of arrival to nine sow farms and then started on the altrenogest regimen within two days of arrival depending on the farm protocol. For these gilts, animal ID, PTS results and BCS were recorded. Animals that did not breed within 29 days after arrival (14 days after removal of altrenogest) were retested with the PTS. Statistical analysis was done with Medcalc.

Results

Of the 38 P1s enrolled, 22/38 (58%) were negative on the PTS and were given PG600 and 16/38 (42%) were positive on the PTS and therefore were not given PG600. 82% of sows that received PG600 bred and of these, 75% bred within 7 days after treatment with an average of 5.3 days from treatment to breed. There was no difference in BCS (1.96 for sows receiving PG600 and 2.06 for sows not receiving PG600, $P = .70$). There was also no difference in percent bred (82% for sows that received PG600 and 93% for sows that did not receive PG600, $P = .33$) or percent culled (18% for sows that received PG600 and 6.3% for sows that did not receive PG600, $P = .30$). There was a trend towards shorter wean to breed interval (21.2 days for sows that received PG600 vs 23.9 days for sows that did not receive PG600, $P = .17$). Of the 225 replacement gilts enrolled, 65% ± 0.063 (95% confidence interval) were positive using the PTS at arrival (Table 1). 86% of the gilts enrolled bred within 29 days, 10% were not bred and were either positive at arrival or at retest and 4% were not bred and were negative at both arrival and retest (Table 1).

Discussion

For the weaned P1s, the progesterone test strip saved 16 doses of PG600, which at $11.00/ dose, saved $176.00. The decreased wean to breed interval on the sows receiving PG600 was worth $129.60 (2.7 days * $3.00 per non-productive day* 16 head). The benefit of the testing program was $305.60 or $8.04 per head tested, not including the labor to collect sera. In a randomly cycling population of replacement gilts 14/21 (67%) should be in the luteal phase and positive on the PTS, which was observed in this study (58%-71% positive (95% CI)). This demonstrates that the PTS can be successfully used to determine if a group of animals has reached puberty and is eligible for treatment with altrenogest as an alternative to heat-no-serve in production systems.

PG600 is a registered trademark of Merck Animal Health.

Table 1: Gilt enrollment and reproductive outcomes by farm

Farm	Gilts enrolled	Percent gilts progesterone positive at arrival	Percent bred	Percent pubertal + gilts "missed" by breeding department	Percent non-pubertal gilts
1	25	52	88	4	8
2	25	60	96	0	4
3	25	48	96	4	0
4	25	64	100	0	0
5	25	72	84	4	12
6	25	68	72	20	8
7	25	75	92	8	0
8	25	80	64	36	0
9	25	68	80	16	4
Total	225	147	193	23	9
Percent		65	86	10	4

https://doi.org/10.54846/am2025/84

Can swine transport trailers be used to evaluate the *Salmonella* status of a farm?

Casondra Snow[1]; Melissa Farber Billing[2], DVM, MVS; Zoe Wolfe[3], DVM; Joshua Duff [3], DVM; Fernando Leite[2], DVM, PhD
[1]College of Veterinary Medicine, University of Minnesota, St. Paul, Minnesota; [2]Boehringer Ingelheim Animal Health USA Inc., Duluth, Georgia; [3]Country View Family Farms, Middletown, Pennsylvania

Introduction

Salmonella infections are often subclinical with intermittent shedding in finishing pigs, but the stress of transportation has been shown to induce more *Salmonella* shedding.[1,2] This increased shedding in the transport trailers and in lairage at the packing plant may lead to an increased likelihood of contamination of the carcasses. The objective of this study was to evaluate if swine transport trailers can be used to determine the *Salmonella* status of a farm at the time of marketing.

Methods

Within three hours of load-out for marketing, six environmental samples were collected from five finishing farms using EnviroBooties® for a total of thirty samples.[3] Pens selected for sampling were spatially distributed throughout the barn and were walked in the same pattern every time. To control for *Salmonella* not associated with the pigs, two samples were collected with EnviroBooties® from one trailer at each farm before the pigs were loaded. Two samples were also taken from the same trailer after all of the pigs were off-loaded at the packing plant. One sample from the lower deck and one from the upper deck of every trailer was collected using the same walking pattern each time. Samples were sent on ice to the Iowa State University Veterinary Diagnostics Laboratory where they were enriched with tetrathionate and Rappaport Vassiliadis broths and cultured for *Salmonella*.

Results

Positive *Salmonella* cultures were only obtained from four of the thirty farm samples, all from farm one. The trailers were 70% (7/10) positive prior to loading and 90% (9/10) positive for *Salmonella* after the pigs were unloaded at the packing plant (Table 1).

Discussion

The use of EnviroBooties® was a successful method to sample and culture *Salmonella* from farms and livestock trailers. However, given the conditions of this study, the transport trailers would not be able to assess the *Salmonella* status of the finishing farms. Interestingly, farm five was only positive for *Salmonella* after the pigs were unloaded from the trailer at the plant indicating that this sampling method may be worth additional exploration. Further investigation into points of trailer contamination is warranted.

References

1. Stege H, Christensen J, et. al. Prevalence of subclinical Salmonella enterica infection in Danish finishing pig herds. *Preventative Veterinary Medicine,* 2000;44:3-4, 175-188. **https://doi.org/10.1016/S0167-5877(00)00103-3**

2. Isaacson R, Firkins L.D, Weigel R.M, et.al. Effect of transportation and feed withdrawal on shedding of Salmonella Typhimurium among experimentally infected pigs. *American Journal of Veterinary Research,* 1999;60:9. **https://doi.org/10.2460/ajvr.1999.60.09.1155**

3. Fangman T, Cline G. Evaluating the use of an Envirobootie to Detect Lawsonia and Salmonella from Known Positive Concrete Surfaces. IPVS 2016. **https://www.aasv.org/swine-information/?item=21244**

Table 1: Percent of positive *Salmonella* cultures for each sample location

Farm	Pen floor culture positive	Pre-loading trailer culture positive	Post-unloading trailer culture positive
1	67% (4/6)	100% (2/2)	100% (2/2)
2	0% (0/6)	100% (2/2)	100% (2/2)
3	0% (0/6)	50% (1/2)	50% (1/2)
4	0% (0/6)	100% (2/2)	100% (2/2)
5	0% (0/6)	0% (0/2)	100% (2/2)

https://doi.org/10.54846/am2025/85

An evaluation of sow feeding behavior and stanchion preference in gestation pens with 48-inch stanchions

B. Rogers[1], BS; R. Stika Jensen[2], DVM; C. Haden[2], DVM; R. Edler[2], MS; R. Retterath[3], BS; C. Cortes[2], BS; D. Stoltz[2], BS
[1]University of Illinois at Urbana-Champaign, Urbana, Illinois; [2]Pipestone Veterinary Services, Pipestone, Minnesota; [3]University of Minnesota, St. Paul, Minnesota

Introduction

In the United States, there is a shift in sow housing moving from individual stalls to pen housing. This changes how sows behave in a group, especially when it comes to group dynamics at the time of feeding. The amount of feed supplied to gestating sows is limited to a set amount and consumed in a short time period. The feed is typically dispensed once per day in barns with pen gestation. Stanchions are in place at the feeding troughs in an effort to reduce feed aggression between sows. With this housing change, there is an increased demand to better understand sow behavior and interactions within the pen hierarchy. A previous study showed that sow aggressive actions were lessened in stanchions that were 48 inches long when compared to shorter lengths.[1] The primary objective of this study was to evaluate behavioral feeding patterns of sows in group housing in a modern pen design with 48-inch feed stanchions when fed once a day to identify if there is a preference or predictability with the stanchion each sow chooses to consume feed at each day. There are sustainability, cost, and welfare benefits to seeing if sows consume feed at the same stanchion each day.

Materials and methods

This study was conducted at a sow farm that utilized pen gestation post pregnancy check with 48-inch stanchions. Data was collected the week of mixing (approximately 40 days bred), then subsequently, every other week until farrowing. Each collection week, data was recorded for three consecutive days at a time. Five weeks of data was collected totaling 15 time points per sow. Sows were housed in groups of 30, with four pens being used for the study. Out of the 30 sows in each pen, 16 were randomly selected, resulting in 64 total sows on trial. The parities of the sows in the study ranged from parity zero (gilts) to parity eight. Two video cameras were used in each of the four pens, being placed on opposite ends of the pen to capture multiple vantage points. Video footage was collected each morning, starting at the moment feed was dropped; approximately 6:00 am. The video footage was reviewed to see which stanchion each sow on the trial preferred at the time of feed drop, with sow and stanchion recorded each day. Sows placed in "hospital stalls" during the study were removed from the trial at that time sampling. Because of this, 58 sows remained on the trial to the conclusion of the study. To obtain a more concrete understanding of sow feeding behavior in gestating pens, data was analyzed by "quadrant" or area of the pen, rather than a specific stanchion. Quadrant 1 had seven stanchions, quadrant 2 had six stanchions, quadrant 3 had ten stanchions, and quadrant 4 had eight stanchions. Quadrants 1 and 2 shared a side of the pen and quadrants 3 and 4 shared a side of the pen. There were a lesser number of stanchions on the quadrants 1 and 2 side compared to the side with quadrants 3 and 4, due to the hospital stalls being placed on the quadrants 1 and 2 side. Pearson's chi-square tests were constructed for each sow, comparing theorized visits to observed visits in quadrants. Each sow had a 25% theoretical chance of visiting each quadrant. Theorized visits used was a 25% chance of visiting each quadrant.

Results and discussion

Results are shown in Table 1 representing each pen's data. There were a limited number of sows (8, 13.8%) that showed quadrant preference. Among this group, there was no parity predilection. Of the sows that had preference, it appears that quadrant three was first choice. Under the conditions of this study, feeding location was not predictable and cannot be used for sustainability, cost and welfare benefits. Some limitations of this study were that only one pen design was used (48-inch stanchions), as well as one genetic line. Further studies are needed to understand the pen dynamics within stanchion group housed gestation sows.

Reference

1. Pipestone Technical Services. Stanchion Length for Pen Gestation. Pipestone, MN. 2020.

Table 1: Feeding quadrant frequency by pen and individual sow

Pen A		Quadrant					Pen C		Quadrant				
Sow ID	**Parity**	1	2	3	4	***P*-value**	**Sow ID**	**Parity**	1	2	3	4	***P*-value**
1	4	4	3	4	4	.99	1	4	3	1	5	4	.65
2	2	5	2	4	4	.87	2	2	2	3	7	2	.57
3	7	5	4	5	1	.58	3	5	3	5	5	2	.81
4	1	2	0	0	1	Excluded	4	7	2	1	6	6	.38
5	3	5	1	2	0	.31	5	6	5	5	2	2	.72
6	1	3	0	4	6	.25	6	5	3	4	6	2	.77
7	5	5	6	0	3	.21	7	2	2	1	9	3	.22
8	7	4	0	5	5	.25	8	7	5	4	4	2	.87
9	5	5	1	5	4	.58	9	6	2	3	5	4	.86
10	5	5	2	4	3	.86	10	3	3	3	0	0	Excluded
11	3	5	3	6	1	.50	11	8	7	4	4	0	.19
12	2	1	3	9	2	.22	12	3	1	6	4	4	.55
13	2	7	3	1	2	.41	13	4	3	1	7	3	.45
14	2	8	3	1	2	.31	14	8	0	0	11	4	**.01**
15	2	1	0	0	1	Excluded	15	7	6	1	4	4	.55
16	3	2	3	3	7	.64	16	7	5	5	2	3	.81
Pen B		**Quadrant**					**Pen D**		**Quadrant**				
Sow ID	**Parity**	1	2	3	4	***P*-value**	**Sow ID**	**Parity**	1	2	3	4	***P*-value**
1	0	3	1	5	3	.68	1	2	3	9	1	1	.16
2	0	0	0	0	0	Excluded	2	2	2	3	8	0	.13
3	0	7	0	1	0	.07	3	4	0	0	12	3	**.008**
4	0	4	1	4	4	.70	4	6	1	1	9	3	.15
5	0	1	0	4	1	Excluded	5	4	2	1	10	0	**.04**
6	2	3	0	5	6	.22	6	3	0	0	8	6	**.02**
7	3	0	15	0	0	**.0004**	7	5	3	0	0	0	Excluded
8	2	0	3	9	3	.11	8	3	5	4	3	3	.95
9	2	5	3	4	3	.95	9	2	6	3	3	2	.77
10	2	1	3	6	3	.57	10	6	1	0	9	4	.06
11	6	1	1	12	1	**.03**	11	2	0	1	2	0	Excluded
12	6	6	3	2	4	.77	12	2	4	5	2	2	.79
13	2	4	4	4	2	.92	13	8	1	2	6	4	.50
14	2	4	1	8	2	.30	14	2	4	9	0	2	.09
15	2	5	3	1	5	.58	15	4	3	4	5	2	.86
16	2	1	10	3	0	**.04**	16	6	0	0	8	6	**.02**

Color key for Table 1	Number of visits to a quadrant
White	0
Blue	1
Yellow	2
Green	3+

Poster 12 https://doi.org/10.54846/am2025/86

Regulation of periweaning body temperature is improved in piglets given higher doses of injectable iron dextran during the suckling period

M. Jones[1], BS; T. Petznick[2], DVM; E. Pratt[2], DVM; W. Lyons[3], DVM; C. Olsen[3], MS, DVM
[1]North Carolina State University College of Veterinary Medicine, Raleigh, North Carolina; [2]ArkCare, Omaha, Nebraska; [3]Pharmacosmos Inc., Watchung, New Jersey

Introduction

Optimizing injectable iron dosage has been the focus of recent research showing that higher iron doses offer multiple benefits to both pigs and producers. Notably, these studies have shown that pigs receiving two, 200mg doses of iron before 10 days of age have improved average daily gain from wean to market and are approximately 7 pounds heavier at marketing compared to those that receive just one 200mg dose.[1,2] One study reported changes in the expression of numerous genes of interest when different iron doses were given to suckling pigs. Interestingly, the TRPV1 gene had a > 4.0-fold increase in expression in piglets receiving two, 200mg doses of iron.[3] This gene encodes for Transient Receptor Potential Vanilloid-1 which is a sensory receptor involved with the detection and regulation of body temperature. These findings prompted the primary objective of this project, which was to evaluate the relationship between injectable iron dose and piglet body temperature during and after weaning.

Materials and methods

On a 4,500-head commercial sow farm in Nebraska, 16 crossbred gilts of similar size and weight were enrolled within 24 hours of birth and evenly allotted to the control (200mg iron) or treatment (400mg iron) group. At enrollment, birth weights and hemoglobin levels were recorded, and ear tags were placed for individual identification. At three days of age, all piglets received a 200mg dose of Uniferon® iron dextran, and treatment pigs received an additional 200mg dose three days later. To monitor body temperature without repeated handling, a novel device (Thermochron® DS1921G) was programmed to record a temperature reading every five minutes. The data loggers were attached to the skin in the axillary region to obtain 1,621 readings from each pig over the span of 12 hours pre-weaning through five days post-weaning. One day prior to weaning, body weight and hemoglobin were measured, and data loggers were affixed to the piglets. To attach the loggers, piglets were anesthetized with an intra-muscular tiletamine/zolazepam/ketamine/xylazine injection, and axillary skin was shaved and cleaned with surgical scrub. Loggers were secured with an adhesive medical patch then sutured around the circumference of the patch with 2-0 Vicryl® using aseptic technique. Once recovered from anesthesia, piglets were returned to their respective litters before weaning the next morning. Data collection began once all pigs had recovered from anesthesia and returned to a normal body temperature. Times for all pig movement events and environmental conditions at the nursery site were recorded. The ambient temperatures in the nursery throughout the rest of the week remained within expected ranges for summertime in Nebraska. At the conclusion of the data collection period, patches and sutures were removed to retrieve the data loggers.

Results and discussion

Throughout the collection period, there are clear daily patterns during which control pigs' temperatures consistently drop lower than the treatment pigs, even in a warm, summer environment (Figure 1). The treatment group exhibited an overall mean temperature that was 0.5°F higher than the control group, but the *P*-value is just .12. It is possible, however, that a significant difference could have been found with a larger sample size. There was a noticeable spike in the temperature of both groups post-placement that lasted 10 hours. The control pigs had a higher mean temperature over this period and exhibited greater variability overall, with higher maximums and lower minimums compared to the treatment group. Altogether, the additional 200mg dose of iron resulted in improved thermoregulation, bringing their average temperature closer to normal (101.5°F) during the evaluation period. Thermal stress can negatively affect pig health and divert energy away from growth. Improved thermoregulation with higher iron doses, potentially driven by increased TRPV1 expression, may enhance energy utilization in the post-weaning phase, thus contributing to the improved wean-market growth performance reported in previous studies.

Figure 1: Average temperatures of treatment vs control groups.

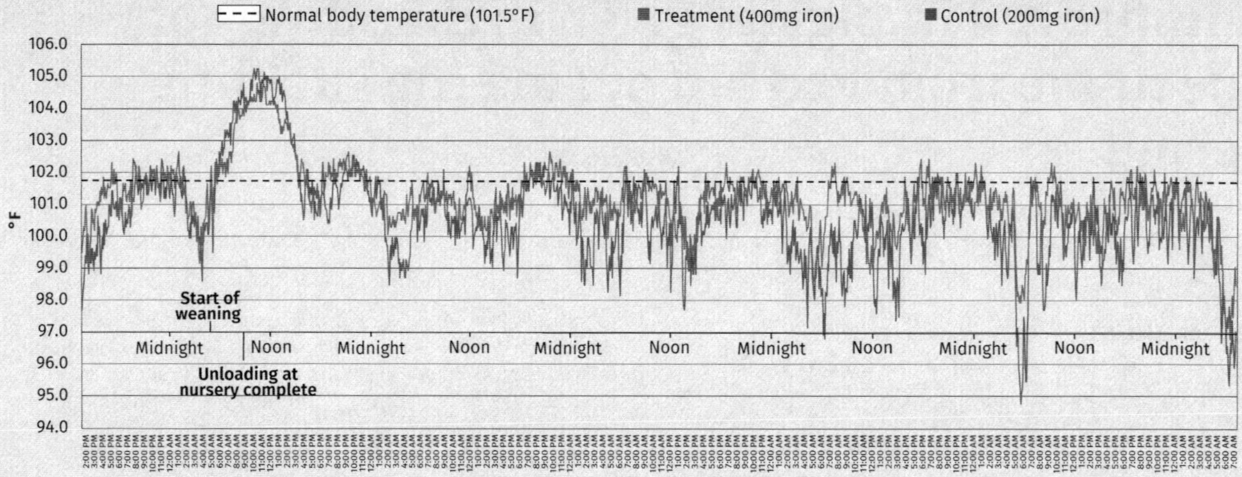

References

1. Gomez Cruz F., Krantz, S., Lyons, W. Pierce, J. L. (2024). Comparison of 200 mg injectable iron dextran vs 400 mg in either a single timepoint or split-dosed into two timepoints on growth and hemoglobin in pigs from birth to end of finishing. *American Association of Swine Veterinarians Annual Meeting.* **https://doi.org/10.54846/am2024/19**

2. Chevalier, T., Lyons, W., Paczosa, D., Rentfrow, G., Lindemann, M. (2023). A second iron injection administered to piglets during lactation improves hemoglobin concentration, growth performance, and carcass characteristics at slaughter. *Journal of Animal Science.* **https://doi.org/10.1093/jas/skad270**

3. Pierce, J., Lyons, W., Chevalier, T., Lindemann, M. (2024). Effects of a second iron-dextran injection administered to piglets during lactation on differential gene expression in liver and duodenum at weaning. *Journal of Animal Science.* **https://doi.org/10.1093/jas/skae005**

https://doi.org/10.54846/am2025/87

An evaluation of the use of artificial intelligence in lung lesions scoring

Sadie Sims[1]; Shuo Wang[2]; Dongyue Xin[2]; Heather Kittrell[3]
[1]College of Veterinary Medicine, University of Tennessee, Knoxville, Tennessee; [2]Boehringer Ingelheim, USA; [3]Boehringer Ingelheim Animal Health USA Inc.

Introduction

The percentage of lung lobe pathology (PLP) is currently used to evaluate lung lesions associated with swine respiratory diseases.[1] The availability of severe different methods can contribute to subject and variable scores by evaluators.[2-5] Artificial intelligence (AI) tools have been used to accurately identify and score pneumonia and pleurisy lesions in images collected from a swine abattoir.[2] The use of AI as a tool for scoring lung lesions is still underdeveloped and needs further exploration.[4,6,7] The objective of this study is to evaluate the use of AI technologies as a tool for scoring lung lesions associated with swine respiratory diseases when compared to scores provided by veterinarians.

Methods

A total of 825 images were used in this study. There were 729 and 96 images collected from previous studies in which animals were challenged with either *Mycoplasma hyopneumoniae* (MHP) or porcine reproductive and respiratory syndrome virus (PRRSV), respectively. These images were scored by either one (PRRSV) or two (MHP) veterinarians during image collection. The images were collected and annotated by the Subject Matter Expert (SME, veterinary student) using LabelMe version 5.5.0 (Labelme, open source, **https://github.com/labelmeai/labelme**). The SME annotated images were used to train deep convolutional neural networks (CNNs) within an internally developed image analysis platform.[8] To train the AI model, the data was split into 80% (662) for training and 20% (165) for testing the model. After training, the model was presented an image for segmentation. The model segmented the image into either background, healthy lung, or lesion. Lesions were then scored using pixel counting, with each pixel labeled as either 'healthy', 'lesion', or 'background'.

Results

Analysis showed that the Intersection over Union (IoU) is 0.99 for the background, 0.91 for the healthy, and 0.38 for the lesion areas. The median scores for veterinarian #1 (VS#1) and veterinarian #2 (VS#2) were similar, and the median scores for AI and SME were higher than the vet scores (Table 1).

Discussion

The AI was able to distinguish between background and healthy lung with high accuracy, however it had lower accuracy in distinguishing lesion areas. The median scores for both veterinarians were similar, and the AI and SME median scores were similar. However, there was a higher difference when veterinary scores were compared to AI or SME.

Table 1: Comparison between the minimum, median, and maximum lung lesion scores between veterinarian #1 (VS#1), veterinarian #2 (VS#2), artificial intelligence (AI), and subject matter expert (SME)

Group	Number	Min	Median	Max
VS#1	84	0.00	4.35	49.50
VS#2	72	0.00	2.25	42.50
AI	84	0.86	9.60	15.17
SME	84	0.00	10.07	25.59

Conclusion

An increased sample size and consistency in quality of photographs of lungs could improve the accuracy of the AI and may be able to provide more similarity in median veterinary scores compared to AI in future studies. Improvement of the AI model that has the ability to accurately score lungs could provide a more subjective tool for veterinarians in clinical and research settings.

References

1. Thacker, BJ, Strait, E., Kesl, L., Comparison of mycoplasmal lung lesion scoring methods. *AASV Annual Meeting.*

2. Bonicelli L, Trachtman AR, Rosamilia A, Liuzzo G, Hattab J, Mira Alcaraz E, Del Negro E, Vincenzi S, Capobianco Dondona A, Calderara S, et al. Training Convolutional Neural Networks to Score Pneumonia in Slaughtered Pigs. *Animals.* 2021; 11(11):3290. **https://doi.org/10.3390/ani11113290**

3. Garcia-Morante, B., Segalés, J., Fraile, L., Pérez de Rozas, A., Maiti, H., Coll, T., & Sibila, M. (2016). Assessment of *Mycoplasma hyopneumoniae*-induced Pneumonia using Different Lung Lesion Scoring Systems: a Comparative Review. *Journal of comparative pathology, 154* (2-3), 125-134. **https://doi.org/10.1016/j.jcpa.2015.11.003**

4. Maes, D., Sibila, M., Pieters, M. *et al.* Review on the methodology to assess respiratory tract lesions in pigs and their production impact. *Vet Res* **54**, 8 (2023). **https://doi.org/10.1186/s13567-023-01136-2**

5. Zimmerman, J. et. al., (2006). Disease of Swine (9th ed.) "Respiratory System", Table 7.4.

6. Hattab, J., Porrello, A., Romano, A., Rosamilia, A., Ghidini, S., Bernabò, N., Capobianco Dondona, A., Corradi, A., & Marruchella, G. (2023). Scoring Enzootic Pneumonia-like Lesions in Slaughtered Pigs: Traditional vs. Artificial-Intelligence-Based Methods. *Pathogens (Basel, Switzerland), 12*(12),1460. **https://doi.org/10.3390/pathogens12121460**

7. Steinmann, T., Blaha, T. & Meemken, D. A simplified evaluation system of surface-related lung lesions of pigs for official meat inspection under industrial slaughter conditions in Germany. *BMC Vet Res* **10**, 98 (2014). **https://doi.org/10.1186/1746-6148-10-98**

8. Zhao, H., Shi, J., Qi, X., Wang, X., & Jia, J. (2017, April 27). *Pyramid Scene Parsing Network.* arXiv.org. **https://arxiv.org/abs/1612.01105**

https://doi.org/10.54846/am2025/88

Effect of antioxidant supplementation in sows pre-farrow and during lactation on sow and piglet performance

A. Bohr[1]; A. Reever[2], MS, DVM; B. Burton[2], DVM; J. E. Guzman[3]; D. McKilligan[3]; K. Coulson[3]; V. Utrera[3], DVM
[1]Iowa State University, College of Veterinary Medicine, Ames, Iowa; [2]Suidae Health and Production, Algona, Iowa;
[3]TechMix, LLC, Stewart, Minnesota

Introduction

Previous studies have shown that the physiological and nutritional demands for a sow during late pregnancy and lactation impose an aggressive metabolic burden that eventually leads to systemic oxidative stress.[1] In a high stress event such as farrowing or weaning, the sow's total antioxidant capacity declines, resulting in detrimental effects on piglet survivability and sow health.[2] Therefore, investigation of antioxidant strategies to mitigate the effects of oxidative stress during this period is warranted. ImmuFend® is a combination of micronutrients and antioxidants designed to minimize the negative effects of oxidative stress and to fortify the immune system in pigs. The objective of this study was to evaluate the differences in sow and piglet farrowing and weaning metrics with two ImmuFend supplementation protocols under a natural porcine reproductive and respiratory syndrome (PRRS) challenge.

Materials and methods

A randomized controlled trial design was used at a 6,500 head AASV category 1[3] PRRSV positive sow farm with no piglet movement. Upon loading (3-12 days prior to farrowing) into farrowing crates, sows were randomly allotted, blocking by parity (P1, P2, P3), to one of three treatment groups (n = 275). Treatment group A received ImmuFend treatment upon loading and continued daily through weaning. Group B received ImmuFend treatment upon loading and continued daily through 5 days post farrow. Group C served as the control and received no treatment. ImmuFend was supplemented to each sow individually as a top dress with a target dose of 2.5 grams per day. Individual piglet weights were recorded at processing (3-5 days of age) and at weaning (19-25 days of age) (n = 680). All key farrowing performance metrics were extracted from the farm's record-keeping software. Statistical analysis was performed using Generalized Linear Mixed Model and MANOVA Repeated Measures functions of JMP PRO 18. Litter was the experimental unit. Results were considered significant if $P < .05$.

Results and discussion

ImmuFend supplementation had a significant effect on number of pigs weaned per sow and piglet average daily gain through lactation (Table 1). Numerical differences were only observed for litter characteristics and piglet weights. Therefore, vitamin and antioxidant strategies can provide an opportunity to mitigate oxidative stress for sows in a high disease state. This study suggests that supplementing sows with this vitamin and antioxidant strategy at the time of a health challenge may result in an improvement in piglet performance, more pigs out the door, and the potential to reduce the economic burden of these challenges for the producer. Due to the significant difference found between the number of weaned pigs in the control group and group A, an economic analysis was conducted. Given the current weaned pig price and cost of ImmuFend implementation ($0.11/sow/day), there was a favorable return on investments of 18:1 and 26:1 for each of the treatment group A and group B respectively. Additional labor costs of implementation and possible variable cost changes due to higher pigs weaned per sow were not included in the economic analysis. Further research should be conducted to validate findings from this study with a larger number sample size to provide insight for further supplementation protocols as well as economic analyses.

References

1. Al-Gubory KH, Fowler PA, Garrel C. The roles of cellular reactive oxygen species, oxidative stress and antioxidants in pregnancy outcomes. *Int J Biochem Cell Biol*. 2010;42(10):1634-1650. **doi:10.1016/j.biocel.2010.06.001**

2. Li Q, Yang S, Chen F, Guan W, Zhang S, Nutritional strategies to alleviate oxidative stress in sows. *Anim Nutr*. 2021;9:60-73. Published 2021 Nov 3. **doi:10.1016/j.aninu.2021.10.006**

3. Holtkamp DJ, Polson DD, Torremorell M, et al. Terminology for classifying swine herds by porcine reproductive and respiratory syndrome virus status. J Swine Health Prod. 2011;19(1):44–56.

Table 1: Effect of ImmuFend treatment on sow and piglet performance

Litter characteristics	A	B	C	SEM	P <
Born alive %	69.96	68.03	65.98	1.764	0.259
Mortality %	20.65	23.22	22.64	1.139	0.243
Total weaned, n	7.14[a]	6.55[ab]	6.10[b]	0.280	**0.025**
Piglet weights, lbs					
Processing	3.85	3.76	3.78	0.092	0.752
Weaning	11.81	12.18	11.49	0.354	0.368
ADG	0.497[ab]	0.509[a]	0.474[b]	0.007	**0.002**

ADG = average daily gain

https://doi.org/10.54846/am2025/89

Effect of ceftiofur sodium and stimbiotic interventions in the periparturient period on sow mortality

A. M. Senn[1], BS; A. Joy[2], BS; B. Kitting[3], VMD, MS; E. A. Jablonski[4], DVM, MS; D. Pyburn[4], DVM; M. Squires[4], BS; D. Amodie[4]; L. Gomez Paredes[3], MS; R. Schmitt[3], PhD; MS; K. A. Vonnahme[4], PhD
[1]Kansas State University, Manhattan, Kansas; [2]Oklahoma State University, Stillwater, Oklahoma; [3]Seaboard Foods, Guymon, Oklahoma; [4]Zoetis, Parsippany, New Jersey

Introduction

Sow mortality is a rising concern across the swine industry. According to MetaFarms, sow mortality has steadily increased since 2014, with a record high rate of 15.8% reported in 2023.[1] Sow treatment protocols vary greatly, and while veterinarians encourage appropriate use of anti-inflammatory or antibiotic products when needed, treatment decisions may occasionally be based on cost instead of clinical outcome studies. Globally, sows are the pigs least treated with antimicrobials.[2,3] Solutions to the ever-growing sow mortality rise are of significant interest to the industry. Recently published work in humans has shown a single oral dose of the antibiotic azithromycin can reduce the risk of postpartum sepsis and death by one-third among women who deliver vaginally, according to a large multi-country clinical trial funded by the National Institutes of Health. Only 1.6% of women in the study who received azithromycin during labor developed sepsis or died within six weeks after delivery, compared to 2.4% of those who received placebo ($P < .001$).[4] Stimbiotics are microbiome modulators that promote dietary fiber fermentation, which may influence the vaginal microbiota to improve clinical outcomes for sows post-farrow.[5] The objective of this study was to compare injectable ceftiofur sodium (NAXCEL®; Zoetis, Parsippany, NJ) and stimbiotic (SIGNIS®; AB Vista, Marlborough, Wiltshire, UK) interventions at the time of farrowing, alongside both interventions and a negative control, on subsequent sow clinical outcomes and mortality.

Materials and methods

Pre-trial diagnostic yielded numerous instances of *Escherichia coli*, *Staphylococcus hyicus*, *Streptococcus dysgalactiae* and *Streptococcus suis* isolates from both clinical sows (n = 8; sample type = systemic tissues) and healthy cohort sows (n = 48; sample type = vaginal swabs). Antimicrobial susceptibility results supported the use of ceftiofur. The study was conducted on a 10,000-sow, porcine reproductive and respiratory syndrome (PRRS)-positive, low prevalence farrow-to-wean farm. Sows were randomly assigned to four treatment groups when loaded into farrowing rooms. T01 (n = 555) received a NAXCEL (ceftiofur sodium sterile powder; NAX) injection within 24 hours after farrowing (3-5 mg/kg intramuscularly). T02 (n = 561) received SIGNIS (stimbiotic; SIG) top dress upon being loaded into farrowing and continuing through lactation (0.2 lbs/day; average duration = 21.88 days). T03 (n = 555) received both (BOTH) interventions as described above. T04 (n = 559) served as a negative control (CON). Sows were enrolled over a five-week period. Pigs were cross fostered across treatment groups to teat count within 24 hours. Sow was the experimental unit. The primary variable of interest was sow mortality from farrow to wean and from wean to subsequent pregnancy check. Secondary variables included the total number pigs born, pre-weaning mortality, subsequent pregnancy rate, and sows requiring additional antibiotic treatment post-farrow. Binary variables were analyzed by a generalized linear mixed model approach using SAS Proc Glimmix (SAS 9.4, Cary NC). The model considered treatment as a fixed effect and room and residual error as random effects. Treatment Least Square Means (LSMeans) were compared by the two-sided Student's t-test. Continuous variables, including transformed percent pre-weaning mortality, used the Proc Mixed procedure. Pre-weaning mortality LSMeans were back-transformed and presented as geometric means. All effects were tested at the 5% alpha level.

Results and discussion

NAX sows tended to have lower mortality from farrowing to pregnancy check and higher subsequent pregnancy rate compared to SIG (Table 1). Reproduction is allowed only after metabolic and immunologic demands are met; therefore, we hypothesize that NAX sows were in a healthy state allowing for successful pregnancy to be established. While there was no statistical difference in mortality in sows receiving BOTH, a decreased subsequent need for antibiotic intervention post-farrow compared to CON was observed. Overall, lowered sow mortality, improved pregnancy rate, and fewer individual sows requiring antibiotic treatment may be of clinical significance to practitioners. Antibiotic interventions should always be supported by diagnostic and clinical evidence to support responsible use.

Table 1: Effect of periparturient sow intervention on sow mortality, subsequent pregnancy rate, need for additional antibiotic treatment and litter characteristics

	T01 (NAX)	T02 (SIG)	T03 (BOTH)	T04 (CON)	SEM	*P*-value
Sows, n	555	561	555	559		
Sow mortality and performance						
Farrow to wean mort, %	2.66	4.02	2.84	3.18	0.01	.5721
Wean to preg check mort, %	0.90	2.14	1.08	2.33	0.004	.1575
Overall mort, %	3.07^{x}	5.71^{y}	3.80^{xy}	4.84^{xy}	0.01	.1516
Sow subsequent pregnancy rate, %	87.4^{x}	82.5^{y}	86.5^{xy}	83.6^{xy}	0.02	.0931
Sow additional antibiotic treatment, %	29.0^{xy}	28.2^{xy}	24.0^{y}	30.8^{x}	0.02	.0738
Litter characteristics						
Avg total born,† n	16.05	16.09	16.02	15.84	0.18	.7328
Piglet pre-weaning mortality, %*	14.16	13.17	13.59	13.07	0.01	.5987

NAX = injectable ceftiofur sodium; SIG = stimbiotic; CON = control

xy LSMeans ± SEM within a row with different superscripts differ; $P \leq .05$; unprotected F test

† Total born = live born + stillborn + mummified fetuses

* Geometric mean ± SEM reported

References

1. Eckberg B. U.S. sow mortality trends continue to climb. Available at: **https://www.nationalhogfarmer.com/livestock-management/u-s-sow-mortality-trends-continue-to-climb**. Accessed October 28, 2024.

2. Jensen VF, Emborg H-D, Aarestrup FM. Indications and patterns of therapeutic use of antimicrobial agents in the Danish pig production from 2002 to 2008. *J Vet Pharmacol Ther.* 2012;35(1):33-46. **https://doi.org/10.1111/j.1365-2885.2011.01291.x3**

3. Sjölund M, Postma A, Collineau L et al. Quantitative and qualitative antimicrobial usage patterns in farrow-to-finish pig herds in Belgium, France, Germany, and Sweden. *Prev Vet Med.* 2016;130:41-50. **https://doi.org/10.1016/j.prevetmed.2016.06.003**

4. Tita ATN, Carlo WA, et al. Azithromycin to prevent sepsis or death in women planning a vaginal birth. *N Engl J Med.* 2023 Mar 30;388(13):1161-1170. **https://doi.org/10.1056/NEJMoa2212111**

5. Monteiro MS, Poor AP, Muro BBD et al. The sow microbiome: Current and future perspectives to maximize the productivity in swine herds. *J Swine Health Prod.* 2022;30(4):238-250. **https://doi.org/10.54846/jshap/1277**

https://doi.org/10.54846/am2025/90

Evaluating postpartum prostaglandin F2 alpha (Lutalyse®) administration on preweaning growth in modern swine production

J. Bell[1], BS; T. Novak[2], DVM; N. Garbes[3], DVM; D. Amodie[3], MS; K. Vonnahme[3], PhD
[1]Iowa State University College of Veterinary Medicine, Ames, Iowa; [2]Veterinary Medical Center, Williamsburg, Iowa; [3]Zoetis, Parsippany, New Jersey

Introduction

Historical studies administering prostaglandin F2 alpha (PGF2α) postpartum report increased weaning weights when looking at a select parity group (sows and gilts < P5) but individual parity information was not reported.[1,2] Given the genetic advancements in swine breeding over the past 25 years, it is crucial to reassess whether modern sows respond similarly to postpartum PGF2α administration. Lutalyse® (dinoprost tromethamine injection; Zoetis, Parsippany, New Jersey) is a PGF2α approved for use in swine since 1979. The objective of this study was to evaluate adjusted weaning weights from litters of sows administered Lutalyse (LUT) 24-48 hours postpartum compared to control sows of modern swine practices.

Materials and methods

Sows from two commercial farms in Iowa were randomly assigned to treatment (LUT 2 mL IM 24-48 hours post-delivery of the first pig vs negative control (CON)) within a parity group (ie, 1, 2-3, 4+). On Farm A, a filtered, batch farrow farm, 360 sows from 3 batches were enrolled (LUT, n = 180; CON, n = 180); and Farm B, non-filtered where all sows enrolled came from two consecutive breed weeks, 316 sows (LUT, n = 158; CON, n = 158). Cross-fostering within treatment was allowed within the first 2 days of life. On day 2 and at weaning, pig number and litter weight were recorded. Adjusted weaning weight (AWW) and average daily gain (ADG) was calculated. The effects of treatment, parity group, and their interaction were analyzed by farm using a generalized mixed model approach (SAS 9.4, Cary, NC). Sow was the experimental unit.

Results and discussion

During initial data analysis, Farm A (LUT, n = 138; CON, n = 141 and Farm B (LUT, n = 132; CON n = 147) failed to show a difference in AWW and/or ADG by treatment or treatment by parity. Because cross fostering occurred outside the boundaries set for this project, a decision was made to look at litters with no foster on events. Table 1 shows data after removal of 372 sows from the analysis. Within a farm, the number of pigs on day 2 was similar ($P \geq .12$) across treatment groups. On Farm A pigs from LUT sows tended ($P = .08$) to be heavier than CON (3.6 vs 3.3 ± 0.2 lbs) on day 2, with no differences ($P \geq .65$) in weight for Farm B. Within a farm, the number of pigs weaned was similar ($P \geq .15$). On Farm A, AWW and ADG were greater in P1 litters whose dams received LUT compared to CON, with treatment not impacting the larger parity groups. On Farm B, AWW was reduced ($P = .03$) in P1 vs older parity dams, however, ADG tended to be affected by treatment and parity with pigs from P1 dams treated with PG having a greater ADG compared to CON. Use of PG in first parity dams shortly after farrowing may augment lochia clearance via increased myometrial contractions allowing for greater milk production, thus heavier AWW and improved ADG. Additional data should be collected to confirm these findings.

Table 1: Adjusted weaning weights (AWW) and average daily gain (ADG) for litters evaluated for the effects of treatment (control; CON or Lutalyse (LUT), parity group (1, 2-3, 4+) and their interaction

Variable	Parity 1		Parity 2-3		Parity 4+			*P* values		
	CON	LUT	CON	LUT	CON	LUT	SEM	Tx	Par	Tx*Par
Farm A, # sows	**4**	**13**	**17**	**17**	**5**	**5**				
Day 2 # piglets	13.8	13.5	14.1	13.7	13.8	13.2	0.5	.12	.40	.85
Day 2 wt, lb	2.8	2.9	3.6	3.9	3.5	4.0	0.3	.08	< .01	.77
Wean age, d	18.0	18.3	17.5	17.5	17.2	18.2	0.9	.44	.56	.70
Wean wt, lb	10.1	12.1	13.3	13.9	14.7	14.6	0.7	.06	< .01	.24
Wean, #	13.7	13.5	13.6	13.5	13.8	12.8	0.5	.16	.69	.46
AWW, lb	12.1^{a}	14.3^{b}	15.9^{cd}	16.7^{cd}	18.1^{e}	17.0^{de}	0.8	.16	< .01	.04
ADG, lb	0.44^{a}	0.54^{b}	0.59^{c}	0.61^{c}	0.69^{d}	0.62^{cd}	0.03	.34	< .01	< .01
Mortality, %	0.4	0.2	1.3	0.1	0.0	1.3	1.3	.90	.90	.12
Farm B, # sows	**8**	**3**	**26**	**21**	**32**	**37**				
Day 2, # piglets	14.6	14.7	14.7	14.5	14.3	14.4	0.5	.90	.30	.63
Day 2 wt, lb	3.0	2.8	3.2	3.3	3.1	3.1	0.3	.75	.07	.65
Wean age, d	23.6^{x}	22.4^{x}	$22.3y^{y}$	22.6^{xy}	22.3^{y}	21.7^{y}	1.0	.19	.09	.14
Wean wt, lb	13.0^{x}	14.1^{xy}	14.4^{y}	14.7^{y}	14.9^{y}	14.3^{y}	1.0	.52	.16	.14
Weaned, #	13.7	12.3	13.4	13.1	13.0	13.1	0.8	.15	.71	.31
AWW, lb	12.3	13.3	14.1	14.2	14.5	14.3	0.9	.41	.03	.34
ADG, lb	0.44^{x}	0.51^{y}	0.51^{y}	0.51^{y}	0.54^{y}	0.53^{y}	0.04	.18	.03	.12
Mortality, %	3.9	15.6	6.1	7.7	6.7	6.4	7.7	.10	.82	.27

AWW = ([Wean weight – day 2 weight]/[wean age – 2]) × [21+day 2 weight])
ADG = (weaning weight – day 2 weight)/(wean age – 2)
SEM = standard error of the mean; Tx = treatment; Par = Parity; largest SEM presented
abcde LSMean ± SEM within a row with differing superscripts differ; $P \leq .05$
xy LSMean ± SEM within a row with different superscripts differ; $P \leq .05$; unprotected F test

References

1. Morrow M, Britt J, Belschner A, et al.; Effect of injecting sows with prostaglandin F2 alpha immediately postpartum on subsequent reproductive performance. *J Swine Health and Production*. 1996; 4(2):73-78.

2. Kurt J, Dudley S, Bosch G, et al. Injection of prostaglandin F2 alpha in postpartum sows: Effects on lactation and fertility. *Proc AASP* 2000, 73-75.

https://doi.org/10.54846/am2025/91

Evaluating porcine circovirus 2 maternal immunity decline via indirect immunofluorescent antibody assay titers in wean pigs

I. Black[1], BS; E. Mahan-Riggs[2], DVM; N. Lauterbach[3], DVM; J. Okones[3], DVM
[1]North Carolina State University, Raleigh, North Carolina; [2]Passel Farms; [3]Pharmgate

Introduction

Porcine circovirus type 2 (PCV2) is a highly prevalent virus in swine production. PCV2 infection in growing pigs is commonly characterized by postweaning multisystemic wasting syndrome (PMWS) presented as slow growth, reduction in average daily gain, and increased mortality.[1] To minimize loss associated with PCV2, vaccinations are routinely administered to both sow and growing pig populations. Maternal immunity has been proven to interfere with a vaccine's establishment of immunity postweaning.[2] The objective of this study was to evaluate the decline in maternal immunity of PCV2 in growing pigs, via indirect immunofluorescent antibody assay titers (IFA) to maximize the immune response of a single dose PCV2 vaccination regimen.

Materials and methods

Three groups of piglets were bled to measure PCV2 antibody titers via IFA levels at three weeks, six weeks, and nine weeks of age. Specific PCV2b and PCV2d genotypes were targeted in IFA assays performed on serum of both piglets and sows. Group 1 (Gp1) consisted of 11 gilt litters and five parity one (P1) litters with three barrows selected randomly at weaning. Gp 1 sows did not receive a PCV2 vaccination prior to farrowing or gestation. A whole herd PCV2 vaccination was administered to sows and gilts 1 week before Gp 2 farrowed and 3 weeks before Gp 3 farrowed. Gp 2 and 3 consisted of seven gilt litters and six P1 litters. Three barrows and two gilts were tagged within 24 hours of parturition from each litter. Cross-fostering of tagged litters was not allowed during the trial. Following phlebotomy at six weeks, all pigs were vaccinated with a commercial PCV2/*Mycoplasma hyopneumoniae* (MHP) vaccine. Trial pigs were placed at four different wean to finish sites and located in pens central to each barn with other ear-tagged pigs from the same Gp.

Results

A positive IFA titer result was > 1:40. Gp 2 and 3 sows were 100% positive for both PCV2b and PCV2d IFA's at weaning. Comparatively, Gp 1 sows were 93.8%. Of the 23 and 21 pigs in Gp 1 positive for PCV2b and PCV2d IFA respectively at six weeks, 57% of the PCV2b and 76% of the PCV2d positive pigs presented negative (< 1:40) or suspect (1:40) at nine weeks. Figure 1 displays a significant difference ($P < .05$) of Gp 1 piglets compared to Gp 3 piglets for both PCV2b and PCV2d at weaning. Gp 2 pigs with positive titers at six weeks became negative or suspect by 68% for PCV2b and by 91% for PCV2d titers. Gp 3 positive titers at six weeks became negative or suspect by 24% for PCV2b and 51% for PCV2d at nine weeks. Titers in Gp 3 at 9 weeks differed ($P < .05$) from the other two Gps.

Discussion

Sows from all groups had positive PCV2b and PCV2d IFA titers at weaning. Results demonstrated piglets from whole herd PCV2/MHP vaccinated sows presented higher percent positive titers for both PCV2b and PCV2d at weaning when compared with piglets from Gp 1 sows. The high percentage of positive IFA titers at weaning indicate that vaccination at this stage may have challenges due to the potential of maternal antibody interference. All Gps presenting with positive IFA titers at six weeks, following vaccination, observed an increase in negative or suspect IFA titers of both PCV2b and PCV2d at nine weeks. Further evaluation of IFA titers should be conducted in future studies post-vaccination and with the addition of a PCV2 challenge. The results from this study indicate the need for further evaluation of single dose PCV2 vaccination regimens in the presence of high maternal immunity.

Figure 1: Positive IFA titers.

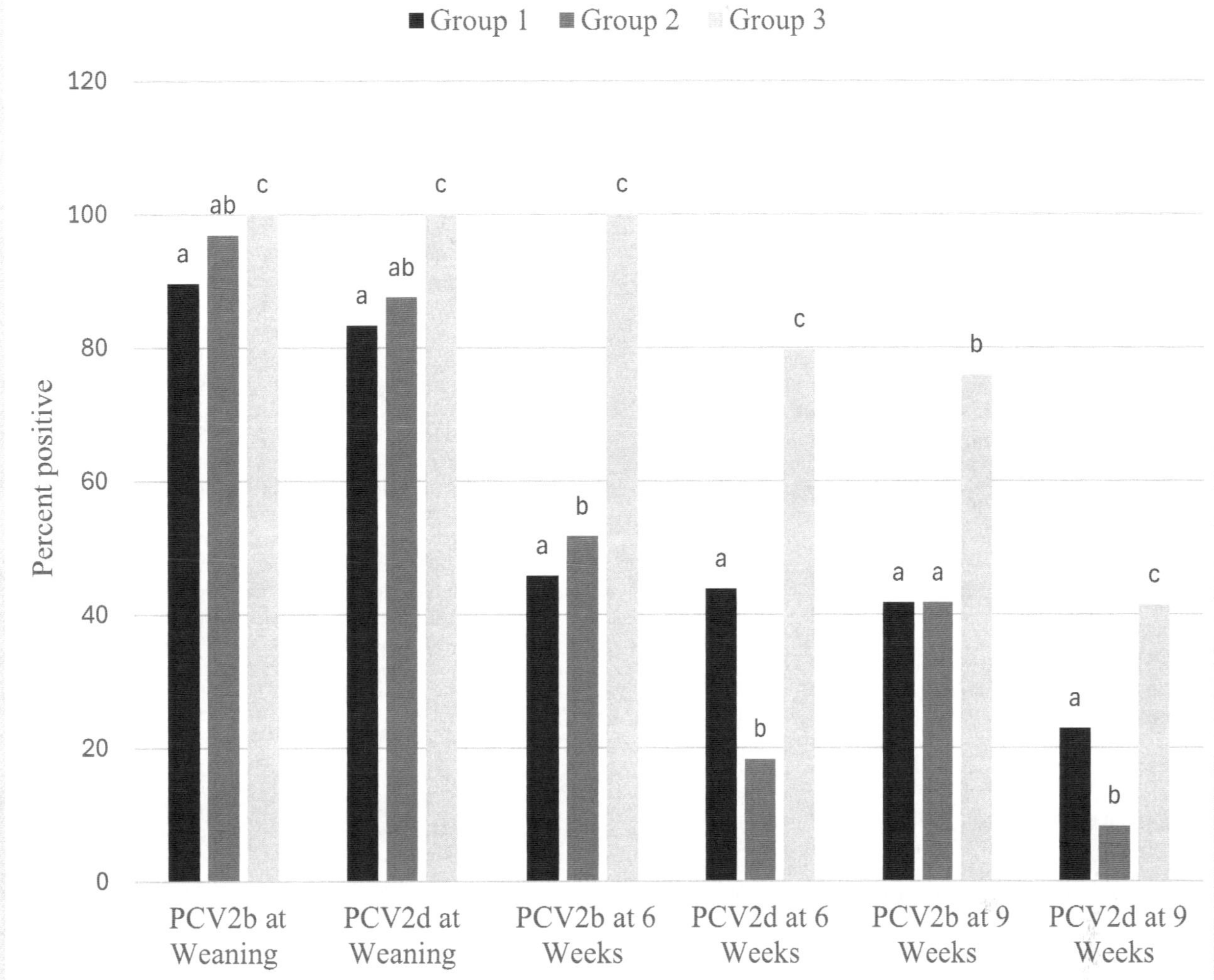

References

1. Zimmerman J, Karriker L, Ramirez A, et. al. Porcine Circoviruses. *Diseases of Swine.* 2012.

2. Figueras-Gourgues S, Fraile L, Segalés J, Hernández-Caravaca I, et al. Effect of Porcine circovirus 2 (PCV-2) maternally derived antibodies on performance and PCV-2 viremia in vaccinated piglets under field conditions. *Porcine Health Management.* 2019 Sep 5;5(1).

https://doi.org/10.54846/am2025/92

Sample type comparison for PRRSV detection in individually housed gestating populations

Cassidy Cordon[1]; Erin Kettelkamp[2]; Isadora Machado[3]; Amanda Sponheim[4]; Daniel Linhares[3]; Marjorie Schleper[5]
[1]College of Veterinary Medicine, University of Minnesota, St. Paul, Minnesota; [2]Swine Vet Center, St. Peter, Minnesota; [3]Iowa State University, Ames, Iowa; [4]Boehringer Ingelheim Animal Health USA Inc., Duluth, Georgia; [5]Christensen Farms, Sleepy Eye, Minnesota

Introduction

Porcine reproductive and respiratory syndrome virus (PRRSV) is a highly impactful pathogen in the swine industry, significantly affecting both production economics and animal welfare.[1] Effective management of PRRSV outbreaks relies heavily on early detection, which is essential for controlling the movement of pigs and personnel to limit disease spread.[2] Traditional methods of detecting PRRSV in individually housed gestation herds, such as serum or tonsil scrape sampling, require animal restraint via snare.[3] Alternative sample types that do not require restraint, could facilitate more efficient sampling in these herds.[4] Here we assess the detection of PRRSV positivity rates across various sample types, including environmental trough swabs, modified tonsil oral scraping (TOSc), and serum, in individually housed gestating swine populations.

Methods

Two commercial sow farms undergoing a PRRSV elimination program were sampled for this study. Farm A was 13 weeks post-PRRSV outbreak detection, six weeks post-oral planned exposure, two weeks post-live virus inoculation (LVI). It was producing piglets of variable quality, with piglet seroconversion observed post weaning, and had not begun monitoring processing fluids. Farm B was 40 weeks post-PRRSV outbreak detection, 36 weeks post-oral planned exposure, 33 weeks post-LVI, and had observed two sequential weeks of PCR-negative processing fluid samples. From each farm, groups of six sequentially housed animals were randomly selected from 20 regions, blocked by location, within the gestation and breeding barns. After feed drop, one dry Swiffer® sweeping cloth (trough swab) was used to collect saliva from the feed trough from each group of dams. Additionally, another trough swab was collected from the full length of each row in the barn. A modified TOSc tool was created using a post cervical artificial insemination rod covered with a 4"x4" piece of square cotton gauze, held in place with a rubber band.[4] TOSc and blood samples were collected from all animals in the study and pooled by group. To prevent cross-contamination, personnel changed gloves and sampling tools between groups. All sample types were tested for PRRSV via rt-PCR.

Results and discussion

PRRSV RNA was not detected via any sample type during this study. These findings provide insight into the limitations of detection at the specific time points sampled post-outbreak in gestating populations on these farms. It is important to consider potential limitations that despite following established procedures to minimize contamination and degradation, the potential for undetected errors in sample handling cannot be entirely ruled out. In a study performed by Peng Li et al,[4] the TOSc sample successfully detected PRRSV in a herd during the early phase of PRRSV infection. However, the results from this study differ, as the TOSc did not detect PRRSV in these herds. The differences are likely contributed to differences in sampling timing relative to the stages of PRRSV infection in sows, highlighting the importance of considering infection stages when evaluating the effectiveness of sampling methods. Additional research is needed to assess these sample types for PRRSV detection across various stages of infection.

References

1. Zimmerman, J. J. (Ed.). (2019). *Diseases of swine* (Eleventh edition.). John Wiley & Sons, Inc.

2. Mil-Homens, M.P., Jayaraman, S., Rupasinghe, K., Wang, C., Trevisan, G., Dórea, F., Linhares, D., Holtkamp, D., Silva, G. Early Detection of PRRSV Outbreaks in Breeding Herds by Monitoring Productivity and Electronic Sow Feed Data Using Univariate and Multivariate Statistical Process Control Methods. Transboundary and Emerging Disease. Volume 2024, Issue 1.

3. Lopez, W., Almeida, M., Silva, G., et al. Evolution of PRRSV monitoring: 3 decades of advance and improvement. *50th Annual Meeting of the AASV.* 2019.

4. Peng Li, Silva APSP, Moraes DCA, Yeske P, Osemeke OH, Magalhães ES, De Sousa E Silva G, Linhares DCL. Comparison of a novel rapid sampling method to serum and tonsil scraping to detect PRRSV in acutely infected sows. Prev Vet Med. 2024 Feb;223:106082.

https://doi.org/10.54846/am2025/93

Transition to single tier vaccine label claims: General overview

E. Larsen, MS; E. Russell, MS; B. Thacker, DVM, PhD, MBA
College of Veterinary Medicine, Lincoln Memorial University, Harrogate, Tennessee

Introduction

USDA's Center for Veterinary Biologics (CVB) has transitioned vaccine label claims from a multi-tiered scheme where vaccines were assigned claims of infection prevention (highest), disease prevention, aid in disease prevention and aid in disease control/reduction (lowest) to a single tier claim that states "this product has been shown to be effective for the vaccination against (a disease)." Firms were required to provide information including the raw study data to CVB for preparing individual Product Summaries that are available on CVB's website.[1] Raw data was not required for products approved prior to 2007. Our objective was to provide an overview of the study data that is available in the Product Summaries, provide insights for interpreting the study data, and identify potential limitations with regard to study design and data presentation.

Methods

From each Product Summary, the following information was obtained: **1)** Study type-efficacy (short term- "immuno" or duration of immunity- DOI) or field safety; **2)** Study design - number of doses, route of administration, age at vaccination, days post-vaccination when challenged, challenge pathogen and days post-challenge when necropsied or study ended; and **3)** Study results - metrics used for evaluation including macroscopic and microscopic lesions and pathogen level as determined by immunohistochemistry (IHC) or PCR. The variable type for each metric was classified as binary (positive/negative), categorical (0 to 2-4 score) or continuous (percent affected, qPCR). Whether raw study data was presented for each study within a Product Summary was also recorded. Excel was used for data management. Data analysis was limited to descriptive summaries and no inferential analyses were done.

Results

Product Summaries were available for 119 of the 120 products listed. Study design and results of "immuno" studies were available for 35.8% of the products; DOI studies, 19.2% and safety studies, 41.7%. Nine Product Summaries, mainly for platform vaccines (RNA particle, DNA, baculovirus vectored), contained only safety studies. The Product Summaries that contained no raw study information were primarily older vaccines that are used in breeding herds for reproductive failure and sow health (parvovirus, leptospirosis, erysipelas) or baby pig scours (*E coli*, Clostridium, TGE, rotavirus). For contemporary diseases, efficacy study data was available by disease: Lawsonia - immuno studies, 4 out of 6 products and DOI studies, 4 out of 6 products; *Mycoplasma hyopneumoniae* - immuno, 11 of 23 and DOI, 6 of 23; PCV2 - immuno, 13 of 13 and DOI, 11 of 13 (two DOI studies were not listed); PRRS, immuno, 12 of 14 and DOI, 5 of 13 (DOI studies not performed for reproductive failure); Salmonella - immuno, 2 of 7 and DOI, 0 of 7 (no DOI studies performed); and swine influenza - immuno, 13 of 16 (many product update studies) and DOI, 3 of 16.

With regard to interpreting the raw data, variable type was the main concern, especially with regard to understanding the severity of the challenge in the non-vaccinated control group. Presenting lesion (gross or microscopic) or IHC data as positive or negative precludes any assessment of disease severity or level of infection. This was observed with several PCV2 and lawsonia containing products. Other products assigned categories such as none, mild, moderate or severe enabled severity assessment although the categories were not always explained.

Discussion

This information provides a broad overview of licensing study data and emphasizes the limitation of using a binary variable approach for characterizing lesions in vaccination-challenge studies. CVB provides a precautionary statement in the Product Summaries to avoid comparing one product to another due to differences in study design and execution. Swine practitioners and producers may find this information useful when discussing and evaluating vaccination protocols recommended by vaccine suppliers.

Reference

1. USDA Center for Veterinary Biologics Product Summaries. Accessed multiple times in 2023 and 2024. **https://www.aphis.usda.gov/veterinary-biologics/product-summaries**

https://doi.org/10.54846/am2025/94

Could viral vaccines be used as alternatives for antimicrobials? Insights from US swine veterinarians

Danqin Li[1], Xirui Zhang[1,4]; Michael D. Apley[2]; Locke Karriker[5]; Joseph F. Connor[6]; Corinne Bromfield[4]; Brian Lubbers[2]; Hatem Kittana[3]; Dustin Pendell[7]; Jishu Shi[1]

[1]Center on Biologics Development & Evaluation and Department of Anatomy and Physiology; [2]Clinical Sciences; [3]Diagnostic Medicine/Pathobiology, CVM, Kansas State University, Manhattan, Kansas; [4]CVM, University of Missouri, Columbia, Missouri; [5]Department of Veterinary Diagnostic and Production Animal Medicine, CVM, Iowa State University, Ames, Iowa; [6]Carthage Veterinary Service, LTD., Carthage, Illinois; [7]Department of Agricultural Economics, COA, Kansas State University, Manhattan, Kansas

Introduction

According to the FDA, in 2023, swine accounted for the highest percentage (44%) of medically important antimicrobial sales among all food-producing animals as measured in kilograms. Additionally, swine was also the only category to have experienced four consecutive years of rising antimicrobial sales from 2020 to 2024.[1] Swine viral diseases often result in severe clinical signs due to secondary bacterial complications, necessitating the use of antimicrobials.[2] Vaccines can effectively reduce the infectious diseases burden and greatly decrease the demand for antimicrobials in swine health management. Veterinarians play a vital role in guiding producers' vaccine use; however, reports about their insights on current vaccines are lacking. This study seeks insights from US swine veterinarians on the safety, efficacy, availability, and cost of current viral disease vaccines, aiming to identify limitations and explore opportunities for better vaccines.

Methods

Data collection included two phases: an online cross-sectional survey of US swine veterinarians, followed by interviews with those who consented to provide further insights. This study sought their insights on the safety, efficacy, availability, and cost of current viral disease vaccines. The survey was based on their expert opinions and practical experiences. Data were initially exported from Qualtrics to Microsoft Excel for preliminary statistical management, where key information was extracted and summarized. The summarized data were then imported into GraphPad Prism (version 10) for advanced statistical analysis and visualization.

Results

Between June and August 2024, survey invitations were sent to 202 veterinarians. 23% (46/202) of veterinarians responded, 19 completed the survey and 5 participated in the interviews. The 19 respondents had an average of over 24 years of experience. Most practiced in the Midwest, which has the highest density of swine in the US, with an average responsibility of 482,736 pigs per veterinarian. Porcine reproductive and respiratory syndrome (PRRS) (84%), swine influenza (SIV) (79%), and rotaviral enteritis (79%) are the top three viral diseases urgently needing better vaccines (Figure 1). Over 25% of veterinarians reported using antimicrobials to manage secondary bacterial complications for these three diseases. Our study also found out that current viral disease vaccines share common challenges but also have vaccine-specific limitations. A major issue affecting all three viral vaccines is the lack of cross-protection against emerging variants. Specific challenges include the risk of virulence reversion in PRRS vaccines, the interference of maternally derived antibodies (MDA) in SIV vaccines, and the difficulties in adapting type C rotavirus (RVC) to cell culture. Notably, while veterinarians generally view the current vaccines as moderately effective and somewhat expensive, they are still willing to pay over 30% more for better vaccines, which shows their desire to have better vaccines in their toolbox.

Discussion

By identifying the common and vaccine-specific challenges, we were able to propose targeted improvement strategies for each vaccine. For PRRS, needle-free intradermal administration shows promise, as it offers faster immune responses, lower vaccine viremia, and reduces the risk of recombination.[3] For SIV, adenovirus-vectored vaccines can resist MDA interference and provide long-lasting immunity.[4] For RVC, customized prescription vaccines can bypass propagation issues while maintaining a similar cost. Our study suggested that using vaccines as alternatives to antimicrobials will be a long-term process, requiring the insights from veterinarians in exploring effective and sustainable solutions.

Figure 1: Swine diseases urgently need vaccines or better vaccines. The top 3 viral diseases in urgent need are: Porcine reproductive and respiratory syndrome (PRRS), swine influenza virus (SIV) and rotavirus, highlighted in purple boxes.

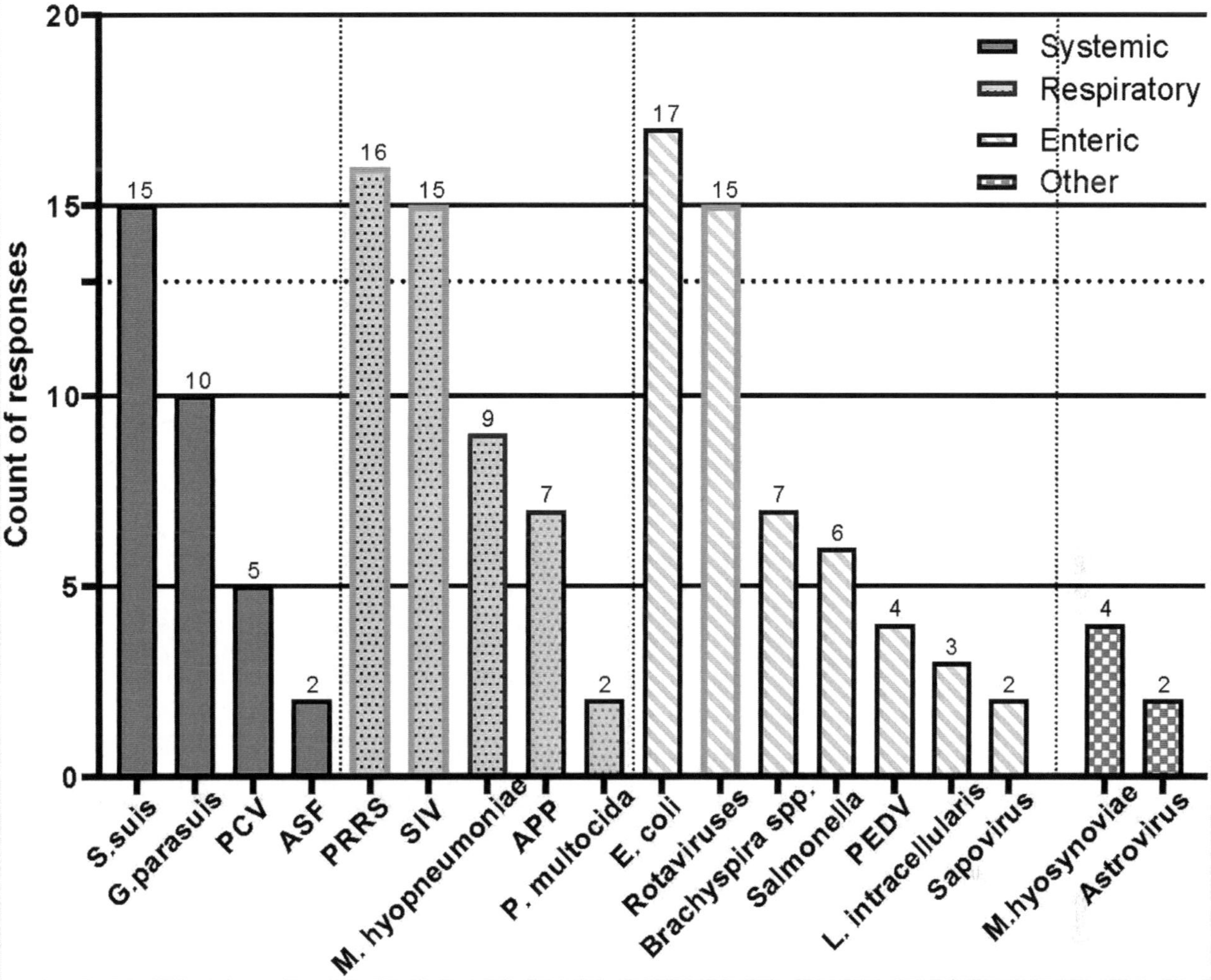

PCV – porcine circovirus; ASF – African swine fever; APP – *Actinobacillus pleuropneumoniae*; PEDV – porcine epidemic diarrhea virus

References

1. 2023 Summary Report On Antimicrobials Sold or Distributed for Use in Food-Producing Animals. FDA. October 10, 2024. Accessed October 26, 2024. **https://www.fda.gov/animal-veterinary/antimicrobial-resistance/2023-summary-report-antimicrobials-sold-or-distributed-use-food-producing-animals**

2. Machado I, Petznick T, Silva A P S P, et al. Assessment of changes in antibiotic use in grow-finish pigs after the introduction of PRRSV in a naïve farrow-to-finish system. Preventive Veterinary Medicine, 2024: 106350. **https://doi.org/10.1016/j.prevetmed.2024.106350**

3. Renson P, Mahé S, Andraud M, et al. Effect of vaccination route (intradermal vs. intramuscular) against porcine reproductive and respiratory syndrome using a modified live vaccine on systemic and mucosal immune response and virus transmission in pigs. BMC Veterinary Research, 2024, 20(1): 5. **https://link.springer.com/article/10.1186/s12917-023-03853-4**

4. Petro-Turnquist E, Pekarek M J, Weaver E A. Swine influenza A virus: challenges and novel vaccine strategies. Frontiers in Cellular and Infection Microbiology, 2024, 14: 1336013. **https://www.frontiersin.org/journals/cellular-and-infection-microbiology/articles/10.3389/fcimb.2024.1336013/full**

https://doi.org/10.54846/am2025/95

Transitioning to single tier vaccine label claims: Swine influenza vaccines

E. Russell, MS; E. Larsen, MS; B. Thacker, DVM, PhD, MBA
College of Veterinary Medicine, Lincoln Memorial University, Harrogate, Tennessee

Introduction

Recently, USDA's Center for Veterinary Biologics (CVB) has transitioned vaccine label claims from a four-tiered scheme to a single tier that states "this product has been shown to be effective for the vaccination against (a disease)." Firms were required to provide the study design and raw data for products approved since 2007 from which individual Product Summaries were prepared and made available on CVB's website.[1] We previously reported our work regarding *Mycoplasma hyopneumoniae* vaccines where we found that the percent reduction in pneumonia in efficacy studies was significantly lower with products containing other antigens compared to monovalent products. In this paper, we report our review of swine influenza virus (SIV) vaccine efficacy studies. Our interest in swine influenza virus (SIV) vaccines was due to 1) SIV is most often used as the antigen for the "proof of concept" licensing study that is required for platform based, prescription vaccines; and 2) the recent approval of an RNA particle (RP) vaccine that is directed against the neuraminidase (NA) antigen of the virus. Previous platform SIV vaccines (RNA particle, baculovirus vectored, DNA) targeted the hemagglutinin (HA) antigen. Likewise, the HA is the target antigen for strain matching for both traditional commercial and autogenous killed virus vaccines. Our objective was to provide an overview of SIV vaccine efficacy study data and identify potential limitations of the data.

Methods

Eight Product Summaries representing 5 firms and containing 14 efficacy studies were available. For all studies, the metric for determining efficacy was a lung score that was based on the percentage of lung tissue or surface that exhibited pneumonia. Within each study, a percent reduction in pneumonia was calculated ([control-vaccine]/control). Statistical analyses were performed using Statistix 10 and inferential analyses were performed by Kruskal-Wallis non-parametric one-way ANOVA.

Results

Of the 8 vaccines, 3 were killed virus vaccines (5 studies), 2 were RNA particle vaccines (6 studies), 2 were baculovirus vectored vaccines (2 studies) and 1 was a DNA vaccine (1 study). One of the killed vaccines used a one dose regimen while the other seven vaccines used a two-dose regimen. Age at vaccination was approximately 3 weeks except for 3 days of age with the RP-NA vaccine. The time interval between vaccination and challenge ranged from 10 to 84 days. The time interval between challenge and necropsy was 5 days. The method for assigning lung scores appeared to be similar across all studies. In all 14 studies, the level of pneumonia was significantly lower in the vaccinated group compared to controls with P-values < .003. The range of percent reduction in pneumonia was 62.5% to 99.7%. With regard to duration of immunity, one of the 5 RP-NA studies used a vaccination-challenge interval of 84 days. All three studies with the one dose, killed vaccine used a vaccination-challenge interval of 70 days. The average percent pneumonia in the control groups ranged from 2.9% to 30.5% and in three studies, ranged from 2.9% to 3.6%. Two of these studies (out of 2 total) were with the baculovirus vaccines and one (out of 3 total) was with the killed, 1 dose vaccine. The group average percent pneumonia in the control groups of these three studies was significantly lower ($P < .005$) compared to the other 11 studies.

Discussion

The information reported here provides a broad overview of and potential limitations with SIV vaccine licensing study data that should be considered when formulating SIV vaccination programs. CVB provides a precautionary statement in the Product Summaries to avoid comparing one product to another due to differences in study design and execution. Swine practitioners and producers may find this information useful when discussing and evaluating vaccination protocols recommended by vaccine suppliers.

Reference

1. USDA Center for Veterinary Biologics Product Summaries. Accessed multiple times in 2023 and 2024. **https://www.aphis.usda.gov/veterinary-biologics/product-summaries**

https://doi.org/10.54846/am2025/96

Mapping the environmental impacts of ASF outbreaks in the Philippines to support ASF preparedness in the US and globally

Mikayla G. Schlosser[1]; Rachel A. Schambow[1]; Amit K. Pradhananga[2]; Amy C. Kinsley[3]; Ruth Miclat-Sonaco[4]; Andres M. Perez[1,3]
[1]Center for Animal Health and Food Safety, College of Veterinary Medicine, University of Minnesota, St. Paul, Minnesota; [2] Department of Forest Resources, Center for Changing Landscapes, University of Minnesota, St. Paul, Minnesota; [3] Department of Veterinary Population Medicine, College of Veterinary Medicine, University of Minnesota, St. Paul, Minnesota; [4]National Livestock Program, Department of Agriculture, Quezon City, Philippines

Introduction

Since its emergence and significant global spread beginning in 2007, African swine fever (ASF) has had devastating effects on swine populations and economies worldwide. Although ASF is well-recognized for its severe impact on trade, socioeconomics, and animal health, little research has focused on its environmental impacts from practices such as mass depopulations and carcass disposal. The Philippines has been affected by ASF since its initial outbreak in 2019, resulting in the loss of an estimated three million pigs in the first year alone.[1] As they decompose, animal carcasses may pose pollution risks such as nitrogen, chemical and biological leachate components, and hazardous gasses, which would be exacerbated at scales used for disease response. Control activities for ASF could lead to risks in land, water, and air quality. ASF has also put endangered swine species at risk, such as the Bornean bearded pig, Pygmy hog in India, and warty pigs of southeast Asia. However, the impact of ongoing, global ASF outbreaks on environmental health is relatively unknown.

Methods

To explore ASF's environmental impacts, we conducted a scoping literature review and an exploratory systems mapping exercise with 18 private and public swine veterinarians in the Philippines, where ASF has been present since 2019. The workshop, held in June 2024, involved participants experienced in ASF prevention and response who were divided into three groups representing the "pre-outbreak," "immediate response," and "endemic/long-term management" phases. Each group created a "rich picture" map illustrating key stakeholders and connections relevant to ASF's environmental impact in the Philippines.[2] After presenting the maps, workshop data were transcribed and analyzed. The literature review highlighted a significant research gap in understanding ASF's environmental effects.

Results

The scoping literature review found minimal relevant literature that directly addresses the environmental impacts of ASF only focusing on water and soil pollution, highlighting a significant research gap in understanding how ASF affects environmental factors.[3] The system maps identified many important stakeholders that have decision-making power and impact on environmental health in the Philippines. Concerns raised by the group included the threat of ASF to native, endangered Visayan Warty Pigs and other swine species, and the potential for water contamination from carcass burial because of the high-water table.

Discussion

Findings from this study will guide future qualitative and quantitative studies to better elucidate ASF's environmental impacts and, ultimately, be used in policy development to safeguard environmental and agricultural sectors from potential ASF outbreaks. This study underscores the need for future studies to address ASF's environmental impact comprehensively, filling a critical research gap that could improve response strategies for current and future outbreaks in the Philippines and US preparedness.

References

1. Hsu CH, Schambow R, Montenegro M, Miclat-Sonaco R, Perez A. Factors Affecting the Spread, Diagnosis, and Control of African Swine Fever in the Philippines. *Pathogens.* 2023;12(8):1068. **DOI: 10.3390/pathogens12081068**

2. Siau K, Wang Y. Cognitive evaluation of information modeling methods. *Information and Software Technology.* 2007;49(5). **https://doi.org/10.1016/j.infsof.2006.07.001**

3. Cooper TL, Smith D, Gonzales MJC, Maghanay MT, Sanderson S, Cornejo MRJC, Pineda LL, Sagun RAA, Salvacion OP. Beyond Numbers: Determining the Socioeconomic and Livelihood Impacts of African Swine Fever and Its Control in the Philippines. *Frontiers in Veterinary Science.* 2022;8. **https://doi.org/10.3389/fvets.2021.734236**

https://doi.org/10.54846/am2025/97

Evaluation of the impact of a pre-farrow novel astrovirus 4 vaccine protocol on piglet performance pre- and post-weaning

C. Ulmer[1], BS; A. Hartman[2], PhD; E. Stephenson[2], MS; C. Mainquist-Whigham[2], DVM, MS; J. Madigan[2], DVM
[1]Iowa State University, Ames, Iowa; [2]Pillen Family Farms, Columbus, Nebraska

Introduction

Porcine astrovirus 4 (PASTV-4) is a new and emerging respiratory disease within swine medicine. Often, positive clinical cases present similar to influenza, with a hacking cough, lethargy, and poor performance seen early in the nursery. Ongoing research is focused on understanding specific pathogenesis, but currently, there is no information on effective vaccines or vaccination strategies for the management of this virus. The lack of information related to vaccines or vaccination strategies has created a need for better scientific understanding. Thus, the objective of this study was to determine the pre- and post-weaning growth and survivability impact in piglets following a pre-farrowing vaccination protocol.

Methods

Our study was conducted at a 1650-head, farrow-to-wean farm. The site had an ongoing history of piglets presenting with a hacking cough starting at about 12 days of age, and subsequent challenges in the nursery. The site consistently tested negative on PCR for influenza and PRRSV, and positive for PASTV-4. Sows were randomly allocated into 2 treatment groups, vaccinated (VX) and non-vaccinated (NVX), and blocked by parity. VX treatment group received 2mL of PASTV-4 vaccine at 6- and 3-weeks pre-farrow. All piglets were weighed and tagged within 24 hours of birth. Cross-fostering was allowed for 24 hours within treatment. At 3-5 days post-farrowing, 30 VX and 30 NVX sows and their largest piglet were selected for diagnostic sampling. Sows were nasal swabbed and udder wiped, while piglets were double tagged. The day prior to weaning all piglets were individually weighed again. The subset of sample sows were udder wiped again the day prior to weaning and the previously identified largest piglet was nasal swabbed. Piglet mortality and mortality reasons were recorded. All pigs were weaned into one room on a 12000-head nursery and were penned by treatment in a checkerboard pattern. At 6 and 9 weeks of age, all 60 sample pigs were nasal swabbed. Individual exit weights were collected at the end of the nursery prior to shipment to the finisher. Mortality and mortality reasons were recorded. All sow nasal swabs, udder wipes, and piglet nasal swabs were pooled by 5 within the treatment group and tested by PASTV-4 PCR.

Results

All sample sow nasal swabs post-farrowing tested PASTV-4 PCR negative, but all udder wipes tested PCR positive. All piglet nasal swabs tested PASTV-4 PCR positive prior to weaning and at both time points in the nursery. Pre-wean mortality for the VX and NVX groups were 11.03% and 9.81% respectively. Nursery mortality for the VX and NVX groups were 3.80% and 4.44% respectively. Growth data were analyzed using a MIXED model procedures in SAS (version 9.4) where treatment served as a fixed effect and covariates were included in each model when appropriate. Birth weights in the VX treatment were lower *($P = .0195$)* than the NVX pigs (2.84lb and 3.05lb, respectively). Similar observations were observed in the end of nursery weights, where VX (51.84lb) had lower *($P = .0523$)* body weights than NVX (54.11lb).

Discussion

In summary, PASTV-4 vaccination of sows prior to farrowing had a negative impact on both birth weight and end-of-nursery weights, as well as on pre-weaning mortality. There was a numerically lower nursery mortality for piglets coming from a VX compared to NVX. Overall, the current analysis would suggest that this pre-farrow vaccination protocol may not be the ideal management strategy, and further research on PASTV-4 and vaccination strategies is warranted.

https://doi.org/10.54846/am2025/98

Developing a *Lawsonia intracellularis* surveillance program to define prevalence in a commercial swine operation

H. Walker[1]; N. Lauterbach[2], DVM; K. Castevens[3], DVM
[1]Carlson College of Veterinary Medicine, Oregon State University, Corvallis, Oregon; [2]Pharmgate Animal Health, Wilmington, North Carolina; [3]Country View Family Farms, Middletown, Pennsylvania

Introduction

Lawsonia intracellularis (Li), a known obligate intracellular bacterium that replicates in the cytoplasm of immature enterocytes is the cause of proliferative enteropathy, more commonly referred to as ileitis.[1] Ileitis is an endemic disease within herds, with an estimated 80-100% of herds infected.[2] An endemic infection can lead to poor production performance, higher costs, and has a significant impact on overall herd health and productivity.[3] Understanding prevalence of ileitis can help with determining specific treatment and vaccination strategies. The objective of this study is to develop and execute a surveillance program for Li to determine herd-level prevalence within a commercial swine operation and whether the sampling method is sufficient.

Methods

The treatment group was defined as the "flow", which includes gilts entering the sow farm, sows within the farm, and single source finisher pigs downstream from that sow farm. There were five flows. Flows 1, 4, and 5 vaccinated weaned pigs against Li, while Flows 2 and 3 were Li unvaccinated flows. Flow 1 had recently vaccinated all sows against Li due to an outbreak. Flows 1 and 5 were no antibiotics ever (NAE) herds, with no feed or water meds during study. Flows 2, 3, and 4 were able to receive water meds as needed, but none were used during the study period. Those same flows had chlortetracycline (CTC) and denagard (DEN) in the nursery feed only. No additional feed medications were given in the finisher feeds. From each flow, blood and fecal samples were obtained from gilts entering the sow farm, and parity 1, 2, and 3 sows (P1-3). Fecal samples were pooled by three at the lab and tested via Li PCR. Blood samples were tested individually for antibodies via Li blocking ELISA. For finishers, there were 3 oral fluid (OF) sampling periods: 16 weeks of age (WOA), 20 WOA, and 24 WOA. The same groups of pigs were followed throughout the sampling periods. Samples were collected using cotton ropes, each containing samples from multiple pigs. The OF were tested via Li PCR.

Results

A negative fecal Li PCR is a CT level > 35. Flows 1, 2 and 3 reported negative fecal PCRs across all tested groups of gilts and sows. Flow 4 and Flow 5 each reported a single positive fecal PCR pool (3 pigs), a 31.8 CT value from P1, and a 34.4 CT value from gilts, respectively. A positive Li ELISA is a percent inhibition ≥ 30% while results ≤ 20% are negative for Li antibodies, and between 20% and 30% are considered suspect. The percentage of positives for each flow are as follows: Flow 1 – 59.1% (8 suspects), Flow 2 – 88.3% (6 suspects), Flow 3 – 58.3% (8 suspects), Flow 4 - 50% (5 suspects), Flow 5 - 58.3% (7 suspects). ELISA result breakdown by parity and flow, and PCR results can be seen in Table 1. All OF collections from all finisher sites showed negative PCRs. Clinical observations supported these diagnostic results, with little to no diarrhea observed during sampling within any of the populations.

Table 1: Percentage positive ELISA results for *Lawsonia intracellularis* for each flow and parity group and PCR results

	Flow 1	Flow 2	Flow 3	Flow 4	Flow 5
Gilts	83%	97%	83%	100%	100%
P1	63%	27%	67%	70%	77%
P2	50%	83%	37%	27%	27%
P3	40%	77%	50%	3%	30%
Overall % positive for all groups	**59.1%**	**88.3%**	**58.3%**	**50%**	**58.3%**
Total suspects	**8**	**6**	**8**	**5**	**7**
Fecal PCR sow farms	All collections negative	All collections negative	All collections negative	1 positive P1 pool - 31.8 CT	1 positive gilt pool - 34.4 CT
OF PCR finisher	All collections negative	All collections negative	All collections negative	All collections negative	All collections negative

P = parity; OF = oral fluids

Conclusion

This study aimed to develop a method of sampling to determine herd-level prevalence for Li. For this study, prevalence was expected to be low due to lack of a history of clinical signs, no production impact, and the vaccination history of the flows. This was supported by the levels of antibodies seen in the breeding herd, with flow 2 being much higher than expected. The high levels in Flow 2 is suspected to be related to previous vaccination history in the barn or a previous undiagnosed outbreak. Unfortunately, there are no set guidelines in the current literature comparing the level of detectable Li antibodies in vaccinated pigs versus pigs infected in Li. Overall lack of clinical signs and lack of positive diagnostic tests during this study, even at the age of expected Li prevalence in finishing pigs, would indicate a low prevalence of Li in this system. Various factors may influence the presence of Li in finishing pigs, such as maternal antibodies, rodent control, and feed or water medications. Following the same finisher barns over a period of time would be valuable in understanding potential impacts of environmental build-up, particularly in unvaccinated herds. Future work should be done to reinforce this testing strategy in a low Li prevalence environment for the use of making decisions on Li vaccination and treatment programs.

References

1. McOrist S, Gebhart CJ, Boid R, et al. Characterization of Lawsonia intracellularis gen. nov., sp. nov., the obligately intracellular bacterium of porcine proliferative enteropathy. *Int J Syst Bacteriol.* 1995;45(4):820–825.

2. Collins A, Collins C. Epidemiology Tools to Evaluate the Control of Proliferative Enteropathy in Commercial Pig Herds. *Animals (Basel).* 2024;14(9):1357.

3. Vannucci FA, et al. Proliferative enteropathy. In: Zimmerman JJ, et al, eds. *Diseases of Swine,* 10th ed. John Wiley and Sons, Inc; 2019: 811-820., Hoboken, NJ.

https://doi.org/10.54846/am2025/99

Adjuvant evaluation of combining *Streptococcus suis* (*S suis*) and influenza A virus (IAV-S) in pigs

Jinnan Xiao[1]; Edgar Diaz-Estrada[2]; Maria Clavijo[1]; Tyler Holck[1]
[1]College of Veterinary Medicine, Iowa State University, Ames, Iowa;
[2]Swine Marketing and Veterinary Services North America, Ceva Animal Health, Kansas City, Kansas

Introduction

Streptococcus suis (*S suis*) is a significant pathogen in swine, causing considerable economic losses and posing zoonotic risks to humans.[1] It is considered one of the most important bacterial pathogens in pigs globally, including in the United States.[2] The prevalence of *S suis* is notable in pig farms, with serotype 2 being the most frequently isolated in clinical cases from North America.[3] Efforts to control *S suis* through vaccination have been challenging due to the pathogen's genetic diversity and the lack of commercial vaccines specifically licensed for *S suis* in many regions.[2] Autogenous vaccines, tailored to specific strains from individual farms, offer a potential solution.[4] This study aims to evaluate the efficacy of two new adjuvants, Adjuvant A and Adjuvant B, developed by Ceva Animal Health LLC, in combination with *S suis* and Influenza A virus (IAV-S) antigens, to enhance immune responses and protection in pigs.

Methods

This study aims to assess the immunogenicity and protective efficacy of two different adjuvants (Adjuvant A and Adjuvant B) when combined with *S suis* and/or IAV-S antigens. Seventy-six clinically healthy three-week-old pigs from twelve litters, free from IAV-S MDA and with low *S suis* presence, were allocated to six treatment groups of 12 pigs each, each coming from a different litter. This ensures that all 12 litters are represented in each group. An additional group of five pigs served as the sentinel control group. Pigs received their first dose of the vaccine at 18-24 days of age, followed by a booster dose 14 days later. Two weeks post-booster, pigs were challenged with 2 mL each of *S suis* serotype 2 intramuscularly and intraperitoneally. Clinical signs were monitored for seven days post-challenge daily, and any moribund or deceased pigs were subjected to brain swabbing to confirm the presence of *S suis*. Observed parameters include clinical observations and serological responses. Any animal shows severe signs will be humanely euthanized, and a necropsy and brain swab will be performed. Any animals found dead will be brain swabbed. Blood samples were collected at study day (SD) 0, 14, 34/35, to evaluate seroconversion rates for IAV-S. Upon euthanasia/death or seven days after the challenge procedure (SD 35), the pigs will be humanely euthanized by barbiturate overdose following AVMA-approved methods and a brain swab taken. Brain swab samples will be operated for pathogen isolation.

Results and analysis

The results indicate that the groups that received a vaccine containing either of the two adjuvants evaluated, used alone with *S suis* or with the influenza virus, demonstrated a significant reduction in mortality and lesions. Combining both antigens did not affect the serological response in the groups vaccinated with *S suis* and influenza.

Discussion

The challenge models for *Streptococcus suis* showed their effectiveness in the test regarding the development of clinical symptoms and their effect on mortality. The tested adjuvants demonstrated their effectiveness in the evaluated formulations for *S suis*, as they did not produce a negative impact when combined with the swine influenza virus.

Acknowledgment

College of Veterinary Medicine, Iowa State University; Ceva Animal Health, Kansas City

References

1. Hlebowicz, M., et al. (2019). Prevalence of *Streptococcus suis* in the USA.
2. Gottschalk, M., Segura, M. (2000). Diseases of Swine.
3. Gebhart, C.J., et al. (2020). Epidemiology and control of *S. suis*.
4. Hopkins, D., et al. (2019). Field studies evaluating the efficacy of *Streptococcus suis* autogenous vaccine.

Poster 26

https://doi.org/10.54846/am2025/100

Challenges and opportunities of bacterial vaccines as alternatives to antimicrobials in swine health management: Insights from US veterinarians

Xirui Zhang[1,2]; Danqin Li[1]; Michael D. Apley[3], DVM, PhD, DACVCP; Locke Karriker[4], DVM, MS, ACVPM; Joseph F. Connor[5], DVM, MS; Corinne Bromfield[2], DVM; Brian Lubbers[3], DVM, PhD, ACVCP; Hatem Kittana[6], DVM, PhD; Dustin Pendell[7], PhD; Jishu Shi[1], DVM, PhD

[1]Department of Anatomy and Physiology, CVM, Kansas State University, Manhattan, Kansas; [2]CVM, University of Missouri, Columbia, Missouri; [3]Department of Clinical Sciences, CVM, Kansas State University, Manhattan, Kansas; [4]Department of Veterinary Diagnostic and Production Animal Medicine, CVM, Iowa State University, Ames, Iowa; [5]Carthage Veterinary Service, LTD., Carthage, Illinois; [6]Department of Diagnostic Medicine/Pathobiology, CVM, Kansas State University, Manhattan, Kansas; [7]Department of Agricultural Economics, College of Agriculture, Kansas State University, Manhattan, Kansas

Introduction

Antimicrobial resistance (AMR) poses an escalating threat to human and animal health and welfare. According to an FDA report, medically important antimicrobials are sold more for swine than for any other livestock species.[1] While vaccines are widely recognized as a promising tool to curb AMR, the efficacy of vaccines against several key swine pathogens still requires improvement. To effectively reduce antibiotic use and mitigate production losses, identifying the bacterial diseases most urgently in need of improved vaccines is essential. In addition to evaluating the vaccines' potential impact on AMR, this study also considers a range of practical factors affecting vaccine application in real-world settings, offering a more comprehensive perspective on this critical issue.

Methods

This study was a survey-based cross-sectional study, designed and distributed via the Qualtrics software. The survey was disseminated through email to licensed swine veterinarians across the US from mid-May to mid-August. A total of 202 veterinarians were contacted, yielding 46 responses, of which 19 were fully completed. Only completed responses were deemed valid for further analysis. The comparison of vaccine efficacy between current and ideal vaccines was conducted using Two-way ANOVA analysis, whereas vaccine price comparisons utilized a two-tailed t-test. All assessments were conducted at a 95% confidence interval (CI).

Results

Among the 11 prioritized candidate bacteria, *Escherichia coli* (*E coli*), *Streptococcus suis* (*S suis*), *Glaesserella parasuis* (*G parasuis*), and *Mycoplasma hyopneumoniae* (*M hyopneumoniae*) were identified as the top four diseases requiring improved vaccines. Respondents reported the highest incidence rates for *E coli* (average nursery morbidity of 25.0% and mortality of 11.9%) and the lowest for *G parasuis* (morbidity of 11.3% and mortality of 3.7%). Significant efficacy gaps between current and ideal vaccines were observed in all four diseases. Regarding economic impact, *M hyopneumoniae* incurred the highest average loss per pig at $2.23, while *G parasuis* had the lowest at $0.96, with *E coli* and *S suis* falling in between, $1.37 and $1.25, respectively. When examining the ratio of willingness to pay for an ideal vaccine relative to the current vaccine price, *M hyopneumoniae* displayed the highest increase, with an ideal-to-current price ratio of 4.2 ($1.03/$0.25;*). This was followed by *E coli* at 1.88 ($0.79/$0.42;*), *S suis* at 1.79 ($0.83/$0.47;*), and *G parasuis* at the lowest, with a ratio of 1.26 ($0.72/$0.57;ns). This ranking aligns positively with the estimated economic loss per head.

Discussion

M hyopneumoniae stands out with intriguing contrasts: respondents identified it as the most manageable disease under current protocols, yet it also exhibits the poorest vaccine efficacy, the highest economic losses per head, and the largest ideal-to-current vaccine price ratio. This combination suggests a strong determination to accelerate the eradication of *M hyopneumoniae* by introducing a more effective vaccine. As sole *M hyopneumoniae* infections are rarely treated, an effective vaccine might become one of the most cost-effective strategies to mitigate their impact. A deeper understanding of the virulence mechanisms of *S suis* and *E coli*, two prevalent and challenging pathogens in production, is crucial for developing more effective vaccines.[2,3] Beyond enhancing cross-protection, further research on reducing production timelines for *S suis* farm-tailored vaccines and overcoming timing challenges for *E coli* vaccines will be critical for the time being. For *G parasuis*, challenges likely stem not only from the specific immune response mechanisms but also from the need to improve vaccine efficacy while controlling costs to meet pricing expectations.

References

1. U.S. Food and Drug Administration. 2022 Summary Report On Antimicrobials Sold or Distributed for Use in Food-Producing Animals. FDA.gov. July 12, 2023. Accessed October 28, 2024. **https://www.fda.gov/animal-veterinary/antimicrobial-resistance/2022-summary-report-antimicrobials-sold-or-distributed-use-food-producing-animals**

2. Gottschalk M, Segura M. Streptococcosis. Diseases of Swine2019. p. 934-50. **https://doi.org/10.1002/9781119350927.ch61**

3. Fairbrother JM, Nadeau É. Colibacillosis. Diseases of Swine2019. p. 807-34. **https://doi.org/10.1002/9781119350927.ch52**

https://doi.org/10.54846/am2025/101

Comprehensive evaluation of the economic and epidemiological benefits of market haul trailer decontamination in the US swine industry

Benjamin Blair; James Lowe
University of Illinois College of Veterinary Medicine

Introduction

The transmission of porcine epidemic diarrhea virus (PEDV) within the US swine industry remains a pressing challenge,[1,2] particularly through market haul trailers that transport pigs between wean-to-finish farms and slaughter facilities. Given the highly contagious nature of PEDV and its severe impact on swine health and productivity,[2] effective biosecurity measures,[3] including the decontamination of these trailers, are essential. This study aims to determine the optimal decontamination threshold (DT) for market haul trailers, balancing the reduction of PEDV transmission with economic feasibility.

Methods

Our approach involved simulation modeling of swine production systems under various conditions. We conducted simulations across two primary scenarios: A single production system consisting of 24,000 sows distributed across eight sites, and a region with 24,000 sows spread across four geographically related but otherwise unconnected systems. Each scenario was evaluated under varying PEDV prevalence levels - low (5%), moderate (10%), and high (20%) - to assess how different trailer washing practices affect disease spread and costs. The decontamination threshold (DT) was defined as the product of truck washing efficacy and the proportion of trailers washed.

Results

Key findings include:

- In single production systems, washing 60% of trailers (DT of 48%) emerged as a cost-effective strategy, reducing PEDV transmission to 23.13 infected premises at an approximate cost of $32,956 per farm.
- For geographically related systems, washing 0% of trailers was found to be the most economical, costing $25,664 per farm, yet still providing adequate disease control.
- High PEDV prevalence scenarios required a higher DT (64%), achievable by washing 80% of trailers, costing $48,957 per farm.
- In low-prevalence conditions, no trailer washing was necessary, minimizing costs to $26,862 per farm.

Conclusion

The findings of this study underscore the importance of strategic trailer decontamination in mitigating PEDV transmission within the swine industry. Trailer decontamination strategies should be tailored to specific production systems and regional disease prevalence. By doing so, they can optimize biosecurity measures while managing operational costs effectively.

This research provides a practical framework for swine producers, enabling them to make informed decisions about trailer decontamination practices. The integration of epidemiological and economic analyses offers a comprehensive approach to enhancing swine health and productivity, ensuring that biosecurity measures are both effective and financially sustainable in diverse production scenarios. The implications of this study extend beyond PEDV control, offering insights into the broader challenge of disease management within livestock production systems while managing resource allocation to ensure optimum returns.

References

1. VanderWaal K, Deen J. Global trends in infectious diseases of swine. *Proc Natl Acad Sci*. 2018;115(45):11495-11500. **doi:10.1073/pnas.1806068115**

2. Zhang X, Rong X, Li J, et al. Modeling the outbreak and control of African swine fever virus in large-scale pig farms. *J Theor Biol*. 2021;526:110798. **doi:10.1016/j.jtbi.2021.110798**

3. Dee SA, Deen J, Otake S, Pijoan C. An experimental model to evaluate the role of transport vehicles as a source of transmission of porcine reproductive and respiratory syndrome virus to susceptible pigs.

4. Xu Y, Sun M, Wang Z, Li X, Du Y, Jiang P. The prevalence and shedding of porcine epidemic diarrhea virus in intensive swine farms of China from 2022 to 2023. *Vet Microbiol*. 2024;298:110273. **doi:10.1016/j.vetmic.2024.110273**

5. Martínez-López B, Ivorra B, Ramos AM, Sánchez-Vizcaíno JM. A novel spatial and stochastic model to evaluate the within- and between-farm transmission of classical swine fever virus. I. General concepts and description of the model. *Vet Microbiol*. 2011;147(3-4):300-309. **doi:10.1016/j.vetmic.2010.07.009**

6. Rajkowski KT, Eblen S, Laubauch C. Efficacy of Washing and Sanitizing Trailers Used for Swine Transport in Reduction of Salmonella and Escherichia coli. *J Food Prot*. 1998;61(1):31-35. **doi:10.4315/0362-028X-61.1.31**

7. Dee S, Torremorell M, Thompson B, Deen J, Pijoan C. An evaluation of thermo-assisted drying and decontamination for the elimination of porcine reproductive and respiratory syndrome virus from contaminated livestock transport vehicles.

8. Kim Y, Krishna V, Torremorell M, Goyal S, Cheeran M. Stability of Porcine Epidemic Diarrhea Virus on Fomite Materials at Different Temperatures. *Vet Sci*. 2018;5(1):21. **doi:10.3390/vetsci5010021**

9. Perri AM, Poljak Z, Dewey C, Harding JCS, O'Sullivan TL. Factors Associated With Time to Elimination of Porcine Epidemic Diarrhea Virus in Individual Ontario Swine Herds Based on Surveillance Data. *Front Vet Sci*. 2019;6:139. **doi:10.3389/fvets.2019.00139**

10. Meurer A, Smith CP, Paprocki M, et al. SymPy: symbolic computing in Python. *PeerJ Comput Sci*. 2017;3:e103. **doi:10.7717/peerj-cs.103**

11. McKinney W. Data Structures for Statistical Computing in Python. In: 2010:56-61. **doi:10.25080/Majora-92bf1922-00a**

12. Harris CR, Millman KJ, Van Der Walt SJ, et al. Array programming with NumPy. *Nature*. 2020;585(7825):357-362. **doi:10.1038/s41586-020-2649-2**

13. Virtanen P, Gommers R, Oliphant TE, et al. SciPy 1.0: fundamental algorithms for scientific computing in Python. *Nat Methods*. 2020;17(3):261-272. **doi:10.1038/s41592-019-0686-2**

14. Hunter JD. Matplotlib: A 2D Graphics Environment. *Comput Sci Eng*. 2007;9(3):90-95. **doi:10.1109/MCSE.2007.55**

https://doi.org/10.54846/am2025/102

Comparative pathogenesis of a Chinese highly pathogenic porcine reproductive and respiratory syndrome virus (PRRSV) to a contemporary L1C variant from the United States

A. Buckley[1], DVM, PhD; B. Arruda[1], DVM, PhD; E. Zanella[1], DVM, PhD; L. Hanson[1], BS; J. Zhang[2], MD, MS, PhD; J. Wiarda[1], PhD
[1]Virus and Prion Research Unit, National Animal Disease Center, ARS, USDA, Ames, Iowa;
[2]Vet Diagnostic and Production Animal Medicine, CVM, ISU, Ames, Iowa

Introduction

Highly pathogenic PRRSV (HP-PRRSV) strains were documented circulating in China around 2006 and were characterized by high fevers and morbidity/mortality. In 2020, reports of high mortality PRRSV isolates came from farms in the Midwest of the United States. These isolates clustered in the sublineage L1C (specifically L1C.5) and many had 1-4-4 RFLP patterns. Subsequently, L1C.5 isolates have spread through the United States swine industry resulting in significant pig mortality and economic losses. The objective of this study was to compare the pathogenesis of a Chinese HP-PRRSV strain to a contemporary L1C.5 strain in a weaned pig model.

Materials and methods

Three PRRSV isolates were utilized for challenge. A Chinese HP-PRRSV isolate, rJXwn06, that was rescued from a full-length cDNA clone, an L1C.5 isolate USA/MN/01775GA/2021, and a commonly utilized challenge strain MN184. Three groups of weaned pigs (n = 20/group) were challenged with 2 mL of 1x10^5 TCID50/mL intranasally (rJXwn06, L1C.5, and MN184) with a fourth group receiving 2 mL of sham cell culture media (sham). Severity of clinical disease was evaluated by scoring lethargy and respiratory distress, measuring febrile response, weighing pigs, and comparing mortality rates. Viral pathogenesis was compared by analyzing acute viremia and nasal shedding via PCR, gross and histologic lung scores, and viral load in selected tissues. Five pigs were necropsied on 2 days post inoculation (dpi), 6 dpi, and 10 dpi for tissue collection and evaluation. The study was ceased at 12 dpi due to no rJXwn06 or LC1.5 animals remaining.

Results

Both the rJXwn06 and L1C.5 groups had significantly lower average daily gains and higher body temperatures than those pigs in the MN184 and sham groups though not significantly different than one another. Temperatures for these groups peaked at 8 and 9 dpi respectively. The rJXwn06 group had multiple lethargic pigs by 6 dpi and had higher lethargy scores than all other groups through 10 dpi. Lethargy in the L1C.5 group was delayed compared to the rJXwn06 group by approximately 2 days. Respiratory signs were observed following signs of lethargy and were observed in the rJXwn06 and L1C.5 groups at 10 dpi. The MN184 and sham groups did not have pigs with clinical scores above zero. Both the rJXwn06 and L1C.5 groups had comparable levels of genomic copies of PRRSV in serum and nasal swabs from 1 to 10 dpi and were significantly higher than the MN184 group. Starting at 7 dpi, animals in the rJXwv06 group had to be euthanized due to progression of clinical signs with the remainder of the animals euthanized on 10 dpi. Mortality in the L1C.5 group was delayed compared to the rJXwn06 group with animals euthanized starting on 9 dpi and all remaining animals euthanized by 12 dpi. The thymus, lung, and tracheobronchial lymph nodes (TBLN) were PCR positive in most pigs from both the rJXwn06 and L1C.5 groups at necropsy in addition to the brain at 10 dpi. Only a couple pigs in the MN184 group had PCR positive TBLN.

Conclusion

This study has demonstrated that this L1C.5 PRRSV isolate from the United States in 2021 had a similar level of pathogenicity to a highly pathogenic Chinese PRRSV strain from 2006. It also supported previous research demonstrating HP-PRRSV localization in central nervous system tissue. A better understanding of these isolates can help guide intervention strategies and control measures to reduce the spread of PRRSV in the swine industry.

 https://doi.org/10.54846/am2025/103

A comparative assessment of oral toltrazuril and injectable iron administration vs combined toltrazuril + iron injectable treatment on weaning weight, clinical signs and mortality in piglets

Cesar Caballero[1], DVM, DVSc; Nicole Seebach[2], BSc; Ryan Tenbergen[3], DVM, MSc;
Enrique Moncada-Fernandez[4], DVM, PhD; Daniel Sperling[5], DVM, PhD
[1]Ceva Animal Health Canada, Guelph, Ontario, Canada; [2]University of Guelph, Guelph, Ontario, Canada;
[3]Demeter Veterinary Services PC, Woodstock, Ontario, Canada; [4]Ceva Animal Health, Lenexa, Kansas;
[5]Ceva Santé Animale, Libourne, France

Introduction

Cystoisospora suis is a parasite which causes cystoisosporosis in piglets, with either clinical and/or subclinical presentations in the farm.[1] Prevalence of coccidiosis has been reported as high as 70%, leading to economic losses caused by the impact of diarrhea, mortality, and stunted growth in affected farms.[2] Standard program to prevent coccidiosis and anaemia in piglets is based in the administration of toltrazuril and iron.[1] The aim of this study was to compare the effects of two preventive programs: a combination of an oral toltrazuril and injectable iron vs a new injectable combination of toltrazuril and iron for parenteral application (Forceris®). The outcomes measured in this study include disease control, weight gain, and mortality in piglets that were treated with either of the two treatments or were left untreated with only iron.

Materials and methods

In a farrow-to-finish operation (400 sows), clinically affected by pre-weaning diarrhea, *Cystoisospora suis* infection was confirmed in fecal samples via autofluorescence technique. 35 litters were randomly selected and assigned to one of three groups: 12 litters (141 piglets) were injected with 1.5mL of Forceris® 1-3 day of age (doa) (group F); 11 litters (132 piglets) received oral toltrazuril and injectable iron 3-4 doa (group OT), and 12 litters (141 piglets) received only Iron 1-3 doa (group C). Piglets in each group were individually tagged and weighed at processing and weaning age (21 days). Score of diarrhea was recorded at three different times: Period 1: Processing 1-3 doa; Period 2: 10-12 doa; Period 3: 21 doa; (0 solid, 1 pasty, 2 semi-liquid, 3 liquid).

Results

Average daily weight gain (ADWG) in the group F with the injectable combination of toltrazuril and iron for parenteral application (Forceris®) was significantly higher compared to group OT and group C as indicated in the Table 1. No differences were found among treatments on mortality. In terms of scores of diarrhea, there were no differences among treatments at period 1 (processing 1-3 doa) and period 3 (18-21 doa). There was a significant difference between treatments and control group at period 2 (10-12 doa).

Table 1: Mean ± standard deviation on ADWG during the lactation period, per treatment

Treatment	ADWG (kg)
Forceris® (Group F)	0.228 ± 0.052[a]
Oral toltrazuril and injectable iron (Group OT)	0.199 ± 0.058[b]
Iron (Group C)	0.195 ± 0.057[b]

ADWG: Average daily weight gain
[a,b] Means with different superscripts differ ($P < .05$)

Discussion

Under the clinical neonatal coccidiosis in this farm, results in the trial suggested that the use of a combined injectable combination of toltrazuril and iron for parenteral application (Forceris®) in piglets at processing (1-3 doa) improves growth with near to 30 grams more of average daily weight gain (ADWG) vs a combination of an oral toltrazuril and injectable iron and near to 33 grams more of ADWG vs untreated pigs with only iron.

References

1. Joachim, A., et al. Comparison of an injectable toltrazuril gleptoferron (Forceris®) and an oral toltrazuril (Baycox®) + injectable iron dextran for the control of experimentally induced piglet cystoisosporosis. Parasites & Vectors. 2018. 11:206. **https://doi.org/10.1186/s13071-018-2797-5**

2. Aliaga-Leyton, A., Webster, E., Friendship, R., Dewey, C., Vilaça, K., & Peregrine, A. S. An observational study on the prevalence and impact of Isospora suis in suckling piglets in southwestern Ontario, and risk factors for shedding oocysts. *The Canadian veterinary journal* 2011; 52(2), 184–188.

https://doi.org/10.54846/am2025/104

Probability of detecting *Lawsonia intracellularis* DNA by PCR in oral fluids and pooled fecal samples based on pen prevalence

G. A. Cezar[1]; F. Leite[2]; J. Morgan[2]; A. Vegter[1]; G. Duhe[2]; D. C. A. Moraes[1]; I. Machado[1]; M. Musskpof[1]; E. Kirwa[1]; T. T. P. Tien[1]; C. Yang[1]; A. H. Gris[1]; E. Braga[1]; E. M. S. Silva[1]; R. Nicolino[1]; P. Gauger[1]; G. Trevisan[1]; D. C. L. Linhares[1]; G. Silva[1]
[1]Iowa State University, Ames Iowa
[2]Boehringer Ingelheim Animal Health USA Inc., Duluth, Georgia

Introduction

Lawsonia intracellularis (*L intracellularis*) is a major enteric pathogen causing clinical and subclinical infections in pig populations, resulting in significant economic losses.[1] Oral fluids have been reported for *L intracellularis* DNA detection by quantitative polymerase chain reaction (qPCR), but there are limited reports on herd sensitivity of using this sample type.[2,3] This study assessed the probability of detecting *L intracellularis*, in oral fluids compared to pooled pen fecal samples based on the pen prevalence.

Methods

Three finishing sites with a previous clinical history of *L intracellularis* and pigs at 20 weeks of age were selected for this cross-sectional study. Oral fluids were collected following a spatial distribution, and from the same pen, individual fecal and pooled pen fecal samples (targeting feces in multiple locations) were collected. All the samples were tested by qPCR and considered positive for Ct values below 35. Pen prevalence was calculated as the proportion of individual qPCR-positive results relative to the number of pigs within a pen. A logistic regression assessed the probability of detecting *Lawsonia* in oral fluids and pooled pen fecal samples using the R program v.4.2.3. Also, agreement among oral fluids and pen fecal samples qPCR results was assessed through Cohen's Kappa statistic.

Results

The study included a total of 21 pens and 573 pigs, where 293 pigs were positive (51%) in individual fecal samples. For the oral fluids, 12 of 21 pens were positive, and 16 of 21 were positive for pooled fecal samples by qPCR. Pen prevalence based on individual fecal qPCR ranged from 0-100%. In scenarios of low prevalence (Figure 1), the pooled pen fecal samples had a higher probability of detecting *L intracellularis* DNA. However, in scenarios of high prevalence (> 80%), the oral fluids had a similar likelihood of detecting *L intracellularis* as pooled pen fecal samples. The Kappa agreement among oral fluids and pen fecal samples was 0.43 (CI: 0.27, 0.58).

Discussion and conclusion

Oral fluids had a ≥ 80% probability of detecting *Lawsonia* when the pen prevalence was above 65%. Pen feces had a higher detection probability, which was expected since *L intracellularis* is an enteric pathogen. However, it is essential to highlight that oral fluids can be tested for multiple pathogens, making it a practical choice for the surveillance of *L intracellularis* and other pathogens. In addition, the Kappa value represented a moderate agreement between oral fluids and pooled fecal sample results.

Figure 1: Probability of detecting *Lawsonia* in oral fluid and pen fecal samples by qPCR.

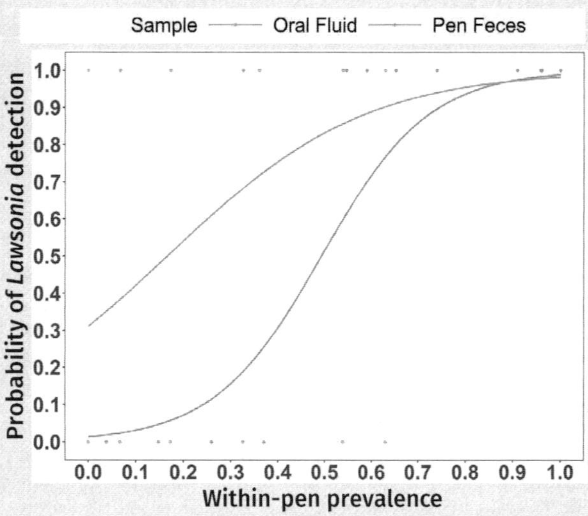

Primarily, this agreement occurred in high pen prevalence scenarios. In conclusion, this study demonstrated that *L intracellularis* DNA was detected in oral fluid, mainly in scenarios with high within-pen prevalence where the detection probability increases.

References

1. Vannucci FA, Gebhart CJ, McOrist S. Proliferative Enteropathy. In: Diseases of Swine. John Wiley & Sons, Ltd; 2019. p. 898–911. **https://doi.org/10.1002/9781119350927.ch58**

2. Burrough ER, Rotolo ML, Gauger PC, et al. Correlation of *Lawsonia intracellularis* semi-quantitative fecal polymerase chain reaction assay results with the presence of histologic lesions of proliferative enteropathy and positive immunohistochemical staining. J Swine Health Prod. 2015;23(4):204–207. Available at: **http://www.aasv.org/shap.html**

3. Campler M, Cheng T-Y, Angulo J, Van De Weyer L, Arruda AG. Detection of *Lawsonia intracellularis* by oral fluids and fecal samples in Canadian swine. J Swine Health Prod. 2024;32(4):156-163. **https://doi.org/10.54846/jshap/1388**

https://doi.org/10.54846/am2025/105

Evaluation of a comprehensive platform (Pork MultiPath™) for efficient and high throughput swine pathogen surveillance and health management

J. F. Connor[1], DVM, MS; B. Genz[2], PhD; B. Payne[3], DVM; J. Gerszon[2], PhD; M. J. Sellars[2], PhD; R. J. Moser[2], PhD
[1]Carthage Veterinary Service, Ltd, Carthage, Illinois; [2]Genics Pty Ltd, Brisbane, Queensland, Australia;
[3]Veterinary Pharmaceutical Solutions, Saint Peter, Minnesota

Introduction

Bacterial and viral co-infections pose significant challenges in pig health management, leading to severe clinical signs and prolonged illness. Molecular diagnostics are restricted to specific tests requested, often limited by available veterinary health budget. Testing a single sample against multiple pathogens in a multiplex can reveal otherwise missed pathogens. The PCR-based multi-pathogen detection platform Pork MultiPath™ (PMP) was designed following WOAH's validation pathway,[1] offering a broad spectrum of pathogen detection (12 + 2 respiratory [respiratory panel detects 12 pathogen targets quantitatively plus 2 R&D targets qualitatively] or 19 enteric targets), scalable throughput, and cost-effectiveness.

Implementing PMP respiratory (PMP1) and enteric (PMP2) pathogen detection panels, Gerszon et al,[2] found strong associations between clinical symptoms and pathogen thresholds for *Glaesserella parasuis, Mycoplasma hyorhinis,* and *Brachyspira pilosicoli*. Additionally, markers such as *Escherichia coli* F4, F5, F18, LT, ST1, and ST2 were linked to higher disease likelihood. In this study, both panels were utilized to understand pathogen status and co-infection prevalence of respiratory and enteric pathogens in US and Latin American pig productions. Further, panel performance was compared to a US-based diagnostic laboratory.

Methods

For concordance testing six respiratory samples (pooled oral fluids, lung and tonsil tissue, brain/heart and nasal swab) and four enteric samples (pooled feces, ileum, jejunum) were collected from a nursery farm in the Midwest, US, and split samples were send for PMP analysis (Genics Pty Ltd, Brisbane, Australia) and equivalent pathogen testing at a US-based diagnostic laboratory. Additionally, a cross-sectional surveillance sampling study was performed on two commercial pig farms with a throughput of 1100 pigs per week and an all-in-all-out production flow in Latin America. Pooled oral fluid and pooled feces sock samples were collected (Farm A: 7 batches, 3 – 9 weeks of age; Farm B: 11 batches, 9 – 20 weeks of age) (n = 8 samples per batch and panel, total: n = 140 samples per panel). Extracted TNAs (RNA / DNA) were tested on PMP panels including control assays and GLP processing. The presence of pathogen markers was reported as copies per reaction

Results

M hyorhinis (6/6 samples), *G parasuis* (5/6 samples) and *Streptococcus suis* (5/6 samples) were most prevalent in US respiratory samples tested on PMP1. PMP2 detected *E coli* F4 (4/4 samples), F18 (4/4 samples), and ST2 (4/4 samples) as most abundant pathogen markers in fecal samples. Additional low prevalence pathogens were detected in individual sample types with both panels. In general, PMP results were highly concordant with real-time PCR results provided by the commercial US laboratory. Costs for PMP analysis (including shipping, diagnostics and reporting) were approximately 80% less, and the turn-around time was approximately 1 week shorter than the complete PCR request from the US provider. PMP1 surveillance testing in Latin America revealed *S suis* (100% of samples positive), *G parasuis* (Farm A: 69%; Farm B: 70%), PCV-2 (Farm A: 56%; Farm B: 78%), and *Klebsiella pneumoniae* (Farm A: 23%; Farm B: 13%) as the most frequently detected respiratory pathogens (Figure 1, left). Further, on Farm B *M hyorhinis* (38%) and *Bordetella bronchiseptica* (13%) were detected with high frequency. The PMP2 enteric panel showed most frequent detection of *E coli* EAE (Farm A: 96%; Farm B: 22%), ST2 (Farm A: 78%; Farm B: 73%), and ST1 (Farm A: 52%; Farm B: 35%), RVC (Farm A: 62%; Farm B: 49%) and PCV-2 (Farm A: 44%; Farm B: 95%) (Figure 1, right). Additionally, *E coli* F4 (36%), F18 (32%) and STX2e (23%), as well as RVA (30%) were present in samples from Farm A.

Discussion

This study underlines the power of PMP panels in surveillance testing and identifying co-infections with pooled and individual samples. Clearly, a customized panel with desired pathogens can be a powerful, cost-effective, and statistically sound screening tool for the US pork industry. Current PMP panels are capable of monitoring multiple markers in a single test from a wide variety of sample types. Both panels can be used to evaluate the efficacy of antibiotic usage, variation across populations, determine co-infection correlations and monitor efficacy of vaccine strategies. Moreover, these tools are particularly advantageous for detecting pathogens at early stages of enteric and respiratory clinical disease, enabling farmers and veterinarians to intervene before symptoms occur. PMP technology could be deployed as a holistic surveillance tool whose results might find its use for a more balanced and targeted antibiotic drug strategy.

Figure 1: Pathogen counts per age group as analysed by Pork MultiPath™. Pooled oral fluid samples (n = 138) (left graph) and pooled faeces from sock samples (n = 140) (right graph) from two Latin American pig farms (Farm A and B) were processed on Pork MulitPath™ respiratory and enteric panels. The cumulative pathogen count per age group is illustrated for every pathogen as a stacked graph. Data was processed with the pandas library, and graphs were generated using matplotlib.pyplot in a Python environment.

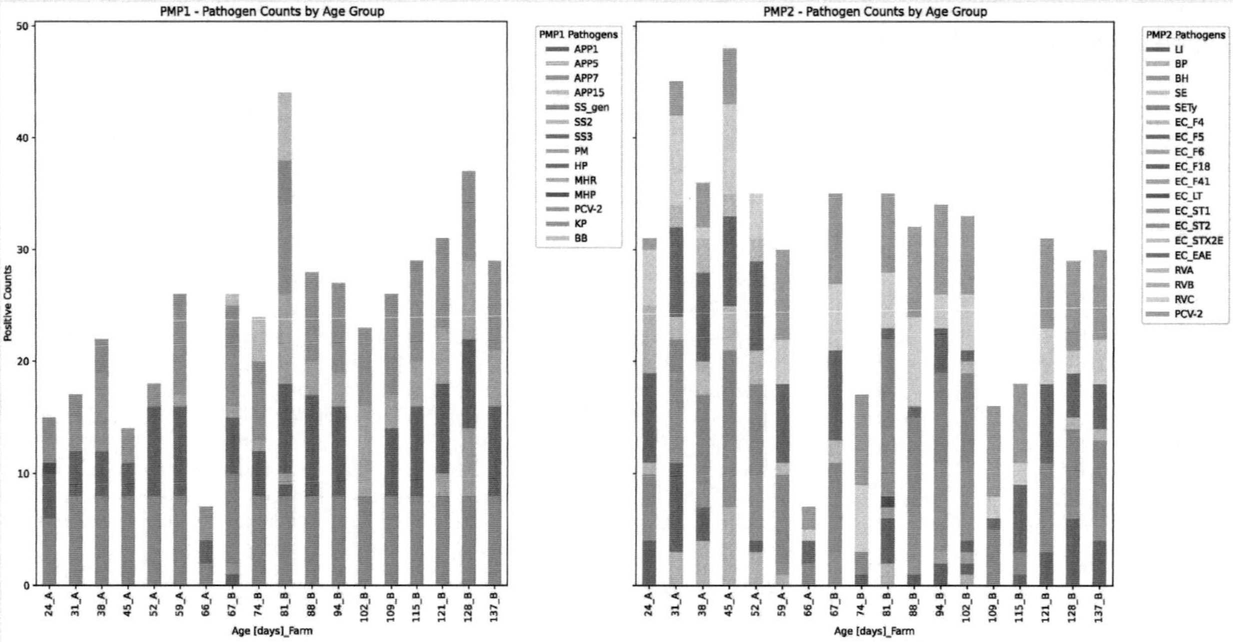

APP – *Actinobacillus pleuropneumoniae*; SS – *Streptococcus suis*, gen – generic; PM – *Pasteurella pneumoniae*; HP – *Glaesserella* (former *Haemophilus*) *parasuis*; MHR – *Mycoplasma hyorhinis*; MHP – *Mycoplasma hyopneumoniae*; PCV-2 – porcine circovirus type 2; KP – *Klebsiella pneumoniae*; BB – *Bordetella bronchiseptica*; LI – *Lawsonia intracellularis*; BP – *Brachyspira pilosicoli*; BH – *Brachyspira hyodysenteriae*; SE – *Salmonella enterica*; SETy – *Salmonella Typhimurium*; EC – *Escherichia coli*, RVA – rotavirus A; RVB – rotavirus B; RVC – rotavirus C.

References

1. Gardner IA, Colling A, Caraguel C, et al. Introduction - Validation of tests for OIE-listed diseases as fit-for-purpose in a world of evolving diagnostic technologies. Rev Sci Tech. 2021;40(1):19-28. **doi.org/10.20506/rst.40.1.3207**

2. Gerszon J, Büchse A, Genz B, et al. The use of oral fluids and sock samples for monitoring key pathogens in pig populations for surveillance purposes. Prev Vet Med. 2024;228:106237. **doi.org/10.1016/j.prevetmed.2024.1062**

 https://doi.org/10.54846/am2025/106

The PRRS-resistant pig and response to vaccination

Lucina Galina Pantoja; Justin Holl; Perry Harms; Matt Culbertson
PIC, Hendersonville, Tennessee

Introduction

The deletion of CD163 Exon 7, which codes for domain 5, confers resistance to porcine reproductive and respiratory syndrome virus (PRRSV); pigs carrying two copies of the deletion are homozygous (HOM) ($CD163^{\Delta E7/\Delta E7}$) and are fully resistant to disease, while heterozygous (HET) ($CD163^{\Delta E7/+}$) or null pigs (NULL) ($CD163^{+/+}$) are susceptible.[1-3] We recently reported resistance to PRRSV type I and type 2 most prevalent contemporary lineages *in vivo* and *in vitro*.[2,3] This time, we wanted to understand *in vivo* responses to a field isolate with high homology to a modified live virus (MLV) vaccine (vaccine-like isolate) and a commercial MLV vaccine. The aim was to characterize the responses of pigs with different CD163 genotypes to this field isolate and vaccine and increase the lineages tested. We hypothesized that the response should be similar to the wildtype isolates tested earlier.[3]

Materials and methods

In two studies, 3-week-old PRRSV-negative pigs were exposed to i) a PRRSV lineage 5, RFLP 2-5-2, MLV vaccine-like field isolate. This was intranasally inoculated, with 5 mL of $1x10^5$ $TCID_{50}$, ½ in each nostril, and ii) a commercial lineage 1D MLV vaccine, 1 mL intramuscular injection, following label instructions. One room was used in each separate study, and each room housed pigs with mixed CD163 genotypes (5 HOM and 5 NULL/HET), ensuring that pigs with different genotypes shared the same environment. Following 1 week of acclimation, serum was obtained before exposure (day 0) and at 3-21 post-exposure (Table 1) and sent to the ISU VDL for PRRSV PCR and ELISA.

Table 1: Pigs with various CD163 zygosities and their response to the lineage 5, "vaccine-like" field isolate, RFLP 2-5-2, and lineage 1D, commercial "MLV vaccine." PCR responses are negative when the Ct is < 37, and ELISA responses are negative when ≥ 0.4 and shown in days post inoculation (dpi). Negative responses are shown in green and positive in red.

Isolate	ID	Zygosity	PCR (Ct < 37) dpi						ELISA (S/P ≥ 0.4) dpi					
			0	3	7	14	10	21	0	3	7	10	14	21
Vaccine like	511	HOM	< 37	< 37	< 37	< 37	< 37	< 37	< 0.4	< 0.4	< 0.4	< 0.4	< 0.4	< 0.4
	541	HOM	< 37	< 37	< 37	< 37	< 37	< 37	< 0.4	< 0.4	< 0.4	< 0.4	< 0.4	< 0.4
	546	HOM	< 37	< 37	< 37	< 37	< 37	< 37	< 0.4	< 0.4	< 0.4	< 0.4	< 0.4	< 0.4
	588	HOM	< 37	< 37	< 37	< 37	< 37	< 37	< 0.4	< 0.4	< 0.4	< 0.4	< 0.4	< 0.4
	597	HOM	< 37	< 37	< 37	< 37	< 37	< 37	< 0.4	< 0.4	< 0.4	< 0.4	< 0.4	< 0.4
	504	HET	< 37	20.1	17.4	18.4	21.9	26.7	< 0.4	< 0.4	< 0.4	1.20	1.53	1.49
	510	HET	< 37	18.8	17.3	17.8	23.5	26.7	< 0.4	< 0.4	< 0.4	1.27	1.83	1.74
	545	HET	< 37	24.2	19.9	21	22.5	25.5	< 0.4	< 0.4	< 0.4	0.74	1.20	1.35
	592	HET	< 37	22.7	17.9	19	21.5	26.2	< 0.4	< 0.4	< 0.4	1.03	1.17	1.15
	599	HET	< 37	20.1	19.6	20.9	22.8	24.4	< 0.4	< 0.4	< 0.4	0.86	1.11	1.23
MLV vaccine	543	HOM	< 37	< 37	< 37	< 37	< 37	< 37	< 0.4	< 0.4	< 0.4	< 0.4	< 0.4	< 0.4
	538	HOM	< 37	< 37	< 37	< 37	< 37	< 37	< 0.4	< 0.4	< 0.4	< 0.4	< 0.4	< 0.4
	551	HOM	< 37	< 37	< 37	< 37	< 37	< 37	< 0.4	< 0.4	< 0.4	< 0.4	< 0.4	< 0.4
	555	HOM	< 37	< 37	< 37	< 37	< 37	< 37	< 0.4	< 0.4	< 0.4	< 0.4	< 0.4	< 0.4
	549	HOM	< 37	< 37	< 37	< 37	< 37	< 37	< 0.4	< 0.4	< 0.4	< 0.4	< 0.4	< 0.4
	682	NULL	< 37	28.4	26.6	27.8	31.1	28.1	< 0.4	< 0.4	0.45	0.45	0.85	0.92
	601	NULL	< 37	28.1	26.7	27.4	26.9	30.5	< 0.4	< 0.4	0.83	0.83	1.19	1.48
	672	NULL	< 37	26.1	27.4	30	31.8	<37	< 0.4	< 0.4	0.73	0.73	1.50	1.32
	668	NULL	< 37	28.6	25.1	29.6	29.1	32.5	< 0.4	< 0.4	0.39	0.39	0.60	0.97
	610	NULL	< 37	27.1	25	28.5	25.9	32.6	< 0.4	< 0.4	0.43	0.43	0.74	1.03

Results and conclusion

As expected, all pigs remained PRRSV-negative before exposure, corroborating the lack of infection during transport and acclimation. HOM pigs ($CD163^{\Delta E7/\Delta E7}$) remained negative after exposure to the 2-5-2 MLV-like isolate and MLV L1D vaccine (Table 1). In contrast, HET and NULL pigs showed evidence of infection by detecting genomic material and immune recognition post-infection. These results concur previous studies and together demonstrate that HOM pigs are resistant against multiple isolates from the most prevalent types and lineages in the US, including PRRSV type 1 (SD13-15, SD01-08) and type 2, lineage L1A (1-7-4), L1C (144), L1D (commercial MLV vaccine), L1H (1-8-4), L1E (1-4-2), L5 (2-5-2 vaccine-like) and L8.

References

1. Burkard et al., Journal Virology, 2019: 92 (16). **doi: 10.1128/JVI.00951-20**
2. Burger et al., The CRISPR Journal, 2024, 7(1) 12:28. **doi: 10.1089/crispr.2023.0061**
3. Nesbitt et al., Frontiers in Genome Editing, 2024, open access. **https://www.frontiersin.org/journals/genome-editing/articles/10.3389/fgeed.2024.1322012/full. doi: 10.3389/fgeed.2024.1322012**

Poster 33 https://doi.org/10.54846/am2025/107

Evaluation of deployable fan coverings for biocontainment of airborne swine pathogens

E. Kettelkamp, DVM; C. Schroeder, BS; P. Yeske, DVM, MS; A. M. Betlach, DVM, PhD
Swine Vet Center, Saint Peter, Minnesota

Introduction

With rising disease outbreaks across the Midwest, controlling swine respiratory disease pathogens has become crucial.[1] While air filtration reduces pathogen introduction via aerosols, biocontainment remains underexplored. Pathogens such as porcine reproductive and respiratory syndrome virus (PRRSV), porcine epidemic diarrhea virus (PEDV), and influenza A virus (IAV) are spread by respiratory secretions and other microscopic particles.[2,3] Studies have shown virus detectability and viability to be influenced by particle size, with PRRSV and IAV isolated from particles larger than 2.1 µm.[2,4] There is demand for ready-to-use biocontainment tools that can be quickly implemented on swine farms to reduce aerosol transmission pressure. This study evaluated the efficacy of rapidly deployable exhaust fan coverings for aerosol biocontainment by measuring total air particle quantity reductions.

Materials and methods

One wean-to-finish site stocked with mid-finishing pigs was selected, featuring various fan sizes (24″ pit fans, 36″ wall fans, 50″ wall fans). Three materials were compared to a fan with no covering (negative control): PolyKlean™ synthetic media (Blue Poly), Nylon, tear-resistant fan sock (Fan Sock) staked to the ground to redirect airflow, and Polyethylene privacy screen (Black Screen). Airborne particle counts were collected using an optical particle counter (OPC) measuring particles from 0.3 to < 5 µm. Sampling points were mapped based on the diameter of the fan cone. Five air particle measurements were collected 1 meter from the interior of the fan shutters and 1, 2, and 3 meters from the exterior of the fan. Three consecutive air particle measurements were performed per distance and location to calculate an average particle count, with three replicates completed for each treatment and fan. Minimum ventilation pit fans and stage two pit fans ran continuously at 100% power during sample collection. Each enrolled fan ran individually at 100% power while treatments were applied. Weather conditions were recorded at all sample locations using a portable weather station. Data were analyzed using a two-way repeated measures ANOVA to evaluate the effects of fan cover treatment and fan size on air particle quantity measurements with log transformation. Analysis was conducted using the lmer function in R studio, with fan identification as a random effect. Results were considered significant with $P \leq .05$.

Results and discussion

At 1 meter outside the fan, the fan sock treatment had a lower average 5.0 µm particle count than no cover or the blue poly treatments ($P = .046$). At 1 meter in front of the fan on the inside of the barn, the 50″ fan had significantly lower particle counts of particles measuring 1.0 µm ($P = .002$) and 5.0 µm ($P = .003$), and overall ($P < .001$). Differences in air particle counts between the 24″, 36″ and 50″ fan size were not observed ($P > .10$). No effects were observed based on weather conditions during the sampling period. Under the conditions of this study, the fan sock treatment had the greatest impact of reducing air particles at 1 meter outside of a fan compared to the other fan cover treatments, especially for 5.0 µm particles. However, as the distance from the fan increased, there was no difference in air particle quantity across the fan cover treatments. As a result, significant differences in the total number of air particles detected were not observed across the fan cover treatments. This pilot study suggests that rapidly deployable fan coverings may reduce air particle quantity, thus questioning its role in improving the biocontainment of aerosolized pathogens.

References

1. Linhares, D., & Trevisan, G. (2024). (rep.). *Swine Disease Reporting System* (Vol. 76, pp. 2–8). Ames, IA: SDRS. Retrieved 2024.

2. Alonso C, Raynor PC, Davies PR, Torremorell M. *Concentration, size distribution, and infectivity of airborne particles carrying swine viruses.* PloS one. **https://pubmed.ncbi.nlm.nih.gov/26287616/**

3. Otake S., Dee S., Corzo C., Oliveira S. & Deen J.; (2010). Long-distance airborne transport of infectious PRRSV and mycoplasma hyopneumoniae from a swine population infected with multiple viral variants. Veterinary microbiology. **https://pubmed.ncbi.nlm.nih.gov/20418029/**

4. Zuo, Z. , Kuehn, T.H. , Verma, H. , Kumar, S. , Goyal, S.M. , Appert, J. , Raynor, P.C., Ge, S. et al (2013) Association of airborne virus infectivity and survivability with its carrier particle size. Aerosol Sci Technol 47, 373–382.

https://doi.org/10.54846/am2025/108

Refining Senecavirus A disease occurrence estimates in the US breeding herd

M. Kikuti, DVM, MPH, PhD; C. M. Melini, DVM, MA; X. Yue, BSc, MSc, PhD;
S. Vadnais, BS, MPH; C. A. Corzo, DVM, MS, PhD
Department of Veterinary Population Medicine, University of Minnesota, Saint Paul, Minnesota

Introduction

Once clinical signs and lesions caused by Senecavirus A (SVA) are present, a foreign animal disease (FAD) investigation is needed to differentiate it from other vesicular diseases of concern. However, the data needed for a comprehensive understanding of its national occurrence in the US are not readily available. Most reports of SVA occurrence rely on seroprevalence or case counts from submissions to Veterinary Diagnostic Laboratories (VDL). While informative, these may include environmental sampling or multiple submissions from the same outbreak, which can potentially lead to an overestimation of disease occurrence. Furthermore, since laboratory submissions are likely driven by clinical suspicion, the positivity rate might be higher than expected in the general population. Therefore, we aimed to estimate SVA incidence in breeding herds in the US to better characterize the disease burden and begin outlining prevention and control measures.

Methods

The diagnostic data analyzed in this study were provided by the Veterinary Diagnostic Laboratories of University of Minnesota and Iowa State University from January 2015 to May 2024, representing 60% of the US breeding herd. Data including specimen submissions and Premises Ids were linked to SVA PCR submissions, excluding environmental samples, isolates, and miscellaneous specimens. Positive submissions from the same site within nine weeks of each other were considered to be representative of the same outbreak. An outbreak was also considered if no positive submission was detected but the site reported a SVA outbreak to the Morrison Swine Health Monitoring Project (MSHMP). Yearly cumulative incidence was calculated using the number of breeding herds reporting statuses to MSHMP for either PRRS, PED, or SVA (ie, sites sharing information for at least one of the primary diseases monitored by MSHMP) as the denominator, and the number of positive SVA submissions as the numerator.

Results

From 2015 to May 2024, 63 SVA outbreaks were detected in 55 sites from 13 production systems across 12 states. Five sites experienced two SVA outbreaks each, while three sites had three SVA outbreaks during the study period. For sites that experienced multiple outbreaks, the median time interval between SVA outbreaks was 98 days (interquartile range: 80.5 – 147 days). Most outbreaks occurred in the third and fourth quarters of the year (28 and 14, respectively) followed by the first and second quarters (12 and 9, respectively). The majority of outbreaks detected originated in the Midwest (82.54%), followed by the South (12.70%) and the Northeast (4.76%). The yearly cumulative incidence ranged from 0.09% in 2015 and 2021 to 1.93% in 2023.

Discussion

The findings indicate that SVA is significantly present in the US breeding herd, with notable regional concentration and periodic outbreaks. Most cases occurring in the Midwest highlights this region as a hotspot, likely due to the high density of swine production, although efforts to better assess disease distribution are still needed. Temporal patterns indicate a higher frequency of outbreaks in the third and fourth quarters of the year, suggesting possible seasonal effects. Yearly variability in incidence rates may reflect differences in surveillance, reporting, or actual disease fluctuations. The higher incidence in recent years may suggest either an increase in the true prevalence of SVA or improvements in detection and reporting mechanisms. Enhanced surveillance systems that integrate on-farm observations and other diagnostic tools could provide a more accurate picture of SVA dynamics, which is crucial for developing more effective control measures.

https://doi.org/10.54846/am2025/109

Comparison of pre-castration and post-castration tongue fluids to processing fluids to monitor PRRSV in suckling pig populations

I. Machado[1], DVM, MS; T. Petznick[1,2], DVM; A. Silva[1], DVM, MS, PhD; E. Magalhaes[1], DVM, MS, PhD; R. Nicolino[3], DVM, MS, PhD; E. Pratt[2], DVM; G. Silva[1], DVM, MS, PhD; D. Linhares[1], DVM, MBA, PhD
[1]VDPAM, Iowa State University, Ames, Iowa; [2]ArkCare, Omaha, Nebraska;
[3]Universidade Federal de Minas Gerais, Minas Gerais, Brazil

Introduction

Detecting porcine reproductive and respiratory syndrome virus (PRRSV) cost-effectively is crucial to breeding herd systems undergoing virus stabilization. While processing fluids (PF) are sensitive and commonly used for PRRSV monitoring in the US, questions persist about virus circulation among different piglet ages. Additionally, not all herds perform physical castration, hampering the use of PF for monitoring. Recently, tongue fluids (TF) from dead animals have emerged as a risk-based approach, with similar PRRSV RNA positivity compared to serum, PF, and family oral fluids.[1,2] Thus, this study aimed to compare PRRSV RNA detection in PF and TF in breeding herds undergoing PRRSV-2 stabilization.

Methods

Three breeding herds (A, B, C) undergoing PRRSV stabilization through herd closure and whole-herd live virus exposure were identified for the study. Tongue tips were collected daily from dead piglets of two groups: TF-pre and TF-post castration (0-3 and 4-21 days old, respectively), and PFs were collected daily from ~3-day-old piglets. Samples were tested for PRRSV RNA detection by RT-qPCR: TF were pooled per age group by week prior testing and PF were individually tested. The herds were combined and analyzed as herd-weeks using a mixed-effect logistic regression model. The outcome was the PCR positivity; the sample type was included as a fixed effect, and the breeding herd as a random effect. The Ct values were compared between sample types using non-parametric analysis (Kruskal-Wallis and Dunn tests). The Cohen Kappa agreement test was performed between the sample types. All statistical analyses were performed in R program software, using a significance level of 0.05.

Results

Data from 74 herd-weeks were analyzed, with results displayed in Table 1. Herd C was withdrawn from the statistical analysis as it had a single positive PF during the project. TF-pre and TF-post castration presented a substantial level of agreement (Kappa = 0.62), while both TF-pre and PF (Kappa = 0.46) and TF-post and PF (Kappa = 0.47) had moderate levels of agreement (P-value < .0001). No statistical differences were detected between TF-pre, TF-post, and PF for RT-qPCR detection (Table 1) (P-value > .05). Additionally, TF-pre (P-value = .01) and TF-post (P-value = .0007) groups had significantly lower Ct values (≤ 3.4) compared to PF.

Conclusion

The results from this study demonstrated that TF is an effective sample type for monitoring PRRSV-RNA in breeding herds, with lower Ct values than PF. TF can be collected from piglets of any age group in the farrowing room, and its collection was reported as practical within the routine, with sampling occurring on all of the days in the project.

Acknowledgment

This material is based upon work supported by American Association of Swine Veterinarians Foundation.

References

1. Machado, I. F., et al. (2022). Porcine reproductive and respiratory syndrome virus RNA detection in tongue tips from dead animals. Frontiers in Veterinary Science, 9, 993442.

2. Jansen, T., et al. (2023). Comparison of tongue tips, serum, and processing fluids for PRRSV monitoring in neonates. *54th AASV Annual Meeting*. 2023.

Table 1: Porcine reproductive and respiratory syndrome virus RNA detection by sample type combined as herd-weeks for herds A and B

	Sample type		
RT-qPCR	**Tongue fluids pre-castration**	**Processing fluids**	**Tongue fluids post-castration**
Positivity (%)	47.3% (35/74)[a]	44.5% (33/74)[a]	58.1% (43/74)[a]
Median Ct value and ranges	25.4 (20.5; 36.5)[b]	28.2 (22.5; 35.0)[a]	24.8 (18.2; 35.8)[b]

[a,b] = Different letter indicates statistical differences (P-value < .05).

https://doi.org/10.54846/am2025/110

Evaluating stillborn presence and litter size as indicators of PRRSV detection in live piglets, and the use of stillborn tongue fluids as risk-based samples for PRRSV monitoring

I. Machado, DVM, MS; L. Peng, DVM, MS; T. Petznick, DVM; O. Osemeke, DVM, MS, PhD; E. Kirwa, DVM, MS; G. Silva, DVM, MS, PhD; D. Linhares, DVM, MBA, PhD
VDPAM, Iowa State University, Ames, Iowa

Introduction

A risk-based approach to animal selection for sampling can enhance porcine reproductive and respiratory syndrome virus (PRRSV) detection, as it significantly improves the probability of selecting an animal harboring the pathogen while requiring a much smaller sample size. Recently, aggregated tongue fluids (TF) have been proposed as a risk-based approach, with a similar PRRSV RNA positivity rate as serum, PF, and family oral fluids.[1] The TF composition is complex, especially in stillborns, consisting of blood, saliva, respiratory mucus, amniotic fluid, meconium, and environmental contaminants, potentially reflecting the presence of PRRSV within the litter. Since TF samples are obtained from dead animals, they are hypothetically risk-based.[2] Thus, we determined the association of stillborn presence on the probability of having viremic live piglets in the litter, and compared the probability of PRRSV detection in stillborn TF to that of liveborn littermates.

Methods

From two PRRSV-positive breeding herds, samples were collected from 130 litters (1,830 piglets) within 12 hours after the farrow to assess the piglet's PRRSV status. TF and intracardiac blood (IB) were collected from stillborn piglets, and tail blood swabs (BS) were collected from all liveborn littermates. Samples were individually tested for PRRSV RNA by RT-qPCR. Generalized linear regression models were used to evaluate the effect of stillborn presence, TF PRRSV-RT-qPCR result, and litter size on the probability that a litter and/or ≥ 1 live littermate would test PRRSV positive. Litters with ≤ 11 liveborn piglets were defined as small litters (based on the first quantile for liveborn piglets). The Cohen Kappa agreement test was performed between TF and IB at a litter level. Analyses were conducted in R program software, using a *P*-value < .05 to determine statistical significance.

Results

TF and IB demonstrated a substantial agreement (Cohen's Kappa = 0.706, *P*-value < .0001). The mean positivity of live pigs within the litter was 5% (0 to 85.7%), while the total born was 4.6% (0 to 76.4%). In litters with at least one stillborn compared to those without stillborn, the odds of having a PRRSV-positive result (IB, TF, or BS) increased 5.6 times (95% CI: 2.21–16.28, *P*-value < .001), and the odds of having at least one viremic liveborn littermate (BS) increased 2.8 times (95% CI: 1.04–8.40, *P*-value = .049). When the stillborn TF was positive, the odds of having a viremic liveborn littermate (BS) increased 20.8 times (95% CI: 6.90–69.56, *P*-value < .001). Results are summarized in Table 1.

Additionally, small litters had 5.3 times higher odds (95% CI: 2.21–13.13, *P*-value = .002) of yielding a positive result (IB, TF, or BS) than litters with more than 11 liveborn piglets. The odds were 7.4 times higher (95 % CI: 2.72–21.91, *P*-value < .001) for having at least one positive liveborn piglet, and 2.7 times higher (95 % CI: 1.05–7.19, *P*-value = .03) for having at least one positive stillborn TF (Table 1).

Conclusion

In this study, positive stillborn TFs were a reliable predictor of the liveborn littermates' PRRSV status. Therefore, veterinarians and pig producers are encouraged to collect TFs, targeting stillborn piglets and small litters ($\leq$ 11 liveborn piglets) to increase the odds of yielding positive PRRSV RT-qPCR results. This targeted strategy in the first hours after farrow enhances the effectiveness of PRRSV monitoring programs in breeding herds. It can help guide better biomanagement practices, such as restricting piglet movements between sows. Lastly, although the TFs were individually collected to test the hypothesis in this study, their use as an aggregated sample, similar to processing fluids, is recommended.[1]

Table 1: Logistic regression analysis of predictors for PRRSV-RNA detection in piglets and litters

Outcome	Predictor	Odds ratio	95% CI*	*P*-value
Positive litter: at least 1 positive BS, IB, or/and TF	No stillborn	1		
	At least 1 stillborn piglet in the litter	5.6	2.21, 16.28	.0005
Positive liveborn littermate: BS	No stillborn	1		
	At least 1 stillborn piglet in the litter	2.8	1.04, 8.40	.049
Positive liveborn littermate: BS	Negative stillborn: IB and TF	1		
	Positive stillborn: IB or/and TF	20.8	6.90, 69.56	< .0001
Positive liveborn littermate: BS	Negative stillborn: TF	1		
	Positive stillborn: TF	20.8	6.90, 69.56	< .0001
Positive liveborn littermate: BS	Negative stillborn: IB	1		
	Positive stillborn: IB	21.7	6.06, 91.66	< .0001
Positive litter: at least 1 positive BS, IB, or/and TF	Litter size ≥ 12 liveborn piglets	1		
	Litter size ≤ 11 liveborn piglets	5.31	2.21, 13.13	.0002
Positive liveborn piglet: BS	Litter size ≥ 12 liveborn piglets	1		
	Litter size ≤ 11 liveborn piglets	7.42	2.72, 21.91	.0001
Positive stillborn: TF	Litter size ≥ 12 liveborn piglets	1		
	Litter size ≤ 11 liveborn piglets	2.76	1.05, 7.19	.035
Positive stillborn: IB	Litter size ≥ 12 liveborn piglets	1		
	Litter size ≤ 11 liveborn piglets	3.70	1.14, 12.41	.028

PRRSV = porcine reproductive and respiratory syndrome virus; BS = blood swab; IB = intracardiac blood; TF = tongue fluids

* 95% CI = 95% confidence interval.

Acknowledgment

Funding, wholly or in part, was provided by the Swine Health Information Center.

References

1. Machado, I. F., et al. (2022). Porcine reproductive and respiratory syndrome virus RNA detection in tongue tips from dead animals. *Frontiers in Veterinary Science*, 9, 993442.

2. Dürlinger, S., et al. (2024). Detection of PRRSV-1 in tongue fluids under experimental and field conditions and comparison of different sampling material for PRRSV sow herd monitoring. *Porcine Health Management*, 10(1), p.18.

https://doi.org/10.54846/am2025/111

Use of tongue tip fluids to detect swine pathogens in growing pigs

C. M. Melini[1]; M. Kikuti[1]; X. Yue[1]; I. Paploski[1]; A. Canturri[2]; S. Rossow[2]; C. A. Corzo[1]
[1]Department of Veterinary Population Medicine, University of Minnesota, Saint Paul, Minnesota;
[2]Veterinary Diagnostic Laboratory, University of Minnesota, Saint Paul, Minnesota

Introduction

Most post-weaning sampling relies on oral fluid samples since jugular venipuncture can be time-consuming and requires skilled personnel. Assessing whether easy-to-collect post-mortem samples, such as tongue tip fluids (TTF), can provide values for diagnosis. Monitoring during the post-weaning stages is necessary as this methodology can be adopted by the industry when resources are scarce. Therefore, the objectives of this study are **1)** to assess the sensitivity and specificity of TTF, and other specimens including intracardiac blood (IC), oral/nasal swabs (ONS), rectal swabs (RS), and superficial inguinal lymph nodes (SILN) in growing pigs for the detection of porcine reproductive and respiratory syndrome virus (PRRSV); and, **2)** to characterize the detection of porcine circovirus type 2 and 3 (PCV2, PCV3), porcine parvovirus type 1 and 2 (PPV1, PPV2), *Lawsonia intracellularis* (Li), and influenza A virus (IAV) in TTF, TTF, ONS, RS, and SILN.

Methods

Two farms were included in this study. One 2,400-head wean-to-finish farm undergoing a PRRS outbreak was sampled when pigs were 6 and 12 weeks of age (WOA). The second farm was a 3,300-pig finishing site undergoing a similar health status and was sampled when pigs were 15 WOA. During each sampling event, 30 dead pigs were sampled, resulting in a total sample of 90 pigs. All TTF samples were tested individually by RT-PCR for specific pathogens. All specimens were tested individually for PRRSV, while additionally ONS was individually tested for IAV; RS for PPV1, PPV2, and Li; while SILN were individually tested for PCV2 and PCV3. The sensitivity (Se), specificity (Sp), positive predictive value (PPV), and negative predictive value (NPV) were calculated for TTF for PRRS detection using IC as the gold standard. Descriptive statistics compared the proportion of RT-PCR positive results from specimens tested for other pathogens.

Results and discussion

Most pathogens were detected at least once in TTF with Ct values ranging from 11.6 to 39.8. For PPRSV the best results for all specimens were at 11 WOA, when TTF had Se = 84%, Sp = 9%, PPV = 62%, NPV = 25%, ONS had Se = 74%, Sp = 73%, PPV = 82%, NPV = 62%, and SILN had Se = 100%, Sp = 9%, PPV = 66%, NPV = 100%. PCV2 was detected in 43% of TTF and 11% of SILN samples, PCV3 was not detected in any sample. PPV1 was detected in 1% of TTF and 0% of RS, PPV2 was detected in 97% of TTF and 61% of the RS samples. Li was detected in 6% of TTF and 0% of the RS samples. IAV was detected in 38% of TTF and 38% of the ONS samples.

The majority of pathogens were detected on TTF samples at three different ages during the nursery, grower and finishing phases indicating that this specimen can provide value for post-mortem diagnosis. However, the overall diagnostic performance of the used specimens still requires further investigation. Indeed, complete and exhaustive collection of multiple clinical specimens from different body systems remains the standard in diagnostic investigations.

https://doi.org/10.54846/am2025/112

Practices implemented for live virus inoculation (LVI) material preparation and pathogen diversity presence in the sample

Mariamawit Mohammed; Michael Zeller; Daniel Linhares; Gustavo Silva; Christopher Rademacher; Christina Peterson; Giovani Trevisan
Iowa State University, Ames Iowa

Introduction

PRRS (porcine reproductive and respiratory syndrome) is a devastating viral disease that wreaks havoc on the swine industry, causing respiratory distress in growing pigs and reproductive failure in breeding animals.[1,2] A common practice for managing PRRS virus outbreaks is the use of live viral inoculum (LVI). This research employed an observational study to thoroughly characterize the PRRSV strain composition and identify practices in LVI preparation.

Methods

Samples used for LVI preparation were collected from multiple farms across ten swine production systems. Veterinarians provided survey data on LVI preparation and administration practices. Samples were tested by RT-qPCR to quantify PRRSV genomic copies and calculate expected viral particles based on a dilution formula provided by the veterinarians. Next-generation sequencing (NGS) followed by metagenomic analysis was conducted on each sample for detailed genomic profiling, characterizing the presence and number of PRRSV and the presence of other swine pathogens.

Results

Thirty-nine (n = 39) LVI serum samples were included in this study. Samples were collected, primarily from multiple piglet donors (94.87%, median = 4, max = 49). LVI was administered to the entire breeding herd. Saline was the most common diluent, followed by PBS. Antibiotics were not reported as added to LVI in 76.92% (n = 30) of cases. LVI was used alone in half of the cases and combined with MLV or killed PRRS vaccines in the rest. Intramuscular injection was the sole exposure method, with single administration in 76.92% (n = 30) of cases, twice in 20.51% (n = 8), and three times in 2.57% (n = 1). PRRSV concentration ranged from 49 to 88,350,183 particles/mL (median: 11,438 particles/mL and mean: 3,512,612). In 15.38% (n = 6) of LVI samples more than one unique PRRS virus was recovered. Other pathogens like *Salmonella enterica, Escherichia coli, Streptococcus suis,* and *Staphylococcus aureus* were also recovered.

Discussion

This study highlights the diversity of practices in LVI preparation. Pathogens detected by NGS provide valuable insights for veterinarians on potential clinical risks in the herd. To ensure consistency in LVI material, performing qPCR to quantify PRRSV and adjusting calculations accordingly, along with NGS to characterize pathogen diversity are recommended practices.

References

1. Lunney, J. K. et al. Porcine reproductive and respiratory syndrome virus (PRRSV): Pathogenesis and interaction with the immune system. Annual Review of Animal Biosciences, 2016, vol. 4 129 - 154. Preprint at **https://doi.org/10.1146/annurev-animal-022114-111025**

2. Trevisan, G. et al. Macroepidemiological aspects of porcine reproductive and respiratory syndrome virus detection by major United States veterinary diagnostic laboratories over time, age group, and specimen. PLoS One 14, (2019).

https://doi.org/10.54846/am2025/113

Assessment of PRRSV and PEDV detection in wean-to-finish pig barns before and after manure pumping

D. C. A. Moraes[1], DVM, MBA, MS; A. P. Poeta Silva[1], DVM, MS, PhD; M. N. Musskopf[1], DVM, MBA; E. K. Kirwa[1], DVM; M. P. Mil-Homens[1], DVM, MS; T. T. P. Tien[1], DVM, MS; Peng Li[1], DVM, MS; G. A. Cezar[1], DVM, MS; I. F. Machado[1], DVM, MS; H. Tong[1], DVM; K. Will[1], DVM, MS, PhD; E. Magalhaes[1], DVM, MS, PhD; S. R. Chagas[2], DVM, MS, PhD; C. Rademacher[1], BS, DVM; J. Zhang[1], MD, MS, PhD; G. Trevisan[1], DVM, MBA, PhD; G. S. Silva[1], DVM, MS, DVM; D. C. L. Linhares[1], DVM, MBA, PhD
[1]Iowa State University, Ames Iowa; [2]Universidade Federal de Goiás, Brazil

Introduction

Manure is commonly pumped out from many pig sites twice yearly, during the fall and spring seasons. Manure pumping may risk animal health due to the gases and pathogens recirculating during manure agitation and spreading in the crop fields.[1] This study aimed to assess the detection of porcine reproductive and respiratory syndrome virus (PRRSV) and porcine epidemic diarrhea virus (PEDV) in oral fluids and manure pit samples in wean-to-finish (W2F) barns before and after manure pumping.

Materials and method

A prospective longitudinal study investigated the frequency of PRRSV RNA and PEDV RNA detection in W2F populations before and after manure pumping. Eighty-four barns were selected for collecting oral fluid and manure pit samples before and two weeks after manure pumping. Six oral fluids and four manure samples were collected from each barn and tested for PRRSV by reverse transcription real-time polymerase chain reaction (RT-rtPCR), and PEDV was tested by a multiplex RT-rtPCR, which can also detect porcine delta coronavirus (PDCoV), and transmissible gastroenteritis virus (TGEV). Manure samples were collected using a disposable 9-foot-long pole with a 200 mL collection cup. Descriptive statistics were used for frequencies of PRRSV and PEDV detections before and after manure pumping in the R program.[2]

Results

From the eighty-four grow-finish barns initially enrolled, seven barns were not sampled after manure pumping because the animals were sent to the slaughterhouse. Thus, 77 barns were monitored for PRRSV and PEDV RNA by RT-rtPCR in oral fluids and manure pit samples collected before and after manure pumping. Among the 77 barns, 25 and 74 were eligible for PRRSV and PEDV evaluations, respectively. For PRRSV, 88% (22/25) of oral fluids tested PRRSV-positive after manure pumping, and 52% (13/25) of manure samples tested PRRSV-positive after manure pumping (Table 1). Two of 74 (2.7%) oral fluid samples tested PEDV-positive after manure pumping, while 97.3% (72/74) remained PEDV-negative after pumping (Table 1). PDCoV was detected in 1.1% of oral fluids before manure pumping, 23.8% of manure samples before manure pumping, and 25.9% of manure samples after manure pumping. TGEV was not detected in any samples before and after manure pumping.

Table 1: Eligible barns tested for PRRSV and PEDV using RT-rtPCR in oral fluids and manure before and after manure pumping

Sample type	PRRSV before pumping	PRRSV after pumping	
	PCR-negative	PCR-negative	PCR-positive
Oral fluid	25	3/25 (12%)	22/25 (88%)
Manure	25	12/25 (48%)	13/25 (52%)
Sample type	**PEDV before pumping**	**PEDV after pumping**	
	PCR-negative	**PCR-negative**	**PCR-positive**
Oral fluid	74	72/74 (97.3%)	2/74 (2.7%)
Manure	7*	6/7 (85.7%)	1/7 (14.3%)

PRRSV = porcine reproductive and respiratory syndrome virus; PEDV = porcine epidemic diarrhea virus

* 7 barns tested negative for manure samples from the 74 eligible barns for PEDV.

Discussion and conclusion

There was an increase in PRRSV detection following pumping, supporting the hypothesis that pumping is a risk factor for PRRSV spread in growing pig populations. PEDV positivity in manure was higher than that of PRRSV, and PEDV positivity in oral fluids was relatively low, with no apparent increase after pumping. In summary, this study highlights the need to understand better the impact of manure pumping on growing pig health and production in the Midwestern US.

References

1. Banks; D., Battrell, M. 2022. Porcine epidemic diarrhea virus survivability and infectivity in hog earthen manure storages. AASV Annual Meeting 2022 Defining Our Future Student Seminar **https://doi.org/10.54846/am2022/31**

2. Team, R.C., 2021. R: A language and environment for statistical computing. Vienna, Austria: R Foundation for Statistical Computing; 2020.

https://doi.org/10.54846/am2025/114

Effect of sampling interval on the time-to-detection of an FAD introduced into the US

B. Munguía-Ramírez[1], DVM, MS; P. Morris[2], PhD; G. Trevisan[1], DVM, PhD; G. Silva[1], DVM, PhD; D. Zhang[2], MS; C. Wang[2], BS, PhD; R. Main[1], DVM, PhD; J. Zimmerman[1], DVM, PhD
[1]Veterinary Diagnostic and Production Animal Medicine, Iowa State University, Ames, Iowa;
[2]Department of Statistics, Iowa State University, Ames, Iowa

Introduction

The extensive movement of animals, personnel, and materials facilitates the spread of infectious agents between and within countries, eg, porcine epidemic diarrhea virus, African swine fever virus, ASFV, and others. Thus, implementing an effective, sustainable, and affordable surveillance system capable of early detection is crucial in the event of a foreign animal disease (FAD) incursion. Previously, we showed that "active participatory regional surveillance", ie, surveillance based on targeted sampling/testing of 10 poor-doing pigs per site could achieve detection at 0.01% herd level prevalence.[1] If participatory surveillance were to be implemented in contiguous 48 states, how often should sampling be performed? That is, what is the effect of sampling interval on the time to detection of an FAD introduced into the US? We addressed this question by simulating the spread of a pathogen in US swine farms using the USDA's Animal Disease Spread Model (ADSM) software (v3.5.10.0) and then calculating the probability of detecting ≥ 1 positive farms for different sampling intervals as a function of percentage (%) of participating herds, herd-level sensitivity, and herd-level prevalence.

Methods

Using the 2017 USDA Census of Agriculture, we created a dataset of 66,637 farms (71,555,218 pigs) in the lower 48 states (8,080,470 km^2). Each farm was identified by geolocation, production type (breeder, feeder, breeder/feeder), and inventory. The spread of a "generic" pathogen was simulated for 180 days over a range of spread parameter values for direct contact (movement of infected animals), indirect contact (movement of people, vehicles, etc), and local area spread (probability of infecting a farm at ≥ 1 km). Infection was initiated in an index herd randomly selected from four "pig density" categories, ie, < 1 pig, 1 to 10 pigs, 10 to 100 pigs, and > 100 pigs per km^2. Each of the 6,075 scenarios was replicated 1,000 times to account for the stochastic nature of ADSM. The output for each scenario provided the infection status of each farm by day post-outbreak. From these data, the probability (*P*) of detecting ≥ 1 positive farm was calculated as a function of the percentage (%) of herds participating in the surveillance program, herd-level sensitivity, and disease prevalence for specific days post-outbreak (day 14, 28, 42, and 56). In addition, we estimated the cumulative *P*, ie, the probability of detecting ≥ 1 positive farm if sampling were performed each 14 days. The cumulative *P* was calculated using the complementary probability formula (Equation 1).[2] In Equation 1, "*p*" represents the probability of not detecting any positive herds, and "*n*" is the (%) of positive participating herds.

(Equation 1) $P = 1 - (1 - p)^n$

Results

The complementary probability formula showed that at a prevalence of 0.03% (17 infected herds, day 28 of the outbreak), with 80% participating herds submitting samples for surveillance and 10% herd-level sensitivity, we can achieve a 90% probability of detecting ≥ 1 positive farm if the sampling is done every 14 days vs a 77% probability of detection if sampling every 28 days (Table 1).

Conclusions

The results of the simulations suggested that a two-week sampling interval in a national participatory surveillance program would detect a fast-spreading pathogen at a 0.03% prevalence. Such a system could be achieved through national surveillance programs (US Swine Health Improvement Plan) and diagnostic testing performed at NAHLN laboratories.

Table 1: Probability of detecting ≥ 1 positive herd as a function of percentage (%) of participating herds, herd-level sensitivity (%), and disease prevalence (%)

		Day 14 (0.01% prevalence)	Day 28 (0.03% prevalence)		Day 42 (0.04% prevalence)		Day 56 (0.07% prevalence)	
Producer participation (No. herds)	Herd-level sensitivity	P[1] (point in time)	P[1] (point in time)	Cumulative P[2]	P[1] (point in time)	Cumulative P[2]	P[1] (point in time)	Cumulative P[2]
20% (13,327)	0.1	0.19	0.31	0.44	0.45	0.69	0.64	0.89
40% (26,654)	0.1	0.34	0.52	0.68	0.70	0.91	0.87	0.99
60% (39,982)	0.1	0.47	0.67	0.82	0.84	0.97	0.95	1.00
80% (53,310)	0.1	0.57	0.77	0.90	0.91	0.99	0.98	1.00
100% (66,637)	0.1	0.65	0.84	0.94	0.95	1.00	0.99	1.00
20% (13,327)	0.3	0.51	0.71	0.86	0.87	0.98	0.97	1.00
40% (26,654)	0.3	0.76	0.92	0.98	0.98	1.00	1.00	1.00
60% (39,982)	0.3	0.88	0.98	1.00	1.00	1.00	1.00	1.00
80% (53,310)	0.3	0.94	0.99	1.00	1.00	1.00	1.00	1.00
100% (66,637)	0.3	0.97	1.00	1.00	1.00	1.00	1.00	1.00
20% (13,327)	0.5	0.75	0.91	0.98	0.98	1.00	1.00	1.00
40% (26,654)	0.5	0.94	0.99	1.00	1.00	1.00	1.00	1.00
60% (39,982)	0.5	0.98	1.00	1.00	1.00	1.00	1.00	1.00
80% (53,310)	0.5	1.00	1.00	1.00	1.00	1.00	1.00	1.00
100% (66,637)	0.5	1.00	1.00	1.00	1.00	1.00	1.00	1.00

1 Probability of detecting ≥ 1 positive herd during a single sampling event calculated as the complement of the probability of detecting zero positive herds among participating farms: $P = 1 - (1 - p)^k$.

2 Cumulative probability of detecting ≥ 1 positive herd across multiple sampling events (assuming independence among all sampling time points), calculated as the complement of the probability of zero detection of positive herds across all events: Cumulative $P = [1- (1-P_1) \times (1-P_2)\times \ldots (1- P_n)]$.

References

1. Trevisan G, et al. 2024. Active participatory regional surveillance for notifiable swine pathogens. *Animals* 14:233.

2. Rotolo M, et al. 2017. Sampling guidelines for oral fluid-based surveys of group-housed animals. *Vet Microbiol* 209:20-29.

https://doi.org/10.54846/am2025/115

Analysis of PRRSV ORF5 lung and serum samples reveals higher rtPCR Ct values in vaccine-like strains compared to wild-type strains

R. Naseer; M. Zeller, PhD; P. Gauger, DVM, PhD; G. Trevisan, DVM, PhD
Department of Veterinary Diagnostic & Production Animal Medicine, College of Veterinary Medicine,
Iowa State University, Ames, Iowa

Introduction

Various strains of porcine reproductive and respiratory syndrome virus (PRRSV) may cause differences in the severity of clinical signs, adverse animal health impacts, and economic losses in swine populations. Understanding the relationship of cycle threshold (Ct) values between wild-type and vaccine-like strains is essential for assessing viral dynamics, and identifying samples that may pose future clinical risk.

Methods

Data was sourced from the Iowa State University Veterinary Diagnostic Laboratory (ISU VDL), encompassing years of PRRSV testing records. This repository offers access to historical and current PRRSV diagnostic testing results, including specimen type, Ct value, collection date, sequence lineage, and associated metadata. The dataset included only lineage 5A (L5A) and lineage 1D (L1D) sequences from 2018 to 2023, as these lineages possess robust diagnostic data and associated vaccine strains; Ingelvac MLV (modified live virus) and Prevacent PRRS, respectively. The data was divided by specimen type, with an emphasis on lung and serum samples. Analyses focused on positive Ct values, defined by a cutoff of < 38. L5A had over 1000 sequences recovered from lung and over 950 from serum. L1D had over 250 sequences recovered from lung and over 200 from serum. Alignments and nucleotide identity were utilized to classify sequences as vaccine-like or wild-type based on a 98% threshold comparison to the relevant vaccine strains. Statistical analyses performed in R compared Ct value distributions and their correlation regarding classification as wild-type or vaccine-like and lineage.

Results

Statistically significant differences in median Ct values between vaccine-like and wild-type samples were identified at a 0.05 significance level. For wild-type samples, median Ct values were 20.8 (lung) and 25.2 (serum) in L5A, and 21.2 (lung) and 23.3 (serum) in L1D. Conversely, vaccine-like samples exhibited median Ct values of 21.9 (lung) and 28.7 (serum) in L5A, and 23.2 (lung) and 29.6 (serum) in L1D. In L5A, mean Ct differences between vaccine-like and wild-type strains were 1.4 (lung) and 2.9 (serum), while in L1D, they were 1.7 (lung) and 4.2 (serum). Annotated phylogenetic trees showed distinct clusters for wild-type and vaccine-like sequences, indicating phylogenetic differentiation. Wild-type sequences had lower Ct values, suggesting higher viral loads. However, although unexpected, some vaccine-like strains also had low Ct values, warranting further investigation and surveillance. Next steps include whole genome analysis of samples with vaccine-like strains and a review of relevant clinical data to evaluate differences in disease severity and identify potential recombination events.

Discussion

Understanding the implications of Ct value distribution could help inform the early detection of strains with more severe clinical implications and viral behavior, enabling the rapid development of effective biosecurity and management strategies against PRRSV in swine.

https://doi.org/10.54846/am2025/116

Bio-exclusion practices in US wean-to-market farms: A factor analysis across 18 production systems

M. Musskopf[1]; R. Nicolino[2]; T. T. P. Tien[1]; D. Holtkamp[1]; G. Silva[1]
[1]VDPAM, Iowa State University, Ames, Iowa; [2] Epidemiologyi, Universidade Federal de Minas Gerais, Belo Horizonte, MG, Brazil

Introduction

Bio-exclusion practices are the first barrier to avoiding pathogen introduction in swine farms. Despite their importance, there is a lack of information regarding the practices implemented in wean-to-market pig populations across the United States. Measuring and identifying distinct factors of biosecurity practices is crucial for understanding variations in implementation and developing targeted interventions to prevent pathogen introduction. This study aims to identify the profiles of bio-exclusion practices implemented by production systems at wean-to-market sites and to describe the relationship between the distinct practices.

Materials and methods

The bio-exclusion information was obtained through a questionnaire fulfilled by the herd veterinarians from 18 swine production systems. A factor analysis (FA) using a principal components method was used to generate unobserved underlying factors. Initially, 36 variables related to bio-exclusion practices were tested. Variables with no variability between production systems were removed from the FA. The uniqueness criteria was used to keep variables with values below 0.5, and the Kaiser–Meyer–Olkin (KMO) test was the statistical measure to determine the fitting of the factor analysis. Using the Kaiser criteria, factors with eigenvalues greater than one were maintained.

Results

Among 36 variables, 7 variables divided into 2 factors were identified, explaining 76.8% of their variability. The distribuition of production systems on factor adoption is displayed in Figure 1. Factor 1 highlighted an inverse relationship between on-site composting of dead animals and central or direct rendering disposal. Factor 2 combined shower-in and shower-out practices with unidirectional flow and defined clean/dirty lines. For factor 1, 50% of production systems showed high values, meaning they adopt more on-site composting, 11% had values near 0.15, and 39% had negative values indicating more off-site disposal. For factor 2, 44% had high adoption of showers, 50% had negative values indicating lower adoption, and 6% showed balanced adoption. Systems were categorized into four groups based on their adoption of these factors: high adoption of both, high adoption of factor 1 and low for factor 2, high adoption of factor 2 and low for factor 1, and low adoption of both factors. The results highlight the variability between and within production systems to rendering and site entry/exit.

Figure 1: Production system distribution on factor adoption.

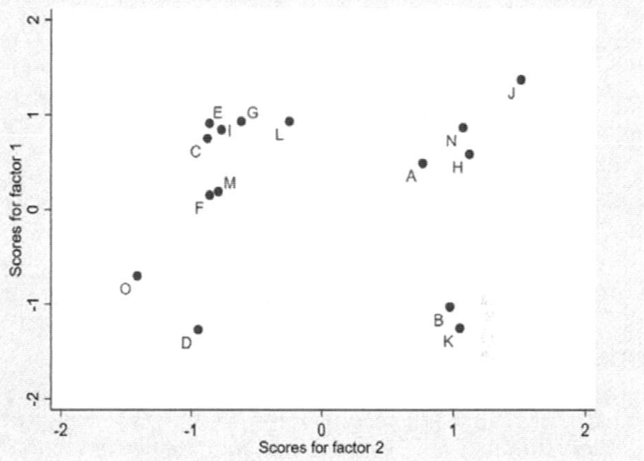

Potential causes for the variability could be the cost to adopt the practices, practicality, labor resources, region, and weather conditions. Adopting specific bio-exclusion practices must consider what better fits the system's necessities. However, it is expected that more layers of biosecurity will decrease the risk of pathogen introduction, given that the risk may increase with the frequency of the events.

Discussion

Among all variables, the practices related to rendering disposal and farm entry were the factors that had higher variability in the production system. These results highlight that while some production systems perceive site entry as a high priority, others may not have the resources to implement these practices. Production systems with low adoption of both factors could have a higher risk for pathogen introduction.

Poster 43
https://doi.org/10.54846/am2025/117

Enhancing PRRSV isolation from postmortem tongue fluids: Evaluation of collection protocols and cell lines

Onyekachukwu Henry Osemeke[1], DVM, MS, PhD; Isadora Machado[1], DVM, MS; Elisa De Conti[1], DVM; Mariah Musskopf[1]; Mafalda Mil-homens[1], VM, MS; Samuel Stutzman[2]; Baoqing Guo[1], BVSc, MS, PhD; Thomas Petznick[1], DVM; Phil Gauger[1], DVM, MS, PhD; Gustavo Silva[1], DVM, MS, PhD; Jianqiang Zhang[1], MD, MS, PhD; Daniel Linhares[1], DVM, MBA, PhD
[1]Veterinary Diagnostic and Production Animal Medicine, Iowa State University College of Veterinary Medicine;
[2]Ronnfeldt Farms, Nebraska

Introduction

Porcine reproductive and respiratory syndrome virus (PRRSV) remains a significant threat to swine health globally. The postmortem tongue tissue fluids (TF) sample has been demonstrated to be a cost-effective option for investigating PRRSV presence in swine herds by reverse transcription real time polymerase chain reaction (RT-rtPCR),[1,2] however, effective isolation of PRRSV is vital for confirming infectious virus presence. Using different sample collection protocols and cell lines, this study evaluated the feasibility of using tongue tissue fluids (TF) from perinatal piglet mortalities as a sample source for PRRSV isolation.

Materials and methods

Four collection protocols were assessed: 1) TF extraction in phosphate buffered saline (PBS), 2) TF extraction in virus transportation medium (VTM), 3) TF extraction in PBS after one freeze-thaw cycle (Freeze-thaw), and 4) preparation of tongue tissue homogenate (Homogenate). Two cell lines (ZMAC and MARC-145) and one batch of primary alveolar macrophages (PAM) were also tested for their effect on successful PRRSV isolation.

Over four days, samples were collected from 597 perinatal mortalities from a 5000-head PRRSV-positive breeding herd. The tongues were collected in 20 batches (~ 30 mortalities per batch). Within each batch, each tongue tissue was cut into 4 quarters with each quarter randomly assigned to one of four collection protocols. In total, 80 bags of samples (20 batches × 4 protocols) were sent for reverse transcription real-time polymerase chain reaction (RT-rtPCR) analysis. The RT-rtPCR cycle threshold (Ct) values were averaged across protocols for each of the 20 batches, and the 10 batches with the lowest mean Ct were selected for virus isolation (VI). Statistical models built using the R statistical software characterized statistical differences in RT-rtPCR Ct between the collection protocols and the statistical differences in the probability of successful PRRSV isolation between the collection protocol groups and between the cell types.

Results

The PBS, VTM, Freeze-thaw, and Homogenate groups had mean Cts of 21.9, 21.8, 22.6, and 24.8, respectively, with the mean Ct in the homogenate group statistically higher than the means of other groups (Figure 1). The probabilities of successful VI (and 95% confidence intervals) were 22.6% (8.1%, 49.1%) in the PBS group, 12.1% (3.4%, 39.1%) in the VTM group, and 2.8% (0.4%, 16.4%) for both the Homogenate and Freeze-thaw groups. There were no statistical differences ($P > .1$) in the probability of successful VI between collection protocols. The probabilities of successful VI in MARC-145 and ZMAC cell lines were 3.1% (0.59%, 14.9%) and 21.0% (8.4, 43.5%), respectively, and 4.8% (1.1%, 19.1%) in PAM cells. The probability of successful VI was statistically higher ($P < .1$) in the ZMAC cells. ORF5 sequencing indicated that all 10 batches of TF samples used for VI and the obtained cell culture isolates all belonged to L1C.5 with 99-100% nucleotide identity to each other.

Figure 1: Distribution of RT-rtPCR Ct values for the phosphate-buffered saline (PBS), tissue homogenate group (Homogenate), virus transport medium (VTM), and freeze-thaw (Freeze-thaw) groups. Groups with different letters are statistically different ($P > .1$). The central triangles within the boxplots represent the mean Ct value for the collection protocol.

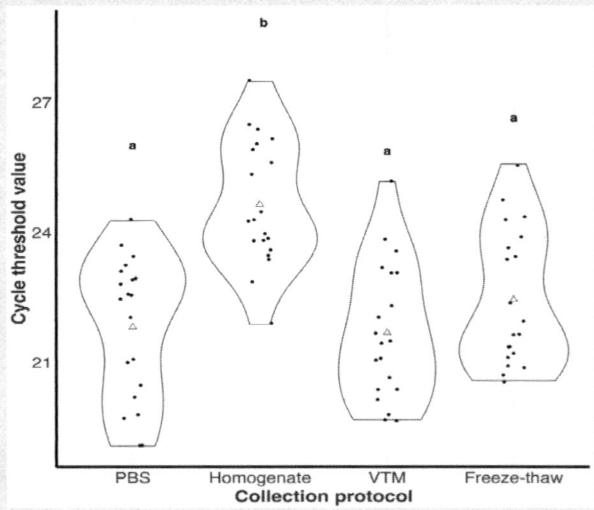

Discussion

It was possible to isolate PRRSV from TF from perinatal mortalities. However, the VI success rate varied among different sample collection protocols and cell culture systems. Successful virus isolation from TF allows practitioners to obtain a more herd-representative sample for surveillance and whole genome sequencing.

References

1. Machado, I. F. *et al.* Porcine reproductive and respiratory syndrome virus RNA detection in tongue tips from dead animals. *Front Vet Sci* 9, 1356 (2022).

2. Baliellas, J., Novell, E., Enric-Tarancón, V., Vilalta, C. & Fraile, L. Porcine Reproductive and Respiratory Syndrome Surveillance in breeding Herds and Nurseries Using Tongue Tips from Dead Animals. *Veterinary Sciences 2021, Vol. 8, Page 259* 8, 259 (2021).

 https://doi.org/10.54846/am2025/118

Evaluating postmortem tongue fluids as a tool for monitoring PRRSV and IAV in the nursery and grow-finish phases of swine production

Onyekachukwu Henry Osemeke[1], DVM, MS, PhD; Isadora Machado[1], DVM, MS; Mafalda Mil-Homens[1], DVM, MS; Micheal Paustian[3], BS, PhD; Grant Allison[2], DVM; Daniel Linhares[1], DVM, MBA, PhD; Gustavo Silva[1], DVM, MS, PhD
[1]Veterinary Diagnostic and Production Animal Medicine, Iowa State University College of Veterinary Medicine; [2]Walcott Veterinary Clinic, Iowa; [3]Paustian Enterprises, Walcott, Iowa

Introduction

Postmortem tongue fluids (TF) sample has been demonstrated to be an efficient, herd-sensitive tool for monitoring the porcine reproductive and respiratory syndrome virus (PRRSV) in swine herds.[1,2] However, the utility of TF for longitudinal monitoring of PRRSV and influenza A virus (IAV) in wean-to-market (WTM) phases remains underexplored. Oral fluids are currently the primary surveillance sample type submitted from growing pig populations. The objectives of this study were 1) to compare the PRRSV and IAV reverse transcription-real time polymerase chain reaction (RT-rtPCR) detection in weekly oral fluids (OF) and TF obtained from WTM pigs, 2) to compare the probability of ORF-5 sequencing success between sample types (OF vs TF).

Methods

Three groups of growing pigs from two production systems were monitored weekly for PRRSV and IAV from nursery placement until market. Two groups (A and B) were sourced from PRRSV-positive stable sow farms. The third group (C) was sourced from a PRRSV positive unstable farm. Groups B and C were given PRRSV modified live virus (MLV) vaccine at weaning. OF (n = 6) were collected weekly, tested individually for PRRSV, and tested in 2 pools of 3 OF for IAV. TF was collected daily and tested as one weekly pool for PRRSV and IAV. All samples were tested by RT-rtPCR, and 22 sets of TF and OF across all 3 groups, paired by week, were sent for Sanger sequencing of the ORF-5 region. Cohen's Kappa analysis was used to assess agreement in RT-qPCR detection rates of both pathogens using both sample types. All statistical analyses were done using the R statistical software.

Results

There were 60 weeks across all groups with at least one sample type tested (OF was submitted in all 60 weeks); 43 weeks had both TF and TT submitted. Of the 43 matched weeks, IAV was detected in OF and TF in 15 (34.9%) and 13 (30.2%) weeks, respectively, while PRRSV was detected in OF in 29 (67.4%) OF and TF in 23 (53.5%) weeks. The Cohen's kappa values that assessed agreement beyond chance was 0.26 (fair agreement) for PRRSV and 0.24 (fair agreement) for IAV. The marginal mean (mm) PRRSV RT-rtPCR Ct (Figure 1) in TF (29.13 (22.37, 35.89)) was statistically lower ($P < .05$) than in OF (32.8 (24.74, 40.87), but not different for IAV (30.88, 25.11, 36.64) in TF samples and 29.66 (23.26, 36.07) in OF samples). Of the 22 week-matched TF and OF sent for PRRSV sequencing, 10 (45.5%) OF and 14 (63.6%) TF were successfully sequenced.

Figure 1: The distribution of RT-rtPCR cycle threshold values for influenza A virus in oral fluids (IAV_OF) and postmortem tongue fluids (IAV_TF), porcine reproductive and respiratory syndrome virus (PRRSV) in oral fluids (PRRSV_OF), and PRRSV in postmortem tongue fluids (PRRSV_TF). Distributions with different letters are statistically different ($P < .05$).

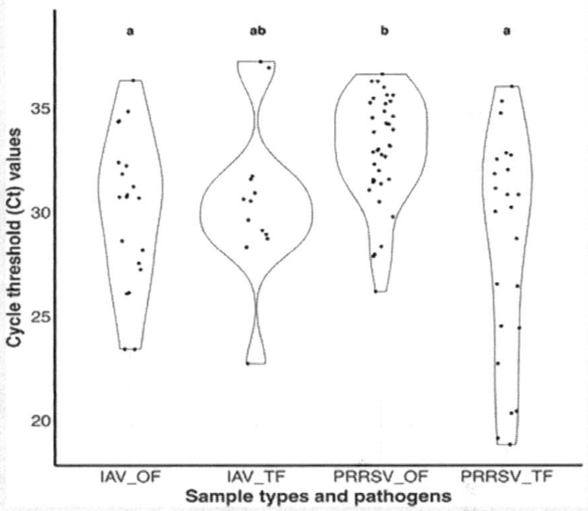

Discussion

TF samples present a practical, cost-efficient alternative for monitoring growing pigs for PRRSV and IAV when mortalities are available. The absence of mortalities in certain weeks made it impossible to use this sample type, presenting a potential challenge for routine monitoring. Nonetheless, the presence of dead pigs is a common trigger for disease investigation in wean-to-market barns, making the sample type useful in such scenarios. Although swine pathogens were detected more frequently in OF in weeks where TF was also obtained, 6 samples were tested for OF while only one test was done per week for TF. The increased PRRSV ORF-5 sequencing success can be attributed to the relatively lower Ct range inherent with TF samples.

References

1. Machado, I. F. *et al.* Porcine reproductive and respiratory syndrome virus RNA detection in tongue tips from dead animals. *Front Vet Sci* 9, 1356 (2022).

2. Baliellas, J., Novell, E., Enric-Tarancón, V., Vilalta, C. & Fraile, L. Porcine Reproductive and Respiratory Syndrome Surveillance in breeding Herds and Nurseries Using Tongue Tips from Dead Animals. *Veterinary Sciences 2021, Vol. 8, Page 259* 8, 259 (2021).

Poster 45

https://doi.org/10.54846/am2025/119

Whole genome sequencing of *E coli* isolates associated with colibacillosis cases in pigs from 2015 to 2022

Rodrigo C. Paiva[1], DVM, MBA, MS; Maud de Lagarde[2], DVM, PhD; John Morris Fairbrother[2], BVSc, PhD; Eric Burrough[1], DVM, PhD, DACVP; Nubia Macedo[1], DVM, MS, PhD; Pablo Pineyro[1], DVM, MVSc, DVSc, PhD; Marcelo Almeida[1], DVM, MS, PhD

[1]Department of Veterinary Diagnostic and Production Animal Medicine, Veterinary Diagnostic Laboratory, Iowa State University, Ames, Iowa

[2]Faculty of Veterinary Medicine, Université de Montréal, Saint-Hyacinthe, Canada

Introduction

Whole genome sequencing (WGS) surpasses traditional methods like genotyping and broth microdilution to characterize critical pathogenic *E coli* isolates.[1] This study characterized *E coli* isolates submitted to the Iowa State University Veterinary Diagnostic Laboratory (ISU-VDL) and the world organisation for animal health reference laboratory for *E coli*, Faculty of Veterinary Medicine, Université de Montréal, using WGS. Temporal trends in pathotypes, antimicrobial resistance mechanisms, and the emergence of a putative high-risk clone associated with colibacillosis in swine were described.

Materials and methods

FastQC and MultiQC were used to evaluate the quality of raw data from the 71 sequenced isolates. Short-read sequences were assembled using SPAdes; the presence of virulence and resistance genes and replicons was determined using the Center for Genomic Epidemiology (CGE) platform. The SAS.9.4 software (SAS, SAS Institute Inc., Cary, NC, USA) was used to create heatmaps, filter the data, and generate the Interactive Tree of Life (iTOL) datasets. The presence of chromosomal point mutations was determined using PointFinder on the CGE platform. Then, the phylogeny was generated through an "in-house" pipeline available on GitHub, and the tree was visualized with iTOL.

Results

Two isolates possessed F4, and sixty-nine possessed F18 fimbrial type. 21% of isolates were phylogroup A, 32% were phylogroup B1, and 47% were phylogroup C. Concerning the pathotype, 67% were hybrid ETEC:STEC, and 33% were ETEC. All isolates of the serotype O116:H11 (44%) were hybrid ETEC:STEC and belonged to phylogroup C. The serotype O141:H4 represented 17% of the isolates, of which 92% were hybrid ETEC:STEC belonging to the phylogroup A. Serotype O147:H2 represented 17% of the isolates, of which 83% were ETEC and 17% ETEC:STEC, all isolates being phylogroup B1. Heat-labile (LT) and heat-stable (ST) toxin genes were present in 93% and 100% of the isolates, respectively. Genes associated with colicin resistance, hemolysin modulators, serum resistance, attachment and invasion, and apoptosis induction were detected in a large proportion of the isolates. All isolates carried multiple antimicrobial-resistance genes and were thus classified as multidrug-resistant (MDR). 69% of isolates carried a mutation in the gene *gyrA*, which confers fluoroquinolone resistance, and 17% possessed the gene $bla_{CMY\text{-}2}$, conferring cephalosporin resistance. Genes coding for aminoglycoside, beta-lactam, fosfomycin, glycopeptide, macrolide, phenicol, rifampicin, sulphonamide, tetracycline, and trimethoprim resistance were commonly observed.

Conclusions

Most *E coli* isolates in this study demonstrated a hybrid pathotype and possessed multiple genes associated with pathogenicity, putatively causing more severe disease in post-weaning pigs. All isolates were multi-drug resistant (MDR). Our data suggest that an ETEC:STEC O116:H11 clone non-susceptible to fluoroquinolones is emerging and has a high putative virulence capacity.

Reference

1. Fairbrother, J.M., Gyles, C.L., 2022. John Wiley & Sons, Ltd, pp. 117–161. **https://doi.org/10.1002/9781119754862.ch7**

https://doi.org/10.54846/am2025/120

Which *E coli* specimen and morphology are more associated with post-weaning colibacillosis in pigs in the US?

Rodrigo C. Paiva, DVM, MBA, MS; Eric Burrough, DVM, PhD, DACVP; Nubia Macedo, DVM, MS, PhD; Pablo Pineyro, DVM, MVSc, DVSc, PhD; Marcelo Almeida, DVM, MS, PhD
Department of Veterinary Diagnostic and Production Animal Medicine, Veterinary Diagnostic Laboratory, Iowa State University, Ames, Iowa

Introduction

The *E coli* microbiota is highly structured, and different gut compartments are preferentially colonized by distinct phylogenetic clones.[1,2] This study aimed to identify trends in *E coli* specimens and morphologies associated with colibacillosis from 2014 to 2023 in the United States.

Materials and methods

Data from 5,784 cases submitted to Iowa State University Veterinary Diagnostic Laboratory (ISU-VDL) were retrospectively assessed and included. Case-individual information about histological evaluation, *E coli* colonization and colony morphology, specimen, age, and disease code were collated and evaluated. *E coli* disease code (Dx-code) considered histopathological evaluation from the ISU-VDL diagnosticians, which consider clinical signs, histopathological tissue evaluation, and ancillary test (genotyping) in the final diagnostic results. For this study, all selected cases had a histological evaluation, specific age (weaning age, 3-6 weeks), absence of tissue autolysis, and an *E coli* Dx-code. SAS.9.4 software (SAS, SAS Institute Inc, Cary, NC, USA) was used to calculate the risk ratio considering a diagnostic of post-weaning colibacillosis (PWC) versus non-confirmed cases.

Results

Smooth mucoid morphologies were likely associated with colibacillosis comparing all other morphologies and greater than smooth, intermediate, and rough, respectively (Table 1). Small intestine specimens were more likely (*P*-value < .001) to be associated with colibacillosis than colon and fecal swabs, respectively. The specimens' risk estimates and 95% confidence intervals (CI) were 0.822 (0.809-0.836) for the small intestine, 0.033 (0.026-0.039) for the colon, and 0.019 (0.014-0.023) for fecal swab, respectively.

Discussion and conclusions

Colibacillosis is still an important worldwide disease for the swine industry, and information to help enhance its diagnosis is fundamental to reducing losses due to this condition. This study demonstrated the likelihood of specific *E coli* morphologies more associated with colibacillosis and the greater chances of using specific specimens to diagnose the disease. These results might help veterinarians and producers optimize the diagnosis of colibacillosis in cases where this condition is concerned and consequently reduce losses associated with potential new outbreaks.

References

1. Fairbrother, J.M., Gyles, C.L., 2022. Escherichia coli, in: Pathogenesis of Bacterial Infections in Animals. John Wiley & Sons, Ltd, pp. 117–161. **https://doi.org/10.1002/9781119754862.ch7**

2. Khan, I., Bai, Y., Zha, L., Ullah, N., Ullah, H., Shah, S.R.H., Sun, H., Zhang, C., 2021. Mechanism of the Gut Microbiota Colonization Resistance and Enteric Pathogen Infection. 2021.

Table 1: Comparison of morphologies and the likelihood of colibacillosis in pigs

Morphologies		Risk estimates	95% CI	*P*-value
Smooth mucoid	Rough	0.993	0.987-0.998	< .001
Smooth mucoid	Intermediate	0.968	0.956-0.979	
Smooth mucoid	Smooth	0.758	0.733-0.782	
Rough	Smooth	0.021	0.004-0.037	
Rough	Smooth mucoid	0.006	0.001-0.012	
Smooth mucoid	All others	0.282	0.266-0.298	

https://doi.org/10.54846/am2025/121

Evaluation of colostrum intake in suckling piglets in Spain

V. Rodriguez-Vega, DVM; S. Gaviria; E. Sánchez, DVM; F. Gonzalvo, DVM
Boehringer Ingelheim España, S.A. Spain

Background and objectives

Colostrum is essential for the survival of newborn piglets due to its energy contribution and the passive immunity transmission from their mother. It has been estimated that the minimum amount of colostrum needed for a piglet is 250g,[1] and the average colostrum production per sow and per litter is 3.67 kg.[2] Therefore, a homogeneous distribution of colostrum among all piglets in a litter is crucial. However, litter size and weight dispersion have been increasing, making this distribution challenging. The present study aims to evaluate colostrum intake by estimating the amount of IgGs in serum from piglets between 24-36 hours of life.

Materials and methods

A total of 4,104 piglets from 114 farms were included in the study. In each farm, six litters were randomly selected at 24-36 hours of life. In each litter, 6 piglets were chosen and weighed estimating on visible weight differences (2 heavy-weighed, 2 middle-weighted and 2 light-weighted piglets). Blood samples were taken and. 1,547 piglets were weighed and followed until weaning.

The IgG level in piglet serum was determined using refractometers that convert the refractive index into %Brix that using the formula y = 7,1823x-33,945 (R^2) can estimate the concentration of IgG in the serum of piglets. We consider a piglet to have ingested an adequate amount of colostrum when the quantity of IgGs in serum is greater than 20 mg/mL.[3]

Results

The average concentration of IgG was 27.12 mg/mL with a coefficient of variation of 40.69%, the percentage of piglets with lower serum concentration than 20mg/mL was 27.13%.

Depending on piglets' size, 69.5% of light-weighted piglets, 73.5% of the middle-weighted piglets and the 76.4% of heavy weighted piglets ingested enough colostrum with an average of IgG of 26.16 mg/mL, 27.22mg/mL 27.43 mg/mL respectively (Figure 1A).

Depending on the piglet's weight, piglets with less than 1 kg (n = 462) had an average of IgG of 22.88 and only the 45.7% of them had more than 20 mg/mL.

Regarding the parity number of sows, the offspring born from gilts show the worst results (68.3%), progressively improving from the second birth (71.5%) until the fourth (81.8%) and gradually decreasing after the fifth birth (75.9%) until those with more than 7 farrows (75.1%). (Figure 1B).

Depending on the litter size, piglets born to litters with more than 17 born alive piglets (BAP) (n = 1,111) had an average of IgG of 27.17 and the 30% had less than 20 mg/mL.

Figure 1: Colostrum intake evaluation **A)** per piglet size and **B)** per parity of sows.

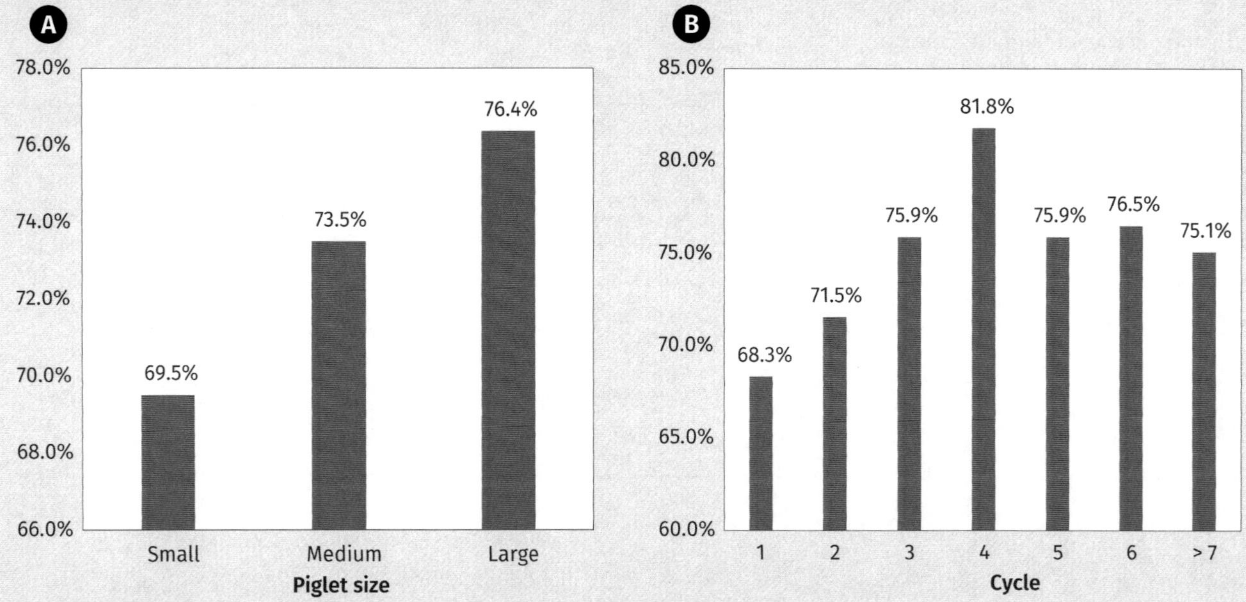

Discussion and conclusions

The concentration of IgG is generally very variable between farms and also within the same farm. 22.365% of piglets have less than 20 mg/mL, which has been shown to be correlated with lower growth[3] and the mortality of them was higher.

The smaller piglets, the offspring of gilts and piglets from litters with more than 17 BAP ingest less colostrum than the rest of the piglets.

References

1. Quesnel H, Farmer C, Devillers N, et al. Colostrum intake: Influence on piglet performance and factors of variation. Livestock Science. 2012; 146:105-114. **doi:10.1016/j.livsci.2012.03.010**

2. Devillers N, Farmer C, Le Dividich J, Prunier, A. Variability of colostrum yield and colostrum intake in pigs: Animal, 2007; (1):1033-1041 **doi: 10.1017/S175173110700016X**

3. Jansen R, Bonekamp R, Marchal L. Impact of inadequate colostrum intake in piglets 2015; 7th ESPHM Symposium OP 019.

Poster 48 https://doi.org/10.54846/am2025/122

How to reduce sow mortality in gestating sows

Joel Sparks[1]; Lauren Wilson[1]; Amy Woods[1]; Melissa Farber Billing[2]; Chris Rademacher[3]
[1]AMVC; [2]Boehringer Ingelheim Animal Health USA Inc.; [3]Iowa Pork Industry Center

Introduction

Sow mortality has significantly increased over the past 10 years in the US swine industry. One database suggests that there was a 50% increase in annualized sow mortality between 2015 and 2020.[1] The objective of this project was to investigate if comparable results could be achieved as reported in one production system during the Improving Pig Survivability project. The protocol for the successful application titled "A Practical Approach to Early Intervention to Reduce Sow Mortality" was used for this evaluation.[2] The protocol (SOP) listed on the website was modified to make it feasible for an extremely large breeding herd with numerous barns of assorted sizes.

Materials and methods

A 10,000 head, commercial breeding herd located in Indiana was used in this study. This farm has fourteen breeding/gestation barns with a combination of individual and group sow housing. Breeding females were bred in stalls and then moved into group pens after pregnancy confirmation. Sows were fed once a day at 5:00am via an automated feed system. To improve early identification and treatment of at-risk sows, the feed drop was changed to 7:15am to coincide with the beginning of barn employees' shift. One team member would walk through the gestation pens or stalls during feeding time and identify sows that were either off-feed, lame, or both. When a sow was found, a bright yellow treatment card would be filled out with her ID number, location, and reason for being identified. The card was hung above her crate or pen. After the daily walk through was complete, designated staff would return at their convenience to evaluate each flagged sow, consult the treatment protocol, and administer the appropriate treatment. The sow treatments were recorded on the sow treatment card, barn clipboard, and in the farm electronic data capture system.

Results

The number of sow treatments and weekly annualized sow mortality were the primary metrics measured to determine the impact of the new identification and treatment protocol. A 5.02% decrease in weekly annualized sow mortality was observed as the average of the 22 weeks prior was 19.96% and the average of the 10 weeks post intervention was 14.94%. The number of sow treatments per week was increased by nearly one hundred treatments as 154 sows were treated prior to the intervention and 246 treatments were given post-implementation of the new protocol.

Discussion

This sow farm was able to reduce weekly annualized sow mortality 5.02% by changing basic practices related to current animal husbandry. Adjusting the feeding time to drop when workers were present gave better opportunity to identify at-risk sows since not eating is often an early indication that the female is needing extra attention. Implementing a highly visible treatment card was an easy method to flag and return to further evaluate the sows for treatment when it is convenient for the barn staff. Emphasis on early identification of at-risk females as a priority and consistent sharing of results motivated farm staff to continue the new protocol. These interventions led to a re-focus on sow husbandry, an increase in sow treatments, and resulted in a decrease in sow mortality on this large sow farm. These results also re-affirmed the results seen in the production system referenced in the Improving Pig Survivability project.[2]

References

1. PigCHAMP Benchmarking Summaries. Accessed July 12, 2024. **https://www.pigchamp.com/benchmarking/benchmarking-summaries**

2. Rademacher, C.J. A Practical Approach to Early Intervention to Reduce Sow Mortality. Improving Pig Survivability Project. January 2024. Accessed July 15, 2024. **https://piglivability.org**

https://doi.org/10.54846/am2025/123

Use of face masks to decrease exposure to swine influenza A virus in swine farm workers

J. Alvarez-Norambuena; M. Yang; M. Culhane; M. Torremorell
College of Veterinary Medicine, University of Minnesota, St. Paul, Minnesota

Introduction

Influenza A virus (IAV) transmission between pigs and swine farm workers is a risk to public health. Farm workers can get exposed to IAV by direct contact with infected animals, indirectly by touching contaminated surfaces, and by the airborne route from breathing infectious aerosols containing viral particles. Personal protective equipment (PPE), such as face masks can be used to decrease aerosol exposure. Unfortunately, there is substantial variation in the levels of comfort, convenience, and protection face masks offer. In this study, we evaluated and quantified the reduction of IAV exposure imparted by various face masks worn by swine farm workers (workers).

Methods

Twenty-four workers from three Midwestern swine farms were included in the study. The selected farms had confirmed IAV infection in the pigs at the time of obtaining workers' consent. Participation of workers in the study was voluntary and participants could withdraw from the study at any time. Participants were asked to wear one of three face masks - an N95 mask, dust mask, or surgical mask – or to not wear a mask (control), according to a set schedule. Each participant wore each mask type for a total of five days for at least two hours/day. Each participant was asked to self-collect a nasal swab sample before wearing the mask at the beginning of the workday and after removing the mask. Participants that did not wear a mask collected swabs at a similar timeframe as those wearing the masks. All nasal swabs were tested by multiplex RT-PCR for RNase P and IAV matrix genes. Differences between treatments were compared using the pairwise Wilcoxon test with correction for multiple comparisons and test the proportion of positive samples between treatments. This study was conducted according to the guidelines and approved protocols of the University of Minnesota IACUC, IBC, and IRB.

Results

A total of 79 out of 832 (9.50%) nasal swabs tested IAV RT-PCR positive. Twenty-eight out of 416 (6.73%) samples, collected before using the face masks, tested positive while 51 out of 416 (12.26%) tested positive after wearing face masks. Before wearing the masks, there were no differences between treatments ($P > .05$). The participants not wearing masks had the most IAV detections (24.27%), which was significantly higher ($P < .001$) than all the detection rates for all three mask types (10.89% - surgical mask, 8.33% - dust mask, and 5.77% - N95 mask; Figure 1). However, there were no statistically significant differences in the IAV detection rate between the three mask types ($P > .05$).

Figure 1: Summary plot of pairwise test comparisons of proportions of positive samples for each of the treatments after wearing the masks.

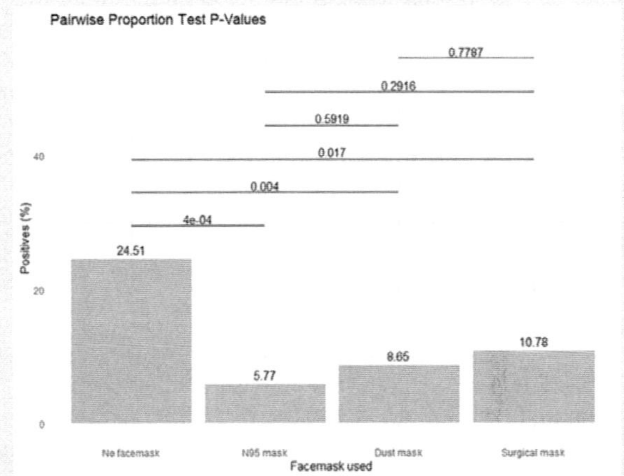

Conclusions

Our results add to the evidence that IAV exposure in swine farms is common and supports the use of face masks to decrease the risk of transmission at the human/animal interface.

Acknowledgment

This work was funded by the National Pork Board.

https://doi.org/10.54846/am2025/124

Evaluation of gaseous ozone technology for decontaminating PRRSV and PEDV from non-porous surface in truck cabins

T. Tran-Pham-Tien; J. Zhang; M. P. Mil-Homens; B. Guo; K. Dion; E. Aljets; H. Tong; M. N. Musskopf; E. R. DeConti; P. Li; E. K. Kirwa; D. C. A. Moraes; G. A. Cezar; I. F. Machado; O. H. Osemeke; G. S. Silva; G. Trevisan; D. C. L. Linhares; D. J. Holtkamp
Iowa State University, Ames Iowa

Introduction

Porcine reproductive and respiratory syndrome virus (PRRSV) and porcine epidemic diarrhea virus (PEDV), both economically detrimental to the swine industry, can spread through contaminated truck cabins, posing significant transport biosecurity risks[1-3]. This study evaluated the effectiveness of gaseous ozone technology in decontaminating PRRSV and PEDV on non-porous surfaces in truck cabins.

Methods

An incomplete factorial study comprising 26 treatment groups, including controls and ozone treatments using three ozone generator capacities (30, 38, and 68 grams of ozone per hour), was conducted over varied exposure times (0.5, 1, or 2 hours) in a climate-controlled truck cabin, where virus-contaminated rubber coupons were subjected to treatments under continuous environmental monitoring, followed by virus elution and titration in cell cultures (MARC-145 cells for PRRSV and Vero cells for PEDV). Statistical analyses encompassed linear mixed models, ANOVA with Tukey's pairwise comparisons of least-square means for virus titers ($\log_{10}$ $TCID_{50}$/mL), standard deviations (SD) for ozone machines, and Pearson correlations between ozone concentration and humidity, ozone and temperature, and ozone concentration and virus titer.

Results

Ozone treatments showed variable results. While some combinations reduced viral titers for both viruses, none reduced viral titers by more than two logs, which was the minimum reduction considered effective. Ozone concentrations varied across identical setups (SD: 4.75 ppm for 30 g/h, 5.43 ppm for 38 g/h machines) during the first 30 minutes. With temperature (19.4-24.7°C, SD < 0.77°C) and humidity (29.9-40.4%, SD 0.23-3.70%) across treatments, the Pearson correlation revealed a weak negative correlation between virus titer and ozone concentration ($P > 0.05$), a moderate positive correlation between ozone and temperature ($r = 0.5$, $P < 0.0001$), and a moderate negative correlation between ozone and humidity ($r = -0.56$, $P < 0.0001$).

Conclusion

Given the conditions of this study, ozone treatments showed inconsistent and limited effectiveness in PRRSV and PEDV decontamination.

References

1. Osemeke O.H., Holtkamp D. PRRSV in U.S. Swine Herds: Updated Economic Costs Attributable to Lost Productivity (2016-2020). In: Proceedings of the 2024 ISU James D. Mckean Swine Disease Conference, Iowa State University, Ames, Iowa; 2024.

2. Holtkamp, D.J., Kliebenstein, J.B., Neumann, E.J., Zimmerman, J.J., Rotto, H.F., Yoder, T.K., Wang, C., Yeske, P.E., Mowrer, C.L., Haley, C.A. 2013. Assessment of the economic impact of porcine reproductive and respiratory syndrome virus on United States pork producers. J. Swine Health Prod. 21(2):72–84.

3. Schulz, L.L., Tonsor, G.T. 2015. Assessment of the economic impact of porcine epidemic diarrhea virus in the United States. J. Anim. Sci. 93(11):5111–5118.

https://doi.org/10.54846/am2025/125

Experimental evidence of vaccine-driven evolution of PRRSV-2 in pigs-to-pig infection chains

N. Pamornchainvakul[1]; I. A. D. Paploski[1]; D. N. Makau[1,2]; J. Baker[1]; J. Huang[1]; C. P. Ferreira[1]; C. A. Corzo[1]; A. Rovira[1]; M. C. J. Cheeran[1]; S. Lycett[3]; A. Doeschl-Wilson[3]; D. C. Schroeder[1]; K. VanderWaal[1]
[1]College of Veterinary Medicine, University of Minnesota, Twin Cities, Minnesota;
[2]Institute of Agriculture, University of Tennessee, Knoxville, Tennessee;
[3]Roslin Institute, University of Edinburgh, Edinburgh, United Kingdom

Introduction

Immune escape mechanisms of porcine reproductive and respiratory syndrome virus type 2 (PRRSV-2) are believed to drive the emergence of genetic variants despite extensive vaccination in the United States.[1] The direct effect of vaccine-induced immune pressure on PRRSV-2 evolution is rarely demonstrated. This study analyzed PRRSV-2 genomes from pig-to-pig infection chains with varying immune backgrounds to investigate this in a controlled environment. We hypothesized that genomes from vaccinated pigs would show greater genetic divergence from the challenge virus than those from non-vaccinated groups, with changes more frequent in antigenic regions.

Methods

An in vivo experiment compared PRRSV-2 micro-evolution across pig-to-pig transmission chains. Weaned pigs were divided into two vaccinated groups and one unvaccinated group. Each group had seven 3-month-old pigs challenged 64 days post-vaccination with PRRSV-2 sub-lineage 1A (L1A). Serum samples were collected over 14 days post-challenge, and samples with the highest virus titer were used as inoculum for the next batch of pigs within the same treatment, repeated for six batches, involving 126 pigs over 84 days. Serum samples with a virus PCR Ct-value of < 30 (n = 184) were sequenced using Oxford Nanopore technology. Genetic relationships were analyzed via pairwise nucleotide distance comparison and phylogenetic approaches. Deep sequencing data were used to quantify and compare viral quasispecies dynamics across samples.

Results

Over the 84-day pig-to-pig infection chain, challenge virus genomes from the last batch of non-immunized pigs remained genetically similar to that from the first batch (mean distance to the first batch's = 0.11%) and to each other (mean within-group distance = 0.12%). In contrast, vaccinated groups showed greater genetic variation (mean within-group distance = 0.24 - 0.4%) and were more genetically different from the first batch's (mean distance to the first batch's = 0.34 - 0.38%). This genetic distinctiveness in vaccinated groups was also seen at the protein level, with non-synonymous mutations altering amino acids. On average, 36.5% of these mutations occurred in immunologic epitopes, including the gp5-M protein ectodomain. Phylogenetic analysis showed samples from non-immunized pigs clustered near the tree's root (the first batch samples), while those from vaccinated groups formed ladder-like structures that were progressively more distant from their first batch samples, indicating directional evolution under selection pressure. This selection was also evident in vaccinated groups at the viral quasispecies level, where previously minor haplotypes dominated over others at the end of the experiment.

Discussion

In summary, these preliminary findings confirm that immune pressure influences PRRSV-2 evolution. Viruses in vaccinated animals were more diversified and genetically distinct from the parent virus, with many changes occurring in immunogenic regions. This study is the first to demonstrate that continuous vaccination impacts PRRSV-2 evolution in controlled setting and may contribute to the emergence of new virus variants. Further research is needed to understand the phenotypic consequences of these differences and the complexity of virus population dynamics within the host.

Reference

1. Paploski, I.A.D.; Pamornchainavakul, N.; Makau, D.N.; Rovira, A.; Corzo, C.A.; Schroeder, D.C.; Cheeran, M.C.J.; Doeschl-Wilson, A.; Kao, R.R.; Lycett, S.; et al. Phylogenetic structure and sequential dominance of sub-lineages of prrsv type-2 lineage 1 in the United States. Vaccines 2021, 9, 608.

https://doi.org/10.54846/am2025/126

Assessment of efficacy of a porcine circovirus vaccine, type 1 – type 2 chimera (PCV2), killed virus, *Mycoplasma hyopneumoniae* (MHP) bacterin administered via a needle-free injection device (NFID) at 3 weeks of age against a relevant PCV2d challenge

C. Venegas Vargas; J. Marx; Y. Diamondidis; L. Taylor; M. Dickson; M. Bandrick
Zoetis Inc. Veterinary Medicine Research and Development, Kalamazoo, Michigan

Objective

Of the at least 9 PCV2 genotypes, PCV2a, PCV2b, and PCV2d are the most clinically important. Vaccination is a common practice to help avoid PCV2-related losses and needle-free intramuscular (IM) vaccine delivery presents further benefits to the pig, the pig caregiver, and the consumer.[1] The goal of this study was to assess efficacy of an experimental PCV2a/PCV2b/MHP vaccine when administered IM via a needle-free injection device (NFID) against a relevant PCV2d challenge.

Methods

Sixty pigs, clinically healthy, and negative for PCV2 antibodies and viremia were blocked by litter and randomly assigned to one of two treatment groups at 21 ± 3 days of age. Animals were administered 2 mL of the inactivated adjuvanted PCV2a/PCV2b/MHP vaccine (T02) or control (T01) containing only MHP fraction and adjuvant IM via a NFID. All pigs were challenged with virulent PCV2d 3 weeks post vaccination. Following challenge, blood and fecal samples were collected twice weekly for PCV2 viremia and fecal shedding assessment via quantitative PCR. A sample/animal was assigned positive or negative viremia and fecal shedding status based on test endpoint limits. Pigs were euthanized and necropsied 21 days post challenge. At necropsy, lymphoid tissues were collected and examined for typical PCV2 lesions (lymphoid depletion and histiocytic replacement) and evaluated for PCV2 colonization by immunohistochemistry. The protocol was approved by the Zoetis IACUC.

Results

All pigs were viro- and seronegative at enrollment and all control animals remained seronegative until challenge. The pigs remained free of PCV2 DNA prior to challenge.

Vaccine efficacy was assessed through comparisons of vaccinates (T02) and controls (T01) for viremia, fecal shedding, PCV2 colonization in lymphoid tissues, lymphoid depletion, and/or histiocytic replacement, with significance determined as $P \leq .05$. Vaccinates had a significantly lower percent of animals with PCV2 colonization in lymphoid tissues, lymphoid depletion, and histiocytic replacement as well as had a significantly lower percent of animals ever positive for PCV2 viremia or fecal shedding (Table 1). When comparing treatments by day, viremia and fecal shedding were significantly less in vaccinates (T02) compared to controls (T01) on each day post-challenge, except for fecal shedding on day 24.

Conclusions

The PCV2a/PCV2b/MHP vaccine induced significant protection when administered IM via a needle-free injection device against PCV2d challenge 3 weeks post vaccination. Vaccination via IM needle-free injection device is a convenient and effective method for administration.

Table 1: Vaccine efficacy results of a PCV2a/PCV2b/MHP vaccine administered via a needle-free injection device at 3 weeks of age against a relevant PCV2d challenge

Protection endpoint	Ever positive		Contrast: T01 (control) vs T02 (vaccinated) Significance
	T01 n/N (%)	T02 n/N (%)	*P*-value*
PCV2 colonization	26/29 (89.7)	3/29 (10.3)	< .0001
Lymphoid depletion	19/29 (65.5)	6/29 (20.7)	.0013
Histiocytic replacement	22/29 (75.9)	6/29 (20.7)	.0002
Lymphoid depletion and/or histiocytic replacement	23/29 (79.3)	8/29 (27.6)	.0004
PCV2 (ever) viremic	30/30 (100)	13/29 (44.8)	< .0001
PCV2 (ever) fecal shedding	30/30 (100)	3/29 (10.3)	< .0001

* Significance determined as $P \leq .05$

n= num

N= tot

Reference

1. Cho H, Ahn Y, Oh T, Suh J, Chae C. Non-inferiority Field Study Comparing the Administrations by Conventional Needle-Syringe and Needle-Free Injectors of a Trivalent Vaccine Containing Porcine Circovirus Types 2a/2b and *Mycoplasma hyopneumoniae*. *Vaccines*. 2022; 10(3):358. **https://doi.org/10.3390/vaccines10030358**

Poster 53

https://doi.org/10.54846/am2025/127

Influenza A vaccines and maternally derived antibodies in piglets

K. Vilaca, DVM, MSc; J. Buchan, DVM; G. Greaves, MSc; C. Werth, MSc
South West Ontario Veterinary Services, Stratford, Ontario, Canada

Introduction

Vaccinating prior to farrowing remains the most common strategy for protecting sows and their offspring from influenza A virus (IAV).[1] Whole inactivated virus (WIV) IAV vaccines are known for inducing high hemagglutinin inhibition (HI) titres in sows, who then pass maternally derived antibodies (MDA) to offspring.[2] When faced with an IAV challenge during lactation or shortly after weaning, piglets with high levels of strain specific MDA had a lower prevalence of IAV at weaning and overall circulation in the nursery compared to piglets with low levels of MDA.[3] More recently, alphavirus replicon particle (RP) IAV vaccines have become available; however, their ability to generate MDA in piglets compared to WIV vaccines is not well understood. The study objective was to understand the difference in MDA levels in piglets between WIV and RP IAV vaccines.

Methods

The herd selected housed > 1500 sows, was naïve to the 6 most prevalent IAV strains in Ontario, did not use IAV vaccines, and was negative for PRRS and Mhp. Sows were vaccinated with either a WIV (n = 30) or RP vaccine (n = 30) at 6- and 3-weeks pre-farrow, each containing the H1-pandemic IAV strain (pH1N1). Control sows (n = 6) were enrolled as sentinels to detect any natural pH1N1 circulating during the study. Serum samples were collected from sows pre-vaccination and post-farrowing, and from piglets at 1-day of age, weaning, and 3-weeks into nursery. All samples were tested for pH1N1 HI titres.

Results

Prior to vaccination, all sows had zero HI titres, including control sows. Post-farrowing, sows vaccinated with the WIV vaccine had a median pH1N1 HI titre that was 8-fold higher than the median titre of sows vaccinated with the RP vaccine. In addition, 80% of WIV vaccinated sows had titres ≥ 40 post farrowing compared to 7% of RP vaccinated sows. Piglets from WIV vaccinated sows had significantly higher median HI titres at 1-day of age, weaning, and early nursery timepoints in comparison to piglets from RP vaccinated sows ($P < .001$). Overall, 96% of piglets from WIV vaccinated sows had HI titres ≥ 40 at 1- day of age, followed by 15% at weaning, and 4% at early nursery. In contrast, only 37% of piglets from RP vaccinated sows had HI titres ≥ 40 at 1-day of age, with no animals displaying titres at weaning or early nursery timepoints. In all sampling timepoints, control piglets had zero-value HI titres.

Discussion

Based on HI titres, the WIV vaccine demonstrated effective seroconversion and antibody production for the pH1N1 IAV strain in majority of sows and a more robust transfer of MDA to piglets compared to the RP vaccine. However, to fully evaluate efficacy differences between vaccine platforms and the correlation between level of MDA and piglet protection, future studies should employ a challenge model at weaning or early nursery.

References

1. Keay, Sheila, Zvonimir Poljak, Famke Alberts, Annette O'Connor, Robert Friendship, Terri L. O'Sullivan, and Jan M. Sargeant. Does Vaccine-Induced Maternally-Derived Immunity Protect Swine Offspring against Influenza a Viruses? A Systematic Review and Meta-Analysis of Challenge Trials from 1990 to May 2021. Animals: An Open Access Journal from MDPI 13, no. 19 (October 3, 2023): 3085.

2. Petro-Turnquist, Erika, Matthew J. Pekarek, and Eric A. Weaver. Swine Influenza A Virus: Challenges and Novel Vaccine Strategies. Frontiers in Cellular and Infection Microbiology 14 (April 3, 2024).

3. Chamba Pardo FO, Wayne S, Culhane MR, Perez A, Allerson M, Tormorell M. Effect of strain-specific maternally-derived antibodies on influenza A virus infection dynamics in nursery pigs. PLoS One. 2019 Jan 14;14(1).

https://doi.org/10.54846/am2025/128

Effect of an electrostatic precipitator on mitigating the transmission of airborne viruses in pigs under experimental conditions

Lan Wang[1]; José Morán[2]; My Yang[1]; Bernard Olson[2]; Chris Hogan[2]; Montserrat Torremorell[1]
[1]Department of Veterinary Population Medicine, University of Minnesota;
[2]Department of Mechanical Engineering, University of Minnesota

Introduction

Airborne viruses are costly and difficult to contain and protect against. Although air filters are used to reduce incidence of airborne infections in swine farms, there are still shortcomings, such as high costs and pressure drop due to resistance to the airflow. Electrostatic precipitators (ESP) are air cleaning devices that utilize high voltage electrodes to ionize airborne particles and remove them. ESPs have a much lower pressure drop, and potentially lower energy consumption compared to air filtration systems. ESPs tend to have high particle collection efficiencies across a wide range of particle sizes. In this study, we assessed the effect of a single-stage, wire-plate ESP on mitigating airborne transmission of influenza A virus (IAV) and porcine reproductive and respiratory syndrome virus (PRRSV) using experimentally infected pigs.

Methods

In each test, four IAV or PRRSV inoculated pigs were placed in two connected isolators (2 pigs per isolator) upstream of the ESP, and two uninoculated sentinel pigs were placed downstream in one isolator. An airflow of 51 m^3/h moved unidirectionally from inoculated pigs, through the ESP, then to sentinel pigs. For each virus, a positive control test with the ESP turned off was performed, followed by three tests with the ESP operating at 12 kV or 14 kV. Inoculated and sentinel pigs were sampled daily by collecting nasal swabs from 0 to 6 DPI (days post inoculation) for IAV and 0 to 10 DPI for PRRSV. Air samples were collected daily, with two Andersen cascade impactors simultaneously sampling the air from upstream and downstream of the ESP, respectively. Samples were tested by RT-qPCR to quantify viral RNA. PCR positive air samples were inoculated in cell culture to assess viral viability.

Results

IAV and PRRSV inoculated pigs shed virus at 1 DPI. When the ESP was turned OFF, sentinel pigs tested positive within 1 day of exposure to the IAV aerosols generated by inoculated pigs and 2 days to the PRRSV aerosols. In contrast, with the ESP turned ON, sentinel pigs did not test positive for IAV until 5 - 6 days of exposure to the ESP-treated aerosols, and 7 - 8 days for PRRSV.

When the ESP was turned OFF, airborne IAV RNA was detected both upstream and downstream of the ESP in particles ranging from 0.22 µm to > 8 µm. When the ESP was ON, limited levels of IAV RNA were detected in air samples collected in the downstream isolator before sentinel pigs tested positive. However, almost no or only low concentration of PRRSV RNA was detected in air samples over the course of the tests. No viable IAV or PRRSV were isolated from the air samples collected in this study.

Conclusions

Under the conditions of this study, the ESP delayed the onset of IAV and PRRSV infections in the sentinel pigs, indicating the ESP was effective at removing viruses from the aerosols generated by infected pigs, and likely resulted in a lower airborne infectious dose to sentinel pigs.

Acknowledgment

This project was supported by the Agriculture and Food Research Initiative Competitive Grant no. 2021-68014-33655 from the USDA's National Institute of Food and Agriculture.

https://doi.org/10.54846/am2025/129

Detections of human seasonal H3N2 influenza A virus reverse-zoonoses in swine during the 2022-2024 influenza season

M. Zeller[1], PhD; D. C. A. Moraes[1], DVM; G. C. Zanella[2,3], DVM; C. K. Souza[2], PhD; T. K. Anderson[2], PhD; A. L. Baker[2], DVM, PhD; P. C. Gauger[1], DVM, PhD
[1]Department of Veterinary Diagnostic & Production Animal Medicine, Iowa State University, Ames, Iowa;
[2]Virus and Prion Research Unit, National Animal Disease Center, USDA-ARS, Ames, Iowa;
[3]Department of Veterinary Microbiology & Preventive Medicine, Iowa State University, Ames, Iowa

Introduction

The Iowa State University Veterinary Diagnostic Laboratory (ISU VDL) detected 32 human-to-swine H3N2 transmissions from November 7, 2022 to July 17, 2024. Sixteen H3N2 detections occurred during the 2022-2023 human influenza season, and 16 detections during the 2023-2024 season.

Methods

Samples were submitted from Colorado, Illinois, Indiana, Ohio, Oklahoma, Missouri, Nebraska, North Carolina, and Pennsylvania, as well as Chile and Mexico. Samples included 23 oral fluids, 6 nasal swabs, 1 lung, and 2 virus isolates. The spillover H3N2 viruses originated from cocirculating human seasonal H3 lineages. The hemagglutinin gene was sequenced using the Sanger method and classified through Nextclade.[1] Sixteen samples were submitted for additional whole-genome sequencing. Maximum-likelihood phylogenetic inference was employed to understand the relationship of the viruses on a genetic and geographic level.

Results

During 2022-2023, lineages 3C.2a1b.2a.2a1 and 3C.2a1b.2a.2b were detected. In the 2023-2024 season, 3C.2a1b.2a.2a1 and 3C.2a1b.2a.2b lineages continued to be detected in swine by genetic sequencing, in addition to the 3C.2a1b.2a.2a.3a.1 lineage. Eighteen neuraminidase (NA) genes were sequenced. Eleven were derived from the human seasonal H3N2 lineage, while the remaining sequences were mixed contemporary swine NA lineages. Whole-genome sequencing performed on 16 samples yielded 6 human-seasonal genomes, and 7 cases had mixed infections with gene segments reflecting the evolutionary origins of endemic swine influenza A virus (IAV) (TRIG, classical, and/or pandemic lineages). Maximum-likelihood phylogenetic inference indicated multiple independent human-to-swine transmissions occurred ($n \geq 8$). Following spillover, phylogenetic evidence supported the transmission of human seasonal IAV in pig populations in North Carolina and Illinois for at least 12 months. Representative human seasonal spillover virus was antigenically distinct when measured using hemagglutination inhibition assays that included a panel of endemic swine lineages.

Discussion

Human seasonal H3 continued to spill into swine at regular intervals, with onward transmission occurring between swine. Whole genome sequencing indicates co-circulation and a high likelihood of reassortment between human spillover IAV and endemic swine virus. The use of personal protective equipment and routine vaccination of workers with the human seasonal vaccine can reduce the risk of virus transmission into swine. Additional measures such as appropriate biosecurity practices and swine vaccinations with homologous strains are necessary to limit human IAV spillovers and prevent the endemic establishment of a new swine H3 lineage.

Reference

1. Aksamentov I, Roemer C, Hodcroft EB, Neher RA. Nextclade: clade assignment, mutation calling and quality control for viral genomes. *J. Open Source Softw.* 2021;6(67):3773. **https://doi.org/10.21105/joss.03773**

https://doi.org/10.54846/am2025/130

Evaluation of the efficacy of oral vaccination with Enterisol Salmonella T/C® by gel

Joel Nerem[1], DVM; Amanda Sponheim[2], DVM; Daniel Hanson[3]; Erin Little[3], MS; Justin Rustvold[2]; Roy Edler[3], MS; Fernando Leite[2], DVM, MS, PhD
[1]Pipestone Veterinary Services, Pipestone, Minnesota; [2]Boehringer Ingelheim Animal Health USA Inc., Duluth, Georgia; [3]Pipestone Research, Pipestone, Minnesota

Introduction

Salmonella can present a significant challenge to swine health and serve as a foodborne human pathogen.[1] Vaccination is an important tool to help prevent this pathogen.[1,2] In this study, we evaluated a novel method of *Salmonella* vaccination using an edible gel. Advantageously, the gel allows pigs to be administered the oral vaccine without being individually handled, saving time and labor. This study had the objective of evaluating the efficacy of the oral administration of Enterisol Salmonella T/C® vaccine by gel and to compare its efficacy with traditional vaccination by oral drench.

Material and methods

A randomized, controlled trial was conducted with three different treatment groups containing 24 pigs each: 1) non-vaccinated; 2) *Salmonella* vaccinated by gel; 3) *Salmonella* vaccinated by oral drench. Pigs in groups 2 and 3 and were vaccinated at 14-16 days of age. For the administration by gel, the live attenuated *Salmonella* vaccine was mixed with a gel (Underline® Gel) and applied to the farrowing crate mat to allow for its consumption. All pigs were individually weighed and challenged with a virulent strain of *Salmonella* Typhimurium at 28 days post vaccination. The study ended ten days post challenge when intestinal lesions were evaluated, and pigs were individually weighed.

Results

Treatment group 1 (non-vaccinated) had a significantly higher level of colon lesions with a mean of 0.40 than both group 2 (0.09) and 3 (0.04) ($P < .001$). The average daily weight gain measured during the ten-day challenge period was significantly lower in group 1 (0.27 lbs) compared to group 2 (0.82 lbs) and 3 (0.83 lbs) ($P < .01$, Figure 1). While both vaccinated groups performed significantly superior to the non-vaccinated challenged group, no significant differences were observed between the gel and oral drench vaccinated groups.

Discussion and conclusion

This study demonstrates that the administration of Enterisol Salmonella T/C® vaccine can be successfully performed with the use of an edible gel. This importantly saves time, labor and minimizes any stress associated with individual piglet handling.

Figure 1: Average daily weight gain measured from time of challenge to necropsy. Error bars represent standard error. Different superscript letters indicate statistical significance ($P < .05$).

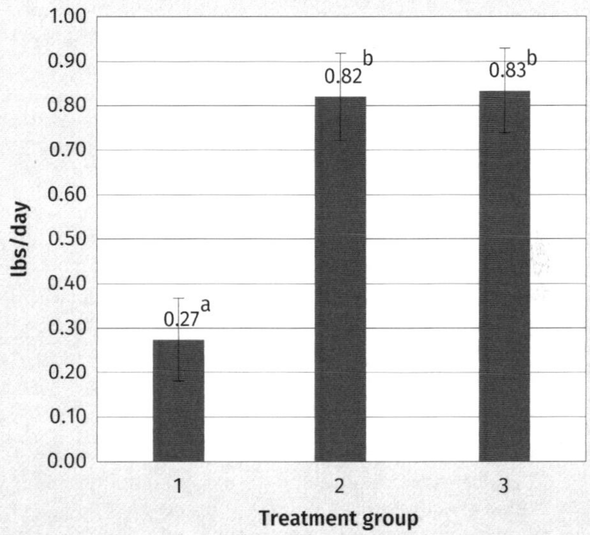

References

1. Griffith RA, Carlson SA, Krull AC. Salmonellosis. Diseases of Swine. 2019.

2. Leite F, Arruda P, Jordan P, Bearson S. Efficacy of Enterisol® Salmonella T/C in protecting against monophasic *Salmonella* I 4,[5],12:i:-. AASV 2022. https://doi.org/10.54846/am2022/52

https://doi.org/10.54846/am2025/131

Effect of lactation creep feeding on pre and postweaning performance

John Sonderman[1], PhD; Brady McNeil[1], MS; Amanda Cross[1], PhD, MBA; Christine Mainquist-Whigham[1,2], DVM, MS; Ethan Stephenson[2], MS; Ashley Hartman[1,2], PhD; Tom Rathje[1], PhD, MBA
[1]DNA Swine Genetics, Columbus Nebraska; [2]Pillen Family Farms, Columbus, Nebraska

Introduction

Setting weaned pigs up for success at weaning can be impacted by many things including health and management practices. Weaning is a time of severe stress, and feed disruption should be limited as much as possible. One potential way to decrease off feed events is through creep feeding pigs while they are still nursing. Over the last 15 years, creep feeding pigs has shown mixed results.[1,2] However, as genetic companies have selected for lean gain to heavier end points, starting pigs post weaning has become more challenging.[3] Therefore, the need to reevaluate the value of creep feeding pigs in late lactation needs to be evaluated, specifically to understand the transition time frame of the first week post weaning.

Methods

A total of 98 litters (DNA L600 x DNA L241) with an average parity of 3.5 were used from a 4-week batch farm. Treatments were assigned randomly at birth with 49 of the litters assigned to the creep feed treatment and 49 assigned to not receive any creep feed. Pigs (n = 1506) were individually tagged and weighed at birth. Cross-fostering was completed within 24 hours of birth within treatment. The creep feed treatment litters were provided a complex early nursery creep diet 7 days prior to weaning at a rate of 0.23 kg per litter per day. Pigs were weaned at approximately 21 days of age and wean weights were recorded. At nursery placement, pigs were randomly assigned to pen by treatment. Individual mortality and morbidity date and reason along with pen feed intake were recorded until nursery exit at approximately 9 weeks of age. Pigs were weighed one week post placement and at nursery exit.

Results and discussion

Wean weights, weight 1-week post placement, and nursery exit weights were different ($P < .05$) with the creep feed treatment weighing 1.40 kg more at 9 weeks of age (Table 1). Feed intake was 0.026 kg/day higher ($P < .05$) for the first week post weaning in the creep feed treatment and 0.02 kg/day ($P < .10$) for the entire nursery period. Nursery mortality was not different between groups. Overall, utilizing creep feed in lactation improves consumption in the first week post weaning and results in a heavier pig at the end of the nursery phase.

References

1. Sulabo R, Jacela J, Tokach M, Dritz S, Goodband R, DeRouchey J, Nelssen J. Effects of lactation feed intake and creep feeding on sow and piglet performance. *J Anim Sci.* 2010;88(9):3145–3153. **https://doi.org/10.2527/jas.2009-2131**

2. Wensley M, Woodworth J, Tokach M, Goodband R, DeRouchey J, Gebhardt J. Effect of early vs. late maturing sire lines and creep feeding on the cortisol response, intestinal permeability, and growth performance of nursery and finishing pigs. *J Anim Sci,* 2023;101:skad169. **https://doi.org/10.1093/jas/skad169**

3. Wensley M, Tokach M, Woodworth J, Goodband R, Gebhardt J, DeRouchey J, McKilligan D. Maintaining continuity of nutrient intake after weaning. I. Review of pre-weaning strategies. *Transl. Anim. Sci.* 2021;5:1–15. **https://doi.org/10.1093/tas/txab021**

Table 1: Creep feed vs no creep feed in lactation weights pre and postweaning

	No creep feed	Yes creep feed	Difference (yes – no)	*P*-value
Birth weight (kg)	1.26	1.31	0.05	.2152
Wean weight (kg)	6.03	6.51	0.48	.0053
1-week post placement weight (kg)	6.62	7.04	0.42	.0042
6-weeks post placement weight (kg)	20.8	22.2	1.4	.0018

https://doi.org/10.54846/am2025/132

Mycotoxicosis and swine inflammation and necrosis syndrome

S. Hough[1], DVM; L. Zheng[1], PhD; P. Gott[1], PhD; J. Bergstrom[1], PhD, M. Thomas, PhD[2], M. Lechner[3]
[1]dsm-firmenich, Plainsboro, New Jersey; [2] Professional Animal Auditor Certification Organization, Kearney, Missouri; [3]UEG Hohenlohe, Franken, Germany

Introduction

Mycotoxins are toxic secondary metabolic products of molds, which reduce crop quality. With feed accounting for 70% of the cost of production, mycotoxin contamination can cause huge economic losses, both in losses in feed quality and animal performance. Diseases related to consuming contaminated feed are labeled as mycotoxicosis, and the systems affected vary by toxin, with potential impacts on liver or kidney function, gastrointestinal and/or respiratory health, and reproductive performance (Figure 1). Mycotoxins can also have subclinical implications and can play as predisposing factors of disease. With their potential influence on pig health and welfare, veterinarians need to be aware of clinical expression when present.

Survey

Biomin, now part of dsm-firmenich, has conducted an annual global mycotoxin survey since 2004.[1] In the United States, samples are screened at Romer Labs (Union, MO) via liquid chromatography with tandem mass spectrometry (LC-MS/MS) method for six major mycotoxin groups. The 2024 corn crop year data is just becoming available (n = 48) as harvest is still in progress in the writing of this abstract. To date, 94% of the samples evaluated have contained at least one mycotoxin, which is similar to findings from 2023 (94%). The percentage of Fumonisin contaminated samples in 2024 corn is lower compared to 2023 (67 vs 82%), but both B-trichothecenes and Zearalenone are higher (69 vs 64% and 81% vs 73%). The percentage of Aflatoxin, A-Trichothecenes and Ochratoxin contaminated samples have all been below 10%. The preliminary results of the 2024 US corn crop survey indicate mycotoxin occurrence and contamination levels will continue to challenge animal health and performance as these ingredients are fed out over the coming months.

Figure 1: Mycotoxicosis clinical signs in pigs.

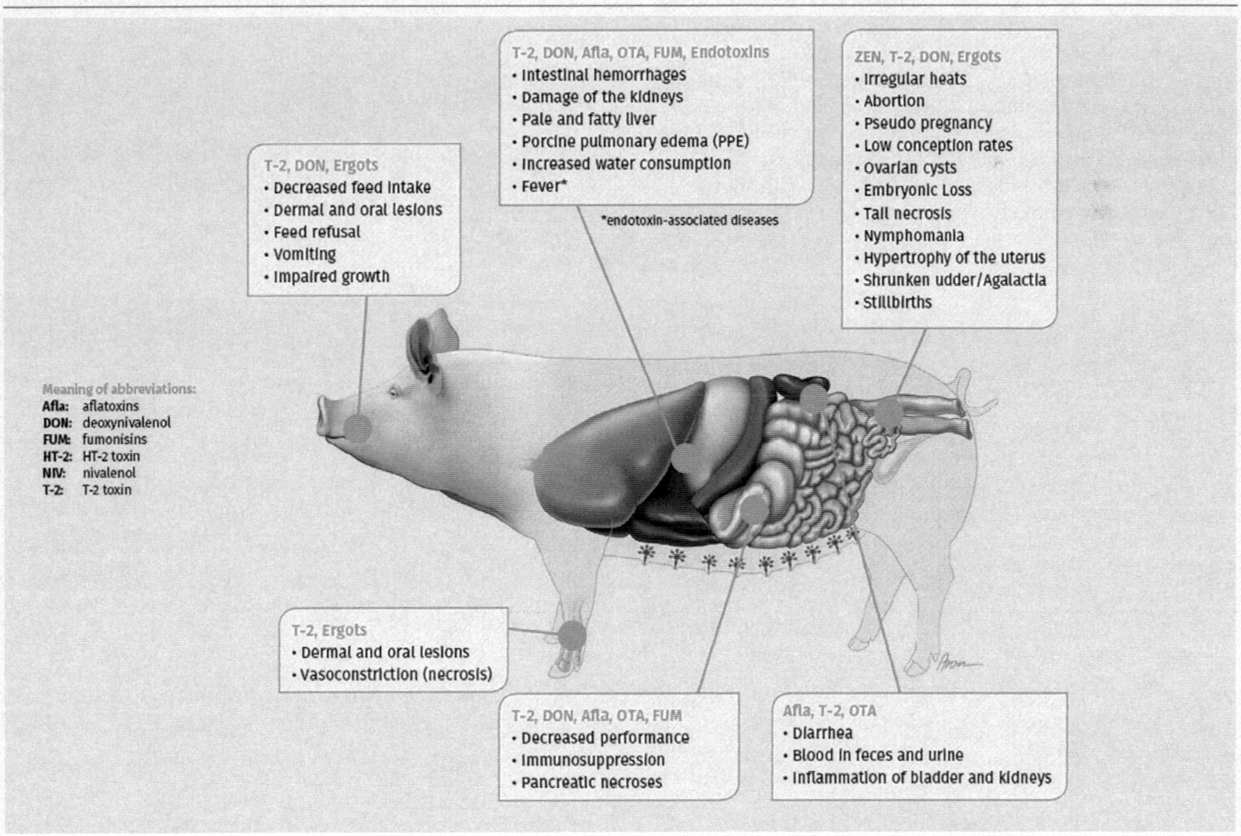

Clinical expression of mycotoxicosis

Several factors affect the clinical signs and severity of mycotoxin toxicity in pigs, like type of mycotoxin, level in the feed, duration of exposure, age of the pig and production phase.[2] Young pigs and breeding swine are the most susceptible and combinations of certain mycotoxins may be more harmful than individual mycotoxins.[3,4] Consumption of mycotoxins causes clinical signs such as feed refusal and diarrhea, impacts the immune system, and makes pigs more susceptible to disease.[5]

Swine inflammation and necrosis syndrome (SINS) is a new term that has been used to explain the occurrence of inflammation and subsequent dead tissue at the tail, ears, coronary bands, teats, navel, and face of various ages of pigs.[6] Considerations of water availability, thermal stress, endotoxins, mycotoxins, and others have been mentioned as contributing factors.[7] Mycotoxins, particularly DON, has also been considered a potential stress factor for necrosis of extremities, like the tail, especially when clinical signs are appreciated shortly after birth in offspring from affected dams.[8] Ergot alkaloids cause vasoconstriction and terminal necrosis in various locations, like the tail, ears, and hoof as well.[9] Dr Mirjam Lechner and Dr Gerald Reiner have been researching this phenomenon since 2015 and they have developed a scoring system to better understand the severity of the syndrome. The focus has mostly been around suckling piglet scoring as they could be indicators of sow status. Case studies and pictures will be shared within the poster.

Discussion

Considering the frequent occurrence of mycotoxins in feeds and the potential negative impact they can have on animal production following exposure, mitigation options should be considered. Veterinarians need to work with their nutrition team to determine the best program for their herds. Strategies for mitigation must consider the chemical structure of each mycotoxin and the potential to be bound with conventional products at the relatively harsh pH of the gastrointestinal tract. Biotransformation or degradation is an alternative detoxification method which can address mycotoxins that have limited binding potential.

Acknowledgments

Authors would like to acknowledge Romer Labs for the mycotoxin testing and thank those willing to share case examples and details regarding clinical significance.

References

1. Biomin Mycotoxin Compendium. 2018. Biomin Holding GmbH, Erber Campus 1, Getzersdorf, Austria.

2. Whitney, M. University of Minnesota Swine Extension Website. Swine Nutrition: Mycotoxins harm swine health and reproductive performance. Accessed November 24, 2024. **https://extension.umn.edu/swine-nutrition/mycotoxins-harm-swine-health-and-reproductive-performance**

3. Andretta, I. M. Kipper, L. Hauschild, C.R. Lehnen, A. Remus and R. Melchior. Meta-analysis of individual and combined effects of mycotoxins on growing pigs. *Scientia Agricola*. 2016. 73:4, 328-331.

4. Pierron, A., I. Alassane-Kpembi, I.P. Oswald. Impact of mycotoxin on immune response and consequences for pig health *Animal Nutrition*. 2016. 63-68.

5. Iowa State University College of Veterinary Medicine Veterinary Diagnostic and Production Animal Medicine Disease Topics: Mycotoxins. Accessed November 14, 2024. **https://vetmed.iastate.edu/vdpam/research/disease-topics/toxicology/mycotoxins**

6. Reiner, G.; Kuehling, J.; Loewenstein, F.; Lechner, M.; Becker, S. Swine Inflammation and Necrosis Syndrome (SINS). *Animals* 2021. 11, 1670.

7. Langbein, F., M. Lechner and G. Reiner. Swine Inflammation and Necrosis Syndrome (SINS) – a new syndrome related to tail biting n pigs. IPVS 2016 – 343.

8. Van Limbergen, T., M. Devreese, S. Croubels, N. Broekaert, A. Michiels, S. De Saeger, D. Maes. Role of Mycotoxins in herds with and without problems with tail necrosis in neonatal pigs. *Vet Record*. 2017. 10, 1136.

9. Merck Veterinary Manual. Overview of Mycotoxicosis. Topic: Ergotism in animals. Accessed November 12, 2024. **https://www.msdvetmanual.com/toxicology/mycotoxicoses/ergotism-in-animals**

https://doi.org/10.54846/am2025/133

Management through PRRSV with Swine Awake™ and Pulmotil® AC

B. Mason[1], DVM; P. Chang[2], PhD; S. Weiland[2], MS; J. Risser[2], DVM
[1]Country View Family Farms, Hagerstown, Indiana;
[2]Elanco Animal Health, Greenfield, Indiana

Introduction

Clinical implications of porcine reproductive and respiratory syndrome virus (PRRSV) are often contributed to secondary bacterial pathogens endemic in the herd. The industry continues to seek practices that mitigate the clinical manifestations of both PRRSV and secondary bacterial pathogens. Tilmicosin (Pulmotil® AC; Elanco Animal Health) assists neutrophils and concentrates in macrophages in the fight against respiratory bacterial pathogens and in coinfections of *Mycoplasma hyopneumoniae* and PRRSV.[1,2] Studies have shown that tilmicosin influences cytokines involved in the inflammatory process.[3] Swine Awake™ (Elanco Animal Health) is a naturally occurring *Bacillus licheniformis* and *Lactobacillus casei* administered through the water. Pigs administered Swine Awake prior to challenge with PRRSV have shown significantly decreased neutrophil and increased lymphocytes levels in the serum, with a trend of increased IL-1β and IL-12.[4] Both tools act through interaction with the innate immune system. This field observation aimed to determine the impact of Swine Awake and Pulmotil AC administered together on nursery performance and an inflammation biomarker in control of swine respiratory bacterial and PRRSV challenges.

Methods

Over 20 weeks between March 2023 and July 2023, 150,000 wean pigs, placed in nursery sites single sourced from Herd B and Herd N, were rotated by week into one of four treatment groups: Control (CTRL), Pulmotil AC (PAC), Swine Awake (SA), Swine Awake plus Pulmotil AC (SA+PAC). The CTRL pigs received Tiamulin/CTC. The PAC groups received Pulmotil AC according to label instructions for 5 days. The SA groups received Swine Awake according to label instructions for 21 days. The SA+PAC groups received Swine Awake and Pulmotil AC administered together for 5 days, followed by 16 days of Swine Awake administered alone. The Pulmotil AC stock solution was mixed following label instructions with the respective quantity of Swine Awake mixed into it. Nursery close-out production metrics were evaluated by treatment group. In each of the first four weeks of the project, sixty pigs from each herd were selected prior to treatment for individual identification (ear tags), weight, and serum (3 mL) collection for systemic inflammation biomarker analysis. Weights and biomarker were repeated 6 weeks later. All serum samples were collected two days prior to or after any vaccinations and sent overnight on ice packs to Elanco Animal Health for analysis. Data were analyzed using the JMP (SAS Inst. Inc, Cary, NC) software.

Results

Results are presented in Table 1 for both individual pig and group data.

Conclusion

PAC+SA tended to improve average daily gain (ADG) in individual pigs compared to other treatments. Biomarker levels improved the most post-treatment for the PAC+SA treatment, which indicates lower inflammatory status of those pigs. Whole lot data showed improvements in mortality when using PAC or PAC+SA treatment. Performance for the whole lots tended to improve with administration of the treatments and ADG was numerically higher in groups that received Pulmotil AC.

References

1. Ivers D, Puls C, Weber T, et al. Effectiveness of tilmicosin phosphate (Pulmotil AC) for controlling respiratory disease associated with Mycoplasma Hyopneumoniae in swine in the presence of porcine reproductive and respiratory syndrome virus (PRRSV). AASV Annual Meeting. 2016.
2. Buret AG, Nerland EM, Leblanc JM, et al. Pro-apoptotic and anti-inflammatory effects of oral tilmicosin in the Actinobacillus pleuropneumoniae-infected porcine lung. AASV. 135-138. 2004.
3. Cao X, Dong M, Shen J, et al. Tilmicosin and tylosin have anti-inflammatory properties via modulation of COX-2 and iNOS gene expression and production of cytokines in LPS-induced macrophages and monocytes. International Journal of Antimicrobial Agents. 27:431-438. 2006.
4. Elanco data on file (from Nutriquest).

Table 1: Results for individual pig and group performance

Individual pig performance and biomarker levels*

		All			Herd B only			Herd N only		
		LSMean	StdErr	*P*-value	LSMean	StdErr	*P*-value	LSMean	StdErr	*P*-value
ADG, lb/day	CTRL	0.75^{b}	0.02	< .0001	0.60^{b}	0.02	< .0001	0.91	0.03	.4117
	PAC	0.78^{b}	0.02		0.62^{b}	0.02		0.90	0.03	
	SA	0.70^{b}	0.02		0.48^{c}	0.02		0.94	0.03	
	PAC + SA	0.85^{a}	0.02		0.74^{a}	0.02		0.96	0.03	
Change in biomarker levels, %	CTRL	28.5^{c}	7.1	< .0001	11.3^{c}	4.5	< .0001	43.9^{c}	11.1	< .0001
	PAC	51.8^{a}	7.1		24.7^{bc}	4.4		78.8^{bc}	11.4	
	SA	72.7^{ab}	7.4		33.8^{ab}	4.4		121.3^{ab}	12.7	
	PAC + SA	96.7^{a}	7.2		42.9^{a}	4.4		155.8^{a}	11.8	

Group performance and mortality**

		All groups			Herd B groups only			Herd N groups only			Whole groups		
		LSMean	StdErr	*P*-value	LSMean	StdErr	*P*-value	LSMean	StdErr	*P*-value	LSMean	StdErr	*P*-value
ADG, lb/day	CTRL	0.86	0.02	.5746	0.87	0.3	.9691	0.82	0.04	.1874	0.76	0.05	.0971
	PAC	0.90	0.02		0.85	0.3		0.95	0.04		1.05	0.05	
	SA	0.89	0.02		0.87	0.3		0.95	0.04		0.93	0.05	
	PAC + SA	0.87	0.02		0.86	0.3		0.88	0.04		0.95	0.06	
Mortality %	CTRL	5.5	0.6	.3563	4.9	0.7	.4194	6.8	1.1	.2212	8.3a	0.5	.0111
	PAC	5.1	0.6		4.3	0.8		5.2	0.9		2.3bc	0.5	
	SA	5.5	0.6		5.7	0.8		4.4	0.9		5.4ab	0.4	
	PAC + SA	4.2	0.6		4.0	0.7		3.7	0.9		1.4c	0.5	

Least square mean (LSMean), standard error (StdErr), average daily gain (ADG), control (CTRL), Pulmotil AC (PAC), Swine Awake (SA), Swine Awake plus Pulmotil AC (SA+PAC).

* Individual pig performance and biomarker levels, the model included the fixed effect of treatment, and starting weight was used as a covariate for growth performance analysis. Sow was included in the model as a random effect for the biomarker analysis of the B group. Tukey's HSD test was used to compare means between treatment groups. Statistical separation of treatments was determined using an unadjusted alpha = 0.05. Trends will be identified as $P < .10$

** Group performance and mortality analyses, the model included the fixed effect of treatment and days on feed was used as a covariate. A total of 39 wean groups were included in the analysis: 24 groups from Herd B with 6 groups in each treatment; 15 groups from Herd N with 3-4 groups in each treatment. Several of the lots included wean groups that divided across multiple treatments. Only 8 wean groups corresponded to complete or whole lots, giving 2 whole lot data points in each of the treatments. Tukey's HSD test was used to compare means between treatment groups. Statistical separation of treatments was determined using an unadjusted alpha = 0.05. Trends will be identified as $P < .10$

 https://doi.org/10.54846/am2025/134

A dynamic network evaluation of human, animal, and truck movement data across a swine farm system

Tara Prezioso[1], DVM, MPH; Alicia Boakes[2], DAGR; Jeff Wrathall[2]; Suman Bhowmick[1], MS, MSc, PhD; Rebecca Lee Smith[1], DVM, MS, PhD
[1]University of Illinois Champaign/Urbana Department of Pathobiology
[2]Farm Health Guardian

Objective

A healthy North American swine industry is important to agriculture, related industries, the country's gross domestic product, and international relations and trade. Many movement network evaluations in North America, which are used to identify disease spread risks and design response plans, focus only on animal or truck movements.[1,2] Very few include human movements. We previously examined a network of movements among three North American swine farms over the course of a year and found that when human movements are included, the network properties change, leading to a change in inference.[3] We now break down those networks dynamically to evaluate how addition of the human movement changes the dynamic network structure and the impact of those changes on disease transmission between farms.

Methods

Farm visit data were collected from a private database over the period from December 22nd, 2021, to April 29th, 2024, representing three swine management companies. The data includes property and group IDs, location, and user/truck IDs, all of which were anonymized. Other variables include the property type, vehicle type, entry type, and dates of visits. Observations without a property ID, user/truck ID, or at least two farm visits were removed from the data. Porcine epidemic diarrhea virus (PEDV) outbreak data was collected during the same time frame for one of the management companies. Three dynamic networks were built using the *network* and *networkDynamic* packages in R. One included all the movement data, one included just the animal and truck movement data, and a third included the human movement data. Each time a truck entered a truck wash, it was assumed cleaned and free of disease, and was provided a new vehicle ID. Each level in the network represents a break timepoint (60d) of PEDV disease transmission. The final full dynamic network includes 28261 unique edges (or 200913 total trips), 182 properties, 17 different property types and 10 unique vehicle types.

Results

Measures of network structure include density, the indegree and outdegree of each node, betweenness and global transitivity at each timepoint. Visualizations show a cyclic pattern to edge formations over time. Both in- and out-degree tend to be quite stable for most properties at each timepoint. Betweenness is relatively stable but still appears to have a cyclic pattern present. Simulations on forward reaching paths show a highly connected network over time and highlight the underlying hub structure.

Discussion

The cyclic pattern seen during edge formations may be related to an all-in, all-out production system. Disease transmission changes with network structure, therefore disease patterns may also have a cyclic pattern reflective of the edge metrics. High degree properties, which are potential network hubs, stay hubs throughout the two years. Forward reaching path analysis shows that an outbreak can potentially reach every property in the system within a short number of movements. The addition of human movement into the network allows for more accurate disease transmission models.

References

1. Black, N. J., Cheng, T.-Y. & Arruda, A. G. (2022). Characterizing the connection between swine production sites by personnel movements using a mobile application-based geofencing platform. *Preventive Veterinary Medicine, 208*, 105753. **https://doi.org:10.1016/j.prevetmed.2022.105753**

2. Galvis, J. A., Corzo, C. A. & Machado, G. (2022a). Modelling and assessing additional transmission routes for porcine reproductive and respiratory syndrome virus: Vehicle movements and feed ingredients. *Transboundary and Emerging Diseases.* **https://doi.org:10.1111/TBED.14488**

3. Prezioso, T., et.al. (2024). A network evaluation of human and animal movement data across multiple swine farm systems in North America. *Preventative Veterinary Medicine* (In review).

https://doi.org/10.54846/am2025/135

Impact of water delivery of monoglycerides and natural extracts (Furst Strike Direct™) on lateral PRRS 144 L1C challenge measured by mortality, morbidity, and injectable treatments

W. Kwon, MS, PhD; G. Hartsook; P. Aparachita, MS, PhD; K. Soltwedel, MS, PhD; R. Fischer, MS, PhD; F. Sandberg, PhD
Furst-McNess Company, Rockford, Illinois

Introduction

The cost of porcine reproductive and respiratory syndrome virus (PRRSV) on the US swine industry is large with an annual cost estimate of $664 million, and infections in growing pigs accounting for over half of these losses.[1] Most recent research revealed that wild-type PRRSV was present in 42% of herds, predominantly during the mid to late grow-finish phases, averaging 20 weeks after placement.[2] Furst Strike Direct™ (FSD) is a water-soluble nutritional supplement and flavoring that contains monoglycerides, natural extracts, and emulsifiers. FSD is delivered through a standard proportioner (ie, water medicator) that allows for high levels of longer chain monoglycerides to be provided with drinking water to support overall health and performance. Previous research has shown that FSD supplementation via drinking water reduced the mortality of nursery pigs within the lower health environment group in a large-scale commercial research project.[3]

Methods

The objective of this experiment was to test the hypothesis that FSD could be beneficial when a combined lateral influenza and PRRS 144 L1C challenge occurred in a wean-to-finish commercial scale research barn. Saliva fluids collected from the barn were tested positive with a wild type PRRSV and an influenza A virus at the same time (tested at Veterinary Diagnostic Laboratory in Iowa State University). A total of 500 DNA-sired growing pigs (initial BW: 78.4 ± 3.7 lb) were randomly assigned to 2 water treatments in a completely randomized design. There were 19-21 pigs in each pen and 12 replicate pens per treatment. The 2 water treatments were **1)** Control (water only) and **2)** 8 oz. of FSD in 5 gallons of water (1:128 dilution rate) for 5 days supplementation. Body weight and feed disappearance were recorded by pen on d 0, 14, 28, and 72 and used to calculate average daily gain, average daily feed intake, and gain to feed ratio. After d 72, an equal number of pigs from each pen were marketed, and marketed weight was recorded based on the 4 individual cuts. Morbidity, mortality, and medical injectable treatments were measured for the entire experimental period. Data were used to calculate all parameters and were analyzed using the GLM procedure in Minitab with Fisher's t-test determining differences between treatments.

Results

There were no differences in growth performance parameters between the 2 water treatment groups throughout the entire period. And there was no difference in mortality and morbidity until d 28 regardless of water treatments during the acute phase of PRRSV (Table 1). However, Control group had higher ($P < .05$) mortality and morbidity compared to the FSD group from d 28 until d 72. Similarly, both injectable treatments of Enrofloxacin and Dexamethasone were greater ($P < .05$, respectively) in the Control group compared to the FSD group during d 28 to 72.

Discussion

Nutritional supplementation with FSD through drinking water under PRRSV and influenza A challenges seemed to have long term benefits on the survivability of pigs after the acute phase of challenges. Using FSD may reduce subsequent fall out and overall mortality and could lead to a reduced need for medical treatments.

References

1. Holtkamp D, Kliebenstein J, Neumann E, et al. Assessment of the economic impact of porcine reproductive and respiratory syndrome virus on United States Pork producers. *J Swine Health Prod.* 2013;21(2):72-84.

2. Angulo J, Yang M, Rovira A, et al. Infection dynamics and incidence of wild-type porcine reproductive and respiratory syndrome virus in growing pig herds in the US Midwest. Prev Vet Med. 2023;217:105976.

3. Sandberg F, Soltwedel K, Fischer R, et al. Effects of Furst Strike Direct™on mortality and injectable treatments in nursery pigs. *In: Proceedings of the AASV 54th Annual Meeting, Denver, CO.* 2023.

Table 1: Impact of water delivery of monoglycerides and natural extracts (Furst Strike Direct™) on removals and injectable treatments of pigs under lateral porcine reproductive and respiratory syndrome 144 L1C and influenza A challenges

Water treatment[1]:	Control	FSD	SEM	*P*-value
Replicate pen	12	12		
Mortality, %				
D0-14	0.8	0.8	0.56	1.000
D14-28	–	–	–	–
D28-72	0.4	–	0.30	.328
D72-MKT	0.5	0.4	0.47	.938
Overall	1.7	1.3	0.92	.752
Morbidity, %				
D0-14	–	–	–	–
D14-28	0.8	0.9	0.56	.959
D28-72	2.6	0.0	0.70	.016
D72-MKT	0.4	-	0.30	.328
Overall	3.7	0.8	0.87	.027
M + M[2] %				
D0-14	0.8	0.8	0.56	1.000
D14-28	0.8	0.9	0.56	.959
D28-72	3.0	0.0	0.82	.017
D72-MKT	0.9	0.4	0.53	.541
Overall	5.4	2.1	1.37	.101
Enrofloxacin, injections/pen				
D0-14	7.0	7.4	1.02	.776
D14-28	2.8	3.1	0.68	.733
D28-72	4.2	1.6	0.79	.030
Overall	13.9	12.1	1.68	.448
Dexamethasone, injections/pen				
D0-14	6.3	6.3	0.95	1.000
D14-28	2.8	2.3	0.64	.651
D28-72	3.0	0.9	0.60	.022
Overall	12.0	9.5	1.50	.252

[1] Control = water only; FSD = Furst Strike Direct™ with drinking water (8 oz. of FSD in 5 gallons of water; 1:128 dilution rate).

[2] M + M = a total of mortality and morbidity.

https://doi.org/10.54846/am2025/136

Synergistic effects of atrophic rhinitis and Glässer disease vaccination on pig performance

T. C. G. D. Souza[1]; L. De Lucas[2]; I. Ballara[2]; O. Boix[2]
[1]HIPRA Brazil; [2]Laboratorios HIPRA S.A., Amer, Spain

Introduction

Swine respiratory pathogens often interact synergistically, resulting in worsened clinical outcomes, increased morbidity and mortality rates, and significant economic losses for producers compared to single infections.[1,2] *Bordetella bronchiseptica* (Bb) is particularly notable for causing necro-hemorrhagic bronchopneumonia and facilitating the opportunistic colonization of *Streptococcus suis* (*S suis*) and *Glässerella parasuis* (Gp).[3,4] This study aimed to assess the effectiveness and productivity impact of comprehensive vaccination strategies against Bb, *Pasteurella multocida* type D, and Gp on pig performance and the overall economic benefit of the farm, as an alternative to the prophylactic use of antibiotics during both the nursery and fattening phases.

Materials and methods

This study was conducted with pigs from a commercial herd comprising 2,000 sows in the state of Paraná, Brazil. The existing antibiotic prophylaxis plan included in-feed treatments with florfenicol (administered between 40-70 days of age) and amoxicillin (administered between 70-90 days of age). The farm had a previous diagnosis of *Glässer* disease (GD) caused by serotypes 4, 5, 7, 12, and 13. The experimental design aimed to compare two groups: Group 1 (G1 [n = 3,340]), which included six non-treated batches of sows, vaccinated with Rhiniseng® 6 and 3 weeks before farrowing and Hiprasuis® Glasser, administered as two mass vaccinations plus a cycle vaccination in sows, along with two vaccinations for piglets at 3 and 6 weeks of age. Group 2 (G2 [n = 4,981]) consisted of 12 non-vaccinated batches receiving the described prophylactic plan. Productive parameters (average daily weight gain [ADWG], feed conversion ratio [FCR], and body weight [BW]), respiratory parameters (atrophic rhinitis index and pleurisy), and mortality were recorded in both groups. Animals that died with symptoms suggestive of GD were necropsied, and samples were collected for diagnostic testing of Gp and *S suis* by PCR. Statistical analyses were performed using the R software.

Results

Results were analyzed during both the nursery and fattening periods (Table 1). During the nursery phase, G1 demonstrated improved ADWG (445 g/d ± 16 vs 428 g/d ± 7), better FCR (1.366 ± 0.055 vs 1.445 ± 0.042), and a lower mortality rate (1.07% ± 0.36 vs 2.70% ± 0.43). These benefits persisted into the fattening phase, where ADWG was also higher (949 g/d ± 15 [G1] vs 920 g/d ± 14 [G2]), along with an improved FCR (2.199 ± 0.031 [G1] vs 2.362 ± 0.061 [G2]) and a lower mortality rate (0.57% ± 0.22 [G1] vs 1.72% ± 0.29 [G2]). Additionally, BW at slaughter was significantly greater in G1 compared to G2 (103.67 kg ± 1.51 vs 100.58 kg ± 1.31). Regarding respiratory parameters, pigs in G2 showed 10% pleurisy, while G1 only showed 1.3% ($P < .05$). Furthermore, the incidence of

Table 1: Productive performance of G1 and G2 in the nursery and fattening phases

		Group 1	Group 2	Dif	*P*-value
Nursery	Nº of piglets (n)	3,340	4,981		
	Entry weight (kg)	5.97	6.05	- 0.08	.35
	Exit weight (kg)	27.77	27.02	0.75	.053
	ADWG	0.445	0.428	0.017	.047*
	FCR	1.366	1.445	-0.08	.014*
	Mortality (%)	1.07	2.7	-1.63	< .001*
	IRA[1]	0.13	0.39	-0.26	< .001*
Fattening	Slaughter weight (kg)	103.67	100.58	3.09	.002*
	ADWG	0.949	0.920	0.029	.003*
	FCR	2.199	2.362	-0.163	< .001*
	Mortality (%)	0.57	1.72	-1.15	< .001*
	IRA[1]	0.062	0.375	-0.313	< .001*
	Pleurisy (%)	1.33	10	-8.67	< .001*

* Indicate significant differences ($P < .05$).

[1] IRA: atrophic rhinitis index (levels: 0 – no lesions; 1 – mild atrophy. primarily in the lower/ventral turbinates; 2 – moderate to severe atrophy in the lower/ventral turbinates; and 3 – severe lesions affecting the entire turbinate.)

ADWG = average daily weight gain; FCR = feed conversion ratio

mortality due to *S suis* and Gp decreased by 83% and 100% in G1 pigs, respectively ($P < .05$). The atrophic rhinitis index was also significantly improved in both the nursery and fattening phases in the vaccinated group. The economic return on investment was calculated to be 8.82 when comparing G1 to G2.

Conclusion

The vaccination strategy using Rhiniseng® and Hiprasuis® Glasser significantly enhanced performance, demonstrating superior ADWG, improved FCR, and reduced mortality throughout both the nursery and fattening phases. The lower incidence of pleurisy and the improved atrophic rhinitis index support the efficacy of the vaccination program compared to antibiotic use. These results suggest that vaccination against non-progressive atrophic rhinitis contributes to reducing the incidence of *S suis* in piglets. An economic return on investment of 8.82 indicates that these strategies not only improve animal health but also enhance the profitability of swine production.

References

1. Opriessnig T, Giménez-Lirola LG, Halbur PG. Polymicrobial respiratory disease in pigs. Anim Health Res Rev. 2011 Dec;12(2):133-48. **doi: 10.1017/S1466252311000120**

2. Nielsen DW, Hau SJ, Mou KT, Alt DP and Brockmeier SL (2023) Shifts in the swine nasal microbiota following *Bordetella bronchiseptica* challenge in a longitudinal study. Front. Microbiol. 14:1260465 **doi: 10.3389/fmicb.2023.1260465**

3. Hau SJ, Nielsen DW, Brockmeier SL. Prior infection with *Bordetella bronchiseptica* enhanced colonization but not disease with *Streptococcus suis*. Vet Microbiol. 2023 Sep;284:109841. Epub 2023 Aug 1. **doi: 10.1016/j.vetmic.2023.109841**

4. Brockmeier SL. Prior infection with *Bordetella bronchiseptica* increases nasal colonization by *Haemophilus parasuis* in swine. Vet Microbiol. 2004 Mar 26;99(1):75-8. **doi: 10.1016/j.vetmic.2003.08.013**

https://doi.org/10.54846/am2025/137

Evaluating microbial levels in feed mills with and without a formaldehyde-based feed mitigant

O. Harrison; D. Mellick; H. Vander Veer; V. Iseri; E. Rankin
Kemin Industries, Inc., Des Moines, Iowa

Introduction

Feed and feed ingredients have been shown to be transmission vehicles for pathogens, such as *Salmonella*.[1] This can increase the likelihood of feed mills becoming contaminated and being a continual source of contamination to feed, thus chemical antimicrobials such as formaldehyde are used. Indicator organisms, such as *Enterobacteriaceae* (EBAC), are commonly used to evaluate disinfectants and hygiene programs in food and feed operations. EBAC is a family of gram-negative organisms that includes both non-pathogenic and pathogenic bacteria, such as *Salmonella* and *Escherichia coli*.[2] EBAC is commonly found in feed and feed ingredients, as well as on surfaces in feed mills. Previous feed mill studies have indicated a background level of EBAC between 1.16 and 4.22 log10 colony forming units per gram (CFU/g).[3] The objective of this study was to evaluate EBAC levels in commercial facilities and determine how the use of a formaldehyde-based antimicrobial, Sal CURB® (SC), affects EBAC levels in feed and the feed mill environment.

Methods

EBAC levels in feed samples and environmental swabs were collected weekly at two feed mills representing one Midwest-based cooperative, manufacturing feed for poultry and swine producers. Samples were collected from January 8, 2024, through February 21, 2024. Feed mills were identified as either West or East Feed Mill. West Feed Mill was a multi-species commercial feed mill that does not use SC. East Feed Mill was a multi-species commercial feed mill that uses SC at 1.0 kg/t in all layer diets, representing > 50% of the feed produced. Two types of samples were collected: feed samples and environmental swabs. Each week feed samples were collected in sterile sample cups on two different days. At the West Feed Mill, one swine diet and one layer diet were collected, both did not include SC. At the East Feed Mill, one swine diet with 0.0 kg/t SC and one layer diet with 1.0 kg/t SC were collected. Environmental swab samples were collected at pre-identified locations to include the control room, receiving, processing, and load-out. A total of 42 feed samples and 73 environmental swabs were collected. Samples were analyzed for EBAC level to obtain a final CFU/g result.

Results

EBAC levels in feed samples averaged 3.21 ± 1.16 log10 CFU/g, while environmental swabs averaged 2.61 ± 1.02 log10 CFU/g. As shown in Figure 1A, feed samples containing SC had a lower level of EBAC compared to samples without SC. The level of EBAC at the four sample locations was similar, with the highest level measured in the receiving samples, while the lowest levels were measured in the load-out samples (Figure 1B). In all environmental samples, EBAC levels were similar between the West and East Feed Mills at 2.84 ± 1.08 and 2.37 ± 0.91 log10 CFU/g, respectively. For samples that would be considered feed contact surfaces, the EBAC level was lower for samples collected from the East Feed Mill compared to the West Feed Mill at 3.24 ± 0.60 and 1.66 ± 0.64 log10 CFU/g, respectively. The East Feed Mill applied SC at 1.0 kg/t to > 50% of the total feed production.

Figure 1: *Enterobacteriacea* level in **(A)** feed samples with and without Sal CURB® (n = 42) and **(B)** swab samples collected from different locations at the East Feed Mill or West Feed Mill (n = 73).

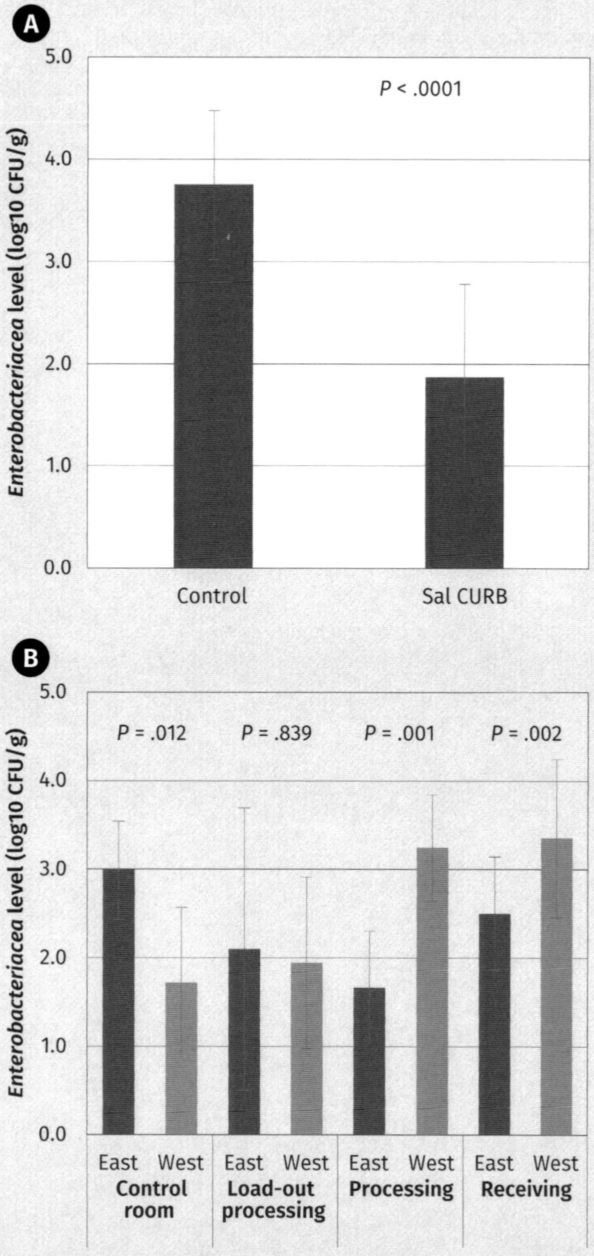

Discussion

EBAC was an effective indicator organism and was present in all non-SC treated samples. SC was able to reduce the level of EBAC in feeds by approximately 2 logs. Additionally, feed contact surfaces coming into contact with SC had a reduction of EBAC by > 1.5 logs. Both reductions were observed with a low-level usage of 1.0 kg/t, compared to the label rate of 3.25 kg/t.

References

1. Crump, J., Griffin, P. and Angulo, F. Bacterial contamination of animal feed and its relationship to human foodborne illness. *Clinical Infectious Diseases*. 2002;35(7):859-65. **https://doi.org/10.1086/342885**

2. Tortorello, M. Indicator organisms for safety and quality—uses and methods for detection: minireview. *Journal of AOAC International*. 2003;86(6):1208-17. **https://doi.org/10.1093/jaoac/86.6.1208**

3. Jones, F. and Richardson, K. *Salmonella* in commercially manufactured feeds. *Poultry Science*. 2004;83(3):384-91. **https://doi.org/10.1093/ps/83.3.384**

https://doi.org/10.54846/am2025/138

Post-cervical insemination in gilts: A future perspective

R. Ausejo-Marcos[1]; F. De Mergelina[1]; J. Miguel[1]; I. Tardío[2]; F. Gonzalvo[2]; P. Martínez[1]; U. Álvarez[1]; A. Climente[1]
[1]R&D Veterinary Department, Magapor S.L., Ejea de los Caballeros;
[2]Veterinary Technical Service at Cincaporc S.A., Fraga, Huesca

Introduction

In contemporary swine reproduction practices, cervical artificial insemination (C) is the primary method chosen for inseminating gilts, while post-cervical insemination (PC) is the preferred technique for sows.[1] The PC technique offers several advantages, including a reduction in the time required for each insemination, decreased backflow in inseminated females, and an enhancement of the genetic progress.

The effectiveness of artificial insemination (AI) is significantly influenced by the precision of heat detection practices implemented on the farm. Therefore, a previous accurate heat detection is crucial for the success of any AI protocols in sows, as it ensures optimal timing for insemination and, thereby maximizing the likelihood of a highest fertility rate.

The main objectives of the trial are therefore to highlight the importance of accurate heat detection and to assess the effectiveness of the PC insemination technique in gilts. This study utilizes the Magaplus N® catheter, specifically designed for gilts to facilitate the work. By focusing on these aspects, the study aims to improve reproductive outcomes and overall productivity in swine farming.

Material and methods

This experience was carried out on a commercial farm with 3500 hyperprolific females. The trial involved 672 gilts, of which 193 were inseminated using C technique, 154 using PC technique and 325 with a mixed technique, allowing the assessment of potential differences between methods.

Results

No significant differences were observed between the PC and C groups in terms of prolificacy, with respective averages of 19.6 and 18.9 total born. The fertility rates assessed via ultrasound were 88.3% for the PC group and 89.1% for the C group, while the farrowing rates were 84.4% and 84.5%, respectively. Regarding the number of doses per gilt per estrus, an increase in fertility was noted in gilts inseminated three times (87.9%) compared to those inseminated twice (83.6%). However, no differences in prolificacy were observed, with a total of 19.3 (3 times AI) and 19.2 (twice AI) total born. In cases where the probe did not cross the cervix completely, traditional insemination (mixed insemination) was employed. It was noted that gilts inseminated first using the C technique and subsequently with two PC inseminations exhibited a lower fertility percentage compared to gilts inseminated in the reverse order, beginning with two PC inseminations followed by one using the C technique (Figure 1).

Figure 1: Effect of mixed insemination techniques by comparing fertility rates.

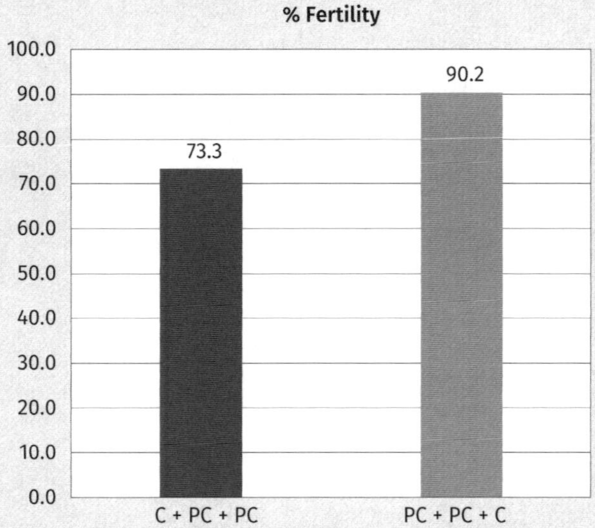

C = cervical artificial insemination; PC = post-cervical insemination

Discussion

The implementation of PC insemination in gilts not only resulted in reproductive parameters comparable to C, but also recorded an average increase of 0.7 piglets born per litter. The PC insemination technique, utilizing a specialized catheter and probe, enhances the operational protocol and accelerates genetic progress.[2] These two benefits are particularly significant in the context of farms employing batch management systems, optimizing farm occupancy, or farms operating at a larger scale, as they contribute to a reduction in insemination duration.

Furthermore, PC technique enables the farm staff members to make informed decisions regarding the timing of the first artificial insemination, contingent upon whether the probe successfully crosses the cervix, indicating that the gilt is nearing ovulation and that an effective heat detection has been achieved.

References

1. Suárez-Usbeck A, Mitjana O, Tejedor MT, Bonastre C, Moll D, Coll J, Ballester C, Falceto MV. Post-cervical compared with cervical insemination in gilts: Reproductive variable assessments. *Anim Reprod Sci.* 2019 Dec; 211:106207. **doi: 10.1016/j.anireprosci.2019.106207**

2. Ausejo R, Mendoza N, Dahmani Y, Mitjana O, Úbeda J, Falceto MV. Effect of incidents associated to post-cervical artificial insemination on reproductive performance of sows. *Bulgarian Journal of Veterinary* Medicine. 2018; 21(2), 198-205. **doi.org/10.15547/bjvm.1031**

https://doi.org/10.54846/am2025/139

Field experience with Circumvent® CML

James Lehman, DVM, MS; Robyn Fleck, DVM; Brett O'Brien VMD, MS
Merck Animal Health, Lenexa, Kansas

Introduction

Circumvent® CML (CML) was licensed in 2022 as the first-ready-to-use, single-dose vaccine effective against disease caused by porcine circovirus (PCV2) Types 2a and 2d, *Mycoplasma hyopneumoniae* (Mhp) and *Lawsonia intracellularis* (LI) in pigs three weeks of age or older. It is given as a single 2 mL intramuscular (IM) dose and commercial sales began in 2023. CML has been integrated into the health protocols of multiple US production systems and overall use in those systems has steadily increased following its release. Systems report positive feedback from farm personnel regarding its ease of use due to less on-farm cold storage requirements, low dose volume, less empty plastic bottle waste and no mixing required at time of vaccination. The 500-dose bottle size also saves time during vaccination by reducing the number of required bottle changes. Production team members value CML's ability to protect against multiple pathogens with a single ready-to-use product that also allows for serologic compliance testing when needed. Co-infections of porcine reproductive and respiratory virus (PRRSV) and PCV2 significantly increase the impact of PCV2 infection to the point where PCV associated disease (PCVAD) is created. Two post-licensing studies evaluating CML are described below.

PCV2 efficacy evaluation with a PCV2d/PRRSV co-challenge

Two piglets from 20 litters were enrolled one week prior to weaning to either a CML or placebo cohort. At weaning (study day 0), piglets were given either 2 mLs of CML or sterile saline IM. On study day 45, piglets were co-challenged with PCV2d and wild type PRRSV intranasally. Serum collected on day 52 confirmed a successful PRRSV challenge. Body weights were taken on study days 0, 28 and 73. Fixed lymphoid tissues were scored by a pathologist at Iowa State University Veterinary Diagnostic Laboratory who was masked to treatment group.

Compared to the control group, CML vaccinated pigs demonstrated solid PCV2 protection as evidenced by reduced lymphoid depletion, reduced lymphoid inflammation, reduced tissue viral loads and reduced viremia post-challenge ($P < .05$, see Table 1). Mean final body weights on study day 73 were 118.4 lbs for the controls and 127.4 for CML vaccinated pigs ($P = .06$).

Post-vaccination serology study

Three piglets from 20 litters were enrolled at three weeks of age and were given either a 2 mL dose of CML IM, or a 2 mL dose each of Circumvent® PCVM G2 (CV) and Porcilis® Ileitis (PI) IM administered on opposite sides of the neck, or 2 mLs sterile saline. Sequential sera were collected at study days 21 and 52 and tested with the Mhp and LI ELISAs.

The probability of testing positive on Mhp ELISA was 44% for CML and 32 % for CV + LI on day 21 and 32% and 61% on day 52 (vaccinates vs controls $P = .46$ for day 21 and .12 for day 52). The probability of testing positive on LI ELISA was 60% for CML and 93% for CV+LI on day 21 and 83% and 98% on day 52 (vaccinates vs controls $P = .003$ for day 21 and .001 for day 52). Mean titers for Mhp were 0.288 for CML and 0.318 for CV+LI on day 21 and 0.261 for CML and 0.393 for CV+LI on day 52 ($P < .05$). Mean titers for LI were 41 for CML and 53 for CV+LI on day 21 and 53 for CML and 72 for CV+LI on day 52 ($P < .05$).

Conclusion

Circumvent CML studies published post-licensure have demonstrated protection against PCV2 in a robust disease co-challenge model (PCV2d/PRRSV). Post-vaccination seroconversion rates for CML on Mhp and LI ELISAs are similar to the parent vaccines given separately. One injection for three pathogens reduces stress on pigs and their caretakers as well as reducing the number of bottles required to handle. Reports from the field indicate that swine producers are content with the product's efficacy and appreciate the convenience of having all three antigens in one bottle, resulting in fewer injections for the pigs and improved animal welfare.

Table 1: Mean PCV2 qPCR copies per mL and percent positive PCV2 PCR

	Mean PCV2 qPCR copies		% PCV2 positive	
Study Day	**Controls**	**Circumvent CML**	**Controls**	**Circumvent CML**
45 (challenge)	0	0	0	0
52	18^{a}	2^{b}	35	5
59	$245,8103^{a}$	124^{b}	100	60
66	$10,957,210^{a}$	5^{b}	100	20
73 (necropsy)	$2,460,934^{a}$	1^{b}	100	5

[a,b] Different superscripts within row are significantly different from one another at $P < .05$

https://doi.org/10.54846/am2025/140

Fecal shedding of *Lawsonia intracellularis* in finishing pigs treated with water-soluble Aivlosin or lincomycin after an ileitis challenge and the resulting correlation on average daily gain

N. Winkelman[1], DVM; A. Mueller[1], DVM; H. Jung[2], BS
[1]Swine Services Unlimited, Inc., Rice, Minnesota; [2]The Ohio State University, Colombus, Ohio

Introduction

Lawsonia intracellularis (Li; ileitis) is an obligate intracellular bacterium that is widespread throughout commercial swine finishing operations and carries a significant economic burden, costing swine producers $5.98 to $17.34 per head,[1] due to losses in average daily gain (ADG), feed conversion rate, and, in severe cases, mortality. The objective of this study was to examine the fecal shedding of Li in finishing pigs treated with water-soluble Aivlosin (62.5% w/w tylvalosin as tylvalosin tartrate; ECO Animal Health) or lincomycin hydrochloride soluble powder (lincomycin hydrochloride; Huvepharma) after an ileitis challenge and the resulting correlation on ADG.

Materials and methods

One hundred seventy, 10-week-old commercial pigs were placed into a finisher on Day -5 (D-5) and allocated to one of three treatment groups: non-medicated control (n = 50), Aivlosin (n = 60) or Lincomycin (n = 60). Pigs were confirmed to be negative for Li at placement into the finisher by IPMA antibody and fecal Li PCR analyses. On D0, all pigs were orally inoculated with Li mucosal homogenate (1.85×10^{10}). Water medication began in the two treated groups on D13 when > 10% of the pigs showed clinical signs of ileitis. Aivlosin was administered at 50 ppm,[2] while Lincomycin hydrochloride soluble powder was administered at 66 ppm for five consecutive days.[3] The study concluded on D35, when all animals were euthanized. Fecal samples were collected from 30 pigs in the control group and 36 pigs in each of the treated groups on D7, D13, D21, and D35 and tested individually for Li polymerase chain reaction (PCR). Individual pig weights were collected on D-5, D0, D13, D21, and D35. Pearson correlation and linear regression models were built to estimate the strength of the correlation and impacts of the treatments on PPE gross lesion Severity Index (SI),[4] fecal PCR, and ADG.

Results and discussion

Only the Aivlosin treatment showed statistically significant improvements in ADG over all time periods after D13. Mean fecal PCR ct values (40 cycles) showed no positives on D7, but by D13, all pigs were shedding high levels of Li with a mean ct of 26.3. On D21, Aivlosin pigs had 8.3 times decrease in Li/gram of feces vs the Lincomycin pigs, based on U of MN VDL qPCR standard curves. Day 35 ct values were 36-37 for all treatments.

The Figure 1 scatter plot has a moderate correlation (0.51) with a high P value < .001. For each ct unit decrease, there is a corresponding ADG decrease of 0.083 lbs. Ct values < 30 begin to show dramatic drops in ADG.

Figure 1: Correlation of D13 fecal PCR ct and average daily gain (ADG) from D13 to 35.

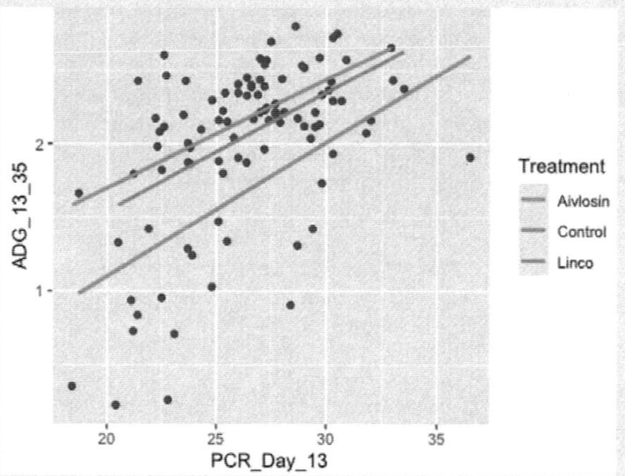

Implications

Li PCRs from fecal and/or oral fluid samples are the gold-standard ante-mortem diagnostic tool for ileitis. Interpreting PCR samples and early intervention with water medications to mitigate performance loss is cost-effective whenever clinical ileitis is present and/or PCR ct values are < 30ct of 40. There was a statistical ADG advantage for Aivlosin over the challenged controls in this trial, and only a numerical advantage for lincomycin.

References

1. Holtkamp D. 2019. Economic losses associated with ileitis. MSD Animal Health.

2. Aivlosin water soluble granules label information. Pharmgate Animal Health, Wilmington, Nc. Accessed 11/01/2024. **https://pharmgate.com/usa/aivlosin-water-soluble/**

3. Lincomycin hydrochloride soluble powder label information. Huvepharma, Peachtree City, GA. Accessed 11/01/2024. **https://www.huvepharma.us/product/lincomycin-hcl/**

4. Jung, H., Comparative efficacy of water soluble tylvalosin and lincomycin in finishing pigs administered a Lawsonia intracellularis (Li of ileitis) oral challenge. AASV 2025 Oral Student Presentation.

https://doi.org/10.54846/am2025/141

Impact of dietary sodium diformate in sows on suckling piglets under poor sanitary conditions

C. Lückstädt; S. Petrovic
ADDCON, Bitterfeld-Wolfen, Germany

Introduction

The application of organic acids and their salts to diets for pigs has been studied extensively under varying conditions.[1] Despite well-documented effects of diformates on growing pigs, data on their impact when fed to sows during late gestation and lactation, and the subsequent effects on their suckling piglets under commercial conditions are not available. In this trial, a lower dosage (5 kg/t) was used than that normally recommended, to investigate whether further economic improvement could be achieved.

Material and methods

This study tested the effects of sodium diformate (Formi® NDF, ADDCON) fed to sows on their suckling piglets and was conducted under veterinary supervision. Fourteen sows (DanBred) on a commercial farm (Northern Serbia) were equally divided into two groups, each containing 7 sows and fed a typical lactation diet (corn-wheat-soy based), containing either 0.5% of NDF, or a negative control without the additive, from one week before farrowing till the end of weaning (day 28). The lactation diet was fed ad libitum and daily feed intake measured. The following parameters were monitored at farrowing: number of piglets born alive, individual weight of new-born piglets and litter weight of new-born piglets. At weaning the number of weaned piglets, the individual body weight of weaned piglets and litter weight during weaning was recorded. Data were analyzed using the t-test with a significance level of .05.

Results

Feed was well accepted by both groups. Feed intake is only available as pooled data - sows fed the NDF-diet had a numerically higher feed intake compared to the negative control group (+560 g/day, Table 1). At farrowing, the weight and number of piglets born alive increased significantly ($P < .05$) due to the addition of the additive: live births increased by 1.5 piglets per sow. Litter weight of piglets differed significantly by more than 2.3 kg, for the NDF-fed sows. No difference was observed for the number of weaned piglets per sow although the weaning weight of the piglets at day 28 was highly significantly ($P < .001$) improved (520 g heavier piglets in the NDF-group), leading to a numerically ($P = 0.15$) increased litter weight at weaning of more than 8.5 kg in NDF-fed sows. The uniformity of overall litter weight was also improved.

Conclusion

As previously reported,[2,3] lower dosages of sodium diformate (Formi NDF) in sow diets during the suckling period will indirectly support piglet production.

References

1. Lückstädt C and Mellor S 2011. Recent Advances in Animal Nutrition – Australia. 18,123-130.

2. Øverland M, Bikker, P and Fledderus J. 2009. Livestock Science 122, 241-247.

3. Lückstädt C and Petrovic S 2019. Advances in Animal Biosciences, 226.

Table 1: Performance parameters of piglets from sows fed with or without sodium diformate (NDF)

Parameter	Control (n = 7)	0.5% NDF (n = 7)	*P*-value
Daily feed intake sow [kg]	5.18	5.74	[+560 g]
Piglets born alive, per sow [n]	13.6 ± 1.7	15.1 ± 1.4	.049
Weight of new-born piglets [kg]	1.41 ± 0.07	1.42 ± 0.07	.46
Litter weight at birth [kg]	19.11 ± 1.90	21.45 ± 1.99	.03
Piglets weaned, per sow [n]	10.6 ± 2.4	11.0 ± 1.6	.36
Weight of weaned piglets [kg]	6.74 ± 0.16	7.26 ± 0.23	.0003
Litter weight at weaning [kg]	71.17 ± 15.95	79.68 ± 10.59	.15

https://doi.org/10.54846/am2025/142

A milk-based bioactive (FxP®) mitigates a severe and natural enteric disease challenge

J. D. Spencer[1]; C. L. Puls[1]; M. McCallum[1]; A. Gaines[2]; N. Horn[1]
[1]United Animal Health, Sheridan, Indiana; [2]AniTek, Shelbina, Missouri

Introduction

Getting pigs off to a good start at weaning can be challenging, especially when health challenges occur. Factors such as wean age, environment, management, nutrition, and health all interact to influence how pigs grow and mitigate disease during this critical transition period. Spray-dried animal plasma has been used in swine diets for decades to assist with weaned pig health and growth,[1] but more recently a novel source of milk-based bioactives has been shown to improve performance and reduce post-weaning mortality in pigs fed commercially relevant plasma free diets.[2]

Methods

Sixty-four weaned pigs (DNA 610E × DNA 241 F1) were transported from the UAH research farm to a contract research facility to conduct a planned controlled *E coli* challenge trial. The single room facility used raised decks that were approximately six feet apart and contained 4 pens per deck and 4 pigs per pen, allowing for 8 replicate pens per treatment (32 pigs per treatment). Pigs were allotted by weight and fed one of two dietary treatments for the 23-d duration of the study that consisted of 1) control diet or 2) control diet + 0.6% FxP.

The designed oral *E coli* challenge model was cancelled due to severe diarrhea starting on d 5 post weaning. Diagnostics on d 7 revealed rotavirus A present in both treatments with pig mortalities starting on d 10. Since the natural health challenge was equal between treatments and pigs were allotted, housed, and fed dietary treatments for a designed disease challenge model, treatments continued and the planned metrics were collected. Routine diagnostic swabs were collected on d 7, 14, 23 and live pigs were sent to the Purdue diagnostic lab at the end of the study to assess the natural disease challenge. Individual weights were collected on d 0, 14, and 23. Further, pen fecal samples and pig serum samples (pooled by pen) were collected on d 14 and 23 to measure total coliforms and serum haptoglobin, respectively. At the end of the study following euthanasia, the mid ileum was excised from 7 pigs per treatment for morphological measurements.

Results and discussion

A significant multi-pathogen challenge was present in all pens throughout the study that included high prevalence of Rotavirus A, F18 *E coli, S enterica* and *S suis* around d 7,14 and 23, respectively. All mortalities were diagnosed as enteritis by the consulting veterinarian at time of necropsy. Overall mortality was high with pigs fed FxP having a lower mortality on d 14 (4 vs 38%, $P = .07$) and end of trial (43 vs 50%; $P = .11$) (Table 1). Throughout the trial, pigs fed FxP were heavier ($P < .05$) and had reduced serum haptoglobin and total coliform measurements ($P < .10$). The change in population body weight variation during the trial was less for pigs fed FxP (weight CV change: -1.84% vs 11.34%; not statistically analyzed) and the ileal morphology was also improved with the feeding of FxP (villus height: crypt depth ratio 2.13 vs 1.51; $P = .07$).

These results show that FxP can reduce mortality and improve pig nursery growth during a severe and natural enteric challenge. Reductions in coliforms, serum haptoglobin and improved intestinal morphology suggest that FxP reduces the severity of the complex disease event allowing for improved survivability, performance, and gut health.

Table 1: Impact of FxP on piglet growth, mortality, and health metrics during a natural challenge in a controlled research trial

	Control	FxP	SE	*P*-value
Weight, lbs				
Day 0	12.5	12.5	0.46	NS
Day 14	13.2	15.5	0.74	.01
Day 23	18.9	23.8	1.43	.04
Weight CV change,* %				
Day 0 - 14	7.46	1.72		
Day 14 -23	3.87	-3.56		
Day 0 - 23	11.34	-1.84		
Serum haptoglobin, ng/mL				
Day 14	3509	1378	383	.04
Day 23	682	432	95	.08
Total coliforms, log CFU/g				
Day 14	8.13	6.53	0.588	.09
Day 23	5.00	2.70	0.804	.04
Mortality, %				
Day 14	38	4	2.54	.07
Day 23	50	43	1.24	.11
Ileal villus:crypt ratio				
Day 23	1.51	2.13	0.18	.07

* Change in treatment population weight variation around the treatment body weight mean was not statistically analyzed.

References

1. Coffey, R.D., and G.L. Cromwell. 1995. The impact of environment and antimicrobial agents on the growth response of early -weaned pigs to spray-dried porcine plasma. *J. Anim. Sci.* 73:2532-2539.

2. Horn, N., K. Ajuwon, E. Li, A. Gaines, J. Goodwin. (2022). Novel whey protein concentrate improves nursery pig growth and intestinal morphology. *Proceedings of 2022 Digestive Physiology of Pigs Conference:* P33. Rotterdam, Netherlands.

https://doi.org/10.54846/am2025/143

Compounded iron supplement used to address hurdles and needs in the field

P. Smith, PharmD, MBA; K. Bretey, DVM; B. Payne, DVM
Veterinary Pharmaceutical Solutions (VPS), St. Peter, Minnesota

Introduction

To prevent iron deficiency anemia in pigs, it is routine practice to inject iron newborn piglets intramuscularly. A common recommendation is to provide 200mg in a single dose. The impact on missing a dose or partially dosing a pig with iron can have negative health and production benefits to the pig.[1] Recent research in pigs representing modern genetics has shown production and carcass benefits for higher dosing, up to 400mg/pig.[2] Compounded iron into larger fluid bags provide fewer opportunities to have missed doses or partial doses.

Administration and dosing

Bottle mount syringes, draw-off syringes and needle-free devises are common instruments to administer iron directly from the bottle. Ergonomics of bottle-mount syringes limit the size of bottles that can be comfortably used. All three delivery types allow air to enter the tube as the bottles are not vacuum sealed once they are punctured. Frequent changing of small bottle sizes allows air into the line. Partial or even missed dosing occurs due to air being in the line or in the barrel of the syringe. Compounded iron fluid bags use a draw-off tube to extend the bag to a syringe. Iron flows through the line and the iron bag decompresses, pushing all of the iron into the spike and tube. The fluid in the bag is not replaced with air unlike vented bottles, therefore greatly reducing the risk of air getting into the lines. This reduces the risk of air being injected into pigs which would lead to partial or missed dosing.

Legal concerns and contamination prevention

Pharmacists and veterinarians have the right to compound for livestock, but they must follow 21CFR.530.13 and USP797 for sterile preparations. This requires proper clean-room facilities, training and SOPs. To monitor compliance, both environmental and product testing for sterility and endotoxins are conducted routinely at VPS. The rigorous internal audits and external Boards of Pharmacy inspections provide a check on the VPS processes. If compounding is conducted by the veterinarian while he/she is on the farm under non-sterile conditions, it must be used immediately. A contamination survey of four independent farms with on-farm iron compounding found 100% of swabs from internal bottles and stoppers were contaminated. There were 49 unique bacteria isolates from 17 genera (Table 1). Fungus was cultured from 90% of the swabs. Compounding pharmacies may compound sterile iron from a veterinarian prescription that can legally be used for longer periods as they are achieving sterile transfer that can be verified. Compounding pharmacies can also advise to which antibiotics can legally be admixed and those which should not be compounded due to potency concerns (ie, penicillin).

Conclusion

Sterile compounding of iron for baby pigs by a livestock compounding pharmacy can provide a legal alternative to immediate use compounding, reducing contamination and ultimately, the risk of missed or partial doses.

Table 1: List of contaminants, bacteria and fungus identified in a diagnostic survey for contamination of four farms' iron that was compounded on farm under the direction of veterinarians

Bacterial genus and species identified	Additional bacterial genera identified
Clostridium perfringens	*Staphylococcus* species
Escherichia coli	*Vibrio* species
Myroides odoratus	*Aeromonas* species
Providencia rettgeri	*Aerococcus* species
Kluyvera cryocrescens	*Pseudomonas* species
Klebsiella oxytoca	**Fungus identified**
Proteus vulgaris	*Aspergillus terreus*
Enterococcus casseliflavus	*Trichosporon asahii*
Enterobacter hormaechei	Additional fungus cultured but not identified
Enterococcus faecium	
Acinetobacter lwoffi	
Brevundimonas diminuta	
Pseudomonas stutzeri	
Acinetobacter johnsonii	
Morganella morganii	
Aerococcus viridans	

References

1. Chevalier, T, Monegue J, Lindemann, M. Effects of iron dosage administered to newborn piglets on hematological measures, preweaning and postweaning growth performance, and postweaning tissue mineral content. *J. of Swine Health and Production.* Jul 2021. v29n4p189. **doi.org/10.54846/jshap/1207**

2. Chevalier T, Lyons W, Paczosa D, et al. A second iron injection administered to piglets during lactation improves hemoglobin concentration, growth performance, and carcass characteristics at slaughter. *J. of Anim. Sci.*, 2023; 101, 1-8. **doi.org/10.1093/jas/skad270**

https://doi.org/10.54846/am2025/144

PCV2 prevalence and genotype diversity at processing and potential correlation with downstream performance in Canadian production systems

R. Tenbergen[1], DVM, MS; M. P. Labrecque[2], DVM; J. Angulo[3], DVM, MS; L. Rosengren[4], DVM, MAB, PhD
[1]Demeter Veterinary Services P.C., Swine Veterinary Partners, Canada; [2]Zoetis Canada Inc., Canada;
[3]Zoetis Inc., United States; [4]Rosengren Epidemiology Consulting Ltd., Canada

Introduction

This prospective observational epidemiology study carried out in 15 farms across Canada in 2022 using cohort groups and was designed to measure porcine circovirus type 2 (PCV2) prevalence in processing fluids and potential correlation with downstream virus status and performance in late nursery and mid finisher. A secondary objective was to identify PCV2 genotypes at processing, late nursery, and finisher and assess the virus genetic diversity within cohort groups at different production stages.

Methods

At each of 15 farms enrolled, two consecutive cohorts per farm were analyzed for a total of 30 cohort groups. Using longitudinal sampling, 1 processing fluids sample from all piglets processed was collected at 3-5 days of age (n = 30) and 5 pens from the same cohort of pigs was sampled using fixed spatial sampling method in late-nursery (8-10 weeks of age) and in mid-finisher (16-17 weeks of age). 1 pool of 6 serums and 1 oral fluid were collected from each of the 5 pens. All pooled serum (n = 317) and oral fluid samples (n = 267) were tested for PCV2 PCR and genotyped (a, b, d or e). Eligible positive samples were ORF2 sequenced to obtain 1 sequence per production phase for each farm (n = 38). In addition, herd PCV2 vaccination protocol, health history in regards of PCV2 and PRRSV and closeout data information were collected. The primary level of analysis was the cohort. Associations were obtained using logistic regression with robust standard errors to account for clustering within herd (Stata/SE 16.1); statistically significant associations were reported at $P \leq .05$. For sequences, epitope content analysis (EpiCC) was done using EpiVax computational tool.

Results

53% of processing fluids were positive for PCV2 and 12 farms out of 15 (80%) had at least 1 cohort positive for PCV2 on processing fluids during the study. PCV2 was detected in 35% of oral fluids and 8% of pooled serum in late nursery and 55% of oral fluids and 34% of pooled serum in late finisher. Considering both serum and oral fluid samples, 12 (80%) and 13 (87%) farms out of 15 had at least 1 cohort positive for PCV2 in late nursery and late finisher respectively. Although we were unable to show a correlation between PCV2 processing fluid status and downstream performance (limited sample size on performance), higher prevalence of PCV2 was observed in cohorts coming from herds with positive processing fluids (Table 1), finding a significant relationship between processing fluids PCV2 status (+/-) and nursery serum status (prevalence). The odds of a cohort having one or more pooled serum samples test positive for PCV2 in nursery were 10 times higher (95% CI 1.08 – 94.8, $P = .04$) when the processing fluids was positive. Additionally, we found a 100% correlation between having at least one positive pooled serum in nursery and one positive pooled serum in finisher. When analysed at the pen level, a positive oral fluids sample was a significant predictor of a positive pooled serum sample in nursery pens (OR 8.5, 95% CI 1.98 – 36.2, $P < .01$) and finisher pens (OR 17.5, 95% CI 5.0 – 61.3, $P < .001$). The following genotypes were identified in descending order of frequency: 2a, 2d, 2e and 2b. Interestingly, 30% of cohorts had detection of different genotypes at each production phase. EpiCC scores were obtained for the 38 samples sequenced in the project revealing that Fostera® Gold (PCV2ab) epitope coverage average is 82% against vaccine A at 68%, vaccine B at 72% and vaccine C at 68%. For example, Fostera® Gold epitope coverage average against 2d samples is 19% superior (79%) in comparison to vaccine A (60%), vaccine B (62%) and vaccine C (66%).

Table 1: Processing fluids status and downstream prevalence

Suckling piglets	Specimen	Nursery	Finishing
Processing fluids -	Serum	1%	27%
	Oral fluids	17%	24%
Processing fluids +	Serum	13%	39%
	Oral fluids	49%	65%

Discussion and conclusions

PCV2 status using processing fluids may serve as predictive factor for downstream prevalence in the nursery and as potential measure to help define and monitoring breeding herd PCV2 stability. PCV2 is prevalent at all production stages but no conclusions could be made about the association of processing fluids status and the impact on downstream performance due to the limited sample size and missing data. Multiple genotypes can be found in the same group of pigs throughout growing phase suggesting multiple infections and creating an environment conducive to recombination. Interestingly, the genotype identified at processing isn't necessarily predictive of the one(s) identified downstream indicating the need of including PCV2 as part of the regular monitoring plan, furthermore, the potential effects on performances should be investigated having the appropriate data, although we should not underestimate the potential impact of subclinical infection and the potential economical impact of having multiple PCV2 genotypes infections.

https://doi.org/10.54846/am2025/145

Earning the role of the pig's champion

Daryl Olsen, DVM
AMVC, Audubon Iowa

Introduction

Knowledge is gained in many ways. I believe knowledge starts with evidence and then we fill in the gaps by watching, studying, learning from mentors, continuing education and many other ways. As veterinarians, we start with extensive training and then we continue to gain knowledge in so many ways. None of us ever want to grow older but one of the huge advantages of growing older is experience. Experience can be one of the greatest ways to gain knowledge IF we learn from our mistakes and continue to learn every day. My role this morning is to take 40 years of working hard to be the pig's champion and prove that veterinarians more than anyone deserve the right to be the pig's champion. The only problem is that deserving the right and earning the right might be two different things.

Self-promotion is difficult for many of us. I am often reminded of my mother's strict words when I find myself bragging or hear someone bragging. "Self-praise stinkith." It rings in my mind probably more than it should. As I watch many professional athletes, entertainers, and social media enthusiasts today, I find myself wanting my mom there to give them the old "Self-praise stinkith" speech. I agree with my mom and am embarrassed when I hear someone or find myself bragging but as veterinarians, I believe we need to forget what my mom promoted and start telling the world and our industry why we deserve and have earned the right to be the pig's champion. Silence on this subject will just continue to erode our role. We need to be vocal, based on evidence and present our case in the proper way. Seems simple.

After graduating from Iowa State University College of Veterinary Medicine in 1982 I accepted a position in Audubon, Iowa. One of the reasons I accepted the position was the fact that this practice was one of the most progressive practices that I visited. They even had a clinic Apple computer and were starting to keep production records for producers and attempting to understand diets and diet formulation. I fit right in with the three other veterinarians already on staff. I was the male baby boomer, working 80 hours a week, as a large animal veterinarian that dominated the swine industry for 30 years. Dr Chris Rademacher's Alex Hogg presentation last year was an eye-opening challenge to our profession. The demographics and the needs of swine veterinarians are changing rapidly.[1] I believe I am now the history of swine veterinary medicine and not the future. That fact again is based in evidence and helps us develop the knowledge we need to make decisions moving forward. Fast forward to today where social media and artificial intelligence dominate decisions, every eight-year-old is more capable on a smartphone than their grandparents and it is not hard to understand that what we've done in the past does not work today. We must change our message. We must change our approach.

Why swine veterinarians deserve the right to be the pig's champion: The evidence

I absolutely believe that swine veterinarians deserve the right to be the pig's champion. Let's review the many reasons that support this statement. Every individual started their journey to be a veterinarian for one reason: A love for animals. Every generation, male or female, from the farm kid who grew up working with animals to the urban kid who was raised with only a pet, we all started wanting to be veterinarians because we loved animals. It's in our DNA and we just don't tell people enough. Who else has dedicated their entire life, their entire profession based on the love of animals and for many of us, the love of pigs? The number one reason we deserve the right to be the pig's champion is because all of us started our careers and have continued our careers because we love animals. Not many other professions can make that claim. We need to start every conversation and end every conversation with that message and that fact helps build credible evidence.

As swine veterinarians, we take our professional oath seriously and we live by this oath. Our Veterinary Oath only further supports the fact that veterinarians deserve the right to be the pig's champion.

> *Being admitted to the profession of veterinary medicine, I solemnly swear to use my scientific knowledge and skills for the benefit of society through the protection of animal health and welfare, the prevention and relief of animal suffering, the conservation of animal resources, the promotion of public health, and the advancement of medical knowledge.*
>
> *I will practice my profession conscientiously, with dignity, and in keeping with the principles of veterinary medical ethics.*
>
> *I accept as a lifelong obligation the continual improvement of my professional knowledge and competence.*[2]

As I am writing this presentation, I realize I must take a break for an appointment today with my dermatologist. I hate going to the dermatologist, but I trust him totally primarily because he has the intensive formal training to evaluate my concerns. The right to be the pig's champion must also include intensive formal training and who else other than veterinarians have the training to understand and promote what is best for the pig. Today's DVM graduates are leaving school with more knowledge than ever before. Many in the audience today have additional degrees beyond their DVM degree, often specializing in areas that provide even greater support. Building the foundation that we are the best equipped to be the pig's champion starts with a love of the animal but is strongly supported by intensive formal training.

The reason that every veterinarian is in attendance today supports the fact that veterinarians are lifelong learners and committed to enhancing their knowledge through continuing education. The American Association of Swine Veterinarians (AASV) has done an amazing job of providing continuing education that is second to none. Our annual meeting is considered one of the best in all veterinary medicine. Add the various regional meetings such as the Leman Conference, the ISU James D. McKean Swine Disease Conference and the quality of continuing education opportunities is just outstanding. Most swine veterinarians don't look at how many hours they need for licensing but instead how much information they can gain to be the best.

The AASV continues to be an association that leads by example. Mentoring veterinary students is so important and many individual veterinarians have mentored literally hundreds of students. The connection between these students and veterinarians remains strong today. Although individual veterinarians and veterinary clinics continue to provide opportunities for students to be exposed to our profession and our industry, the AASV takes it to the next level. With the support of the AASV Foundation and multiple invested companies, students are welcomed to our annual AASV meeting and get the opportunity to display their talents in many ways including oral presentations and research posters. The AASV Foundation also supports veterinarians who pursue additional education post-graduation. The evidence that swine veterinarians and the AASV are training the next generation of swine veterinarians supports our role of the pig's champion.

As a profession, we must promote the fact that we have led improvements in disease elimination that have changed the industry forever. Diseases that plagued the industry for years (such as pseudorabies, atrophic rhinitis, *Mycoplasma hyopneumoniae*) have either been eliminated or can be eliminated from specific herds due to work done by swine veterinarians. As swine veterinarians, we continue to lead efforts to eliminate disease from North American swine herds but maybe even more important, swine veterinarians have led the charge to keep new diseases out of our swine populations since the introduction of porcine epidemic diarrhea virus (PEDV) in 2013. In response to this outbreak, the formation of the Swine Health Information Center (SHIC) by the National Pork Board was initially promoted by members of the AASV. This organization continues to lead the charge to protect the North American swine population and is further support that as a profession, we deserve the right to be the pig's champion.

Forty years in the profession gives one the opportunity to look back on technological advances in the swine industry that have changed pig well-being and health forever. A must read for all swine veterinarians is Multi-Site Pig Production by D. L. (Hank) Harris.[3] Released in 2000, Hank gives credit to many veterinarians who at that time led many improvements and changes in the industry. The book outlines the backbone of our industry today. Many of the advancements in swine production technology over the past forty years have been led by swine veterinarians attempting to make improvements to the health and production of the swine industry. I was fortunate enough to serve on Boehringer Ingelheim's porcine reproductive and respiratory syndrome (PRRS) Research Committee for many years. I was always impressed with the quality and originality of the proposals. The brilliant thought processes of both practicing veterinarians and research dedicated veterinarians were on display. Technologies including the isowean pig principle, modern truck washes and bakers, modern boar studs to biofiltered swine facilities were principles championed by swine veterinarians. Veterinarians will continue to lead the charge in adapting new technologies that advance swine health.

SHIC is just one example of organizations that have been supported by swine veterinarians. The National Pork Board (NPB), the National Pork Producers Council (NPPC), the US Swine Health Improvement Plan (SHIP), Professional Animal Auditor Certification Organization (PAACO), and the list can go on and on, are all entities that many swine veterinarians influenced and have been instrumental in making positive changes for the pig. Our collaborative work with the industry is so very important to improving pig welfare and pig health. The time and effort that veterinarians spend, without compensation, in support of the pig makes me so proud to be a swine veterinarian. Almost every swine veterinarian has supported allied organizations throughout their careers.

As swine veterinarians, we look forward to attending the annual meeting to gain knowledge but probably just as exciting is the chance to spend time with colleagues, now friends. I am not alone in viewing everyone in attendance as a colleague and not a competitor. In one way or another, we are all providing services to swine producers and are competitors, but I consider every swine veterinarian a colleague, not a competitor. By definition a colleague is a person with whom one works in a profession or business. Synonyms include teammate, coworker, and ally. I continue to admire our profession for the willingness to share information, to work with colleagues to improve the health and well-being of pigs under our care. From sharing PRRS sequences to working through area disease control and elimination efforts, our profession leads the effort to control disease through a network of shared information. When PEDV first decimated our industry, swine veterinarians were searching for answers. The information that was shared throughout the industry to help solve questions that at that time didn't have answers was amazing. Again, strong support that swine veterinarians deserve the right to be the pig's champion.

As a profession, I am not sure we always appreciate the support we receive from the finest diagnostic laboratories in the world. With amazing laboratories at Veterinary Colleges across the United States (Iowa State University, South Dakota State University, University of Minnesota, Kansas State University, Purdue University, and North Carolina State University) that work closely together, sharing, comparing and compiling diagnostic information to the new National Bio and AgroDefense Facility in Manhattan, Kansas, our profession has unmatched support. This unbelievable diagnostic support system is in place because many dedicated swine veterinarians wanted a better system and worked diligently to make it happen. Just one more reason swine veterinarians deserve the role of the pig's champion.

We are approaching the five-year anniversary of one of the greatest disasters our industry has ever experienced. Yes, it was a health concern but not a disease that affected the pig. The diagnosis in January, 2020 of a novel coronavirus (Covid-19) in a Washington man soon escalated to the World Health Organization declaring this novel coronavirus a global pandemic in March, 2020. The sequence of events that followed led to the greatest supply chain disruption the world and our country has ever seen. Pork producers were left with multiple problems from how to keep workers engaged without sacrificing their health to a supply chain that had nowhere to go with a just-in-time and growing supply of pigs. Add to the fact that although our human health professionals are unbelievable at individual health care, their knowledge in large population viral epidemiology was limited. Swine veterinarians were forced to help producers, packers and our country with many serious and difficult decisions. I am always proud of our profession but in my entire professional career, I have never been prouder of how swine veterinarians worked tirelessly to work through a crisis I hope none of us have to experience again. We deserve the right to be the pig's champion with our efforts through the Covid 19 crisis.

Conclusion

The evidence is strong and extremely supportive of the fact that swine veterinarians deserve the right to be the pig's champion. From our love of animals and especially pigs, to our Veterinary Oath, intensive formal training, being lifelong learners, mentoring the next generation of swine veterinarians, a history of driving improvements such as adaptation of new technologies and disease elimination, the advancements made through our involvement in allied entities (NPB, NPPC, SHIP, SHIC), our willingness to share information, definitely the greatest diagnostic capacity in the world and finally our response to the Covid crisis, all of these reasons are strong support why we deserve the role of the pig's champion. But deserving the right and earning the right to be the pig's champion are two huge differences. As a profession, we all deserve the right but we all individually and collectively must earn the right.

We know the AASV, and our profession is changing as supported by Dr Rachemacher's presentation from just one year ago. My generation attempted to earn the pig's champion role by in fact "trying to outwork the problem". The positive to working tireless hours was that we in fact became almost irreplaceable to many producers. We placed ourselves in their circle of trust and thus had a huge influence on health and production. But as Dr Clayton Johnson presented in this year's Howard Dunn presentation, we failed in many ways.[4] And I agree with Dr Cara Haden in her 2025 Alex Hogg Memorial Lecture that the new generation of veterinarians have different professional and lifestyle goals, probably learning from our mistakes.[5] As the AASV and our profession changes, how can we re-establish ourselves again as the pig's champion?

Individually we must continue to take every opportunity to be the pig's champion and tell everyone we can why we not only deserve to be the pig's champion, but we have also earned it. Unfortunately, I don't think this will be enough to move the needle. As veterinarians, we have never collectively worked at establishing or re-establishing our profession as the pig's champion. If swine veterinarians are not considered to be the pig's champion, then what role will we fill? We could quickly fall into a mere technician role and that scares me beyond words. NPB's We Care initiative establishes the producer as the pig's champion but there are issues with self-promotion. I am certainly not encouraging us to distance ourselves from producers or producer groups with the goal of establishing our profession as the pig's champion, but we must do something different. Doing the same thing and expecting different results has never worked.

I believe that as a profession, we deserve, and we have earned the right to be the pig's champion. In the future, to maintain that role, we must be more organized and more vocal. Advocacy is not something our association has the staff or the time to address. I believe it is time we became more organized and focused on re-establishing our profession as the expert, the leader in pig health and welfare. I don't know what the structure of this new entity should be BUT it must be led by the AASV and its members. If it is indeed the AASV that leads this effort, we must be prepared to provide the necessary resources, both financial and personnel. I want our profession to be clearly identified as the pig's champion. We've earned it.

References

1. Rademacher C. 18th Alex Hogg Memorial Lecture. Past, present, and future challenges for the swine veterinary profession. AASV Annual Conference Proceedings. 2024. **https://www.aasv.org/swine-information/?item=27190**

2. AVMA website. Veterinary Oath. **https://www.avma.org**

3. Harris D. Multi-Site Pig Production. Iowa State University Press, 2000.

4. Johnson C. 2025 Howard Dunne Memorial Lecture. The challenges and opportunities of becoming the pig's champion. AASV Annual Proceedings. 2025.

5. Haden C. Alex Hogg Memorial Lecture. Who get's to be the pig's champion? AASV Annual Proceedings. 2025.

https://doi.org/10.54846/am2025/146

The path toward PEDV elimination in Manitoba

Jenelle Hamblin, MSc
Manitoba Pork, Winnipeg, Manitoba, Canada

Background

Elimination has been the chosen strategy for management of porcine epidemic diarrhea virus (PEDV) in Manitoba since its first case in 2014. Regions of the province have since experienced significant PEDV outbreaks in 2017, 2019 and 2021-2022. Beginning in 2022, the industry led the development of a regional PEDV elimination plan for Manitoba. This five-year plan outlines objectives and interventions to reach the sector's elimination goal. Implemented in fall 2023, the elimination plan is a testament of commitment to action and collaboration amongst all parties within the sector.

Manitoba Pork and Manitoba's Chief Veterinary Office have and continue to work together to operationalize the plan and to navigate PEDV risk in the province, all with ongoing guidance from sector stakeholders. This model of industry/government collaboration is unique. Manitoba prides itself on the technical elements of the plan but also the steps that were taken to develop and reach consensus. Regional disease control takes commitment from all stakeholders, and this plan serves as an example.

Planning for the future of PEDV in Manitoba

Despite the sector's commitment and best efforts on PEDV elimination, it became apparent during the 2021-2022 outbreak that the PEDV management efforts needed review. Dealing with a major PEDV outbreak every other year is not sustainable for the industry and an in-depth evaluation of PEDV management was required. Manitoba Pork convened a working group of industry stakeholders in early 2022 that included representation from producers, veterinarians, Manitoba's Chief Veterinary Office and Manitoba Pork.

The mandate of the working group was to:

1. Outline both short- and long-term objectives for managing PEDV in Manitoba.
2. Assess possible strategies that could be employed to manage PEDV.
3. Determine the impacts of the strategies used in the high-risk area on other areas of Manitoba.
4. Provide outreach and messaging to the sector on PEDV management.

As a first step, the working group commissioned a review of PEDV management strategies. Completed by the Western College of Veterinary Medicine in April 2022, this review focuses on PEDV strategies employed in different jurisdictions across North America with examples including elimination, gilt exposure, endemicity, use of vaccines, and/or running positive flows. Based on this review, the working group decided that PEDV elimination remained the best option for Manitoba.

Development of a PEDV elimination plan

In 2022-2023, the working group collaborated with sector partners to develop an elimination framework titled Manitoba's PEDV Elimination Plan – A Sector Approach to Disease Management. The plan outlines an overall goal with associated objectives and prevention/intervention actions to manage PEDV in Manitoba. Spanning five years (2022-2027), the plan is a living document that will be consistently reviewed to adjust strategies based on new experiences and learnings.

The goal is to eliminate 96% of PEDV infections in the high-risk area of southeastern Manitoba by 2027, done in a stepwise process by reducing the number of infections each year through prevention and intervention actions. These farm-level actions focus on surveillance, enhanced biosecurity, and rapid and aggressive elimination protocols. The plan also outlines areas of future consideration where industry can focus risk-mitigation efforts such as managing PEDV infected manure, risk from high traffic facilities, and/or exploring the use of vaccines should an option become available.

In fall 2023, the working group members convened and endorsed the plan for implementation across the industry. Manitoba Pork shared the components of the plan broadly throughout its membership and sector stakeholders through targeted communications and presentations at numerous provincial producer meetings and industry seminars. Regular communication with the swine veterinarians in Manitoba was also integral to the understanding of and implementation of the elements of the PEDV elimination plan on farm.

Implementation of the PEDV elimination plan

Throughout 2024, many elements of the elimination plan saw meaningful progress made. From prevention and enhanced biosecurity to disease modelling and communication, PEDV was top of mind for producers, veterinarians and industry stakeholders.

Based on previous experience, the next large-scale outbreak of PEDV was anticipated to occur in 2024. With that in mind, a widespread prevention and preparedness campaign began in late 2023/early 2024. Producers, veterinarians, sector stakeholders, Manitoba Agriculture and Manitoba Pork enhanced field level biosecurity and communication efforts to actively prevent PEDV infection from taking hold in the high-risk area.

As part of the prevention portion of the plan, biosecurity practices were reviewed and enhanced. Specific focus was given to hands-on training, to ensure that what was communicated in a biosecurity protocol was well understood. Farms were and continue to be empowered to become their own biosecurity champions and call out potential breaches. Training and communication are people focused, which helps staff understand the why and the how behind biosecurity. The commitment to biosecurity from the farms, sector stakeholders and veterinarians is evident in the strides that have been made in 2024, with known PEDV risk still present in Manitoba.

Regular on-farm surveillance for PEDV across the high-risk area is an element of the elimination plan that continues to be explored. Farms are independently performing regular surveillance as a valuable early detection tool to limit the potential for downstream spreading. Expanding on-farm surveillance plan will continue to be worked on in 2025. Regular surveillance at high-traffic facilities such as assembly yards and abattoirs is also currently underway. This surveillance acts as an alert system and reinforces the risk of PEDV to the sector.

Manitoba broadly implemented the use of on-farm biosecurity software. This software provides traceability to people, vehicles, equipment, etc, through geo-fencing technology. The communication component of the software can also broadly alert its farms of potential disease threats and limit access to farms. This has helped the farm to react quickly to a possible disease event, implement biocontainment actions quickly and therefore reduce the risk of downstream spread.

As outlined in the plan, a 2024 objective was to develop a disease spread model for PEDV in Manitoba. The goal of the modelling work was to evaluate the actions and interventions outlined in the plan against the known disease transmission for PEDV in the province. Working with a veterinary epidemiology consultant, the Manitoba swine sector's PEDV disease spread model was created. The results of the model showed enhanced biosecurity both on-farm and during transport as the most effective action to be taken to minimize disease transmission. Applying multiple actions at once, such as enhanced biosecurity, reducing risk from high traffic facilities and limiting movements, did see a greater combined impact on reducing disease transmission. Overall, the results of this project did reinforce the actions outlined in the plan as key measures to take as Manitoba works towards PEDV elimination.

Communication remains an overarching objective of the PEDV elimination plan. All the efforts made to further the elements of the plan in enhanced biosecurity, modelling, surveillance, etc has been strategically communicated to maintain momentum and awareness of the work that is being done and the rationale behind it. Our Manitoba swine veterinarians have been the main conduit of on the ground information to producers while Manitoba Pork has taken the role of preparing and extending larger sector wide updates.

Manitoba collaboration

The Manitoba approach to disease management has been one of collaboration. The key to success in a regional disease response is consistency, rapid response and information sharing among all parties. The Manitoba swine sector recognizes the importance of regional management of PEDV and the need to work together to achieve our goal. The sector's leadership in coming together, openly discussing and problem solving, is a testament to the commitment to the plan and to the greater good of our industry.

Information sharing was one of the first elements the sector led, even prior to the development of the PEDV elimination plan. The Manitoba Coordinated Disease Response (MCDR) was developed in 2017 by the Manitoba swine sector in response to the first major outbreak to PEDV. The MCDR is a password protected online platform where producers can sign up to agree to share their PEDV health status with other producers who have also signed on. This confidentiality agreement between the producer, Manitoba Agriculture and Manitoba Pork, allows for on-farm disease information as well as information on disease risk in the province to be shared and has greatly assisted in the regional disease response which Manitoba Pork and Manitoba Agriculture lead.

The commitment to collaboration has been evident in the sector's continued engagement on the discussion and advancement of PEDV elimination plan but also in how it can be expanded. The established working group is now taking the model we have built for creating the PEDV plan and discussing how we can apply this approach to other disease issues we face in Manitoba. The sector sees value in this not only for production diseases but also for planning and preparation activities related to potential foreign animal disease.

Looking to 2025

The Future of PEDV Working Group continues to meet to review progress and provide guidance on the implementation of the PEDV elimination plan. As immunity in the sow herds continues to decrease and the risk of PEDV to the sector remains, 2025 will be another test for the sector. Manitoba's commitment to eliminate PEDV in the province remains strong. Prevention of disease with enhanced biosecurity, especially in high-risk seasons will continue. Should PEDV cases arise, the plan will not be considered a failure, but a test of the approach that we have developed. Maintaining the step-wise elimination approach coupled with the dedication to continuous learning and improvement will help the sector navigate through to 2027 and beyond.

https://doi.org/10.54846/am2025/147

PEDV elimination: Estimating prevalence in the US herd

G. Trevisan[1]; C. A. Corzo[2]; G. Cezar[1]; M. Melini[2]; M. Kikuti[2]; X. Yue[2]; A. Moore[1]; S. Vadnais[2]; H. Naikare[2]; R. Main[1]; M. Prarat[3]; A. Pillatzki[4]; F. Matias Ferreyra[5]; C. Bowen[6]; D. Linhares[1]
[1]Iowa State University; [2]University of Minnesota; [3]Ohio Animal Disease and Diagnostic Laboratory;
[4]South Dakota State University; [5]Kansas State Veterinary Diagnostic Laboratory;
[6]Purdue University Animal Disease Diagnostic Laboratory

Background

Since its emergence in 2013, the porcine epidemic diarrhea virus (PEDV) has become endemic in the United States and continues to pose an animal health threat to the industry. In naïve breeding herds, PEDV causes pre-weaning mortality that can reach up to 100% for several weeks causing the loss of millions of pigs and dollars to the US swine industry. Fortunately, no zoonotic potential has been described for this virus.

Discussions around the opportunity to eliminate PEDV from the US herd have gained traction in the last few years. At the 2024 AASV meeting, a PEDV Elimination Task Force was assembled to assess what is needed to eliminate this virus and whether it is achievable. At the 2024 US SHIP House of Delegates and the US Animal Health Association meeting, a resolution was passed supporting the PEDV Elimination Task Force initiatives.

The current developments towards PEDV elimination highlight the need to better understand what is the current prevalence and incidence of such pathogen in the swine industry. To estimate the prevalence, a cross-sectional study could be designed to collect samples across multiple farms throughout the US states and use specific diagnostic tests. Farms included in this study would need to be selected randomly. Limitations to develop such a project would be the lack of a robust farm location database, consent of all producers for participation, and budget for testing all the farms. An additional potential testing approach could follow the model used during pseudorabies eradication and survey incoming pigs at the packing plant to determine the number of pigs that would have had previous exposure to the PEDV. In both testing scenarios, seasonality would need to be considered during the study design, determining a need to design distinct cross-sectional studies during the year.

Meanwhile, another approach is to use existing swine health databases that capture diagnostics test results and herd classification. The US is fortunate to have two distinct databases providing public information about PEDV activity, the Morrison Swine Health and Monitoring Project (MSHMP) monitors the incidence and prevalence of PEDV outbreaks in 1000+ participating breeding herds, representing ~3.6M sows, or ~55% of the US swine breeding herd, and, the Swine Disease Reporting System (SDRS) that aggregates and provides real-time information for PEDV RT-PCR test and testing results from six Veterinary Diagnostic Laboratories, providing insights from PEDV activity in all age groups and environmental samples.

Current PEDV status of the US breeding herd

Since the emergence of PEDV in 2013, the MSHMP network of participants has been voluntarily sharing their breeding herd status on a weekly basis. Given that MSHMP uses the number of participating farms as a denominator, disease metrics such as prevalence and cumulative incidence can be calculated weekly, providing the industry with measures of occurrence and elimination over time.

Breeding herd prevalence

Figure 1 summarizes the prevalence of PEDV in the US breeding herd since its emergence by categorizing breeding farms into different statuses. When a herd reports an outbreak, it is automatically placed in Status 1, indicating that the herd is undergoing an outbreak and it is weaning PEDV RT-PCR positive pigs. Once this herd has implemented intervention measures (eg, herd closure, feedback, cleaning and disinfection, wean down) to control the virus and related clinical signs are absent, and no virus is found via RT-PCR, the herd is considered to be in Status 2 or stable. In some cases, herds that have achieved Status 2 choose to maintain a certain level of immunity by exposing naïve gilts to PEDV before they are introduced into the sow herd; in these cases, herds are classified as Status 2FVI (field virus inoculation). Status 3 herds are those that have reopened to gilts and these have remained negative for at least 90 days. If the herd remains free of clinical signs for the following 6 months, it can be classified as Status 4.

Over time, the proportion of herds in each status has varied, but the proportion of herds in Status 4 has consistently increased, indicating that herds continue to remain naïve, and those that undergo a PEDV outbreak reestablish their negative status. Furthermore, the proportion of farms conducting FVI has also decreased over time.

Figure 1: Breeding herd PEDV status in the US between May 2013 and Oct 2024.

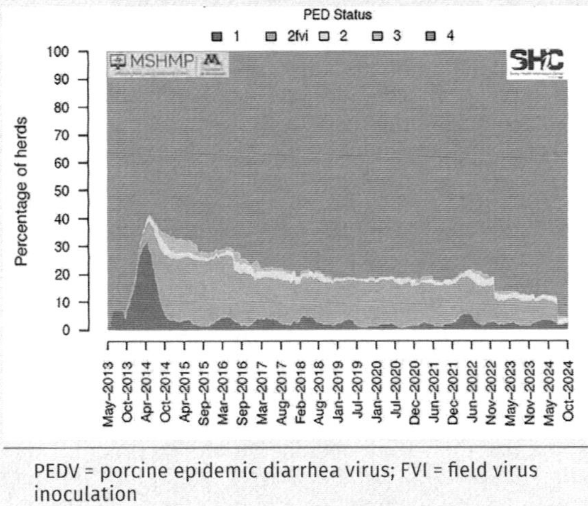

PEDV = porcine epidemic diarrhea virus; FVI = field virus inoculation

The proportion of Status 1 - herds undergoing an outbreak - follows a cyclical pattern, with most outbreaks occurring during the fall and winter seasons. It is important to note that the proportion of these herds remains low, typically varying between 0% and 5% depending on the time of the year. The proportion of actively shedding breeding herds remains low because of two main reasons, effective biosecurity and elimination of the virus. We recently reported on the speed at which breeding herds eliminate the virus as it has significantly improved; on average it took 23 weeks for an unstable herd to reach stability during the epidemic phase, whereas during the endemic phase, herds are achieving stability within 14 weeks.

Breeding herd cumulative incidence

One indicator of PEDV activity in the industry is the number of breeding herd cases reported to MSHMP. The metric MSHMP uses to measure this activity is the cumulative incidence, which is essentially the number of cases of a disease divided by the at-risk population within a specified time frame - in this case, a full year. This metric allows for the assessment of disease trends over time, facilitating data interpretation and aiding in the evaluation of progress.

Figure 2 shows the cumulative incidence since the virus emerged over 10 years ago, clearly illustrating the significant shift between the epidemic (ie, 2012-2013, 2013-2014) and the endemic phase (ie, 2014-2015 and 2024-2025). The US swine industry responded quickly during the early stages of the epidemic by establishing protocols to rapidly eliminate the virus and reduce risk of introduction to herds. This coordinated industry effort has successfully kept the virus at bay, with the cumulative incidence remaining under 10%, and in some years, below 5%. These metrics highlight the success of companies and practitioners in controlling and eliminating this virus.

Another important metric assessed is the frequency of outbreaks in herds over time. This metric helps us understand whether certain herds are experiencing frequent outbreaks, which may suggest they are exposed to a constant high risk of virus introduction. Overall, 46% of the herds in the MSHMP database have never reported an outbreak, while 37% have reported a PEDV outbreak once, and the remaining 17% have reported two or more breaks (Figure 3).

Figure 2: Breeding herd cumulative incidence in the US breeding herd between the 2012-2013 and 2024-2025 MSHMP years.

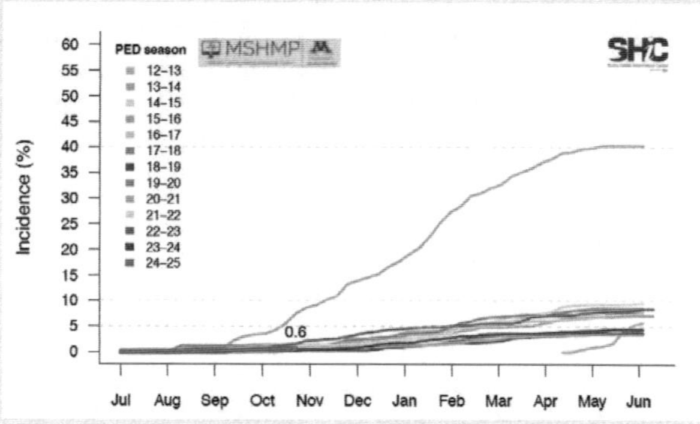

PED = porcine epidemic diarrhea; MSHMP = Morrison Swine Health and Monitoring Project

Figure 3: Frequency of PEDV outbreaks by herd in the 1,096 MSHMP participating breeding herds between 2013 and 2024.

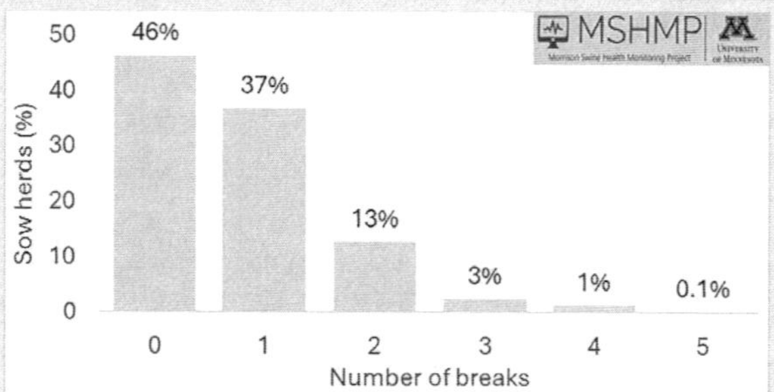

MSHMP = Morrison Swine Health and Monitoring Project; PEDV = porcine epidemic diarrhea virus

Real-time PEDV detection information

The SDRS data covers north of 98% of the porcine molecular testing done on porcine samples received at the NAHLN-accredited laboratories network. The SDRS information is dynamically presented by farm type, ie, boar stud, breeding herds, growing, etc, specimens, and regions, ie, state level. Additionally, SDRS has information about confirmed disease diagnoses from tissue submissions (analyzed by diagnosticians) at the Iowa State University VDL. One of the limitations of the SDRS data is that data is anonymized and no farm location beyond state-level is available for incidence or prevalence estimates. Another SDRS limitation is that some production systems perform in-house PCR testing for PEDV, and this information has not been shared with the SDRS yet.

SDRS information is made publicly available for stakeholder usage through curated monthly PDF reports, audio and video reports, and real-time updated interactive online dashboards. The SDRS also counts on an advisory board composed of key personnel from each participating laboratory, and veterinarians and producers across the US. The advisory board offers monthly input on the disease activity in their respective regions, which are included in the monthly reports.

After the PEDV epidemic years of 2013 and 2014, the percentage of positivity submission had a trend for decreased detection until 2021, when it reached the lowest overall detection at 8.44%. At the end of 2021, the activity started to go up in certain regions, and the detection of PEDV was at 11.73% in 2022, followed by another trend of decrease during 2023. It has been a pattern that every two or three years, the PEDV detection goes up.

In terms of specimens used for molecular testing of PEDV RNA, oral fluids, fecal swabs, feces, and environmental samples make up most of PEDV RT-PCR testing (Figure 4). The specimens with the highest positivity for PEDV are labeled as environmental, swabs, fecal swabs, and feces (Figure 5). Illinois have been having detection between 5 to 10%. Iowa, Missouri, Oklahoma, Kansas, and North Carolina have been having consistent PEDV detection above 10%. Worthily, Missouri and Iowa haven't been able to keep PEDV detection lower than 15% since 2021 (Figure 6). Interestingly PEDV detection in Kansas has been driven by positivity in wean-to-market with consistent spikes every other year. Kansas PEDV positivity in sow farms has been below 5% since 2019.

Figure 4: Ten most frequently used specimens for porcine epidemic diarrhea virus RT-PCR testing over time. The number of submissions by specimen is presented in thousands.

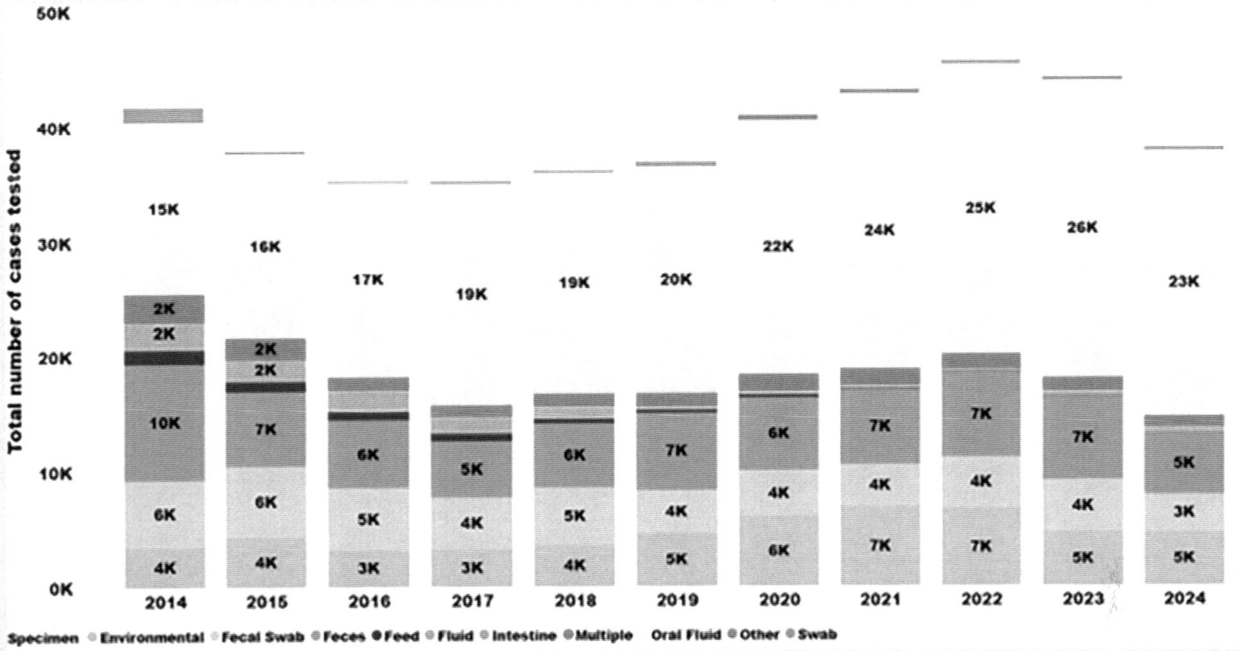

Figure 5: Porcine epidemic diarrhea virus RT-PCR testing percentage of positive results by specimen. Results were aggregated at the case level, which may contain more than one sample by specimens at a submission.

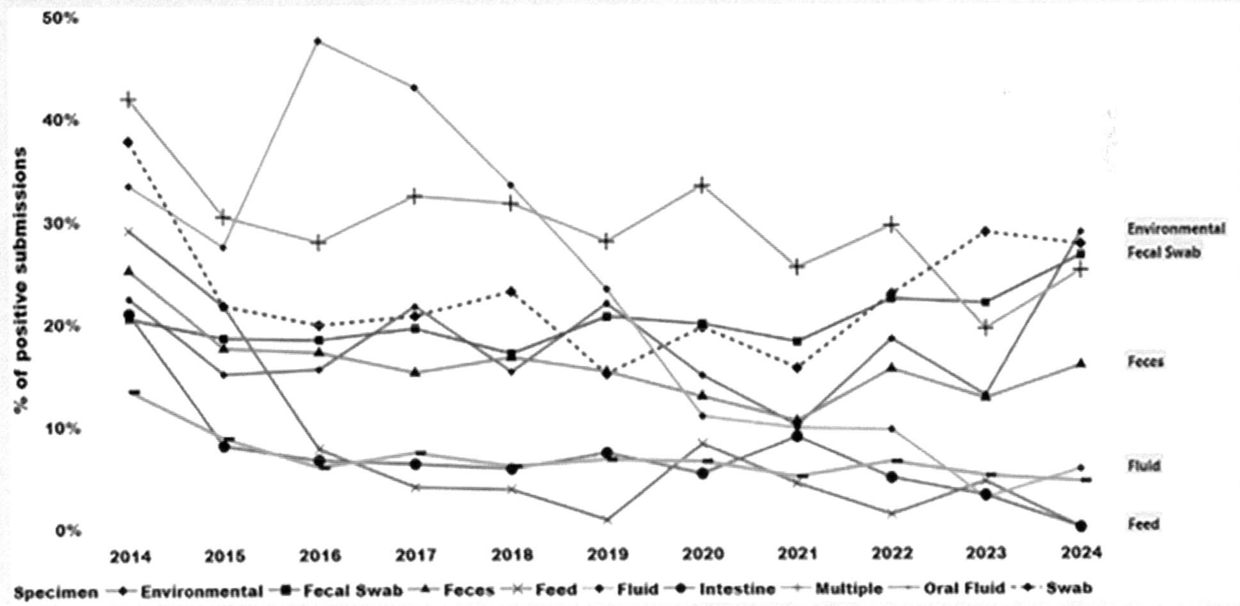

Figure 6: Porcine epidemic diarrhea virus RT-PCR percentage of positive submissions by state.

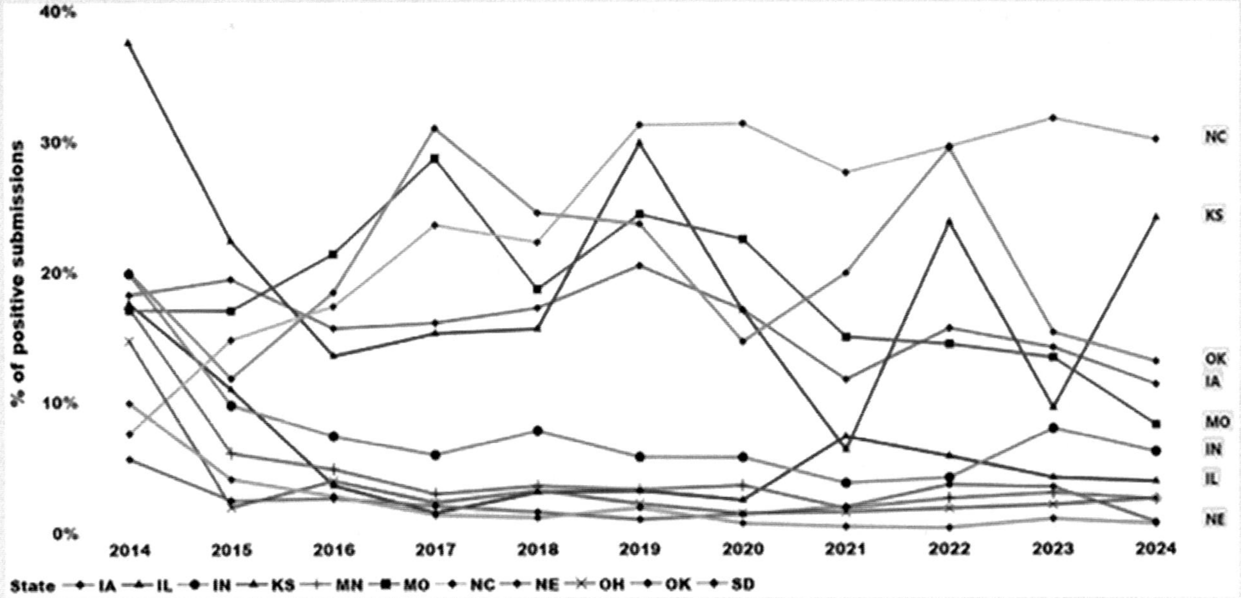

Another strategy that helps understand how much PEDV testing is being performed across states is by normalizing the number of submissions tested by the state swine inventory. The USDA NASS annual report is a source of pig inventory data by state. Using the USDA NASS total pig inventory as the denominator, a rate for the number of submissions by 100,000 pigs was calculated. Oklahoma has been the state with the highest PEDV testing per 100,000 pigs. Ohio, South Dakota, Kansas, Minnesota, and Missouri test between 75 and 100 submissions per 100,000 pigs. Another group of states, ie, Indiana, Illinois, Nebraska, and North Carolina, test between 20 and 45 submissions per 100,000 pigs (Figure 7).

Differences in terms of testing across states can be a factor of large inventory, eg, Iowa, the presence of large breeding or finishing sites, and North Carolina by the usage of in-house testing. Nevertheless, it is important to acknowledge that in-house testing for PEDV performed by production systems is not included in this analysis, which might influence the rate of testing and positivity rate in some states.

In sum, SDRS provides real-time information on PEDV detection across different regions, farm types or age groups, and specimens. Notably, there is a regional difference in terms of PEDV detection, which can also be translated into potential differences in prevalence across US states. States that have consistently lower PEDV positivity are very likely to have a lower prevalence when compared with states with higher PEDV positivity rates. Other states have been repetitively having high levels of PEDV detection with potentially higher prevalence in those areas. Notably, and pending industry interest and support, a capacity to monitor PEDV sequencing data can be included in the SDRS following a model already present for PRRSV ORF5 sequences. PEDV genetic sequencing data would open the opportunity to track PEDV genetic evolution over time and regions.

The regional differences and the seasonal patterns of PEDV RNA positivity across states, age groups, and over time have been shown to follow a consistent behavior over the years. This allows the SDRS to not only describe the detection patterns but also build forecasting algorithms and rapidly detect and communicate with stakeholders relevant, non-expected spikes in detection in respective regions, age groups, farm types, or specimens. Rapid detection of increased PEDV activity is a critical first step toward rapidly implementing appropriate contingency practices for response and recovery at the farm, truck wash, or other relevant premises.

MSHMP and SDRS future collaboration towards a national PEDV elimination

At the time of writing, both SDRS and MSHMP teams were developing a research plan to provide the US swine industry with data focused on national herd prevalence for PEDV. In addition, plans to use both databases to better understand national occurrence trends were being developed with the goal of enhancing our epidemiologic understanding of this virus and thus provide key data that can aid in the future as national elimination programs are assembled.

Both teams have drafted the following objectives:

1. Conduct an exploratory data analysis to determine the prevalence of PEDV in growing pig herds in the US using the MSHMP pig database
2. Explore PEDV positivity-rate and cumulative incidence relationships between SDRS and MSHMP
3. Enhance SDRS PDF reports to provide PEDV levels of detection across different states and specimens
4. Perform a case study on the macroepidemiological factors associated with the disappearance of transmissible gastroenteritis virus from the US swine industry

Figure 7: Number of submissions tested for porcine epidemic diarrhea virus RNA by RT-PCR by 100,000 pigs in each monitored US state.

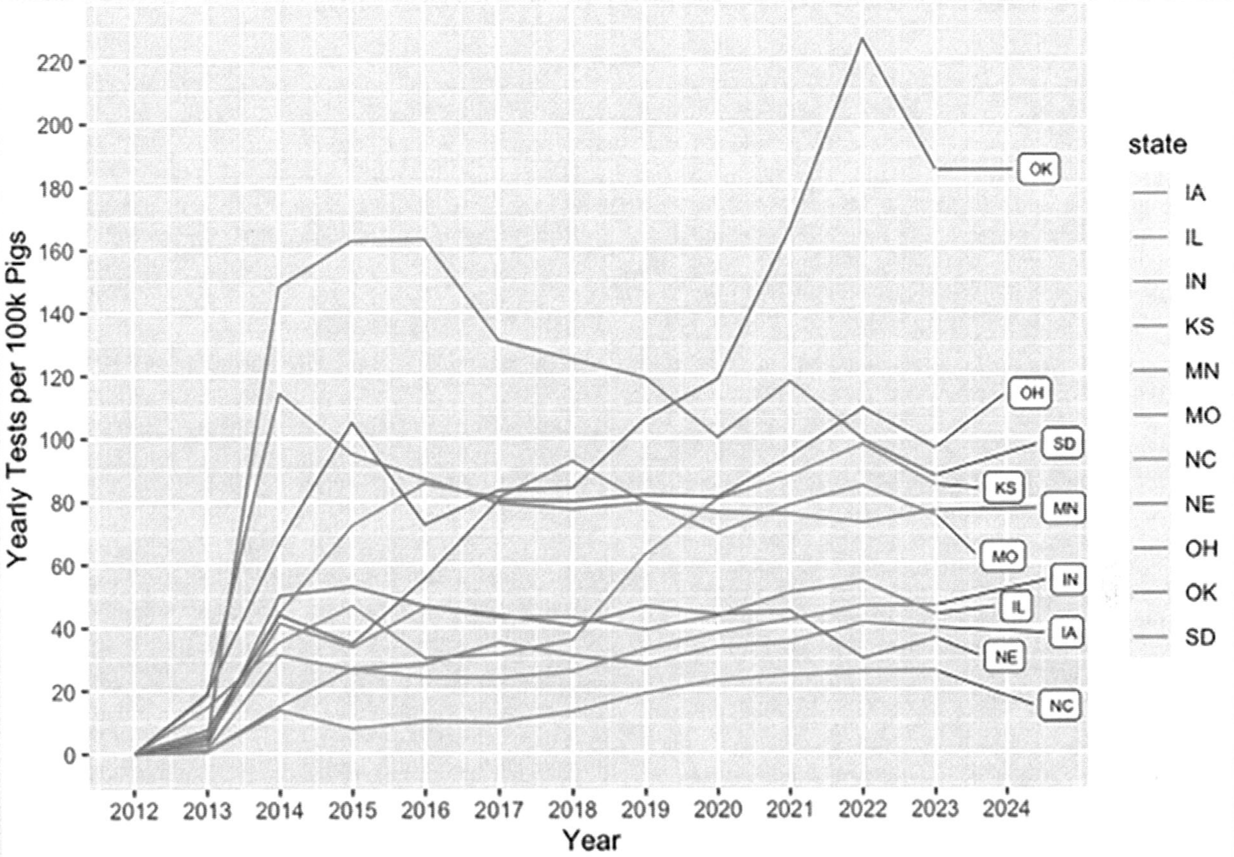

Together, SDRS and MHSMP will support a regional or national PEDV elimination programs, by providing real-time information on where PEDV is detected/reported (by state) and where activity is increasing or decreasing. Regions with lower viral activity levels could be initial targets for PEDV elimination, while regions with higher viral activity could be prioritized for gap analyses to identify the factors contributing to higher virus circulation in those regions.

https://doi.org/10.54846/am2025/149

Comparison of elimination protocols for PEDV, PRRRS and *Mycoplasma hyopneumoniae*

Paul Yeske
Swine Vet Center, St Peter MN

Introduction

There is a renewed interest in pathogen elimination programs by industry. Especially now for porcine epidemic diarrhea virus (PEDV) with the formation of the AASV and NPB PEDV elimination task forces, resolution by US Swine Health Improvement Plan to explore certification of PEDV and move towards elimination and resolution at USHA supporting the exploration of PEDV elimination. Also, in looking at recent SDRS reports there appears to be a trend in *Mycoplasma hyopneumoniae* elimination as well Figure 1.[1] Porcine reproductive and respiratory syndrome (PRRS) elimination continues to be done on a regular basis to control the virus in herds. These pathogen elimination programs are being done on a regular if not daily basis in the industry.

The first goal is to have bioexclusion and keep these pathogens and diseases from ever entering the herds. Bioexclusion is not just one thing but several layers trying to cover the weakness of each intervention (Swiss Chess Model).[2] A newer consideration is biocontainment which looks at the positive farms to help with control as well, planning this ahead of an outbreak to reduce this as potential source of pathogens going forward (Canadian PEDV Elimination Plan).[3] As with a disease outbreak there is a progression that herds work through. Starting first and foremost confirming the diagnosis, then moving to control the disease as much as possible and then deciding if the herd wants to move to continue with control measures or move on to an elimination strategy. There are costs both real and opportunity that can be calculated to better understand the impact of these decisions going forward.

There are good elimination procedures and resources available (ie, *Mycoplasma hyopneumoniae* road to success)[4] that have been done that have good track records of success that can also be a guideline for herds that may be experiencing these for the first time.

Basics of elimination

There are several different elimination strategies[5] that can be used for these pathogens.

- Depopulation and repopulation
- Herd closure
- Test and removal
- Medication
- Vaccination

Depopulation and repopulation

This is the tried-and-true method of pathogen elimination allowing for multiple pathogens to be eliminated at the same time. This method historically has the highest success rate and is nearly 100% effective for all 3 of these pathogens if procedures are followed correctly. The downside to this procedure is that it is usually the most expensive approach. This can be reduced by incorporating an off-site breeding project to reduce the number of down days in the system.

Figure 1: Trends supporting effective *Mycoplasma hyopneumoniae* control in breeding herds.

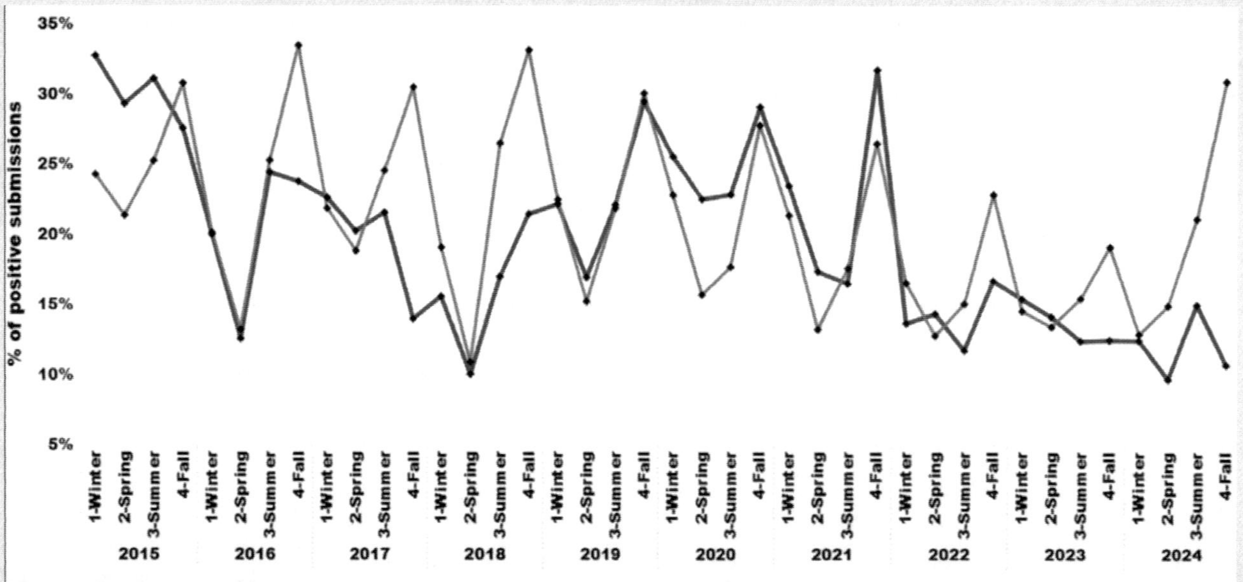

Herd closure

Commonly known as load, close and homogenize. Allowing the herd's immune system to help with stabilizing the population and to clear the pathogen to allow negative replacements to be brought back into the herd. It is most successfully used on farrow to wean farms. For farrow to finish farms this is more difficult to achieve and generally requires a partial if not total depopulation of nursery and finishing space. This is an effective tool for all 3 diseases and can be done simultaneously.

Loading is making sure that there is adequate inventory to maintain the herd in the for the time of closure. There are times when compromises must be made due to limitation in facilities for numbers or in the case of *Mycoplasma hyopneumoniae* some of the replacements haven't even been born yet.

Time of closure is the key difference with the different pathogens. PEDV is the shortest with 90-120 days, PRRS a minimum a minimum of 210 days (more recently with the new viruses circulating many going up to 300 days or longer) and *Mycoplasma hyopneumoniae* at minimum of 240 days. The day of the last animal exposed is day 0 and the counting goes forward from there.

Closure refers to no additions to the inventory gilts or boars from the time of herd exposure to the pathogen until diagnostics testing determines that herd is ready to be opened to new additions.

Homogenization depends on the pathogen and on how it is to be done. PEDV is typically done through feedback to sow herd, PRRS serum exposure and *Mycoplasma hyopneumoniae* through aerosol exposure. The goal of this step is to make sure all animals are exposed to the pathogen and allow the immune system the time it needs to clear the pathogen and not shed to negative contact animals.

Having a detailed plan for the process is very important to communicate with everyone connected to the farm so that areas don't get missed and the projects can be successful. Figure 2[5] details herd closure for *Mycoplasma hyopneumoniae.*

Test and removal

This procedure was used successfully and widely with the pseudorabies (PRV) elimination programs. With PRV there were good vaccines that reduced shedding and where differentiable tests to allow for this to be successful. In looking at PEDV at the current time these vaccines aren't available so really isn't an option. PRRS there may be an opportunity with the tonsil-oral scrubbing (TOSc) sampling to be done in combination with herd closure to shorten the window. *Mycoplasma hyopneumoniae* would be a similar case to augment the herd closure plans and reduce the time of closure by removing shedding animals.

Medication

This procedure has been done with *Mycoplasma hyopneumoniae* successfully, just not as high of success rate as other methods with success approximately 50% of the time. For PEDV and PRRS currently there are not effective treatment options to do these pathogens.

Vaccinations

PRV vaccination was an effective tool and just had to have the rigor to control the shedding and had the differential test as well. Although vaccination can be effective tools in the herd closure process but as a stand-alone program currently for PEDV, PRRS and *Mycoplasma hyopneumoniae* are not good enough. There may be potential in the future as there are breakthroughs in vaccinations.

Diagnostics

This is the important part of all these programs especially the herd closures. The use of these multiple tests helps to determine if the process has been successful and if there needs to an extension to the program to be successful. Depending on the pathogen will change the testing methods used as well as testing results to answer the desired questions. The use of rectal swabs is helpful for PEDV due to less environmental contamination. For PRRS, may need to answer the question, is the virus still circulating in the sow herd or in the farrowing, the use of tongue tips on stillborns can be helpful as well as TOSc samples from sows.

Conclusions

There are many similarities as we look at these pathogens PEDV, PRRS and *Mycoplasma hyopneumoniae* that exist. Depopulation and repopulation can work for many different pathogens at the same time but certainly across these 3 example pathogens. Herd closure is the most popular for these 3 pathogens as well. Not every elimination plan will go as planned so we must remain flexible and use the diagnostic tools to help determine the progress or lack thereof to help in finding the directions necessary to move forward to success. Understanding the principles is the most important part of the process and learning to be adaptable are good skills as we look at elimination of any pathogens as we move forward.

References

1. Trevisan G. Swine Disease Reporting System monthly report 81.
2. Reason, James (1990-04-12). "The Contribution of Latent Human Failures to the Breakdown of Complex Systems". Philosophical Transactions of the Royal Society of London. Series B, Biological Sciences. **327** (1241): 475–84.
3. Hamblin J. Manitoba Pork PEDV Elimination Plan. US SHIP house of delegates.
4. Sponheim A., Fano E., Pieters M. Roadmap to Mhp Eradication.
5. Yeske P. Eradication of Mycoplasma hyopneumoniae from pig herds; Mycoplasmas in Swine, 247-266, 2021.

Figure 2: *Mycoplasma* eradication protocol.

Mycoplasma eradication herd closure protocol

Farm		Example
Date		12/27/2020

Date	WK	Activity	Product	Dosage	Withdrawal	Completion date	INT
12/21/2020	52	Herd closure	N/A	N/A	N/A		
12/28/2020	53	Gestation barn exposure via aerosol	N/A	NA	N/A		
1/18/2021	4	Whole herd vaccination (sow unit & gilt developmen	M + Pac	2 cc	21 days		
1/22/2021	4	Wash gestation barn	N/A	N/A	N/A		
1/22/2021	4	Laryngeal swabs / tracheal swabs and blood samples of sows in herd and GDU to confirm exposure	N/A	N/A	N/A		
1/22/2021	4	Measure sows for whole herd injection with Draxxin	N/A	N/A	N/A		
1/27/2021	5	Myco exposure complete (when confirmed)	N/A	N/A	N/A		
1/29/2021	5	1st whole herd injection with Draxxin (all pigs on site sows, boars, gilts growing gilts & suckling piglets)	Draxxin	1 mL per 88 lbs	33 days		
1/30/2021	5	Start injecting piglets day 1 with Draxxin 25 at (0.2 mLs dose per piglet)	Draxxin 25	0.2 mL per piglet	33 days		
2/12/2021	7	2nd whole herd injection with Draxxin (all pigs on site sows, boars, gilts growing gilts & suckling piglets)	Draxxin	1 mL per 88 lbs	33 days		
2/13/2021	7	Start injecting piglets day 14 with Draxxin 25 at (0.5 mL dose per piglet)	Draxxin 25	0.5 mL per piglet	33 days		
2/26/2021	9	Stop day 1 injections to piglets with Draxxin 25	N/A	N/A	N/A		
3/12/2021	11	Stop day 14 injections to piglets with Draxxin 25	N/A	N/A	N/A		
3/19/2021	12	Whole Draxxin injection elimination completed	N/A	N/A	N/A		
3/19/2021	12	Herd continues to be closed as precaution	N/A	N/A	N/A		
3/26/2021	13	Testing tracheal swabs to see if Draxxin program was successful	N/A	N/A	N/A		
4/19/2021	17	Whole herd vaccination (sow unit & GDU)	M + Pac	2 cc	21 days		
6/25/2021	26	Laryngeal swabs / tracheal swabs gilts in GDU test all pens and keep track of pigs so that they can be retested	N/A	N/A	N/A		
7/19/2021	30	Whole herd vaccination (sow unit & GDU)	M + Pac	2 cc	21 days		
7/30/2021	31	Wash the gestation	N/A	N/A	N/A		
8/6/2021	32	1st whole herd injection with Draxxin (all pigs on site sows, boars, gilts growing gilts & suckling piglets)	Draxxin	1 mL per 88 lbs	33 days		
8/7/2021	32	Start injecting piglets day 1 with Draxxin 25 at (0.2 mLs dose per piglet)	Draxxin 25	0.2 mL per piglet	33 days		
8/20/2021	34	2nd Whole herd injection with Draxxin (all pigs on site sows, boars, gilts growing gilts & suckling piglets > 2 weeks of age)	Draxxin	1 mL per 88 lbs	33 days		
8/20/2021	34	Start injecting piglets day 14 with Draxxin 25 at (0.5 mL dose per piglet)	Draxxin 25	0.5 mL per piglet	33 days		
8/20/2021	34	Start chlortetracycline in gestation and lactation			14 days		
9/3/2021	36	Last piglets treated at birth	N/A	N/A	N/A		
9/10/2021	37	Stop chlortetracycline in gestation and lactation			14 days		
9/17/2021	38	Last piglet treatment at 14 days of age	N/A	N/A	N/A		
9/24/2021	39	Elimination completed	N/A	N/A	N/A		
9/25/2021	39	Negative gilts enter breeding herd	N/A	N/A	N/A		
9/25/2021	39	Date pigs are expected negative	N/A	N/A	N/A		

240	Days since the herd was confirmed (+)		
271	Days since the herd was exposed		
278	Days since closure		
374	Last gilts birth date	9/16/2020	2 months at closure
304	Gilts 10 months of age	7/17/2021	

https://doi.org/10.54846/am2025/150

Advancing trailer sanitation and biosecurity in swine transportation

E. S. Magalhães[1], DVM, MS, PhD; D. C. L. Linhares[2], DVM, MBA, PhD; R. Main[2]
[1] Department of Animal Science, Iowa State University, Ames, Iowa;
[2] Veterinary Diagnostic and Production Animal Medicine Department, Iowa State University, Ames, Iowa

Introduction

Biosecurity in the swine industry is a critical concern, particularly during transportation, where livestock trailers serve as a significant risk factor for transmitting diseases such as porcine epidemic diarrhea virus (PEDV). This research introduced innovative monitoring truck wash tools to enhance sanitation compliance, and evaluated alternative sanitation methods for washing PEDV-positive market trailers. Altogether, these efforts align with biosecurity improvements toward PEDV control and elimination.

The process of building datasets

The first study introduced the truck automated sanitation classification (TASC) platform, a technology-driven system designed to automate compliance monitoring and data collection in trailer sanitation. The TASC platform compared three different methods for recording truck wash events and deliveries of pigs at the packing plant: GPS geofencing records, app-based tools to document wash events, and in-site check-in records. From this data, algorithms were built to combine, standardize, and analyze the raw data provided from each tool to verify sanitation effectiveness and generate compliance reports. By integrating GPS data with photographic evidence of trailer cleanliness, the platform enhanced accuracy in identifying gaps in sanitation practices and improved transparency in compliance monitoring. The findings revealed that while all three methods had their strengths and limitations, GPS-based tracking showed higher accuracy in documenting truck wash events and deliveries at packing plants compared to the other methods. The platform's ability to generate detailed reports on wash frequency, time, and effectiveness demonstrated its potential as a valuable tool for decision-making and scaling biosecurity practices across production systems.

In addition to evaluating the TASC platform, the second research tested alternative trailer wash methods in a separate study. Five trailer cleaning treatments were tested in model-size trailers contaminated with PED-positive feces, where back and forward foot-traffic from inside the truck to the farm-site area was simulated. After foot-traffic simulation, viral presence was quantified using qPCR to measure the viral load on the trailer and farm-site surfaces, and bioassays were conducted by inoculating PEDV-naïve pigs with material from the farm-site areas to assess infection risk. Each treatment contained five replicates, and the five treatments were: Positive control (no wash-trailer inoculated with PEDV-positive material), scrape and bake (thermo-assisted drying and decontamination, or TADD), volume hose wash and disinfect, power wash and disinfect, and negative control (no wash, trailer inoculated with PEDV-negative material). This study simulated real-world conditions by conducting foot traffic between the trailer and the farm site.

Among the methods evaluated, the power wash and disinfect approach demonstrated high efficacy, achieving a 99.99% reduction in viral genomic copies on trailer and farm-site surfaces. Similarly, the volume hose wash and disinfect method achieved high reductions in viral load, with a 99.83% decrease in genomic copies. These water-based methods prevented PEDV transmission in pigs through bioassay for all replicates, which highlights its potential as a scalable and effective sanitation strategy. In contrast, the scrape and bake method was not effective. Even though there was 98.46% reduction in viral presence, scrape and bake was ineffective in inactivating PEDV, as evidenced by infections in bioassay pigs for four of five replicates. All negative control replicates were negative in qPCR and Bioassay, and the positive control group had consistent positive results on both trailer and farm-site surfaces, and all pigs were positive in bioassay, thus reinforcing the necessity of implementing robust cleaning protocols.

Conclusions

This research highlights the importance of effective trailer cleaning methods and technology integration in reducing the risk of inter-premises disease transmission during swine transport. Water-based cleaning methods, particularly power wash and disinfect and volume hose and disinfect, proved to be the most effective in inactivating PEDV contamination and preventing transmission. Also, the TASC platform showcased the potential for automating and standardizing information collected from multiple technologies recording log-key events concerning truck wash compliance. Combining the application of these strategies and technologies provides an opportunity to elevate biosecurity standards for market haul transportation, which reduces disease-associated costs related to endemic pathogens (eg, PEDV) and enhances the swine industry preparedness against foreign animal disease (FAD). Future efforts should focus on industry-wide adoption of these practices and further investment in sanitation infrastructure and automated compliance systems to improve efficiency and scalability.

https://doi.org//10.54846/am2025/151

Reduction of porcine epidemic diarrhea virus (PEDV) transmission: Transportation case study

Pete Thomas, DVM, MS
Iowa Select Farms, Iowa Falls, Iowa

Introduction

Disease prevention is a critical component of successful swine production operations. Porcine epidemic diarrhea virus (PEDV) is one of the most impactful diseases at the sow farm level. In commercial wean-to-finish, the impact of disease has been less characterized. Keeping pigs negative in commercial wean-to-finish production settings can be not only difficult, but also hard to understand the production impacts and associated financial implications. Disease prevention, particularly with PEDV, relies on good biosecurity practices and associated investments.

Iowa Select Farms (ISF) has most of its farms located in swine dense regions of Iowa. In fact, many of the sow farms are in some of the most swine dense counties in the United States. Commercial wean-to-finish farms across the state see large seasonal increases in PEDV outbreaks during the colder months. Investigating and characterizing outbreaks is critical to understanding how disease is being spread and what interventions are impactful.

Contaminated market trailers have been implicated as a source of PED contamination. One study demonstrated an increase in trailers that tested negative at arrival to the packing plant testing positive after unloading.[1] In addition, many producers have experienced the largest incidence of PEDV in their production systems occurring during the marketing phase.

System background

During the time period of October-April 2018-2022, we had 30% of our groups break with PEDV during marketing.

In addition, we have seen re-break rates in new groups of wean-pigs as high as 58%.[2] This has been primarily attributed to groups that broke during marketing and failure to effectively clean and sanitize the premise before the next pigs arrive.

Our system developed two different models to look at the impact of PEDV introduction into wean-to-finish populations. The first model looks at the cost of PED introduction during interim cuts due to dirty trailer or load-out crew contamination of a site. Prior to using this model for PED, it was used to evaluate the impact of ractopamine fed to pigs that were light on first cuts. During the winter of 2016/2017, the model found that pigs infected with PED during marketing weighed 1.98kg less at market and took an average of 4.6 days longer to reach market.[2]

The second model looks at the impact of re-breaking a site of wean-pigs with PEDV. We were able to calculate the re-break rate of wean-pig sites that placed pigs in a barn that had broken during marketing in 2018/2019 at 58%. Mortality of groups placing from 2018-2022 demonstrate a 3.1% increased mortality when breaking during the early wean-to-finish period versus groups that did not.

In order to reduce the impact of PEDV on our system, biosecurity investments were considered. These included: Washing all trailers prior to final cuts, hiring a crew supervisor, and whitewashing all sites. After reviewing the costs and return on investment expected with each intervention at the time, we decided to focus on preventing PED re-breaks in wean pigs by whitewashing all positive facilities and hiring a crew supervisor to oversee load-crew biosecurity and wash inspections of positive sites. In the first year, we saw a reduction in re-break rates from 58% to 19.6%. In the winter of 2021-2022 we saw a re-break rate of 11.5%.[3] The overall PEDV break rate in pigs during the nursery phase is represented in Figure 1.

Transportation case study

Despite the improvement in prevention of wean-pig breaks, we continued to deal with a high percent of our sites breaking during marketing. Recorded break-rates ran between 23%-38% from 2018-2023 (Table 1). As we entered the fall of 2023, we discussed washing market trailers after every load. We anticipated multiple benefits: 1) Substantial reduction in PEDV breaks in our commercial wean-to-market system. 2) Enhance our credibility as we took additional steps in improving wean-to-finish biosecurity. 3) Reduce transmission of other diseases. 4) Reduce the area pressure of PEDV on our breeding stock farms in the same geography.

As we explored options to accomplish washing and disinfecting all trailers between loads, we came to the conclusion that the most practical option was to do a high-volume wash and disinfect at truck-wash locations closest to the plants. We also utilized our internal market wash location with high pressure washing in about 25% of the total washes. It took several weeks during Q4 of 2023 to hit 100% implementation, which occurred on week 46.

Figure 2 demonstrates the week-by-week case rate during the winter of 2022/2023 compared to the winter of 2023/2024. Overall, there was a reduction from 35.6% to 5.8% PEDV break rate in groups during marketing. There were no differences observed prior to week 46, However, we saw an 11X reduction in the PEDV break rate from week 46 of 2023 to week 23 of 2024. This occurred after trailer sanitation had been 100% implemented. In addition, we only had 2 wean-pig sites out of a total of 352 placed during the time period, that re-broke with PEDV after being positive during marketing.

Discussion

Biosecurity is an essential part of disease mitigation and production efficiency. There are not any vaccines currently on the market that are effective at preventing infection and clinical disease in naïve pigs. Biosecurity is effective and critical to PEDV prevention and potential elimination from a production system or geography. In our system, contaminated market trucks were by far the most common way that PEDV was being introduced into our production sites. The benefits of utilizing clean trailers have been discussed, and each production system needs to weigh these benefits against the cost.

Figure 1: Porcine epidemic diarrhea virus break rates in pigs during the first 60 days placed on feed.

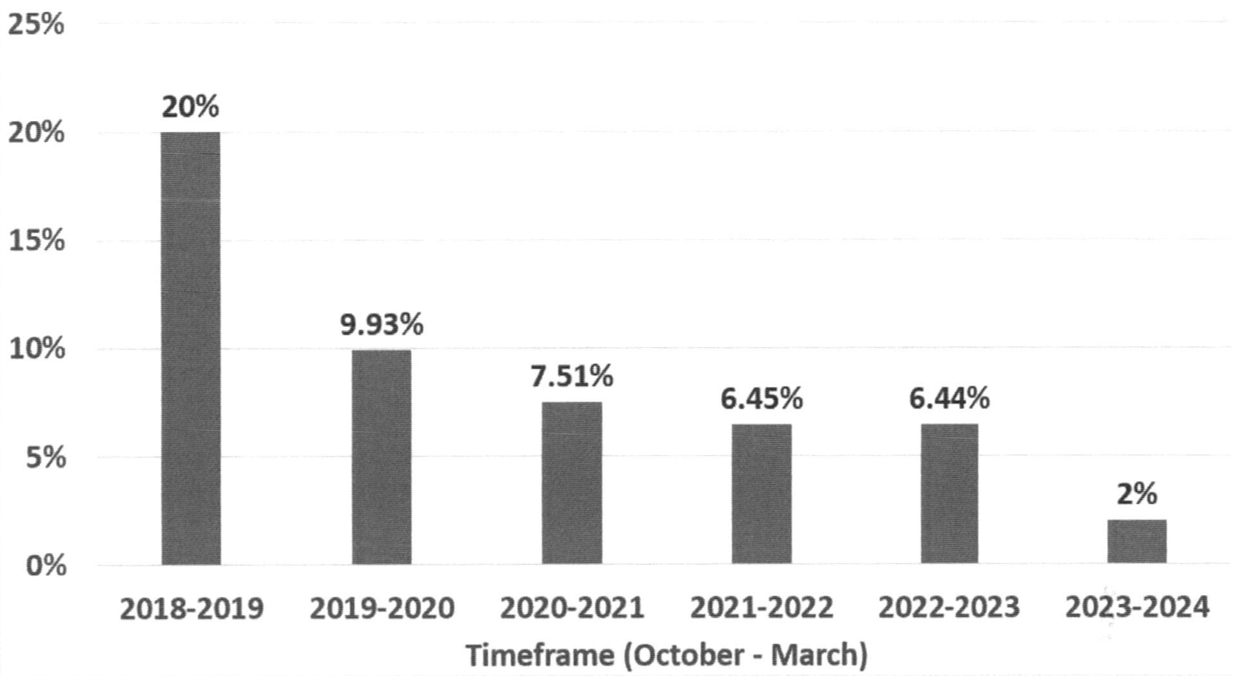

Table 1: Historical porcine epidemic diarrhea virus break rates by category

	2018-2019	2019-2020	2020-2021	2021-2022	2022-2023	2023-2024
Total Sites Coronavirus Positive	198	325	283	238	289	80
Marketing break-rate	23% recorded	38.32%	33.85%	26.86%	35.6%	5.8%
Sites placed break-rate (total)		12.57%	7.23%	9.25%	9.0%	2.8%
Wean sites placed break-rate	20%	9.93%	7.51%	6.45%	6.44%	2%
Feeder pig sites placed break-rate		14.16%	6.98%	14.04%	11.8%	3.5%
Protected sites break-rate	18%	20.51%	26.00%	15.29%	14.6%	—
Positive at prior turn, re-filled with wean pigs break-rate			21.43%	11.40%	17%	8.7%
Positive at marketing, had a wash audit break-rate		23.85%	15.00%	16.67%	4.4%	7.5%

Figure 2: Enteric coronavirus breaks during marketing by calendar week.

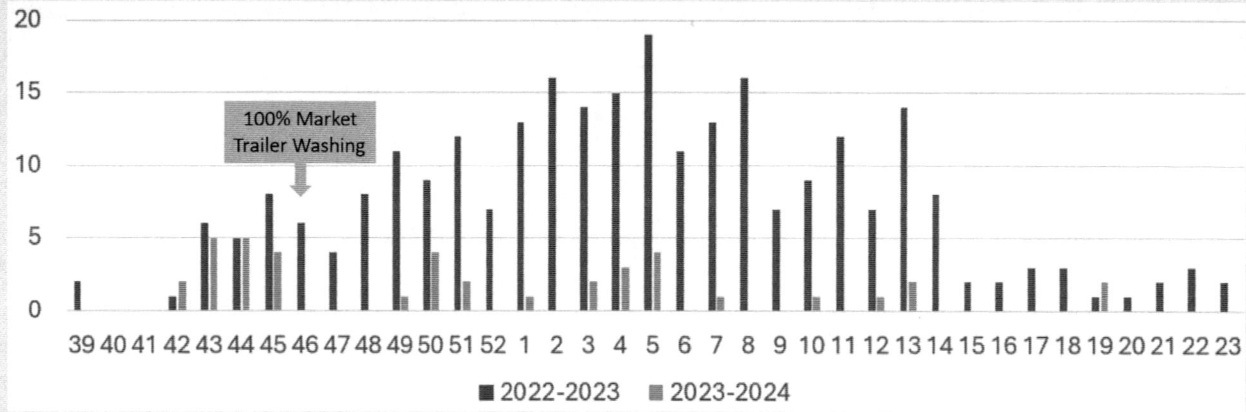

References

1. James Lowe, Phillip Gauger. Role of Transportation in Spread of Porcine Epidemic Diarrhea Virus Infection, United States. *Emerg Infect Dis.* 2014 May;20(5):872–874.

2. Pavlovic, L. Experiences managing wean-to-finish PEDV in a production system. *Proc of the American Association of Swine Veterinarians annual meeting.* 2018:485-487.

3. Thomas, P, Coleman K. Impactful investments: measuring return on investment for biosecurity. *Proc of the American Association of Swine Veterinarians annual meeting.* 2023:344-345.

https://doi.org/10.54846/am2025/152

Swine pharmacology update for the individual pig

J. S. Smith
University of Tennessee, Knoxville, Tennessee

Introduction

Presentation of pig patients to large animal clinicians is increasing in numbers of production, companion as well as sanctuary/rescue animals. While not as covered by most veterinary training programs as goats, sheep, cattle, llamas, alpacas and horses, these swine patients are often in need of specific care beyond production diseases. With both production and companion animal aspects to this species, there are challenges for their care. The goal of this presentation is to review pharmacology for pigs with respect to species issues, regulatory considerations as well as new potential therapies.

Species-specific considerations for swine

Despite the common use of pigs as production, companion, and research animals, the nature of drug metabolism in the species is not completely known. Gastrointestinal physiology is similar to humans with respect to gastric pH and emptying, and the skin of the flank and behind the skin has similarities to human skin.[1] While renal metabolism in the pig has not been exhaustively studied, it is recognized that the renal tubules contain high concentrations of cytochrome p450 enzymes. Phase II hepatic metabolism is where recognized differences exist for the pig with respect to sulfate conjugation. While deficiencies exist here with sulfation, the pig makes up for these by having high activities of glucuronidation and acetylation.[2] While the potential for genotype-specific variation exists, there is currently a paucity of descriptions of drug classes that present challenges for metabolism in the pig.

Drug administration

Drug administration can be challenging in individual pigs. While most injection techniques are described for commercial pig breeds these principles can be extrapolated to individual patient pigs in large animal referral settings. For intramuscular injections needle should be 18-20 G in size and 5/8-1/2" in length for piglets; 16-18 G and 5/8-3/4" for nursery pigs; 16 G and 1" for finishing pigs; and 14-16 G and 1-1 1/2" for adult pigs (**https://swine.extension.org/injection-reference-for-swine/**). Subcutaneous injection needle selection is based on needle length with ½" recommended for finishing pigs, ¾" recommended for finishing pigs, and 1" for sows.

The National Pork Board has described a triangle on the neck region as ideal for injection. The author is aware of references describing injections in pigs in the epaxial and other muscles of the posterior chain, but has also observed abscesses for injections given in those regions. For small pigs being given a subcutaneous injection, loose flaps of skin behind the elbow and flank have also been described. In the hospital setting the author has had success utilizing various slings for restraint prior to injection. The Panepinto sling (Panepinto & Associates) and Feet First Chute (ZinPro) are 2 options for swine restraint that can allow for ease of injections.

Labelled medications

While there are multiple classes of drugs labelled for use in swine in the United States, as displayed in Table 1, there are multiple limitations. For example, only one approved drug has any analgesic properties (flunixin meglumine) although, it is not labelled as an analgesic. Additionally, several of these approved medications are no longer actively produced in the United States. An additional consideration is that the majority of label approvals are for herd-focused conditions, such as respiratory disease, and not necessarily for all conditions encountered for individual pigs. Swine clinicians should be familiar with provisions of extralabel drug usage in the United States.

Pharmacology update for the individual pig

Several drugs have recently been investigated that would have potential for individual pig patients in a large animal specialty hospital setting. These include: Transdermal flunixin meglumine (Banamine Transdermal™), maropitant (Cerenia™), grapriprant (Galliprant™), clonazepam, and afoxolaner. All of these drugs would be considered extralabel drug usage in pigs in the United States.

Flunixin meglumine is a nonsteroidal anti-inflammatory currently labelled for use in pigs by intramuscular injection. While the pharmacokinetics of the injectable formulation are well known, and intramuscular administration is more bioavailable than oral administration (76% vs 22%),[3] the presence of a new transdermal formulation of flunixin raised potential applications for individual swine patients. Unfortunately, the transdermal flunixin formulation was shown to be poorly bioavailable with 1.55% bioavailability in sows[4] and 7.8% in piglets.[5] At this time because of this low bioavailability transdermal flunixin does not appear to be an efficacious route of delivery in pigs.

Maropitant is a synthetic neurokinin type-1 receptor antagonist which inhibits the central effect of Substance P.[6] Used heavily as an anti-emetic in canine practice it has also been explored for equine use,[7,8] and anecdotal perioperative use for swine has been discussed.[9] Preliminary investigation of maropitant in pigs has determined that an intramuscular administration of 1.0 mg/kg achieved concentrations higher than reported for the effective use in dogs and cats. However, dosing with the oral tablet in 2 pigs only achieved bioavailability of approximately 15 and 27%, indicating that higher doses may be necessary when administered orally to pigs compared to dogs.[10,11] At the author's hospital maropitant is used frequently for the perioperative management of anesthetic cases, particularly when adequate time for pre-anesthetic fasting is not possible.

Table 1: Labelled medication for swine in the United States (Source: **FARAD.org**)

Number	Ingredient	Route	Example
2	ampicillin trihydrate	IM	Princillin Injection
1	azaperone	IM	Stresnil Injection
6	ceftiofur crystalline free acid	IM	Excede™ For Swine
1	dihydrostreptomycin sulfate	IM	Pfizer-Strep
2	dinoprost tromethamine	IM	Dinoprost Tromethamine Injection
1	doramectin	IM	Dectomax® Injectable Solution
5	enrofloxacin	IM/SC	Baytril® 100 Injectable Solution
1	fenprostalene	SC	Porcilene® Solution
1	ferric oxide	IM	Felac
1	florfenicol	IM	Nuflor® / Nuflor®-S
3	flunixin meglumine	IM	Banamine® Injectable Solution
1	gelatin	IV	Intragel
2	gentamicin sulfate	IM	Garacin® Piglet Injection
2	gleptoferron	IM	FerroForte®
2	GnRF - DT conjugate	SC/IV	Improvest®
4	gonadorelin	IV	P.L.H. Injectable
10	iron dextran	IM	Ferrextran 100
2	isoflupredone acetate	IM/ISYN	Predef® 2x Sterile Aqueous Suspension
5	ivermectin	SC	Bimectin® Injection for Cattle and Swine
5	lincomycin hydrochloride	IM	Lincocin® Sterile Solution (Rx)
1	neostigmine methylsulfate	IM	Stiglyn 1:500
25	oxytetracycline	IM	Bio-Mycin® 200 Injectable Solution
6	oxytocin	IM/IV/SC	Oxytocin (Label Under Distributors)
4	penicillin G (procaine)	IM	Agricillin Pen Aqueous
2	selenite sodium	IM/SC	BO-SE, L-SE
1	sodium chloride	IV	Intragel
1	tetracycline	IM	Tetracycline
1	thialbarbital sodium	IV	Kemithal L.A.
3	thiamylal sodium	IV	Anestatal™
11	tulathromycin	IM	Arovyn™
3	tylosin	IM	BiloVet®
1	Vitamin E	IM/SC	BO-SE, L-SE

IM = intramuscular; SC = subcutaneous; IV = intravenous; ISYM = intrasynovial; L.A. = long acting

Grapiprant is a prostaglandin receptor blocker marketed for management of osteoarthritis for dogs in the United States. By acting on the prostaglandin receptor, and not the arachidonic acid pathway itself, there is the potential for the benefit of an NSAID without the adverse effects of reduction of concentrations of the prostaglandins necessary for physiologic function. Recently Kleine, et al has demonstrated that grapiprant can achieve similar plasma concentrations in pigs associated with clinical efficacy in dogs, however a higher dosage was required compared to canine applications.[12] The author's hospital is in the early stages of trying this drug in pigs with elbow osteoarthritis, but the cost due to the increased in dose comparted to canines has been a barrier to thorough evaluation.

Clonazepam is an orally administered benzodiazepine used for seizures, anxiety, muscle disorders, with adverse effects including sedation in people. After oral administration of 0.5 mg/kg in three-month old, healthy commercial pigs, sedation was noted for 2-4 hours and serum cortisol concentrations were also decreased.[13,14] As such, oral clonazepam may provide an effective sedation tool with an ease of administration compared to parenteral sedatives in pigs. At the author's hospital we are routinely using clonazepam at a 0.35-0.5 mg/kg oral dose several hours prior to surgical procedures as well as for general sedation. The responses range from light sedation to being able to intubate for anesthesia with this dosing range. For fractious pigs we are using 2 doses 8 hours apart prior to the need for sedation.

Afoxalaner is a member of the isooxazoline class of insecticide/acaricides. Used primarily for external parasite management in dogs, this class of drugs acts as an antagonist on ligand-gated chloride channels, specifically gamma-aminobutyric acid receptors. With high efficacy and low toxicity concerns, these medications have been used in individual pig settings, namely treatment for *sarcoptes* spp infection that was resistant to ivermectin[15] as well as tick prevention in a miniature companion pig.[16] While this medication is not labelled for pigs in the United States, a similar product is labelled for control of the poultry mites in Europe, and contains a short withdrawal time in that application,[15] so there is potential for withdrawal information to be generated in the future.

Extralabel drug usage concerns

It can be challenging to determine appropriate drugs for clinical applications in pigs with a role that likely does not include food production. Additionally, individual pigs presented to referral centers may have needs, such as prophylactic antimicrobial administration, for which there are no labelled products.[17] In the US both AMDUCA and the Code of Federal Regulations (21 CFR part 530)[1] state that additional prohibitions can occur if: (I) a drug has no acceptable analytical method, and can't be detected; or (II) the extralabel use of a drug presents a public health risk. With these additional prohibitions in mind, it is important to note that the absence of a drug on the AMDUCA list would not guarantee that it is legal to use in a food animal species such as a pig. These situations can have different implications, as the risk of getting caught using an impermissible drug in a companion pig may be low, but not zero. For example, the author has been advised from University council, that as a university employee, and as such an employee of the state of Tennessee, they will not be covered for an "illegal act" such as using an impermissible drug in a food animal species. Even though the risk of companion and sanctuary pigs entering the food chain is low, it has been reported,[18] and at the author's hospital when we are asked to consider an illegal drug in a companion or sanctuary pig we decline and cite the legal nature of the prohibitions.

Regardless of the role of the patient, the US FDA recognizes pigs as a major food animal species. From a legal perspective it does not matter what the animal's role is (companion vs production), a pig is considered a food animal based on species alone. Practitioners should be aware, that while not common, there are documented instances of companion pigs entering the food chain.[18] As such a withdrawal time should be provided to the owner and documented in the medical record, regardless of the animal's intention. The author has worked with multiple clients and sanctuary organizations that would find the suggestion of the consumption of their companion pig offensive, and has provided a withdrawal recommendation in the format of this example:

"While we recognize that Piggie Smalls, the notorious P.I.G., is a companion animal, we are legally obligated to provide you a withdrawal time of <XX> days."

This format can diffuse client emotion of the suggestion of the consumption of their animal, as well as to legally document that a withdrawal time was advised.

The author is aware of the challenges of not using therapies when a client requests it for a companion food animal. While it may be unlikely that a purely companion sheep or goat will enter the food chain, there have been documented cases of pet pigs entering the human food supply, these being Vietnamese-Potbellied pigs in California.[18] FARAD can be contacted for withdrawal recommendations for these patients.

References

1. Helke KL, Nelson KN, Sargeant AM, et al. Pigs in Toxicology:Breed Differences in Metabolism and Background Findings. *Toxicologic Pathology* 2016;44:575-590.

2. Toutain PL, Ferran A, Bousquet-Mélou A. Species differences in pharmacokinetics and pharmacodynamics. *Handb Exp Pharmacol* 2010:19-48.

3. Pairis-Garcia MD, Karriker LA, Johnson AK, et al. Pharmacokinetics of flunixin meglumine in mature swine after intravenous, intramuscular and oral administration. *BMC Vet Res* 2013;9:165.

4. Cramer MC, Pairis-Garcia MD, Bowman AS, et al. Pharmacokinetics of transdermal flunixin in sows. *J Vet Pharmacol Ther* 2019;42:492-495.

5. Kittrell HC, Mochel JP, Brown JT, et al. Pharmacokinetics of Intravenous, Intramuscular, Oral, and Transdermal Administration of Flunixin Meglumine in Pre-wean Piglets. *Front Vet Sci* 2020;7:586.

6. Hay Kraus BL. Spotlight on the perioperative use of maropitant citrate. *Vet Med (Auckl)* 2017;8:41-51.

7. Berryhill EH, Knych H, Edman JM, et al. Pharmacokinetics of single doses of maropitant citrate in adult horses. *J Vet Pharmacol Ther* 2019;42:487-491.

8. Berryhill EH, Knych H, Chigerwe M, et al. Pharmacokinetics of maropitant citrate after oral administration of multiple doses in adult horses. *J Vet Pharmacol Ther* 2020;43:282-287.

9. Smith JS, Seddighi R. Miniature Companion Pig Sedation and Anesthesia. *The veterinary clinics of North America Exotic animal practice* 2022;25:297-319.

10. Gebert J, Smith J, Ebner L, et al. Pharmacokinetics of extravascular maropitant citrate in swine. In: *AASV Annual Meeting 2023*.

11. Smith JS, Gebert JE, Ebner LS, et al. Pharmacokinetics of intramuscular maropitant in pigs (Sus scrofa domesticus). *J Vet Pharmacol Ther* 2023;46:158-164.

12. Kleine S, Hampton CE, Smith C, et al. Pharmacokinetics of a single oral dose of grapiprant in juvenile pigs (Sus scrofa domestica). *J Vet Pharmacol Ther* 2023;46:269-275.

13. Shanks GAH, Chiara E; Kleine, Stephanie; Smith, Joseph; Xiaojuan, Zhu; Mulon, Pierre-Yves; Smith, Christopher; Giori, Luca; Cox, Sherry; Seddighi, Reza Mohammed. Pharmacokinetics of oral clonazepam and its effect on selected stress markers in growing commercial pigs (Sus Scrofa Domesticus). In: *54th Annual Meeting of the American Association of Swine Veterinarians. Aurora (Denver), Colorado: AASV*; 2023.

14. Hampton CE, Kleine SA, Smith JS, et al. Pharmacokinetics of oral clonazepam in growing commercial pigs (Sus scrofa domestica). *J Vet Pharmacol Ther* 2024.

15. Smith JS, Berger DJ, Hoff SE, et al. Afoxolaner as a Treatment for a Novel Sarcoptes scabiei Infestation in a Juvenile Potbelly Pig. *Frontiers in Veterinary Science* 2020;7.

16. Needleman A, Videla R. Urolithiasis in a female miniature potbellied pig. *Veterinary Record Case Reports* 2019;7:e000809.

17. Smith J, Merkatoris P, Breuer RM, et al. Letter regarding "Gastrointestinal foreign bodies in pet pigs: 17 cases". *Journal of Veterinary Internal Medicine* 2022;36:1199-1200.

18. Lord LK, Wittum TE. Survey of humane organizations and slaughter plants regarding experiences with Vietnamese potbellied pigs. *J Am Vet Med Assoc* 1997;211:562-565.

https://doi.org/10.54846/am2025/153

Preventive medicine for miniature companion pigs

J. S. Smith
University of Tennessee, Knoxville, Tennessee

Introduction

Currently an increase of pig patients to large animal clinicians is being noted for all types of pigs, including companion, as well as sanctuary/rescue animals. While not as extensively covered by most veterinary training programs as other large animals, these porcine patients are often in need of care for conditions other than common production diseases. Since individual pigs include aspects of production and companion animal medicine there are many nuances to their care. The goal of this presentation is to review preventive medicine as well as husbandry for the individual pig.

Breeds

There are numerous breeds under the umbrella of miniature companion pigs (MCPs). The initial breed that was utilized as a pet was the Vietnamese-potbellied pig, but more recently Kune Kune pigs, Juliana pigs, Ossabow Island pigs and other pig breeds such as the Gottingham, which originated as a miniature pig for laboratory models. Clients should be counseled that terms such as "micro" and "teacup" are just marketing terms, and no specific predictions about the size of an animal are possible exclusively based on breed. While some MCPs can be registered, such as the American Miniature Pig, the author has noted quite a bit of variability in the reporting of breeds by clients as well as the appearance of the animal.

Nutrition

Numerous commercial diets are available for MCPs. The author's institution stocks the Mazuri brand. Clients should be counseled to avoid commercial pig fed, as while it may seem economical, it is designed for rapid growth of larger animals, and this could be problematic for maintain a healthy body condition score (BCS) in an MCP (Figure 1). Clients should also be advised to reduce the overall calories being fed by ~15% for animals that are castrated or spayed, as most feeding recommendations are based on intact animals. Calories should be minimized for treats. Vegetables can be useful, but fruit juices or other high calorie foods should be used sparingly and limited. Some breeds, such as the Kune Kune have grazing behaviors and access to pasture or hay can be incorporated into their planning. Pigs can enjoy rooting behavior, which can be challenging to provide in some companion environments, and this can be mimicked by the providing of feed pellets in a Buster Cube or similar small animal enrichment device. Table 1 adapts previous nutritional recommendations[1,2] for commercial miniature pig diet components.

While these recommendations are specific for life stages of the pig, the author's institutions have seen better BCS scores in animals moved and kept on the maintenance recommendations as soon as possible. An example of this diet commercially available is the Mazuri Mini Pig: Mature Maintenance product.

Vaccination

Several recommendations exist for "core" vaccinations for MCPs.[3] It is important to consider the animal's role, environment and risk factors when considering what is or is not a core vaccine. Typically, coverage for tetanus, erysipelas, and rabies are recommended, with additional considerations given for leptospirosis and respiratory disease, depending on environment. The author has noted vaccine recommendations written by exotic animal veterinarians that do not always align with the risks of MCPs, so evaluation of risks and benefit for each individual case is encouraged by the veterinary team. The agents for tetanus and erysipelas are commonly encountered in soil, and while rabies may be required by various local organizations it is worth noting that currently no rabies vaccines are approved for pigs, and additional permission for extralabel use of the rabies vaccine may be necessary in certain states. Clients should be counseled that in the event of an exposure that even though their pig is vaccinated, their animal may be considered non-vaccinated due to the non-approved vaccine.

Leptospirosis coverage should be considered a core vaccine for breeding animals as well as pigs with potential risks from contact with water or wildlife vectors. *Actinobacillus* and *Mycoplasma* coverage may be useful in multi-pig environments or for pigs that travel, however the efficacy of the vaccine is debated for individual pigs. Parvovirus should be considered for breeding sows, but clinicians should be aware that the swine and canine parvoviruses are different and vaccination with the canine vaccine is not thought to provide protection. The following table is adapted from a Carpenter's exotic animal formulary[3] (Table 2).

Depending on the situation, other vaccinations can be employed for enteritis, circovirus (not appreciated much clinically, but required by some boarding operations), and if injection of the erysipelas vaccine is irritating to the patient, an oral formulation is available for administration. Clinicians should be prepared for misinformation by MCP owners, as several online message boards describe MCPs as being immune to rabies and tetanus, with no evidence existing for those claims.

Table 1: Miniature pig diet component recommendations from Tynes et al[1]

	Starter	Grower	Breeder	Maintenance
Age	Weaning	Weaning at 4-6 months	Adult	Majority
Protein	20 %	> 20 %	12 %	12 %
Fat	~5 %	> 5 %	2 %	2 %
Fiber	~2.5 %	~ 10%	~11-15%	12-15%

Table 2: Vaccination recommendations for MCPs from Carpenter's formulary[3]

Vaccine	Recommendation	Timeline
Erysipelas	Core for all	At 8-12 weeks, repeated in 3-4 weeks afterwards, then 1-2 yearly.
Tetanus	Core for all	At least annually.
Rabies	Core based on risk and evaluation	Annually. With the caveat that this is extralabel use and may require additional permissions from the state veterinarian.
Leptospirosis	Optional based on risk andbreeding status	At 8-12 weeks, repeated in 3-4 weeks afterwards, then 1-2 yearly. If used for breeding, boostering 3 weeks prior to breeding.
Pneumonia (*Actinobacillus pleuropneumoniae*)	Optional based on risk (travel and other pig contact)	At 8-12 weeks, repeated in 3-4 weeks afterwards, then 1-2 yearly.
Parvovirus	Optional based on breeding status	If used for breeding: At 8-12 weeks, repeated in 3-4 weeks afterwards, then 1-2 yearly, boostering 3 weeks prior to breeding.

MCPs = miniature companion pigs

Parasitism

Internal parasitism is not a common cause for MCPs to be hospitalized at the UTCVM, as opposed to our small ruminant patients, but it is encouraged to perform regular fecal exams to make deworming decisions in an evidenced-based manner. Our practice routinely identifies zoonotic parasites such as *Balatidium coli, Ascaris suum,* and Strongyle spp in our MCP patients, and as such advises owners on the risks of finding such parasites in their animals.

Body condition

Body condition should frequently be monitored, as heavier-conditioned pigs appear more likely to develop osteoarthritis and similar musculoskeletal conditions. It is recommended that MCPs maintain a body condition of 3/5 or less. The American Mini Pig Association has adapted a swine body condition system from the commercial pig system developed by the University of Kentucky. This system is more user friendly for MCPs because it's based on a single observation made from the posterior of the patient. Clients should be counseled about the importance of maintaining an appropriate body condition, because it can be incredibly challenging to encourage weight loss once the MCP becomes overconditioned.[1]

Figure 1: Body condition scoring guide. (Reprinted with permission from the American Mini Pig Association.)

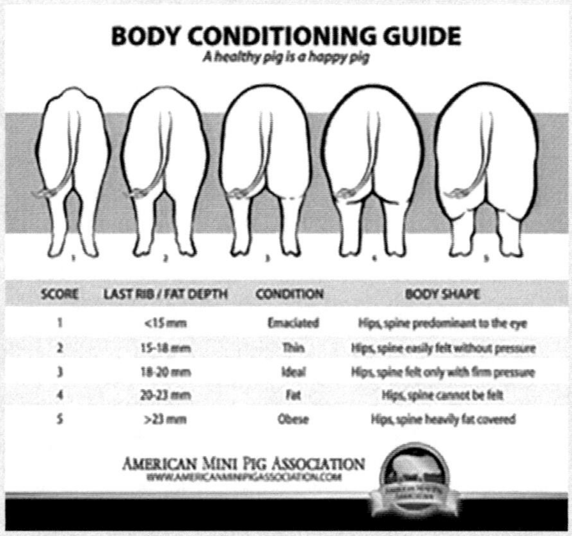

At the author's institution we have notices severe elbow osteoarthritis in over conditioned MCPs. One study identified orthopedic injuries in 21% of MCPs presented to a veterinary teaching hospital.[4] We have also notices that it can be incredibly challenging to remove excess weight from these animals due to their sedentary nature. It is strongly recommended to include exercise and caloric monitoring to maintain a low BCS in these animals to avoid chronic health conditions.

Tusk and hoof care

Yearly or more frequently as needed sessions for tusk trimming and foot trimming are recommended. These can be incorporated into an annual preventive medicine appointment, as the sedation will enhance physical examination. Hooves should be trimmed with vertical cracks dremeled out to encourage growth of health hoof material. At the author's practice excessive material is first removed with a hoof nipper, and then the hooves are shaped with a Dremel. For pigs with minimal pigmentation of the hoof a light can be shined through the hoof from the solar surface to identify the margins of the quick. Dewclaws should be trimmed at this time as well as they are typically minimally worn.

Tusks should be routinely evaluated for occlusion and the gumline examined at the same time. It is not uncommon for older MCPs to have periodontal disease and other dental pathology.[5] The mandibular tusk of the MCP extends almost to or to the angle of the mandible, so tusk infections are incredibly challenging to manage. Extraction of the tusk itself had been historically attempted, but required so much bone removal from the mandible that the pigs were prone to pathological fractures afterwards, although newer attempts at this technique are reported.[6] Tusks should be trimmed with a Dremel tool or gigli wire, as angled nippers have been shown to have an extreme risk of fracturing the tooth. Water should be squirted onto the cut surface of the tusk to reduce the friction and heat buildup, as it is thought that this can potentially cause a thermal necrosis to the inner structures of the tooth. Osteomyelitis or tusk root infection of the mandibular tusk in MCPs can be incredible obstacles for management. Regular tusk inspection and maintenance is important for MCPs as they can develop severe pathology in this region that may go unnoticed until it is too late for therapy.[7]

Restraint and sedation

Restraint of MCPs can be challenging, and there are several techniques that can be utilized. The Panepinto sling (**https://www.panepinto.com/**) provides a method for restraint that can be used to facilitate how trims and venipuncture while reducing excitement for the pig patient. Multiple clients have trained their animals to present in lateral recumbency for belly scratches or forking. Forking is a practice where tactile stimulation along the dorsum or ventrum with a back scratcher or fork will induce recumbency.[8] Care should be taken to not use a hog snare or similar tool for production pigs on MCPs due to anatomical differences in the shape of the head, as well as the stressful nature of snaring. An additional reference for restraint, sedation, and anesthesia[8] for miniature companion pigs can be found here at this link (**https://trace.tennessee.edu/utk_largpubs/54/**) or the QR code below:

At the author's institution most sedation is based on multimodal approaches, such as midazolam (0.2-0.5 mg/kg, IM), butorphanol (~0.2 mg/kg, IM), with potential additions of ketamine and/or xylazine.[8] We have recently evaluated clonazepam as an oral sedative.[9] After oral administration of 0.5 mg/kg in three-month old, healthy commercial pigs, sedation was noted for 2-4 hours and serum cortisol concentrations were also decreased.13,14 As such, oral clonazepam may provide an effective sedation tool with an ease of administration compared to parenteral sedatives in pigs. At the author's hospital we are routinely using clonazepam at a 0.35-0.5 mg/kg oral dose several hours prior to surgical procedures as well as for general sedation. The responses range from light sedation to being able to intubate for anesthesia with this dosing range. For fractious pigs we are using 2 doses 8 hours apart prior to the need for sedation.

Extralabel drug usage concerns

It can be challenging to determine appropriate drugs for clinical applications in pigs with a role that likely does not include food production. Additionally, individual pigs presented to referral centers may have needs, such as prophylactic antimicrobial administration, for which there are no labelled products.[10] In the US both AMDUCA and the Code of Federal Regulations (21 CFR part 530)[1] state that additional prohibitions can occur if: (I) a drug has no acceptable analytical method, and can't be detected; or (II) the extralabel use of a drug presents a public health risk. With these additional prohibitions in mind, it is important to note that the absence of a drug on the AMDUCA list would not guarantee that it is legal to use in a food animal species such as a pig. These situations can have different implications, as the risk of getting caught using an impermissible drug in a companion pig may be low, but not zero. For example, the author has been advised from University council, that as a university employee, and as such a public employee of the state of Tennessee, they will not be covered for an "illegal act" such as using an impermissible drug in a food animal species.

Regardless of the role of the patient, the US FDA recognizes pigs as a major food animal species. From a legal perspective it does not matter what the animal's role is (companion vs production), a pig is considered a food animal based on species alone. Practitioners should be aware, that while not common, there are documented instances of companion pigs entering the food chain.[11] As such a withdrawal time should be provided to the owner and documented in the medical record, regardless of the animal's intention. The author has worked with multiple clients and sanctuary organizations that would find the suggestion of the consumption of their companion pig offensive, and has provided a withdrawal recommendation in the format of this example:

"While we recognize that Piggie Smalls, the notorious P.I.G., is a companion animal, we are legally obligated to provide you a withdrawal time of <XX> days."

This format can diffuse client emotion of the suggestion of the consumption of their animal, as well as to legally document that a withdrawal time was advised.

The author is aware of the challenges of not using therapies when a client requests it for a companion food animal. While it may be unlikely that a purely companion sheep or goat will enter the food chain, there have been documented cases of pet pigs entering the human food supply.[11] FARAD can be contacted for withdrawal recommendations for these patients.

References

1. Tynes VV. Potbellied pig husbandry and nutrition. *Vet Clin North Am Exot Anim Pract* 1999;2:193-208.

2. Tynes VV. Preventive health care for pet potbellied pigs. *Veterinary Clinics of North America: Exotic Animal Practice* 1999;2:495-510.

3. Carpenter JW. *Exotic Animal Formulary*. 5 ed. St. Louis, Missouri: Elsevier Health Sciences, 2018.

4. Sipos W, Schmoll F, Stumpf I. Minipigs and potbellied pigs as pets in the veterinary practice--a retrospective study. *J Vet Med A Physiol Pathol Clin Med* 2007;54:504-511.

5. Smith MA, Rao S, Rawlinson JE. Dental Pathology of the Domestic Pig (Sus Scrofa Domesticus). *J Vet Dent* 2020;37:192-200.

6. Savard C, Dumais Y, Pouillevet H, et al. Extraction of aberrant tusk and mandibular osteomyelitis treatment in a pot-bellied pig (Sus scrofa domesticus). *Journal of Exotic Pet Medicine* 2021;39:24-31.

7. Smith J, Griffin C, Jones P, et al. Severe mandibular osteomyelitis and exfoliation of a mandibular canine tooth in a Vietnamese potbellied pig. *Veterinary Record Case Reports* 2022;10:e348.

8. Smith JS, Seddighi R. Miniature Companion Pig Sedation and Anesthesia. *The veterinary clinics of North America Exotic animal practice* 2022;25:297-319.

9. Hampton CE, Kleine SA, Smith JS, et al. Pharmacokinetics of oral clonazepam in growing commercial pigs (Sus scrofa domestica). *J Vet Pharmacol Ther* 2024.

10. Smith J, Merkatoris P, Breuer RM, et al. Letter regarding “Gastrointestinal foreign bodies in pet pigs: 17 cases”. *Journal of Veterinary Internal Medicine* 2022;36:1199-1200.

11. Lord LK, Wittum TE. Survey of humane organizations and slaughter plants regarding experiences with Vietnamese potbellied pigs. *J Am Vet Med Assoc* 1997;211:562-565.

https://doi.org/10.54846/am2025/154

We sent Wilbur to medical school: Describing the challenges of using swine in biomedical research

C. Vonderohe[1], DVM, PhD; J. McNew[2], DVM; S. Elefson[1], PhD; A. Antonson[3], PhD
[1]USDA/ARS Children's Nutrition Research Center, Department of Pediatrics, Baylor College of Medicine, Houston, Texas;
[2]Center for Comparative Medicine, Baylor College of Medicine, Houston, Texas;
[3]Department of Animal Science, University of Illinois, Urbana-Champaign, Illinois

How and why pigs are used in biomedical research

Animal models of human physiology and pathology are vital facets of biomedical research. Although there are limitations to the degree to which different species can recapitulate human pathophysiology, animal models allow investigators to ask biomedical research questions that would otherwise be impossible to answer in human subjects. The overwhelming majority of animal models used in biomedical research are rodents because they are economic to house and manage, there is a plethora of mouse-specific research tools, as well as have genetically modified strains that can be leveraged to identify the role of specific proteins *in vivo*.[1] However, rodent size, genetics, and physiologic responses to stimuli limit their utility in translational research. Pigs, however, are an excellent translational animal model; they recapitulate human physiology and pathology, they are large enough to test devices, pharmaceuticals and surgical techniques, and have the additional benefits of being less expensive and more socially acceptable than canines or non-human primates.[1-3]

Pigs are used as biomedical models for a wide range of human ages, from preterm neonates to the elderly, as well as a great breadth of fields, including immunology, ophthalmology, orthopedics, cardiology, neonatology, neurology, toxicology, and nutrition.[2] Recent developments in genome editing have significantly increased the utility of swine models in translational oncology and xenotransplantation.[4] Additionally, greater availability of swine-specific reagents, swine-friendly animal housing and swine-specific expertise for collaboration has resulted in significantly more interest in leveraging the pig's size and physiology as a biomedical model.[5] There has also been a great deal of interest in different breeds of pigs, particularly mini-pig breeds because they are easier and more economical to raise and manage at older ages; the small size means that investigators can house and manage multiple aged animals rather than a single pig derived from commercial production genetics.[6]

Regulation of swine in biomedical research

Pigs used for biomedical research are subject to greater institutional and federal regulation compared to animals entering the food chain. Additionally, the use of a pig as a research model shifts the perspective of seeing each animal as an individual rather than as part of the herd. The care of pigs in research is regulated by the government depending on the intended use and funding source for the studies. Pigs used in biomedical research are "covered" by the Animal Welfare Act (AWA),[7] which was initially proposed to protect companion animals (ie, dogs and cats). Laws under this act were enacted for breeders, carriers, and exhibitors to establish minimum standards of care for animals intended for research. This act now covers many species, including swine. Confusingly, for swine used in research activities intended to improve animal nutrition, breeding, management, production efficiency, or the quality of food, they are not "covered" by the AWA.

Any swine research funded by National Institute of Health (NIH), the Centers for Disease Control and Prevention (CDC), Food and Drug Administration (FDA), and the Biomedical Advanced Research and Development Authority (BARDA) must follow the public health service policy (PHS) and therefore the Guide for the Care and Use of Laboratory Animals (Guide).[8] In addition to these organizations, there is a memorandum of understanding of which the PHS policy also applies to research funded by the National Aeronautics and Space Administration (NASA), National Science Foundation (NSF), and the Department of Veteran Affairs (VA). Regulations set forth in the AWA and the PHS policy overlap. Some noteworthy requirements are for a veterinarian to oversee all animal care for the research activities and for an Institutional Animal Care and Use Committee (IACUC). The IACUC reviews and approves all studies performed with animals by the organization. Biomedical research in swine must always follow the AWA, but the guide may or may not apply depending on the funding source. However, institutions may elect to follow these guidelines, and others, through an accreditation with AAALAC International. AAALAC is a voluntary program, and their accreditation demonstrates a program's commitment to the humane care and use of animals in science. In general, these regulations set minimum standards for pen sizes, housing conditions, sanitization protocols, anesthesia, and analgesia.

Challenges of swine use in biomedical research

However, challenges remain that limit the ability of investigators to use swine model to its greatest potential, including the availability of swine-specific expertise in medical schools, the price and availability of pigs in certain parts of the country, and limited capacity to house and manage pigs, regardless of size and genetics.

The biomedical pig's welfare is heavily regulated, but the interpretation of those regulations is at the behest of attending veterinarians and IACUCs. A vast majority of veterinarians working with pigs in biomedical research are trained as laboratory animal veterinarians. Many of these veterinarians complete an American College of Laboratory Animal Medicine (ACLAM)-recognized residency. Residencies vary across institutions, but most of the programs are rodent or non-human primate focused, both which emphasize large animal models to produce a well-rounded laboratory animal veterinarian. Pigs are extensively used in biomedical research but the total populations at an institution may be low at a given time. Taken together, the typical lab animal veterinarian is highly trained across a breadth of species, but their swine-specific knowledge may be limited. This may result in lab animal veterinarians having limited practical familiarity with swine-specific diseases, procedures, nutrition, and management.

Animal care staff, who take care of these pigs daily, may come from a mixed background and, in many instances, are expected to take care of a diverse group of species at an institution. The culmination of these factors leads to an environment where the care of pigs is different in a research setting than an agricultural production setting. Common swine diseases are mitigated by the procurement of pigs from specific pathogen free vendors. Knowledge in production is typically not necessary since breeding is often limited to production facilities. And as with the knowledge and skill of working around and handling animals, more practice tends to improve a person's proficiency and while many individuals greatly enjoy working with pigs, at times, their skills and knowledge could be improved.

The typical biomedical researcher does not have experience working with swine. Most laboratories that use pigs rely on institutional knowledge possessed by the principal investigator, lab manager, or technical staff along with veterinary support staff at the institution for swine handling and management. Cogent to the field of biomedical research, these differences in approach can limit the reproducibility of data. Additionally, laboratories are often unaware of swine-specific physiology and behavior and may treat pigs like dogs or a comparable companion animal species. While this may work for young piglets or mini- or micro- pig breeds, it is less effective when handling commercially bred sows or grower pigs. For example, many researchers require convincing that it is futile to attempt to physically "wrestle" a mature, 350 kg sow. There is a need for people with swine experience to be willing to train and communicate with these individuals to ensure handler safety. Generally, experts in their respective fields of biomedical research may be uncomfortable working with their chosen animal model, but utilize the model given its high translatability to humans and it is imperative that these researchers obtain support from trained individuals to ensure proper care.

Another major challenge to swine use in biomedical research is animal availability. While we recognize that this is not a pressing issue in top pig-producing regions of the world, this issue limits the capacity for some biomedical researchers to complete projects. Pigs used for biomedical research also need to have a very high health status, which may further limit the availability of animals for experimentation. Additionally, institutions which sell pigs for biomedical research need to be licensed under the AWA and are therefore inspected by the USDA, and AAALAC if accredited. As a result, researchers may pay premiums for commercially bred market pigs. For example, investigators in Texas have paid $500 for weaned pigs, and up to $3000 for pregnant sows. Pigs of other breeds, particularly for proprietary phenotypic-specific breeds or transgenic animals are even more limited and expensive than market pigs, regardless of geographic location.

Finally, location and facility design may limit the use of pigs as models for biomedical research. Many medical schools or biomedical research laboratories are located in metropolitan areas, and therefore have a limited geographic footprint. Although using mini-pig breeds or neonatal pigs reduces the space necessary for animal management and housing, that does not change the fact that most laboratory equipment may need to be retrofitted to suit pig needs. For example, most enclosures used for pigs in biomedical research are raised floor pens. Animal enclosures needed for special conditions (such as preterm neonatal pigs) or those for large pigs need to be purpose-built and may require more space in animal facilities.

Opportunities for the swine veterinarian

The use of pigs in biomedical research is increasing.[2] As a reflection of the breadth of the popularity of the pig as a model, a recent conference focused solely on swine in biomedical research had almost 250 registrants and 117 submitted abstracts that were presented as posters or in one of eighteen plenary sessions that represented fields from neonatology to xenotransplantation. While the popularity of the swine model continues to grow, the challenges that face investigators and veterinarians remain, and these challenges represent opportunities for veterinarians who have swine experience. Swine veterinarians can be invaluable resources for biomedical research institutions. Specifically, they can provide consultation to the IACUC when they are reviewing swine protocols. Swine veterinarian's knowledge about controlling and treating disease outbreaks could also be instrumental for research institutions encountering a novel disease challenge or unfamiliar pathology. Additionally, they can also serve as trainers and resources for the proper handling of pigs and common veterinary skills, for investigators, veterinarians, lab animal veterinary residents, and technical staff.

References

1. Beck, A. P. & Meyerholz, D. K. (2020). Evolving challenges to model human diseases for translational research. *Cell Tissue Res* **380**, 305-311, **doi:10.1007/s00441-019-03134-3**

2. Lunney, J. K. *et al.* (2021). Importance of the pig as a human biomedical model. *Science Translational Medicine* **13**, eabd5758, **doi:doi:10.1126/scitranslmed.abd5758**

3. National Academies of Sciences, E. *et al.* in *Necessity, Use, and Care of Laboratory Dogs at the U.S. Department of Veterans Affairs* (National Academies Press (US) Copyright 2020 by the National Academy of Sciences. All rights reserved., 2020).

4. Rogers, C. S. Genetically engineered livestock for biomedical models. *Transgenic Res* **25**, 345-359, doi:10.1007/s11248-016-9928-6 (2016).

5. Meyerholz, D. K., Burrough, E. R., Kirchhof, N., Anderson, D. J. & Helke, K. L. (2024). Swine models in translational research and medicine. *Veterinary Pathology* **61**, 512-523, **doi:10.1177/03009858231222235**

6. Stricker-Krongrad, A., Shoemake, C. R. & Bouchard, G. F. (2016). The Miniature Swine as a Model in Experimental and Translational Medicine. *Toxicol Pathol* **44**, 612-623, **doi:10.1177/0192623316641784**

7. Animal Welfare Act as Amended 2019. 7 U.S.C. 2131-2160.

8. National Research Council (US) Committee for the Update of the Guide for the Care and Use of Laboratory Animals. Guide for the Care and Use of Laboratory Animals. 8th edition. Washington (DC): National Academies Press (US); 2011. Available from: **https://www.ncbi.nlm.nih.gov/books/NBK54050/ doi: 10.17226/12910**

https://doi.org/10.54846/am2025/155

Niche production: It's not for everyone

P. Schneider, DVM, MS; T. Schwartz, DVM

Niche pork production

Niche pork production represents a small but significant segment of the United States swine industry. It is estimated that less than 10% of pigs in the US are raised for niche pork marketing programs, which include initiatives such as California Proposition 12, Massachusetts Question 3, USDA Organic Certification, Certified Humane, and private label no antibiotic ever (NAE), among others. Many of these programs impose restrictions on practices common in conventional pig farming, requiring farmers and veterinarians to find new approaches to address health and production concerns that meet these program standards.

Health considerations in antibiotic-free programs

One major hurdle in niche pork production is the restriction on the use of antibiotics. Programs that require NAE or antibiotic-free (ABF) pigs forbid the use of antibiotics throughout the pig's life, eliminating the option to use them at routine stages such as processing or early nursery phases. The easiest way to mitigate the need for antibiotics is by selecting high-health-flow pigs to enroll in these programs. Farms with persistent porcine reproductive and respiratory syndrome (PRRS), *Mycoplasma* instability, or well-documented systemic bacterial concerns are poor candidates for NAE programs due to the need for antibiotics as a health and welfare intervention. Farms should be aware of potential systemic bacterial pathogens that may be introduced during population transitions or genetic changes. Conventional thought that bacterial challenges, such as *Glaesserella parasuis,* are ubiquitous should be challenged and addressed by elimination and/or vaccination.

Biosecurity and disease prevention

Preventing pathogen exposure is critical on both sow farms and growing pig sites, making biosecurity a must. Proper biosecurity procedures for people, vehicles, and supplies should be regularly audited and reinforced through ongoing training. For farms operating on a continuous-flow system, managing biosecurity during transportation and dead animal removal is especially important. Examples include using only cleaned and disinfected trailers and working with drivers on their biosecurity habits. Onsite dead disposal is preferred, with options such as composting, incineration, or burial (where allowed) being suitable options given they are managed appropriately.

Farm cleanliness to maintain good hygiene should also be a priority. Farms that require bedding must understand the importance of daily scraping and bedding. It's important to maintain a good environment to ensure a healthy pig. Bedded systems in particular need to implement rodent reduction programs to eliminate rodent populations.

Maintaining hygiene between litters is also essential. Some farms may adopt "McRebel"-type practices to reduce pathogen movement between litters. Often, less movements and good colostrum management decrease the prevalence of scour, lameness, and neurologic Strep infections. Additionally, certain management practices that cause inflammation can decrease immune function. For example, sharp scalpel blades have an advantage over other techniques that may cause more tissue damage, increase the propensity for infections and lengthen recovery time.

Vaccination and immunity

A strong vaccination program is essential to build robust immunity and maintain disease stability. Autogenous vaccines are often a critical component of NAE vaccination programs due to their protection for herd-specific bacterial challenges. Veterinarians should build libraries of bacterial isolates and periodically sample the herd to ensure that the isolates included in autogenous vaccines accurately reflect the bacterial challenges present. Timing of vaccine administration is also important for maximizing efficacy and ensuring prolonged immunity. Additionally, veterinarians should consider other strategies to enhance protection and reduce exposure to specific pathogens in growing pigs. Diseases such as *E coli, Salmonella*, and *Lawsonia* may be carried by the sow herd, and vaccinating sows or suckling pigs may be the most effective way to address these concerns.

Feed program

Feed is a critical component of any production program, but in antibiotic-free production it becomes even more important. Feed rations are formulated to optimize livestock performance while meeting the essential nutritional and maintenance needs of the animal. In antibiotic-free production, however, the most efficient feed is not always the best choice. Feed must be carefully formulated to meet and exceed the pig's nutritional requirements. The two key areas to focus on for antibiotic-free pigs are the weaned pig feed program and a growth and maintenance program that is robust enough to support the pig's needs in a more rugged (often outdoor) environment. For weaned pigs, using a fiber source, such as steam-rolled oats, and keeping protein levels lower helps promote a healthy gut flora and reduces the risk of enteric outbreaks. A key element of a successful feed program is having a knowledgeable nutritionist who understands the unique needs of antibiotic-free pigs.

Parasite control

In programs that use solid floors and bedding, managing internal and external parasites is crucial. Regular deworming programs are essential to limit the exposure of suckling piglets to parasite eggs and to ensure the effectiveness of treatment in growing pigs. Deworming programs should be tailored to the types of parasites present and their life cycles. Multiple deworming products are available; it is important to understand the specific stages of the parasites they address and the overall cost-effectiveness of the program.

Animal care and management

In niche pork production, individual animal care must be more intensive than in conventional systems. Early identification of sick or off-feed pigs is critical. In many cases, non-pharmacological interventions, such as improving the barn environment and reducing competition among pigs by moving to individual care pens, can support recovery. In other cases, early identification allows for treatment with anti-inflammatory or non-antibiotic products that are effective in managing the condition. Farmers must recognize that some health challenges can only be treated with antibiotics, and they should be prepared to make decisions that prioritize the welfare of the pigs.

Conclusion

While niche pork programs impose certain requirements that may be difficult for some farmers to meet, the health management strategies for niche pig farming are not significantly different from those in conventional systems. However, farmers raising pigs for niche markets must be able to execute health plans and daily management tasks at a very high level to prevent disease outbreaks and production challenges that could ultimately require the use of antibiotics. Veterinarians and farmers should have open discussions about their ability to implement health protocols, conduct daily observations, and manage interventions effectively before committing to niche pork production.

https://doi.org/10.54846/am2025/156

African swine fever control and management challenges in the Philippines

Angel Tomas Antonio C. Manabat, DVM, FPCSP; Maximino M. Montenegro DVM, MSc
Philippine College of Swine Practitioners

Introduction

The first reported case of African swine fever (ASF) in the Philippines was in August 2019, in Rizal province in the island of Luzon, by mid 2023 it has spread to 17 regions and 78 out of 82 provinces, ASF entered many commercial and small hold farms, a number of which have had several breaks and rebreaks in the past 5 years. It is estimated that around half of the country's swine population was lost and recent trends show very little recovery. As of November 2024, there are still active cases in 6 regions and 20 provinces.

Contingency planning and government initiatives

After ASF hit China in August 2018, the Philippine Bureau of Animal Industry (BAI) organized the ASF technical working group (TWG) composed of government veterinarians, private swine practitioners & swine industry stakeholders in September 2018 to prevent the entry of ASF into the country. The result was an ASF preparedness plan that was put into action in February 2019. The plan covers areas such as risk analysis, prevention strategies, early warning plans, control and eradication strategies, command structure and responsibilities, and support plans. It aims to outline the measures that will be carried out in the event of an ASF outbreak in the Philippines, considering phases of disaster management and in line with international standards from organizations such as OIE and FAO.

After the ASF outbreak in August 2019, a series of administrative orders (DAO) & administrative circulars (AC) were put into effect nationwide to control and stamp out the disease. The significant DA orders and circulars are 1) AC12 s.2019 Natl Zoning and Movement Plan for the Prevention and Control of ASF; 2) AO7 s.2019 Veterinary Quarantine Movement Protocol During Animal Disease Outbreaks/Emergencies (1-7-10); 3) AO7 s.2021 Implementing Guidelines for the "Bantay ASF sa Barangay" Program, 4) AO6 s.2021 Guidelines on the Implementation of Recovery, Rehabilitation and Repopulation Assistance Program for ASF affected and non-ASF affected areas.

Challenges

Despite these stringent measures, the virus continued to spread throughout the country except for a few island provinces that have remained unaffected. Private swine practitioners viewed the failure to control the spread due to:

1. Failure to implement the ASF preparedness plan led to the entry of disease to the country by way of uncontrolled disposal of food wastes that were eventually fed to small holder swine farms in July 2019.
2. Training and knowledge dissemination regarding ASF, on how to prevent, identify and control the disease etc, did not reach many small hold farmers, leading to delayed reaction and reporting of cases.
3. Failure to monitor and restrict the movement of middlemen (viajeros) who control the swine industry supply chain (AC 12 s.2019 National Zoning & Movement Plan).
4. Failure to properly communicate and implement the surveillance and monitoring provisions of BABAY ASF (AO7s.2021 Bantay ASF sa Barangay Program or Prevent ASF in the Barrios).
5. Unsustainable indemnification policy and economic realities in the rural areas where small hold farmers tried to hide their pigs to avoid being culled. (AO7 s.2019 Veterinary Quarantine Movement Protocol During Animal Disease Outbreaks/Emergencies [1-7-10 protocol]).
6. Regional Animal Disease Diagnostic Laboratories were ill equipped, overwhelmed by the volume of submissions, lacked trained staff and funds needed to purchase ASF test kits.
7. Failure of border control and ineffective animal quarantine outposts – lack of personnel, security risks, aging animal extension workers.

These are the major challenges that contributed to failure to control ASF from the point of view of private swine veterinarians. Government have given notice recently that the policies relating to ASF will be thoroughly reviewed by 2026 and if needed be revised or removed.

ASF vaccines

With the continued difficulty to control the spread of ASF, government has entertained alternative approaches to help in managing the problem, from unofficially supporting selective depopulation instead of total depopulation of farms or areas affected by ASF, to the evaluation and use of ASF vaccines. A couple of ASF DNA subunit vaccines have undergone testing in 2022, with mixed results and did not advance to any further evaluation. In 2023 the government entertained a MLV ASF vaccine that has been developed and approved for use in Vietnam, (AVAC), for further safety and efficacy testing in the country. Results from the series of tests conducted by BAI was found satisfactory and the vaccine was forwarded to the Philippine Food and Drug Administration (FDA) for final evaluation and issuance of a Certificate of Product Registration (CPR). As of this time the FDA has only issued a CPR under monitored release good for 2 years, this will allow government to provide vaccines to farmers under monitored vaccination and post vaccination evaluation, to further gather data to show the safety and efficacy of the vaccine. A number of ASF vaccines are applying for testing in the country as of this moment.

https://doi.org/10.54846/am2025/157

The problem of African swine fever in pig production and the impact of the pig and meat trade

R. Niemyjski, DVM
Agri Plus Sp. z o.o. Poland

Introduction

African swine fever (ASF) is a very dangerous, contagious and infectious pig disease caused by African swine fever virus (ASFV). The virus is extremely resistant to low temperatures and retains infectious properties in blood, in faces and tissues for a very long time. It originally occurred in Africa, from where it spread to other continents. Its presence causes very severe financial losses for pig farmers and the entire pork processing industry.

ASF in Poland

First ASF outbreak in Poland occurred in February 2014 in wild boars in the immediate vicinity of the border with Belarus. The virus started to spread very fast among the wild boar's population due to their large number. Subsequently, it appeared in the pig population on small farms keeping a pair of pigs. Unfortunately, in the following years, it also made its way into large pig farms. First outbreaks in pigs caused great panic among the entire pig industry. ASF virus undoubtedly is one of the main reasons for the decrease in the number of pig farms in Poland, which results in a decrease in the population. (Table 1).[1]

Moving pigs in the ASF zones

Due to the occurrence of ASF in Poland, areas subject to various restrictions have been separated. Red-colored area (zone III) is an area where ASF occurred in pig population. This is the area with the highest restrictions. All animal movements are supervised by the Poviat Veterinarian. Pink-colored area (zone II) is an area where ASF occurred in wild boar population. Blue-colored area (zone I) is an ASF-free area, which borders directly on the area of ASF occurrence in pig or wild boar population. In the initial period of ASF occurrence in Poland, moving pigs between the zones was almost impossible. (Figure 1).[2]

The official veterinarians did not consent to such movements. If pigs were to be transported, it could have only been done within one zone (one color). Today, when ASF has been present in Poland for over 10 years, there is a consent to move pigs one zone down, eg, from zone III to zone II, however never to zone I. There is no major problem with moving in the other direction. The exception is zone III, where the official veterinarian must agree to the import of animals regardless of the zone they come from.

Table 1: History of African swine fever in Poland. Number of outbreaks in pigs and wild boars in all years. Current status as of November 12, 2024.

	Outbreaks in pigs	Wild boars cases
2014	2	30
2015	1	53
2016	20	80
2017	81	314
2018	109	286
2019	48	2477
2020	103	3820
2021	113	2232
2022	14	2113
2023	30	2673
2024	44	1617

Trade in meat in and from the country where ASF occurs

Trade in meat originating from ASF zones is as follows: Meat and processed meat from pigs originating from zone III may only be used in a country, from which the animals come from. Meat and processed meat from pigs originating from zone II and I may be used in the EU and cannot be exported to countries outside the EU. Only meat and processed meat from pigs originating from outside the ASF zones may be exported to countries outside the EU, unless a given country introduced embargoes on pork products originating from a country where ASF virus is present. With the increase in the range of ASF, export opportunities are significantly decreasing.

Prevention of ASF in pigs

Over the years, it seems that there are a few effective tools to use in the fight against ASF. There is one condition: They must be performed in a coordinated manner. Their use at the same time significantly reduces the spread of the virus and its entry into pig farms. First one is biosecurity. It is not only limited to having the right equipment such as mats, disinfection baths and gates, tight fencing, separation of the clean and dirty zones through a sanitary lock in a farm. The most important thing in terms of biosecurity is awareness. All animal care personnel must understand the risk of transmitting the virus to the barn. Only through continuous training, reminding of safety rules and protocols are we able to minimize the risk of ASF on a pig farm. Another tool is significant reduction of wild boars as they are the vector of virus in fields and forests. The virus spreads in the environment through migration of wild boars. It is important to work well with hunters and reduce wild boars' population. The third, and probably the most important tool, is searching through fields and forests for collection of dead wild boars and their disposal. In Poland, wild boars are delivered to disposal facilities (are burned). It was observed that in places where the virus occurred and the above three tools were implemented and coordinated at a high level, ASF did not return.

Figure 1: Areas in Poland where African swine fever occurs and outbreaks in pigs (red dots). Status as of November 12, 2024.

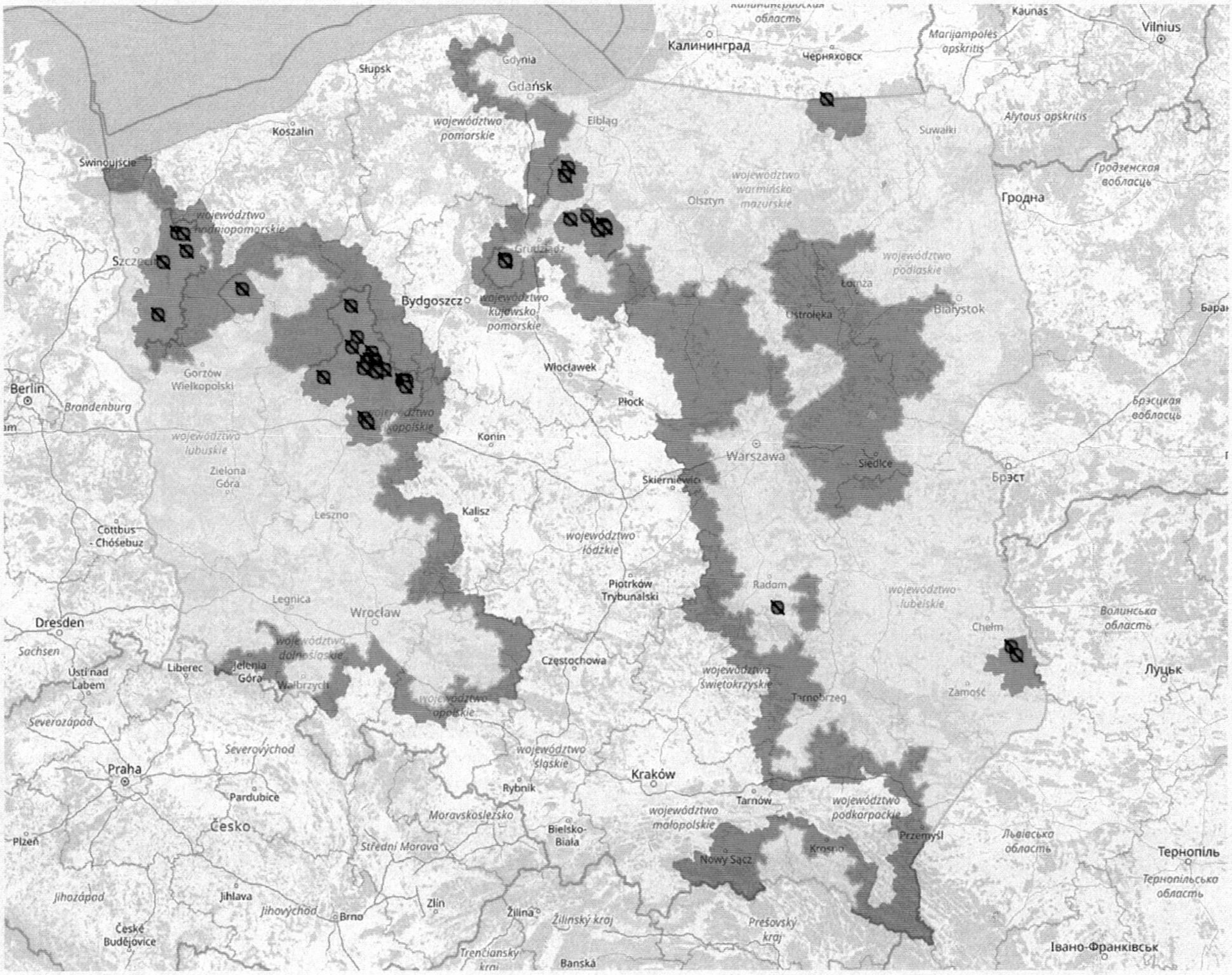

Summary

ASF is undoubtedly one of the largest challenges in pig production in Poland and Europe. Losses caused by the virus are huge and include difficulties in selling pigs and meat from pigs coming from ASF zones, disposal of sick herds and their re-stocking, as well as expenses incurred for rigorous biosecurity rules. The above factors have a big impact on the significant reduction of farms growing pigs, which impacts the significant decrease in the number of pigs in Poland and other European countries where the virus occurs.

References

1. https://www.wetgiw.gov.pl/nadzor-weterynaryjny/asf-w-polsce
2. https://bip.wetgiw.gov.pl/asf/mapa/

https://doi.org/10.54846/am2025/158

African swine fever in Germany: Lessons learned

Timothy Snider, DVM, MPH
Pig Improvement Company, Health Assurance – European Region

Introduction

On September 10, 2020, the first case of African swine fever (ASF) was confirmed in a feral swine carcass that was discovered in the district of Spree-Neiße in the federal state of Brandenburg approximately 7 kilometers from the Polish border in eastern Germany. Since that time and at the writing of this paper, in early December 2024, ASF virus (ASFV) has been found in 6382 feral swine and 35 domestic herds across 7 federal German states (see Figure 1 and Figure 2).[1]

From the onset of the outbreak in Germany, the government and the swine industry have clearly stated their intention to eradicate ASFV from within the borders of Germany. Although there have been many setbacks in the years and months since the first case was detected, the country has a made great strides towards its ultimate goal of eradicating the disease from both the feral swine and domestic herds. Many lessons have been learned in Germany and throughout the European region during this time. The lessons outlined in this paper are drawn from observations and experiences from Germany and across the European region (including the European region of Russia) from the perspective of the author as a swine industry stakeholder in the region (ie, PIC Europe Health Assurance) and as a resident of Germany.

Discussion

In order to provide a systematic framework, the lessons learned are listed below using the following subcategories: Regulatory preparedness and response, industry preparedness and response, epidemiology, and society.

Regulatory preparedness and response

- Legislation outlining the regulatory framework of the European Union (EU) that directs the response to ASF across the EU is online and publicly available. It has been systematically updated since ASFV entered the EU in 2014 to reflect improved knowledge gained from experience and research.
- While member states (MS) of the EU must follow the minimum standards set by the EU legislation, individual MS may incorporate additional, stricter, country level regulations.
- Within Germany, outbreaks are managed at the local district level, but a system of crisis management has evolved that provides a framework of resources for local preparedness and response resources that originate from both the national and state levels.
- Regionally based rapid response teams have developed to train and facilitate more rapid, coordinated, and effective district level responses.
- Regulations guiding the response provide competent officials with the authority to make individual risk-based decisions related to movement controls and this can result in a wide

Figure 1[1]: Feral swine cases per German Federal State September 10, 2020 to December 9, 2024.

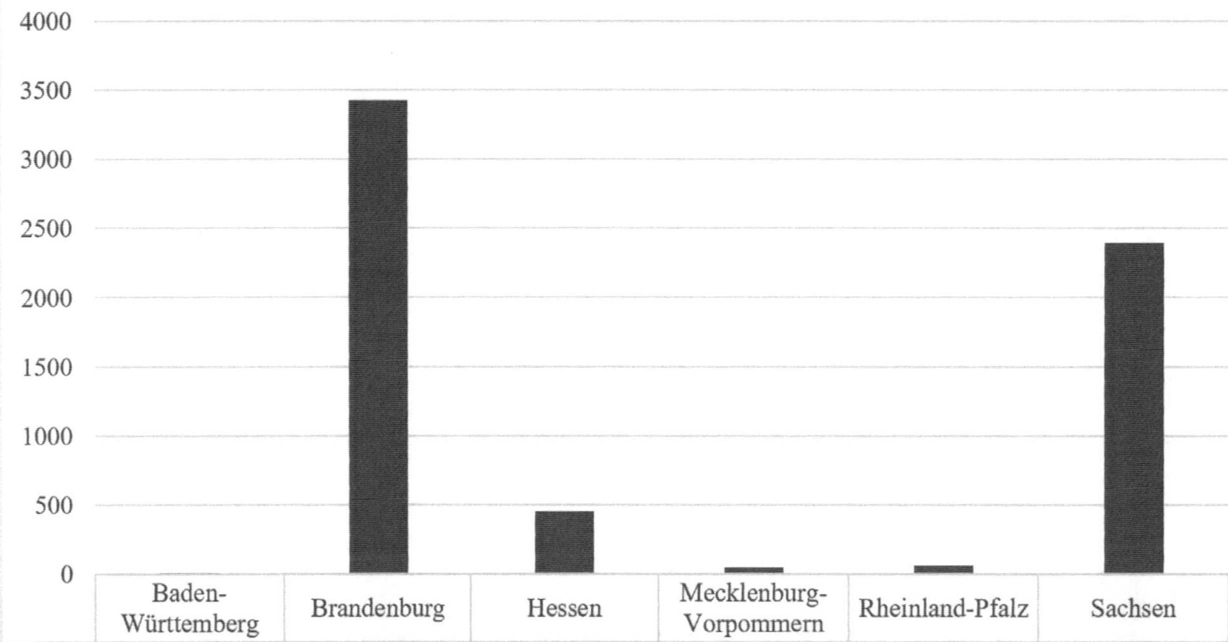

	Baden-Württemberg	Brandenburg	Hessen	Mecklenburg-Vorpommern	Rheinland-Pfalz	Sachsen
■ Total	1	3429	450	47	58	2397

Figure 2[1]: Domestic herd cases per German Federal State September 10, 2020 to December 9, 2024.

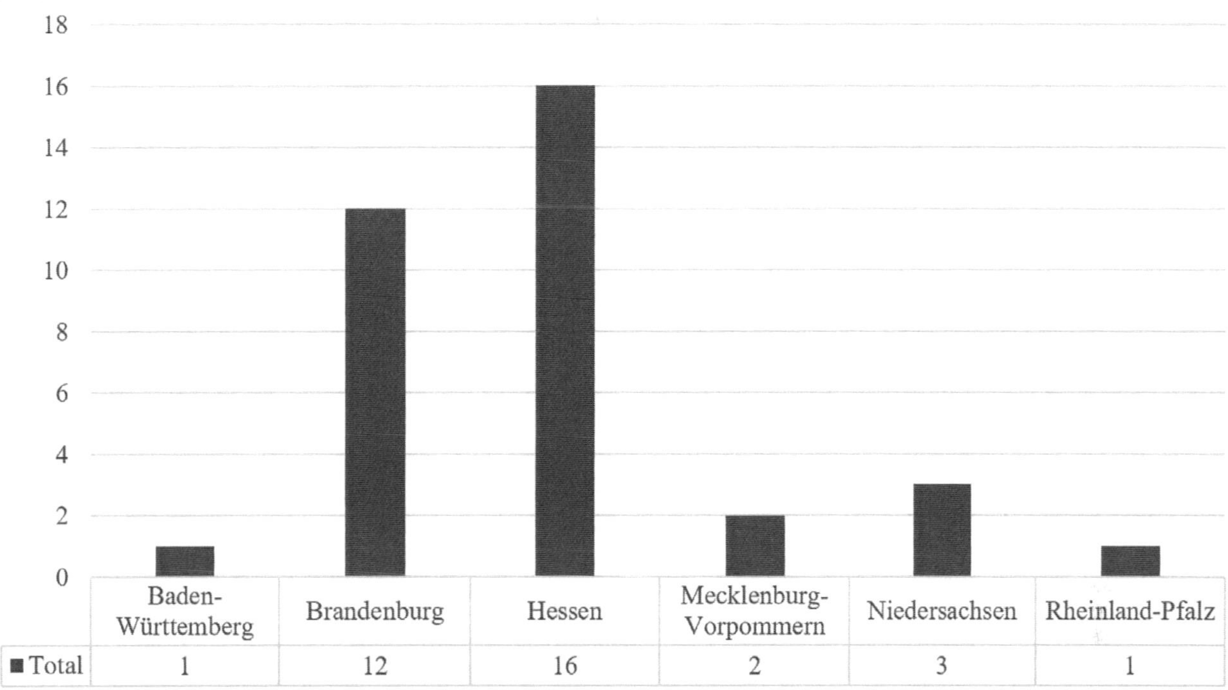

variation of regulatory directives depending on interpretation, risk aversion, and other influential factors.

- Biocontainment and eradication are expensive and not all countries have the financial resources and/or the willingness to pursue to the same level of preparedness and response (eg, Russia, Belarussia, Romania, East Ukraine). On several levels, Germany has the resources and willingness to effectively respond.

Industry preparedness and response

- As with all foreign animal disease (FAD) outbreaks, the industry largely looks to the regulated system of preparedness and response for resources. The industry also looks to governments for prevention awareness information and activities. However, due to real life experiences, industry associations and companies have developed their own well-defined programs of prevention awareness and 'preparedness and response' that include implemented mitigation plans in the hopes of increasing resilience should ASFV enter their regions of operations.
- Even when movements of animals, semen, and products from regulated zones can legally occur, experience has shown that the marketplace ultimately decides if there is a real market for these commodities. German scenarios experienced over the past several years have shown that typically, the precautionary principle prevails in the short term, while 'supply and demand' and habituation prevails in the long term. It can be very difficult to predict if ASFV infected zone will completely stop commercial activity related to swine commercial and breed stock sales.

Epidemiology

- Germany provides food and habitat that is ideal for a large, robust and growing feral swine population which increases challenges associated with the ability to effectively eradicate ASFV from the country.
- Germany is one of Europe's strongest economies, situated in the center of the EU. It has a complex transportation network that provides for the free movement of a large volumes of people and goods across borders within the EU borders. This creates a system under which the movement of ASFV infected products can occur rapidly over long distances creating an ongoing risk of human mitigate transmission opportunities.
- Several different outbreak patterns have occurred in Germany that exemplify the various types of transmission pathways of ASFV. Examples include:
  1. Long distance people facilitated transmission into the feral swine population followed by ongoing local transmission amongst feral swine that results in a larger infected zone that is usually as a result of going undetected for an extended period of time before controls are implemented. This type of outbreak has often resulted in domestic herd outbreaks within the infected zone, most commonly in small holdings, due to low biosecurity combined with increased infrastructure contamination with ASFV;
  2. Long distance people facilitated transmission into the feral swine population followed by very limited local transmission within the feral swine population as a result of rapid detection and control implementation;
  3. Long distance jump into single domestic herds, usually small holding with low biosecurity, which result in no transmission to feral swine due to eradication and decontamination.

- The epidemiological curves of ASFV outbreaks are often misinterpreted. Rapid rises in case numbers may actually reflect the improved efficiency of the response in finding carcasses and not necessarily an increased incidence.
- An effective response must include a multidisciplinary team, including wildlife biologists and hunters, that has a very good understanding of feral swine behaviour.

Society

- Public opinion is not always supportive of the eradication of transboundary diseases such as ASFV and this phenomenon can be exemplified at the local and national level within Germany.

Reference

1. TierSeuchenInformationsSystem - Friedrich-Loeffler-Institut website ASFV database. **https://tsis.fli.de/cadenza/**

https://doi.org/10.54846/am2025/159

Hot topics in European swine production

Vincent ter Beek, MA
Editor, *Pig Progress* at Misset International Publishing, the Netherlands

Introduction

A good start for every paper is to explore its boundaries. Exploring "hot topics in European swine production" also defines what this paper will not be about. It will not cover slow developments. It will not cover general topics in agriculture. And it will not talk about Australia, China or for that matter, the United States.

Perhaps "Europe" requires a little more narrowing down. Because technically this could refer to an entire continent, also including Iceland, the UK, Ukraine, Georgia or Turkey. Seeing that this paper more than once will relate to developments initiated at the headquarters of the European Union, this paper will focus on developments happening in the EU. For most non-EU members the EU is a major trade partner, which is why developments, regulations and standards developed within the EU have a substantial impact on those just outside. The paper will focus on four different areas of swine production: The environment, pig welfare, health issues and nutrition.

Environment

Many developments in swine production relate to one another and that can certainly be said about this hot topic: The environment. What is being discussed in this chapter – nitrogen emissions and zinc oxide – have its repercussions on nutrition as well.

To some, the words "sustainability," "global warming," and "carbon footprint" mean a tremendous lot, others may shrug or others again may be of the opinion that all this "green" attention is merely hogwash. In the EU, environmental policies certainly have found their way into European politics when the ambitious "Green Deal" was introduced to make Europe carbon neutral by 2050. Nevertheless, the announcement of many "green" regulations did not always receive unanimous positive acclaim. The most recent developments in various European capitals give reason to believe that Brussels has understood that a more moderate policy might be the way ahead for the next few years.

One way how environmental legislation has recently affected the swine industry in North Western Europe is related to nitrogen emissions. Particularly, this has become a major problem in the Netherlands, Belgium as well as Denmark in recent years, countries with a thriving swine industry. Bottom line of the problem is that under the European Natura 2000 Directive, every EU member state promised to look well after national parks and nature in general. Now that protection can become jeopardized when nature and agriculture occur close to one another. After all, an excess of nitrogen being emitted can affect biodiversity. Certain plants can thrive whereas others may vanish.

In the Netherlands, the issue has led to extremely fierce debate, with tractors rolling down highways and public controversy impacted at least one and perhaps even two elections. Currently, a government more favorable to agriculture's course is in power but even they cannot make the conflict of interest go away. In Belgium, an approach was taken to engage on a road to reduce livestock numbers and the Danes have also agreed to bring down the number of livestock, including swine.

Another Europe-wide environmental development with implications for the swine industry has been the restriction of the use of zinc oxide in weaner feeds. This certainly has a strong environmental component, as excess use of zinc can end up in the environment – at least that is the thought behind it. As from mid-2022, producers in the EU can only use 150 ppm zinc in weaner feeds, this should be to meet the pigs' nutritious needs. Pharmacological levels of zinc oxide are not to be given any longer to overcome weaning diarrhea. The EU is not the only one doing this – also in Canada and in China laws exist to limit the use of zinc oxide.

Welfare

Many pig welfare initiatives are being researched in the European Union, but when looking to legislation that is likely to become introduced in the next few years, one would be looking at tail docking, farrowing and lactation management, and weaning age.

An end to tail docking is described well in a Scientific Opinion of August 2022, written at the request of the European Commission, published by the European Food Safety Authority (EFSA). This Scientific Opinion is part of the Farm-to-Fork Strategy, a component of the aforementioned Green Deal. The idea of a scientific opinion is to evaluate whether existing practices need a reviewing. Generally, a Scientific Opinion paves the way for further regulations or directives.

In 2008, an EU Council Directive already stated that tail docking should not be performed routinely, but in practice that did not stop producers from tail docking. In case the 2022 Scientific Opinion will be adopted, tail docking of pigs is going to become a thing of the past.

Currently in the EU, tail docking is not performed in Sweden and Finland, and the practice is also banned in non-EU member states Iceland, Norway and Switzerland. Especially Finland has been quite vocal about its ban on tail docking, which was introduced two decades ago. To make the rest of the EU follow, scientific studies are being conducted to figure out what complex multifactorial mechanisms lie behind why tail biting happens and how it can be prevented. Various industry players have also started developing initiatives – worth mentioning is France's largest pig cooperative Cooperl which aims to market pork of pigs whose tails have been left intact.

A second important development is how sows should be kept during farrowing and lactation. The same Scientific Opinion of August 2022 openly discusses that free farrowing should become the standard, with an enlarged area of the total farrowing pen. This will have far-reaching consequences throughout the EU, as it will have an impact on the sizes of farrowing houses. Ahead of new legislation, both Austria and Germany have already adopted new legislation – each with their own specifications. Sweden, as well as non-EU member states Switzerland and Norway, already banned the use of farrowing crates.

Good to mention in this context is recent legislation in Germany with regard to sows during the breeding phase. In 2021, Germany introduced legislation which is pretty similar to what was introduced in California with the Prop 12 initiative – at artificial insemination, sows cannot be confined very long. Germany followed Austria, which initiated legislation along these lines earlier.

A third point in the 2022 EFSA Scientific Opinion relates to the current weaning age. Currently, under EU directives, an age of 28 days is recommended, yet exceptions are allowed in specific circumstances. In practice, the specific circumstances often tend to be the norm. The Scientific Opinion advised to reconsider this, meaning that the weaning age of 28 days would become the standard.

Health issues

Two viral diseases dominate the headlines in Europe these days – African swine fever virus (ASFV) and porcine reproductive and respiratory syndrome virus (PRRSV), each for their own reasons.

For over a decade now, ASFV has been top of mind for the swine industry in Europe. The virus arrived on the continent via Georgia, Armenia and Azerbaijan in 2007 and then gradually made its way into Russia, Eastern Europe and finally the European Union (2014). Late 2024 most of Middle and Eastern Europe had had a visit by the virus, as well as a few countries in Western Europe, including Sweden, Belgium, Germany and Italy.

To date, the Czech Republic (2017), Belgium (2018) and Sweden (2023) are the only countries that have managed to eradicate the virus – with the unfortunate note that the virus returned to a different part of the Czech Republic in 2022. In these successful cases, there happened to be a one-off infection in feral pig populations, which could be controlled when combining the right veterinary policies with cooperation and transparency.

In Germany, the situation is somewhat different, as the country shares a long border with Poland, where the virus is out of control. As a result, there is a constant risk of renewed waves of infected feral pigs migrating across the border. In addition, in summer 2024, the virus emerged near Frankfurt-am-Main, in the country's south-west. So far, this outbreak has formed a challenge, leading to various farms being infected as well.

Italy is also having difficulties getting ASF under control. In 2022, the virus emerged at five different locations in Italy. Four out of five could be controlled, but the fifth – in the north – has led to serious animal health issues. Due to an abundance of feral pigs as well as pig farms in the region around Milan and Genoa, the virus managed to spread quickly, leading to at least 120,000 domestic pigs having to be culled over the last two years.

Health-wise, also PRRSV provides an entirely different challenge. Spain has been fighting the PRRS-variant Rosalía for several years now, which essentially is a combination of the American and European subtype. In comparison to earlier varieties of PRRS, Rosalía causes higher levels of mortality and abortions in sows, more stillborn piglets and higher mortality rates in growers and finishers. As Spain is not exporting live pigs, Rosalía is not creating problems elsewhere.

Not surprisingly, various countries have expressed the desire to get the virus under control once and for all. Hungary, building a centralized nationwide program, says its swine herds are ready to receive the status free from PRRS. As per this year, vaccination against PRRS is forbidden. Denmark, on the other hand, has embarked on a different route, to entirely eradicate the virus. In an initiative supported by many industry stakeholders, the Danish decided to include PRRSV in the national specific-pathogen-free breeding program based on serology tests. This program has been underway now for two years and again – transparency and cooperation are the basis. It is expected it will take some time for the virus to flare out.

Nutrition

The removal of pharmacological levels of zinc oxide from weaner pig diets as per mid-2022 has led to a re-thinking of weaner pig management. Ways to prevent the occurrence of post-weaning diarrhea involve solutions from the areas of genetics, vaccination as well as management – but the majority of suggested solutions undoubtedly come from the field of pig nutrition.

Even though none of them can be considered to be the silver bullet, it shows a market eager to find different solutions. Some ideas for feed-related solutions involve the usage of wheat bran in the diets, or the inclusion of probiotics, benzoic acids or yeast products. In addition, a lot of attention is being paid to train piglets how to consume solid feed prior to weaning, to make them well-prepared.

A more fundamental route of exploration is the search for lowering crude protein levels in weaner pig diets. After all, high levels of proteins in weaner diets can contribute to the occurrence of post-weaning diarrhea. When taking this approach, various other issues are also being tackled. Protein sources are costly anyway, often protein sources need to come from far which adds to pig production's CO_2 footprint, and high protein levels can also lead to higher levels of nitrogen being emitted into the environment.

Reducing levels of crude protein, however, is a balancing act, as it can also impact performance levels, which opens the door to the need for precision technologies. A lot of research is spent to the inclusion of synthetically created limiting amino acids, for example methionine and lysine. Should those be added to the feed separately, crude protein levels can be lowered.

Conclusion

When reviewing the current hot topics in swine production, one might conclude that the future may not look too bright for swine production in the EU, with challenges coming from many sides. Experience showed, however, with the ban on antibiotic growth promoters (2006) or the ban on sow stalls (2013), that faced with new challenges, the European swine industry has also continued to reinvent itself, with the most optimistic, skilled and forward-looking producers emerging to become the leaders of the future.

Swine production in the future will continue to evolve, so much is clear. If there would be one thing the US swine industry could pick up from this paper, it would be that hot topics in Europe's swine industry matter for the US. Perhaps in 2025, Europe's hot topics may lead to a chuckle, a frown or a shrug for professionals in the US swine industry. But give it a decade or perhaps two, and certain topics have crossed the Atlantic as well.

In other words: Forewarned is forearmed.

https://doi.org/10.54846/am2025/160

Unlocking the secrets of *Streptococcus zooepidemicus*

LeeAnn Peters, BSc(Ag), PAg
Director, Technical Services
Maple Leaf Foods Inc, Landmark, Manitoba

Introduction

Until 2019, *Streptococcus equi* subspecies *zooepidemicus* was thought to be a commensal bacterium in swine in North America. It was not known to cause extreme impacts to hog production through significant sudden deaths and abortions. This paper will describe the extensive experience of a large commercial system with *Streptococcus equi* subspecies *zooepidemicus*, (strep zoo) and provide evidence of long-term human asymptomatic carrier status and subsequent zoonotic transmission causing infection in sows. Considering this case of amphixenosis, it will discuss potential protocol changes to consider for the entire hog industry at large.

Background

The bacterium is a β-hemolytic, Lancefield group C streptococcus first isolated by PR Edwards in 1943. He found it was an encapsulated, gram-positive bacteria that is non-motile.[1] Strep zoo is particularly well documented as an opportunistic bacterial with zoonotic potential in horses[2,3] and in other animals like dogs, cats, cattle, monkeys, sheep, goats and chickens.[4-11] In mammals it can cause severe symptoms with pyrexia, septicemia, meningitis, and pneumonia, that often results in death.[2]

Human cases are rare but documented, usually in elderly or immunocompromised individuals that have contact with infected horses or have consumed unpasteurized dairy products or pork.[12] Prior to 2017, there were thirty-two recorded cases of human infections world-wide, with symptoms including rheumatic fever, kidney failure, meningitis, and arthritis.[2] To the author's knowledge, asymptomatic strep zoo human carriers infecting other mammals does not appear well studied or documented to date. Typically, the literature focuses on animal to human transfer.

The most striking characteristic observed was the speed at which adult sows succumbed to the disease. Animals would appear healthy at morning checks, but over the course of hours would develop high fevers, become lethargic and die with no other symptoms.[13] At the peak of the infections, the herd mortality exceeded 45% and abortions soared above 25%. On post-mortem: edema, lymphadenopathy and severe septicemia were common observations. Serotyping identified the strain as ST-194.[13] Costa's development of a type 194 specific real-time PCR for enabled accurate diagnostics for both animal and environmental samplings supporting the investigations.[13]

In the sow facilities, attempts to control the disease through multiple methods failed. Lab results demonstrated the susceptibility of strep zoo to common antibiotics including tiamulin, tilmicosin, penicillin and ceftiofur. However, broad and heavy application of these antibiotics both through multiple methods did little to slow the progression of disease in the field. The hypothesis was that the infection "walled" itself off deep in fatty tissues so despite proven susceptibility in the lab, field application was unsuccessful. Antibiotics administered multiple times in six separate cases, totaling over six figures per case, produced poor results. For the animals that survived, immunity was short-lived with no predictability on survival over the long term. Three different autogenous vaccines had no impact. Morbidity and mortality in nursery and finishing sites downstream persisted throughout the outbreaks resulting in poor growth, high mortality, and condemnation rates at the slaughter plant. In each of the cases resolution came only through complete depopulation and thorough sanitation. Later breaks in two other unrelated commercial units in Canada proved to require the same actions.

Timeline of infections

In March 2019, high sudden mortality occurred in a group of replacement gilts in a multiplication grow out (Figure#1-Site A). Diagnostics indicated the presence of strep zoo. After communication with experts in various sectors of the field, the consensus was it is a commensal bacterial that should not cause widespread herd issues. The replacement gilts were moved to four commercial sites. (Figure #1-Site B-E) By April 2019 animals in all sites had developed clinical symptoms and were formally diagnosed with strep zoo. Site F followed in January 2020, and at the time had no clear connection to the initial breaks.

In each case, the disease spread quickly by nose-to-nose contact. It dispersed first within pen and then between pens, until it had spread through the entire facility. In the later rebreaks, eliminating contact between pens and vigilant boot sanitation provided better control of the transmission in loose-housed barns. Site F, a stalled barn, with shared water troughs observed more rapid infection and mortality rates.

Sites A-D depopulated and underwent extensive cleaning and disinfection. This resulted in successful eradications post repopulation and the sites remain negative today. Site F depopulated and it was decommissioned. Despite strict adherence to the same protocols, Site E experienced two further re-breaks with similar severity in December 2020 and December 2022. (Figure 1)

In March of 2020, a medical professional was contracted to complete human diagnostics via bacterial cultures. This was driven out of an abundance of caution for the health of the employees. Nasal and throat swabs were collected from many employees that either worked within the positive sites or were support staff that visited those sites regularly. This included veterinarians, maintenance technicians and production management employees. All employees were negative for strep zoo on bacterial cultures.

Site E with its subsequent rebreaks became an obsession for the company. Despite extensive "tear-down" style sanitation, intense inspection, rest time for the barn, and swabbing to ensure eradication, the site broke repeatedly. Koirala et al, demonstrated that the bacteria does not survive beyond 25 days even in a perfect environment without a live host.[14] Several papers supported that work demonstrating survival times of 1 to 3 days on the surfaces of wood, metal, and rubber in warm conditions.[15] Another study demonstrated when substrates (wood, boots, clothing etc) were inoculated with the bacteria, its survival was noted at 2 days in warm, dry conditions and extended to 34 days in cold, wet

Figure 1: Timeline.

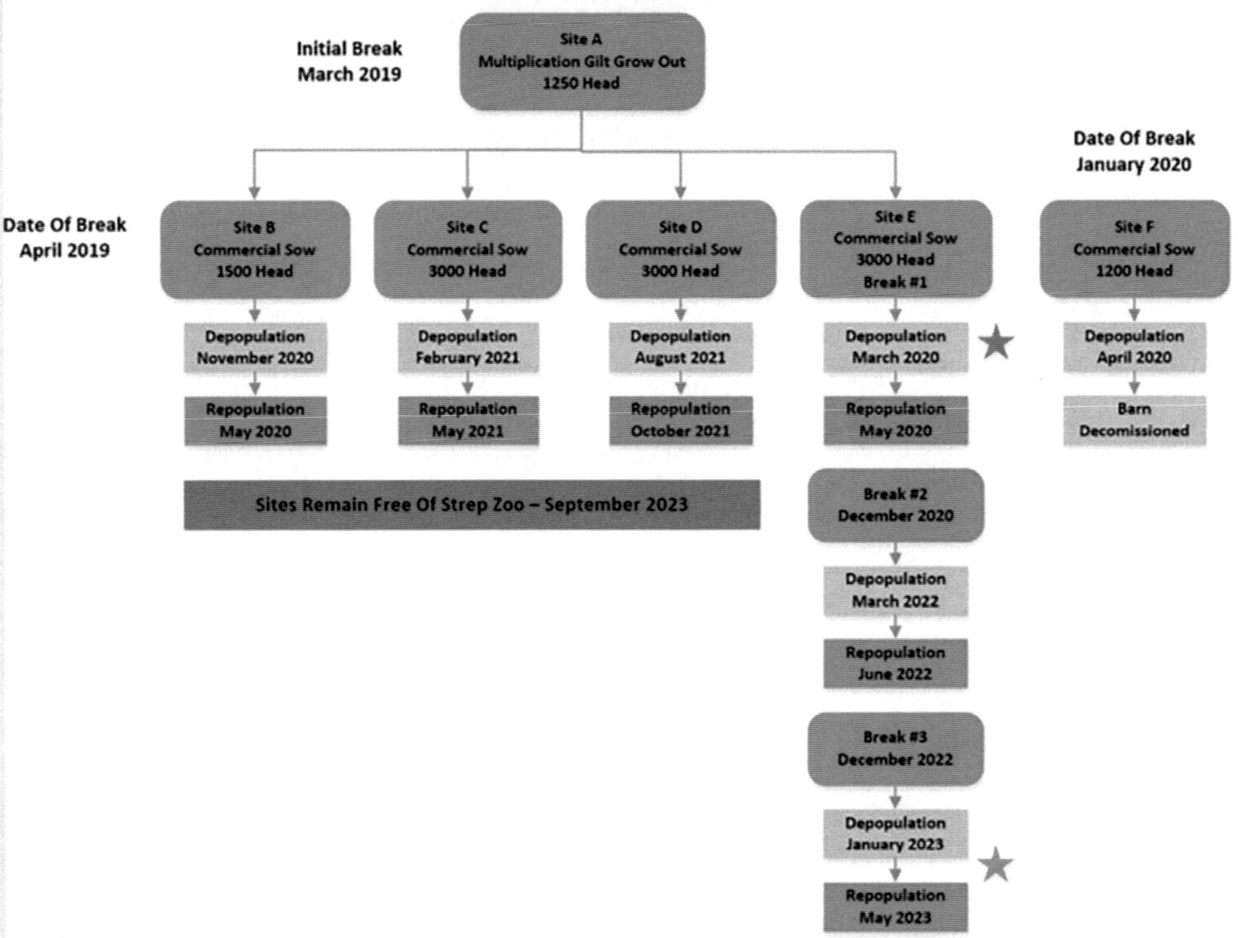

Red star, strep zoo cultures negative; green star, strep zoo positive.

conditions.[16] All eradications exceeded those time periods, which should have resulted in no live bacteria remaining in the environment and therefore an exceptionally low risk of re-infection. In acts of desperation, diagnostics began on all life forms found in or around the facility including voles, mice, and small flying insects looking for live potential hosts that might be causing the repeat infections. All returned negative PCR results.

January 2023 found us at an impasse. The barn depopulated for the third time in four years. Until the source of the breaks was identified the barn would not be repopulated. Another round of strep zoo in this facility was not possible either financially or for the mental health of the staff. Additionally, the cause for the original break in Barn A remained unknown. This left the company feeling very vulnerable to future breaks. These factors combined to trigger a new period of intensive research, discussion and exploring alternative approaches to solving the mystery.

Alternate approach

In early 2023, an un-related large commercial site in Western Canada broke with the same strain of strep zoo. Detailed trace outs demonstrated no points of contact between the sites described above and the new break. The system veterinarian collected and submitted large numbers of environmental samples including employee masks. The nanofiber N95 masks returned positive PCR results. However, as the barn still contained infected animals, it could not be concluded the source of the bacteria was human as opposed to originating from environmental contamination.

Fueled by this information, the exercise was repeated at Site E. By February 2023, the barn had been vacant for 8 weeks, and a full sanitation and disinfection cycle had occurred. In addition, all rodents and insects had been fully eradicated from the facility. One mask returned a positive on PCR tests for both strep zoo and more specifically for ST-194. While promising, it was still not possible to conclude if the bacteria on the mask was from the human wearing it or simply transferred while the employee was working in the environment.

This spawned a new hypothesis: Could humans be asymptomatic carriers detectable through PCR but not through traditional cultures done in the human medical laboratories? Would human tonsil and throat swabs be positive on PCR for strep zoo and specifically ST-194? At the time, the PCR for strep zoo and ST-194 was only available through the veterinary laboratory, which was not permitted to process human diagnostic samples.

In collaboration with veterinary and medical professionals, a novel approach was developed. We would collect tonsil and throat swabs from the employees. Through human medical diagnostic labs, samples would be cultured and then the bacterial DNA would then be extracted. The extracted DNA would then be sent to the veterinary lab for PCR tests for strep zoo and more specifically ST-194.

Results: Laboratory results and investigative findings

While the bacterial cultures again returned negative results, the veterinary lab PCR test results were positive for strep zoo and ST-194 on the tonsil and the throat swabs for one individual. The positive results belonged to the same employee whose mask had assessed positive in the initial mask environmental test. The employee reported being completely asymptomatic for the duration of the infections.

This information culminated in a proposed scenario for the strep zoo breaks: An employee contracts strep zoo from infected sows during the initial break. The employee becomes an asymptomatic carrier of the pathogen for a prolonged period: in this case 4 years. The same human transfers the bacteria back to three separate herds causing full site disease breaks.

Based on this hypothesis, further investigation revealed the following supportive information:

A. **Swine to human transmission:**
For the duration of Site E – break 1, employee A participated in the removal of about 95% of the infected fallen stock. The initial break occurred in the hottest temperatures of the summer. At these temperatures, it is common for employees doing highly physical tasks to work without personal protective equipment like masks and goggles. In the heat, employees wipe perspiration from their faces with their hands and forearms. Doing so while working with infected blood and saliva could present a pathway for infection. The human mucus membranes could come in to contact with the pathogen which could then proceed to colonize the tonsils and nasal passages. This could explain how employee A contracted strep zoo of the same strain from positive sows during break 1. We hypothesized the initial infection in the summer of 2019.

B. **Human to swine transmission:**
The next step was to look at the connection between employee A and subsequent breaks. employee A was working in each site that broke following the initial infection of the four farms infected directly from the gilt multiplier. Figure 2 traces the movement of employee A during the Site E and Site F strep zoo breaks. Three means of transmission of the bacteria back to the animals were identified. First, common respiratory infections result in an increased shedding of the bacteria onto hands and surfaces by the carrier. Secondly, employees removing their masks to facilitate sneezing or spitting within the pens. Finally, wastewater from the showers and bathrooms in the offices, draining through the main manure line in the barn. Animals may be exposed to the bacteria either directly through pit overflow or indirectly through live vectors like flies.

C. **Human to human and other species transmission:**
As part of the disease investigation and to ensure no reservoirs for future infections, members of employee A's family and their pets were evaluated for the bacteria. The results were negative.

Discussion

There are six specific factors surrounding employee A that made detection a reality and support the proposed scenario. There was complete stability of the staff members in Site E for more than 5 years. Employee A had never travelled outside of the province or the country at any point during their lifetime. They are not a consumer of uncooked meats or unpasteurized dairy.

Apart from the swine herds in questions, the employee had no contact with other livestock. They were extremely healthy, reporting a 7-year period since the last treatment with antibiotics for any condition.

Finally, in the weeks, proceeding the third break in Site E, employee A experienced significant respiratory illness, resulting in self-declared breaches in biosecurity and standard operating procedure practices. Under medical supervision, the employee completed a 21-day course of antibiotics to clear the infection. Subsequent testing at regular intervals has returned negative results demonstrating a cure to date.

Although the source of the initial break in the system, has never been formally determined, Barn A had three visitors with animal contact in the weeks proceeding the break: The possibility that one may have been an asymptomatic carrier remains a reasonable conclusion.

Industry implications

As a result of this experience, we believe in encouraging broader discussion to share knowledge of strep zoo and its potential impacts but also to consider changes in practice that might prevent future infections. Suggestions include:

Evaluate the risks from farm employees and support staff that have regular contact with other animals that may be strep zoo carriers. This includes but is not limited to horses, assembly yards, abattoirs, dog kennels and zoos. Prohibit contact with sows by external visitors, with only extremely specific exceptions. When hiring unfamiliar staff, consider the risks and if screening or testing might be prudent.

Execute strong education campaigns on the importance of personal protective equipment when working around animals. Highlight prohibited activities (spitting in pens, sneezing without a mask around animals, poor hand hygiene, urination, and defecation in pig production areas).

Pro-actively screen employees that want to practice the consumption of uncooked meat products and unpasteurized dairy. Impose intensified travel restrictions with high-risk destinations: Including downtimes and prohibited travel activities. Finally, where possible, redesign wastewater systems to isolate human waste to septic tanks.

Conclusion

Site E has been repopulated and is producing live healthy piglets. As of November 2024, it remains negative for strep zoo. All fifteen staff from Site E remain employed, including employee A.

The impact of strep zoo to the business has been staggering. Millions of dollars were spent on diagnostics, treatments, vaccinations, depopulation, sanitation, and repopulation. This does not include the lost opportunity cost on unborn market animals or the cost of vacant barns for extended periods of time. Additionally, the impact on the mental health of the employees should not

Figure 2: Employee movement.

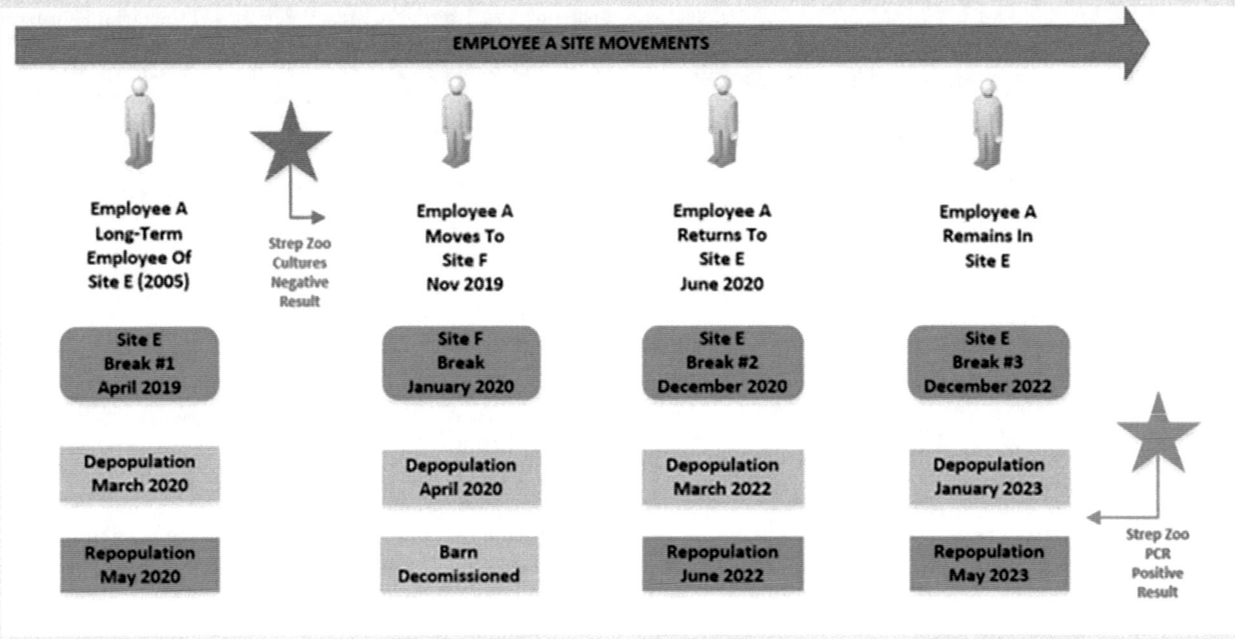

be underestimated: this was a tough battle for all involved. Strep zoo, specifically ST-194 remains a legitimate risk to North American herds and currently prevention remains the most viable management option.

Acknowledgments

Dr Matheus Costa, DVM, PhD, DABVP (Swine Health); Dr Richard Rusk DVM, MD; Dr Frank Marshall, DVM; Dr Jewel White, DVM; Dr Jan Xavier Lumasac, DVM.

References

1. Beres SB, Sesso R, Pinto SW, et al. Genome sequence of a Lancefield group C Streptococcus zooepidemicus strain causing epidemic nephritis: new information about an old disease. *PLoS One.* 2008;3(8):e3026. Published 2008 Aug 21. **doi:10.1371/journal.pone.0003026**

2. Pelkonen S, Lindahl SB, Suomala P, et al. Transmission of Streptococcus equi subspecies zooepidemicus infection from horses to humans. *Emerg Infect Dis.* 2013;19(7):1041-1048. **doi:10.3201/eid1907.121365**

3. Streptococcus Equi: What the CDC Wants Horse Owners to Know About Bacterial Infection Transmission from Horses". *The Horse Owner's Resource.* Retrieved 2018-04-30.

4. Lamm CG, Ferguson AC, Lehenbauer TW, Love BC. Streptococcal infection in dogs: a retrospective study of 393 cases. *Vet Pathol.* 2010;47(3):387-395. **doi:10.1177/0300985809359601**

5. Blum S, Elad D, Zukin N, et al. Outbreak of Streptococcus equi subsp. zooepidemicus infections in cats. *Vet Microbiol.* 2010;144(1-2):236-239. **doi:10.1016/j.vetmic.2009.12.040**

6. Sharp MW, Prince MJ, Gibbens J. S zooepidemicus infection and bovine mastitis. *Vet Rec.* 1995;137(5):128. doi:10.1136/vr.137.5.128-b.

7. Pisoni G, Zadoks RN, Vimercati C, Locatelli C, Zanoni MG, Moroni P. Epidemiological investigation of Streptococcus equi subspecies zooepidemicus involved in clinical mastitis in dairy goats. *J Dairy Sci.* 2009;92(3):943-951. **doi:10.3168/jds.2008-1548.**

8. Dzhikidze ZK, Krylova RI, Voskanian NA, Kukava GG, Dzhindzhiia PIa. Vspyshka infektsii, vyzvannoЀ Streptococcus zooepidemicus, sredi laboratornykh primatov [An outbreak of Streptococcus zooepidemicus-caused infection among laboratory primates]. *Zh Mikrobiol Epidemiol Immunobiol.* 1996;(1):81-84.

9. Priestnall S, Erles K. Streptococcus zooepidemicus: an emerging canine pathogen. *Vet J.* 2011;188(2):142-148. **doi:10.1016/j.tvjl.2010.04.028**

10. Mätz-Rensing K, Winkelmann J, Becker T, et al. Outbreak of Streptococcus equi subsp. zooepidemicus infection in a group of rhesus monkeys (Macaca mulatta). *J Med Primatol.* 2009;38(5):328-334. **doi:10.1111/j.1600-0684.2009.00359**

11. Garmyn A, Van de Velde N, Braeckmans D, Ronsmans S, Boyen F, Verlinden M. An Outbreak Associated with Streptococcus equi Subsp. zooepidemicus in Layers: Evidence of Fecal Transmission. *Avian Dis.* 2020;64(3):343-346. **doi:10.1637/aviandiseases-D-19-00191**

12. Poulin MF, Boivin G. A case of disseminated infection caused by Streptococcus equi subspecies zooepidemicus. *Can J Infect Dis Med Microbiol.* 2009;20(2):59-61. **doi:10.1155/2009/538967**

13. Costa MO, Lage B. Streptococcus equi Subspecies zooepidemicus and Sudden Deaths in Swine, Canada. *Emerg Infect Dis.* 2020;26(10):2522-2524. **doi:10.3201/eid2610.191485**

14. Koirala S, Pantuzza C, Costa MO. Survival of Streptococcus equi subsp. zooepidemicus on environmental samples is affected by material type and incubation temperature [published online ahead of print, 2023 Apr 17]. *Vet Res Commun.*

15. Weese JS, Jarlot C, Morley PS. Survival of Streptococcus equi on surfaces in an outdoor environment. *Can Vet J.* 2009;50(9):968-970.

16. Durham AE, Hall YS, Kulp L, Underwood C. A study of the environmental survival of Streptococcus equi subspecies equi. *Equine Vet J.* 2018;50(6):861-864. **doi:10.1111/evj.12840**

https://doi.org/10.54846/am2025/161

Brazilian swine industry: An update on actual strategies and future perspectives

Glauber S. Machado, DVM, PhD
GFD Consulting / Integrall Solutions, Brazil

Introduction

Brazil is actually the fourth largest exporter of pork in the world (considering European Union as a single entity for comparison purposes), with a total sow inventory of 2.1 million productive sows and 5.4 million tons of carcass equivalent produced in 2024. For many years pork exports were stagnant at volumes ranging between 500 and 600 thousand tons per year, until the 2018 African swine fever (ASF) epidemics in China (and other Asian countries) caused a major positive impact on Brazilian pork demand, which suddenly moved our pork exports to a new level. Since year 2020, Brazil exports more than 1 million tons of in natura and processed pork yearly. Export volumes keep growing every year since then, with a total expected volume of 1.233.330 tons for year 2024 (978.337 tons already exported from January to October 2024, when this article was written). China and the Philippines account nowadays for 35% of Brazilian pork exports, representing therefore the two main destinations. Important to emphasize that Brazilian pork exports are much better distributed worldwide and not anymore concentrated in one or two main markets, making therefore exports to become a very robust and consistent market from Brazilian pork.

Although pork exports have been consistently increasing every year, domestic consumption of pork still accounts for 67% of the national production, with the remaining 33% being exported. In recent years there has been a substantial growth also in the domestic pork consumption in Brazil. Such a growth on domestic consumption can be explained by various factors. Most relevant factors were:

- Rising price of beef, which has been reaching consistently higher prices in the national retail market, making pork a very competitive meat when compared to beef;
- long-term marketing campaigns carried out by several actors from Brazilian pork industry, aiming not only at the consumer but also at decision-making professionals like nutritionists and medical doctors. Training packers and retailers to better present pork meat to the final consumer was also part of such marketing campaigns;
- significant expansion in pork production and slaughter weight averages over the last years has determined greater domestic availability of meat at relatively lower prices to the Brazilian consumer.

Swine production systems in Brazil

Several distinct productive arrangements between packers and swine producers are implemented in Brazil, including classical integration, cooperatives, supply contracts, and other arrangements that allow not only for the implementation and development of a solid swine production chain, but also greater market competitiveness. For easiness of understanding, we could classify existing production arrangements into 7 main categories:

1. Large scale traditional integration
2. Regional partial integration
3. Packers holding their own production system
4. Cooperative systems for joint purchasing and slaughter
5. Intercooperation: Cooperatives of pig producers supplying for a central cooperative
6. Intercooperation: Slaughterhouse/packer shared by several cooperatives
7. Cooperative of pig producers supplying privately owned slaughterhouse/packer

With regard to the consolidation of the activity, as already partially discussed in this article, there is clear tendency for integrated production arrangements to grow, to the detriment of independent production of pigs for the spot market. At each crisis cycle (related to either costs or market prices), significant changes are observed in the productive arrangements of Brazilian swine production, with a reduction in the number of small and medium-sized independent producers, opening space and opportunities for the expansion of large vertical integrated systems.

The Brazilian swine industry has undergone an intense process of professionalization and modernization over the last 20 years, always accompanied by consolidation, grouping, and growth in scale. Didactically, it is possible to establish 4 trends or consolidation models observed over the years in Brazilian swine industry, especially after the cyclical crisis that recently impacted our industry:

a. Gradual disappearance of small independent producers, accompanied by expansion of verticalized systems (integrations and cooperatives): very common trend taking place in the South and Southeast regions.
b. Reorganization of independent producers into cooperatives, joint purchase companies/coops/associations or even informal pools of producers for joint purchasing purposes, with or without verticalization of production.
c. Growth of large independent professional pig farmers, who are already becoming integrators themselves and absorbing small producers into their production systems.
d. Verticalization of very large independent professional pig farmers by investing in the construction of packing plants for slaughtering and processing of pork, thus creating new players in the market.

The modernization observed in Brazilian swine production over the last years has supported the expansion of production, placing Brazil among the largest producers and exporters in the world. Such a modernization process involves not only investments on environmental control, but also in biosecurity standards, waste processing technology and adaptation to animal welfare regulations (including adjustment of facilities and feeding equipment for sows moved to the group-housing).

Productivity and health status of pig farms in Brazil

Brazilian pork sector is among the most efficient and competitive in the world, not only because of lower production costs related to feed and labor, but also due to the high productivity and good average technical indexes of its herds. The overall health status of the Brazilian swine herds is recognized as very privileged when compared to the situation observed in the main swine producing countries. Among several other globally relevant swine viral diseases to which Brazil is free of, emphasis should be given to the Brazilian naïve status for porcine reproductive and respiratory syndrome (PRRS) virus. Brazil is actually one of the few relevant players within global swine industry where this virus has not yet been introduced or disseminated. In most countries, the introduction of this disease has occurred through international trade of swine genetic material. In order to improve genetic level of Brazilian pig sector, most breeding companies established in Brazil also import genetics (semen and live animals) originating from countries which are recognized by their high-level genetic technology. In most countries where PRRSV is widely disseminated, the initial source of infection was the introduction, by importation, of infected swine at the beginning of the 90s, or at the end of the 80s, from countries where PRRS was present. Some other South American countries such as Colombia, Venezuela and Chile have already identified the occurrence of antibodies against PRRSV and typical signs of PRRS in their swine herds. In Brazil, however, serological tests have been jointly and consistently performed by EMBRAPA and MAPA (Ministry of Agriculture) since 1995 in order to monitor Brazilian PRRSV status. There has been no knowledge or report of any outbreak or even clinical diagnosis, viral isolation or detection of PRRSV RNA in Brazil so far. Although results indicate no evidence of PRRS virus in Brazil, which give a positive competitive advantage against other swine producing countries, it is necessary to keep strict surveillance protocols in place to constantly monitor not only Brazilian swine herds but also - and most importantly - every import of live pigs or frozen semen into Brazil to avoid the introduction and dissemination of such virus within the Brazilian herds. Breeding companies are acting together with EMBRAPA and MAPA in order to keep such protocols in place and therefore keep our nationwide PRRSV naïve status as it is.

Key aspects and challenges for long term success of pork business in Brazil

Brazil has a very robust and economically relevant animal protein sector, with global companies well established and a consistently growing share in relevant markets worldwide. The adoption of high technology in the areas of genetics, nutrition, facilities, and management allowed for our national swine industry to reach technical-economic efficiency and high-quality standards. Nevertheless, competing in the international market is a constant challenge that requires strategic positioning and attention to the signs of economical, political, or social related changes. In this context, we can identify some relevant challenges for success and longevity in the pork production and sales business in Brazil. The volatility of the sector is the biggest challenge for the Brazilian swine producer, as it is for the majority of the global pork chain. The tough times experienced by Brazilian swine industry in 2022/23 were a real and recent example of such volatility, after two years of excellent results. The coronavirus pandemic, tensions in the political-economic environment, resumption of the growth of Chinese herds and upward pressure on prices of agricultural commodities were among the main reasons for one of the biggest historical crises in the Brazilian swine industry over the last couple years. For this reason, it is fundamental to have good financial planning, highly professional supply structure, intense monitoring of the grain market, strategic stock capacity of critical raw materials and constant investment in technology, thus ensuring the longevity and growth of the business, even in an environment of cyclical crisis in the sector. Besides the intrinsic challenge related to price and cost volatility, there are also market challenges (to increase domestic per capita consumption of pork), health challenges (keeping the national herd free from diseases that are technically and economically relevant such as PRRS, porcine epidemic diarrhea or ASF), management challenges (to focus on technical performance conciliated with animal welfare premises), and environmental challenges (to seek for the chain's sustainable development). Demand for sustainable development, including environmental compliance, food safety, and animal welfare, is actually mandatory. Brazilian swine farming still keeps its enormous capacity for sustainable production expansion when compared to most other pig producing countries, given the country's territorial extension, climate, and availability of natural resources, without any threat to the preservation of biomes.

We have technical and structural conditions in place to increase production and, at the same time, to reduce the high polluting potential of our activity. In summary, it can be stated that the most critical points for success in the swine production activity in Brazil are the following:

a. Continuous investment in infrastructure and biosecurity culture within the production systems, with a view to preserving sanitary standard of herds, ensuring not only productivity and cost reduction, but also access to the most demanding markets.
b. Proactive management of the supply area, with special emphasis on purchasing strategies, price protection, hedging and stocking strategies for most relevant and costly feedstuffs (corn, soybean meal, sorghum, distiller's dried grains with solubles), once more than 70% of production costs are there.
c. Building solid and safe commercialization channels for your live animals, whether through contracted production, risk dilution among several buyers, integration with industries or coops or even production verticalization.
d. Technical focus on feed conversion (feed efficiency), weight gain, and carcass quality, to the detriment of the exclusive focus on number of piglets produced (PSY). Focus where the money really is.
e. Unequivocal alignment of production system with the major trends and demands of consumers. Among them: Animal welfare, meat quality, food safety, sustainability and governance policies (ESG).
f. Strategic management of human resources associated with meat production, slaughtering, and processing, as there is a trend toward a reduction in available manpower for the primary sector and increasing difficulty for talent retention.
g. Continuous investment in technologies associated with production, especially in segments of environmental control, ventilation, precision feeding, remote management of production indicators, optimization of labor (process automation and/or robotization) and biosecurity associated with the production and transportation of swine.

https://doi.org/10.54846/am2025/162

Human spillover cases in swine and significance to influenza A virus ecology

P. Gauger
Iowa State University, Ames Iowa

Introduction

Influenza A virus (IAV) is a primary cause of respiratory disease in swine and a component of the porcine respiratory disease complex (PRDC).[1] Influenza viruses are an important health and economic concern for swine producers throughout the world. Swine operations may be affected by seasonal infections or endemic circulation of IAV resulting in high morbidity but typically low mortality.[2,3] Infection with IAV manifests as an acute onset of respiratory disease characterized by fever, lethargy, anorexia, respiratory distress, coughing, conjunctivitis and nasal discharge.[4] Reproductive failure associated with IAV-induced fever has also been documented, which may include infertility, abortions and weak-born piglets.[5] Influenza is a zoonotic viral disease that represents a health and economic threat to both humans and animals worldwide.[6]

History of human-to-swine influenza A virus transmission

Transmission of human IAV to swine has occurred often since 1998 and if the human strains become established in swine, they significantly impact the ecology of IAV circulating in swine.[7-10] Human spillovers of IAV to swine often remain unnoticed or undetected if they do not become endemic in swine and actually may occur more often than expected.[11] In spite of the number of human IAV transmitted to swine, the public health concern is the transmission of IAV from swine to humans through zoonotic infections and the threat of a future pandemic virus originating in pigs.[12-15] In reality, humans have played a primary role in transmitting IAV to swine, which has not only contributed to the genetic diversity of IAV that currently circulate in swine populations, but increase the risk of zoonotic infections after the virus has become adapted to swine.[16]

Genetic analysis confirmed the first IAV in swine were similar to the 1918 virus that coincided with the human pandemic known as the Spanish flu (Figure 1).[17] The virus was isolated from pigs in 1930 and belonged to the H1N1 lineage.[18,19] Influenza A viruses in swine that demonstrate similar genetic characteristics to the 1930 virus are currently designated in the classical H1N1 lineage of IAV that still circulates in swine today and globally is referred to as the 1A classical swine lineage.[16] For over 70-80 years, the classical H1N1 virus remained relatively stable at the genetic and antigenic level and was the only subtype responsible for influenza outbreaks of respiratory disease in swine in the United States (US). It is unknown why additional subtypes did not infect pigs during this extended period and become established. Perhaps reassortant influenza viruses did not efficiently transmit among pigs to allow endemic infections to occur.[20] One could also speculate production practices based on smaller groups of pigs, limited movement of swine and restricted exposure to humans may have contributed to this lack of variability as well. The current 1A classical lineage of IAV in swine includes the US clade designations alpha, beta, gamma (Figure 1), and eventually the pandemic H1 viruses after 2009.[21]

In 1998, the introduction of IAVs of the H3N2 subtype into North American swine may be considered one of the most critical events that shaped the future epidemiology of IAV in pigs. At that time, a severe influenza-like disease was observed in swine in North Carolina as well as Texas, Minnesota and Iowa. Genetic analysis of these H3N2 viruses that were isolated from multiple cases demonstrated the presence of two genotypes. A North Carolina isolate was a double reassortant with gene segments similar to those of the classical H1N1 swine lineage (PA, PB2, M, NP, NS) but also included human lineage hemagglutinin (HA), neuraminidase (NA) and PB1 genes. Influenza viruses isolated from swine in Iowa, Minnesota and Texas were triple reassortants containing gene segments from the classical H1N1 lineage (NS, NP, M), and also included segments from avian lineage (PB2, PA) and the same human lineage virus as the double reassortant H3N2 viruses (HA, NA, PB1).[22] Within one year, the triple reassortant influenza viruses in swine became widespread in the US and with time replaced the classical H1N1, suggesting the unique combination of swine, avian and human internal gene segments may have provided a selective advantage over other influenza viruses in swine.[23]

The six internal gene segments of the triple reassortant influenza viruses that includes the PB1 of human lineage, PA and PB2 of avian lineage and NP, M and NS from the swine lineage, which is still present in IAV circulating in swine, is known as the triple reassortant internal gene constellation or 'TRIG' cassette. The vast majority of the contemporary influenza viruses isolated from

Figure 1: Timeline of human-seasonal H1N1, H1N2, and H3N2 influenza A virus spillovers from people to swine since 1918 that have become established in the pig population including the phylogenetic clade designations for the United States and the global clade classification.

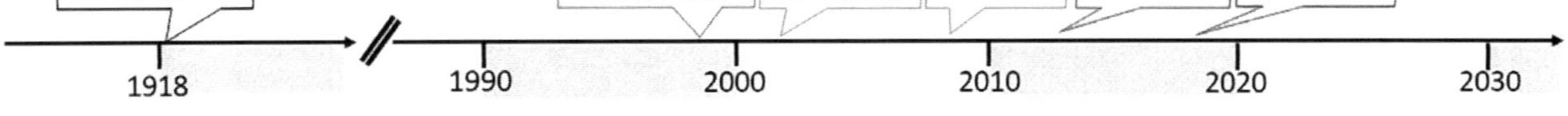

swine in the US and Canada prior to 2009 contained the TRIG regardless of subtype. Shortly after the introduction of H3N2 viruses in 1998, reassortant viruses began to emerge in swine. The classical H1N1 that reassorted with the H3N2 emerged with the HA and NA genes from the classical H1N1 and TRIG from H3N2 viruses.[24-26] Reassortant swine influenza viruses with the HA from the classical H1N1 and NA and TRIG from the H3N2 virus are the contemporary H1N2 viruses.[25,26] All of these subtypes (H1N1, H1N2, H3N2) currently co-circulate in the North American swine population and throughout the world, but globally some IAV have different origins or lineages responsible for establishing the virus in swine.

Further adding to the diversity in H1 influenza viruses in swine was the detection of HA genes derived from human-seasonal IAV in 2005 that became endemic and eventually spread throughout the US pig population.[27] The HA genes of the human-like swine viruses were similar to the human-seasonal H1N1 and H1N2 viruses that were circulating in people during the years prior to that time (2002-2003). The human-like swine virus HA is genetically and antigenically distinct from the classical H1N1 HA, although the TRIG genes of the human-like swine viruses are similar to the TRIG of the contemporary swine triple-reassortant viruses.

The H1 viruses currently circulating in swine have been classified into individual clusters based on branches in a phylogenetic tree and represent the evolutionary diversity demonstrated by the HA genes from each of these groups. The HA from classical H1N1 influenza viruses are known as the 1A clade globally and consist of alpha, beta, and gamma clusters in the US. The 2002 human seasonal spillover HA represents the 1B human seasonal lineage of IAV in swine globally and is known as the delta 1a, delta 1b, and delta 2 clusters of H1 swine viruses in the US (Figure 1).[16] Phylogenetic analysis demonstrated the 1B or delta-cluster contains two subclusters designated 1a and 1b in addition to the delta 2 and likely represents two separate introductions of H1N2 and H1N1 human seasonal influenza viruses into the swine population, respectively.[28] The delta-cluster viruses have a preference for NA genes of human lineage rather than the N1 of classical H1N1 swine lineage although this may vary. The diversity of clusters within H1 viruses circulating in swine and the divergence noted in the delta-cluster group further emphasizes the rapidity with which influenza A viruses are capable of drifting and/or reassorting within the swine population.

In 2009, a novel H1N1 IAV caused severe outbreaks of respiratory disease in humans that quickly spread across the globe at a rate prompting the WHO to declare a pandemic.[29] This novel pandemic H1N1 virus (H1pdm) demonstrated a combination of genes from IAV from North American and Eurasian swine lineages that has never been detected in swine or other species.[30,31] The H1pdm is a triple reassortant with a unique set of six gene segments closely related to the North American triple-reassortant influenza viruses (PB2, PB1, PA, HA, NP and NS) and Eurasian lineage M and NA genes.[32] At this time, no closely related influenza A viruses with this gene combination have been identified in swine in North America or elsewhere in the world as an immediate precursor to the H1pdm.[30,31,33] The closest known HA related to the HA gene of the H1pdm is the North American gamma-cluster H1 lineage. Importantly, the H1pdm forms its own branch within the H1 gamma-cluster viruses confirming its uniqueness among other IAV in North American swine and globally is designated the 1A.3.3.2 clade.[29]

The H3 subtype IAV are also classified according to a clade or cluster designation. Since 1998, three introductions of human H3 subtype viruses have been detected in swine and were classified as clusters I, II and III. The cluster III viruses became dominant in North America and have now evolved into cluster IV which remains one of the common H3 IAV detected in swine.[26,34,35] The H3 cluster IAV in swine are classified according to the global classification system that includes H3.1990.1, 1990.2, 1990.3 and the current 1990.4 clades and a classification scheme that is based on the decade when the virus spilled-over from humans and was detected or became established in swine.[16] Currently, the 1990.4 clade has diverged and includes additional clades 1990.4.a and 1990.4.b in addition to the 1990.4 and each lineage circulates at different frequencies in the swine population. The most recent human-to-swine spillover IAV in the H3 subtype includes 2 different human-seasonal IAV that became established in swine in the 2010 decade.[36-38] The first H3, designated 2010.1, was detected in 2012 and was entirely a human-seasonal IAV that quickly reassorted with swine-lineage N1 neuraminidase genes and emerged as an H3N1 that quickly reassorted with other H3N2 in swine acquiring a swine-lineage N2 neuraminidase to establish the current H3.2010.1 that still circulates in swine.[36,39] The second H3 spillover in the 2010 decade is designated 2010.2 and was detected in 2016. Although this clade of H3 is detected less often compared to the 2010.1, it still persists in the US swine population at low frequency.[37,40] Additional detections of human-seasonal H3 occur at least yearly although many of these do not become established in the swine population.

Why human-seasonal influenza A virus spillover into swine

Known as reverse-zoonosis, the spillover of human-seasonal IAV into swine is common and likely occurs more often than the viruses are actually detected in swine.[41] Humans and swine share several similar characteristics of their respiratory tract as well as sharing the same IAV subtypes that are established in both populations. Thus, bidirectional transmission of IAV between humans and swine can occur.[9] To result in a successful replicative cycle, influenza viruses must efficiently infect the host cell, replicate, and produce functional virus progeny that will be released and infect new cells.[41] The swine respiratory tract contains both α2,3- and α2,6- sialic acids (SA), which are the terminal moieties of glycoprotein receptors on the host cell surface of IAV. Humans also have α2,6-SA sialic acid receptors and research has demonstrated that pigs and people not only share similar IAV receptors but have similar distribution of α2,6-SA receptors in their respiratory tract supporting infection with human or mammalian strains of H1 and H3.[42]

It is well known that swine can serve as mixing vessels of avian, human and swine strains of IAV considering the swine host can support infection and replication of the different lineages of IAV.[43] Therefore, swine are susceptible to human strains of influenza when exposure is adequate to cause infection. Another reason swine support infections with human-lineage IAV that have become endemic is the propensity for reassortment that is a phenomenon of viruses with genetic segments.[44] Reassortment occurs when two different strains of IAV (human and swine) infect the host and replicate in the same respiratory cell allowing the exchange of genetic segments. During reassortment in swine, human strains often acquire the internal gene segments from endemic IAV that circulate in swine. This adaptive process allows IAV with human-lineage surface proteins and internal gene segments from swine-lineage strains to replicate, transmit, and ultimately become established in swine.[41] Preventing human-to-swine transmission of IAV in the workplace has become a necessary measure to preclude the introduction of new human strains into the swine population.

There are other determinants of host range for IAV in swine including binding specificity of the hemagglutinin gene, specific neuraminidase and matrix gene lineages that are compatible with the hemagglutinin of the virus, body temperature, and immunological factors.[41] These additional aspects of IAV may contribute to the potential of a human-lineage strain to become established in swine. However, there are other exposure factors that also play an important role in transmitting IAV from humans-to-swine. These may include the stage of the disease process and magnitude of shedding at the time of exposure to pigs and the extent of contact that occurs between infected/shedding people and swine to allow sufficient transmission of virus. These factors are all key components to successful spillover of IAV to swine.[10,11,16]

How human IAV impact the ecology of H1 and H3 in swine

Human-origin viruses have been repeatedly transmitted to swine worldwide and have had a major impact on the epidemiology of swine IAV.[10,36] Once the H3N2 human-seasonal IAV was transmitted to swine, reassortment with classical strains of H1N1 led to the emergence of genetic and antigenically distinct H3N2, H1N1, and H1N2 strains in swine.[24,26] This also resulted in the emergence of the triple-reassortant internal gene (TRIG) constellation, containing gene segments from a complex reassortment history among swine-, human- and avian-origin IAVs that became the predominant backbone of the viruses circulating in pigs in the US.[45,46] Each introduction of human-seasonal IAV into swine has caused an increase in the genetic pool of IAV circulating in swine, including the surface proteins (HA and NA) as well as the internal gene constellation. However, the spillover of the human pandemic H1pdm into swine has probably had the greatest impact on the genetic ecology of IAV in swine.[47] Shortly after the H1pdm was detected in pigs, reassortant IAV emerged that expanded the genetic diversity of IAV circulating in US swine.[48] This included reassortant IAV with the matrix (M) gene from pandemic strains and a mixture of internal gene segments that originated from either classical strains or the H1pdm. Eventually, this resulted in the emergence of multiple H3 clades that were designated cluster IVA, B, C, D, E, and F that has more recently stabilized to include only the cluster IV, IVA and IVB clades that are currently endemic in pigs (H3.1990.4, 1990.4.a, and 1990.4.b).[48] The H1 subtype of IAV in swine has also been impacted both genetically and antigenically with the introduction of the H1pdm and other human-seasonal IAV, including the HA and NA surface proteins as well as internal gene segments.[44] It is clear human-to-swine transmissions have impacted the current epidemiology of IAV in swine. At least 10 of the current phylogenetic clades in pigs have evolved from a human virus and considering the classical swine H1 virus likely originated from the human 1918 pandemic virus, all IAV clades in swine should be considered of human-origin.[41]

Conclusion

There is sufficient genetic evidence that human viruses are frequently transmitted to pigs, and have had a significant impact on the diversity of viruses that circulate in the US and globally. Additional surveillance is necessary to understand the diversity of IAVs circulating in different regions and the participation of human-origin strains in this overall diversity. The increasing evidence of the important role that human-seasonal strains have played in the genetic and antigenic diversity of IAV in swine suggests control of human IAV may help prevent further spillovers into swine. Understanding the mechanisms behind the reason human strains may or may not spillover into pigs, and the risk factors associated with transmission, requires further research.

References

1. Gracia JCM, Pearce DS, Masic A, et al. Influenza A Virus in Swine: Epidemiology, Challenges and Vaccination Strategies. Frontiers in Veterinary Science 2020;7.

2. Janke BH. Clinicopathological Features of Swine Influenza. Swine Influenza 2013;370:69-83.

3. Janke BH. Influenza A Virus Infections in Swine Pathogenesis and Diagnosis. Veterinary Pathology 2014;51:410-426.

4. Vincent AL, Lager KM, Anderson TK. A Brief Introduction to Influenza A Virus in Swine. Animal Influenza Virus, Second Edition 2014;1161:243-258.

5. Wesley RD. Exposure of sero-positive gilts to swine influenza virus may cause a few stillbirths per litter. Canadian Journal of Veterinary Research-Revue Canadienne De Recherche Veterinaire 2004;68:215-217.

6. Szablewski CM, McBride DS, Trock SC, et al. Evolution of influenza A viruses in exhibition swine and transmission to humans, 2013-2015. Zoonoses and Public Health 2024;71:281-293.

7. Nelson MI, Gramer MR, Vincent AL, et al. Global transmission of influenza viruses from humans to swine. J Gen Virol 2012;93:2195-2203.

8. Nelson MI, Souza CK, Trovão NS, et al. Human-Origin Influenza A(H3N2) Reassortant Viruses in Swine, Southeast Mexico. Emerg Infect Dis 2019;25:691-700.

9. Nelson MI, Vincent AL. Reverse zoonosis of influenza to swine: new perspectives on the human-animal interface. Trends Microbiol 2015;23:142-153.

10. Nelson MI, Wentworth DE, Culhane MR, et al. Introductions and evolution of human-origin seasonal influenza a viruses in multinational swine populations. J Virol 2014;88:10110-10119.

11. Markin A, Zanella GC, Arendsee ZW, et al. Reverse-zoonoses of 2009 H1N1 pandemic influenza A viruses and evolution in United States swine results in viruses with zoonotic potential. Plos Pathogens 2023;19.

12. Bailey ES, Choi JY, Fieldhouse JK, et al. The continual threat of influenza virus infections at the human-animal interface What is new from a one health perspective? Evolution Medicine and Public Health 2018:192-198.

13. Borkenhagen LK, Salman MD, Ma MJ, et al. Animal influenza virus infections in humans: A commentary. International Journal of Infectious Diseases 2019;88:113-119.

14. Bowman AS, Walia RR, Nolting JM, et al. Influenza A(H3N2) Virus in Swine at Agricultural Fairs and Transmission to Humans, Michigan and Ohio, USA, 2016. Emerg Infect Dis 2017;23:1551-1555.

15. Hennig C, Graaf A, Petric PP, et al. Are pigs overestimated as a source of zoonotic influenza viruses? Porcine Health Management 2022;8.

16. Anderson TK, Chang J, Arendsee ZW, et al. Swine Influenza A Viruses and the Tangled Relationship with Humans. Cold Spring Harbor Perspectives in Medicine 2020.

17. Koen J. A practical method for field diagnosis of swine disease. Am J Vet Med 1919;14:468-470.

18. Shope RE. The etiology of swine influenza. Science 1931;73:214-215.

19. Shope RE. Swine influenza: III. Filtration experiments and etiology. The Journal of experimental medicine 1931;54:373-385.

20. Vincent AL, Ma WJ, Lager KM, et al. Swine Influenza Viruses: A North American Perspective. Advances in Virus Research, Vol 72 2008;72:127-154.

21. Lewis NS, Russell CA, Langat P, et al. The global antigenic diversity of swine influenza A viruses. Elife 2016;5.

22. Zhou NN, Senne DA, Landgraf JS, et al. Genetic Reassortment of Avian, Swine, and Human Influenza A Viruses in American Pigs. Journal of Virology 1999;73:8851-8856.

23. Webby RJ, Swenson SL, Krauss SL, et al. Evolution of Swine H3N2 Influenza Viruses in the United States. Journal of Virology 2000;74:8243-8251.

24. Karasin AI, Landgraf J, Swenson S, et al. Genetic characterization of H1N2 influenza A viruses isolated from pigs throughout the United States. Journal of Clinical Microbiology 2002;40:1073-1079.

25. Karasin AI, Olsen CW, Anderson GA. Genetic characterization of an H1N2 influenza virus isolated from a pig in Indiana. Journal of Clinical Microbiology 2000;38:2453-2456.

26. Webby RJ, Rossow K, Erickson G, et al. Multiple lineages of antigenically and genetically diverse influenza A virus co-circulate in the United States swine population. Virus Research 2004;103:67-73.

27. Vincent AL, Ma WJ, Lager KM, et al. Characterization of a newly emerged genetic cluster of H1N1 and H1N2 swine influenza virus in the United States. Virus Genes 2009;39:176-185.

28. Lorusso A, Vincent AL, Harland ML, et al. Genetic and antigenic characterization of H1 influenza viruses from United States swine from 2008. Journal of General Virology 2011;92:919-930.

29. Garten RJ, Davis CT, Russell CA, et al. Antigenic and Genetic Characteristics of Swine-Origin 2009 A(H1N1) Influenza Viruses Circulating in Humans. Science 2009;325:197-201.

30. Smith GJD, Vijaykrishna D, Bahl J, et al. Origins and evolutionary genomics of the 2009 swine-origin H1N1 influenza A epidemic. Nature 2009;459:1122-U1107.

31. Trifonov V, Khiabanian H, Rabadan R. Geographic Dependence, Surveillance, and Origins of the 2009 Influenza A (H1N1) Virus. New England Journal of Medicine 2009;361:115-119.

32. Dawood FS, Jain S, Finelli L, et al. Emergence of a Novel Swine-Origin Influenza A (H1N1) Virus in Humans Novel Swine-Origin Influenza A (H1N1) Virus Investigation Team. New England Journal of Medicine 2009;360:2605-2615.

33. Evseenko VA, Boon ACM, Brockwell-Staats C, et al. Genetic composition of contemporary swine influenza viruses in the West Central region of the United States of America. Influenza and Other Respiratory Viruses 2011;5:188-197.

34. Gramer MR, Lee JH, Choi YK, et al. Serologic and genetic characterization of north American H3N2 swine influenza A viruses. Canadian Journal of Veterinary Research-Revue Canadienne De Recherche Veterinaire 2007;71:201-206.

35. Richt JA, Lager KM, Janke BH, et al. Pathogenic and antigenic properties of phylogenetically distinct reassortant H3N2 swine influenza viruses cocirculating in the United States. Journal of Clinical Microbiology 2003;41:3198-3205.

36. Rajão DS, Gauger PC, Anderson TK, et al. Novel Reassortant Human-Like H3N2 and H3N1 Influenza A Viruses Detected in Pigs Are Virulent and Antigenically Distinct from Swine Viruses Endemic to the United States. Journal of Virology 2015;89:11213-11222.

37. Sharma A, Zeller MA, Souza CK, et al. Characterization of a 2016-2017 Human Seasonal H3 Influenza A Virus Spillover Now Endemic to US Swine. Msphere 2022;7.

38. Zeller MA, Li G, Harmon KM, et al. Complete Genome Sequences of Two Novel Human-Like H3N2 Influenza A Viruses, A/swine/Oklahoma/65980/2017 (H3N2) and A/Swine/Oklahoma/65260/2017 (H3N2), Detected in Swine in the United States. Microbiol Resour Announc 2018;7.

39. Powell JD, Abente EJ, Chang J, et al. Characterization of contemporary 2010.1 H3N2 swine influenza A viruses circulating in United States pigs. Virology 2021;553:94-101.

40. Souza CK, Kimble JB, Anderson TK, et al. Swine-to-Ferret Transmission of Antigenically Drifted Contemporary Swine H3N2 Influenza A Virus Is an Indicator of Zoonotic Risk to Humans. Viruses-Basel 2023;15.

41. Rajao DS, Vincent AL, Perez DR. Adaptation of Human Influenza Viruses to Swine. Frontiers in Veterinary Science 2019;5.

42. Trebbien R, Larsen LE, Viuff BM. Distribution of sialic acid receptors and influenza A virus of avian and swine origin in experimentally infected pigs. Virology Journal 2011;8.

43. Scholtissek C. Pigs as 'mixing vessels' for the creation of new pandemic influenza A viruses. Medical Principles and Practice 1990;2:65-71.

44. Rajao DS, Anderson TK, Kitikoon P, et al. Antigenic and genetic evolution of contemporary swine H1 influenza viruses in the United States. Virology 2018;518:45-54.

45. Anderson TK, Nelson MI, Kitikoon P, et al. Population dynamics of cocirculating swine influenza A viruses in the United States from 2009 to 2012. Influenza and Other Respiratory Viruses 2013;7:42-51.

46. Lorusso A, Vincent AL, Gramer MR, et al. Contemporary Epidemiology of North American Lineage Triple Reassortant Influenza A Viruses in Pigs. Swine Influenza 2013;370:113-131.

47. Howden KJ, Brockhoff EJ, Caya FD, et al. An investigation into human pandemic influenza virus (H1N1) 2009 on an Alberta swine farm. Canadian Veterinary Journal-Revue Veterinaire Canadienne 2009;50:1153-1161.

48. Rajão DS, Walia RR, Campbell B, et al. Reassortment between Swine H3N2 and 2009 Pandemic H1N1 in the United States Resulted in Influenza A Viruses with Diverse Genetic Constellations with Variable Virulence in Pigs. J Virol 2017;91.

https://doi.org/10.54846/am2025/163

Don't be a fomite: The role of personal protective equipment in preventing influenza A virus

Montserrat Torremorell, DVM, PhD
Veterinary Population Medicine Department, College of Veterinary Medicine, University of Minnesota, St. Paul, Minnesota

Introduction

Influenza A virus (IAV) in pigs can be transmitted by direct pig to pig contact, and indirectly, through the contamination of fomites (ie, inanimate objects contaminated with infectious materials) and the air. Secretions and excretions of pigs (eg, feces, urine, respiratory and oral secretions, semen, etc) can contaminate fomites, and farm workers can act as fomites as their clothes, boots and personal objects may be contaminated with IAV. Worker's skin, and also their nares, can become contaminated with IAV and they can be exposed to IAV by handling infected animals, touching contaminated surfaces, and/or breathing contaminated aerosols.

To prevent exposure to IAV in an occupational setting such a farm it is highly recommended to use personal protective equipment (PPE). PPE is considered the last line of defense to prevent infection in an individual but also is the least effective compared to other disease control strategies such as segregation of infected animals, vaccination or use of mechanical controls (eg, ventilation). Part of the challenge resides in how well the PPE is used throughout the exposure period since often wearing PPE is uncomfortable, may not be well fitted and may deteriorate during the duration of the exposure. Examples of PPE include gloves, coveralls, footwear, face shields, and face masks.

Here we will review a few studies that speak of the effectiveness of on-farm PPE with the goal to prevent infection and transmission of IAV.

Face masks

Transmission of IAV via aerosols is of particular interest in swine farms because IAV can be found in the air associated with particles of multiple size ranges able to be inhaled by workers[1] and penetrate the respiratory tract. In a recent study by Lopez-Moreno et al,[2] nasal swabs of farm workers tested positive for IAV when the workers had been working in farms with IAV infected pigs. Forty one percent of workers (24/58) tested positive at the end of a working day during an observation period of 8 weeks. At the sample level, 4.3% (38/887) of nasal swabs tested positive and direct whole genome sequencing from these samples showed evidence of swine-origin IAV on the nasal passages of the workers. These results indicated that farm workers were more likely to test IAV positive at the end of the working day (OR = 1.98, 95% CI 1.14-3.41) compared to testing positive prior to working with the pigs indicating risk of exposure while working with the pigs.

Face masks are recommended for people working with pigs to decrease aerosol exposure. However, PPE use levels vary by farm and activity in part due to lack of comfort wearing them, particularly for face masks. Type N-95 respirators are considered the standard for preventing exposure to viral aerosols since 95% of airborne particles are captured in the respirator. However, a tight fit of the respirators to the face is necessary for particles to be removed and not by-pass the seal. Unfortunately, N-95 respirators are difficult to keep in place when performing common farm chores and environmental conditions with elevated temperatures and humidity make wearing the N-95 respirators particularly uncomfortable for people. Other masks such as dust masks and surgical masks may also be used in farms. However, there is limited information available regarding their effectiveness for protecting workers from IAV at the pig farms.

In a recent study funded by the National Pork Board (Torremorell personal communication), three face masks named N-95 respirators, dust masks, and surgical masks were investigated as a means to decrease the risk of IAV exposure to farm workers and compared the results to not using the masks (control treatment). Twenty four percent (25 out of 102) of samples from workers not wearing masks tested positive while 5.77% (6/104), 8.65% (9/104) and 10.78% (11/102) tested positive for workers wearing N-95, dust mask and surgical masks, respectively. There were statistically significant differences between not wearing masks and wearing masks ($P < .01$) with workers not wearing masks testing positive more often and having lower RT-PCR Ct values indicative of more viral RNA in the worker's nasal passages than any other face mask group. In contrast, there were no differences in the number of positive workers and their Ct values among types of masks ($P > .05$) although N-95 masks had the highest numerical reduction.

These results add to the evidence that IAV exposure through aerosols is common in swine farms and supports the use of face masks as an intervention strategy to decrease the risk of IAV transmission to workers. Understanding which face mask is more comfortable to swine workers should increase compliance and the likelihood of wearing masks.

Clothing and farm activities associated with IAV contamination

There are farm activities that result in the contamination of farmworker's hands and clothing while performing chores at the farm, Lopez-Moreno et al,[3] reported that activities conducted around the weaning period were more likely to result in worker's contamination than activities performed at processing. Of the samples that tested IAV PCR positive, there were 16% (12/76) collected immediately after processing, 96% (45/48) collected after vaccination, and 94% (29/31) collected after weaning. Samples collected immediately after vaccination and weaning, i.e., activities that took place during the peri-weaning period when pigs were about 3 weeks of age and more likely to test IAV positive, had almost 6 times higher risk of IAV detection and had more samples IAV positive (P-value < .0001) than samples collected after processing, ie, an activity that took place in the first few days of life when IAV infected pigs tends to be low. Both, hands and clothes had similar contamination rates (46% and 55% respectively, P-value = .42) and viable virus was isolated from both. These results indicate that contaminated coveralls and hands are not only a source of transmission of IAV between pigs but also are a source of IAV for the workers. Thus, changing coveralls between high risk IAV contamination activities or washing hands or wearing globes should be considered when trying to decrease risk of exposure to workers.

Gloves and hand washing

Detection of IAV on hands of workers handling IAV infected pigs should not come as a surprise. In an experimental setting, 100% of hand wipes tested IAV positive after researchers interacted with the pigs for 15 minutes (Torremorell personal communication). Under farm conditions, a study by Lopez-Moreno et al,[4] reported 46 % of the instruments used when handling piglets and 58 % of farmworkers' hands testing IAV positive.

Properly sanitizing hands is crucial to prevent disease transmission and exposure to workers. In a National Pork Board study, we assessed the effectiveness of four hand sanitation protocols in decreasing or removing IAV from hands in an experimentally IAV infected pig setting. The treatments assessed included: A) hand washing with soap and water, B) hand washing with water only, C) hand sanitizing with alcohol-based hand sanitizer, and D) wearing gloves. The alcohol-based hand sanitizer and wearing disposable gloves were the most effective treatments at reducing a larger amount of virus from hands. In fact, wearing disposable gloves prevented the hands of workers from becoming contaminated in the first place. Hand washing with water only and washing with soap and water reduced the amount of detectable virus but did not eliminate it. Overall, results from this study emphasized the importance of using a hand sanitation protocol to prevent the transmission of influenza between pigs and farm workers.

H5N1 infections and occupational exposure

Occupational exposure of H5N1 2.3.4.4b genotype has been documented in dairy farm workers with direct exposure to milking activities in parlors, or in workers involved in flock depopulation events where the quantity of infected animals and environmental exposure is considered high. The most likely route of infection to workers under these scenarios is the respiratory, oral or conjunctival route of pathogen entry. Effectively wearing PPE such as N-95 masks, coveralls, face shields and gloves under conditions of environmental stress is challenging but necessary. Thus, it will be of the utmost importance to use PPE that protects mouth, nose, eyes and hands should H5N1 be found in pigs. Protecting the health of workers against H5N1 infections is a responsibility that we all should embrace.

Summary

There is more we can do to protect farm workers from IAV infections and most of that entails basic biosecurity practices directed at hand washing, changing clothes after performing high risk IAV activities, wearing gloves or wearing masks. However, wearing PPE well and consistently can be a challenge. Thus, a risk-based approach to use different levels of PPE is recommended. Changing coveralls, wearing disposable gloves and changing them, and washing hands are recommended after performing high risk contamination activities such as loading or vaccinating pigs. Wearing farm dedicated clothing and showering in and out of swine facilities is also recommended to remove IAV-laden particles from the workers' skin. IAV exposure through aerosols and the use of face masks to protect workers from infectious aerosols, should be considered during periods of IAV outbreaks in particular when pigs are housed in enclosed buildings. This may need to be considered specially if H5N1 infections are detected in swine where exposure through aerosols is likely to occur. Ultimately, it is the collection of measures applied comprehensively, including mechanical controls in farms and PPE usage, that will help decrease the risk of IAV transmission to farm workers.

References

1. Alonso C, Raynor PC, Davies PR, Torremorell M (2015). Concentration, size distribution, and infectivity of airborne particles carrying swine viruses. PLoS One. 10(8):e0135675. **doi: 10.1371/journal.pone.0135675. eCollection 2015**

2. Lopez-Moreno G, Davies P, Yang M, Culhane MR, Corzo CA, Li C, Rendahl A, Torremorell M. Evidence of influenza A infection and risk of transmission between pigs and farmworkers. Zoonoses Public Health. 2022 Aug;69(5):560-571. Epub 2022 Apr 20. **doi: 10.1111/zph.12948**

3. Lopez-Moreno G, Culhane MR, Davies P, Corzo C, Allerson MW, Torremorell M. Farm management practices associated with influenza A virus contamination of people working in Midwestern United States swine farms. Porcine Health Manag. 2023 May 15;9(1):13. **doi: 10.1186/s40813-023-00304-2**

4. Lopez-Moreno G, Garrido-Mantilla J, Sanhueza JM, Rendahl A, Davies P, Culhane M, McDowell E, Fano E, Goodell C, Torremorell M. Evaluation of dam parity and internal biosecurity practices in influenza infections in piglets prior to weaning. Prev Vet Med. 2022b, Nov;208:105764. Epub 2022 Sep 21. **doi: 10.1016/j.prevetmed.2022.105764**

https://doi.org/10.54846/am2025/164

Influenza A virus in non-commercial swine and best practices for control and prevention

A. Bowman, MS, DVM, PhD
Department of Veterinary Preventive Medicine, The College of Veterinary Medicine, The Ohio State University, Columbus, Ohio

Introduction

The extended interaction between pigs and humans at agricultural exhibitions fosters conditions that promote transmission of influenza A virus (IAV). This underscores the need to control IAV within these events and the highly mobile population of exhibition swine. Managing IAV spread at agricultural exhibitions is particularly challenging because these events gather pigs from diverse locations and backgrounds, allowing them to intermingle for prolonged periods (3-10 days). This extended comingling amplifies IAV transmission among pigs, leading to higher viral loads and increasing the likelihood of humans being exposed to infectious doses. Notably, most human infections with variant IAV are detected toward the end of exhibitions, when a significant portion of the pigs are actively shedding the virus. Therefore, reducing pig-to-pig transmission is essential for protecting public health in addition to swine health.

Moreover, IAV transmission during exhibitions enables newly infected pigs to spread the virus to other events or home farms after the show. Molecular data confirms that pig movements between exhibition sites have driven recent outbreaks of IAV. Specifically, pigs from national-level jackpot shows have been identified as key vectors for introducing IAV into state-level circuits, which then spread the virus to local county fairs, where human infections occur. Implementing measures to reduce IAV transmission at individual exhibitions not only safeguards the immediate health of attending pigs and humans but also provides broader protection by preventing downstream outbreaks at subsequent shows.

Limit show-to-show movement

The team at Ohio State collected snout wipe or nasal swab samples from 17,009 pigs across 350 national, state, and local swine exhibitions in eight states between 2016 and 2018.[1] Influenza was detected in 13.9% of pigs sampled at jackpot shows, with 76.3% of these events featuring at least one infected pig. Jackpot shows demonstrated 4.3 times higher odds of detecting an influenza-positive pig compared to county fairs. At county fairs where influenza was identified, nearly half of the pigs tested positive, explaining the high risk of zoonotic transmission in these settings. The timing of jackpot shows and the repeated, long-distance movement of individual pigs facilitate the spread of influenza to county fairs. Targeted mitigation strategies to control influenza at jackpot shows could significantly reduce transmission at county fairs and the risk of zoonotic infections.

Shorten shows

We assessed the impact of limiting swine exhibitions to 72 hours or less as a strategy to reduce IAV risk.[2] Our research included daily longitudinal sampling of every pig throughout 16 county fairs during 2014–2015, resulting in 39,768 nasal wipes collected from 6,768 pigs. Additionally, we analyzed IAV prevalence data from 195 fairs held during 2018-2019 to test the hypothesis that shorter exhibition durations would correlate with reduced IAV prevalence. Both studies revealed that shortening swine shows significantly lowers IAV prevalence among exhibition pigs at county fairs. Reducing viral load in exhibition barns is essential to minimize the risk of interspecies IAV transmission and its pandemic potential. Based on these findings, we strongly recommend fair organizers adopt shorter swine show durations to safeguard the health of animals and humans alike.

Modify entry procedures

We collected environmental wipes from surfaces used to corral and guide swine during the weighing and identification process upon their arrival at six fairs, two of which had pigs testing positive for IAV.[3] Since chute systems typically involve multiple stops, three high-contact surfaces were identified at each fair and sampled every 30 minutes. At the fairs with IAV-positive pigs, 30% and 41% of the surface wipes tested positive for IAV via RT-PCR, respectively. Furthermore, IAV isolates were successfully recovered from these environmental samples at both fairs. Detection and isolation of IAV occurred shortly after infected pigs passed through the chute, depositing nasal secretions on these surfaces. This suggests that these surfaces likely serve as fomites, enabling the virus to spread from early-arriving infected pigs to others later in the sequence. To mitigate this risk, the authors recommend mechanically washing and disinfecting the chute surfaces regularly or between animals during weigh-in procedures to reduce the potential for IAV transmission.

Vaccinate

We conducted a study to evaluate viral shedding and transmission in market-age swine vaccinated with live-attenuated influenza virus (LAIV), killed influenza virus (KV), or a sham vaccine (NV) following challenge with IAV.[4] Additionally, naive ferrets were exposed to these pigs in contact scenarios simulating human interactions in field settings to assess post-challenge transmission. Both LAIV and KV vaccination significantly reduced peak nasal viral titers in swine by nearly 100-fold compared to NV swine (LAIV mean: 4.55 log TCID50/mL; KV mean: 4.53 log TCID50/mL; NV mean: 6.40 log TCID50/mL). Air sampling during the post-challenge period showed a decrease in cumulative airborne IAV in study rooms housing LAIV and KV pigs (LAIV, AUC: 57.55; KV, AUC: 24.29) compared to NV pigs (AUC: 86.92). Transmission analysis revealed a delay in infection onset among ferrets exposed to LAIV pigs compared to NV pigs (rate ratio: 0.66; $P = .028$). Ferrets exposed to vaccinated pigs also exhibited lower cumulative nasal wash virus titers (LAIV vs NV, $P < .0001$; KV vs NV, $P = 0.3490$) and experienced milder disease. These findings strongly support pre-exhibition vaccination of swine with LAIV or KV to mitigate the public health risks associated with IAV in agricultural exhibitions.

References

1. McBride DS, Perofsky AC, Nolting JM, Nelson MI, Bowman AS. Tracing the Source of Influenza A Virus Zoonoses in Interconnected Circuits of Swine Exhibitions. J Infect Dis. 2021 Aug 2;224(3):458-468. **doi: 10.1093/infdis/jiab122**

2. McBride DS, Nolting JM, Nelson SW, Spurck MM, Bliss NT, Kenah E, Trock SC, Bowman AS. Shortening Duration of Swine Exhibitions to Reduce Risk for Zoonotic Transmission of Influenza A Virus. Emerg Infect Dis. 2022 Oct;28(10):2035-2042. Epub 2022 Sep 9. **doi: 10.3201/eid2810.220649**

3. Lauterbach SE, Zentkovich MM, Nelson SW, Nolting JM, Bowman AS. Environmental surfaces used in entry-day corralling likely contribute to the spread of influenza A virus in swine at agricultural fairs. Emerg Microbes Infect. 2017 Feb 22;6(2):e10. **doi: 10.1038/emi.2016.138**

4. Lorbach JN, Nelson SW, Lauterbach SE, Nolting JM, Kenah E, McBride DS, Culhane MR, Goodell C, Bowman AS. Influenza Vaccination of Swine Reduces Public Health Risk at the Swine-Human Interface. mSphere. 2021 Jun 30;6(3):e0117020. [Epub ahead of print] **doi: 10.1128/mSphere.01170-20**

https://doi.org//10.54846/am2025/165

Expect the unexpected: Are swine veterinarians ready for H5N1?

C. Scanlon Daniels, DVM, MBA
Circle H Headquarters LLC, Dalhart, Texas

Introduction

On March 25, 2024, the United State Department of Agriculture (USDA) announced that highly pathogenic avian influenza H5N1 (HPAI) clade 2.3.4.4b, genotype B3.13 had been identified in milk and an oropharyngeal swab from two dairy farms in Texas and two dairy farms in Kansas.[1] On April 10, 2024 the USDA issued a federal order requiring additional testing for and reporting of HPAI in livestock and milk.[2] As of January 10, 2025 the total outbreak consisted of 924 confirmed cases in 16 states.[3]

Background

The detection of a known animal and human health pathogen in a new species is a notable event. While bovines had been known to be susceptible to influenza infection and seroconversion to human and swine origin influenza, the recent finding and spread of HPAI in dairy cattle in the US has been an interesting experience. In 1997, the southwest of England recognized a syndrome in dairy cattle they termed "sporadic milk drop syndrome".[4] At that time, the syndrome was associated with seroconversion to seasonal human influenza viruses.[5-7] Description of that clinical syndrome is remarkably like what we have recently seen in the US. In 1998, a retrospective study of 2,345 bovine sera in Minnesota revealed that 58% had some measurable level of antibodies to H1 hemagglutinin protein.[8] Seasonally, prevalence of antibodies peaked during the months of September to November and February to March following the seasonal pattern historically seen in swine. For those that want a deeper dive into the history of influenza in bovines, a relatively recent literature review from 2019 provides a comprehensive summary of past cases and studies occurring globally in the ruminant population.[9]

As swine veterinarians, we can learn much from what has happened in the year since HPAI was identified in dairy cattle in the US. On one hand, we have advantages relative to bovine veterinarians. On the other hand, many of the factors that have caused consternation will continue to be problematic if HPAI is identified in swine.

Swine veterinarian advantages

I believe we have several advantages should HPAI be identified in swine.

1. Strong background and applied experience in diagnostics. Most of us are conducting diagnostics in our clinical practice on a regular basis and are already comfortable in our understanding of strengths and limitations of various diagnostic assays and population/sub-population statistical sampling.
2. Strong working relationships with state and federal animal health officials. Since so many swine moving interstate and many production systems span multiple states, most of us already have routine contact and interaction with multiple state veterinarians.
3. Demonstrated experience with novel and emerging pathogen epidemiology and control. Almost all of us have direct experience working with variant porcine reproductive and respiratory syndrome virus (PRRSV), porcine epidemic diarrhea virus (PEDV), deltacoronavirus, Seneca valley virus, pandemic H1N1 influenza, porcine circovirus type 2 and H3N2 influenza to name a few. We work diligently to elucidate the impact of lessor known agents like parainfluenza virus and porcine circovirus type 3.
4. We have a mindset of disease elimination where possible. PRRSV, *Mycoplasma* and PEDV all have well worked out elimination protocols.
5. We have a history with successful program diseases like pseudorabies.
6. We can direct pork checkoff dollars toward research on disease. The dairy industry does not have this ability.
7. Strong, integrated relationships with producers. Because of the disease risk inherent in swine production, veterinarians have a "seat at the table" for important decisions and our contribution is valued by producers.

Continuing challenges

Some of the challenges that have been identified in the dairy HPAI outbreak are likely to be problematic for veterinarians and producers should HPAI be identified in swine.

1. HPAI in any livestock species is subject to USDA authority. However, this becomes problematic when there are no rules or regulations in place for infections in a new species.
2. HPAI is a select agent.[10] Once samples are known to be positive for H5N1 antigen, many requirements related to storage, handling and testing come into play.
3. Overlapping federal and state jurisdiction of animal health. Federal rules and regulations can promote uniform national requirements while also allowing individual states to promote their own requirements. In the absence of comprehensive federal regulation, there has been a wide variety of state responses to HPAI in dairy cattle.
4. HPAI is of economic significance to dairy producers but has not been great enough to stimulate much independent industry action. For comparison, the relative economic impact of HPAI in dairy is estimated to be approximately 10% that of a PEDV outbreak in swine. We have seen growing interest by veterinarians and producers in eliminating H1 and H3 influenza infections in swine, but the interest and adoption is less than what we see for PRRSV and PEDV. Would we see the same if HPAI was identified in swine?
5. Disproportionate impact to poultry producing neighbors. Currently, identification of HPAI in dairy results in the inability to move lactating cows. Infection of a layer facility results in depopulation.[11] This disparity in response has created consternation.

6. Uncertain impact on public health. We have been fortunate that human infections with the B3.13 strain have been mild and self-limiting and no human deaths have been associated with it to date of this writing. Notwithstanding, in the western pacific region (Australia, Cambodia, China, Lao PDR and Viet Nam), human cases of H5N1 between January 1, 2003 and November 1, 2024 are associated with a human case fatality rate of 54%.[12] We have no way of predicting how or when H5N1 may become more adapted to humans and/or gain additional virulence factors.
7. Confusion regarding the emergence of different influenza clades. More recently, HPAI infections in poultry, swine and/or humans have been identified with D1.113 and D1.2[14] clades. B3.13 continues to be identified in dairy cattle and poultry facilities near affected dairies, but the B3.[13] clade is no longer being identified in migratory and wild bird populations.

Summary

We still have an opportunity to develop the response to HPAI in commercial swine before it occurs. We would be well served to work proactively through the issues identified above.

References

1. United States Department of Agriculture. Federal and state veterinary, public health agencies share update on HPAI detection in Kansas, Texas dairy herds. March 25, 2024. Accessed January 10. 2025. **https://www.aphis.usda.gov/news/agency-announcements/federal-state-veterinary-public-health-agencies-share-update-hpai**

2. United States Department of Agriculture. Federal order requiring additional testing for and reporting of highly pathogenic avian influenza (HPAI) in livestock and milk. April 24, 2024. Accessed January 10, 2025. **https://www.aphis.usda.gov/sites/default/files/dairy-federal-order-eng-sp.pdf**

3. United States Department of Agriculture. HPAI confirmed cases in livestock. January 10, 2025. Accessed January 10, 2025. **https://www.aphis.usda.gov/livestock-poultry-disease/avian/avian-influenza/hpai-detections/hpai-confirmed-cases-livestock**

4. Gunning R.F., Pritchard G.C. Unexplained sporadic milk drop in dairy cows. Vet. Rec. 1997;140:488.

5. Brown I.H., Crawshaw T.R., Harris P.A., Alexander D.J. Detection of antibodies to influenza A virus in cattle in association with respiratory disease and reduced milk yield. Vet. Rec. 1998;143:637–638.

6. Crawshaw T.R., Brown I. Bovine influenza. Vet. Rec. 1998;143:372.

7. Gunning R.F., Brown I.H., Crawshaw T.R. Evidence of influenza A virus infection in dairy cows with sporadic milk drop syndrome. Vet. Rec. 1999;145:556–557.

8. Jones-Lang K., Ernst-Larson M., Lee B., Goyal S.M., Bey R. Prevalence of influenza A virus (H1N1) antibodies in bovine sera. New Microbiol. 1998 Apr; 21(2):153-60.

9. Sreenivasan C.C., Thomas M., Kaushik R.S., Wang D., Li F. Influenza A in bovine species: a narrative literature review. Viruses. 2019 Jun 17;11(6):561

10. Code of Federal Regulations. Title 9, Chapter I, Subchapter E, Part 121-Possession, use, and transfer of select agents and toxins. Amended 12/31/2024. Accessed January 10, 2025. **https://www.ecfr.gov/current/title-9/chapter-I/subchapter-E/part-121**

11. United States Department of Agriculture. Highly pathogenic avian influenza (HPAI). April 2022. Accessed January 10, 2025. **https://www.aphis.usda.gov/sites/default/files/fs-hpai-restocking-your-poultry-flock.508.pdf**

12. World Health Organization. Human infection with avian influenza A(H5) virus. Avian influenza weekly update number 977. December 13, 2024. Accessed January 10, 2025. **https://cdn.who.int/media/docs/default-source/wpro---documents/emergency/surveillance/avian-influenza/ai_20241213.pdf?sfvrsn=5f006f99_147**

13. Centers for Disease Control. Genetic sequences of highly pathogenic avian influenza A(H5N1) viruses identified in a person in Louisiana. December 26, 2024. Accessed January 10, 2025. **https://www.cdc.gov/bird-flu/spotlights/h5n1-response-12232024.html**

14. United States Department of Agriculture. USDA animal and plant health inspection service shares update on H5N1 detection in Oregon swine, bovine vaccine candidate progression. November 6, 2024. Accessed January 10. 2025. **https://www.aphis.usda.gov/news/agency-announcements/usda-animal-plant-health-inspection-service-shares-update-h5n1-detection**

https://doi.org/10.54846/am2025/166

Where to begin when selecting IAV strain(s) for your custom IAV vaccine for swine?

Emily Mahan-Riggs[1], DVM, MVSc; Marie Culhane[2], DVM, PhD
[1]Indianola, Iowa; [2]Shoreview, Minnesota

Introduction

Influenza A virus (IAV or flu) is a common cause of porcine respiratory infections resulting in fever, anorexia, lethargy, nasal discharge, coughing, and sneezing. Flu is detectable at low levels in pig populations year-round. When disease burden due to flu becomes a problem, there are several control methods to explore, including, but not limited to, vaccinations. There is a long history of flu vaccine use in humans. A bivalent (influenza A and influenza B) whole inactivated vaccine was first approved for use in the military in 1945 and then for the general population in 1946.[1] The global influenza program for human influenza vaccines began in 1947 and still continues as the World Health Organization (WHO) Global Influenza Programme (GIP) and the Global Influenza Surveillance and Response System (GISRS).[2] These WHO initiatives support a systematic approach for seasonal influenza vaccine make-up and administration for humans. For pigs in the United States, the history of flu vaccination begins almost a half century later. It was 1994 when the first flu vaccine for pigs became available in North America – a monovalent H1N1 whole inactivated vaccine; however, European flu vaccines for pigs became available in the late 1980s.[3,4] Unfortunately, no global or national programs/systems exist for pig flu vaccinations and therefore the approach to flu vaccination – what strains, when, to what subpopulation of pigs, how, and why – is less well defined. By working together as a practitioner and researcher, we herein attempt to define a systematic approach to flu vaccinations of pigs, with a special focus on strain selection, that we have developed together. We use experiences gained from flu surveillance and vaccinations programs in midwestern United States pig farms and pork production systems to describe our systematic approach.

Approach

Surveillance

When embarking on a flu vaccine program for pigs, the first step is to do conduct flu surveillance, both active and passive. For active surveillance in midwestern US pig farms, the targeted subpopulations for active surveillance are incoming gilts and weaned pigs, since both have been shown to be sources of flu introduction, dissemination, and maintenance on pig farms with ongoing flu problems.[5] For incoming gilts, flu PCR tests are completed on oral fluids collected upon arrival to the gilt developer unit (GDU) and prior to delivery to the sow farms. For weaned pigs, snout wipes, udder wipes, or nasal swabs are collected monthly following published methods.[6] These active surveillance protocols are accomplished by purposely collecting and screening incoming gilts and weaning age pigs regardless of clinical signs. The results of active surveillance can be used to identify newly introduced and emerging flu viruses in a timely manner, often before clinical disease manifests.[7] In addition to active surveillance, passive surveillance for flu in pigs with clinical signs of respiratory disease is conducted. The results of passive surveillance are useful to correlate pathogen detection with respiratory tract lesions; as such, the passive surveillance results from diagnostic case submissions are useful for confirming flu is a disease problem in the herd. Tissues from sick or dead pigs can be submitted for diagnostic testing for flu and are especially helpful for yielding virus isolates that can be subtyped and genetically characterized.

Strain characterization and selection

Data-driven vaccine strain selection for pigs can best be accomplished with the fully characterized flu virus isolates obtained from the passive and active surveillance programs. For the selection of custom IAV vaccines for the midwestern US pig farms under our care and study, we routinely evaluated the flu strains to find a strain that matched with the assumption that matching vaccine strains are the most efficacious.[8] In our case, matching meant finding the flu strains that had the highest similarity in HA antigenic properties, eg, those with an overall hemagglutinin (HA) protein identity of 95% or more and having identical or nearly identical amino acid motifs, to the most flu strains isolated during the previous two years of surveillance. Once a strain with the highest HA protein similarity was identified, the neuraminidase (NA) protein was genetically characterized to ensure that NA mismatch was avoided. The importance of NA matching has been receiving more attention since antibodies inhibiting NA can also reduce disease, virus shedding, and spread.[9] In addition, with pigs having flu strains that are expanding in their diversity[10], attempting to match both the HA and NA may be beneficial. Finally, once HA/NA matched strains were identified, the veterinary and production teams were asked if they thought the strain was a problem strain, present throughout the system, causing disease, or found in a sow farm or GDU that was likely to spread to many destination sites.

Timing of vaccination

In addition to providing the much-needed flu strain characterization data, the diagnostic submissions with flu positive results were often the impetus for production managers and owners to begin a flu vaccination program or change the existing flu program. For example, at the beginning of our shared experience, the flu vaccination program consisted of vaccinating the whole herd within each sow farm twice a year. The goal of this whole herd approach was to stabilize both the progeny from weaning to market at destination production sites and at the origin sow farm sites. The vaccines used were autogenous, adjuvanted, whole inactivated vaccines. Autogenous vaccines were chosen in order to have more current or "new" strains that would match the circulating strains in the farms rather than "old" strains that were part of the commercial "off the shelf" vaccines. This decision is not uncommon and there are criteria available to support this approach.[11] As flu continued to cause problems in pigs from weaning to market age despite the twice yearly, whole sow herd vaccination with autogenous flu vaccine, the decision was made to switch to pre-farrow vaccination of sows (at 3-weeks pre-farrow). Although we both had conducted previous work with herds in the eastern US to control flu by doing both pre-farrow and twice or thrice yearly whole herd vaccines, this approach was not tenable in the midwestern US farms.

Evaluation of custom vaccine usage

As with all interventions on pig farms, there needs to be an evaluation to determine if the intervention was successful or not. For our farms that decided to update their autogenous vaccines and to change the vaccination program to pre-farrow, they continued to perform BOTH passive and active surveillance. This allowed for the monitoring of the number of flu PCR positive tests per months, with the goal of seeing less positives post-vaccination. In addition to monitoring PCR positives, a serosurveillance program was recommended to determine if the pigs mounted a detectable antibody response to the vaccine. Studies on flu vaccine responses have made us believe hemagglutination inhibition (HI) antibody titers of > 40 are protective against respiratory disease.[12,13] In lieu of serology, success of this flu vaccination program was determined by having less disease, lower treatment costs, reduced death loss, and fewer complaints from pig caretakers.

Conclusion

The bottom line is, to be well informed about flu and control it, a systematic approach is needed. It should include routine sampling - both active surveillance of pigs in targeted populations AND passive surveillance of pigs with respiratory disease. The sampling should lead to informed, data-driven decisions regarding vaccine strain selection. The vaccine needs to be given at a time that is biologically/immunologically beneficial for the pig and agreeable to the producers. The benefits of a successful flu vaccination program can be broad and include increased profitability, better production, and happy/healthy people and pigs.

References

1. Barberis I, Myles P, Ault SK, Bragazzi NL, Martini M. History and evolution of influenza control through vaccination: from the first monovalent vaccine to universal vaccines. J Prev Med Hyg. 2016 Sep;57(3):E115-E120. **doi:10.15167/2421-4248/jpmh2016.57.3.652**

2. World Health Organization, "Global Influenza Programme." Accessed 11-01-2024. **https://www.who.int/teams/global-influenza-programme/about-us**

3. Van Reeth, K., Ma, W. (2012). Swine Influenza Virus Vaccines: To Change or Not to Change—That's the Question. In: Richt, J., Webby, R. (eds) Swine Influenza. Current Topics in Microbiology and Immunology, vol 370. Springer, Berlin, Heidelberg. **doi:10.1007/82_2011_142**

4. Petro-Turnquist E, Pekarek MJ, Weaver EA. Swine influenza A virus: challenges and novel vaccine strategies. Front Cell Infect Microbiol. 2024 Apr 3;14:1336013. **doi:10.3389/fcimb.2024.1336013**

5. Diaz A, Perez A, Sreevatsan S, Davies P, Culhane M, Torremorell M. Association between Influenza A Virus Infection and Pigs Subpopulations in Endemically Infected Breeding Herds. PLoS ONE 10(6): e0129213. **doi:10.1371/journal.pone.0129213**

6. Garrido-Mantilla, J., Alvarez, J., Culhane, M. et al. Comparison of individual, group and environmental sampling strategies to conduct influenza surveillance in pigs. BMC Vet Res 15, 61 (2019). **doi:10.1186/s12917-019-1872-4**

7. Corzo CA, Culhane M, Juleen K, Stigger-Rosser E, Ducatez MF, Webby RJ, et al. Active Surveillance for Influenza A Virus among Swine, Midwestern United States, 2009–2011. Emerg Infect Dis. 2013 Jun;19(6):954–60. **doi:10.3201/eid1906**

8. Tricco AC, Chit A, Soobiah C, Hallett D, Meier G, Chen MH, et al. Comparing influenza vaccine efficacy against mismatched and matched strains: a systematic review and meta-analysis. BMC Med. 2013 Jun 25;11:153. **doi:10.1186/1741-7015-11-153**

9. Kaplan BS, Anderson TK, Chang J, et al. Evolution and Antigenic Advancement of N2 Neuraminidase of Swine Influenza A Viruses Circulating in the United States following Two Separate Introductions from Human Seasonal Viruses. J Virol. 2021;95(20):e0063221. **doi:10.1128/JVI.00632-21**

10. Walia RR, Anderson TK, Vincent AL. 2019. Regional patterns of genetic diversity in swine influenza A viruses in the United States from 2010 to 2016. Influenza Other Respir Viruses. 2019;13:262–273. **doi:10.1111/irv.12559**

11. Sandbulte MR, Spickler AR, Zaabel PK, Roth JA. Optimal Use of Vaccines for Control of Influenza A Virus in Swine. Vaccines (Basel). 2015 Jan 30;3(1):22-73. **doi:10.3390/vaccines3010022**

12. Hobson D, Curry RL, Beare AS, Ward-Gardner A. The role of serum haemagglutination-inhibiting antibody in protection against challenge infection with influenza A2 and B viruses. J Hyg (Lond). 1972;70(4):767-777. **doi:10.1017/s0022172400022610**

13. Coudeville L, Bailleux F, Riche B, Megas F, Andre P, Ecochard R. Relationship between haemagglutination-inhibiting antibody titres and clinical protection against influenza: development and application of a bayesian random-effects model. BMC Med Res Methodol. 2010;10:18. Published 2010 Mar 8. **doi:10.1186/1471-2288-10-18**

https://doi.org/10.54846/am2025/167

Ask the experts: How is a herd-specific vaccine produced?

Panel: Erin Strait[1]; Ben Hause[2]; Mark Mogler[3]; Moderator: Rebecca Robbins[4]
[1]Ceva Animal Health, Lenexa, Kansas; [2]Cambridge Technologies; [3]Merck Animal Health; [4]PIC, Hendersonville, Tennessee

We have decades of experience in influenza vaccine manufacturing on our panel. The goal of our panel is to improve the understanding of the manufacturing process for herd-specific influenza vaccines and what constitutes success in developing and manufacturing these vaccines:

Dr Mogler: One area of success in influenza vaccine development has been the use of sequence data to inform vaccine composition and the use of platform technologies to rapidly deploy those decisions in the field.

Dr Hause: Well-produced whole inactivated influenza virus vaccines generate sterilizing immunity to antigenically similar viruses. Effective herd protection to influenza begins with thorough diagnostic surveillance of circulating viruses to ensure strain selection to maximize immune coverage. Nuances of vaccine production, including isolate purification and characterization, antigen production and inactivation, and formulation are critical components of efficacious vaccines.

Dr Strait: Custom influenza vaccines, through autogenous and/or veterinary prescription, can be highly effective when paired with effective surveillance programs and strain selection(s) guided by sequencing and HI, when possible. As we continue to learn more about key antigenic sites as well as look beyond HA sequences to other important gene segments such as NA our ability to predict protection and cross-protection will only improve.

We want to strengthen the practitioners' knowledge of the topic and invite them to come with their own questions. We will use the following 3 common field scenarios and the responses to guide us into a deeper discussion during the panel.

1. My autogenous vaccine has worked great; so well that combined with some simple management practices influenza has been eliminated from the herd. My isolate is going to expire; what are my options if I want to continue administering vaccine?
   a. Isolates must be used within 60 months of isolation.[1]
   b. This is not an issue for platform vaccines which only require a sequence from the gene of interest for manufacture.
   c. Identifying an adjacent or non-adjacent herd from which to obtain the homologous isolate as long as the regulatory definitions are met, Center for Veterinary Biologics is notified, and, when necessary, the State Veterinarian approves it.
2. How many IAV strains/antigens can I put in my vaccine? How do I know if I've put too many antigens in a formula?
   a. This is dependent on
      i. final volume of a dose,
      ii. antigen to adjuvant ratio which varies by adjuvant,
      iii. origin of antigen you want to mix together (different subtypes of IAV or IAV and other antigens like rotavirus or *E coli*)
      iv. immunogenicity of each antigen
      v. for platform vaccines, this may be limited by the license
      vi. clinical outcome desired
   b. Consequences of too many antigens in a single dose of vaccine
      i. Poor protection to all or some of the target strains/pathogens
      ii. Dose volume is too large to effectively administer leading to partial dosing
      iii. Tissue reactions usually due to increased amount of adjuvant necessary to elicit an immune response
3. I am preparing to stock a commercial sow farm and the multiplier source is IAV positive. The multiplier uses an autogenous IAV vaccine. I don't want to leave my commercial herd unvaccinated. What are my options?
   a. Peruse non-adjacent approval for purchase of vaccine from the same manufacturer the multiplier uses;
   b. Obtain the HA sequence and use a platform vaccine;
   c. Use a commercial vaccine;
   d. Collect appropriate samples for obtaining an isolate suitable for vaccine production, which entails sampling pigs with clinical signs and collecting samples with a good track record for successful viral isolation, ie, nasal swabs and lung tissue.

Additional themes and questions we hope to cover during the panel discussion:

A. The typical development and manufacturing process for autogenous and prescription platform influenza vaccine products.
Q: Are commercial vaccines safer?
Q: Explain backbone for RxPP. Is RP different from mRNA?

B. Advantages/disadvantages to using formulas with adjuvant, whole virus, gene of interest, and multivalent?
Q: What does balanced mean when referring to multivalent influenza vaccines? Does it affect efficacy?
Q: How can practitioners measure efficacy of different custom vaccine products? Should it differ from measuring efficacy for commercial licensure?
Q: What delays can occur when making a custom product? Are there vaccine manufacturing processes that take more/less time than others? Can practitioners do anything to prevent delays?

C. Is it possible to create an efficacious universal flu vaccine (like that used for humans)? What are the barriers to making one for swine?

A list of common terms and definitions to custom vaccine manufacture:

Autogenous - custom vaccines that consist of herd specific (homologous) antigens for use by or under the direction of a veterinarian or approved non-veterinarian specialist[2] prepared from cultures of microorganisms which have been inactivated and are nontoxic.[3]

Commercial - must be produced at a USDA-approved establishment, and be demonstrated to be pure, safe, potent, and efficacious.

Efficacy or efficacious - Specific ability or capacity of the biological product to effect the result for which it is offered when used under the conditions recommended by the manufacturer.[4]

Licensed - products produced in a facility, according to an approved Outline of Production. Autogenous, Platform and Commercial vaccines are all produced in licensed facilities.[5] Exempt products are beyond the scope of this discussion.

Serial - total quantity of completed product which has been thoroughly mixed in a single container and identified by a serial number.

Prescription Platform (RxPP) - custom biological product prepared for an individual animal or animals under a prescription written by a licensed veterinarian. Serials are manufactured based upon a licensed production platform.[6]

References

1. VS Memorandum (VSM) 800.69. Guidelines for Autogenous Biologics. Accessed 12/9/2024. **https://www.aphis.usda.gov/sites/default/files/memo800-69.pdf**

2. Autogenous Biologics. Accessed 11/15/2024. **https://www.aphis.usda.gov/sites/default/files/pel_4_16.pdf**

3. Evaluation of Autogenous Biologics Manufacturer's Adherence to Policy During the Inspection Proper. Accessed 11/15/2024. **https://www.aphis.usda.gov/sites/default/files/ICWI0213.pdf**

4. 9 CFR, Subchapter E, VIRUSES, SERUMS, TOXINS, AND ANALOGOUS PRODUCTS; ORGANISMS AND VECTORS. Accessed 11/15/2024. **https://www.govinfo.gov/content/pkg/CFR-2014-title9-vol1/pdf/CFR-2014-title9-vol1-chapI-subchapE.pdf**

5. VS MEMORANDUM NO. 800.50. Basic License Requirements and Guidelines for Submitting Materials in Support of Licensure. Accessed 11/15/2024. **https://www.aphis.usda.gov/sites/default/files/memo_800_50.pdf**

6. VS MEMORANDUM NO. 800.214. Prescription Platform Product Biologics. Accessed 11/15/2024. **https://www.aphis.usda.gov/sites/default/files/memo_800_214.pdf**

American Association of Swine Veterinarians
830 26th Street
Perry, IA 50220-2328
USA

ISBN 979-8-3313-1636-5

9 798331 316365

CURRAN ASSOCIATES INC.
proceedings.com

20th Australasian Weeds Conference 2016

Perth, Western Australia
11-15 September 2016

THE WEEDS SOCIETY OF WA INC